# Materials Science and Engineering of Carbon

## Fundamentals

# Materials Science and Engineering of Carbon
## Fundamentals

**Second Edition**

**Michio Inagaki**

**Feiyu Kang**

TSINGHUA UNIVERSITY PRESS

ELSEVIER

AMSTERDAM • BOSTON • HEIDELBERG • LONDON
NEW YORK • OXFORD • PARIS • SAN DIEGO
SAN FRANCISCO • SINGAPORE • SYDNEY • TOKYO
*Butterworth-Heinemann* is an imprint of Elsevier

*Butterworth-Heinemann* is an imprint of Elsevier
225 Wyman Street, Waltham, MA 02451, USA
The Boulevard, Langford Lane, Kidlington, Oxford, OX5 1GB, UK

First edition 2006
Second edition 2014



**British Library Cataloguing-in-Publication Data**
A catalogue record for this book is available from the British Library

**Library of Congress Cataloging-in-Publication Data**
A catalog record for this book is available from the Library of Congress

ISBN: 978-0-12-800858-4

For information on all Butterworth-Heinemann publications
visit our website at http://store.elsevier.com

Printed and bound in the US

14 15 16 17 18 10 9 8 7 6 5 4 3 2 1

# Contents

# Preface

One of the authors (M. Inagaki) has been emphasizing the importance of nanotexture, as well as structure, to understand carbon materials. In 1985, he proposed the classification of nanotexture of carbon materials on the basis of preferred orientation schema of anisotropic layers of carbon hexagons, planar, axial, point and random orientation. Nanotexture is formed during carbonization of organic precursor, as well as structure, and governs the structure development during heat treatment at high temperatures, which has been understood as graphitizing and non-graphitizing. Nanotextures can explain the reason why fibrous carbon materials exist, such as carbon fibers and tubes, and the spherical carbons, such as carbon blacks and fullerenes, from strongly anisotropic carbon layers consisting mainly of hexagons. In order to convince of the importance of nanotextures in carbon science and engineering, he published a book in Japanese entitled 'Materials Engineering of Carbons' in 1985, and another book in Japanese entitled 'New Carbon Materials – Structure and Functions' with his friend, Y. Hishiyama in 1994. In 2000, the book in English entitled 'New Carbons – Control of Structure and Functions' from Elsevier added the concept of carbon families, diamond, graphite, fullerene and carbyne, to follow the rapid progress in science, engineering and applications of carbon materials. However, he strongly felt that, even though many young scientists and engineers are interested in and working on nanocarbons, such as carbon nanotubes and graphenes, fundamental knowledge on carbon materials is necessary for them. Most basics of carbon materials were already clarified before 1985. The books, which give such fundamental knowledge on carbon materials, are rather few and also it must be handy and easy to buy. Therefore, he discussed with Prof F. Kang and decided to publish the book from Tsinghua University Press, China, entitled 'Carbon Materials Science and Engineering – From Fundamentals to Applications'. The book aims to give comprehensive information firstly on fundamental science on preparation and characterization of various carbon materials, and secondly on engineering and applications of various carbon materials, on the basis of the same basic concept as published before, i.e., classifications based on carbon families and nanotextures.

Since so many copies have been sold mostly in China, the present authors (M. Inagaki and F. Kang) decided to write advanced science and engineering on carbon materials under the corporation of two more authors (M. Toyoda and H. Konno) by taking in recent developments, and it was published in September, 2013, with the title 'Advanced Science and Engineering on Carbon' by two publishers, Tsinghua University Press and Elsevier. At the same time, the publishers asked the present authors to revise and up-date the previous book 'Carbon Materials Science and Engineering – From Fundamentals to Applications'. Here, the revised version is presented.

In this revised version, the content is largely revised and up-dated, for example, a chapter of nanocarbons is newly added, although fundamentals in carbon science and engineering do not change and the basic concepts, carbon families and nanotextures, are still valid. The authors hope to provide fundamental science and engineering on carbon materials, associated with some applications, to young graduate students who are working on various carbon materials and also engineers whose works are more or less related to carbon materials. It will be a great pleasure for the authors if they will always bring this book with them to discuss their results and to read the scientific papers published. They may find out how the data they got and/or those published either agree or disagree with the general information explained in this book, and also what is missing from this book.

# Acknowledgments

The authors would like to express their sincere thanks to the people who kindly provided the data and figures for this book, the names and affiliations of contributed persons being mentioned in the caption of figures and tables. They also thank all of the people who took care of this book in Tsinghua University Press and also in Elsevier.

CHAPTER

# Introduction

# 1

## 1.1 Carbon materials

Carbon C is one of the abundant elements on the Earth, because almost all organics are composed from carbon networks, and it is very familiar in our daily lives, for example, ink for newspapers, lead for pencils, activated carbons in refrigerators, etc. Carbon materials, which consist mainly of carbon atoms, have been used since prehistoric era as charcoal. In Japan, a large amount of charcoal (about 800 tons) was reported to be used for casting a great image of Buddha in Nara from 747–750. Soft graphite has been used for a long time as lead and carbon blacks as black inks. Diamond crystals are fascinating for all human beings not only as jewels but also the hardest materials were found to consist of carbon atoms, the same atoms as lubricating soft graphite in 1799. Nowadays various carbon materials are used in our daily lives, though many of them are inconspicuous; activated carbon produced from coconut shells for a filter of tobacco, carbon fibers for reinforcement of rackets and fishing rods, leads for automatic pencils, activated carbons for deodorization in refrigerators, membrane switches composed of graphite flakes for keyboards of computers and various instruments, etc. Charcoal may be the first carbon material used practically, as it has been used since the pre-historic age. Carbon materials started to be used as electrodes for batteries around 1800. Since 1878, large-sized carbon rods were used as electrodes for iron refining, which were industrially produced by heat treatment at high temperatures (as high as 3000°C) and called graphite electrodes because crystalline graphite structure was well developed in most of them. Later on, various carbon materials having graphitic structure for various applications were developed, which were called graphite materials, even though the development of graphitic structure is not complete. At the same time, carbon materials without noticeable graphite structure, such as charcoal, were also developed and opened new applications.

There was no clear definition and no clear-cut classification on what graphite materials are and what carbon materials are. In the present book, however, we will use the term ‘carbon materials’ for materials composed predominantly of carbon element, irrespective of their structure, so including fullerenes and carbon nanotubes, and also the terms either ‘carbon materials’ or ‘carbons’ for the materials without three-dimensional graphite structure. On the other hand, ‘graphite materials’ and sometimes ‘graphites’ were used for the materials which have three-dimensional graphite structure, even partly. In industry, the term ‘graphite’

Materials Science and Engineering of Carbon: Fundamentals. DOI: http://dx.doi.org/10.1016/B978-0-12-800858-4.00001-2

and 'graphitized' are often used, even though graphite structure is not developed appreciably; for example PAN-based carbon fibers heat-treated at a high temperature used to be called 'graphite fibers', even though almost no graphite structure was developed, as will be explained later in detail.

Polycrystalline graphite materials have been used in various fields of industries using their different properties. Their characteristics can be summarized as follows; (1) high thermal resistance in non-oxidizing atmosphere, (2) high chemical stability, (3) high electrical and thermal conductivities, (4) small thermal expansion coefficient and, as a consequence, high thermal shock resistance, (5) very light weight, (6) high mechanical strength at high temperatures, (7) high lubricity, (8) highly reductive at high temperatures and easily dissolved into iron, (9) non-toxic, (10) radiation resistance, and (11) low absorption cross-section and high moderating efficiency for neutron.

Since all polycrystalline graphite materials consist of parallel stacking of carbon hexagonal layers, like graphite, which are called crystallites, their properties of a bulk material are strongly governed by different factors, such as how large the crystallites are, how these anisotropic crystallites orient in the bulk, to what temperature they were heat-treated, etc. The preferred orientation of crystallites in bulk graphitic materials depends strongly on the condition of forming process and the heat treatment temperature governs the size and perfection of the structure. Therefore, most of the properties of carbon materials distribute in a wide range. In Fig. 1.1, electrical conductivity, bulk density, thermal expansion coefficient (expansivity), and tensile strength are compared for different carbon materials, including natural graphite, various fibrous carbon materials, and graphite intercalation compounds (GICs).

Polycrystalline graphite is a good electric conductor, but its electrical conductivity of roughly $2 \times 10^5$ S/m is inferior to metals. By intercalation of different species into the interlayer spaces of graphite, however, electrical conductivity is much improved and becomes even higher than that of metallic copper. Thermal expansion coefficient of graphite single crystal is very high along the c-axis (perpendicular to the graphite layer plane), but negative (i.e., shrinkage) along the a-axis (parallel to the layer). In polycrystalline graphite materials, this anisotropy in thermal expansion is spaciously averaged, depending strongly on the size and arrangement of crystallites (i.e., structure and texture). In fibrous carbons, it is mainly governed by expansion along the layer planes and so rather small values. In most of physical properties, such as electrical and thermal characteristics, the highest and the lowest values are realized in the directions perpendicular and parallel, respectively, to graphite layers, as shown on the electrical conductivity and thermal expansion coefficient in Fig. 1.1. Mechanical properties, such as tensile strength, and bulk density are texture-sensitive characteristics and so they show a wide range of values, in general. The practical values for various carbon materials including polycrystalline graphite materials (high-density isotropic graphite and graphite electrodes) are inferior to the theoretical values for graphite single crystal, because of their polycrystalline nature.

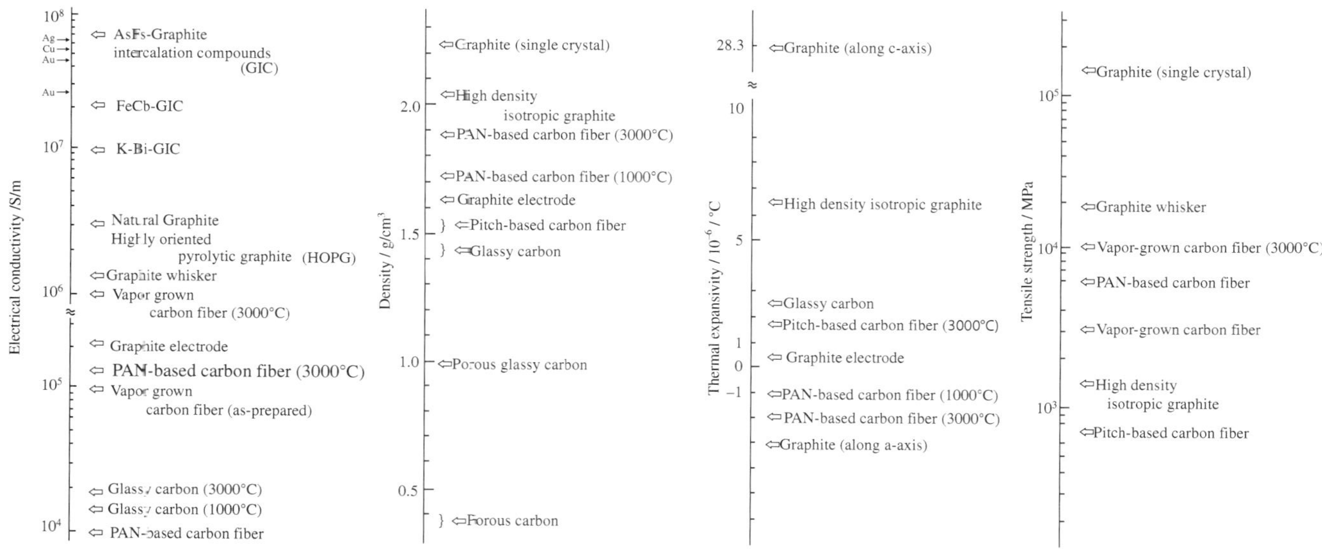


**FIGURE 1.1**

Range of various properties of carbon materials.

## 1.2 Short history of carbon materials

Development of carbon materials was discussed by dividing into three periods, before 1960, between 1960 and 1985 and after 1985, as summarized in Table 1.1. The year 1960 may be said to be the beginning of the era of new carbons, because of the inventions of carbon fibers from poly(acrylonitrile), pyrolytic carbons by CVD process, and glass-like carbons from thermosetting resins, which were completely different from the carbon materials used before 1960.

Up to 1960, four carbon materials were known and had practical applications in various fields of industries; artificial graphite blocks mainly used for steel refining, carbon blacks for ink and reinforcement of rubbers, and activated carbons for water purification, in addition to natural diamond. These carbon materials, except diamond because of its very different appearance and properties, were proposed to be called **classic carbons**.

In 1960, three carbon materials, carbon fibers, glass-like carbons, and pyrolytic carbons were developed, which were completely different from classic carbons in their production processes and also properties. Following these three carbon materials, different kinds of carbon materials had been developed under the modifications in precursors, preparation conditions, etc. So, we called these carbon materials **new carbons**, in the contrast to classic carbons. After the finding of graphite intercalation compounds having a high electrical conductivity, higher

**Table 1.1** Carbon Periods

| Period | Years | Carbon Materials Developed | Remarks |
|---|---|---|---|
| I | Pre-1960 | Artificial graphite blocks<br>Activated carbons<br>Carbon blacks<br>Natural diamond | Mass production<br>Sale in either tons or kilograms |
| II | 1960–1985 | Various carbon fibers<br>Glass-like carbons<br>Pyrolytic carbons<br>High-density isotropic graphites<br>Intercalation compounds<br>Various composites<br>Synthetic diamond<br>Diamond-like carbons | Introduction of various techniques for the production of carbon materials (e.g., CVD, composite with other materials, etc.).<br>Development of new applications<br>Sale mostly in grams |
| III | Post-1985 | Fullerenes<br>Carbon nanotubes<br>Graphene | Nano-sized<br>Sale in milligrams |

than copper, a boom in research on intercalation compounds has arisen in a world-wide scale, although it could not open the practical applications.

The year 1985 was another epoch for carbon materials, where a carbon cage consisting of 60 carbon atoms was found, which was named buckminsterfullerene $C_{60}$ and followed by a series of carbon cages, such as $C_{70}$, $C_{86}$, etc. In 1991, multi-walled carbon nanotubes were reported, which was followed by the finding of single-wall carbon nanotubes. In 2004, a single hexagonal carbon layer was reported. Finding of these novel carbons, **nanocarbons**, attracted pronounced attention to nano-scale science and technology, and accelerated the development of the science related to nanotechnology. In the course of nanotechnology development, the word **nanocarbons** came to be often used. Also, structure and texture of most carbon materials were required to be controlled in nanometer scale for all applications.

## 1.3 Classic carbons, new carbons, and nanocarbons

### 1.3.1 Classic carbons

The fundamental science and technology on classic carbons, artificial graphite blocks, carbon blacks, and activated carbons, were established before 1960, in the period I (Table 1.1). It has to be emphasized, however, that these carbon materials are principal products and principal incomes for carbon industries world-wide currently.

In Fig. 1.2, photographs of these carbon materials are shown. These three-carbon materials have a wide range of sizes; graphite electrodes representing artificial graphite blocks are used in a size of about 700 mm in diameter and about 3 m in length, carbon blacks are spherical particles with the diameter from 10 to a few hundred nanometers, activated carbons are porous materials with irregular

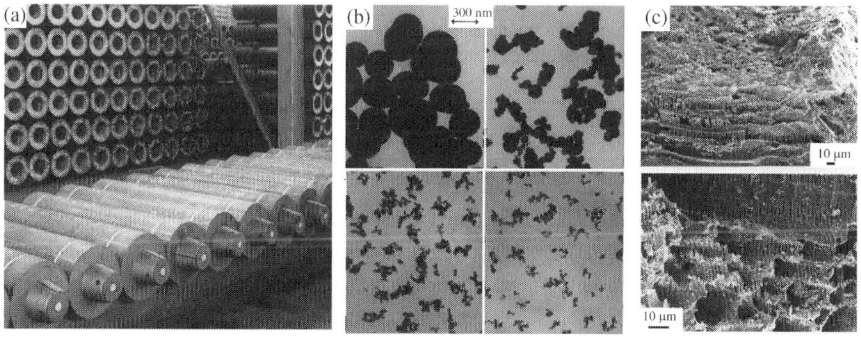


**FIGURE 1.2**

Classic carbons. (a) Graphite electrodes (b) Carbon blacks (c) Activated carbons.

shapes. Diamond is so rare in nature and expensive as to be measured by using the unit of carat, different from other carbon materials (gram).

### 1.3.2 New carbons

The important developments related to carbon materials since 1960, in the periods of II and III, are listed in Table 1.2. The period II started with the developments of PAN-based carbon fiber, pyrolytic carbon, and glass-like carbon, all three of them being completely different from classic carbons.

Carbon fibers [1], which were produced by carbonization of poly(acrylonitrile) fibers after oxidation (PAN-based carbon fibers; Fig. 1.3), fascinated people by their high strength and flexibility and many demonstrative pictures, for example, hanging an automobile by a thin string of carbon fibers, were published in various journals. The developments of other kinds of carbon fiber followed in the 1970s, including pitch-based and vapor-grown carbon fibers. In contrast to carbon fibers, glass-like carbon was very hard and brittle, and its gas impermeability which had never been realized in classic carbons was amazing [2]. It was named from its conchoidal fracture surface, similar to soda-lime glass. Now different products of glass-like carbon were industrially developed, as shown in Fig. 1.4. Pyrolytic

**Table 1.2** Topics Related with Carbon Materials

| Year | Basic Science | Materials Development | Technology Development |
|---|---|---|---|
| 1960 | | PAN-based carbon fibers<br>Pyrolytic carbons<br>Glass-like carbons | Electrodes for electric discharge machining |
| 1965 | Mesophase spheres in pitch | Needle-like cokes<br>Mesophase-pitch-based carbon fibers | |
| 1970 | Biocompatibility of carbon materials | Vapor-grown carbon fibers | Carbon prostheses |
| 1975 | High conductivity of GICs | | Mesocarbon microbeads |
| 1980 | Diamond-like carbon films | Isotropic high-density graphites<br>Carbon-fiber-reinforced concrete | Carbon electrode for fuel cell |
| 1985 | | Buckminsterfullerene $C_{60}$, followed by various fullerenes | First wall for fusion reactor |
| 1990 | Superconductivity of $K_3C_{60}$ | Carbon nanotubes, single-wall and multi-walled | |
| 1995 | | | Carbon anode for lithium ion rechargeable batteries |

**FIGURE 1.3**

Carbon fibers.

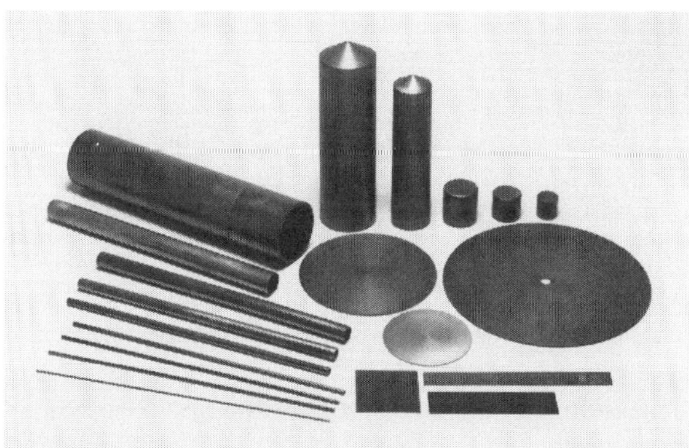

**FIGURE 1.4**

Products of glass-like carbon.

carbons were produced by a completely different technique from conventional ones, chemical vapor deposition (CVD) [3], though it is very common in material production nowadays. Their strong anisotropy in various properties, such as electrical and thermal conductivities, gave a quite new aspect for the application of

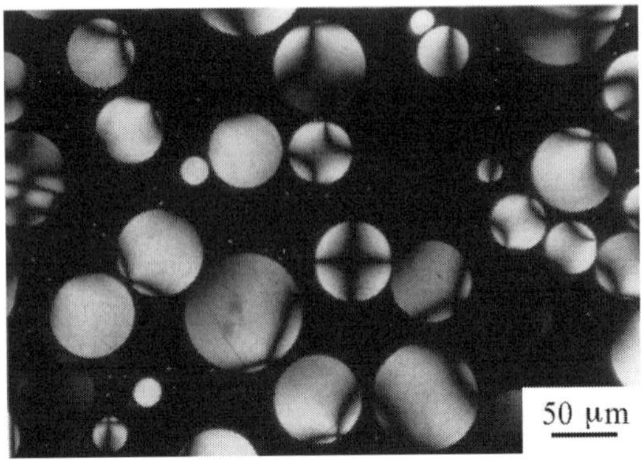


**FIGURE 1.5**

Mesophase spheres formed in a pitch.

carbon materials. Pyrolytic carbons prepared in well-controlled conditions could have very high crystallinity, i.e., well-developed and well-oriented basal planes of graphite, by treatment under high temperature and high pressure, which were called highly oriented pyrolytic graphite (HOPG) and developed new applications as a monochrometer for X-rays and neutrons [4]. This CVD process was successfully applied for carbon coating of nuclear fuel particles [5].

Formation of optically anisotropic spheres in pitches, mesophase spheres, and their coalescence, which was firstly reported in 1964 [6], motivated many fundamental studies, structure of the spheres, growth and coalescence mechanism of spheres and formation of bulk mesophase, and created new carbon products, needle-like cokes which were the essential raw materials for high-power graphite electrodes, mesophase-pitch-based carbon fibers with high performance, and mesocarbon microbeads for different applications [7] (Fig. 1.5).

Thin flakes of natural graphite were successfully used to produce the membrane switches, of which the construction was schematically shown in Fig. 1.6. They contributed to promote the light weight and small size of modern electronics, by being used as keyboards for computers, switching boards for various electric equipments, etc.

Good biocompatibility of carbon materials, found around 1970, led to the development of various prostheses, such as heart valve, tooth root, etc. [8]. Around 1980, industrial technology for producing isotropic high-density graphite blocks by using so-called rubber-press was established and created various applications; reflectors for high-temperature gas-cooled reactors, various jigs for the synthesis of semiconductor crystals and also the electrodes for electric discharge machining. Around 1985, a small amount of mixing of carbon fibers to cement paste was found to result in a pronounced reinforcement of concrete [9]. Its first practical application was the construction of the Arshaheed monument in Iraq and then it was applied in different buildings (Fig. 1.7). Today, not only carbon-fiber-reinforced

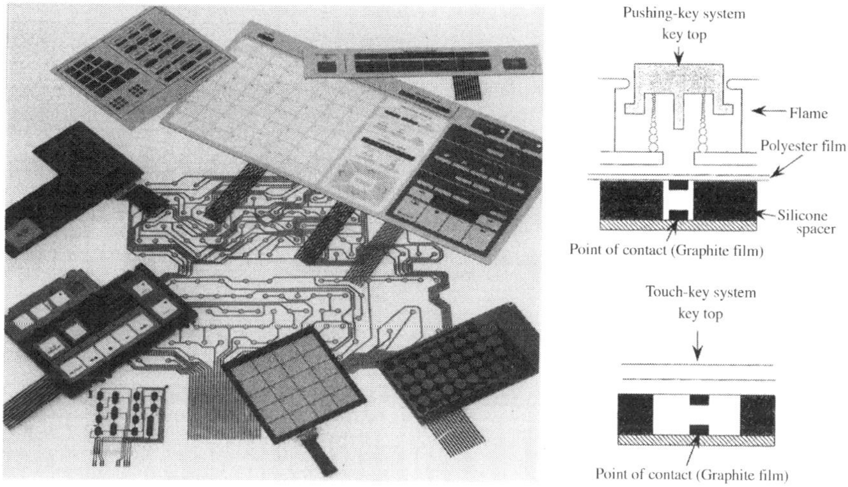


**FIGURE 1.6**

Membrane switches and their construction.

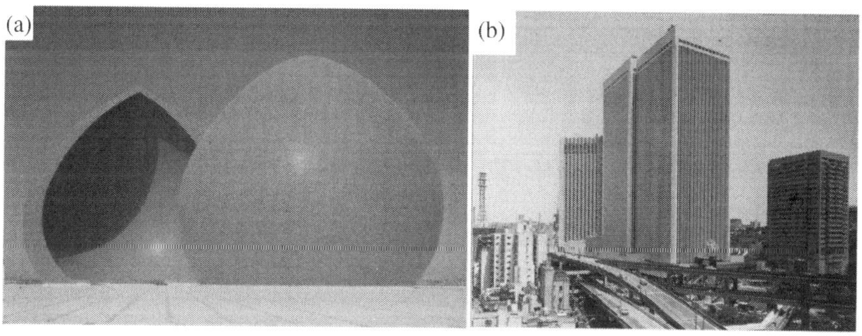


**FIGURE 1.7**

Buildings that used carbon-fiber-reinforced concrete. (a) Arshaheed monument in Iraq and (b) Ark Hills Tower in Tokyo.

concrete but also carbon fibers themselves are used in the field of civil engineering, such as in buildings, bridges, and various constructions [10].

Finding of high electrical conductivity of $AsF_5$−graphite intercalation compound, higher than metallic copper, gave a strong impact to scientists and engineers [11]. The researchers did not give a practical application of these intercalation compounds, mainly because of their poor stability in air. However, practical use of carbon materials as anode for lithium ion rechargeable batteries has led to a great success (Fig. 1.8) and contributed to the development in personal

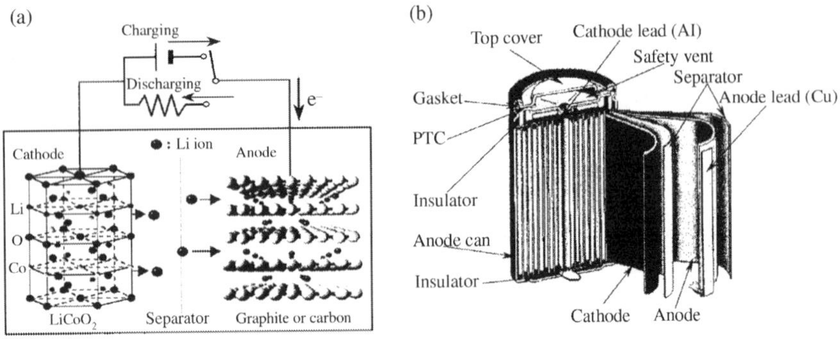


**FIGURE 1.8**

Lithium-ion rechargeable batteries. (a) Principle of charge/discharge (b) An example of construction.

computers and portable telephones, which was based on intercalation and deintercalation of lithium ions into the graphite gallery [12]. As one of the energy storage devices, electrochemical capacitors are developed, they are based on physical adsorption and desorption of ions on porous carbon electrodes to form electric double layers.

### 1.3.3 Nanocarbons

In 1985, a cage (cluster) composed of 60 carbon atoms $C_{60}$ (buckminsterfullerene) was firstly reported to be isolated from the soot obtained by laser irradiation on a graphite block [13], of which the structure consisted of 20 hexagons with 12 pentagons of carbon atoms (Fig. 1.9a). This carbon cluster $C_{60}$ was spherical, in other words, all chemical bonds are closed in the cage. These cages are crystallized to form the face-centered cubic crystal by cubic closest packing. It can be dissolved into some organic solvents, such as benzene, hexane, etc., and behaves as a molecule. Later, cages with different sizes, such as $C_{70}$, $C_{76}$, $C_{82}$, ... and also multi-walled cages were found and isolated. These are called fullerenes. Doping of alkali metals into all interstices of the fullerene crystals (tetrahedral and octahedral sites of cubic closest packing of cages) was found to give superconductivity [14]. Cages containing metal atoms, such as La, Sc, etc., were synthesized [15]. In 1996, The Nobel Prize in Chemistry was awarded jointly to R. F. Curl Jr., Sir H. W. Kroto, and R. E. Smalley for their discovery of fullerenes.

In 1991, carbon nanotubes were reported [16] and later single-wall carbon nanotubes were found [17,18]. In 1960, fibrous carbons had been synthesized by arc discharging between carbon electrodes, which were called graphite whiskers because of their high crystallinity [19]. In 1976, a single-wall carbon nanotube was observed in the first step of the growth of vapor-grown carbon fibers by

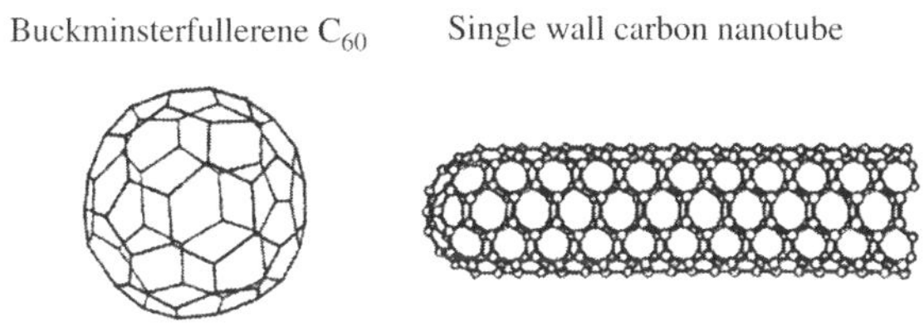


**FIGURE 1.9**

Fullerene and carbon nanotube.

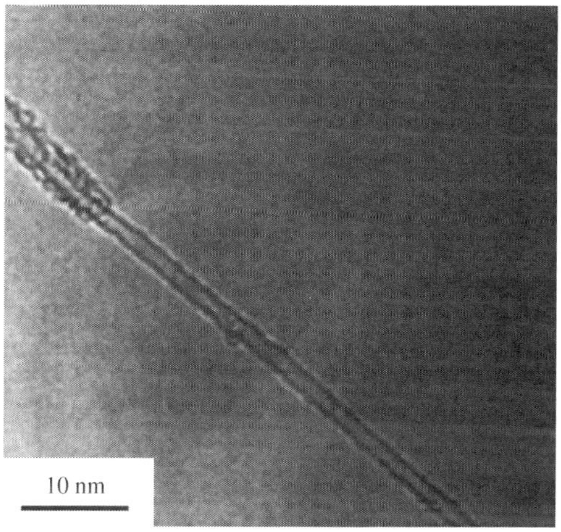


**FIGURE 1.10**

Single-wall carbon nanotube [18].

CVD method using minute particles of catalyst iron [20] (Fig. 1.10). The naming of carbon nanotube [17] was very timely for the start of nanotechnology in various fields.

In 2004, the preparation of graphene, a single two-dimensional sheet of carbon atoms, was firstly reported, as shown in Fig. 1.11 [21], although the term 'graphene' was proposed in 1986 [22]. The Nobel Prize in Physics 2010 was awarded jointly to A. Geim and K. Novoselov for groundbreaking experiments regarding the two-dimensional material graphene.

Nanocarbons were defined as not only their sizes of primary particles are in nanometer scale, but also their structures and/or textures are controlled in nanometer scale [23]. Either nano-size or nano-structure of carbon materials had to be consciously controlled to govern their properties and functions. They were discussed in more detail by emphasizing some novel techniques to produce

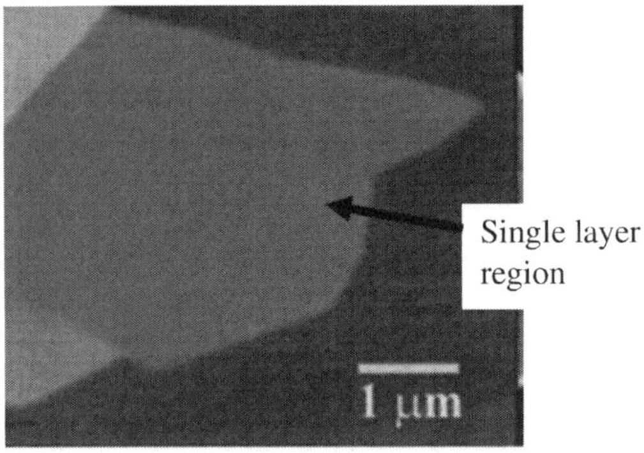


**FIGURE 1.11**

Atomic force microscopic image of single hexagonal carbon layer, graphene [21].

nanocarbons [24]. Nanocarbons were classified mainly based on their preparation processes as follows.

**(I)** *Nano-sized carbons* Carbon materials of which sizes are in nanometers, for example, carbon nanotubes, carbon nanofibers, fullerenes and graphene, were classified into following three:
- **(I-a)** Carbons produced through vaporization of carbon clusters or fragments.
- **(I-b)** Carbons produced through catalytic effects of nano-sized metallic particles.
- **(I-c)** Carbons produced through other processes, such as template, polymer blend, etc.

**(II)** *Nano-structured carbons* Carbon materials of which structure and texture are designed and controlled in nanometer scale were classified into the following four:
- **(II-a)** Carbons produced through controlling nano-size pores.
- **(II-b)** Carbons produced through designing molecular structure in precursors.
- **(II-c)** Carbons produced through controlling the carbonization process of precursors.
- **(II-d)** Carbons consisting of different component carbons and produced through controlling their interfaces in nanometer scale.

Fullerenes and carbon nanotubes were firstly synthesized through carbon vapors produced by arc discharging (I-a) and then expanded to new synthesis method by CVD process using nano-sized metallic particles, such as Fe and Ni (I-b), mainly in order to increase the production efficiency of these nanocarbon materials. Unique processes for the production of nano-sized carbons, which may be classified into (I-c), and also interesting results, have been reported on nano-structured carbons classified into category II [24]. Some of the results are explained in Section 2.4.

## 1.4 Construction and purposes of the present book

The present authors published 'Carbon Materials Science and Engineering – From Fundamentals to Applications' in 2006, and briefly revised and up-dated the version in 2011. We are going to publish an advanced science of carbon materials titled 'Advance Materials Science and Engineering of Carbon' from Tsinghua University Press and Elsevier by focusing on specific items. During editing of the manuscript on advanced science, revision and up-dating of the previous book published from Tsinghua University Press was strongly recommended by the publishers, and the authors also thought it necessary, because of the rapid development of science and engineering on carbon materials, not only in carbon nanotubes and graphenes but also in porous carbons. Therefore, the authors decided to publish the present book under the same title.

The present book is aiming to give basic and thorough understanding of various carbon materials, and to be useful in understanding the advanced carbon science and engineering focuses on different carbon materials. The present book is constructed from two parts, **Fundamental Science of Carbon Materials** and **Engineering and Applications of Carbon Materials**, in addition to this **Introduction**.

In **1 Introduction**, a brief review of carbon materials is presented before getting into the detailed discussion on science and engineering of various carbon materials in the following chapters, by explaining how widely different carbon materials have been developed, which have been called classic carbons, new carbons, and nanocarbons, and how many carbon materials we are using daily.

In **2 Fundamental Science of Carbon Materials**, the concept of carbon families is firstly introduced, and then structural characteristics and the textures, which have arisen from their characteristic structures, are explained. The detailed discussion on texture development in carbon materials (carbonization) is given, by separating novel techniques for carbonization. Structure development in carbon materials with high-temperature treatment (graphitization) is discussed based on their nanotextures and a general view of graphitization process is given. Also, acceleration of the graphitization process is discussed separately. Pores have given characteristic functions to carbon materials and so one section is devoted to the characterization and the control of pore structure in carbon materials. The introduction of foreign species, not only atoms but also molecules, into carbon materials is also explained in an independent section, which has been done through intercalation, substitution, and doping.

In **3 Engineering and Applications of Carbon Materials**, different carbon materials are explained in ten sections by paying particular attention to their preparation and applications. Polycrystalline graphite, highly oriented graphite, glass-like carbons, carbon fibers, nanocarbons, porous carbons including activated carbons, carbon-based composites, and intercalation compounds of natural graphite are explained on their definition, production processes, and applications in separate sections. Two sections are devoted especially to carbon materials for energy storage and environment remediation because of the important roles of carbon materials in these application fields.

The authors aim to give an overview on fundamentals of science and engineering of carbon materials from the point of view on structure and texture. If the readers can get a general view on carbon materials and the fundamental concepts to understand and to study the carbon materials, the authors will have succeeded. In addition, the authors strongly recommend all readers to refer to the original papers and also related papers cited in the present book, in order to understand in more detail. It has to be emphasized here frankly that all published papers are not cited in the present book and many interesting and important papers are omitted here.

One carbon material has different aspects. For example, porous carbons are produced through different processes, from carbonization of thermosetting precursors associated with activation, through template method without activation, from carbon aerogels, from the carbonization of thermoplastic precursors, through exfoliation of graphite via intercalation compounds, etc. The porous carbons thus prepared have a wide range of pores from micropores to macropores, and, as a consequence, they have been applied in various fields, adsorbents of various molecules, molecular sieving, storage of methane and hydrogen, sorption of viscous heavy oils, electrodes of electric double-layer capacitors, etc. As another example, exfoliated graphite has been used as the raw material for flexible graphite sheets, which are applied in various fields of industry, but it has recently been found to be a good sorbent for heavy oils, a support for various catalyst metals, and also a raw material for graphenes. Various carbon materials are explained in different chapters and sections in the present book because they have been used in different fields, as explained on porous carbons and exfoliated graphite. Therefore, the readers are strongly requested to read through whole parts of the present book first, even though they may be interested in a specialized carbon material, and then to visit the sections, which are written on the specified carbon material.

## References


[1] J.B. Donnet, T.K. Wang, S. Rebouillat, J.C.M. Peng, Carbon Fibers (1998) 573.
[2] T. Noda, M. Inagaki, J. Non-Cryst. Solids 1 (1969) 285.
[3] J.C. Bokros, Chem. Phys. Carbon 5 (1969) 1.
[4] A.W. Moore, Chem. Phys. Carbon 11 (1973) 69.
[5] J. Guilleray, R.L.R. Lefevre, M.S.T. Price, Chem. Phys. Carbon 15 (1976) 1.
[6] J.D. Brooks, G.H. Taylor, Symp. Carbon, Tokyo (1964)Paper III-14-1.
[7] J.D. Brooks, G.H. Taylor, Chem. Phys. Carbon 4 (1968) 243.
[8] J.C. Bokros, L.D. LaGrange, F.J. Schoen, Chem. Phys. Carbon 9 (1972) 103.
[9] S. Akihama, T. Suenaga, T. Banno, KICT Report No. 53 (1984).
[10] M. Inagaki, Carbon 29 (1991) 287.

[11] F.L. Vogel, Bull. Am. Phys. Soc. 21 (1976) 263.
[12] G. Pistoia, Lithium Batteries (1994) 483.
[13] E.W. Kroto, J.R. Hearth, S.C. I'Bien, et al., Nature 318 (1985) 162.
[14] R.C. Hadden, A.F. Hebard, M.J. Rosseinsky, et al., Nature 350 (1991) 320.
[15] Y. Chai, T. Guo, C. Jin, et al., J. Phys. Chem. 95 (1991) 7564.
[16] S. Iijima, Nature 354 (1991) 56.
[17] S. Iijima, T. Ichihashi, Nature 363 (1993) 603.
[18] D.S. Bethune, C.H. Kiang, M.S. deVries, et al., Nature 363 (1993) 605.
[19] R. Bacon, J. Appl. Phys. 31 (1960) 283.
[20] A. Oberlin, M. Endo, T. Koyama, J. Cryst. Growth 32 (1976) 335.
[21] K.S. Novoselov, A.K. Geim, S.V. Morozov, et al., Science 306 (2004) 666.
[22] H.P. Boehm, R. Setton, E. Stumpp, Carbon 24 (1986) 241.
[23] M. Inagaki, L.R. Radovic, Carbon 40 (2002) 2279.
[24] M. Inagaki, K. Kaneko, T. Nishizawa, Carbon 42 (2004) 1401.

CHAPTER

# Fundamental Science of Carbon Materials

# 2

## 2.1 Carbon families

### 2.1.1 Carbon–carbon bonds

Carbon atoms can have three different hybrid orbitals, $sp^3$, $sp^2$ and sp, and give a variety of combinations of chemical bonds. Fig. 2.1 illustrates how these three hybrid orbitals of carbon atoms give a large family of organic molecules and how the inorganic carbon materials, diamond, graphite, fullerenes and carbynes, result from the extension to giant molecules of these organic materials. This variety in chemical bonds gives an enormous number of hydrocarbons and constructs an enormous number of organic materials; C–C bond using $sp^2$ hybrid orbitals gives a series of aromatic hydrocarbons, benzene, anthracene, phenanthrene, etc., C–C–bonds using $sp^3$ and sp hybrid orbitals give various aliphatic hydrocarbons, such as methane, ethane, propane, ethylene, acetylene, etc. As schematically shown in Fig. 2.1, infinite repetition of C–C bond using $sp^3$ results in a three-dimensional network of carbon atoms, which is known to be diamond. The atomic positions of carbon in an organic compound named as adamantane are exactly the same as those in diamond. Therefore, if we were able to polymerize adamantane to a giant molecule, we may synthesize diamond through a chemical process, though it has not yet succeeded. A simple repetition of C–C bond with $sp^2$ hybrid orbitals gives flat planes of hexagons of carbon atoms, as benzene, anthracene, ovalene, and finally reaches graphite, where giant flat planes tend to stack with each other due to interaction between $\pi$-electron clouds resonated between neighboring carbon atoms in the same plane. If C–C bonds based on $sp^2$ hybrid orbitals compose pentagons of carbon atoms, a coranulene molecule is formed, being associated with five hexagons, which is somewhat curved. By the polymerization of these coranulene molecules, various clusters of carbon atoms are formed and the resultant carbon materials are called fullerenes, where carbon hexagons have to be located between pentagons, the smallest molecule is $C_{60}$, consisting of 12 pentagons and 20 hexagons of carbon atoms. The infinite repetition of C–C bonds with sp hybrid orbitals gives the carbon materials called carbynes, in which carbon atoms are making linear chains either with double bonds or with the repetition of single and triple bonds.

Materials Science and Engineering of Carbon: Fundamentals. DOI: http://dx.doi.org/10.1016/B978-0-12-800858-4.00002-4

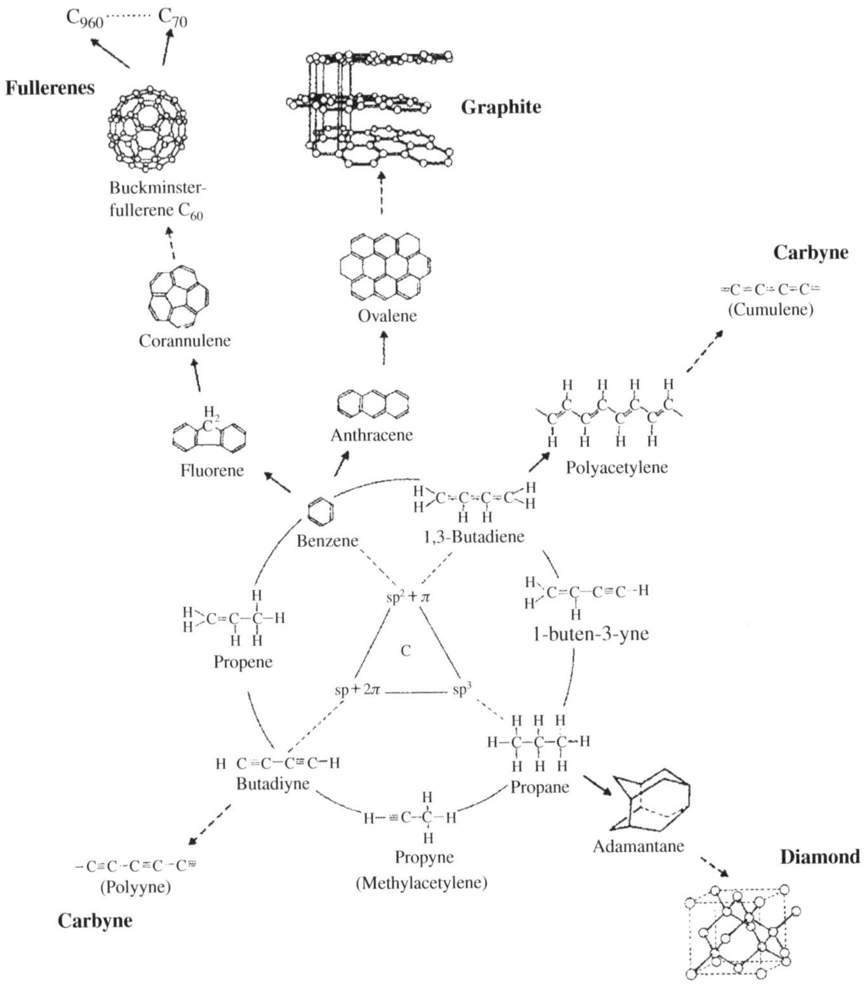


**FIGURE 2.1**

C—C bonds to form a large number of hydrocarbons and their extension to carbon families.

## 2.1.2 Carbon families

Based on the extension by the repetition of three kinds of C—C bonds to infinite molecules, we may define a family of inorganic carbon materials, **carbon family**, consisting of diamond, graphite, fullerene and carbyne. In Fig. 2.2, the structural characteristics and the structural diversities in each family are summarized.

Diamond consists of $sp^3$ orbitals, where chemical bonds extend in a three-dimensional direction and are purely covalent. It is very hard because of covalent

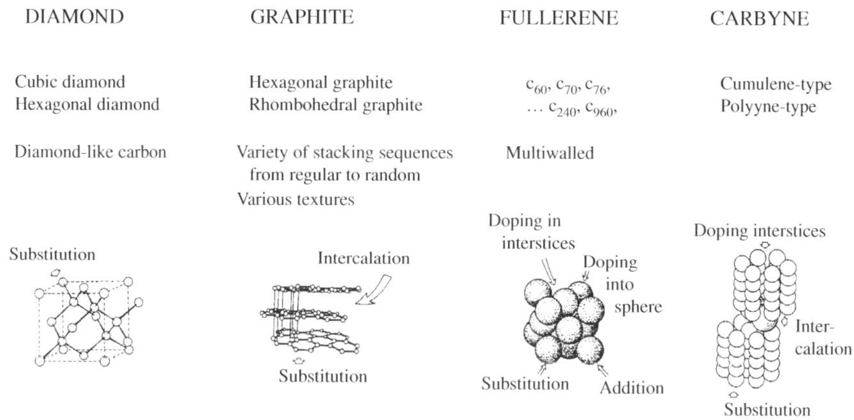


**FIGURE 2.2**

Carbon families and their diversity.

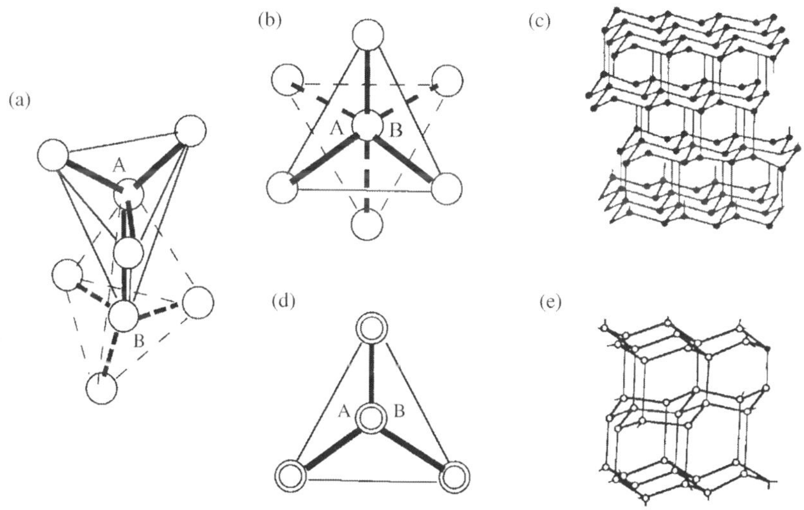


**FIGURE 2.3**

Mutual relations between two tetrahedra of carbon atoms in diamond.

bond and electrical insulator because of high localization of electrons (no $\pi$-electrons exist). In order to construct diamond crystal, periodical and regular repetition of this C—C bond is required in a long range. Let us put our attention on a couple of carbon atoms indicated as A and B in Fig. 2.3a. The carbon atom A has to be connected with four carbon atoms, including B, to make a tetrahedron because of directional sp3 bonds. The atom B has also to be surrounded by four carbon atoms, including A. If we look down these two tetrahedra centered by A and B atoms along their connecting line, there are two possibilities in mutual relation between two basement planes consisting of three carbon atoms, which give two crystal structures. If these two basement planes are rotated with each other by 60°, as shown in Fig. 2.3b, the resultant diamond crystal belongs to cubic crystal system (cubic diamond; Fig. 2.3c). If there is no rotation between these two basement planes (Fig. 2.3d), diamond crystal in hexagonal system is the result (hexagonal diamond, Fig. 2.3e). Most diamond crystals, which are either naturally occurred or synthesized, are cubic.

In the case where long-range periodicity was not attained between interconnected two tetrahedra, in other words, in the case of random rotation between these two basement planes of tetrahedra, an amorphous structure is the result. Because of random rotation between tetrahedra, some carbon atoms cannot make a chemical bond with neighboring carbon atoms, most of which are supposed to be connected with hydrogen atoms to be stabilized. The materials are obtained as thin films, mainly because random repetition of tetrahedra is difficult to keep at a long distance, and called diamond-like carbon (DLC). Some of these are as hard as diamond crystal, because principal C—C bonds are $sp^3$ bonding, and contain a relatively large amount of hydrogen.

Carbon family having $sp^2$ bonding is represented by graphite, where the flat layers of hexagons of carbon atoms bound by using $sp^2$ orbitals are stacked parallel by using $\pi$-electron clouds with a regularity of ABAB..., which belongs to a hexagonal crystal system. A stacking regularity of ABCABC... is also possible, which belongs to a rhombohedral crystal system, but it occurs only locally by introducing stacking faults due to shearing stress during grinding, for example. In addition, the parallel stacking of the layers without any regularity occurs mostly in the carbon materials prepared at low temperatures as 1300°C, where the layers of hexagons are usually small in size and also a few number of layers are stacked in parallel. This random stacking of layers is called 'turbostratic structure'. Since this turbostratic structure can be partly transformed to regular stacking of layers by the heat treatment at high temperatures, a wide range of diversity in structure in the graphite family was caused. The graphite family can have various textures in different scales mainly because the basic structural unit is the stacked flat planes of carbon hexagons which are highly anisotropic. Structure and textures in carbon materials classified into the graphite family will be discussed in more detail in Section 2.2, because most carbon materials, which are used in our lives and also in industries, belong to graphite family.

Bonding nature in fullerene particles is also $sp^2$ hybrid, but different from graphite in the fact that some $sp^2$ bonds are curved to construct pentagons of carbon atoms. The particle of buckminsterfullerene $C_{60}$ is composed of 12 pentagons and 20 hexagons of carbon atoms. The addition of hexagons into $C_{60}$ to make all pentagons apart from each other and to keep closed cluster morphology leads to giant fullerenes, as shown in the upper line of Fig. 2.4. Another way to increase the number of hexagons is to make two groups of six pentagons apart from one another, which results in single-wall carbon nanotube, as shown

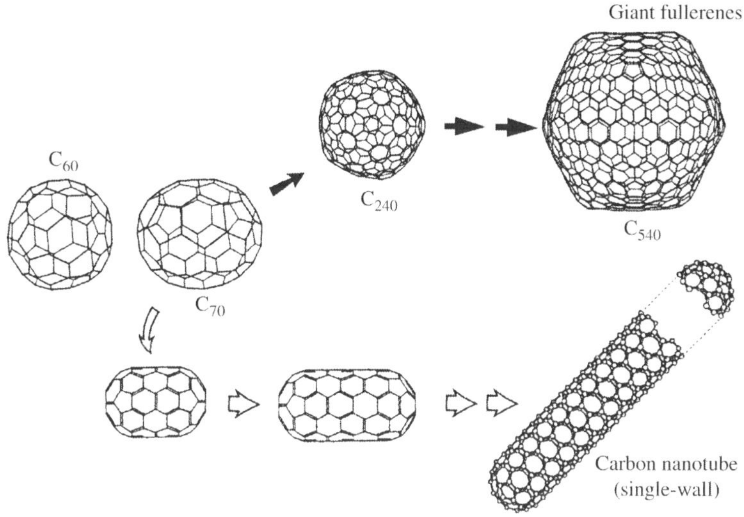


**FIGURE 2.4**

Buckminsterfullerene to either giant fullerenes or single-wall carbon nanotube.

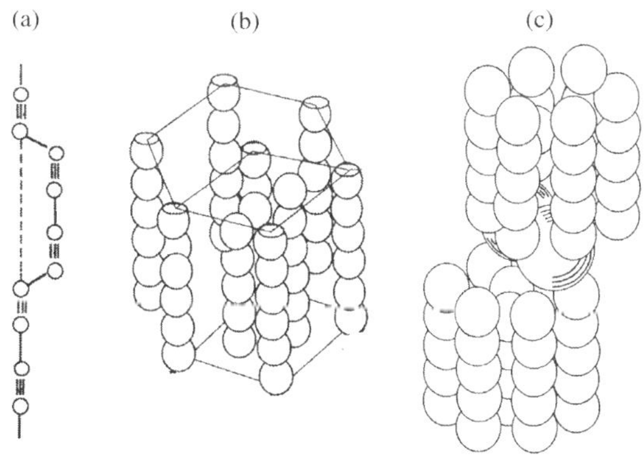


**FIGURE 2.5**

Structural models for carbines.

schematically in the lower line of Fig. 2.4. In this carbon family, a variety of structure is mainly due to the number of carbon atoms consisting of fullerene particles and relative location of 12 pentagons.

Carbynes have been supposed to be carbon atoms bound linearly by *sp* bond, where two $\pi$ electrons have to be resonated, giving two possibilities, i.e., an alternative repetition of single and triple bonds (polyyne) and a simple repetition of double bonds (cumulene) (Fig. 2.1). Its detailed structure is not yet clarified, but some of the proposed structural models are illustrated in Fig. 2.5, where some number of carbon atoms make a line with sp hybrid orbitals and these lines gather by van der Waals interaction between $\pi$-electron clouds to make a layer, and then these layers are stacked. Three models resemble each other and the last model (Fig. 2.5c), where foreign atoms are intercalated, seems to be the most realistic. In this carbyne family, the variety in structure is mainly due to the number of carbon atoms making a line, in other words, the thickness of layers consisting of linear carbon chains, and the density of chains in a layer.

The synthesis of graphdiyne has been reported, which is constructed by replacing one-third of the carbon–carbon bonds in graphite with two acetylenic linkages. Since graphdiyne is supposed to be a flat layer consisting of sp and $sp^2$ bonds, it might be able to be classified to a new carbon family after enough accumulation of experimental evidences of their presence and to be an alternative of graphene.

Each carbon family also shows various possibilities for accepting foreign atoms. Diversities in foreign atom acceptance are summarized in Fig. 2.2. The possibility to accept foreign atoms into diamond structure is restricted to the substitution of carbon atoms by either boron or nitrogen. The possibilities to accept foreign atoms into graphite structure are substitution of carbon atoms in fundamental hexagonal layers and the intercalation into the space between hexagonal layers (gallery). In the fullerene family, various possibilities to accept foreign atoms, i.e., insertion into either the interstices among fullerene particles or the inner space of a particle, and addition of atoms and radicals onto the surface of fullerene particle, in addition to the substitution of carbon atoms by either boron or nitrogen, as occurs in diamond and graphite families. In carbyne family, intercalation among layers of carbon chains, doping into the space between carbon chains in a layer and also substitution of carbon atoms in the chain are considered. The intercalation of either an iron or potassium atom, as shown in Fig. 2.5c, was reported to stabilize the carbyne structure.

### 2.1.3 Structural relation to neighboring atoms

Crystalline structures of the compounds of other atoms neighboring to carbon the Periodic Table, i.e., $B_4C$, $C_3N_4$, SiC, BN, are shown in Fig. 2.6, in relation to the structural modifications of carbon described above. The compound SiC has the same crystal structures as cubic and hexagonal diamond, which have to be called zincblende-type and wurtzite-type structures, respectively, because it is a binary compound of C and Si. From the comparison of these two structures of zincblende and wurtzite of ZnS (Fig. 2.7) to those cubic and hexagonal diamonds (Fig. 2.3), respectively, the similarity is clear; if all constituent atoms in zincblende and wurtzite structures were carbon, the former corresponds to cubic diamond structure and the latter to a hexagonal diamond one. This structural similarity is explained by the rule that an equi-number of electrons in the outermost orbital gives the same crystal structure, in the present case the total number of electron being eight (average four per atom, the same as carbon).

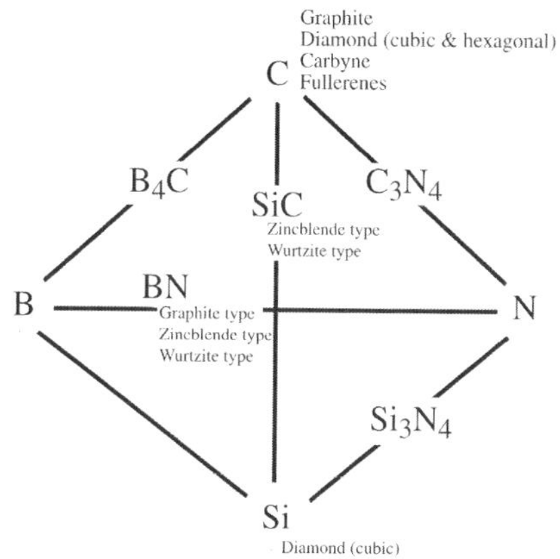


**FIGURE 2.6**

Structural relations to neighboring atoms.

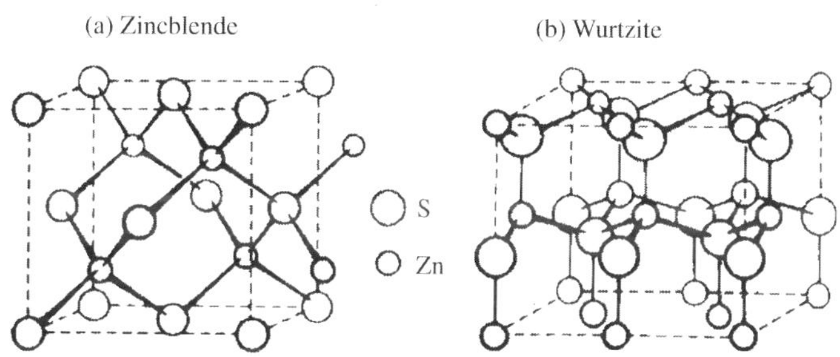


**FIGURE 2.7**

Zincblende and wurtzite type structures of ZnS.

BN, whose constituent atoms, B and N, locate in the neighboring III and V groups in the Periodic Table, respectively, can have a layered structure, similar to graphite, as shown in Fig. 2.8. It consists of a layer composed of hexagons of B and N atoms, a B atom in a layer is neighbored by N atoms in the upper and the lower layers, the stacking sequence being expressed by AA... in the reference to graphite structure as will be shown in Fig. 2.9. Therefore, no structural modification is possible in this layered structure, as hexagonal and rhombohedral modifications in graphite. This layered structure of BN can transform to either zincblende or wurtzite structure under high pressure, corresponding to cubic and hexagonal diamond structures, respectively. BN with zincblende type structure is a super hard material, which has an advantage for cutting iron because no carbon atoms are contained.

The compound between B and C, $B_4C$, is also one of the hard materials. The compound between C and N, $C_3N_4$, was theoretically predicted to be a super hard

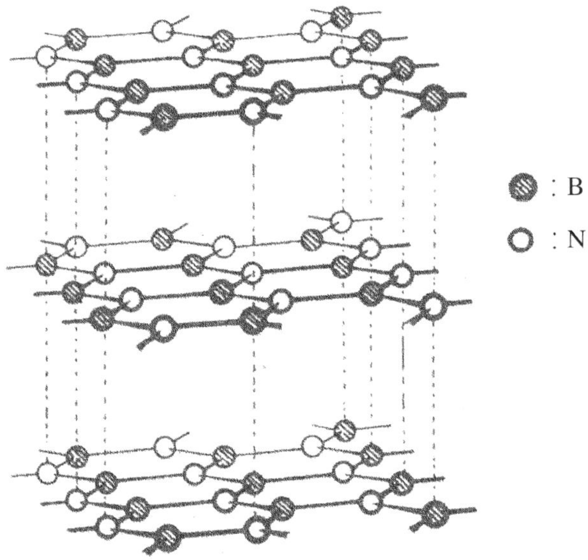


**FIGURE 2.8**

Boron nitride BN with layered structure.

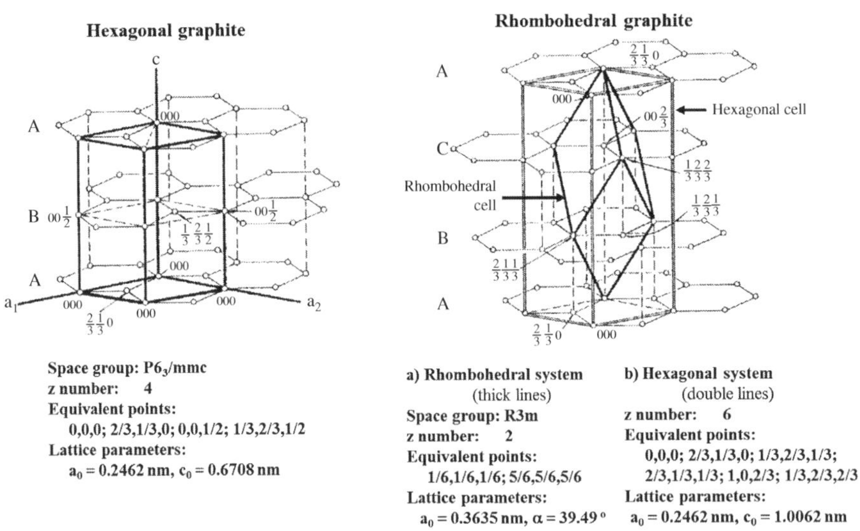


**FIGURE 2.9**

Crystal structures for two modifications of graphite.

material and has been studied on its synthesis. Atoms B and N are known to substitute carbon atoms in the structures of carbon materials. The compound between Si and N, $Si_3N_4$, attracted the attention of ceramists as a high-temperature structure material.

Silicon Si, which locates in the same group as carbon, but in the third row of the Periodic Table, cannot have layered structure, like graphite, and has cubic diamond structure. Such kind of structural anomalies, i.e., only carbon atoms can have layered structure graphite, has often been observed in the atoms belonging to the top row in each group of the Periodic Table.

## 2.2 Structure and texture of carbon materials

### 2.2.1 Structure

The fundamental unit of the structure of carbon materials in graphite family is a hexagonal carbon layer. Regular stacking of these layers give graphite crystal, hexagonal graphite with ABAB... stacking regularity [1,2] and rhombohedral graphite with ABCABC... regularity [3,4]. The unit cells and equivalent points for these two crystal modifications are shown in Fig. 2.9. Rhombohedral graphite is often expressed in a hexagonal system, because of easy comparison with hexagonal graphite, and so two unit cells in rhombohedral and hexagonal systems are shown in Fig. 2.9b, together with equivalent points for each unit cell, where thick lines indicate rhombohedral unit cell and double lines hexagonal unit cell. The distance between neighboring carbon atoms in the layer planes is 0.1412 nm and the spacing between neighboring layers is 0.3354 nm [5].

The structural relation between hexagonal and rhombohedral graphites is explained in Fig. 2.10. In hexagonal graphite, the second layer (denoted as B position) is displaced by (2/3, 1/3) along a1 and a2 axes from the first layer (denoted as A position). With further displacement by (1/3, 2/3) for the third layer, the total displacement becomes unity in two directions, that is, exactly the same position as the first layer A (Fig. 2.10a). Therefore, it is said that AB stacking belongs to the hexagonal crystal system. For the third layer, however, it is possible to displace again by (2/3, 1/3), which is the same displacement between A and B. The third layer is not in the same position as either the first layer A or the second layer B, and so is denoted as layer C. The repetition of the same displacement (2/3, 1/3) in the fourth layer coincides to the first layer A, giving ABC stacking, which belongs to the rhombohedral crystal system.

In addition to these regularly stacked structures, random stacking is possible in many carbon materials, which is called turbostratic structure, as shown in Fig. 2.10b [6]. This turbostratic stacking is often observed in the carbon materials prepared at relatively low temperatures as 1300–1500°C, where the size of layer is small and only a few layers are stacked in parallel. By heating these carbons to high temperatures up to 3000°C, both the size and the number of stacked layers

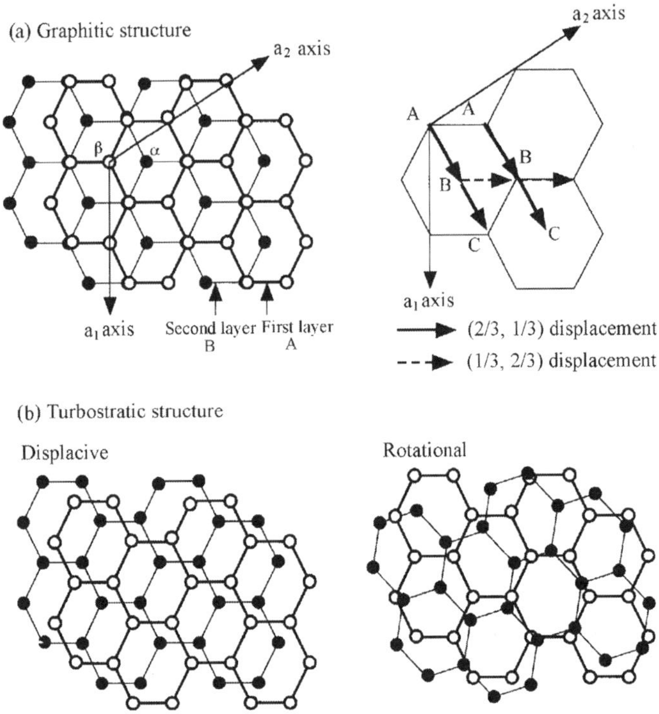


**FIGURE 2.10**

Graphitic and turbostratic structures shown by stacking of two hexagonal carbon layers.

usually increase and also the regularity in stacking is improved. The heat treatment at intermediate temperatures gives partial improvement in stacking regularity. As a consequence, a wide range of structure from completely turbostratic stacking to pure ABAB stacking is possible through the intermediates with variable ratios of these two stacking regularities. The formation of intermediate structure depends primarily on the starting materials (precursors) and the heat treatment temperature (HTT). In turbostratic structure, two kinds of displacement are possible, displacive and rotational, though it is difficult to differentiate them. Recently, the presence of displacive turbostratic stacking was shown through detailed analysis of scanning tunneling microscope (STM) images [7].

X-ray powder diffraction gives useful information to study the structure of carbon materials. In Fig. 2.11a, X-ray powder diffraction pattern for natural graphite with high crystallinity is shown. The diffraction lines for graphite have to be classified into three groups, the lines with 00*l*, *hk*0 and *hkl* indices, mainly because of the strong anisotropy in structure. The lines with 00*l* indices are due to

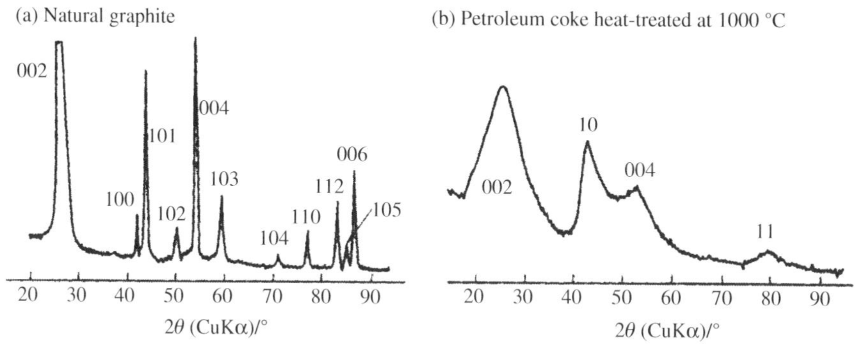


**FIGURE 2.11**

X-ray powder patterns for graphite and low-temperature-treated carbon.

the reflection from basal plane (hexagonal carbon layers), where only even index of *l* is allowed because of the extinction rule due to ABAB stacking sequence of layers. The lines with *hk*0 indices are due to the diffraction from the crystal planes perpendicular to the basal planes and the lines with *hkl* indices come from the planes declined to basal plane, where three-dimensional structure has to be established with graphitic stacking of layers. Therefore, *hkl* lines are called three-dimensional lines.

On the other hand, powder pattern which is given by low-temperature-treated carbons is quite different from natural graphite, as shown on a petroleum coke heat-treated up to 1000°C in Fig. 2.11b because it consists of turbostratic stacking of small layers. The diffraction pattern is characterized by very broad 00l lines, unsymmetrical hk lines, and no hkl lines due to random turbostratic stacking of a limited number of layers. The diffraction lines due to the planes perpendicular to layer planes are missing l index because of no regularity in the direction along the normal to layers, in other words, two-dimensional structure, and so expressed 10 and 11 in Fig. 2.11b.

The rhombohedral structure of graphite was shown to form by applying shearing force, such as grinding [8,9]. It gives additional diffraction lines, because of ABC stacking regularity, different from hexagonal graphite (AB stacking). In Fig. 2.12a, change in diffraction pattern around 44° in 2 with grinding is shown [10]. With increasing grinding time, additional diffraction lines appear because of the introduction of stacking faults, which can be indexed to be 101 and 102 lines based on rhombohedral structure in hexagonal system. The amount of rhombohedral stacking formed, Rh, was calculated from the intensity ratio of 101 line for rhombohedral to that for hexagonal modifications and plotted against the grinding time in Fig. 2.12b. The value of Rh saturates at around 33%, which seems to be reasonable because of the assumption that the rhombohedral stacking is formed due to the introduction of stacking fault in the crystallites with hexagonal stacking [11]. Further increase in grinding time results in increasing structural defects and

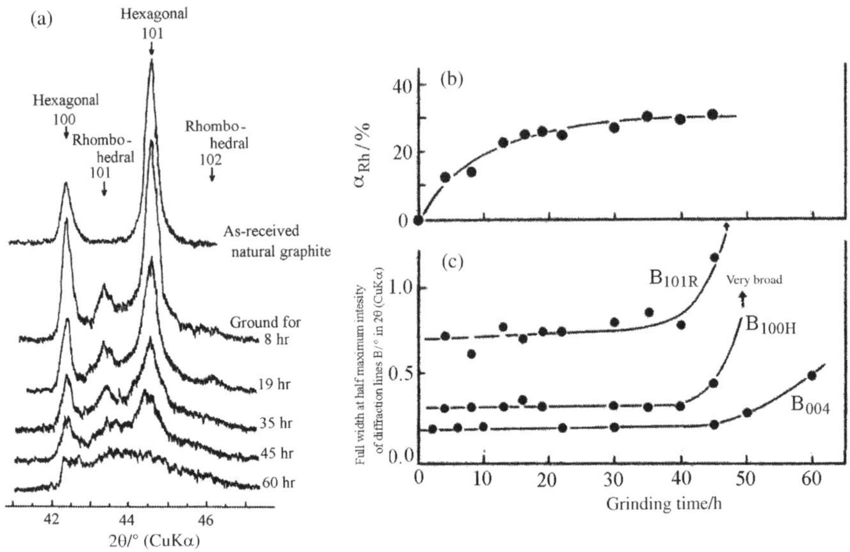


**FIGURE 2.12**

Formation of rhombohedral modification with grinding of natural graphite [10].

so all diffraction lines become broad, as shown by 004 line, together with 100 line for hexagonal modification and 101 line for rhombohedral one, in Fig. 2.12c. After 60 h grinding, 100 line becomes broad and unsymmetrical, and so the introduction of too much amount of stacking faults leads to turbostratic structure.

For graphite, hexagonal carbon layer looks triangular under scanning tunneling microscope (STM), as shown in Fig. 2.13a, in other words, only carbon atoms designated by $\alpha$ in Fig. 2.10a are detected under STM, because of the interaction with lower-lying atoms in the position $\beta$. In turbostratic structure, however, there is no interaction with lower-lying atoms, because of no regular stacking, and so hexagons of carbon layers are observed, as shown in Fig. 2.13b [7].

### 2.2.2 Structure development with heat treatment (carbonization and graphitization)

Structure in carbon materials is known to depend strongly on the temperature that they experienced. In order to get carbon materials, in which carbon atoms are principal constituents, we have to heat-treat some organic polymers, i.e., carbon precursors, in inert atmosphere. The process during the changes from organic precursors to inorganic carbon materials is composed of pyrolysis, cyclization, aromatization, polycondensation and carbonization of the organic precursor, and depends strongly on the carbon precursors and on the conditions of heat treatment (temperature, heating rate, atmosphere, etc.). In addition, these are usually

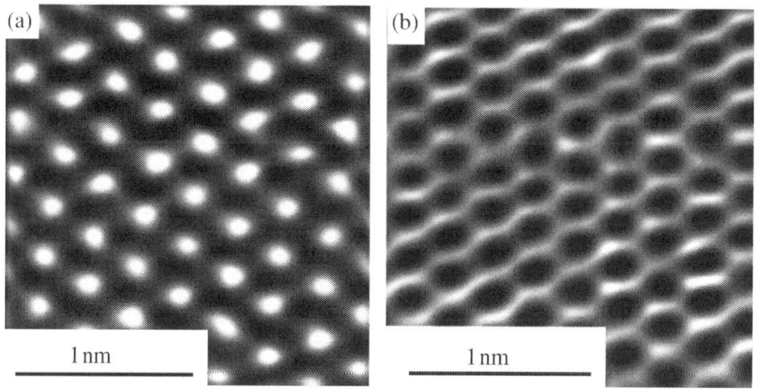


**FIGURE 2.13**

STM image of hexagonal carbon layers in graphitic and turbostratic stackings.

*(Courtesy by Prof. K. Oshida of Nagano Nat. Coll. Tech.)*

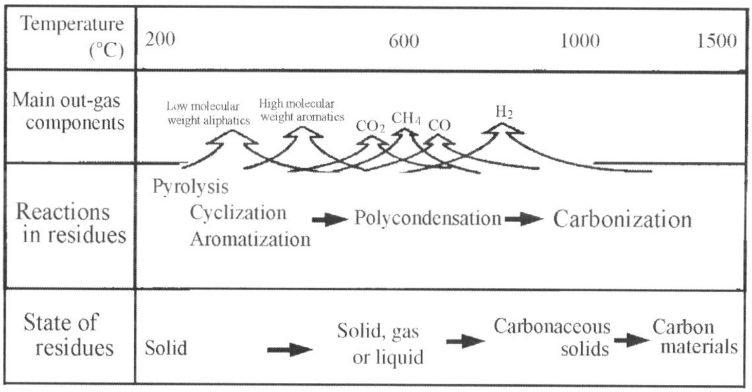


**FIGURE 2.14**

Carbonization process.

overlapped with each other in most carbon precursors. Therefore, often we call the whole process from the precursor to the carbon material '**carbonization**'. The general scheme of the carbonization process is shown by indicating main out-gas components and changes in residues in Fig. 2.14, although it strongly depends on the carbon precursors.

In the beginning of pyrolysis, aliphatic molecules with low molecular weights and then low-molecular-weight aromatics are released as gases, mainly because some of C—C bonds in organic molecules are weaker than C—H bonds. Cyclization and aromatization proceeded in the residues, associated with the release of low-molecular-weight hydrocarbons, and then polycondensation of aromatic molecules occurs. Around 600°C, mainly foreign atoms such as oxygen and nitrogen are released as $CO_2$, CO and $(CN)_2$, together with methane $CH_4$. In this stage the residues are either gas, solid or liquid phases depending on the starting precursors. Above 1000°C, out-gas is mainly $H_2$ because of polycondensation of aromatics and the residues may be called carbonaceous solids, which still contain hydrogen. Above 1300°C, almost all foreign atoms, mainly hydrogen, go out and the residual solids are carbon materials. The details of carbonization process of various precursors are discussed in the Section 2.3.

In this carbonization process, a scheme of preferred orientation of fundamental structural units, i.e., hexagonal carbon layers, is established, which is nanotexture explained in Section 2.2.3. Therefore, this carbonization process is the most important for the preparation of carbon materials. However, carbon layers are still small and in most cases are stacking randomly (turbostratic structure). In order to make these carbon layers grow and stack with graphitic regularity, heat treatment at a high temperature above 2500°C is required. In some cases, graphitic structure develops, and so this process has been called '**graphitization**'. However, it does not occur always, the development of graphitic structure depending strongly on the nanotexture formed during the carbonization process.

During the carbonization and graphitization processes, the structure changes from starting organic precursor through carbon to graphite and, as a consequence, the properties of the materials change markedly. In Fig. 2.15, the change in energy band structure from an aromatic hydrocarbon through coke to polycrystalline graphite is schematically shown as a function of HTT [12,13]. Starting aromatic hydrocarbon is a molecular crystal with a wide band gap $\Delta E$, it being an insulator. The carriers for electrical conduction in carbons heat-treated at low temperatures below 700°C are supposed to be electrons, of which mobility is rather low. Above 1300°C, hydrogen atoms remained in the carbon starts to depart and leave electron traps, most of which may locate at the edge of the hexagonal carbon layer. As a consequence, the Fermi level is lowered and the concentration of positive holes increases which leads to the pronounced changes in

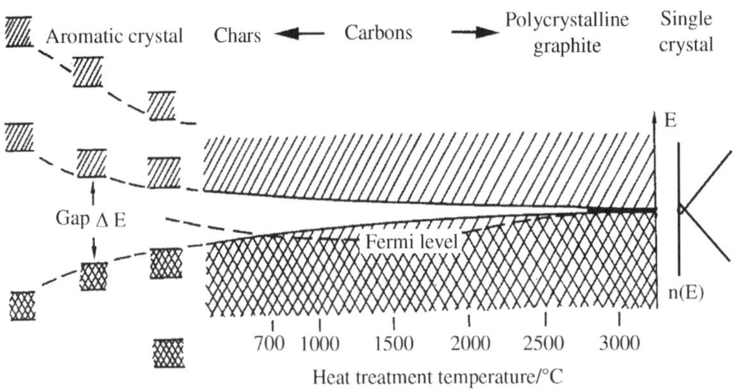


**FIGURE 2.15**

Scheme of changing in energy band structure with HTT [12].

electromagnetic properties of carbon materials. Around 2000°C, Fermi level starts to rise, because of healing of electron traps by high-temperature treatment. Also the band gap between the valence and conduction bands becomes small with the increase in HTT, because of the growth of crystallites (layer size along a-axis and parallel stacking along c-axis) and improvement of stacking regularity, and so the relative concentration of electron increases. In the carbons with highly developed graphite structure, the valence and conduction bands overlap slightly and the concentrations of two carriers, negative electrons and positive holes, are comparable with each other, just as in graphite crystal. Even though the values of electromagnetic parameters approach those of graphite, it cannot reach the exact values measured on a single crystal of graphite.

### 2.2.3 Nanotexture

In the graphite family, the fundamental structural unit is a layer of carbon hexagons, which has a strong anisotropy because the bonds in the layer are covalent and those between the layers are van der Waals-like. As a consequence, these anisotropic layers tend to be oriented during their agglomeration to form certain morphologies as carbon solids. Therefore, the way to agglomerate gives different schemes and different degrees of preferred orientation of layers, and results in a variety of carbon materials, in addition to the mixing ratio of ABAB and turbostratic stacking.

A classification based upon the scheme of preferred orientation of anisotropic layers and its degree has been proposed, as illustrated in Fig. 2.16 [14,15]. Since these are the textures constructed by fundamental structure units in nano-size, they are called **nanotextures**.

Firstly, random and oriented nanotextures are differentiated and then the latter is classified by the scheme of orientation, parallel to the reference plane (planar orientation), along the reference axis (axial orientation) and around the reference point (point orientation).

The extreme case of planar orientation, i.e., perfect orientation with large-sized planes, is a single crystal of graphite. The plates of so-called highly oriented pyrolytic graphite (HOPG) have a very high degree of planar orientation of hexagonal carbon layers, but the size of layers is not so large, in other words, in the direction perpendicular to the plate the structure is close to perfect orientation but in the parallel direction it is polycrystalline. Various intermediates between perfect planar orientation and random orientation are found in pyrolytic carbons and coke particles depending on the preparation conditions and the temperature of heat treatment. Various highly oriented graphite materials, kish graphite, various pyrolytic carbons and carbon films derived from some precursor polymers, were developed (see Section 3.2). In cokes the improvement in planar orientation of layers with HTT was demonstrated through high-resolution transmission electron microscopy (HR-TEM). In Fig. 2.17, the change in 002 lattice fringes observed by HR-TEM with HTT is shown [16].

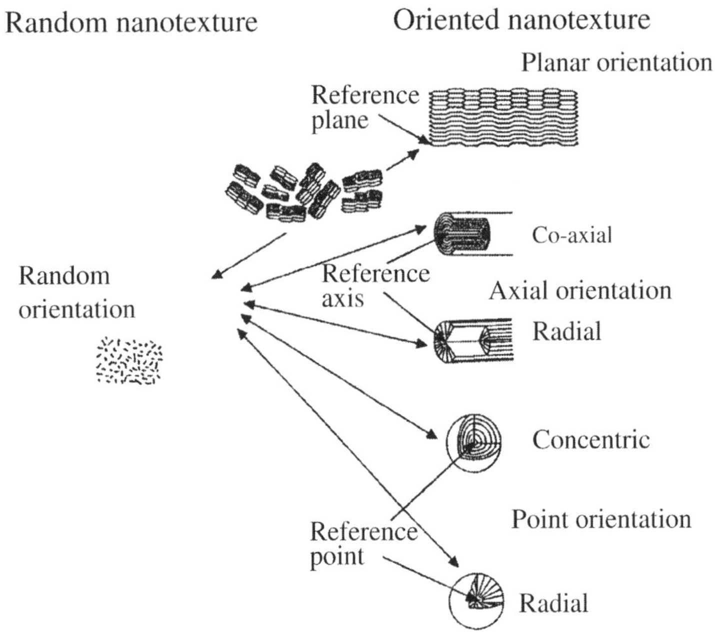


**FIGURE 2.16**

Classification of nanotextures in carbon materials [14].

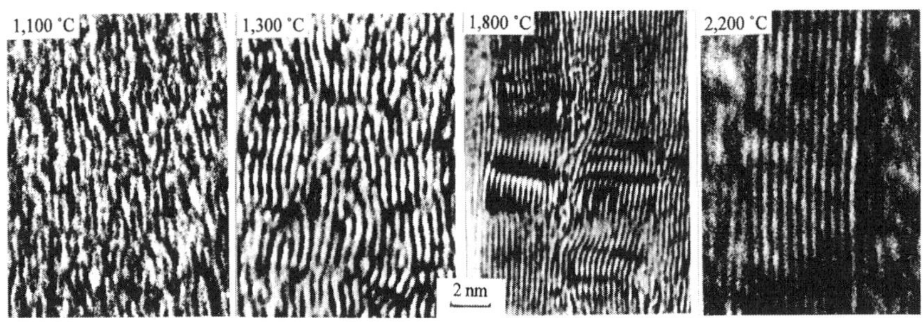


**FIGURE 2.17**

Changes in 002 lattice fringes of a coke with HTT.

*(Courtesy of Mme. A. Oberlin.)*

Axial orientation of layers is found in various fibrous carbon materials, in other words, fibrous morphology of carbon materials is possible because of this axial orientation scheme. In this orientation scheme, the co-axial and radial alignments of layers relative to the reference axis are possible. The co-axial alignment of layers is realized in multi-walled carbon nanotubes and some carbon fibers, typical axial orientation being formed in vapor-grown carbon fibers. The radial alignment of layers can be found in one of mesophase-pitch-based carbon fibers,

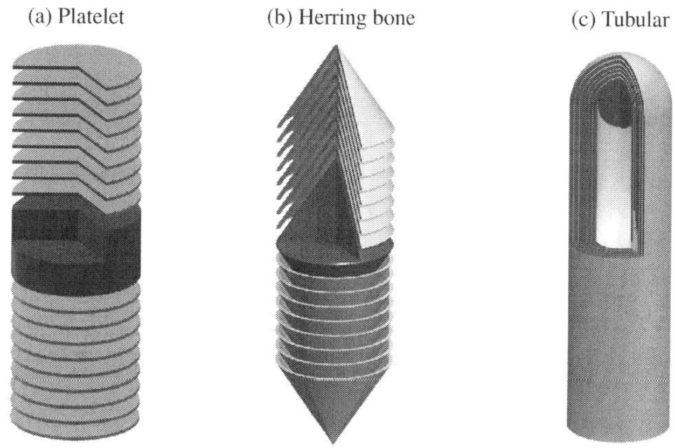


**FIGURE 2.18**

Schematic illustration of nanotextures in carbon nanofibers.

having radial texture in the cross-section of the fibers. In carbon nanofibers, most of which are grown through a chemical vapor deposition process using fine catalyst particles, different orientation schemes along their axis are formed, from parallel (tubular type) to perpendicular (platelet type) through herring-bone type or cup-stacked type, as shown in Fig. 2.18. These fibrous carbon materials will be discussed in more detail in Section 3.4.

In point orientation, concentric and radial alignments have also to be differentiated. The extremes of concentric point orientation are a family of fullerenes, as shown in Fig. 2.4. It can also be found in carbon black particles formed by minute hexagonal carbon layers. Carbon blacks are industrially produced in a large amount, of which sizes are from few tens to a few hundred nanometers, in which minute-layer planes are preferentially oriented along the surface of spherical particles, as shown in Fig. 2.19 by 002 lattice fringe image and shown schematically in Fig. 2.20a. Radial alignment of layers to form spheres is found in the carbon spherules, which are formed from a mixture of poly(ethylene) and poly(vinyl chloride) by pressure carbonization [17], as will be explained in Section 2.4.3.

The structure of mesophase sphere is close to radial point orientation scheme, but in its center the orientation of layers is not radial [18,19]. In Fig. 2.20, the models of nanotexture for carbon black particle [20], mesophase sphere [18] and carbon spherule [21] are compared.

The texture with random orientation occurs in glass-like carbons and carbon materials just after carbonization of precursor polymers (e.g., phenol resin). Fundamental structural units composed of most glass-like carbons are so small as to be difficult to observe under TEM. Therefore, the discussion on their nanotexture was often performed on the basis of the TEM observations on their high-temperature-treated ones, where layers became somewhat larger. Three nanotexture models shown in Fig. 2.21 have been proposed [22–24]. In the shell

**FIGURE 2.19**

002 lattice fringes in the particles of carbon black.

*(Courtesy of Dr. T. Nakata of Tokai Carbon Co., Ltd.)*

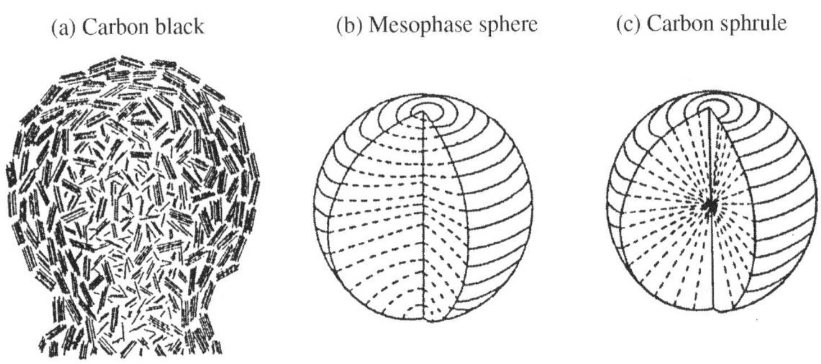


**FIGURE 2.20**

Nanotexture models for spherical carbon materials.

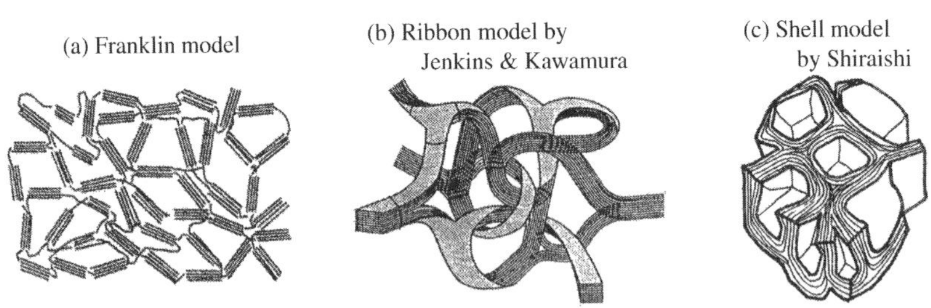


**FIGURE 2.21**

Nanotexture models of glass-like carbon heat-treated at a high temperature.

model [24] (Fig. 2.21), hexagonal layers are locally oriented in concentric scheme to make closed pores. By taking into consideration the presence of a large amount of closed micropores and gas impermeability of glass-like carbons, this model is believed to be realistic, which will be discussed in more detail in Section 3.3.

### 2.2.4 Microtexture (agglomeration)

Most particles with planar and axial orientation, for example, coke particles and fibers, are also anisotropic and so their agglomeration can create further variety in texture. Therefore, it is necessary that the texture formed by the preferred orientation of the constituent particles, particles themselves having anisotropy, has to be taken into consideration and differentiated from nanotexture discussed in the previous section. The texture due to the preferred orientation of anisotropic particles may be called '**microtexture**', because the particles are often in micrometer or millimeter sizes. This microtexture is usually formed during the forming process of bulky carbon materials. For example, in large-sized graphite electrodes for metal refining (Fig. 1.2a) the particles of needle-like coke tend to be oriented along the rod axis during their forming process through extrusion with binder pitch (more detailed explanation will be presented in Section 3.1). To prepare the carbon-fiber-reinforced plastics for fishing rods, for example, constituent carbon fibers have to be oriented along the rod axis in order to get high strength and high Yong's modulus (more details are presented in Section 3.6.

## 2.3 Carbonization (nanotexture development)

### 2.3.1 Formation processes of carbon materials

As discussed above (Section 2.2.2 and Fig. 2.14), the precursors are carbonized to solid carbon materials by heat treatment up to a temperature around 1000–1500°C. During this carbonization process, the precursor passes through either gas, liquid or solid phase. In Table 2.1, carbonization processes are classified on the basis of the intermediate phases and the representative carbon materials formed through the process are listed. The products of carbonization and carbons, are often classified into two categories, graphitizing and non-graphitizing, which means whether the carbon can be converted to graphite through the treatment at high temperature under atmospheric pressure. These terms, graphitizing and non-graphitizing (graphitizable and non-graphitizable or soft and hard are also used), are convenient to predict graphitizability of the carbon, so that they are frequently used to show the characteristics of carbon materials. However, it has to be taken into consideration that carbon materials cannot be divided clearly into these two categories, many carbons showing intermediate behavior, as shown in Section 2.4.

From hydrocarbon gases, different carbon materials have been produced, this process being **gas phase carbonization** (Section 2.3.2), because hydrocarbon

**Table 2.1** Carbonization Processes for the Production of Various Carbon Materials

| Carbonization Process | Precursors | Carbon Materials | Characteristics |
|---|---|---|---|
| Gas phase carbonization | Hydrocarbon gases decomposition in space | Carbon blacks | Fine particles, so-called 'structure' |
| | Hydrocarbon gases, deposition on substrate | Pyrolytic carbons | Various textures, preferred orientation |
| | Hydrocarbon gases, with metal catalysts | Vapor-grown carbon fibers Carbon nanofibers | Fibrous morphology, various nanotextures |
| | Carbon vapor | Carbon nanotubes | Tubular, single-wall & multi-walled |
| | Carbon vapor | Fullerenes | Spherical, molecular nature |
| | Hydrocarbon gases | Diamond-like carbon | Thin film, $sp^3$ bond, amorphous structure |
| Solid phase carbonization | Plants, coals & pitches | Activated carbons | Highly porous adsorptivity |
| | Furfuryl alcohol, phenol resin, cellulose, etc. | Glass-like carbon | Amorphous structure, gas impermeability, conchoidal fracture surface |
| | Poly(acrylonitrile), pitch, cellulose & phenol resin with stabilization | Carbon fibers | Fibrous morphology, high mechanical properties |
| Liquid phase carbonization | Pitch, coal tar | Cokes | |
| | Cokes with binder pitches | Graphite blocks | Various densities, various degrees of orientation |

molecules decompose in gaseous phase to deposit as a solid carbon. Depending mostly on the concentration of hydrocarbon gases, various carbon materials are produced: carbon blacks, pyrolytic carbons, carbon fibers and nanofibers, carbon nanotubes and fullerenes.

On the other hand, most of thermosetting resins, such as phenol, furfuryl alcohol, and celluloses, can be converted to carbon materials without any marked changes in morphology, **solid phase carbonization** (Section 2.3.3). When the carbonization of most of these precursors proceeds rapidly, the resultant carbon materials are porous. If the carbonization is performed so slowly that the resultant carbonaceous solids could shrink, so-called glass-like carbons are produced, which contain a large amount of closed pores.

Polycrystalline graphite blocks (one of the classic carbons; Section 3.1) are industrially produced by forming the coke particles using a pitch as a binder and then carbonizing and graphitizing at high temperatures. The cokes are also produced from pitches. In the case of most pitches, their carbonization process is classified into **liquid phase carbonization** (Section 2.3.4), because all of them transform to viscous liquid during heat treatment to be carbonized even though they are solids at room temperature. To produce graphite blocks, the coke particles are used as fillers in order to avoid a large amount of volatiles, which may introduce some shape distortion and crack in the products, and also may reduce the density. Since these coke particles are not sintered to form blocks, a binder has to be used, pitches having been often selected, which give similar carbon residues after carbonization with relatively high yield. Therefore, the production of graphite blocks is based on liquid phase carbonization in two steps; in the process of the filler coke preparation and in the carbonization of the binder pitch in the formed blocks.

The carbonization process is the most important process for the production of carbon materials because the nanotexture of carbon materials, which has the determining effect on structural change at high temperatures and consequently on their properties, is principally established during this process, as explained below. The carbons in which graphitic structure develops markedly at high temperatures above 2500°C has been called 'graphitizing carbons' and those showing no appreciable development of graphitic structure called 'non-graphitizing carbons', though many carbons have intermediate behavior between graphitizing and non-graphitizing. Graphitization behavior of various carbons is discussed in Section 2.5.

## 2.3.2 Gas phase carbonization

The carbon materials produced by gas phase carbonization depend strongly on the concentration of the precursor hydrocarbon gases. Under a high concentration of the precursors, various carbon blacks, which is one of classic carbons and have the nanotexture of point orientation, are produced in an industrial scale. Under a low precursor concentration with some solid substrates, however, pyrolytic carbons are formed, which is one of new carbons developed in the beginning of Period II of carbon history (Table 1.1) and have the nanotexture of planar orientation. Some fine metallic particles have been found to catalyze the carbonization of hydrocarbon gases to give the fibrous carbon materials, nanofibers, which may be classified as nano-sized carbons and have the axial orientation nanotexture. Some of these nanofibers can be converted to carbon nanotubes by a high-temperature treatment. So-called vapor-grown carbon fibers were developed in the beginning of Period II (Table 1.1) and have different characteristics in comparison with other carbon fibers, such as PAN-based carbon fibers. After the finding of carbon nanotubes, it was recognized that the center of these vapor-grown carbon fibers consists of carbon nanotubes. Through carbon vapor which is formed by arc discharging between carbon electrodes, carbon nanotubes were found to be formed.

### a. Carbon blacks

Carbon blacks are formed through incomplete combustion of either gaseous or mist-like hydrocarbons and are very important industrial products [25]. They were used as a colorant in inks in the third century AD and are used now for reinforcement of rubbers in large amounts. Carbon blacks are classified on the bases of reaction process and raw materials into furnace blacks, channel blacks, lamp blacks, thermal blacks and acetylene blacks, as summarized in Table 2.2. They are characterized by spherical primary particles, which are in different sizes and coalesced into aggregates (secondary particles) more or less, as shown by transmission electron microscope (TEM) images in Fig. 2.22. The aggregation of primary particles into the secondary particles is very important for the performance of carbon blacks, particularly for the reinforcement of rubbers and is often called '*structure*' in industry.

The mechanism of the formation of carbon black particles was discussed in detail in the literature [26]. The formation of oil droplets due to pyrolysis and

**Table 2.2** Production Methods and Raw Materials of Carbon Blacks

| Reaction Process | Carbon Black Formed | Main Raw Materials |
|---|---|---|
| Incomplete combustion | Oil furnace black | Creosote oil |
| | Gas furnace black | Natural gas |
| | Channel black | Natural gas |
| | Lamp black | Heavy oils |
| Pyrolysis | Thermal black | Natural gas |
| | Acetylene black | Acetylene |

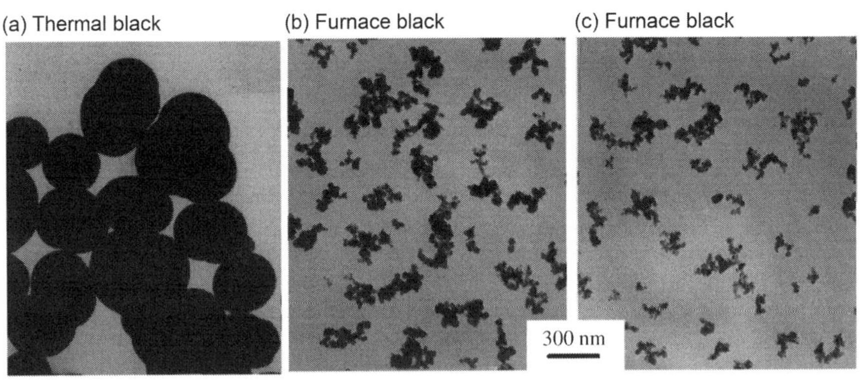


**FIGURE 2.22**

TEM images of carbon blacks.

*(Courtesy of Dr. T. Nakata of Tokai Carbon co. Ltd.)*

polymerization/condensation of organic molecules in the space of the furnace seems to be important in order to interpret *structure*. However, it is very difficult to determine the mechanism because of the formation reaction in a millisecond at a high temperature such as 1300–1500°C.

Furnace blacks are produced from either mists of creosote oil or natural gas consisting mainly of methane by their incomplete combustion in the furnace. Their yield and quality are predominantly governed by the temperature of combustion (usually 1260–1420°C), which is regulated by the mixing ratio of hydrocarbon gas with air and their turbulence conditions. The size of primary particles has a rather broad distribution, the carbon black with the smaller particles having the narrower size distribution, as shown in Fig. 2.23. The aggregation of these primary particles, 'structure', is particularly developed in furnace blacks, this being the main reason why they have been used to reinforcement of rubbers. Fig. 2.24a shows a TEM image of 002 lattice fringes for the coalesced part between primary particles of a furnace black. The concentric point orientation of hexagonal carbon layers is clearly observed on the surface of the particles, but at the center the layers look randomly oriented. Even at the neck between two primary particles, the orientation of layers along the surface is observed. This *structure* of carbon blacks was evaluated by measuring the absorption of either linseed oil or dibutylphthalate (DBP) and also by analyzing their TEM images.

Channel blacks are formed by impingement of natural gas flames on channel irons. Lamp blacks are manufactured by burning aromatic oils in shallow, open pans with limited air supply, which have been used mainly for inks.

Thermal blacks are produced by thermal decomposition of natural gas, while acetylene blacks are formed through an exothermic decomposition of acetylene gas at a relatively high temperature such as 2400°C. In thermal blacks, the size of

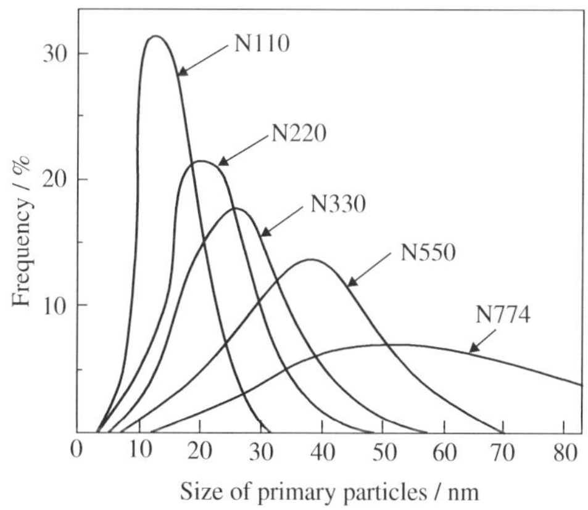


**FIGURE 2.23**

Size distribution of primary particles of furnace blacks.

*(Courtesy of Dr. T. Nakata of Tokai Carbon Co., Ltd.)*

their primary particles is usually large, as shown in Fig. 2.22a and almost no aggregation is observed, i.e., no structure, as shown by TEM image of 002 lattice fringes in Fig. 2.24b. In most of the large particles of thermal blacks, concentric point orientation of hexagonal carbon layers is observed clearly, even at the center of the particles.

Acetylene blacks are produced by pyrolysis of acetylene gas, which have high structure, as shown in Fig. 2.25a. By using similar conditions to furnace blacks, carbon blacks with developed *structure* as comparable with acetylene black (named Ketjenblack), as shown in Fig. 2.25b, are produced [27]. They are often used as electric conductive additives for the electrodes of lithium-ion batteries and various electrochemical capacitors.

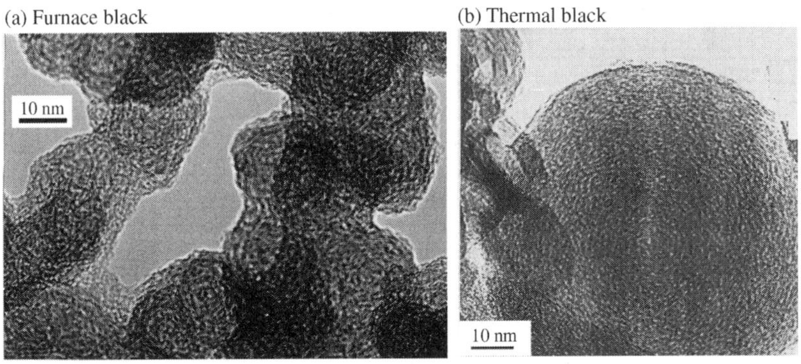


**FIGURE 2.24**

TEM image of 002 lattice fringes of furnace and thermal blacks.

*(Courtesy of Dr. T. Nakata of Tokai Carbon Co., Ltd.)*

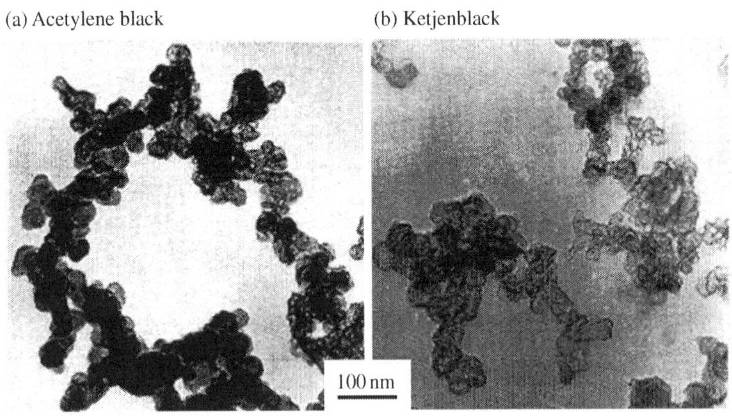


**FIGURE 2.25**

TEM images of acetylene black and Ketjenblack [27].

## b. Pyrolytic carbons

If a substrate is inserted into the combustion flame of hydrocarbon gas, carbon deposition occurs on the surface of the substrate. In order to control the structure of the carbon deposits formed on the substrate, the deposition conditions have to be controlled. Methane and propane have usually been employed as precursor gases. This process is a kind of chemical vapor deposition (CVD) and the products are called pyrolytic carbons.

Practically, the deposition of pyrolytic carbons has been performed on both static and dynamic substrate; in the former case the substrate was placed in the furnace, which is heated either by direct passing of electric currents (cold-wall type apparatus) or from the surroundings (hot-wall type apparatus), and in the latter case the small substrate particles are fluidized in the furnace. When the static substrate is heated by passing electric current through itself, the deposition temperature, in other words the temperature of the surface of deposited carbon, changes with increasing thickness of pyrolytic carbon formed. If the temperature of the surface of the substrate is controlled to be constant, the temperature of the inside becomes higher and even compressive stress arises, which modifies the structure of formed pyrolytic carbons. The factors controlling the structure and properties of pyrolytic carbons are (1) precursor hydrocarbon gas and its concentration, (2) deposition temperature, (3) contacted time between hydrocarbon molecule and substrate heated at a high temperature, which depends strongly on flow rate of gas and size of the furnace, and (4) the geometrical arrangement of the furnace, particularly the physical surface area of the substrate relative to the volume occupied by decomposition gas [28,29]. When the static substrate is heated by passing the electric current, not only the surface of the substrate but also the surrounding space has to reach a temperature enough for the decomposition of hydrocarbon precursor gases (hot sheath). The size of this hot sheath depends strongly on the flowing rate of gases.

In Fig. 2.26, bulk density of pyrolytic carbons prepared from propane and methane is plotted against deposition temperature [30]. Both methane and propane

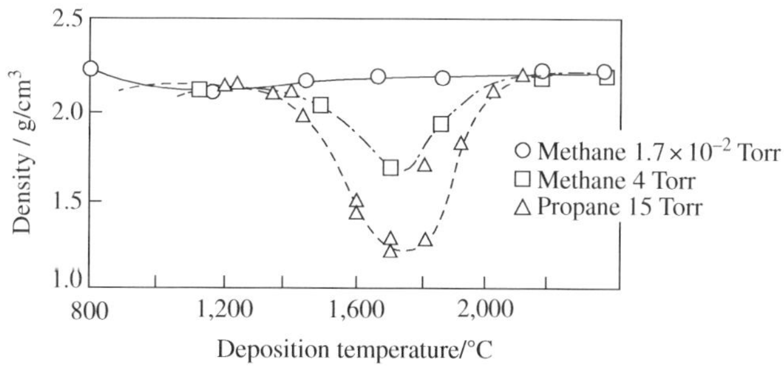


**FIGURE 2.26**

Dependences of bulk density of pyrolytic carbon on deposition temperature as a function of precursor gas and its pressure [30].

gases give relatively low bulk density at the temperature of 1400–2000°C, which is mainly due to the formation of carbon black particles, which are incorporated into the pyrolytic carbon deposited. In order to prepare pyrolytic carbon with high density in this temperature range, the pressure of precursor gas, methane in Fig. 2.26, has to be very low. Below 1400°C and above 2000°C, bulk density does not depend on the pressure of precursor gas, but these temperature ranges are not practical, because the deposition rate is very low, below 1400°C, and high energy consumption and special materials for the furnace construction are required above 2000°C.

In Fig. 2.27, optical micrographs of two different textures are shown on the cross-sections perpendicular to the deposition plane of the pyrolytic carbon, consisting of so-called growth cones [31]. In Fig. 2.27a, growth cones seem to start at the surface of substrate and grow larger with the proceeding of the deposition, which has been called 'singularly nucleated'. The nucleation points for each growth cone are supposed to be due to the roughness of the substrate surface. In Fig. 2.27b, however, growth cones are small and thin, and look to be generated even in the course of the deposition, which has been called 'regeneratively nucleated' or 'continuously nucleated'. In this texture, the nucleation was supposed to occur from the soot-like small particles, which are formed in the gas phase during deposition. The pyrolytic carbons with such regeneratively nucleated texture usually give low bulk density. The decrease in bulk density at certain deposition temperature with relatively high concentration of hydrocarbons, as shown in Fig. 2.26, is mainly due to this co-deposition of soot-like particles into pyrolytic carbon with regeneratively nucleated texture. Reduction of partial pressure of the precursor hydrocarbons is experimentally shown to be effective to suppress the formation of these soot-like particles, resulting in a high bulk density and the change in optical texture to a singularly nucleated one [30], and also effective to shift the temperature giving density minimum to lower [31].

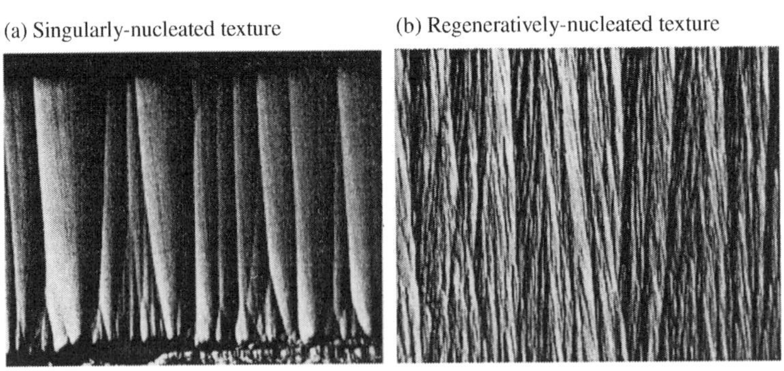


**FIGURE 2.27**

Optical micrographs of the cross-section perpendicular to the deposition plane of pyrolytic carbons [31].

Various optical textures under polarized light were observed on the surface parallel to the deposition plane, i.e., perpendicular to the growth cones. They are usually classified into four groups, granular, columnar, laminar and isotropic. Granular texture corresponds to regeneratively nucleated texture in the cross-section and columnar to singularly nucleated.

Similar optical textures were observed on the pyrolytic carbons deposited in the spaces among carbon fibers to produce carbon fiber/carbon composites (carbon/carbon composites) [32,33] and also on those deposited on small particles of uranium oxides in fluidized bed for the preparation of fuel particles for nuclear reactors [34]. Three optical textures observed in carbon fiber/carbon composites are shown in Fig. 2.28. The textures classified into columnar and laminar are further divided by the size of the regions, which show homogeneous optical nature, such as rough and smooth columnar. Optical texture of the pyrolytic carbons formed in carbon/carbon composites is known to be governed predominantly by the conditions of the deposition, geometry of deposition furnace, area of deposition surface of the substrate, deposition temperature, gas used, gas pressure, flowing rate of gas, etc. [34]. Preparation and properties of carbon/carbon composites are discussed in Section 3.6.2.

This deposition process was successfully applied to the fine particles flowing in a gas stream, in order to produce the carbon-coated nuclear fuel particles of uranium compounds [28]. One example of construction of the cross-section of the particle is illustrated in Fig. 2.29, where the carbon layers with well-oriented and random texture are formed on purpose by controlling deposition conditions. It was successfully employed also to the carbon coating of the substrate with complicated shape; a marked example is a heart valve [29].

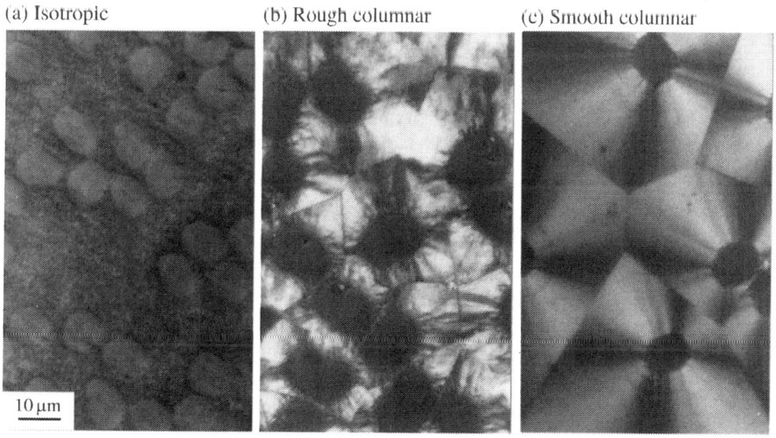


**FIGURE 2.28**

Optical textures of pyrolytic carbon deposited among the fibers.

*(Courtesy of Prof. E. Yasuda of Tokyo Inst. Tech.)*

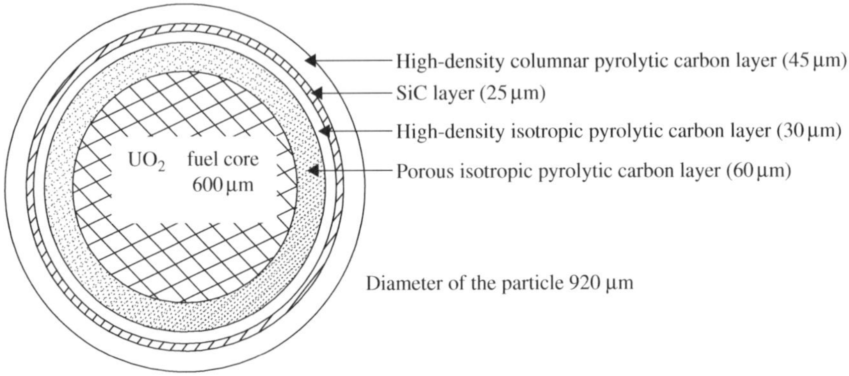


**FIGURE 2.29**

Construction of nuclear fuel particle coated by pyrolytic carbon layers with different optical textures [28].

### c. Vapor-grown carbon fibers and nanofibers

Vapor-grown carbon fibers (VGCFs) were deposited on a substrate from benzene vapor in a flow of high-purity hydrogen gas using a catalyst of fine particle of iron [35–37]. The formation process of the fibers was studied in detail [38].

The growing process of VGCF is divided into three steps, as shown in Fig. 2.30. The first step is the nucleation period, corresponding to the temperature of the furnace of 950–1000°C, where iron compounds (oxides or metal) are reduced to form fine iron particles and their surfaces are covered by carbon produced from benzene vapor. The appearance of the surface of the substrate at the first step is shown in Fig. 2.31a. In this step, the embryo for the carbon fibers is formed, in which fine iron particle is embedded.

In the second step, the length of very thin fibers increases rapidly (elongation step). They are so thin that they look transparent under scanning electron microscope (SEM), as shown in Fig. 2.31b. The detailed TEM observation using different techniques, bright-field, different dark-fields, lattice fringe images and selected-area electron diffraction, showed that in the fibers formed in this step the hollow tubes consisting of a limited number of carbon layers are formed, at the top of which a very fine iron particle is always included [39]. TEM image of the tubes formed in this step is shown in Fig. 2.32a, where even the tube consisting of a single layer (single wall carbon nanotube) is observed. This TEM image also suggests that the deposition of carbon consisting of small layers occurs on the surface of the thin tube.

The third step is the step to increase the thickness of the fiber (thickening step), as shown in Fig. 2.31c. The high-resolution TEM observation shows that the resultant VGCF composes of the hollow tube and oriented small layers (Fig. 2.32b). The detailed analysis shows that the wall of hollow tube consists of few straight layers and that small layers preferentially orient parallel to the tube surface, as

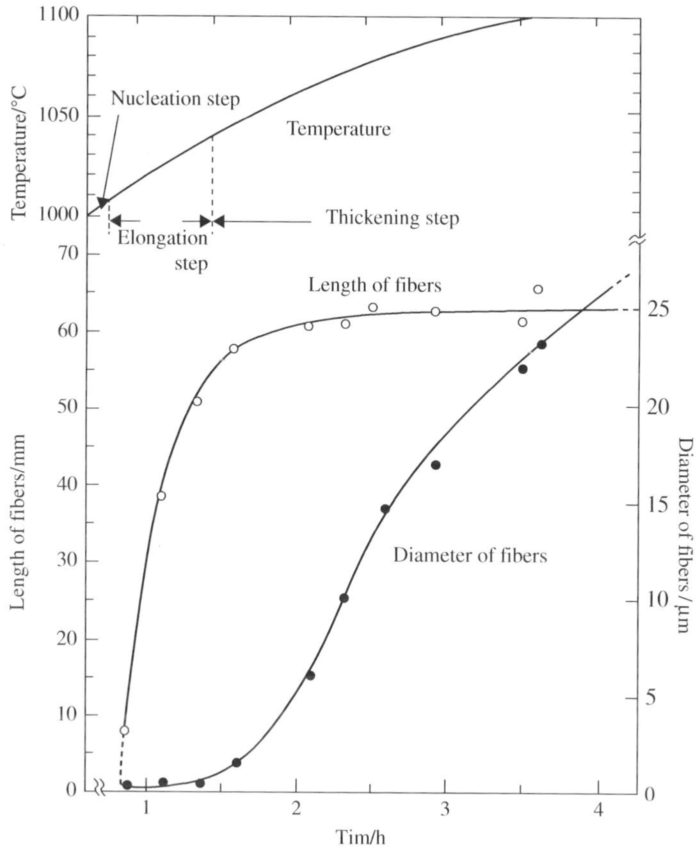


**FIGURE 2.30**

Processes of the growth of vapor-grown carbon fibers.

*(Courtesy of Prof. M. Endo of Shinshu Univ.)*

shown in Fig. 2.33 [39]. On the part of hollow tube wall, ten lattice fringes are observed, because layers are located perpendicular to the electron beams in microscope and are aligning along the fiber axis. On the other hand, layers deposited on the hollow tube wall are small, but preferentially orient along the fiber axis.

The formation of the same VGCFs was reported by using different precursor hydrocarbon gases, carrier gases, catalyst metals and deposition conditions [40–44]. In order to get a high yield of VGCFs, floating catalyst method was developed, instead of the method where the catalysts were seeded on the surface of substrate (seeding catalyst method) [45] (refer to Section 3.4.2.d).

The formation mechanism of VGCFs was also discussed in various references [38,39,46]. In Fig. 2.34, growth models for two methods, seeding and floating catalyst methods, are shown.

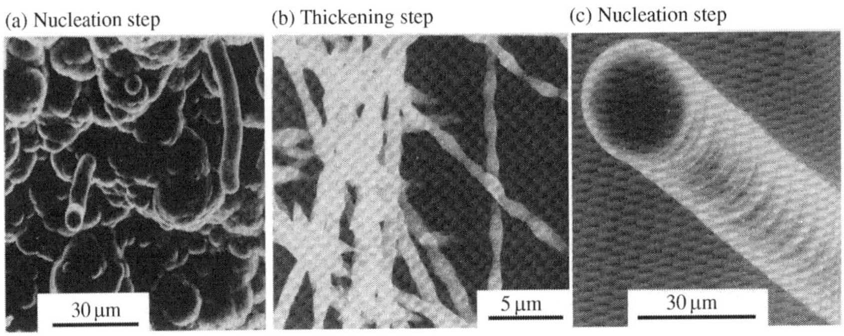


**FIGURE 2.31**

SEM images on different steps for the growth of vapor-grown carbon fibers.

*(Courtesy by Prof. M. Endo of Shinshu Univ.)*

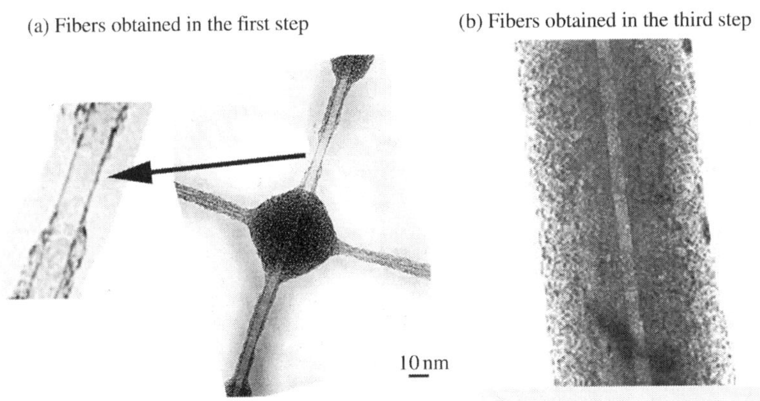


**FIGURE 2.32**

002 lattice fringe images of vapor-grown carbon fibers in the second and third steps.

*(Courtesy by Prof. M. Endo of Shinshu Univ.)*

Since most vapor-grown carbon fibers changed to well-graphitized carbon fibers and they could be prepared without the coexistence of other forms of carbons, like carbon blacks, huge numbers of works to prepare fibrous carbons through the decomposition of various gases, not only hydrocarbons but also CO, under different conditions were performed, aiming to prepare carbon nanotubes in a high purity with a high yield [47–54]. These fibrous carbons are called nanofibers, differentiating from carbon nanotubes and also from vapor-grown carbon fibers. In Fig. 2.35, fibrous carbons formed from CO gas are shown [50]. The nanotexture of these fibrous carbons showed a wide diversity from tubular to platelet, as shown in Fig. 2.18. By selecting the decomposition conditions, carbon helical microcoils with controlled pitch were prepared [55]. The fibrous carbons

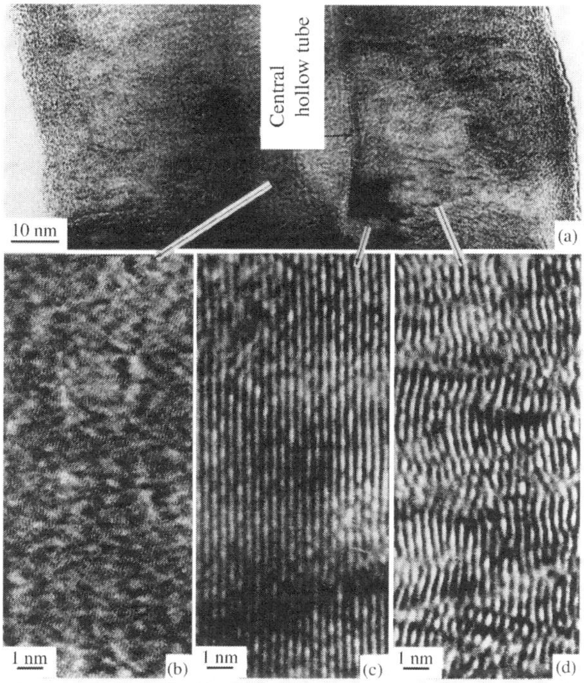


**FIGURE 2.33**

Lattice fringe image of vapor-grown carbon fibers.

*(Courtesy by Prof. M. Endo of Shinshu Univ.)*

with the appearance as many cups are stacked in one direction, cup-stacked nanofibers, as shown in Fig. 1.10, were prepared in a similar process [56]. The carbon materials with fibrous morphologies are summarized in Section 3.4.

## d. Carbon nanotubes

Carbon nanotubes were found in the carbon deposits on the graphite anode during arc discharge in He atmosphere [57]. The temperature at the graphite electrode was estimated to reach up to 2500°C. The hollow tube at the center of carbon nanotubes thus obtained showed a wide range of diameter from 1–50 nm and their wall consisted of different numbers of carbon layers. Most of them are closed at the end [58], the smallest diameter being the same as the size of the smallest fullerene $C_{60}$. The carbon nanotubes of single wall were found later [59,60]. In Fig. 2.36, TEM image of carbon nanotube is shown. By selecting the arc discharging conditions between graphite electrodes, a relatively high yield of carbon nanotubes was reported [61].

Under very similar conditions of arc discharging, fibrous carbons were obtained in 1960, of which the structure was understood to be a scroll of

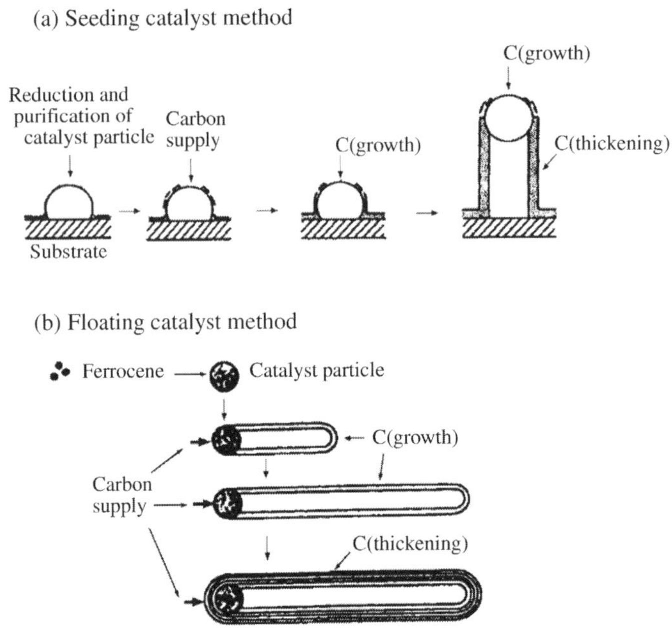


**FIGURE 2.34**

Illustration of the growth of vapor-grown carbon fibers by seeding and floating catalyst methods [46].

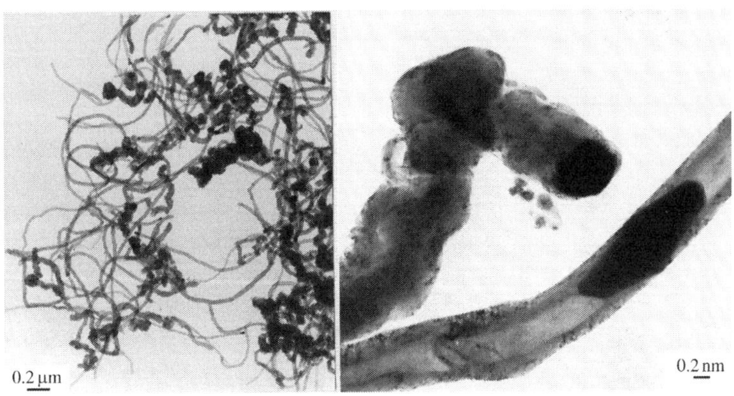


**FIGURE 2.35**

Fibrous carbons obtained from CO at 500°C.

*(Courtesy of Mme A. Oberlin.)*

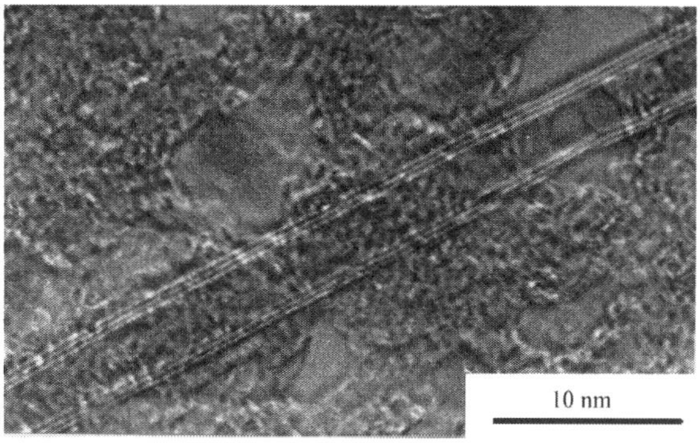


**FIGURE 2.36**

TEM image of carbon nanotube.

*(Courtesy of Prof. M. Endo of Shinshu Univ.)*

hexagonal carbon layers and so-called graphite whisker [62]. Similar carbon nanotubes, even single wall nanotube, were formed at the beginning of the formation of vapor-grown carbon fibers (see Fig. 2.32a) [39].

In the preparation of carbon nanotubes by arc discharge, they are formed together with other forms of carbons, such as carbon blacks and pyrolytic carbons. Therefore, the purification process is unavoidable in order to have nanotubes in high purity. Also the control of diameter and length of carbon nanotubes is difficult in the arc discharge process. By CVD process using catalysts (catalytic CVD), however, carbon nanotubes were successfully grown vertically on the substrate [63–67]. In Fig. 2.37, single-wall carbon nanotubes grown vertically to the substrate (called either 'forest' or 'array') are shown, being synthesized by catalytic CVD process using ethylene gas with a small amount of water vapor [66]. The growth rate of carbon nanotubes was very fast, 2.5 mm length for 10 min CVD without formation of other forms of carbon [65–67]. Arrays of multi-walled carbon nanotubes were also synthesized [63,64]. The conditions of CVD, including precursors, water vapor content, etc., were studied in detail.

### e. Fullerenes

The carbon cluster $C_{60}$ was firstly found in the soot formed by laser ablation in He atmosphere [68,69] and named as buckmisterfullerene. $C_{60}$ was also found in the soot formed through the vaporization of carbon vapor from the graphite rod heated at high temperatures in He atmosphere [70]. Also in the carbon deposits formed by arc discharge between graphite electrodes a relatively high concentration of fullerenes were found [71]. This carbon cluster $C_{60}$ could be isolated using the solvents, such as benzene, carbon disulfide, etc. [72]. By the solvent fractionation, the formation of carbon clusters with different sizes, such as $C_{70}$, $C_{82}$, ... $C_{960}$, is confirmed and this series of clusters is called fullerenes. The cluster $C_{60}$

**FIGURE 2.37**

Carbon nanotubes synthesized by catalytic CVD [66].

can also be separated from the soot by vaporization at around 400°C in vacuum. For the formation of these fullerenes, the atmosphere of different gases is examined, but helium atmosphere resulted in a relatively high yield of fullerenes in most cases.

The formation mechanism of these fullerene clusters was discussed in the literature [73,74]. A model named ring-stacking model is illustrated in Fig. 2.38, $C_{60}$ molecule being formed by the stacking of carbon clusters of different sizes (different number of carbon atoms) [75].

The synthesis of $C_{60}$ through a chemical route has been tried, but not yet succeeded [75]. It was started from trindane $C_{15}H_{18}$, where three pentagons were bonded around the hexagon, as shown as A in Fig. 2.39, and tried to polymerize them. The dimmer (B) and trimmer (C) were successfully synthesized, but no tetrammer (D), i.e., $C_{60}$, was obtained. The trimmer had caged morphology and was said to be very stable. The first chemical synthesis of $C_{60}$ has been done by flush vacuum pyrolysis at 1100°C. However, $C_{70}$ or higher fullerenes are not yet synthesized.

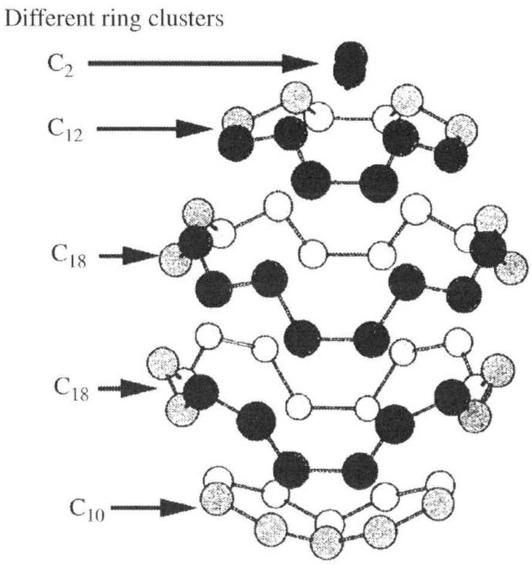


**FIGURE 2.38**

Ring-stacking model for the formation of buckminsterfullerene $C_{60}$ [75].

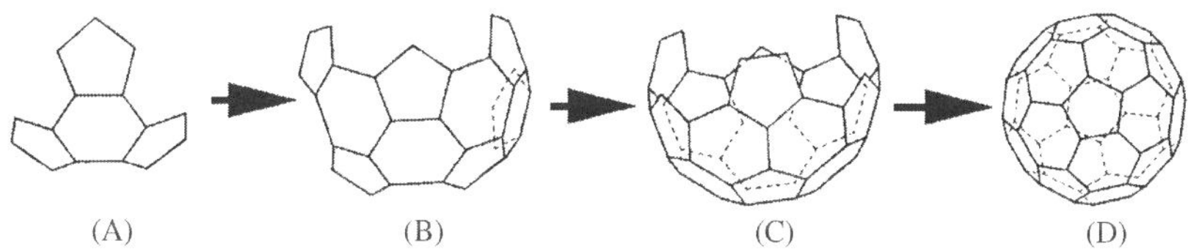


**FIGURE 2.39**

Trial to synthesize $C_{60}$ by polymerization [75].

#### f. Graphenes

The chemical vapor deposition process is one of preparation methods for graphene [76]. Graphene grown on polycrystalline Pt foil by ambient pressure CVD with a low concentration of methane at 1040°C is shown in Fig. 2.40 [77]. The graphene thus synthesized has hexagonal morphology with high crystal perfection and is grown to millimeter-size during 180 min.

### 2.3.3 Solid phase carbonization

In the process of solid phase carbonization, the difference in heating rate gives many different carbon materials. Most thermosetting resins do not pass through liquid state during their pyrolysis and carbonization. When these thermosetting resins are heated very slowly, the carbonized products shrink homogeneously by keeping their forms and give glass-like carbons, which have the nanotexture of random orientation and contain a large amount of closed pores. Glass-like carbons

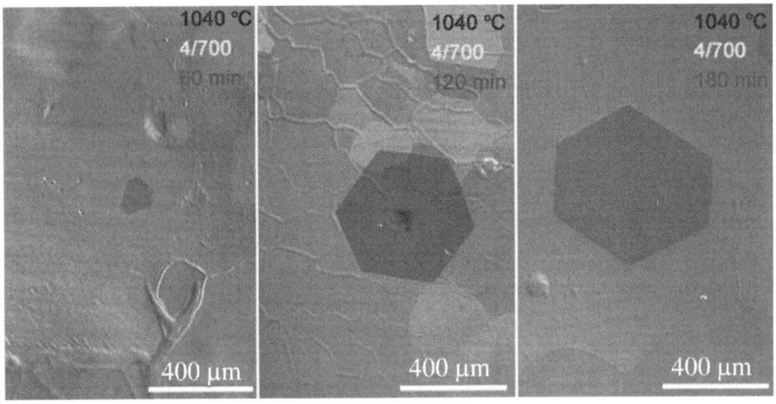


**FIGURE 2.40**

Graphene formed on Pt substrate via CVD of methane [77].

are one of the new carbons developed in the beginning of Period II (refer to Table 1.1). When the thermosetting resins were rapidly heated, a number of cracks were formed in most of the carbonized products and became small fragments, in which many open micropores were easily created by oxidation (activation) to give activated carbons, one of classic carbons. Thin films of polyimides and some other polymers do not pass apparently through liquid phase during carbonization process. Partial oxidation of organic fibers to prevent their softening during pyrolysis and carbonization processes (called stabilization) is an essential process to produce PAN-based and pitch-based carbon fibers. The process for thermosetting resins to form glass-like and porous carbons, including polyimide to form carbon films, may be intrinsic solid phase carbonization. On the other hand, the process for PAN- and pitch-based carbon fibers may be extrinsic solid phase carbonization because these precursors PAN and pitch, which are intrinsically carbonized in liquid state, are changed by the process of stabilization.

## a. Activated carbons

Activated carbons contain a large amount of open nano-sized pores (nanopores), which are usually evaluated by BET surface area. They have been used as adsorbents since prehistoric age, and are now used in a much wider field, from our daily lives to various industries not only for manufacturing some products but also for the treatment of waste. For their predominant applications as adsorbents, pore structure is the most important property to be controlled [78–80]. For the pore development, partial oxidation of carbonized materials under a controlled condition, which has been called '*activation*', was usually carried out. For carbon precursors, not only natural biomasses, such as plants and woods, but also pitches and resins have been used. In Fig. 2.41a, SEM micrograph of a carbon prepared from coconut shell is shown, which are used as a filter in tobacco. Fig. 2.41b is a

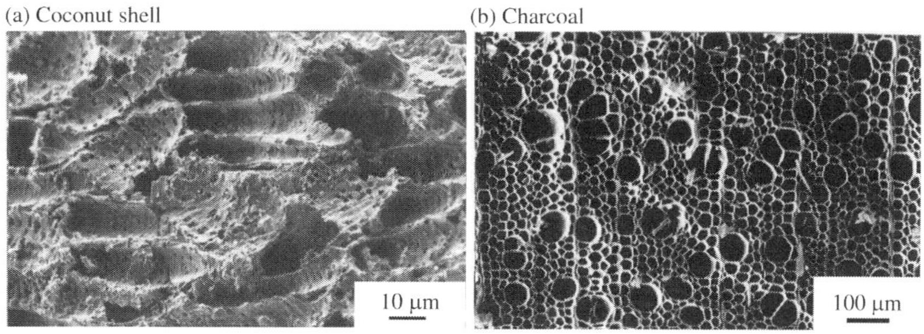


**FIGURE 2.41**

SEM images of a carbon prepared from coconut shell and charcoal.

*(Courtesy of Dr. A. Yoshida of Tokyo City Univ.)*

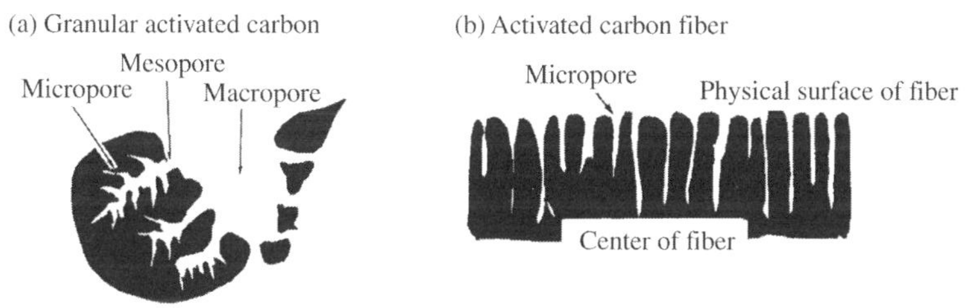


**FIGURE 2.42**

Schematic illustrations of pore structure.

SEM of charcoal, where homogeneous sizes of pores are aligned in a relatively regular way. These pores observed under SEM are macropores, which are originated from the cell structure in the precursor plants, and are not effective for adsorption of various molecules. However, the presence of these macropores before activation is preferable for creating micropores on the wall. Various wasted materials, such as sawdust from woods, wasted cotton, etc., are able to convert to activated carbons.

For activation, different processes were employed, oxidation using diluted oxygen gas, air, water vapor, etc. (physical activation), and oxidation using $ZnCl_2$ and KOH (chemical activation). Recently great success to get a high surface area reaching to about 3600 $m^2/g$ was obtained by using KOH for activation [81]. During the activation process, the creation of micropores of size 0.4–2 nm is the most important. In most carbon materials, however, macropores (the size of >50 nm) and mesopores (2–50 nm) are coexistent with micropores, as shown schematically in Fig. 2.42a. In other words, macropores and mesopores had to be formed during the activation process in order to develop a large amount of micropores, and these pores also play an important role for adsorption as pathways for adsorbates. For the carbons prepared from natural plants, many macropores are already formed during carbonization as a memory of cell structure of the original plants, as shown in Fig. 2.41, which seems to make micropore development by activation easier.

In so-called activated carbon fibers, on the other hand, micropores are exposed to the surface of thin carbon fibers, as schematically shown in Fig. 2.42b. Such a direct exposure of micropores gives an advantage of fast adsorption/desorption [82].

Pore size distribution in activated carbons is important, as well as pore volume (porosity), for practical applications, for example, for the adsorption of liquid materials relatively large pores in a few nanometer size (mesopores), are desired, but for that of gaseous materials small pores (micropores) are effective. The activated carbons for molecular sieving must have sharp pore size distribution at the size suitable for target molecules. In Fig. 2.43, some examples of pore size distributions for different applications.

Recently, the cores of kenaf plant (*Hibiscus cannabinus*) were found to give a high BET surface area, as high as 2700 $m^2/g$, by the carbonization in inert atmosphere, without any activation process [83]. This high surface area was supposed to be due to the departure of metallic impurities (mostly K), which were originally included in the cores, during carbonization.

The development of pores and the control of their sizes in carbon materials will be discussed in detail in Section 2.7 and also in Section 3.5. Also, pore development without activation was carried out by the selection of suitable precursors with different techniques, which are explained in Section 2.4.

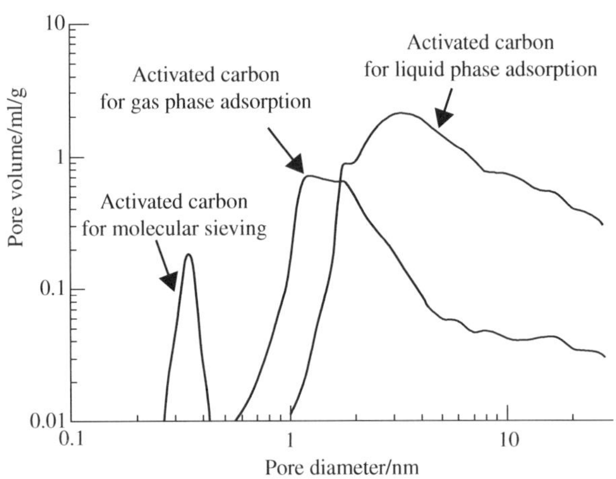


**FIGURE 2.43**

Pore size distributions of activated carbons for different applications.

## b. Glass-like carbons

Glass-like carbons are produced by the carbonization of thermosetting resins, such as phenol-formaldehyde, poly(furfuryl alcohol), cellulose, etc., under an exact control of the heating process and are characterized by amorphous structure. Their properties are very similar to inorganic glasses, such as high hardness, brittle conchoidal fracture and gas impermeability [84]. The most important process for their production is very slow heating, slower than the rate of the shrinkage to compensate the evolution of pores due to the release of decomposition gases from the precursor block [85]. In Fig. 2.44, the changes in weight, volume, BET surface area and adsorption of water vapor with carbonization temperature are shown for poly(furfuryl alcohol) condensates [86]. Rapid decreases in weight and volume up to 700–800°C are due to the decomposition of the precursor and correspondingly both BET surfacc area and adsorption of water vapor increase, with the latter micropores being responsible. Above 800°C, however, BET surface area and water vapor adsorption decrease quickly, suggesting that most pores are closed. Low bulk density at 1.3–1.5 $g/cm^3$ of glass-like carbons commercially available suggests the presence of a large amount of pores, but their low gas permeability as $10^{-12}$ $cm^2/s$, almost gas-impermeable, suggests that all pores are closed. In the process above 800°C, the pores formed during the pyrolysis and carbonization of precursor below 800°C have to be due to the shrinkage of the bulk of the products, so that the heating rate for carbonization is crucial for

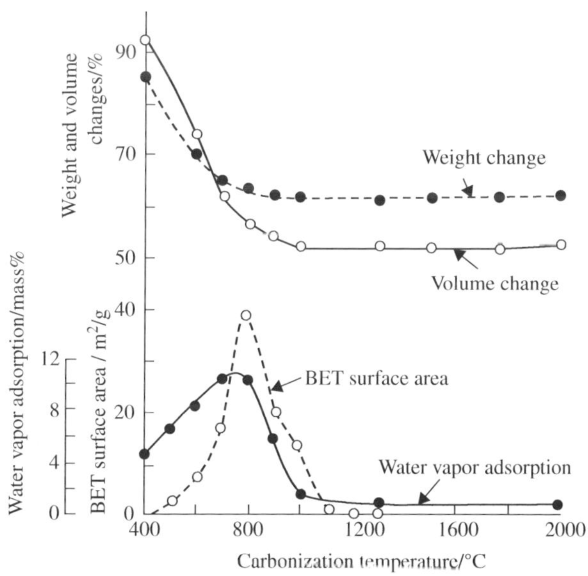


**FIGURE 2.44**

Changes in weight, volume, BET surface area and water vapor adsorption with carbonization temperature for glass-like carbon [86].

the production of glass-like carbons, as slow as completing shrinkage of the matrix carbon. The formation of closed pores with rather homogeneous size of about 5 nm was reported, the volume of which reached around 1/3 of the bulk [87–89].

Most thermosetting resins can give glass-like carbons by selecting appropriate heating rate for carbonization. When the heating rate was high, higher than shrinkage (heating rate conventionally employed is too high for most of thermosetting resins), the pores formed during carbonization were connected with each other and became cracks to give irregular-shaped particles, which contained a large amount of macropores. In convention, the glass-like carbons (sometimes called glassy carbons) indicate the carbon blocks with characteristics mentioned above (refer to Section 3.3), and the carbons produced from the same precursors but consisted of particles with irregular shape are not called glass-like carbons, but their nanotexture can be the same as glass-like carbon.

Glass-like carbons are very hard and brittle, and their machining is so difficult that they have to have a shape near the final product before carbonization. Because of this reason, their carbonization process has also to be controlled exactly by taking account of rather large shrinkage during pyrolysis and carbonization, up to 50% in volume, as shown in Fig. 2.44.

The structure and nanotexture of the carbon materials prepared with rather high heating rate from thermosetting resins are locally very similar to those of glass-like carbons obtained as certain shapes. The structure models consisting of the mixture of diamond-like tetrahedrally coordinated $sp^3$ C—C bonds with graphite-like $sp^2$ C—C bonds [90,91] and consisting of ribbons of carbon layers [92] had been proposed. Based on the experimental results of radial distribution analysis by X-ray [93], high-resolution electron microscopy [94,95], temperature dependence of specific heat [96] and K-emission spectrum measurement [97], however, the closed shell model mentioned in Fig. 2.21c [98] is believed to be the most reliable.

### c. Carbon fibers

In order to produce carbon fibers from poly(acrylonitrile) (PAN) and various pitches, so-called stabilization process after their spinning is essential, which consists of partial oxidation by using different gases, such as air, oxygen, chlorine, hydrochloric acid vapor, etc. The stabilized fibers are then subjected to carbonization in inert atmosphere, which proceeds in solid state (extrinsic solid phase carbonization). The resultant carbon fibers are called PAN-based and pitch-based carbon fibers. When cellulose and phenol fibers were used as precursors, the process for stabilization is not necessary and subjected directly to carbonization (intrinsic solid state carbonization).

The changes in molecular structure from the precursor PAN to hexagonal carbon layer during the processes of stabilization, pyrolysis and carbonization were studied by various authors using various techniques [99]. In the stabilized fibers, the formation of three basic cyclic structures, as shown in Fig. 2.45, was reported from the detailed studies by $^{13}$C-NMR, ESCA and FT-IR [100].

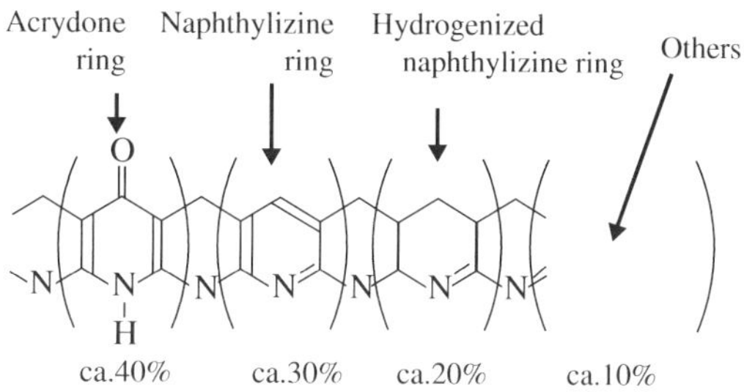


**FIGURE 2.45**

Basic cyclic structures in stabilized PAN-based fiber [100].

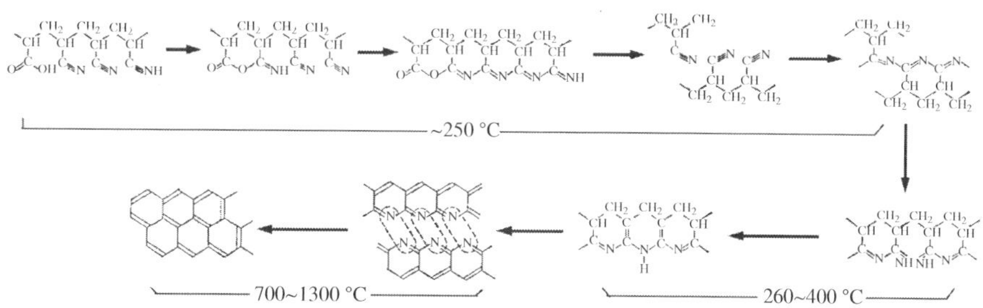


**FIGURE 2.46**

Molecular structure changes during pyrolysis and carbonization of PAN.

The changes in molecular structure in PAN fibers to carbon fibers were estimated to proceed in the steps shown in Fig. 2.46. The cyclization at the position of carboxyl radicals and the formation of ladder polymers occur at up to about 250°C, aromatization and further cyclization occur by being associated with the release of NH radicals as $NH_3$ and HCN gases at a temperature range of 260–400°C, and above 700°C aromatization due to denitrogenation reaction to form fundamental structural units, hexagonal carbon layers, in carbon fibers [99,101–103].

During stabilization and carbonization, the stress was pointed out to be accumulated in the fibers due to different reasons on the course carbonization, such as hindrance of the thermal motion of molecules, shrinkage caused by oxidation reaction during stabilization and also shrinkage by pyrolysis and carbonization accompanied by the departure of hydrocarbon gases. The changes in stress up to 600°C are schematically shown in Fig. 2.47, being divided into four steps [104]. In the first step, stress increases to a maximum and then is released down to a minimum at around 240°C, probably due to the motion of PAN molecules at this temperature range. When the fibers are cooled down from this temperature range, the shrinkage stress becomes zero, as shown by the broken line from the points

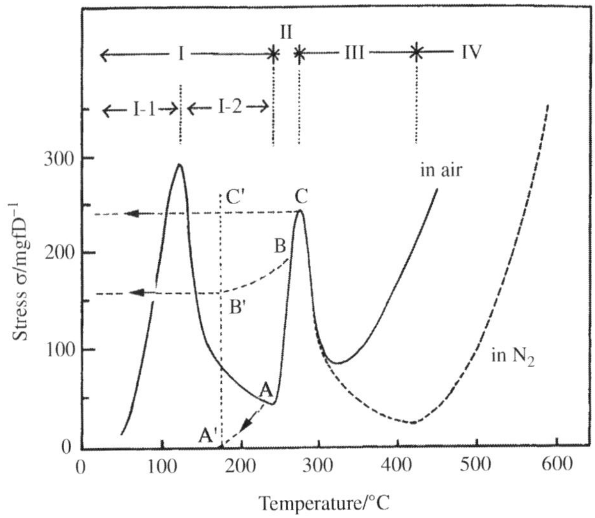


**FIGURE 2.47**

Changes in stress σ during stabilization and pyrolysis of PAN fiber [104].

A to A′ in Fig. 2.47. The second step is the abrupt increase in stress in a narrow temperature range, which is related to the change in chemical structure in the fibers, such as the network formation by oxygen and polymerization of the nitrile side-groups. When the fibers are cooled from this temperature range, some stress remains in the fiber, as shown by the broken line from the points B to B′ and from C to C′ in the figure. In the third and fourth steps, the fibers show different behaviors in stress, depending on atmosphere; stress decreases quickly and reaches a minimum at low temperature in air, but it decreases to a lower minimum at a somewhat higher temperature in nitrogen atmosphere. The third step of stress change is supposed to correspond to further pyrolysis and plasticization of polymers formed in the previous steps. The gradual increase in stress in the fourth step is supposed to be due to the formation and growth of carbon hexagonal layers. From the detailed analysis of this change in stress with temperature, various parameters to produce PAN-based carbon fibers were proposed [104].

To produce the carbon fibers from pitches, the stabilization process is also essential. When optically isotropic pitch (refer to Section 2.3.4) was used as precursor, isotropic-pitch-based carbon fibers were produced through spinning, stabilization and following carbonization, whose cross-sectional nanotexture was random orientation. The precursor pitches for isotropic-pitch-based carbon fibers must have certain homogeneity and appropriate fluidity. In Table 2.3, some examples of optically isotropic pitches used for carbon fiber production are listed [105]. From optically anisotropic pitches (mesophase pitches, refer to Section 2.3.4) mesophase-pitch-based carbon fibers were produced through the

Table 2.3 Pitches used as Precursors for Carbon Fiber Production [105]

| Pitches | | Optically Isotropic | | | Optically Anisotropic | | | | |
|---|---|---|---|---|---|---|---|---|---|
| Origin of pitch | | Coal | Coal | Petroleum | Coal | Coal | Coal | Synth. | Synth. |
| Softening point (°C) | | 267 | 250 | 250 | 304 | 308 | 299 | 257 | 265 |
| Toluene insoluble (mass%) | | 74.6 | 44.1 | 58 | 81 | 86 | 95 | 43 | 74 |
| Anisotropic ratio (%) | | 0 | 0 | – | 94 | 100 | 100 | 100 | 100 |
| Elemental composition (mass%) | C | 94.3 | 93.4 | 93.9 | 94.9 | 94.2 | 95.2 | 94.4 | 95.2 |
| | H | 3.89 | 4.44 | 4.65 | 3.69 | 3.92 | 3.82 | 5.91 | 5.02 |
| | N | 1.01 | 0.85 | 0.2 | 0.68 | 1.05 | 0.89 | <0.1 | 0.1 |
| | S | 0.32 | 0.29 | 0.7 | 0.16 | 0.28 | 0.15 | <0.01 | 0.16 |
| H/C ratio | | 0.50 | 0.57 | 0.59 | 0.47 | 0.50 | 0.48 | 0.75 | 0.63 |

same processes, but various cautions are required. Some raw materials for the production of mesophase-pitch-based carbon fibers are listed in Table 2.3.

### d. Carbon films derived from aromatic polyimides

Polyimides have been developed as thermoresistant polymers and are used in different fields, especially the field of electronics. Polyimide films with different molecular structures have been developed because of their practical and promising applications, which give a wide variety of structure in the carbon films after heat treatment at high temperatures, from highly crystalline graphite to amorphous glass-like carbon films [106]. This is a typical case where the molecular structure of organic precursors and the texture of their polymer films govern the structure and nanotexture of the resultant carbon films, i.e., crystallinity and preferred orientation of hexagonal carbon layers.

As shown in Fig. 2.48, aromatic polyimides (D) are formed by a reaction of tetracarboxylic dianhydride (A) with equimolar aromatic diamine (B) in a solvent of either N-methyl-2-pyrrolidone (NMP) or N,N′-dimethylacetamide (DMAc), through dehydration of the intermediate product of poly(amide acid) (C) by either a thermal or chemical method (imidization). The polyimides, therefore, can be characterized by coupling the names of starting anhydride and diamine, such as PMDA/ODA, which is the polyimide made from pyromellitic dianhydride (PMDA) and 4,4′-oxydianiline (ODA). In Fig. 2.49a, molecular structures of some aromatic polyimides are illustrated with their tradenames.

For a commercially available polyimide film, Kapton, with 25 μm thickness, the changes in weight and shrinkage along the film with carbonization temperature and the changes in the composition of decomposition gases during carbonization are shown in Fig. 2.50a and b, respectively [107]. The results in Fig. 2.50 show that the carbonization of the polyimide Kapton proceeds in two

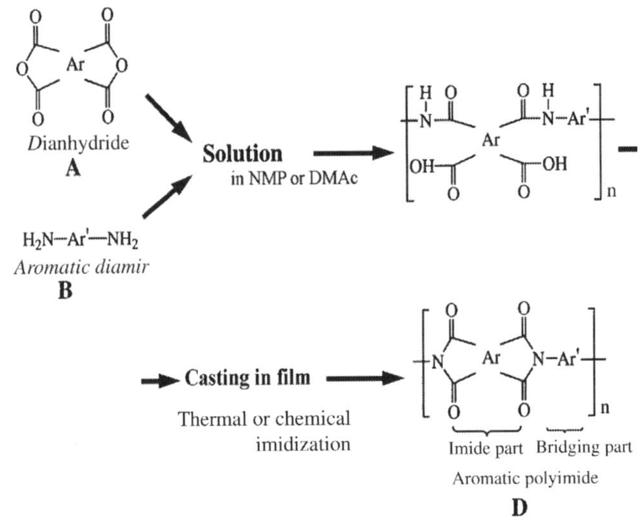


**FIGURE 2.48**

Preparation process of polyimide films.

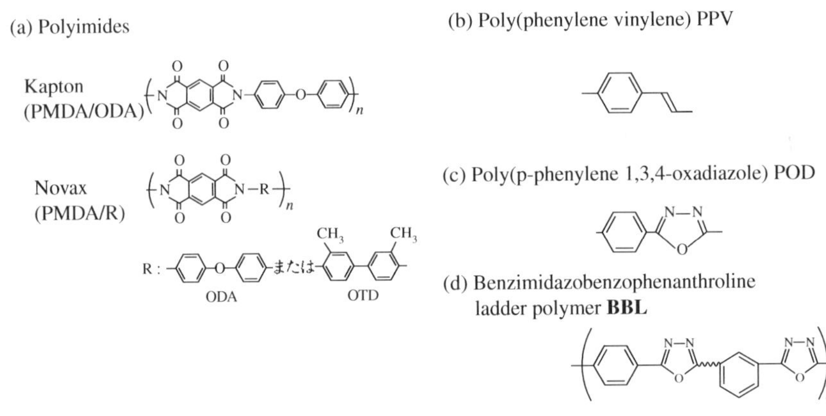


**FIGURE 2.49**

Molecular structures of the precursors for carbon films.

steps, the first step in rather narrow temperature range of 500–600°C and the second step from 700°C to more than 1000°C. In the first step of carbonization, an abrupt weight decrease associated with an evolution of a large amount of carbon monoxide and pronounced shrinkage along the film was observed, which is mainly due to a breakage at carbonyl groups in the imide part. Small weight loss in the second step, associated with a little shrinkage, is due to the evolution of small amounts of methane, hydrogen and nitrogen. The etheric oxygen in

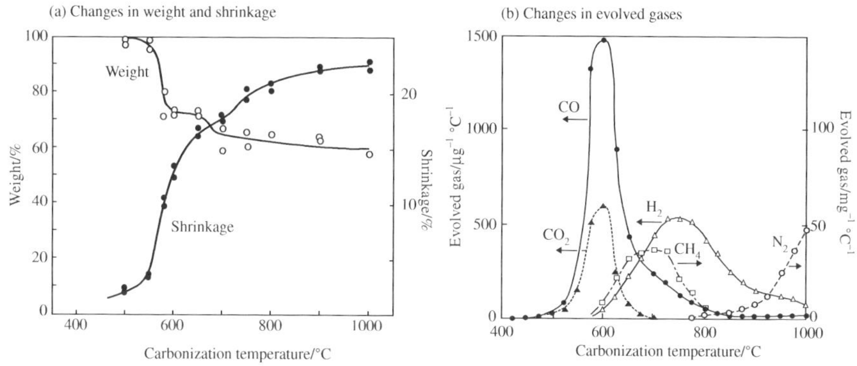


**FIGURE 2.50**

Changes in weight, shrinkage and evolved gases during carbonization of polyimide Kapton film [107].

polyimide molecules was supposed to be released at the end of the first step, from the comparison of the result on the molecule without etheric oxygen. The release of nitrogen in the second step of carbonization was found to continue up to high temperatures above 2000°C, leaving a large amount of pores in the film if it was heated continuously up to 2400°C [108].

The structural change in the second step of carbonization was found to reflect on electrical properties of the film [109]. In Fig. 2.51, the changes in electrical conductivity at room temperature along the film with carbonization temperature are plotted on different laboratory-made polyimide films. All films, including PMDA/ODA (corresponds to Kapton), show pronounced increase in conductivity, more than one order of magnitude, in a narrow temperature range between 700–800°C, although only small weight loss and shrinkage are observed (Fig. 2.50). Further carbonization up to 1100°C gives another increase in conductivity. The difference in electrical conductivity of the carbon films prepared from polyimide films with different molecular structures was explained to be due to the preferred orientation of hexagonal carbon layers along the surface of the film [106]. The facts that methane and hydrogen are released and electrical conductivity increases in the second step suggest that the carbonization process mainly occurs in the second step, and the first step may be pyrolysis of polyimide molecules.

The structure and nanotexture in the carbon films, i.e., preferred orientation of hexagonal carbon layers and graphitizability, were found to be governed by the molecular structure and texture of the precursor polyimides [106]. It is worthwhile to mention that homogeneous and dense carbon films are obtained even though weight loss is as large as 40% and also linear shrinkage as large as 22%. No cracks are observed on the carbonized films even under scanning electron microscope.

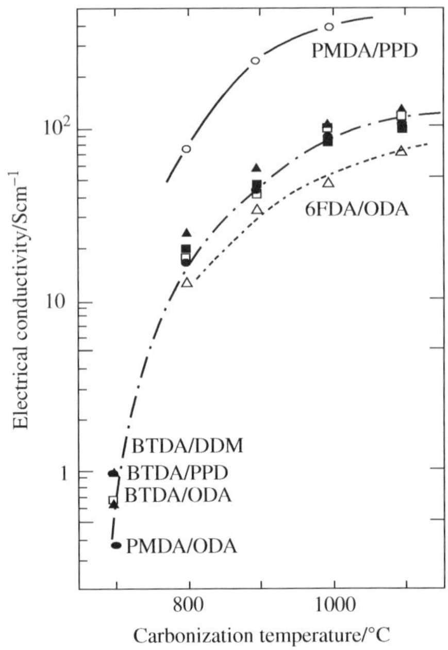


**FIGURE 2.51**

Changes in electrical conductivity along the film at room temperature with carbonization temperature for various polyimides [109].

Carbon films, which are easily converted to graphite films, are prepared from the films of poly(phenylene vinylene) PPV [110], poly(p-phenylene-1,3,4-oxadiazole) POD [111], and benzimidazobenzophenanthroline ladder BBL polymer [112], of which molecular structures are shown in Fig. 2.49b–d. Their carbonization behaviors measured by weight change and linear shrinkage are shown and are similar to polyimide films, Kapton and Upilex [113].

## 2.3.4 Liquid phase carbonization

Some precursors, such as pitches, are carbonized through liquid phase. This liquid phase carbonization is not convenient to give certain forms to the resultant carbons. Therefore, a process was developed for the industrial production of graphite blocks using the cokes, which were the products of liquid phase carbonization of pitches as a filler and a small amount of pitches as a binder. The technique to change the nanotexture in the resultant cokes by applying a shear stress during liquid phase carbonization was invented and gave so-called needle-like cokes. They are now an important raw material to produce large graphite electrodes for metal refining. In the course of liquid phase carbonization, the formation of spheres with optical anisotropy (mesophase) was found, which promoted the developments of some new carbons, such as mesocarbon microbeads and mesophase-pitch-based carbon fibers.

### a. Mesophase in pitches

Pitches, which are the residues of refining of petroleum oils and coals, are the mixtures consisting of condensed aromatic molecules with a wide range of molecular weights. This fact gives certain merits for pitches to use for a binder to prepare carbon blocks, such as high performance for the forming process because of the fluidity in a wide range of temperature, gradual carbonization because of various constituent molecules, relatively high carbonization yield, etc. On the other hand, this fact makes the characterization and specification of pitches difficult.

The characterization of pitches was carried out by the fractionation into different components using different solvents [114]. The fractionation of pitches was performed in industries by using pyridine and benzene (or sometimes quinoline and toluene) as solvents, benzene-soluble (BS) fraction, benzene-insoluble but pyridine-soluble (BI-PS) fraction and then pyridine-insoluble (PI) fraction. The BS fraction usually consists of molecules with relatively low molecular weight (about a few hundred) and has a high H/C ratio of 1.8–0.8, which is often called $\gamma$-resin. The BI-PS fraction consists of the molecules with intermediate molecular weights and called $\beta$-resin. In PI fraction, not only the components with high molecular weight of more than 2000 and low H/C ratio of less than 0.5, but also some impurities like carbon blacks (free carbon) are included and called $\alpha$-resin. The content of $\beta$-resin in the pitch is pointed out to be an important parameter for its use as a binder.

In the course of carbonization, mesophase spheres are firstly formed and by further heating they grow and coalesce with each other to form bulk mesophase and finally create a solid (coke) with different textures [18]. This mesophase, either with spherical or coalesced morphology, is known to be one of liquid crystals (nematic type) [115]. In Fig. 2.52, a series of polarized-light microscopic images show the sequence from the formation of small mesophase spheres to their growth and partial coalescence.

In Fig. 2.53, polarized-light micrographs of mesophase spheres before coalescence and those just at the beginning of the coalescence are compared. The detailed analysis of the changes in extinction counters in the spheres gave some information on the structure in the spheres.

The formation of the mesophase spheres and the texture of bulk mesophase were shown to govern the final texture and properties of the carbonization products, cokes [18,116,117], as will be explained below.

### b. Mesophase spheres

The nanotexture in the mesophase sphere was clarified by polarized-light microscopy [18] and high resolution transmission electron microscopy [118,119]. The nanotexture of the sphere is shown in Fig. 2.54 [119], together with the schemes of aromatic molecules in its various cross-sections. The orientation of anisotropic aromatic molecules is radial around the surface, i.e., perpendicular to the surface

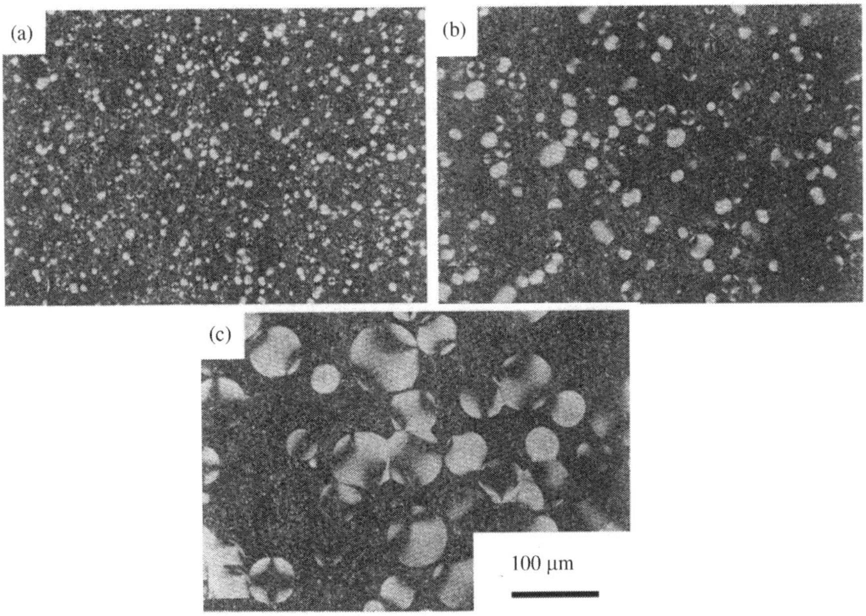


**FIGURE 2.52**

Formation, growth and coalescence of mesophase spheres in a pitch.

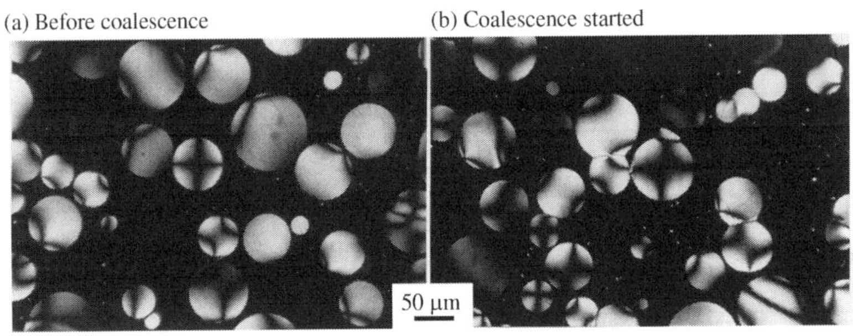


**FIGURE 2.53**

Mesophase spheres and the beginning of their coalescence.

of the sphere, by making two poles, but planar at the center of the sphere. The mesophase spheres formed in most of the pitches have this nanotexture and are named Books-Taylor type, to differentiate them from other spheres having different nanotextures explained below.

The partial orientation of small hexagonal carbon layers, which were formed from aromatic molecules, is kept even after coalescence of mesophase spheres, as shown by the 002 lattice fringes image in Fig. 2.55.

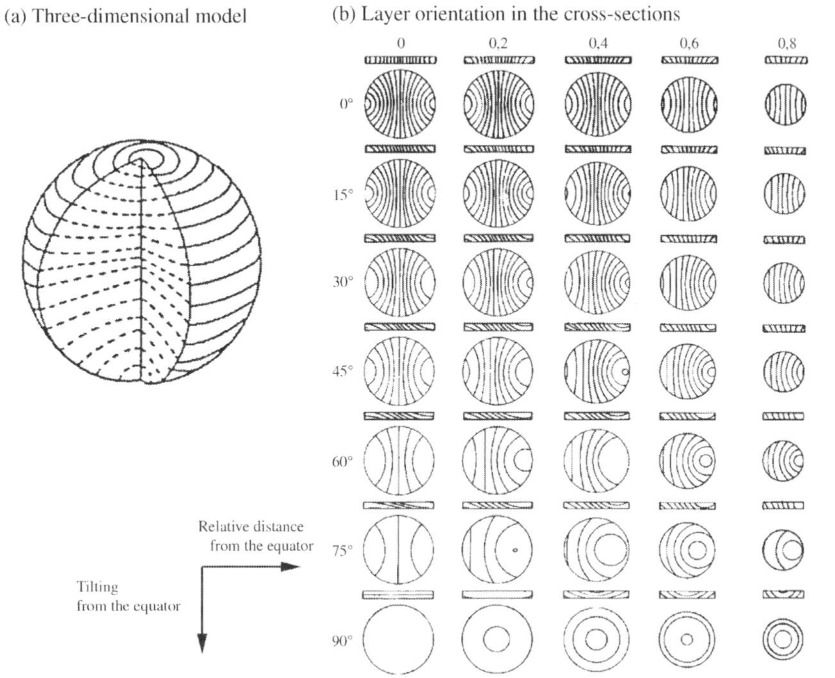


**FIGURE 2.54**

Nanotexture model of Brooks-Taylor type mesophase sphere.

*(Courtesy of Mme A. Oberlin.)*

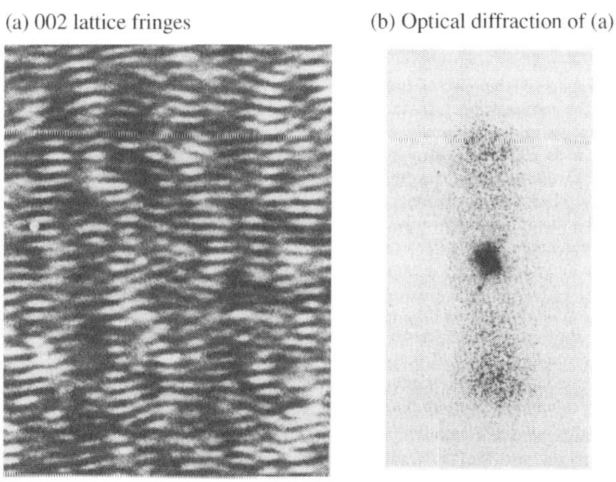


**FIGURE 2.55**

002 lattice fringe image of a part of mesophase sphere and its optical diffraction.

*(Courtesy of Mme. A. Oberlin.)*

Mesophase spheres formed in pitches were separated from isotropic matrix using a solvent, either pyridine or quinoline [120,121]. These separated mesophase spheres were named mesocarbon microbeads (MCMB) [110]. In Table 2.4, the yield, specific gravity and chemical composition of MCMB separated from the starting pitch after different treatments are shown. The type M of MCMB is obtained with a relatively high yield, because of adhesion of fine solid particles (free carbon) presented in advance, and has a homogeneous size of about 10 μm. On the other hand, the yield of type C separated from the quinoline-soluble fraction (QS fraction) of the same coal tar pitch is very low and their sphere sizes distribute in a wide range. Small particles less than 3 μm in this type are not spherical, but have lemon-like appearance. Type P separated from an asphaltene consists of small particles with less than 3 μm size, which also have a lemon-like appearance.

An SEM image of one commercially available MCMB was shown in Fig. 2.56, a relatively homogeneous size of spheres being seen.

Mesophase spheres with different nanotextures have been reported. Some of them are shown in Fig. 2.57b–d, compared to the nanotexture of Brooks-Taylor type spheres (Fig. 2.57a). The mesophase spheres with the nanotexture of Fig. 2.57b were obtained from a coal tar pitch mixed with a carbon black of 3–5 mass% [116] and also from the pitches prepared by heating either naphthalene or anthracene to 300°C with $AlCl_3$ catalyst under a magnetic field [122]. Depending on the amount of carbon blacks mixed and also the strength of magnetic field,

**Table 2.4** Starting Pitch, Yield, Specific Weight and Chemical Composition of Mesocarbon Microbeads [119]

| Mesocarbon Microbeads | Starting Pitch | Yield (Mass %) | Specific Weight (g/cm$^3$) | Chemical Composition (Mass%) | | | | Size (μm) |
|---|---|---|---|---|---|---|---|---|
| | | | | C | H | N | Diff. | |
| Type C | QS from a coal tar pitch, heat-treated at 430°C for 120 min | 6.3 | 1.46 | 91.1 | 3.1 | 0.8 | 5.0 | 1–100 |
| Type M | A coal tar pitch, heat-treated at 430°C for 90 min | 22.4 | 1.48 | 91.8 | 2.8 | 0.7 | 4.7 | ~10 |
| Type P | An asphaltene pitch, heat-treated at 430°C for 60 min | 7.2 | 1.47 | 81.9 | 3.8 | 1.3 | 13.0 | 1–3 |

**FIGURE 2.56**

SEM image of MCMB.

*(Courtesy of Kawasaki Steel Co. Ltd.)*

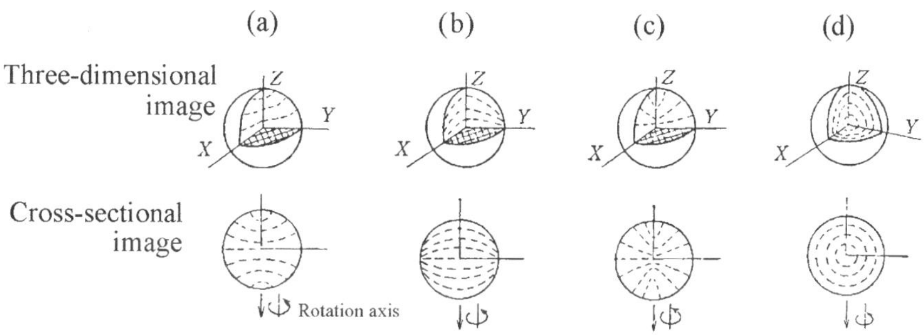


**FIGURE 2.57**

Nanotextures in different mesophase spheres [116].

Brooks-Taylor type mesophase spheres were also formed. The sphere shown in Fig. 2.57c has radial nanotexture even at its center and is formed from a pitch at relative low temperatures of 300°C for a long time [123]. When the pitch containing this type of spheres was heated to 400°C, all mesophase spheres changed to Brooks-Taylor type ones. The nanotexture Fig. 2.57d has the concentric point orientation, being formed by heating decacyclene [124a].

### c. Bulk mesophase

With increasing heat treatment temperature a number of mesophase spheres are formed and then coalesced with each other to form bulk mesophase. The coalescence of mesophase spheres, which is associated with the rearrangement of hexagonal carbon layers, occurs when the concentration of mesophase spheres reached a high level, where the viscosity of the pitch usually became high. Therefore, this rearrangement process of carbon layers proceeded slowly. As the balance between the rate of growth of mesophase and that of rearrangement of carbon layers, both of which strongly depend on the precursors and heating rate, different textures were formed. The optical texture of the bulk mesophase was classified according to the arrangement of extinction contours observed under a polarized-light microscope (optical texture) [124b,125], which depended on the local orientation of hexagonal carbon layers.

The classifications proposed are summarized in Table 2.5. In the classification proposed by the reference [122], the word 'isotropic' has been used, but it may give some confusion with the isotropic matrix where mesophase spheres are formed at a low temperature. Therefore, the word 'mosaic' is considered to be preferable, as employed in the reference [125]. Examples of optical textures of bulk mesophase are shown as polarized-light micrographs in Fig. 2.58. Extinction contours observed on the bulk mesophase under polarized light show the regions where the hexagonal carbon layers are oriented. The crossing of extinction contours gives some

**Table 2.5** Classification of Optical Texture of the Bulk Mesophase

| Optical Texture | Classification from Ref. [122] | Classification from Ref. [125] |
|---|---|---|
| Random arrangement of extinction contours spaced at less than 10 μm | Fine isotropic | Fine mosaic |
| Random arrangement of extinction contours space at more than 10 μm | Coarse isotropic | Coarse mosaic |
| Orientation of contour counters along one direction spaced at less than 10 μm | Fine fibrous | Fine fibrous |
| Orientation of contour counters along one direction spaced at more than 10 μm | Coarse fibrous | Coarse fibrous |
| Aggromeration of spheres without noticeable coalescence | Globular | Globular |
| Mixture of coarse extinction contours with fine ones | Heterogeneous fine isotropic | – |

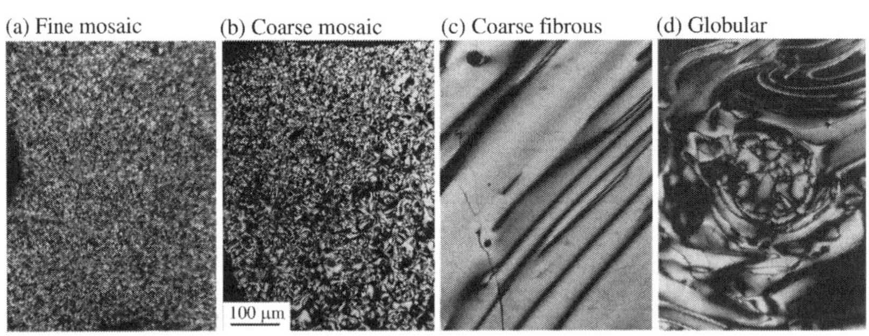


**FIGURE 2.58**

Polarized-light micrographs of the bulk mesophase with different optical textures.

**Table 2.6** Mesophase Formation Parameters $T_{50}$ and $\Delta T_{80}$ for some Pitches [122]

| Pitch | $T_{50}$ (°C) | $\Delta T_{80}$ (°C) | Optical Texture of Bulk Mesophase |
|---|---|---|---|
| Gilsonite | 409 | 13 | Fine mosaic |
| Air-brown asphalt | 414 | 26 | Fine mosaic |
| Propane asphalt | 427 | 26 | Coarse mosaic |
| Reduce-crude mix | 431 | 26 | Fine fibrous |
| Decant oil | 451 | 24 | Coarse fibrous |

information on the disturbance of layer orientation, stacking disturbance, which has been proposed to be explained by using the idea of disclination [126,127].

The formation of mesophase spheres in a pitch depends strongly on the temperature. It was proposed to characterize the mesophase formation behavior quantitatively by using the temperature at which the fraction of mesophase becomes 50%, $T_{50}$, and the temperature range between those for mesophase fraction of 10 and 90%, $\Delta T_{80}$ [122]. In Table 2.6, these parameters are shown for some pitches, together with the final optical texture of bulk mesophase.

Gilsonite pitch had a low $T_{50}$ and very narrow $\Delta T_{80}$, revealing that the formation of mesophase sphere occurred very rapidly at relative low temperature, and, as a consequence, very fine mosaic texture with the spacing of less than 5 μm was obtained. In decant oil, on the other hand, mesophase was formed at a high temperature and grew relatively slowly, resulting in coarse fibrous texture, because of a high $T_{50}$ and relatively wide $\Delta T_{80}$.

### d. Control of optical texture of bulk mesophase

The optical texture of pitches and resultant cokes is determined as a balance between two rates, the rate for the formation and growth of mesophase spheres and that for the rearrangement of fundamental structural units during the coalescence of mesophase spheres to form bulk mesophase. The latter never overcomes the former, i.e., the latter is much slower than the former. Therefore, it is usually difficult to obtain a large area with homogeneous orientation. The former depends strongly on the precursor pitches and the latter on the temperature. Since the control of optical texture of bulk mesophase is a very important process for the manufacturing of cokes and mesophase-pitch-base carbon fibers, it is discussed from two points of view, selecting of carbonization conditions, such as heating rate, pressure and additives, and modifying the starting pitch [128].

One of the techniques to control or change the optical texture in bulk mesophase is the application of a magnetic field [129–131]. Since the aromatic compounds in the pitch are diamagnetic, it is possible to make the mesophase spheres orient and to prepare the sample in which all hexagonal carbon layers are oriented by making their layer planes parallel to the direction of magnetic field. For these highly oriented cokes, high graphitization degree was obtained after a high-temperature treatment [132].

Another technique to control the optical texture of bulk mesophase and resultant coke is to mix some other precursors to the pitch. By mixing a phenol resin into a pitch by using a common solvent, such as pyridine, the former giving an isotropic texture and the latter coarse mosaic texture, optical texture was reported to be controlled, from coarse mosaic through fine mosaic to isotropic with increasing mixing ratio of phenol [133,134]. Polarized-light micrographs of the resultant cokes are shown in Fig. 2.59. On these mixtures, a wider range of texture could be obtained by the carbonization under pressure, as shown in Fig. 2.60 [134]. Intimate mixing of phenol resin with pitch using solvent was essential to obtain a homogeneous texture, otherwise heterogeneous optical textures were obtained, coexistence of coarse mosaic regions with isotropic ones [135].

From the mixtures of pitch and coal, a wide range of optical texture was obtained in the resultant carbons [136]. The results were explained by using

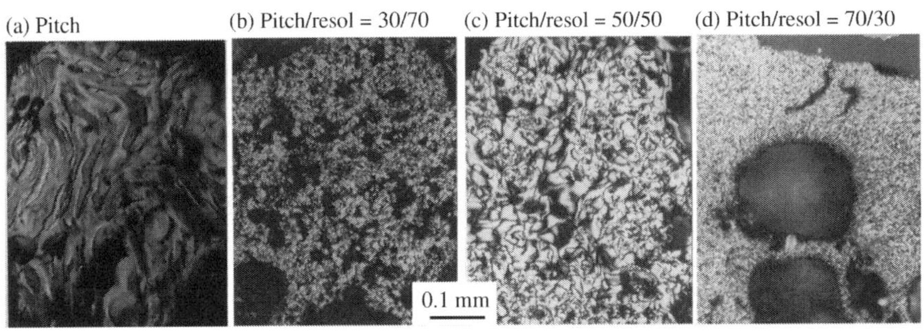


**FIGURE 2.59**

Polarized-light micrographs of the coke prepared from the mixture of pitch and resol-type phenol resin [134].

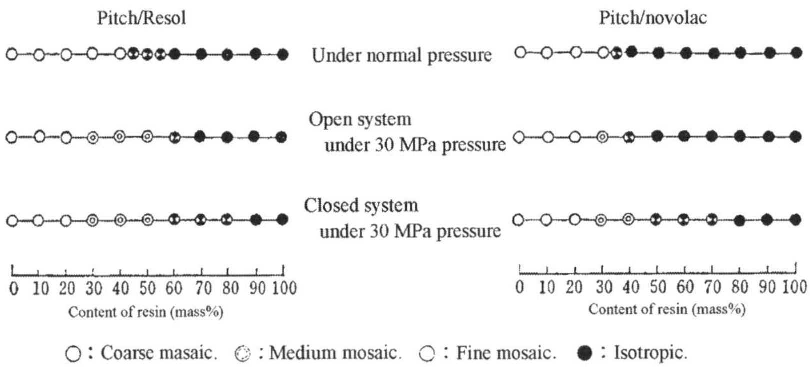


**FIGURE 2.60**

Optical texture of the coke as functions of the mixing ratio of phenol resin to pitch and the pressure during carbonization [134].

hydrogen donability of the pitch and hydrogen acceptability of the coal. When the ratio of donability to acceptability was around 1, the mixture gave a fine mosaic texture. An isotropic texture was obtained from the mixtures with the ratio less than 0.8, and coarse mosaic texture from those with the ratio more than 1.

### e. Fractionation of pitches

More detailed fractionation processes for the pitches were proposed [137–139]. One of the fractionation processes proposed is shown in Fig. 2.61 [138], where BS fraction is separated into five fractions with a much narrower range of molecular weights using mixed solvents of benzene and n-hexane. In Table 2.7, the yield and molecular weight of each fraction are tabulated on two pitches from different origins, together with the carbonization yield at 600°C.

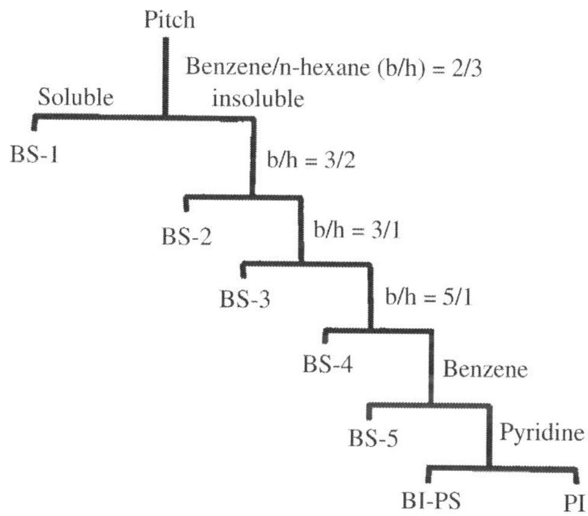


**FIGURE 2.61**

Procedure for the fractionation of the pitch [138].

**Table 2.7** Yield and Number-Averaged Molecular Weight of the Fractions of Two Pitches and the Carbonization Yield at 600°C [138]

| Fraction | Coal Tar Pitch | | | Petroleum Pitch | |
|---|---|---|---|---|---|
| | Fractionation Yield (Mass%) | Molecular Weight | Carbonization Yield (Mass%) | Fractionation Yield (Mass%) | Molecular Weight |
| BS-1 | 17 | 300 | 8 | 28 | 380 |
| BS-2 | 17 | 350 | 13 | 11 | 450 |
| BS 3 | 15 | 390 | 20 | 14 | 590 |
| BS-4 | 11 | 420 | 22 | 3 | 750 |
| BS-5 | 7 | 460 | 26 | 2 | 910 |
| BI-PS | 27 | – | 75 | 27 | – |
| PI | 8 | – | 88 | 14 | – |

By selecting non-polar solvents, benzene and n-hexane, the pitch was fractionated depending on their molecular weight. The molecular weights of the fractions BI-PS and PI, however, were not able to be determined, but were reasonably supposed to be much higher than that of the fraction BS-5. Carbonization yield depends strongly on the number-averaged molecular weight of each fraction. Viscoelastic properties of fractionated pitches were discussed in Section 3.1.3b.

## 2.4 Novel techniques for carbonization

### 2.4.1 Template method

Carbonization using template (template carbonization) is the process to replicate the nanotexture of template into carbon material. It gives a high possibility of controling nanotexture from one to three dimensions by selecting a suitable template; one-dimensional carbon nanofibers were synthesized using anodic aluminum oxide films, two-dimensional graphite layers using layered compounds, clay minerals, and three-dimensional nanoporous carbons using zeolite, silica, MgO, metal-organic framework, etc. These techniques are reviewed from different viewpoints [140–144].

The template carbonization technique has firstly been developed for the preparation of thin oriented graphite films using two-dimensional spaces in layered compounds, such as montmorillonite and taeniolite [145–147]. The fundamental scheme of the template method is shown in Fig. 2.62; carbon precursor such as acrylonitrile molecules is intercalated into the two-dimensional gallery of montmorillonite (MONT), followed by polymerization to polyacrylonitrile (PAN) in the gallery and then carbonized at 700°C. After removal of montmorillonite components by HF and

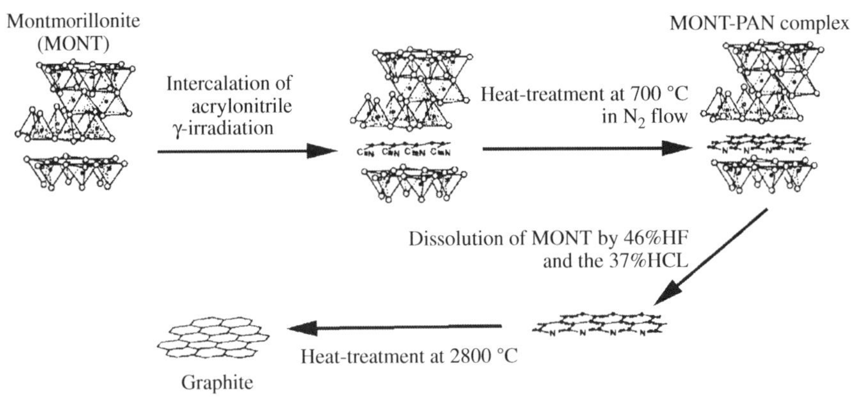


**FIGURE 2.62**

Scheme of template method applied to montmorillonite lamellae.

*(Courtesy of Prof. T. Kyotani of Tohoku Univ.)*

HCl, flaky carbon particles are obtained [146]. The flakes heat-treated up to 2800°C show the 002 lattice fringe image under transmission electron microscope (TEM) similar to graphite and the negligibly small D-band in Raman spectrum.

Carbon nanofibers have also been prepared by carbon deposition from propylene gas at 800°C on the inner walls of nano-sized channels in the anodic aluminum oxide (AAO) film [148,149]. The preparation procedure is schematically shown in Fig. 2.63; carbon deposition on the pore wall results in nanofibers after the dissolution of AAO film either by HF at room temperature or NaOH aqueous solution at 150°C in an autoclave. Scanning electron microscope (SEM) and TEM images of carbon nanofibers prepared by using AAO film with 30 nm diameter channels are shown in Fig. 2.64. In the as-prepared tubes, carbon layers are small and do not yet have a perfect orientation along the tube axis, but they grow markedly and align perfectly along the tube axis after heat treatment at a high temperature as 2800°C. The carbon tubes prepared have very uniform diameter and length, and align along the channels of AAO film (Fig. 2.64a and b), which are perpendicular to the film surface. The diameter and length of nanofibers can be controlled by the diameter of channels and thickness of the AAO film, respectively.

In the as-prepared tubes, carbon layers are small and do not yet have a perfect orientation along the tube axis, as shown in Fig. 2.64e, but they grow markedly and align perfectly along the tube axis after heat treatment at 2800°C, the resultant carbons being comparable to multi-walled carbon nanotubes synthesized by arc discharging.

This preparation technique has different advantages for the control of structure of carbon nanofibers, in addition to high homogeneity in diameter and length of tubes. One of the advantages is the fact that fibers are fixed inside the channel of template aluminum oxide. It gives a possibility to fill the inside of the nanofibers by either metals or metal oxides to prepare nanowires which are expected to be useful for nanotechnology [150–153]. The inner surface of these nanofibers is

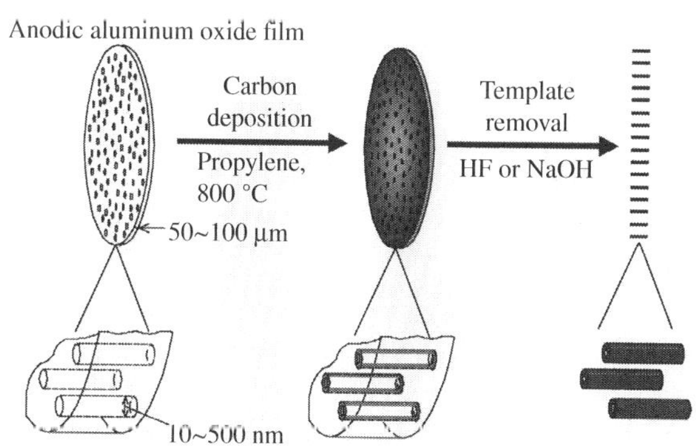


**FIGURE 2.63**

Preparation procedure of carbon nanofibers through template carbonization technique using AAO film.

*(Courtesy of Prof. Kyotani of Tohoku Univ.)*

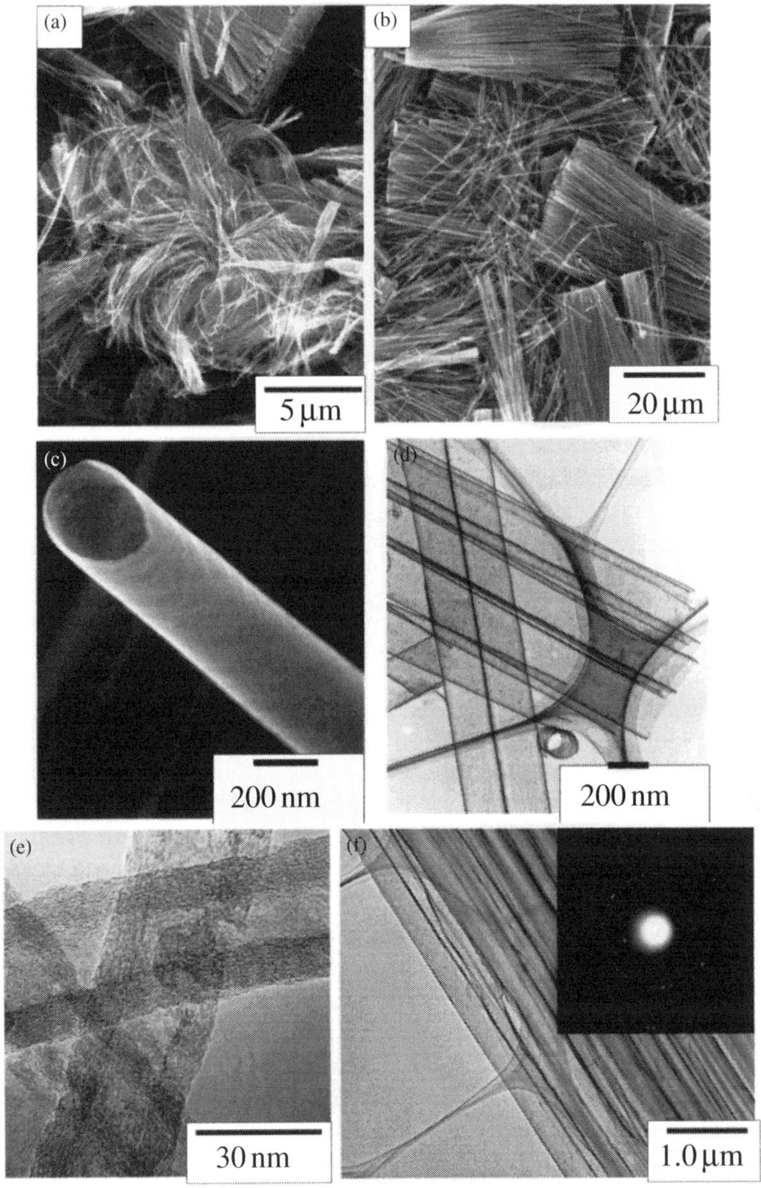


**FIGURE 2.64**

SEM and TEM images of carbon nanofibers prepared. (a)–(c) SEM images with different magnifications, (d)–(f) TEM images, a) to e) as-prepared and f) 2800°C-treated carbon nanofibers.

*(Courtesy of Prof. Kyotani of Tohoku Univ.)*

also able to modify by fluorination [154] and oxidation [155], because only inner surfaces of tubes are able to be exposed to the atmosphere.

By liquid phase carbonization of poly(vinyl chloride) and poly(vinyl alcohol) by using alumina-based templates, anodic aluminum oxide films and tunnel-etched aluminum films with different hole diameters, carbon nanofibers having the same diameter as the template holes are obtained [156]. The nanofiber prepared has platelet nanotexture without a hollow tube at its center. In Fig. 2.65, SEM and TEM images of these nanofibers are shown.

Microporous carbons with a high specific surface area, higher than 3000 $m^2/g$, have been prepared using three-dimensional channels of zeolite, of which size and shape are strictly defined by the crystal structure of zeolite used [157–159]. In Fig. 2.66, SEM images of original zeolite, zeolite after poly(acrylonitrile) (PAN) deposition in its channels and carbon after removing zeolite network are shown. Three types of zeolites, zeolite Y, β and L, which have different pore interconnections, have been used for template carbonization, of which the nanotextural features are preserved on the resultant templated carbons.

By using mesoporous silicas, di- and tri-block copolymers, metal organic frameworks (MOFs) and MgO as templates, mesoporous carbons of different pore geometry have been synthesized. Ordered mesoporous structures of silicas, which are formed by templating a self-assembly of surfactants (MCM-48, MCM-41, SBA-1 and SBA-15, etc.), are successfully inherited into carbons to form ordered mesoporous carbons. The pores (channels) in the template silicas are replicated in the carbons by keeping the pore symmetry via either impregnation or CVI of a carbon precursor, followed by carbonization and removal of the templates [160–164]. The resultant carbon has SBET of 1380 $m^2/g$ and ordered channels of the width of 3.0 nm, together with micropores of 0.5–0.8 nm width; Vmeso of 1.1 $cm^3/g$ and Vmicro of 0.3 $cm^3/g$. The same surfactants used for the preparation of mesoporous silicas have also been used for the solvent evaporation-induced self assembly of the mixture of a carbon precursor with a surfactant block

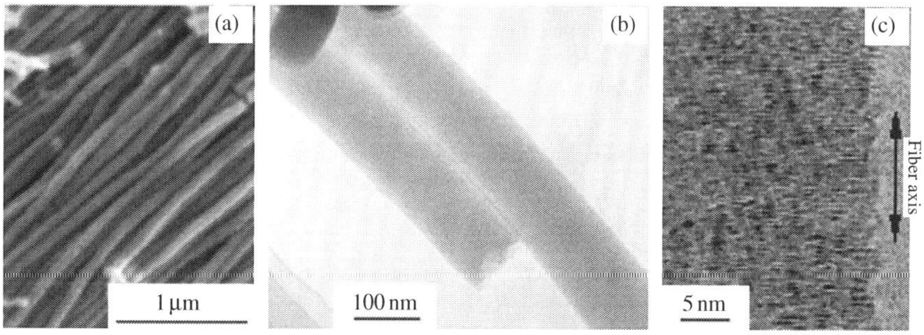


**FIGURE 2.65**

SEM and TEM images of the nanofibers formed from PVC with AAO film and heat-treated at 1500°C [156].

copolymers [165–167]. The channels with a diameter of 34 nm and a wall thickness of 9.0 nm are formed perpendicularly in the carbon film. In Fig. 2.67, SEM images are shown on the carbon prepared from resorcinol/formaldehyde and triethyl orthoacetate by using the triblock copolymer Pluronic F127 as template [166]. Metal-organic frameworks (MOFs) also offer a potential as a template to synthesize nanoporous carbon materials, because they have nano-scaled cavities and open channels [168–171]. By using MOF-5 ($Zn_4O(OOCC_6H_4COO)_3$) and furfuryl alcohol, mesoporous carbon with $S_{BET}$ of 2872 $m^2/g$ and pore volume of 2.06 $cm^3/g$ has been obtained [168].

**FIGURE 2.66**

SEM images of zeolite NaY (a), zeolite/PAN carbon (b) and the carbon recovered (c).

*(Courtesy of Prof. T. Kyotani of Tohoku Univ.)*

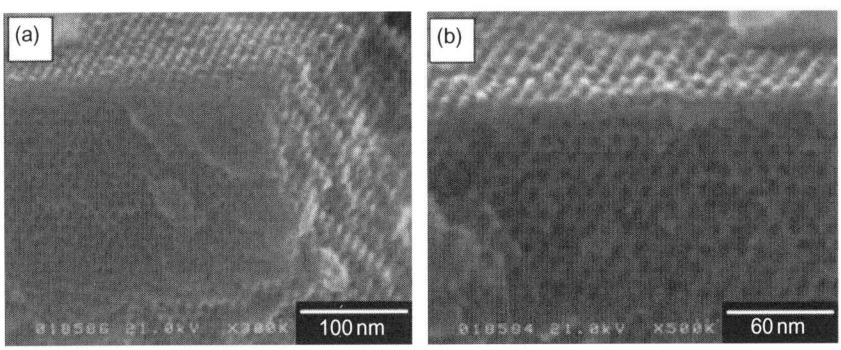


**FIGURE 2.67**

SEM images of mesoporous carbon prepared from resolcinol/formaldehyde via triblock copolymer template at 800°C [166].

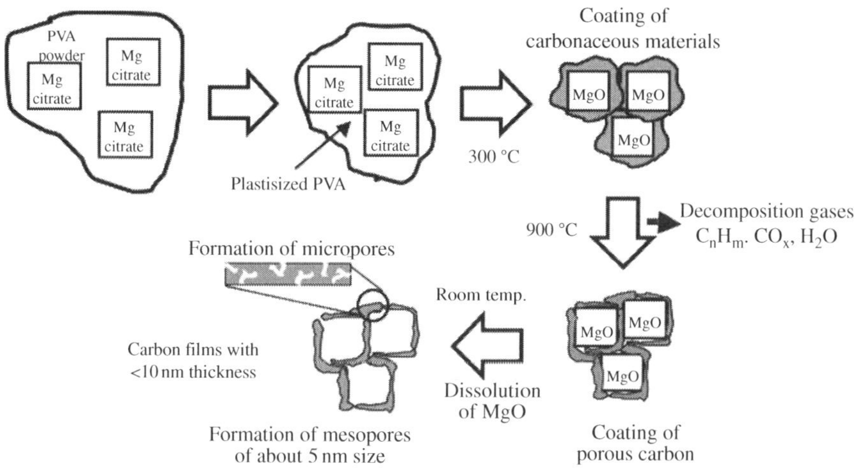


**FIGURE 2.68**

Scheme for the formation of MgO-templated porous carbon from Mg citrate and poly(vinyl alcohol) (PVA).

A mixture of MgO precursor, such as Mg acetate, Mg citrate etc., which gives nano-sized MgO particles after its pyrolysis, with carbon precursor is heat-treated at 900°C in inert atmosphere, and then template MgO is dissolved out using a diluted acid at room temperature to isolate the mesoporous carbon formed [172,173]. The scheme for the formation of mesoporous carbon via MgO-template process is shown in Fig. 2.68. By using Mg citrate as template, the pore size distribution of the resultant carbons shows a sharp maximum at about 5 nm, and $S_{meso}$ and $V_{meso}$ reaches about 1600 $m^2/g$ and 1.7 $cm^3/g$, respectively. MgO is easily dissolved out by citric acid, which can be used as template, in other words, template MgO can be recycled.

A carbon material with honeycomb-type open cell structure (Reticulated Vitreous Carbon, RVC) has been developed by carbonization of the polyurethane (PU) foam impregnated with phenol, furfuryl alcohol or epoxy resin [174,175]. Carbon foams are prepared using PU foam as template with impregnation of poly (amide acid) followed by imidization and carbonization [176]. In this process, there is no need to take off the template because poly(urethane) gives only a small amount of carbon residues after carbonization.

## 2.4.2 Polymer blend method

In order to control the pore structure in carbon materials, blending of two kinds of carbon precursors, one giving a relatively high carbonization yield and the other having a very low yield, has been proposed and called polymer blend method [176]. This idea gives certain success to prepare macroporous carbons through the synthesis of poly(urethane-imide) films [177–181]. Poly(urethane-imide) films are prepared by blending poly(amide acid) and phenol-terminated polyurethane prepolymers. Poly(urethane-imide) films after heating up to 200°C show phase-separation of polyimide (PI) and polyurethane (PU), where the former forms the matrix and the latter forms the small islands. By heat treatment up to 400°C, the PU component is pyrolyzed to gases and results in porous polyimide films. Porous polyimide films thus obtained are able to convert easily to porous carbon films by carbonization in an inert atmosphere. Pore sizes in these polyimide and carbon films are controlled by changing the blending ratio of PI to PU and also the molecular structure of PU. The carbon films prepared are shown to be suitable as a medium for culturing biological cells, macropores in the range of 0.6 to 3.0 μm sizes being shown to be preferable for the cells [182].

By coupling polymer blend method with spinning, carbon nanofibers have successfully been prepared [183,184]. The preparation procedure is schematically shown in Fig. 2.69. The first and the most important step of this procedure is the preparation of fine spheres of polymers with core-shell structure (microcapsules). The polymer consisting of the core is the one which disappears during heat treatment at high temperatures (pore forming polymer, polyethylene PE or poly(methylmethacrylate) PMMA) and that of the shell is the one which gives carbon after carbonization at high temperatures (carbon precursor polymer, phenol-formaldehyde resin PF or poly(acrylonitrile) PAN). As illustrated in Fig. 2.69, the mixture of microcapsules with PMMA is subjected to melt-spinning, followed by heating in dry air at 250°C for 5 h. In the resultant polymer fibers, PMMA forms the matrix of the fibers and inner core of extended thin filaments of PAN, and they are converted to thin carbon tubes, carbon nanofibers, by carbonization at 900–1000°C in an inert atmosphere. Representative TEM images of carbon nanofibers thus prepared are shown in Fig. 2.70.

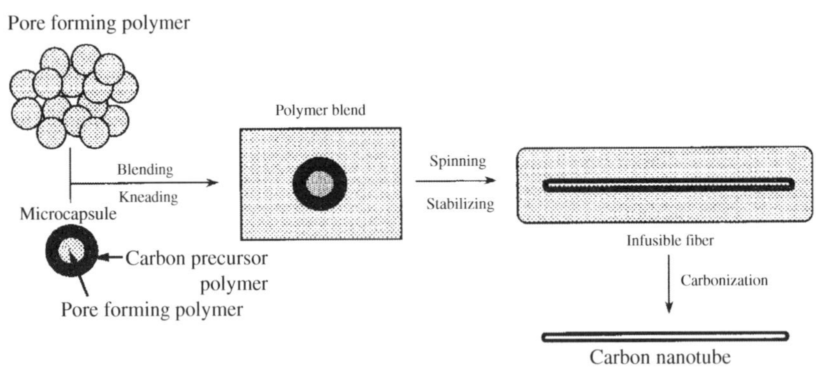


**FIGURE 2.69**

Preparation procedure of carbon nanofibers through polymer blend technique coupled with spinning.

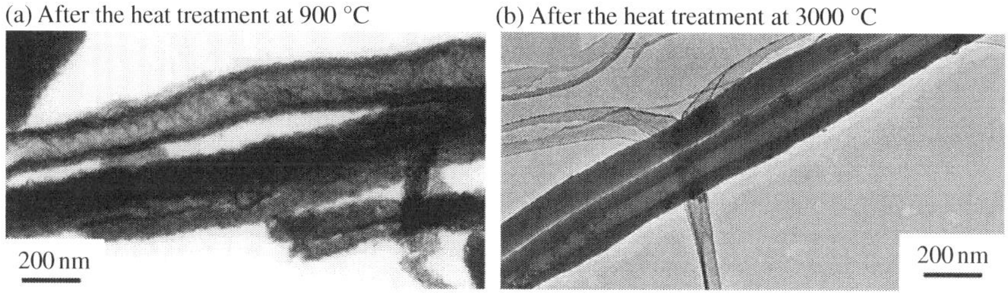


**FIGURE 2.70**

TEM images of carbon nanofibers prepared through polymer blend technique.

*(Courtesy of Prof. A. Oya of Gunma Univ.)*

This technique has been developed for the preparation of very fine carbon fibers with a diameter of about 200 nm by using microspheres of PAN mixed with PMMA spheres [185]. By designing the inside structure of microcapsules, carbon materials with different morphologies can possibly be prepared [186]. When the microspheres of carbon precursor polymer with pore-forming polymer as the matrix are used, the carbon tube containing a large number of thin carbon fibers is possible to be prepared. When the microsphere consists of alternative multilayers of pore-forming and carbon precursor polymers, multi-walled carbon tubes can be prepared.

### 2.4.3 Electrospinning

Electrospinning has been used to produce various polymer fibers with diameters from a few tens of nanometers to a few micrometers as different forms of nonwoven mats (webs), yarns, etc. This technique is reviewed from different viewpoints; focusing on polymer nanofibers [187–192], but it is particularly interesting for the preparation of carbon nanofibers [193,194]. It is a relatively simple and low-cost strategy to produce continuous nanofibers from polymer solutions or melts, as shown in Fig. 2.71. A viscoelastic solution of polymers is charged by a high potential difference between the spinneret and grounded target. The repulsion among the charges on the surface of the droplet at the tip of the spinneret competes with the surface tension, which tends to stabilize the droplet. Once a critical condition is reached at the point where surface charge repulsion dominates, a jet is drawn from the spinneret under a constant flow rate. As the solvent evaporates, the jet solidifies to form thin fibers, which are deposited on the grounded target (or collector).

More than 100 polymers have been used to produce their nanofibers via electrospinning, but the carbonization of electrospun polymer nanofibers to carbon nanofibers is rather limited, as polyacrylonitrile (PAN), polyimide (PI), poly(vinyl alcohol) (PVA), poly(vinyliden fluoride) (PVDF) and pitches [193]. In order to convert electrospun polymer nanofibers to carbon nanofibers, carbonization process at around 1000°C has to be applied. For the carbon precursors, such as PAN

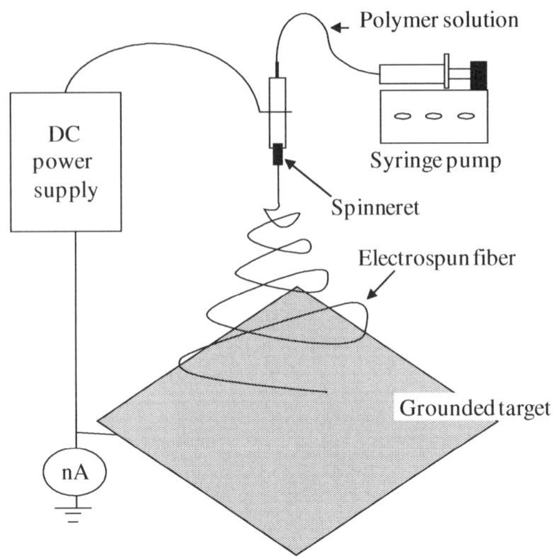


**FIGURE 2.71**

Scheme of fundamental setup for electrospinning.

and pitches, so-called stabilization process before carbonization is essential to keep fibrous morphology, as in the production of PAN-based carbon fibers. During stabilization and carbonization of polymer nanofibers, they showed significant weight loss and shrinkage, resulting in a decrease in fiber diameter.

PAN has been commonly electrospun into high-quality fibers with various diameters and also converted to carbon nanofibers. Structural analysis was performed on carbon nanofibers, which were prepared from PAN/DMF solution by carbonization at 750°C followed by 1100°C [195]. The resultant carbon nanofibers had average diameters of 110 nm and skin-core heterogeneity; the skin carbon layers being oriented predominantly parallel to the fiber surface. The diameter of nanofibers was approximately 330 nm for as-spun, 250 nm for 1000°C-treated and 220 nm for 1800°C-treated [196]. Hollow carbon nanofibers were synthesized by electrospinning of DMF solution of PAN and PMMA in different ratios, followed by carbonization at 1000°C [197,198].

Carbon nanofibers prepared from a PI (PMDA/ODA) with the diameter less than 2–3 μm by electrospinning could give relatively high tensile strength as 74 MPa and electrical conductivity of 5.3 S/cm after the heat treatment at 2200°C [199]. The addition of PAN in poly(amic acid) solution improved spinnability and decreased the diameter of resultant carbon nanofibers [200]. In Fig. 2.72, an SEM image of fibers and diameter distribution is shown for different mixing ratios of PAN/polyimide. From the mixtures of PAN/polyimide = 70/30 and 30/70, carbon nanofibers with the average diameter of 350 and 700 nm, respectively, were obtained, pure PAN and polyimide resulting in the diameter of about 300 and 1000 nm [201].

From novolac- and resol-type phenolic resins, microporous carbon nanofibers were easily prepared via electrospinning and following carbonization [202–204]. The nanofibers carbonized at 1000°C gave large $S_{BET}$ of 860 $m^2/g$ and total pore volume of 0.365 $cm^3/g$ without any activation process [204]. SEM images for the nanofibers carbonized at 1000°C and $N_2$ adsorption isotherms for the nanofibers carbonized at different temperatures are shown in Fig. 2.73.

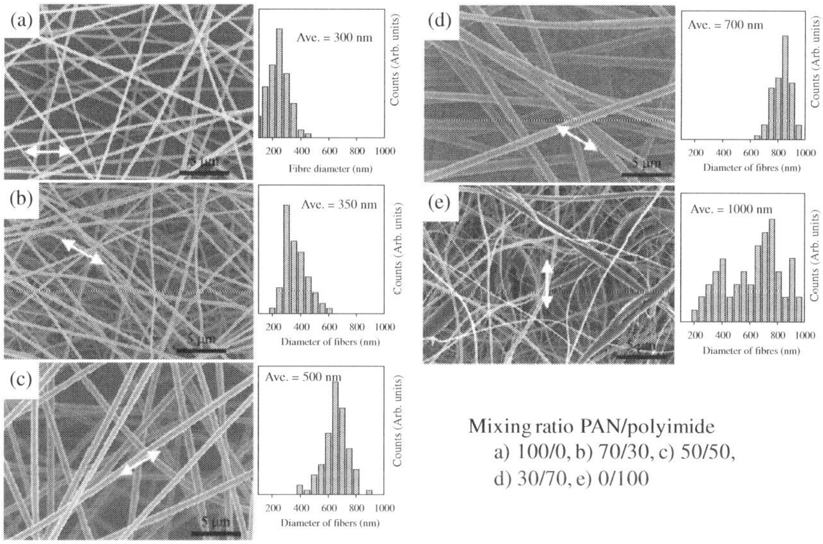


**FIGURE 2.72**

SEM images and diameter distribution of electrospun carbon nanofibers prepared from the mixtures of PAN and polyimide [201].

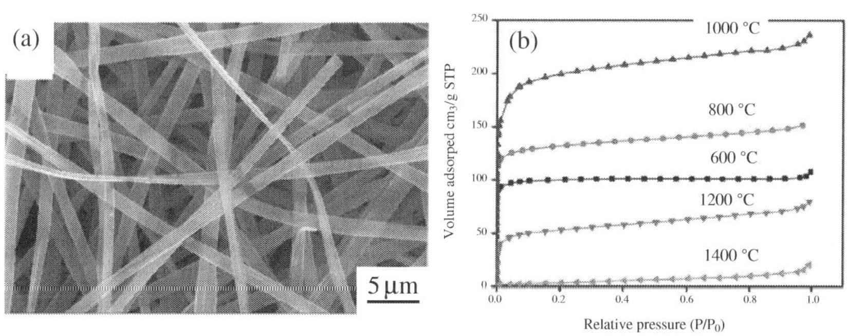


**FIGURE 2.73**

Carbon nanofibers prepared from phenol resin: (a) SEM image and (b) $N_2$ gas adsorption isotherm at 77 K for different carbonization temperatures [204].

### 2.4.4 Pressure carbonization

On the process of carbonization, a part of decomposition products from carbon precursors becomes gaseous phases and various hydrocarbon gases come out, which makes the final carbon yield lower. If the carbonization process is performed under pressure, gaseous decomposition products can be carbonized to give solid carbon and, as a consequence, a high carbon yield is expected. In addition, it gives the resultant carbons a nanotexture different from that obtained under normal and reduced pressures, and results in carbon particles with a characteristic texture and morphology. The pressure carbonization process has been realized either by applying external compression or by build-up of pressure due to gas formation in the closed vessel. Whether the pressure vessel is heated homogeneously or not is a very important factor. If not, a certain part of gaseous decomposition products is deposited on the low temperature place(s) and so excluded from the carbonization system (open system under pressure). If the carbon precursor is included into a gold tube, which is backed up by gas pressure in the autoclave, all gaseous products are kept in the reaction system and are possible to convert to carbonaceous materials (closed system under pressure).

Yield from a pitch increases to more than 90 mass% by the carbonization in a closed system under 10–30 MPa pressure at 700°C [205]. The carbonization yields of fractionated pitches by using benzene and pyridine (refer to Fig. 2.61) are plotted on number-averaged molecular weights of fractions as a function of carbonization conditions in Fig. 2.70 [206]. Under atmospheric pressure, carbonization yield is relatively low, depending strongly on molecular weight. Carbonization under 30 MPa pressure but open system gives a somewhat larger yield. Under 30 MPa pressure in a closed gold tube, all fractions give the yield more than 90 mass%. On benzene-insoluble but pyridine-soluble fraction (BI-PS) and pyridine-insoluble fraction (PI), which are supposed to have high molecular weights, high carbon yields are observed even under atmospheric pressure. Carbonization under pressure in a closed system gives higher yield than under pressure in an open system. For BI-PS fraction, optical texture of the resultant carbon changes markedly, from isotropic under pressure in open system to coarse fibrous texture under pressure in closed system, although the increase in carbon yield is not so pronounced. Benzene-soluble (BS) fraction gives coarse mosaic texture irrespective of carbonization conditions, although the yield depends strongly on the conditions (Fig. 2.74).

Under high pressure, mesophase spheres and their coalescence to bulk mesophase occur on the course of carbonization, as under atmospheric pressure [205]. A distribution in the concentration of mesophase spheres has been observed along the height of the pitch, from only a few mesophase spheres at the top to bulk mesophase at the bottom through a high density of mesophase spheres. It is worthwhile to mention that the density of mesophase spheres is very high at the middle of the sample.

Carbonization under a pressure of 200 MPa at 600°C gives spherical carbons from anthracene [207] and divinylbenznee [208,209]. The carbon spheres formed from anthracene have an anisotropic nanotexture, similar to the one of mesophase spheres, but those from divinylbenzene are isotropic.

Carbonization of the powder mixtures of poly(ethylene) (PE) with either 5–20 mass% poly(vinyl chloride) (PVC) or poly(vinylidene chloride) (PVDC) under a pressure of 10–30 MPa at a temperature above 600°C in a closed gold tube gave carbon spheres with homogeneous size with relatively high yield [210–216]. In Fig. 2.75, SEM images of carbon spheres obtained are shown.

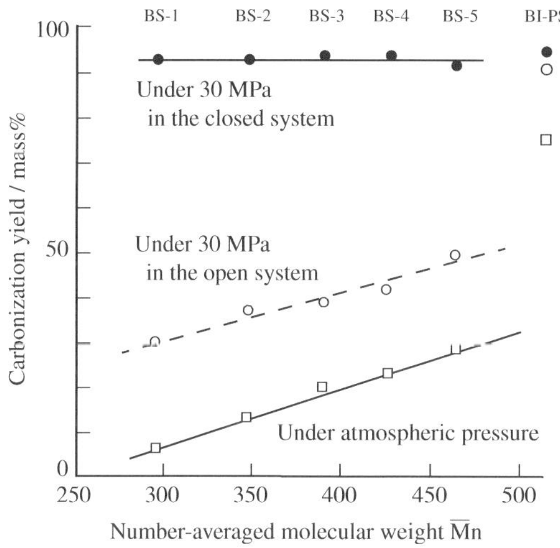


**FIGURE 2.74**

Dependences of carbonization yield at 650°C under different pressure conditions on number-averaged molecular weight of fractionated pitches [206].

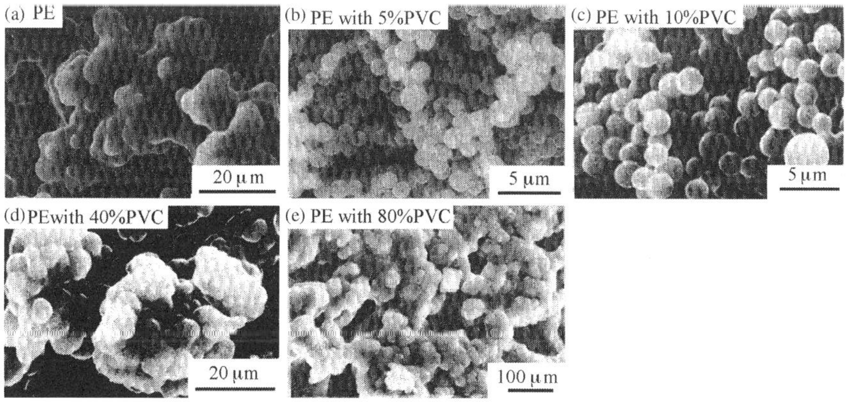


**FIGURE 2.75**

SEM images of carbon spherules obtained from the mixtures of PE with different amounts of PVC by pressure carbonization (30 MPa, 650°C, 1 h) [210].

In Fig. 2.76, changes in the yield with carbonization temperature under 30 MPa pressure are shown on the mixtures of PE and PVC [211]. PVC is converted to carbonaceous materials, which has been known to be very active and is called polyene, by releasing HCl below 300°C (Fig. 2.76e). However, PE starts to decompose at around 540°C and gives low-molecular-weight products, which can volatilize under normal pressure (Fig. 2.76a). In Fig. 2.76, the upper curve shows the yield measured just after opening the gold tube after carbonization under pressure and the lower curve shows the yield after leaving for a long time in air. Therefore, the part hatched in the figure shows the weight loss in air due to vaporization of the carbonization products under pressure, which are reasonably supposed to be low-molecular-weight products formed from PE and the mixture of PE with PVC. From the mixtures of these two precursors, separated spheres with homogeneous size are obtained above 600°C when the mixing ratio of PVC is less than 20 mass% (Fig. 2.75b and c). When PVC is mixed in more than 20 mass%, the relative percentage of spheres decreases with increasing PVC content and the spheres formed coagulated with each other (Fig. 2.75d). Above 80 mass% PVC, porous carbons like coke are obtained, no more spheres being detected (Fig. 2.75e). From these experimental results, it is concluded that the coexistence of active polyene, which is formed by the departure of HCl from PVC, with a large amount of

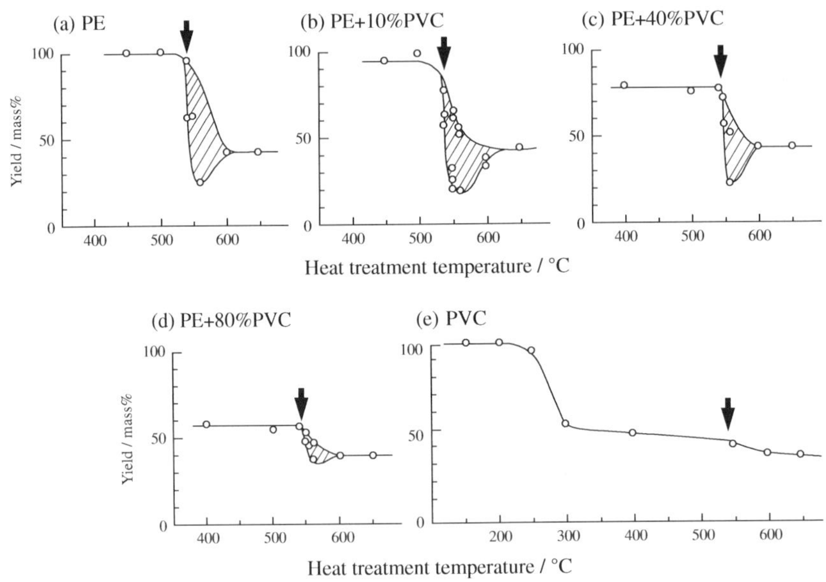


**FIGURE 2.76**

Changes in yield after heat treatment up to different temperatures on the mixtures of PE and PVC [210].

low-molecular-weight carbonaceous materials under pressure is an essential condition to give spherical carbons (carbon spherules). Based on this conclusion, different combinations of two precursors, either poly(propylene) PP and poly(stylene) PSt as the precursor for low-molecular-weight materials and either poly(vinyl chloride) PVC or poly(vinylidene chloride) PVDC as the precursor of polyene, have been proved to give carbon spherules under pressure [212]. Benzene-soluble fraction of a pitch is also used as the precursor for low-molecular-weight materials [215].

By selecting an appropriate condition for the synthesis, carbon spherules with homogeneous size without any coalescence, as shown in Fig. 2.77, are obtained. Nanotexture of these carbon spherules is clarified by detailed analysis of 002 dark-field images on the cross-sections of these spherules [21]. In the figure, the nanotexture of different cross-sections of the spherule, which are assumed from the model, because only cross-sections of the spherules are able to be observed by thin-sectioning of the spherules embedded in a resin. The carbon spherules obtained by pressure carbonization have radial point orientation of hexagonal carbon layers, as shown in Fig. 2.78. The nanotexture around the central part of the sphere is different in the spherules obtained by pressure carbonization from the mesophase spheres, in the former radial but in the latter planar orientation, as shown in Fig. 2.20c and b, respectively. This is the reason why the former is called carbon spherules to differentiate from mesophase spheres.

In supercritical water (above 374°C and 22.1 MPa), single carbon spheres with submicron size and a nanotexture of concentric point orientation have been obtained from the mixture of benzene with 53 mass% of hydrogen peroxide ($H_2O_2$) at 400°C and 48 MPa [217].

(a) From the mixture of PE with 5 mass% PVC

(b) From the mixture of PE with 5 mass% PVDC

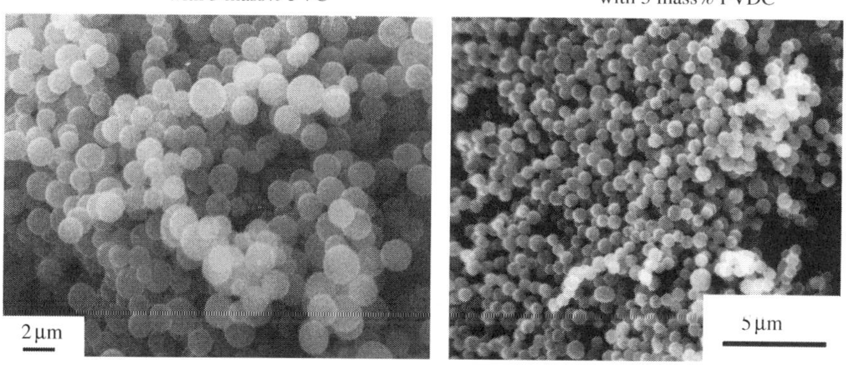


**FIGURE 2.77**

Carbon spherules with homogeneous size synthesized under 30 MPa pressure at 650°C [210].

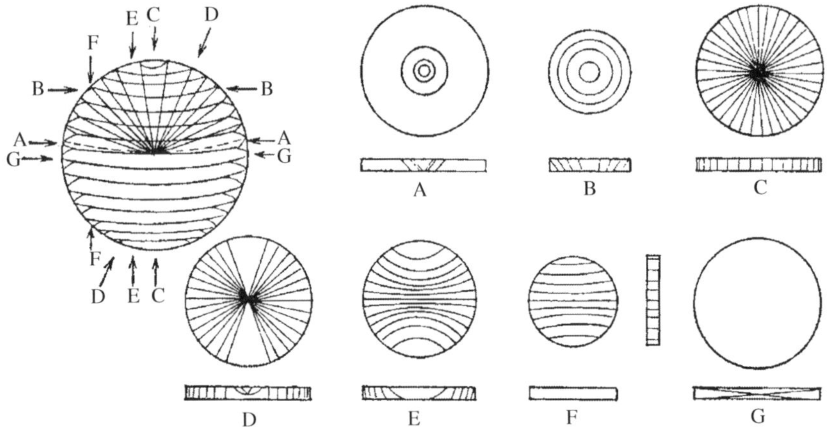


**FIGURE 2.78**

Model of carbon spherules obtained by pressure carbonization [21]. From A to G, textures in different cross-sections are shown on the basis of the nanotexture model.

Hydrothermal carbonization was applied to various biomasses, including sucrose, glucose, fructose, cyclodextrin, starch, etc., and also to biomass derivatives for the synthesis of carbon spheres below 240°C [218–224]. The size of carbon spheres depends strongly on the carbonization conditions, the precursor saccharide used and its concentration, hydrothermal temperature and holding time, etc. Under the hydrothermal conditions at 190°C, the size is about 0.25 μm from 0.15 mol/L solution and about 5 μm from 1.5–3.0 mol/L [218]. The spheres synthesized below 240°C can keep their spherical morphology after heat treatment at 1000°C in argon flow [219].

## 2.4.5 High-yield carbonization

Most of the precursors that transform to carbon materials, such as pitches with different origins, phenol resins of different molecular structure, and various hydrocarbons, such as poly(vinyl chloride) and poly(furfuryl alcohol), have relatively low carbon yields, because some parts of carbon atoms composing precursor molecules are converted to volatile components. Various trials and efforts have been exerted to increase the carbon yield (high-yield carbonization).

Treatment of pitches with iodine vapor at 90°C has been found to affect strongly their carbonization behavior, resulting in high carbon yields, changes in optical texture from flow-type texture to either mosaic-type or isotropic texture, and suppression of graphitization at high temperatures [225–227]. Yield after 800°C carbonization increases with increasing time of iodine treatment, as shown in Fig. 2.79, on a coal tar pitch [228]. Carbon yield increases markedly in the beginning of iodine treatment and then tends to saturate. It is worthwhile to point

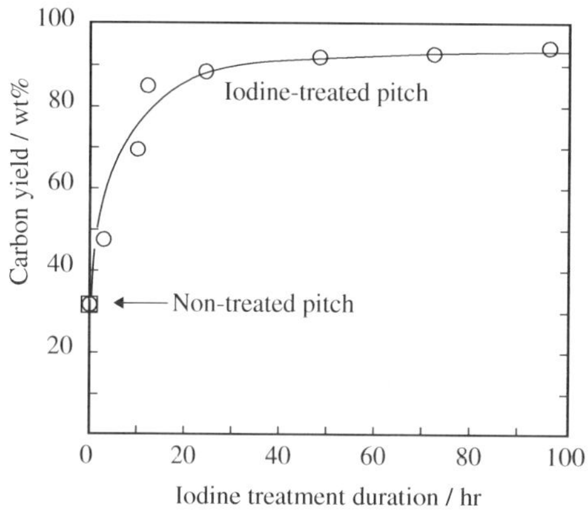


**FIGURE 2.79**

Change in yield after 800°C carbonization with iodine treatment time.

*(Courtesy of Prof. E. Yasuda of Tokyo Inst. Tech.)*

out that the yield obtained is close to 100 mass%. Iodine treatment is found to be mostly effective for BS fractions in pitches by increasing their carbonization yield and changing their optical texture, and no detectable modification is observed for other fractions, BI-PS and PI fractions [227]. Ultrafine particles of pitch after iodine treatment can keep their morphology even after carbonization up to 800°C to give carbon particles of about 50 nm [226].

Mechanism of iodine treatment has been studied by using different techniques, such as $^{127}$I-NMR, EPR, FT-IR, TG-MS, etc. [228,229]. The amount of iodine absorbed into pitch at 100°C, iodine penetrating easily into the pitch tablet, increases with increasing treatment time. Iodine absorbed into pitch can be divided into two; iodine which can be washed out by ethanol and that remains even after ethanol washing. The latter iodine is supposed to be kept in a constant amount in the pitch, irrespective of treatment time, because an almost constant value of atomic ratio I/C of around 0.045 is obtained after ethanol washing, and to be in ionic state, based on $^{127}$I-NMR and EPR measurements. The experimental results on iodine-treated pitches after ethanol washing suggests the formation of charge-transfer complexes of relatively large-sized aromatic components with 10 or more benzene rings with iodine. The former iodine, washable with ethanol, is supposed to be weakly bound to pitch molecules and to be distributed uniformly in the pitch. During heat treatment to high temperatures to carbonize, the former acts as a dehydrogenation reagent below 200°C, even during iodine treatment at 100°C, the latter decomposes above 200°C associating dehydrogenation reaction. The mechanism proposed is illustrated in Fig. 2.80. These dehydrogenation reactions at low temperatures promote polymerization of aromatic molecules in the pitch before going to grow, which leads to isotropic texture and low graphitizability of the resulting carbons.

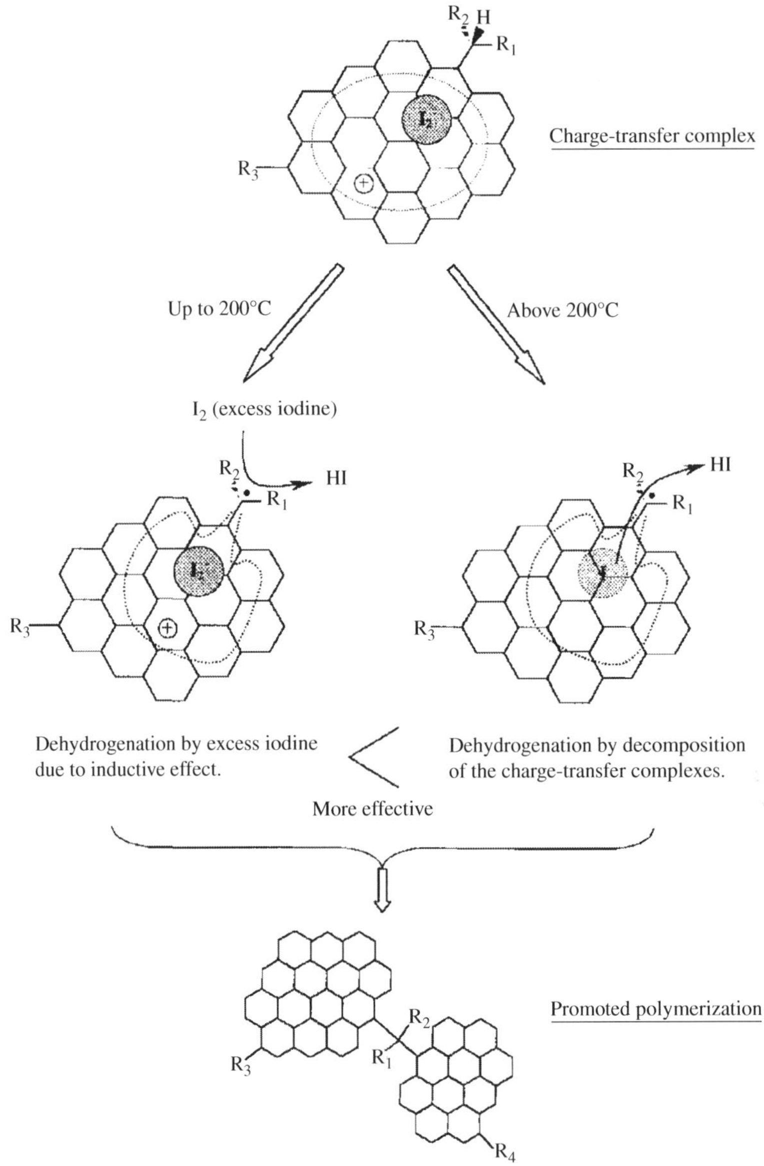


**FIGURE 2.80**

Mechanism of dehydrogenation of pitch by iodine.

*(Courtesy of Prof. E. Yasuda of Tokyo Inst. Tech.)*

The effect of this iodine treatment is confirmed using different model compounds with different molecular structures [230]. Only the compounds containing condensed aromatic rings in their molecules, such as anthracene and perylene, show significant effect, i.e., increase in carbon yield, revealing an agreement with the above-mentioned mechanism, in other words, the formation of charge-transfer complexes with iodine.

This iodine treatment on the production of carbon–carbon composites has successfully been applied to modify the carbon yield and texture of matrix carbon [231].

Acetylene terminated resin and poly(silyleneethylene)s were reported to give high carbon yields above 80%, but the inclusion of hetero-atoms in resultant carbons was inevitable [232,233].

Recently, poly(phenylene butadiynylene)s were found to have very high carbon yields without hetero-atoms [234,235]. The molecular structures of poly(phenylene butadiynylene)s are shown in Fig. 2.81, where carbon yields after 900°C treatment and carbon contents in molecular structures (i.e., theoretical carbon yield) are listed. In these molecules **A**, **B** and **C**, benzene rings are linked through butadiynylene (–C≡C–C≡C–) in *para-*, *metha-* and *ortho*-positions, respectively. They are synthesized from the corresponding diethynylbenzene by oxidative polycondensation using a CuCl-N,N,N′,N′-tetramethylethylen-diamine catalyst. In Fig. 2.82, a representative TG-DTA curve is shown. Their pyrolysis behavior is characterized as follows: a very sharp exothermic peak at around 200°C without any accompanying weight change, and a little but gradual weight loss around 600°C. The exothermic peak at around 200°C is concluded to be due to 1,4-polymerization of butadiynylene moiety and to result in cross-linking between molecules. These materials are so highly and strongly cross-linked that hydrogen atoms remaining are mostly stripped off as hydrogen molecules. Carbon yield from these three molecules after the treatment at 900°C is more than 90 mass%, very close to carbon content (97 mass%). The derivatives with methyl radicals on benzene ring give also high carbon yields, close to carbon content of about 80 mass% [234].

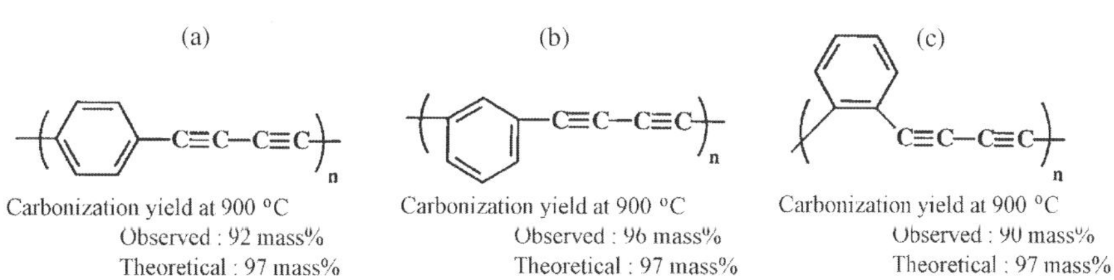


**FIGURE 2.81**

Molecular structures of poly(phenylene butadiynylene).

*(Courtesy by Prof. M. Kijima of Tsukuba Univ.)*

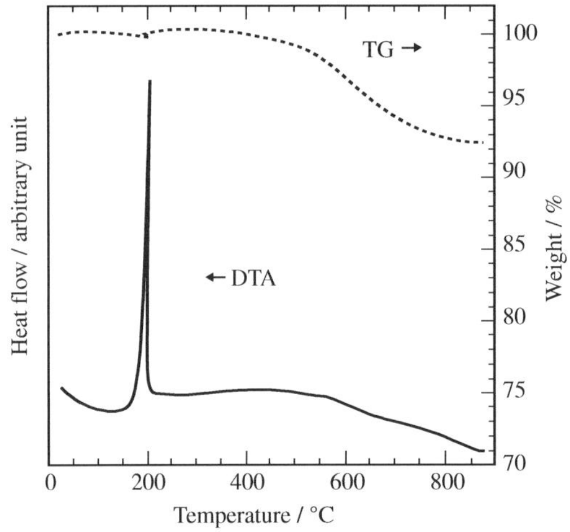


**FIGURE 2.82**

TG-DTA thermograms of poly(phenylene butadiynylene).

*(Courtesy by Prof. M. Kijima of Tsukuba Univ.)*

The resultant carbons are amorphous state and microporous, having total surface area of 1330 $m^2/g$, microporous surface area of 1300 $m^2/g$ and micropore volume of 0.49 $cm^3/g$. They were prepared in a form of either powder or pellet under Ar atmosphere.

## 2.4.6 Low-temperature carbonization

For the carbonization of organic precursors, heating to a high temperature is essential. In order to exclude non-carbon atoms, such as hydrogen and oxygen, from organic precursor molecules, a temperature of more than 400°C is needed and certain loss of carbon atoms can not be avoided because of the departure as small hydrocarbon molecules and carbon oxides. In order to take out hydrogen atoms as hydrogen molecule, heating to a temperature more than 800°C is required. These facts are due to strong chemical bonding of hydrogen and oxygen to carbon. When polyimides are used as precursors, a small amount of nitrogen atoms remains in the resultant carbon, which comes from the precursor imide molecules. A heat treatment above 2500°C is needed for complete elimination of nitrogen [236,237].

If all of the hydrogen atoms in a hydrocarbon molecule can be replaced by a halogen atom, there might be a possibility of eliminating the halogen atoms by using metals. Since the formation of metal halogenide usually occurs at a low temperature, even at room temperature, we may expect the formation of carbon

**Table 2.8** Chemical Composition of the Carbon Prepared Through Defluorination of PTFE by Sodium Metal at 200°C and Heat-Treated at Different Temperatures

| Heat Treatment Temperature | Chemical Composition (Mass %) | | |
|---|---|---|---|
| | C | H | Difference |
| As-prepared (200°C) | 73.82 | 1.53 | 24.66 |
| 400°C | 81.69 | 1.11 | 17.20 |
| 600°C | 90.35 | 1.14 | 8.52 |
| 800°C | 96.47 | 1.05 | 2.49 |
| 1000°C | 92.82 | 0.77 | 6.42 |

*(Courtesy of Dr. O Tanaike of AIST.)*

materials at a low temperature and also high carbon yield, theoretically 100% carbon yield.

Poly(tetrafluoroethylene)(PTFE) has firstly been used as the precursor to react with alkali amalgam for the preparation of porous carbon [238,239]. Then this reaction is applied to synthesize carbynes, because linear structure of the precursor molecules may be retained even after taking off the fluorine atoms [240–242].

Defluorination of PTFE has been studied to prepare the porous carbons [243–245] (Section 3.5). When alkali metals were used for defluorination of PTFE, the reaction proceeded at a low temperature between room temperature and 200°C in a closed vessel. Almost 100% carbon atoms in the precursor PTFE were fixed as a solid carbon, though the carbon content in the starting PTFE is only 24 mass%. It has to be pointed out, however, that the resultant carbon just after defluorination is so reactive that it chemically adsorbs oxygen and water vapor in the air. To eliminate these impurity atoms, it is necessary to heat at a high temperature, because they are chemically bonded to carbon network. In Table 2.8, chemical composition of the carbons obtained through defluorination of PTFE by sodium metal and heat-treated at a high temperature is shown [244]. By heat treatment up to 1000°C, about 30 mass% weight loss is observed.

Defluorination of perfluoropolyimide films by alkali metals was possible to give carbonaceous films [246]. In Fig. 2.83, the changes in Raman spectrum are compared for the carbon prepared by defluorination at 200°C and that prepared by the thermal carbonization at different temperatures above 600°C. The carbon film prepared by defluorination has the structure corresponding to the carbon film prepared by heating to 1000°C. The yield of carbon atoms is almost 100% and the apparent carbonization yield is about 60 mass%. However, it has to be pointed out that the carbon films prepared contain a pretty large amount of impurities, mostly oxygen, after being exposed to air. The carbon films thus obtained at 200°C are porous, rich in micropores, after dissolution of alkalifluoride.

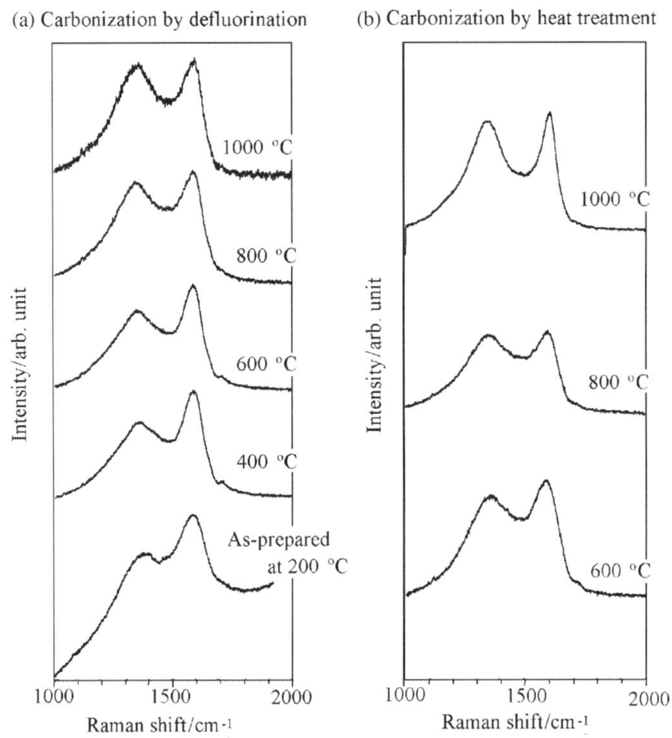


**FIGURE 2.83**

Raman spectra of the carbons prepared from perfluoropolyimide film by defluorination at 200°C and by carbonization at different temperatures.

*(Courtesy of Dr. O. Tanaike of AIST.)*

# 2.5 Graphitization (structure development)

## 2.5.1 Structure parameters

Structure of carbon materials has been characterized by X-ray diffraction, different microscopy techniques, i.e., transmission and scanning electron microscopy, optical microscopy, tunneling scanning microscopy, and Raman spectroscopy. According to the characterization techniques, various structure parameters have been employed. Physical properties, which are structure-sensitive, have also been used for the characterization of structure, magnetoresistance being especially effective for the purpose.

X-ray powder diffraction gives useful information on structure of carbon materials and also on its change with heat treatment [247]. The structure parameters are derived from X-ray diffraction and often used for the characterization of carbon materials in graphite family. The interlayer spacing between neighboring

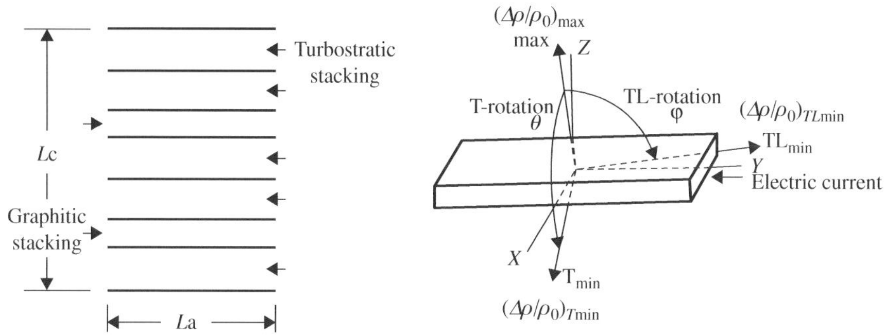


**FIGURE 2.84**

Structure parameters derived from X-ray diffraction (a) and Measurement of magnetoresistance (b).

hexagonal carbon layers, $d_{002}$, corresponds to a half of the c-axis length $c_0$ of the graphite unit cell (Fig. 2.9a) and so is sometimes expressed as $c_0/2$. The a-axis length $a_0$ corresponds to the distance between carbon atoms in the hexagonal carbon layer $d_{C-C}$, i.e., $d_{C-C} = a_0/\sqrt{3}$. The sizes of crystallites, i.e., hexagonal layers stacked in parallel, are also important parameters for the characterization of carbon materials, which have to be differentiated in two directions, parallel and perpendicular to the hexagonal layers, *La* and *Lc*, respectively. The probability for the stacking of two neighboring layers in graphitic relation, $P_1$ [248,249], which are determined through Fourier analysis of diffraction line profiles, is the most exact parameter to define the degree of the development of graphitic structure, graphitization degree. The definitions of some of these parameters are illustrated in Fig. 2.84a. The changes of these X-ray parameters, $d_{002}$, *La*, *Lc* and $P_1$, with heat treatment temperature (HTT) are shown on some carbon materials in Fig. 2.85a–d.

Interlayer spacing for graphitic stacking is known to be 0.3354 nm and that for turbostratic stacking is reported to be about 0.344 nm. Therefore, the observed interlayer spacing $d_{002}$ is an average value depending on the relative ratio of graphitic stacking to the turbostratic one, and so decreases gradually to the value of 0.3354 nm with structure improvement by heat treatment.

The broadening of the diffraction lines is due to the small size of crystallites and the presence of strain in the structure. By ignoring the structural strain, the crystallite size L can be calculated from the full width at the half maximum intensity (FWHM) $\beta$ of each line by Scherrer's equation,

$$L = K\lambda/\beta\cos(\theta), \tag{2.1}$$

where $\lambda$ is the wavelength of X-ray used, K is a constant and $\theta$ is the diffraction angle of the line. The crystallite sizes along the c-axis calculated from

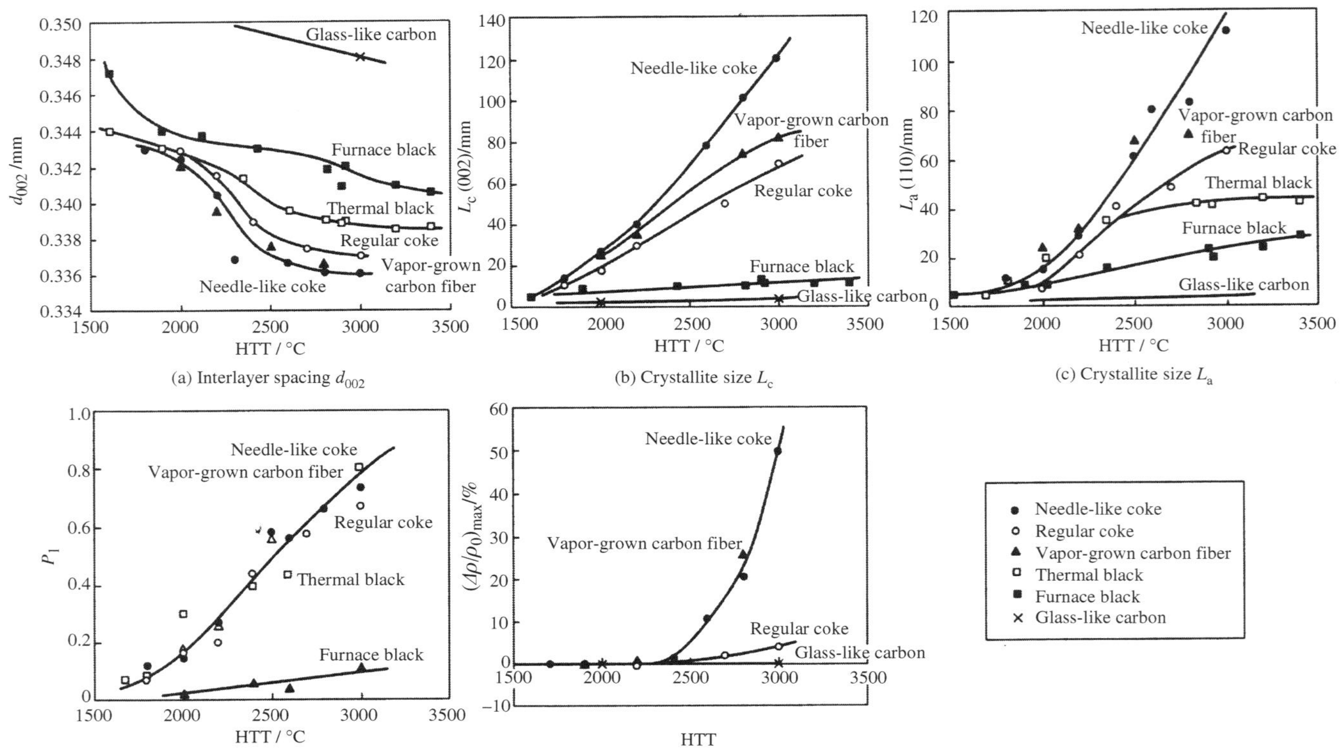


**FIGURE 2.85**

Dependences of various structure parameters on HTT for various carbon materials.

002 and 004 lines of most carbon materials are usually different, the former being larger than the latter. Therefore, it is recommended to measure the crystallite sizes from these two lines and to write the Miller indices of the line used, such as *Lc* (002), *Lc*(004) and *La*(110) [250]. By the assumption that the main reason for lattice strain is due to the random co-existence of graphitic and turbostratic stacking sequences, as schematically illustrated in Fig. 2.84a, the crystallite size *Lc* and lattice strain $\varepsilon_c$ along the c-axis can be separately evaluated from the plot of $\beta_{obs}\cos(\theta)/\lambda$ against $\sin(\theta)/\lambda$ on the basis of the empirical equation [251]:

$$\beta_{obs}\cos(\theta)/\lambda = K/Lc + 2\varepsilon_c\sin(\theta)/\lambda \qquad (2.2)$$

where, $\beta_{obs}$ is FWHM observed on the diffraction lines.

Here, it has to be pointed out that the term 'crystallite' is not exactly the correct terminology in crystallography, though it has been commonly used, because it contains random stacking of layers (not crystalline), as shown in Fig. 2.84a.

For the measurements of these X-ray parameters, a specification was proposed [250]. The principal points of the specification are the two following points: (1) the diffraction profiles must be corrected from the line broadening due to different intensity factors, each of which is a function of diffraction angle, and (2) an internal standard of silicon must be used in order to avoid the shifting and broadening of diffraction profiles mainly due to the instrument used. According to the experiences of the Japanese carbon group, the parameters determined by this specification were much reproducible and reliable [250]. The specification proposed was expected to be used not only in the academic research laboratories but also in the companies that are producing various carbon materials.

Electrical resistivity $\rho$ and galvanomagnetic properties, such as Hall coefficient $R_H$ and transverse magnetoresistance $\Delta\rho/\rho_0$, for carbon materials are structure-sensitive and depend strongly on the precursor used and its heat treatment temperature (HTT), which are due to the change in electronic band structure accompanied by the structural change with HTT. Particularly, magnetoresistance measurement is a powerful tool to characterize the structure and nanotexture of carbon materials [247,252]. Transverse magnetoresistance $\Delta\rho/\rho_0$ is defined as a change in electrical resistivity $\rho$ by the application of magnetic field with the strength B:

$$\Delta\rho/\rho_0 = [\rho(B) - \rho(0)]/\rho(0), \qquad (2.3)$$

where $\rho(0)$ and $\rho(B)$ are the resistivity without and with magnetic field B, respectively.

As seen in the equation, magnetoresistance is a relative change in electrical resistivity and depends only on electrical current for the measurement of resistivity $\rho$ and magnetic field strength $B$ applied, not influenced by specimen geometry, size and shape. However, it depends on the orientation of crystallites in the specimen because of their strong anisotropy in $\rho$ between the direction perpendicular and parallel to the hexagonal carbon layer. Therefore, $\Delta\rho/\rho_0$ values of carbon materials are measured by applying the magnetic field in three orthogonal directions, as shown in Fig. 2.84b. On a specimen with a planar orientation of crystallites along the plane of specimen, a maximum value of $\Delta\rho/\rho_0$, $(\Delta\rho/\rho_0)_{max}$ is measured when the magnetic field is applied from the direction 'max'. After defining the direction 'max', $\Delta\rho/\rho_0$ is measured in two orthogonal directions, $TL_{min}$ and $T_{min}$, the former giving a minimum value of magnetoresistance, $(\Delta\rho/\rho_0)_{TLmin}$, near the direction of current and the latter being perpendicular to these two directions, $(\Delta\rho/\rho_0)_{Tmin}$. Changes in $(\Delta\rho/\rho_0)_{max}$ with heat treatment temperature are shown on different carbon materials in Fig. 2.85e. $(\Delta\rho/\rho_0)_{max}$ gives useful information on the changes in structure and nanotexture in the carbon material with heat treatment. In order to characterize more exactly the structure of a wide range of carbon materials and to compare with each other, crystallite magnetoresistance $(\Delta\rho/\rho_0)_{cr}$ is defined as:

$$(\Delta\rho/\rho_0)_{cr} = (\Delta\rho/\rho_0)_{max} + (\Delta\rho/\rho_0)_{TLmin} + (\Delta\rho/\rho_0)_{Tmin} \quad (2.4)$$

Dependences of $\Delta\rho/\rho_0$ on the angles, $\varphi$ and $\theta$, of $TL$ and $T$ rotations (Fig. 2.84b), respectively, give the orientation of crystallites in the specimen. In order to characterize the nanotexture in various carbon materials, anisotropy ratios $r_{TL}$ and $r_T$, are defined:

$$\begin{aligned} r_{TL} &= (\Delta\rho/\rho_0)_{TLmin}/(\Delta\rho/\rho_0)_{max}, \\ r_T &= (\Delta\rho/\rho_0)_{Tmin}/(\Delta\rho/\rho_0)_{max} \end{aligned} \quad (2.5)$$

For the ideal planar, axial, point and random orientation, the following values of $r_{TL}$ and $r_T$ have to be obtained:

$$\begin{array}{lll} \text{Ideal planar orientation} & : & r_{TL} = 0 \text{ and } r_T = 0 \\ \text{Ideal axial orientation} & : & r_{TL} = 0 \text{ and } r_T = 1 \\ \text{Ideal point orientation} & : & r_{TL} = 1 \text{ and } r_T = 1 \\ \text{Ideal random orientation} & : & r_{TL} = 1 \text{ and } r_T = 1. \end{array} \quad (2.6)$$

In axial orientation scheme, coaxial and radial alignments cannot be differentiated from these $r_{TL}$ and $r_T$ values. Also for point and random orientation schemes, both $r_{TL}$ and $r_T$ are equal to 1 and two schemes cannot be differentiated.

In Fig. 2.86, the dependences of $\Delta\rho/\rho_0$ on the rotation angle $\varphi$ and $\theta$ are shown on typical orientation schema, planar, axial and point orientations, i.e., a highly oriented graphite film prepared from polyimide, a PAN-based carbon fiber heat-treated at 3000°C and a gilsonite coke of spherical particle heat-treated at 3000°C, respectively.

For the study on structure of carbon materials, different techniques of transmission electron microscopy (TEM) have been employed; bright-field image, selected area electron diffraction, dark-field images using different diffraction lines and lattice fringes [247,253]. The bright-field image is the image by electron beam transmitted through specimen and gives an appearance of the specimen particle. The electron diffraction occurs as the case of X-ray, but in a microscopy only the diffraction lines from the crystal planes parallel to the incident beam can

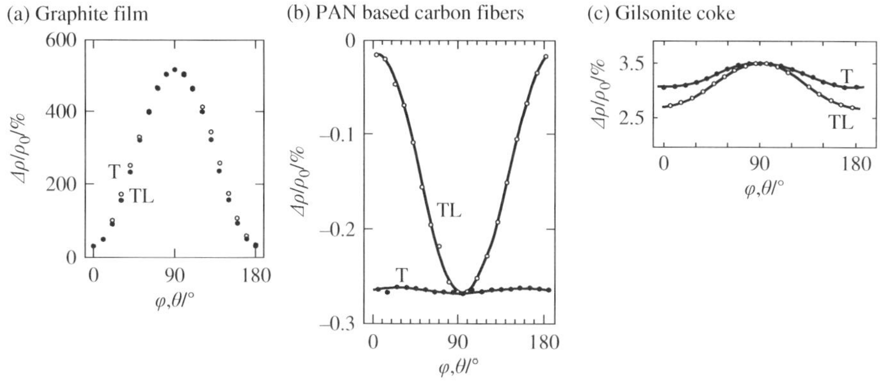


**FIGURE 2.86**

Dependences of $\Delta\rho/\rho_0$ on rotation angles $\varphi$ and $\theta$ on three different orientation schema.

*(Courtesy of Prof. Y. Hishiyama of Tokyo City Univ.)*

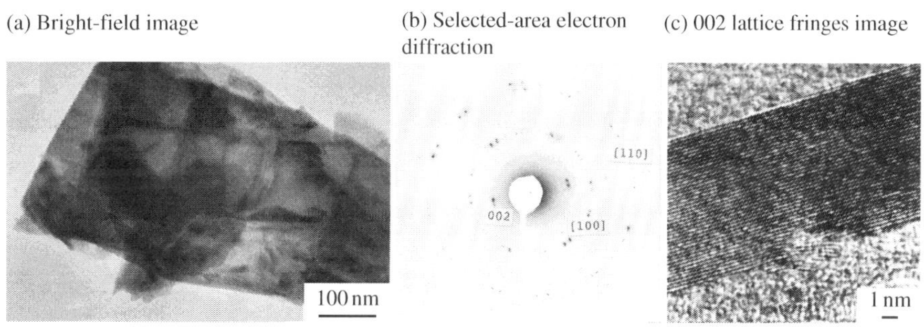


**FIGURE 2.87**

TEM images for a carbon particle.

*(Courtesy of Mme A. Oberlin.)*

be observed because of the very short wavelength of electron beam. The lattice fringes are observed as a result of interference between the transmitted beam and one of diffracted beams. For carbon materials, mostly 002 diffraction line was used to observe lattice fringes (002 lattice fringes). Bright-field image, selected area electron diffraction pattern and 002 lattice fringes image are shown for a carbon particle in Fig. 2.87a–c, respectively.

The dark-field image observation in TEM was also applied successfully to the nanotexture analysis in carbon particles. For the particle shown by the bright-field image of Fig. 2.88a, only the edges are lightened under 002 dark-field images, as shown in Fig. 2.88b and c, and the central part is never lightened under any positions of the aperture on 002 diffraction ring, but lightened under 10 dark-field images, as shown in Fig. 2.88d. By the rotation of the aperture of the TEM on the

002 diffraction ring, lightened parts of the particle edge rotate, as seen from the two images of Fig. 2.88b and c. From the analysis of these dark-field images of 002 and 10 diffraction lines, the nanotexture of the particle can be clarified, the present particle being a shell surrounded by hexagonal layers and so layers are perpendicular to the image plate (edge-on) at the edge of the particle and parallel (lay-down) at the center.

Scanning electron microscopy (SEM) has been commonly used to observe the morphology and the surface of different carbon materials. In order to have correct information on the surface of carbon materials, the use of low acceleration voltage, less than 10 kV, for primary electron beam is strongly recommended [247,254]. It has been pointed out that special care to prepare the fractured surface for SEM observation has to be paid; fracturing the sample carbon at liquid nitrogen temperature by giving a shock through frozen acetone [247,255]. This is particularly important for well-crystallized carbon materials, such as highly

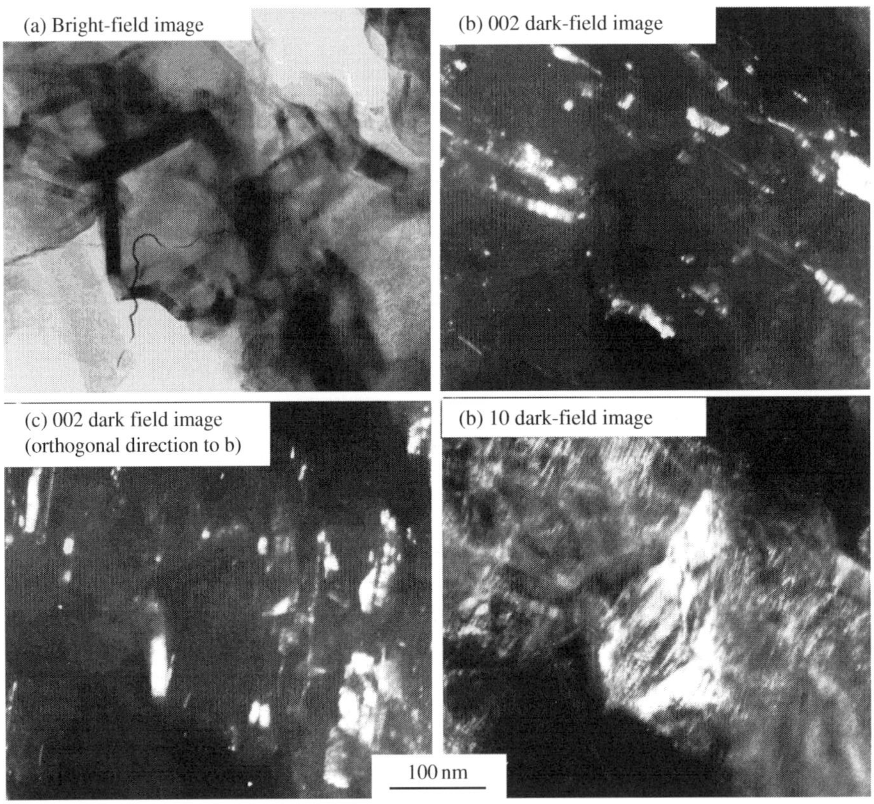


**FIGURE 2.88**

Dark-field images of a carbon particle.

*(Courtesy of Mme. A. Oberlin.)*

oriented graphite and vapor-grown carbon fibers because of easy cleavage. In Fig. 2.89, SEM images of the cross-section of pitch-based carbon fibers with different nanotextures are shown, radial and random alignments of carbon layers being recognized clearly.

Raman spectroscopy is also used for the structure characterization of carbon materials [256–258]. The Stokes rays scattered from the surface of the sample solid are measured as a function of its frequency shift from the frequency of the incident laser beam (Raman shift). In Fig. 2.90, the Raman spectra for different carbon materials are shown. Theoretically, the Raman bands at 1580 and 42 cm$^{-1}$ are possible, which are due to vibration mode of $2E_{2g}$ along the hexagonal carbon layer, the former band being usually used for the characterization of carbon materials and called the G band. Additional bands are observed at around 1360 and 1620 cm$^{-1}$ often, which are called D and D′ bands, respectively, because

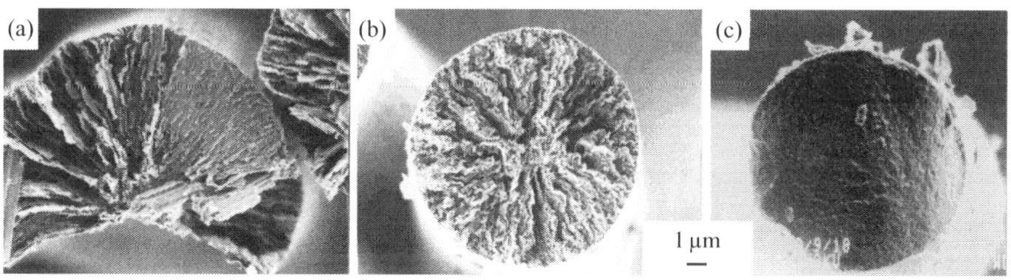


**FIGURE 2.89**

SEM images of the cross-sections of pitch-based carbon fibers with different nanotextures.

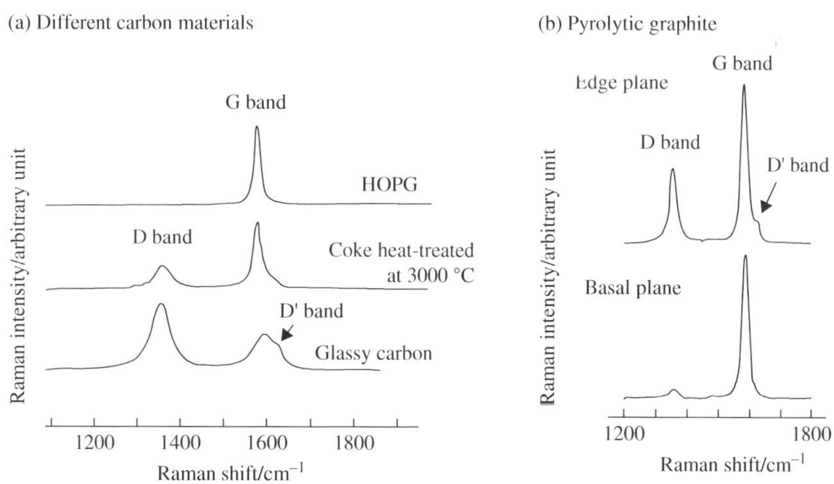


**FIGURE 2.90**

Raman spectra for different carbon materials.

they are supposed to be due to some structural disorders. As shown in Fig. 2.90a, the surface of highly oriented pyrolytic graphite (HOPG) plate does not give D and D′ bands, but a coke heat-treated at 3000°C and a glass-like carbon show marked D and D′ bands, for the glass-like carbon even stronger than the G band. The intensity ratio of D to G band, $I_D/I_G$, was often used as a parameter for structure characterization. However, its usage as common parameter for the characterization of wide range of carbon materials is doubtful, because D band is detectable on the cross-section of HOPG plate [258]. It has to be pointed out here that the structural information obtained by Raman scattering is that on the surface of the sample solid. The origins for the D and D′ bands observed on carbon materials are not clearly understood, but they are partly due to the edge planes of hexagonal carbon layers, as shown in Fig. 2.90b.

### 2.5.2 Graphitization behavior

With the increase in heat treatment temperature (HTT), the development in graphite structure has been pursued mainly by measuring different structure parameters, $d_{002}$, *La*, *Lc* and $(\Delta\rho/\rho_0)_{max}$. The changes of these parameters, including $P_1$ with HTT, are shown on different carbon materials in Fig. 2.85. The changes in these parameters are quite different from the starting carbon materials. The decrease in $d_{002}$ and the increases in *Lc*, *La*, $P_1$ and $(\Delta\rho/\rho_0)_{max}$ indicate the development of graphite structure (graphitization). The graphitization behaviors of needle-like coke and vapor-grown carbon fibers are similar from HTT dependences of $d_{002}$, *La*, $P_1$ and $(\Delta\rho/\rho_0)_{max}$, but the growth of *Lc* is suppressed in vapor-grown carbon fiber, which is probably due to the thin fibrous morphology. Regular coke always gives somewhat larger $d_{002}$ and smaller *Lc* and *La* than needle-like coke, which is due to their nanotextures, needle-like coke having a higher orientation degree along its axis of particles. In both cokes, $\Delta\rho/\rho_0$ changes its sign from negative to positive, indicating the change in charge carriers from a one-carrier system (positive holes) to a two-carriers system (positive holes and negative electrons). Since thermal black consists of much larger primary particles than furnace black, the former gives smaller $d_{002}$ and larger *Lc* and *La* than the latter. These two carbon blacks give quite different graphitization degrees $P_1$. In glass-like carbon, the development of graphite structure is almost completely suppressed $d_{002}$ being 0.348 nm, *La* and *Lc* being only 4 nm and $P_1$ being difficult to be determined even after 3000°C treatment.

The difference in graphitization behavior among carbon materials is predominantly due to their nanotextures. In Fig. 2.91, 002 lattice fringe images are compared between a petroleum coke and a sugar coke, the former having oriented nanotexture and the latter random nanotexture. In the petroleum coke after the heat treatment at 2200°C, already layer planes stack highly in parallel and grow in large size. In the sugar coke, however, layers are much smaller than those in the petroleum coke and oriented randomly, even though it is heat-treated at 2700°C. In the sugar coke, small layers are aggregated randomly and as a

**FIGURE 2.91**

002 lattice fringe images of two carbons heat-treated at different temperatures.

*(Courtesy of Mme. A. Oberlin.)*

consequence the growth of layers is markedly suppressed. In the petroleum coke, however, carbon layers orient already in the low-temperature-treated sample and as a consequence they easily coalesce with each other to result in large layers, which stack in parallel already after 2200°C treatment. The difference in graphitization behavior between thermal and furnace blacks in Fig. 2.85 is also explained by the size of primary particles; particles of thermal black are very large, more than 500 nm, and as a consequence the growth of layers is relatively easy, but those of furnace black are less than 100 nm and layer growth is suppressed even at a high temperature.

As explained briefly, graphitization behavior is strongly governed by the nanotexture of carbon materials, which is predominantly determined by the structure of precursors and is established during the carbonization process. Therefore, the graphitization process was discussed separately on the carbon materials with different nanotextures, planar, axial, point and random orientation, in the following sections.

### a. Carbon materials with planar orientation

Carbon materials, in which hexagonal carbon layers highly orient in the scheme of planar orientation, are represented by pyrolytic carbons, thin carbon films derived from some polyimide films and flexible graphite sheets. In these carbons high graphitization degree is usually obtained by a high temperature treatment. In the particles of needle-like cokes, which have been used as important raw materials for graphite blocks, such as graphite electrodes and high-density isotropic graphites, planar orientation scheme is kept.

Graphitization behavior of the pyrolytic carbons has been discussed through the detailed kinetic studies on $d_{002}$ and total diamagnetic susceptibility $\chi_T$ [259]. In Fig. 2.92, $d_{002}$ and $\chi_T$ measured for a pyrolytic carbon, which has been prepared at 2150°C from methane, are plotted against the logarithm of residence time at different temperatures. The graphitization proceeds in two steps; in the

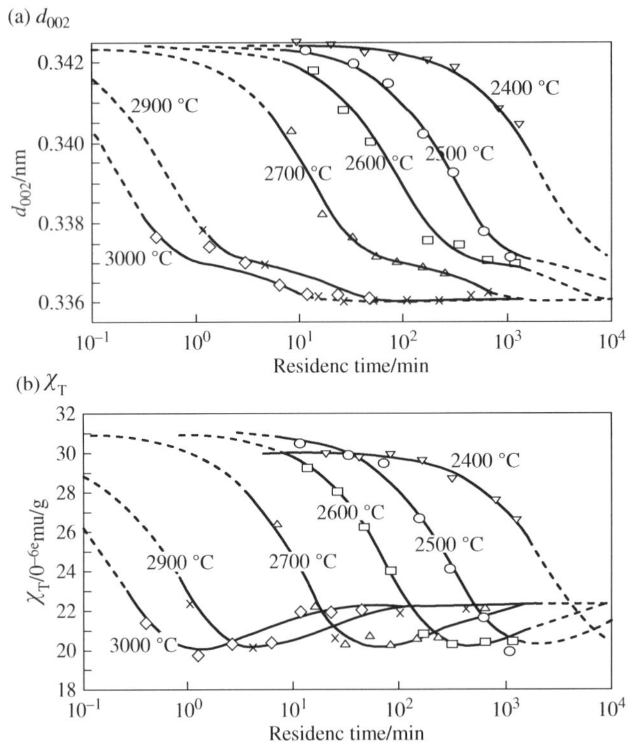


**FIGURE 2.92**

Dependences of $d_{002}$ and $\chi_T$ of a pyrolytic carbon on residence time [259].

first step, $d_{002}$ decreases down to about 0.337 nm and $\chi_T$ to a minimum, $20 \times 10^{-6}$ emu/g, and in the second step $d_{002}$ decreases again and $\chi_T$ increases to approach graphite values, 0.3354 nm and $22 \times 10^{-6}$ emu/g, respectively. It has to be pointed out that the residence time dependences of $d_{002}$ and $\chi_T$ at each heat treatment temperature are very similar with each other and were able to overlap by shifting along the abscissa to get a master curve at a reference temperature for each parameter, $d_{002}$ and $\chi_T$. This shifting reveals that the exchange between heat treatment temperature and residence time for structure change, i.e., for graphitization process, is possible. In Fig. 2.93, the master curves for $d_{002}$ obtained through the kinetic study are compared for four pyrolytic carbons with different preparation conditions [259]. The plot of shift factors to get the master curves against the reciprocal temperature is a kind of Ahrrenius plot and gives apparent activation energy. For different pyrolytic carbons, almost a constant value of about 270 kcal/mol was obtained [259].

The studies on $\rho$, $R_H$ and $(\Delta\rho/\rho_0)_{max}$ using a commercially available pyrolytic carbon have been performed, paying a particular attention on structural change

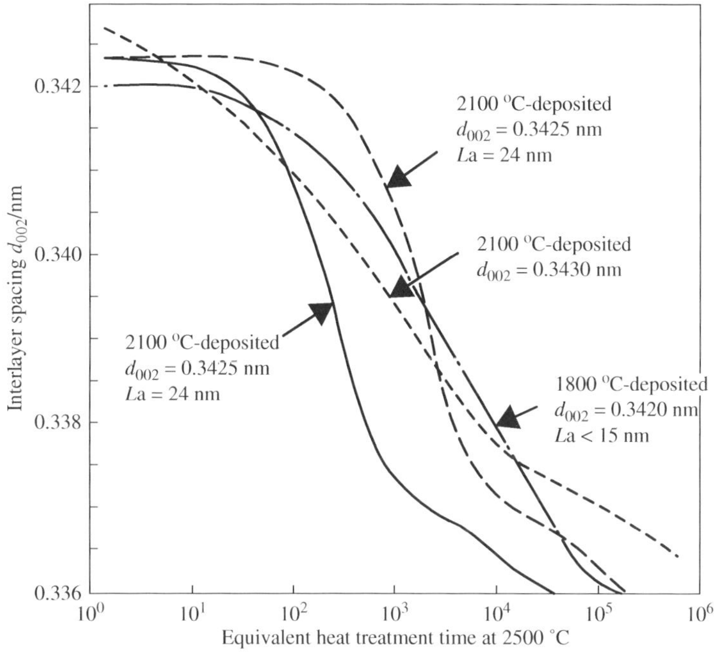


**FIGURE 2.93**

Master curves of $d_{002}$ for different pyrolytic carbons [259].

with HTT [260]. In Fig. 2.94, the dependences of $(\Delta\rho/\rho_0)_{max}$ and $R_H$ on magnetic field B measured at liquid nitrogen temperature are shown with a parameter $d_{002}$. $(\Delta\rho/\rho_0)_{max}$ increases its value with the decrease of $d_{002}$ (i.e., the increase in HTT) and changes its sign from negative to positive on the sample with $d_{002}$ of about 0.338 nm. $R_H$ decreases with $d_{002}$ and its sign changes also at $d_{002}$ of 0.338 nm. The changes in $(\Delta\rho/\rho_0)_{max}$ and $R_H$ with heat treatment are also transferred from the first step to the second at around the same $d_{002}$ value as kinetic studies on $d_{002}$ and $\chi_T$ [261].

Aromatic polyimide films are known to be carbonized in inert gas atmosphere without any cracks and pores, the shape of the film being retained after carbonization and heat treatment at high temperatures [262–264]. Highly crystallized and oriented graphite films are prepared by heat treatments at 3100 and 3200°C from carbonized films of various aromatic polyimide films [265,266].

The changes in various structure parameters, $d_{002}$, $Lc$, $\rho$, $(\Delta\rho/\rho_0)_{max}$ and $R_H$ with HTT are studied on Kapton-derived carbon film [267]. In Fig. 2.95, $d_{002}$ and crystallite sizes $Lc$ determined from 002 and 004 diffraction lines are plotted against HTT. The HTT dependences of $\rho$, $(\Delta\rho/\rho_0)_{max}$ and $R_H$ at liquid nitrogen temperature are shown for the same carbon films in Fig. 2.96.

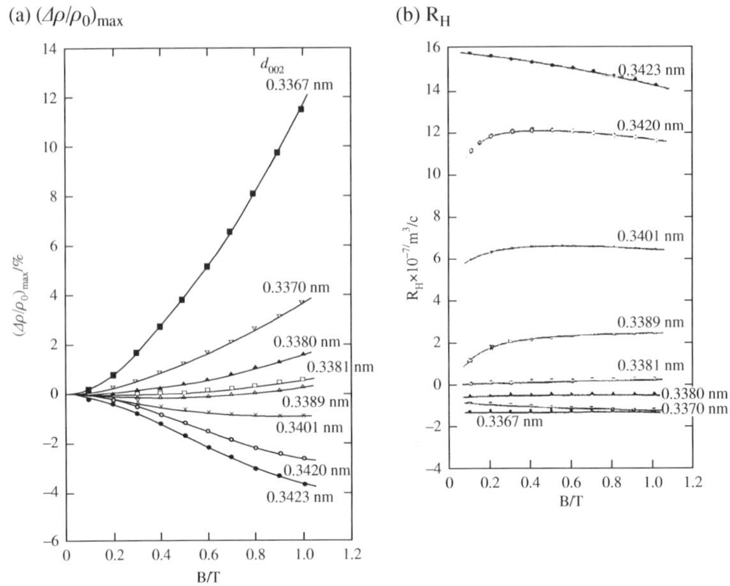


**FIGURE 2.94**

Dependences of $(\Delta\rho/\rho_0)_{max}$ and $R_H$ on magnetic field B at 77 K for a pyrolytic carbon.

*(Courtesy of Prof. Y. Kaburagi of Tokyo City Univ.)*

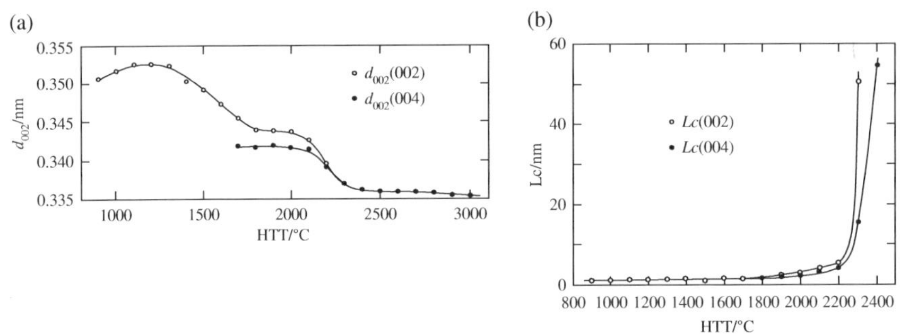


**FIGURE 2.95**

Changes in $d_{002}$ and $Lc$ of carbon film derived from Kapton with HTT.

*(Courtesy of Prof. Y. Hishiyama of Tokyo City Univ.)*

Above 2200°C, $d_{002}$ values derived from both 002 and 004 lines and $\rho$ decrease after plateau, Lc increases suddenly, and RH shows a maximum and changes its sign to negative. For the films heat-treated between 1600 and 2200°C, $(\Delta\rho/\rho_0)_{max}$ is measured as negative, but it suddenly changes to positive above 2300°C. Above 2600°C, crystallite size increases rapidly with HTT and reaches more than 100 nm. The carbon films heat-treated above 2600°C become highly

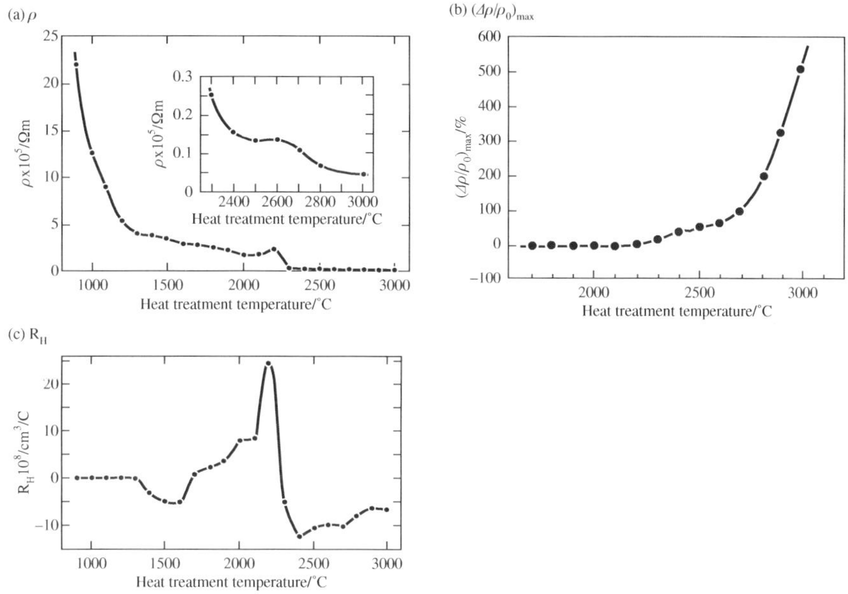


**FIGURE 2.96**

Changes in $\rho$, $(\Delta\rho/\rho_0)_{max}$ and $R_H$ at 77 K and 1 T.

*(Courtesy of Prof. Y. Hishiyama of Tokyo City Univ.)*

anisotropic and have well-oriented texture. These abrupt changes in structure occurred in association with the departure of nitrogen atoms, which were substitutionally doped into hexagonal carbon layers [268,269].

Fig. 2.97 shows the dependences of $R_H$ and $(\Delta\rho/\rho_0)_{max}$ on magnetic field B for the films heat-treated at different temperatures [270]. $R_H$ is negative at low HTTs and becomes positive above 1700°C, showing a maximum at about 2200°C, and then suddenly changes negative again. The films heat-treated above 2300°C show the magnetic field dependence of $R_H$ similar to HOPG. At a constant magnetic field, the absolute value of $(\Delta\rho/\rho_0)_{max}$ increases with increasing HTT. The dependence of $(\Delta\rho/\rho_0)_{max}$ on magnetic field for the 2200°C-treated carbon film is different from others, showing the trend of saturation at high fields. Above 2300°C, the dependence of $(\Delta\rho/\rho_0)_{max}$ on magnetic field becomes pronounced with the increase in HTT.

The word 'coke' has been used to call a wide range of carbons, from the cokes prepared from pitches, like pitch coke, to carbonized sugar, like sugar coke. These two cokes have completely different nanotexture, as can be seen from their TEM images in Fig. 2.90, the pitch coke having a nanotexture based on planar orientation, but the sugar coke on random orientation. Even in the cokes prepared from pitches, different morphologies of particles are produced, from needle-like to spherical. Most of the cokes prepared from pitches have the nanotexture with

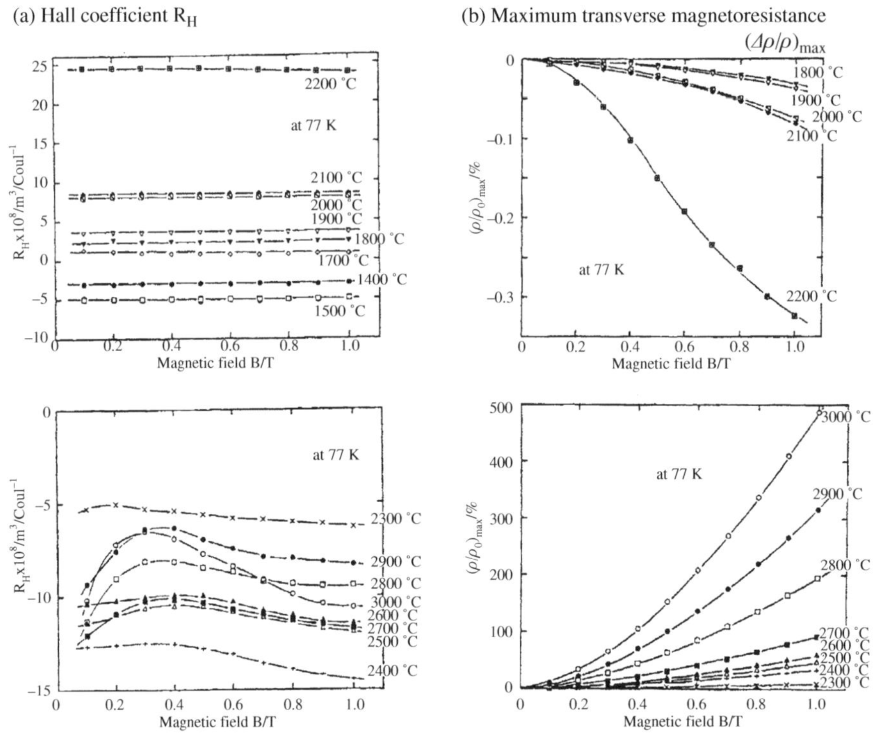


**FIGURE 2.97**

Magnetic field dependences of $R_H$ and $(\Delta\rho/\rho_0)_{max}$ of the carbon films from polyimide.

*(Courtesy of Prof. Y. Hishiyama of Tokyo City Univ.)*

planar orientation, even though the degree of orientation depends strongly on the coke. In Fig. 2.98, $d_{002}$ and $Lc(002)$ are plotted against HTT for different cokes. For fluid and gilsonite cokes whose particles have spherical morphology, the changes in $d_{002}$ and $Lc$ are rather depressed, mainly because their particles have point orientation scheme. The coke derived from poly(vinyl chloride) in the laboratory and two kinds of commercially available needle-like cokes have very similar dependence of $d_{002}$, but the growth of $Lc(002)$ is a little depressed in one of needle-like cokes.

The growth of crystallite size by the high-temperature treatment has been studied as a function of $d_{002}$ and the results are shown in Fig. 2.99 [271]. Both $Lc$ (002) and $Lc(004)$ show plateau at $d_{002}$ from 0.342 to 0.338 nm, a rather narrow range of heat treatment temperature from 1900 to 2200°C. The thickness of three-dimensional graphitic stacking is measured from 112 diffraction line by converting it to the thickness along the c-axis, $Lc(112)$, and plotted in Fig. 2.99b against $d_{002}$. The diffraction line 112 can be detected on the sample which had the $d_{002}$ value below 0.342 nm, though its *FWHM* is very broad. In the range of $d_{002}$

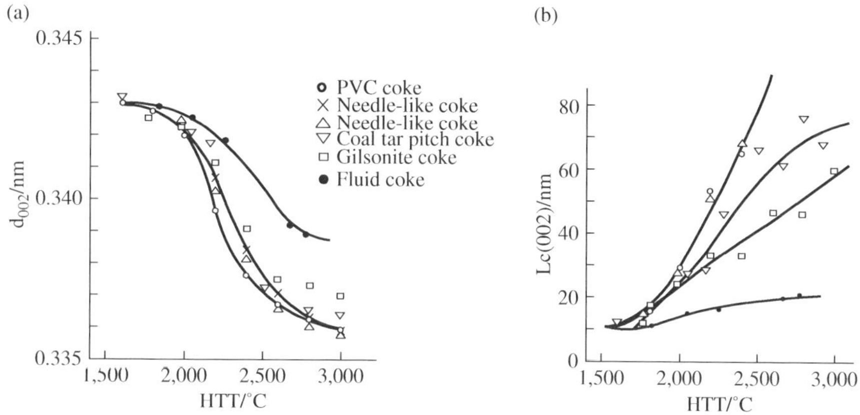


**FIGURE 2.98**

Dependences of $d_{002}$ and $Lc(002)$ of different cokes on HTT.

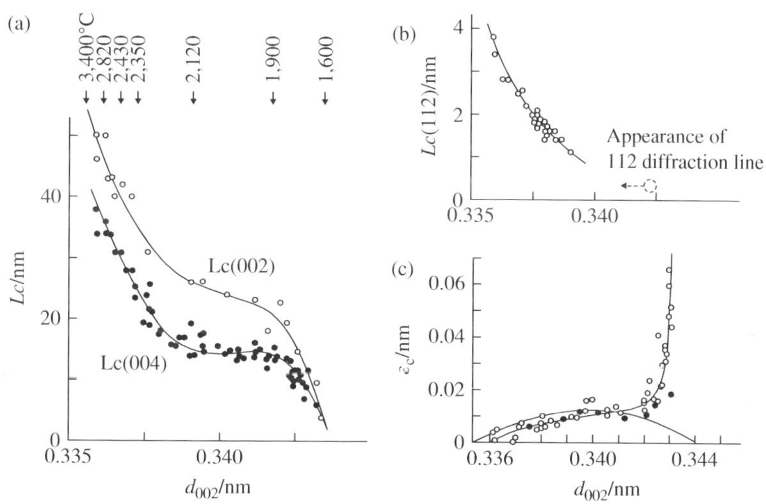


**FIGURE 2.99**

Relations of $Lc(002)$ and $Lc(004)$ (a), $Lc(112)$ (b) and $\varepsilon_c$ (c) to $d_{002}$ of the cokes [271].

below 0.339 nm, $Lc(112)$ can be determined and grows quickly with decreasing $d_{002}$, i.e., increasing HTT.

The lattice strain $\varepsilon_c$, which is calculated from *FWHMs* of 002 and 004 lines, is plotted in Fig. 2.99c against $d_{002}$. With decreasing $d_{002}$ to about 0.342 nm, $\varepsilon_c$ decreases rapidly and then decreases gradually. The change in $\varepsilon_c$ with $d_{002}$ below 0.342 nm was tried to be explained by the coexistence of graphitic stacking with $d_{002}$ of 0.3354 nm with turbostratic stacking with 0.344 nm, the calculated strain

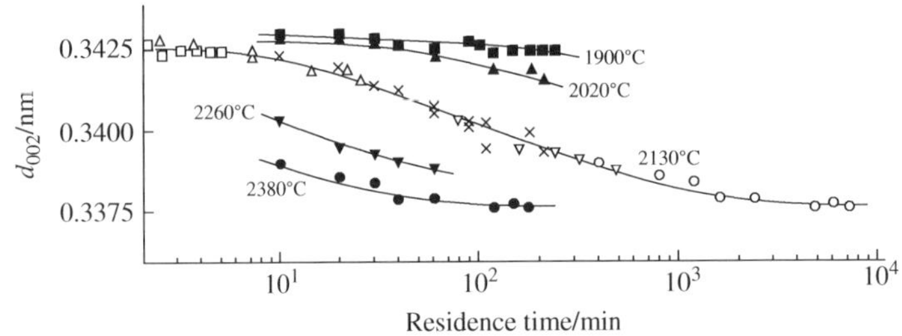


**FIGURE 2.100**

Changes in $d_{002}$ with residence time at different HTTs under air pressure of 0.1 Torr for a petroleum coke [274].

due to mixing of two different interlayer spacings being shown by the broken line in Fig. 2.99c [272]. Above 0.342 nm, the strain observed is much larger than predicted, suggesting another kind of strain, which was discussed in relation to other experimental results in Fig. 2.126.

Graphitization kinetics has also been studied on different cokes [273]. Fig. 2.100 shows the changes in $d_{002}$ with residence time at different HTTs for a petroleum coke. Applying the same analysis procedure for pyrolytic carbons (Fig. 2.92), the master curve at 2130°C is obtained, as shown in Fig. 2.100 and the apparent activation energy for graphitization was calculated to be 750 kJ/mol. A marked effect of the pressure of atmosphere is observed; under air pressure of 2 and 4 Torr the obtained activation energy was about 310 kJ/mol, which is much smaller than that obtained for the same coke under the pressure of 0.1 Torr and also the value of 1130 kJ/mol for the pyrolytic carbons under inert gas flow [274]. The graphitization behavior of the cokes under reduced pressures was discussed as the acceleration due to the catalytic effect of atmospheric oxygen gas [273,274]. Carbonization conditions of the cokes have been reported to give certain influence on graphitization kinetics by using the coke prepared from poly (vinyl chloride) in the laboratory; low-temperature carbonized coke gives low apparent activation energy as 590 kJ/mol but high-temperature carbonized coke high activation energy as 1090 kJ/mol [275].

In Fig. 2.101, the changes in galvanomagnetic properties, electrical resistivity $\rho$, maximum transverse magnetoresistance $(\Delta\rho/\rho_0)_{max}$ and Hall coefficient $R_H$, with heat treatment temperature are shown for a petroleum coke, which represents general scheme of changes in these properties in the cokes [276]. $\rho$ exhibits a small hump around 1800°C and then decreases gradually with the increase in HTT. $R_H$ is negative; which is independent of magnetic field strength at low HTT, becomes positive above 1800°C, reaches a maximum around 2200°C and then decreases to be negative again with a weak field dependence above 2800°C. $(\Delta\rho/\rho_0)_{max}$ cannot be measured on the cokes heat-treated below 1600°C, but above this temperature it shows negative value. Through a minimum around 2200°C it starts to increase with the increase in HTT and becomes positive. The

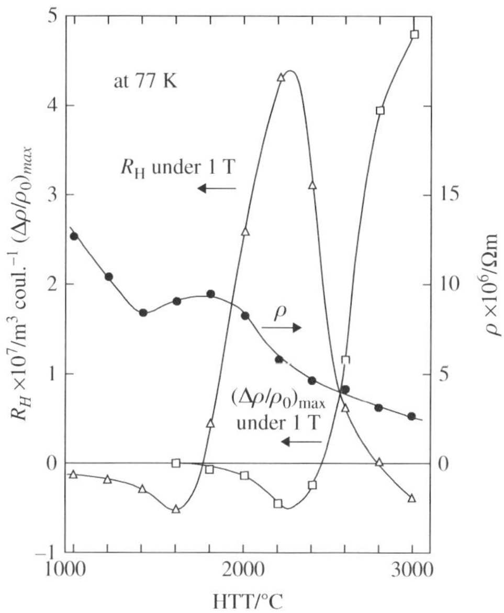


**FIGURE 2.101**

Changes of galvanomagnetic properties, $\rho$, $(\Delta\rho/\rho_0)_{max}$ and $R_H$, at 77 K for a petroleum coke with HTT.

*(Courtesy of Prof. Y. Hishiyama of Tokyo City Univ.)*

transition behavior from negative to positive values in $(\Delta\rho/\rho_0)_{max}$ at HTT between 2400 and 2600°C is studied in detail by measuring its magnetic field dependence [277]. The results are shown in Fig. 2.102. The dependence of $(\Delta\rho/\rho_0)_{max}$ on magnetic field changes markedly with the change in HTT from 2450 to 2570°C.

These experimental results are qualitatively explained using a model on the change in energy band structure for carbon materials [12,13]. One of the models was schematically shown in Fig. 2.16. For the carbon materials with the nanotexture of planar orientation, the carriers for electrical conduction after the heat treatment at low temperatures below 1600°C are supposed to be electrons of which mobility is rather low. Above 1600°C, hydrogen atoms remained in the precursor starts to depart and leave electron traps, most of which may locate at the edge of the hexagonal carbon layers. As a consequence, Fermi level is lowered and the concentration of positive hole increases, which leads to the increase in $R_H$ to be positive and to the decrease in $(\Delta\rho/\rho_0)_{max}$, as shown in Fig. 2.101. Passing through the maximum in $R_H$ and the minimum in $(\Delta\rho/\rho_0)_{max}$ around 2200°C, Fermi level starts to rise, because of healing of electron traps by high-temperature treatment and also the valence band touches with conduction band. After high-temperature treatment around 3000°C, valence and conduction bands overlap slightly, because of the growth of crystallites and improvement of stacking

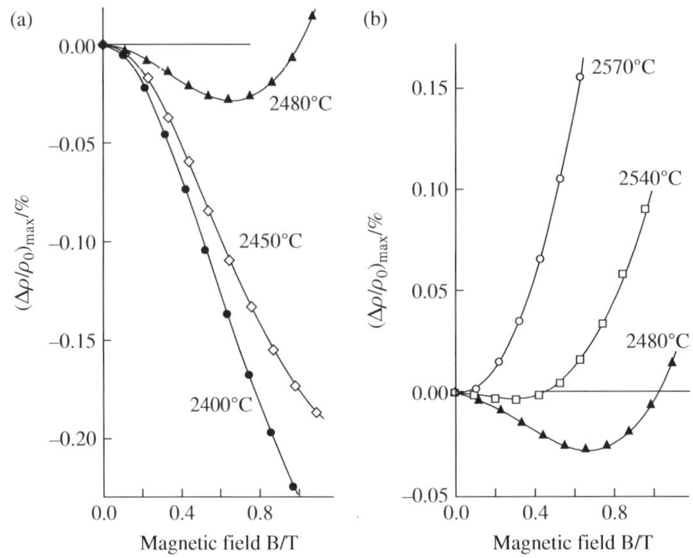


**FIGURE 2.102**

Magnetic field dependences of $(\Delta\rho/\rho_0)_{max}$ for a petroleum coke pieces heat-treated at different temperatures.

*(Courtesy of Prof. Y. Hishiyama of Tokyo City Univ.)*

regularity, which results in the increase in relative concentration of electron. In the highly graphitized carbons, the concentration of two carriers, negative electrons and positive holes, are comparable, which gives the high value of $(\Delta\rho/\rho_0)_{max}$, and a little higher mobility of electrons than that of holes gives negative $R_H$, as in graphite. Even though the values of these parameters approach those of graphite, they cannot reach the exact values which have been measured on single crystals of graphite.

### b. Carbon materials with axial orientation

Different carbon fibers have been developed, vapor-grown, isotropic-pitch-based and mesophase-pitch-based carbon fibers as well as poly(acrylonitrile)-based (PAN-based) carbon fibers. In these fibers, structural development with high-temperature treatment depends strongly on the nanotexture in their cross-sections which is governed mainly by precursors and spinning conditions and, as a consequence, the changes in various physical properties are also different from each other.

In Fig. 2.103, $d_{002}$ of four different kinds of carbon fibers is plotted against HTT [278]. The change observed on vapor-grown carbon fibers is very similar to needle-like coke, as shown in Fig. 2.85a, and its graphitization proceeds with increasing HTT. For isotropic-pitch-based carbon fibers, $d_{002}$ value saturates at

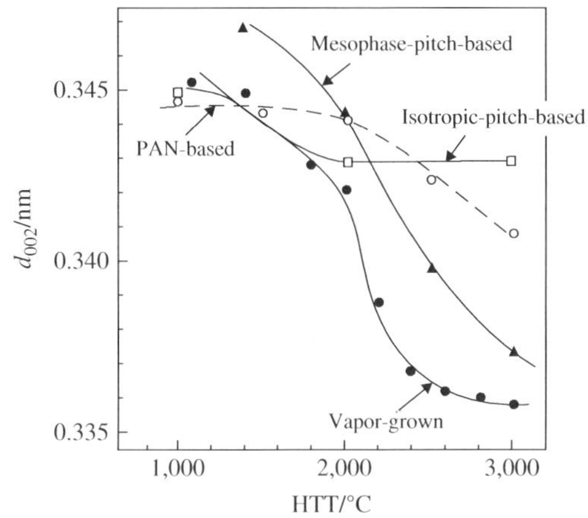


**FIGURE 2.103**

Changes in $d_{002}$ with HTT for different carbon fibers [278].

about 0.343 nm and almost no graphitic structure is developed. The mesophase-pitch-based carbon fibers show a relatively marked decrease in $d_{002}$, mainly because they have straight radial nanotexture in their cross-sections, but its change is obviously retarded from that for vapor-grown carbon fibers.

Fig. 2.104 shows the dependences of the maximum transverse magnetoresistance $(\Delta\rho/\rho_0)_{max}$ along the fiber axis on magnetic field B measured at liquid nitrogen temperature for four kinds of carbon fibers heat-treated at high temperatures. Vapor-grown carbon fibers show large positive values of $(\Delta\rho/\rho_0)_{max}$, but the other three show negative values even after high-temperature treatment, revealing a big difference between vapor-grown and the other three carbon fibers in the development of graphite structure by high-temperature treatment (graphitizability). Isotropic-pitch-based carbon fibers give a very low absolute value of negative $(\Delta\rho/\rho_0)_{max}$ and mesophase-pitch-based carbon fibers a relatively large absolute value, even though still negative. This figure shows clearly that the graphitizability is very much different on carbon fibers.

In Fig. 2.105a, $(\Delta\rho/\rho_0)_{cr}$ determined at liquid nitrogen temperature in a magnetic field of 1 T for various carbon fibers are plotted against HTT [279]. In vapor-grown carbon fibers $(\Delta\rho/\rho_0)_{cr}$ increases markedly with increasing temperature. In PAN-based and isotropic-pitch-based carbon fibers, on the other hand, $(\Delta\rho/\rho_0)_{cr}$ does not increase and is negative even after 3000°C treatment. PAN-based carbon fibers heat-treated under stretching show an increase in $(\Delta\rho/\rho_0)_{cr}$ above 2500°C and mesophase-pitch-based ones give a small value but positive $(\Delta\rho/\rho_0)_{cr}$. This result suggests that the $(\Delta\rho/\rho_0)_{cr}$ development, i.e., structure development, retards in PAN-based and isotropic-pitch-based carbon fibers along with HTT axis, in other words, structure development in these fibers even after 3000°C corresponds to that in vapor-grown carbon fibers heat-treated below

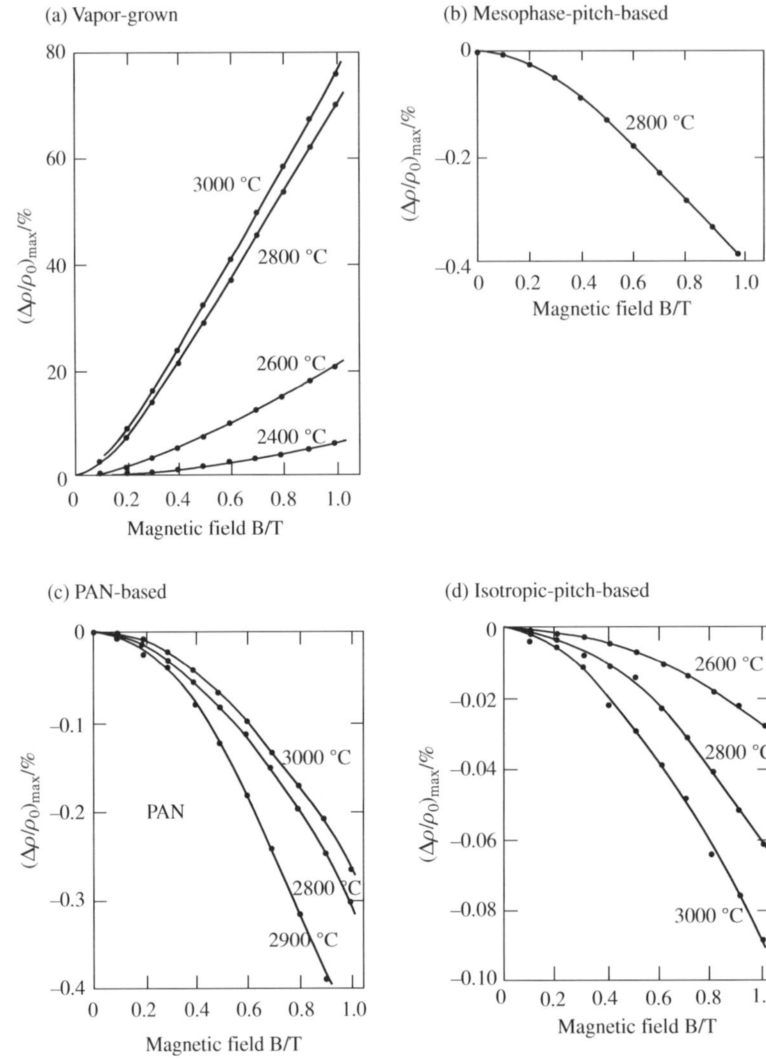


**FIGURE 2.104**

Magnetic field dependences of $(\Delta\rho/\rho_0)_{max}$ of different carbon fibers heat-treated at different temperatures.

*(Courtesy of Prof. Y. Hishiyama of Tokyo City Univ.)*

2200°C. In Fig. 2.105b, therefore, $(\Delta\rho/\rho_0)_{cr}$ values are plotted in logarithmic scale on the curve obtained by the vapor-grown carbon fiber [279]. In this figure, the actual HTT is indicated at the end of each sample code. This figure shows that the change in magnetoresistance, in other words structural development, with heat treatment follows a common scheme; $(\Delta\rho/\rho_0)_{cr}$

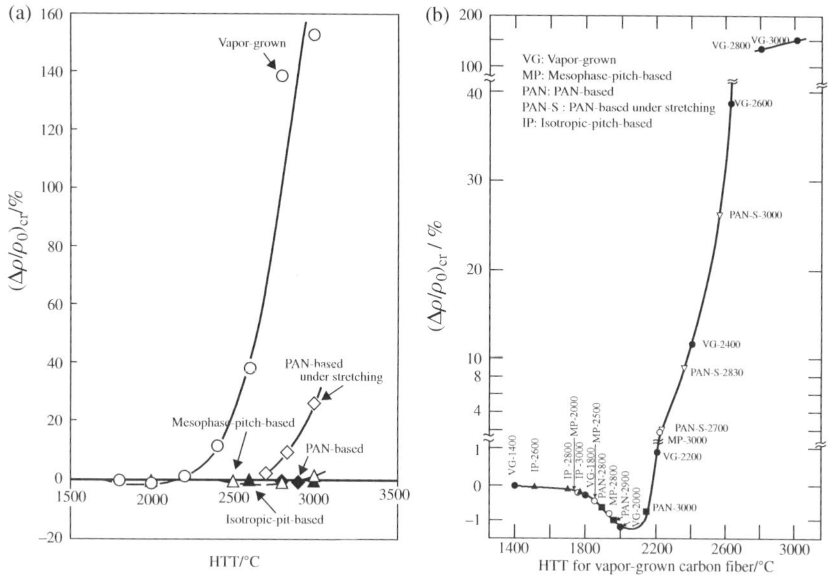


**FIGURE 2.105**

Change in $(\Delta\rho/\rho_0)_{cr}$ of various carbon fibers with HTT.

*(Courtesy of Prof. Y. Hishiyama of Tokyo City Univ.)*

decreases gradually and after passing a minimum value it increases rapidly to change the sign from negative to positive. However, this structural development is retarded more or less in each carbon fiber; for example, the structure development evaluated by $(\Delta\rho/\rho_0)_{cr}$ of isotropic-pitch-based carbon fibers heat-treated up to 3000°C corresponds to that of vapor-grown carbon fibers heat-treated at about 1800°C, 3000°C-treated PAN-based carbon fibers to that of 2000°C-treated vapor-grown carbon fibers, and 3000°C-treated mesophase-pitch-based carbon fiber to 2200°C-treated vapor-grown one. Stretching during carbonization and graphitization processes is very effective for increasing $(\Delta\rho/\rho_0)_{cr}$, i.e., for the structure improvement.

For this pronounced difference in graphitizability in carbon fibers, nanotexture in their cross-sections is predominantly responsible. In Fig. 2.106, SEM images of the cross-sections of these carbon fibers were heat-treated at around 1200–1800°C (after carbonization) and at 3000°C. Carbon layers in the vapor-grown carbon fibers can grow relatively easily at high temperatures, probably because of their co-axial nanotexture, and finally the cross-section of the fiber was polygonized, as shown in Fig. 2.106a. For the radial type of axial orientation shown on mesophase-pitch-based carbon fibers in Fig. 2.106b, the growth of layers is not so difficult, in comparison with the isotropic-pitch-based carbon fibers with random orientation in their cross-section (Fig. 2.106c). In the case of isotropic-pitch-based carbon fibers, the orientation is almost random not only in

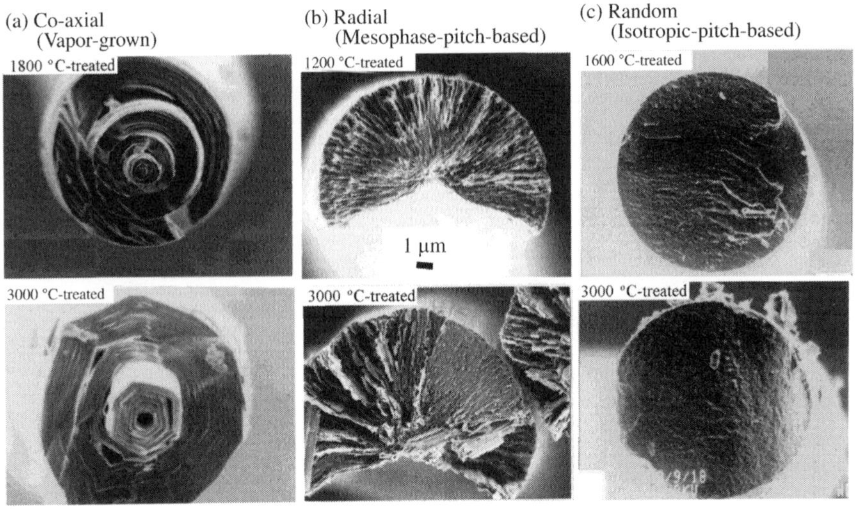


**FIGURE 2.106**

SEM images of the carbon fibers with different nanotextures in their cross-sections.

*(Courtesy of Prof. A. Yoshida of Tokyo City Univ.)*

their cross-sections but also along the fiber axis, which is reasonably supposed to hinder the growth of crystallites.

A marked effect of nanotexture on the crystallite growth can also be seen in nanofibers. In Fig. 2.107, 002 lattice fringe images of two nanofibers with different nanotextures, herring-bone and tubular ones (refer to Fig. 2.18) are compared between as-prepared fiber and high-temperature-treated fibers [280].

By high-temperature treatment, hexagonal carbon layers grow larger in both fibers. In the herring-bone nanotexture, however, the growth of crystallites is seen to be limited within each packet. In the tubular-type nanotexture, the layers grow along the fiber axis and become very long, and the resultant nanofibers may be called multi-wall carbon nanotubes. In vapor-grown carbon fibers, the same marked crystallite growth as in Fig. 2.107b occurs as explained in detail in Section 3.4.2d, and a high graphitization degree is attained by a high-temperature treatment, as shown in Fig. 2.105a.

### c. Carbon materials with point orientation

Carbon blacks, which have the nanotexture with concentric point orientation, consist of the primary particles with different sizes from 15 to a few hundred nm and, as a consequence, their graphitization behaviors are different mainly depending on their sizes. In Fig. 2.108, interlayer spacing $d_{002}$ is plotted against HTT for different carbon blacks, channel and furnace blacks consisting of the aggregated primary particles with small sizes around 20 nm, but thermal and lamp blacks of mostly separated particles with large sizes as a few hundred nm. The latter two

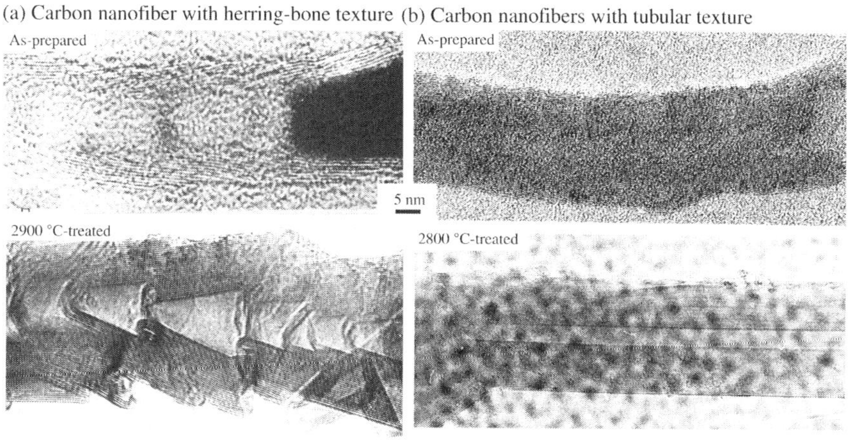


**FIGURE 2.107**

002 lattice fringe images of two nanofibers with different nanotextures.

*(Courtesy of Mme S. Bonnamy of CRMD-CNRS, France.)*

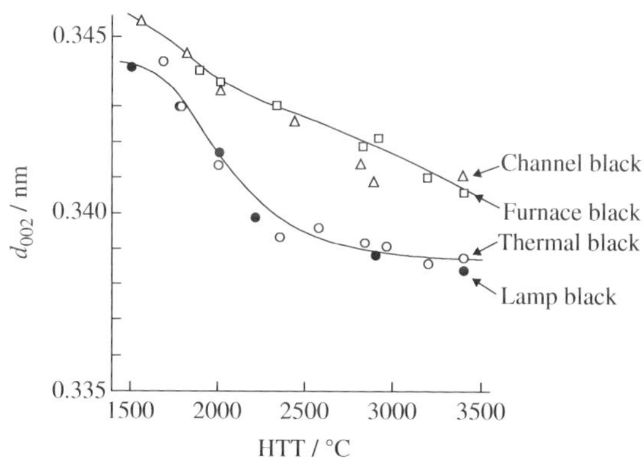


**FIGURE 2.108**

Changes in $d_{002}$ with HTT for carbon blacks.

give smaller $d_{002}$ than the former two, but their $d_{002}$ values decrease only to 0.339 nm even after the heat treatment to 3400°C, far from the value for graphite 0.3354 nm.

The growth of crystallites in carbon black particles has been pointed out to be constrained mainly due to the size of the primary particles [281–284]. In Fig. 2.109, 002 lattice fringe image is shown for furnace blacks heat-treated at 3000°C, whose $d_{002}$ change retards as shown in Fig. 2.108. Lattice fringes corresponding to the size of La become larger by the heat treatment, but still spherical

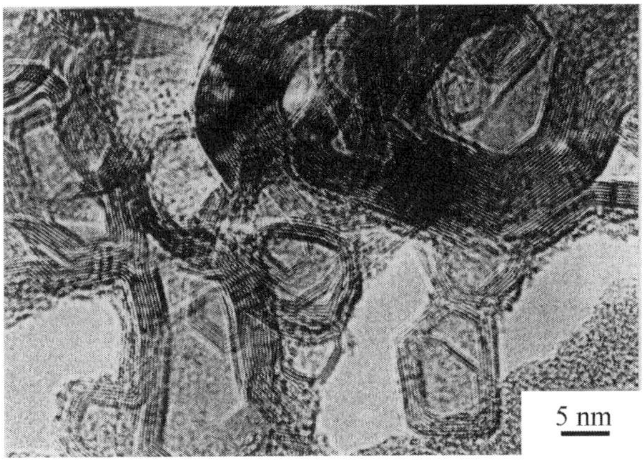


FIGURE 2.109

TEM images of channel black after the heat treatment near 3000°C.

morphology is kept even after 3000°C treatment. It has been experimentally proven that crystallite growth depends on the size of particle size, revealing that spherical morphology constrains the growth of crystallites in the particles. Some particles are broken into a few fragments by the heat treatment at around 3000°C, probably in order to release the constraint stress accumulated due to crystallite growth [282], but no acceleration of graphitization (crystallite growth) has been detected even after the heat treatment at 2000°C under a pressure as high as 0.5 MPa [285].

Carbon spherules obtained from the mixture of poly(ethylene) with poly(vinyl chloride) by pressure carbonization have radial point orientation, as shown in Section 2.4.3 and in Fig. 2.74. These spherules with the size of about a few μm show the $d_{002}$ value of 0.337 nm, close to graphite, but the layer size La is only around 6 nm, revealing that the growth of crystallites is also depressed in these spherules [286]. In Fig. 2.110, SEM images of these spherules heat-treated at 1000 and 2800°C are shown. The flattening of the particles is supposed to be due to the shrinkage along the line connecting two poles of the sphere, and the formation of large cracks around the equator and concentric formation of small cracks around the poles are consistent with the nanotexture in these spherules (refer to Fig. 2.74).

Mesophase spheres formed in pitches are separated from isotropic matrix using a solvent and studied their graphitization behavior [287,288]. These separated mesophase spheres are named as mesocarbon microbeads (MCMB) (refer to Section 2.3.4a and b) [288]. Because they also have a nanotexture with radial point orientation at least near the surface (refer to Fig. 2.52), crack formation after high-temperature treatment is very similar to carbon spherules shown in Fig. 2.110 [288]. Changes in $d_{002}$, $La(110)$ and $Lc(002)$ with HTT are studied in detail for MCMBs prepared from different precursor pitches [289].

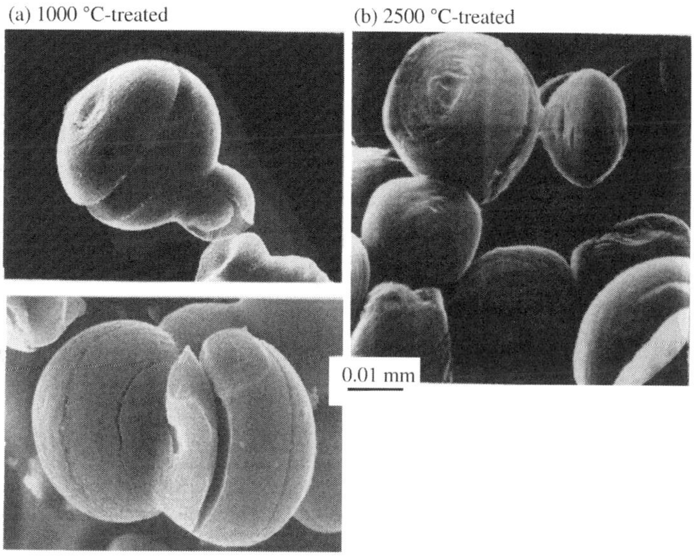


**FIGURE 2.110**

SEM images of carbon spherules after the heat treatment at 2800°C [285].

The cokes of spherical particles, fluid coke and gilsonite coke, also have the nanotexture of point orientation. The appearance and polarized-light microscopic images of the particles of a fluid coke are shown in Fig. 2.111a and b, and the structure models of the particles are shown in Fig. 2.111c [290]. It consists of two kinds of the particles, denoted as A and B in Fig. 2.111, which give quite different appearance and optical micrograph. The former has the nanotexture of a concentric point orientation, but the nanotexture of the latter is similar to the flower of chrysanthemum. It is shown in Fig. 2.98 that the changes in $d_{002}$ and $Lc(002)$ with HTT for fluid coke are retarded much from the needle-like cokes with planar orientation. Even after 2800°C treatment, $d_{002}$ and $Lc(002)$ become 0.339 and 20 nm, respectively, the former being much larger and the latter much smaller than those for needle-like cokes. Gilsonite coke has a similar nanotexture to the particle A of fluid coke and as a consequence a similar structure change at high temperatures [291].

Various experimental results suggest that crystallite growth is markedly depressed in the particles with spherical morphology consisting of nanotexture of point orientation. In Fig. 2.112, therefore, La is plotted against the logarithm of average particles size for carbons consisting of spherical particles, showing clearly that in the smaller particles crystallite size becomes the smaller [14].

### d. Carbon materials with random orientation

In glass-like carbons, which have the nanotexture of random orientation, the crystallites do not grow even by heat treatment above 3000°C, having pretty large

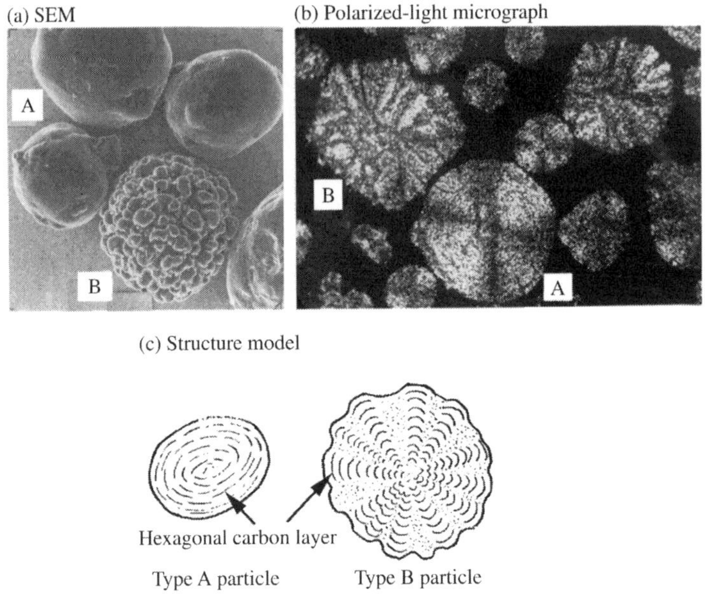


**FIGURE 2.111**

Fluid cokes [290].

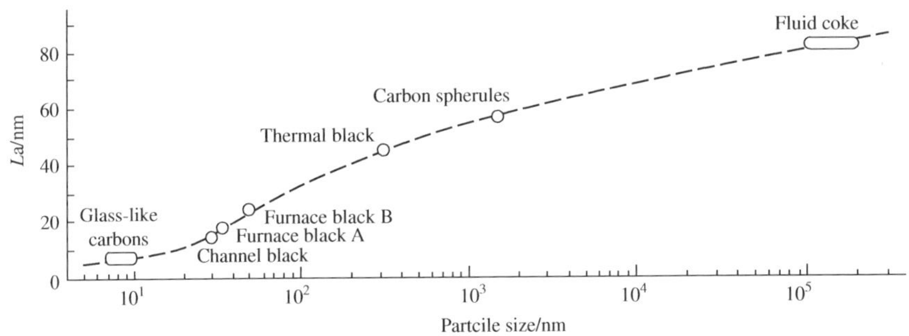


**FIGURE 2.112**

Dependence of *La* on average particles size in spherical carbon particles [14].

$d_{002}$ and very small crystallite sizes *Lc*(002) and *La*(110), as shown in Fig. 2.85. The following two experimental results show how strongly the graphitization of these carbons is depressed. In order to develop the graphitic structure in these carbons, their nanotexture of random orientation had to be destroyed, the heat treatment under as a high pressure as 30 MPa being necessary to graphitize (refer to Section 2.6.2) [292,293]. When a rod of glass-like carbon was melted by directly passing the electrical current, a ball with graphite structure was obtained in a

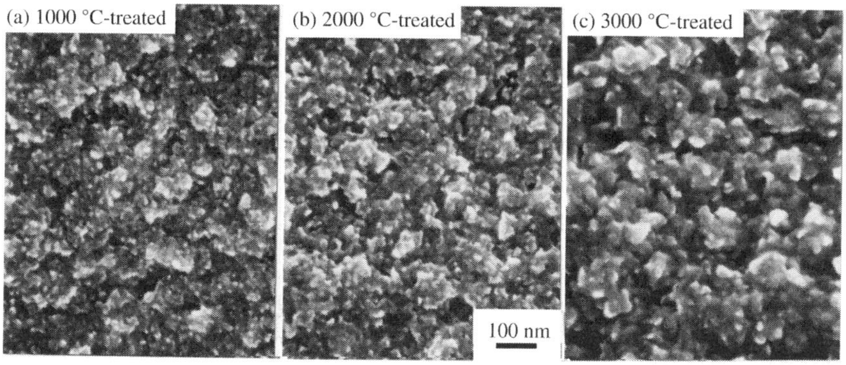


**FIGURE 2.113**

SEM images of fractured surface of glass-like carbons heat-treated at different temperatures.

*(Courtesy of Prof. A. Yoshida of Tokyo City Univ.)*

**Table 2.9** Grain size *D* Determined from FE-SEM, X-ray Parameters and Magnetoresistance Parameters on Glass-Like Carbons Heat-Treated at Different Temperatures [295]

| HTT (°C) | Grain size *D* (nm) | X-ray Parameters (nm) | | | Magnetoresistance Parameters | | |
|---|---|---|---|---|---|---|---|
| | | $d_{002}$ | *Lc* (002) | *La* (110) | $(\Delta\rho/\rho_0)_{cr}$ (%) | $r_T$ | $r_{TL}$ |
| 1000 | 7.0 | 0.3468 | 1.9 | 2.5 | – | – | – |
| 2000 | 10.0 | 0.3442 | 3.3 | 3.1 | −0.085 | 0.77 | 0.89 |
| 3000 | 13.1 | 0.3436 | 3.6 | 3.5 | −0.182 | 0.96 | 0.86 |

crater-type cavity in the middle of the rod, but the wall of crater kept the characteristics of glass-like carbon even though it had been heated to a temperature very near the melting point of carbon [294].

SEM images of the fractured surface of the glass-like carbons heat-treated at 1000, 2000 and 3000°C are shown in Fig. 2.113 [295]. The sizes of constituent grains measured on these micrographs are listed in Table 2.9, together with X-ray and magnetoresistance parameters. The grain size in the glass-like carbon increases slightly with increasing heat treatment temperature, associated with only slight changes in $d_{002}$, $Lc(002)$, $La(110)$ and $(\Delta\rho/\rho_0)_{cr}$. However, all of these parameters are very far from the values for graphite. Anisotropic ratios $r_T$ and $r_{TL}$ are close to 1, indicating isotropy in structure.

For one of the carbons with the nanotexture of random orientation, i.e., the carbon prepared from sugar, the change in 002 lattice fringes image with HTT is

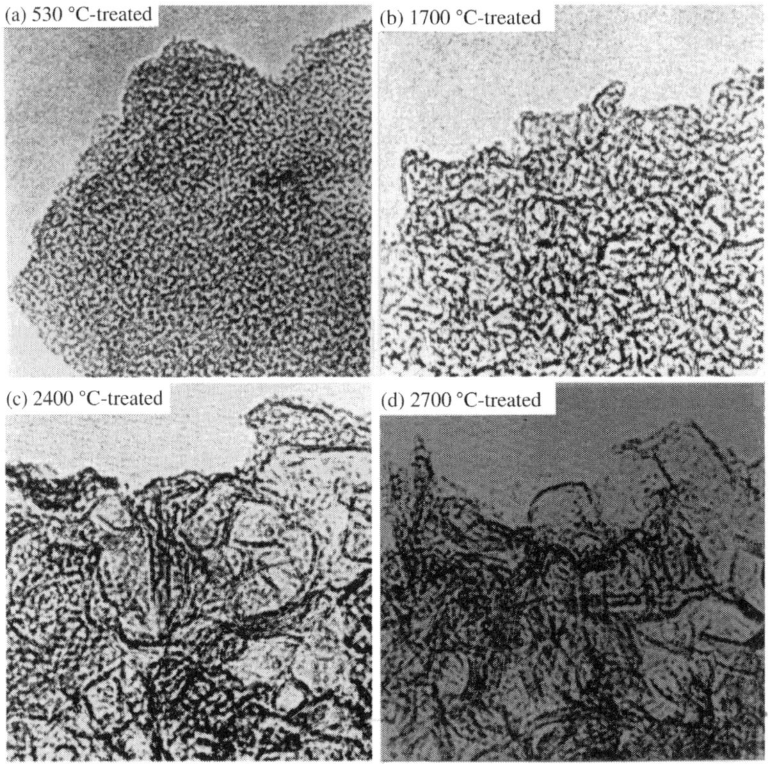


**FIGURE 2.114**

002 lattice fringe images of the carbon prepared from sugar and heat-treated at different temperatures.

*(Courtesy of Mme. A. Oberlin.)*

shown in Fig. 2.114 [94]. It is clearly seen that the size of fringes (the size of crystallites) increases slightly, but randomness in their orientation is not improved even after treatment at 2700°C. The random orientation even after the high-temperature treatment in these carbons is demonstrated by measuring selected area electron diffraction pattern and 110 dark-field images, in addition to 002 lattice fringes for the carbons prepared from different precursors [253,296].

The HTT dependences of $R_H$ and $(\Delta\rho/\rho_0)_{max}$ measured for two glass-like carbons at different temperatures in a field of 0.65 T are shown in Fig. 2.115a and b, respectively [297,298]. Positive $R_H$ and negative $(\Delta\rho/\rho_0)_{max}$ observed for the sample heat-treated at 3200°C are the characteristics of these carbons. $R_H$ for glass-like carbon is independent of magnetic field and insensitive to temperature of measurement in the range from 2.8 K to 300 K, but it depends strongly on HTT. In contrast to the HTT dependences of these parameters for the coke in Fig. 2.102, $R_H$ is positive and $(\Delta\rho/\rho_0)_{max}$ is negative even after heat treatment at

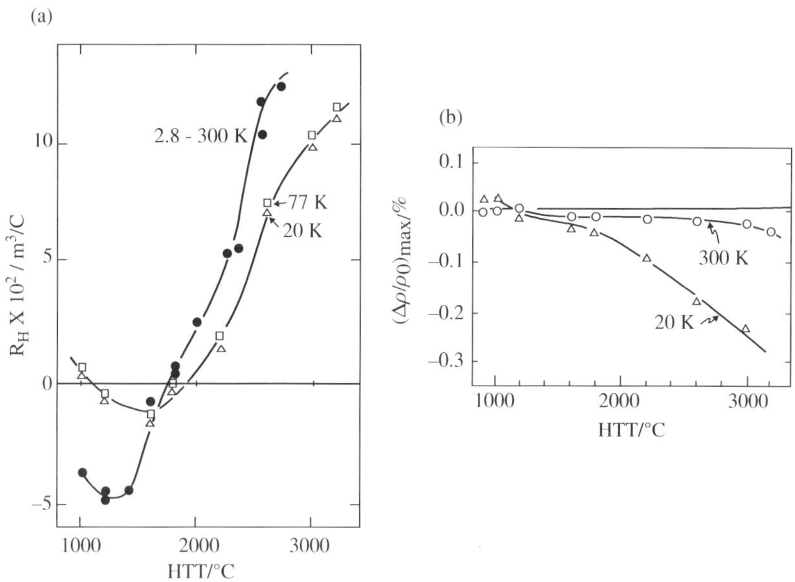


**FIGURE 2.115**

Dependences of $R_H$ and $(\Delta\rho/\rho_0)_{max}$ on HTT for glass-like carbon [297,298].

3000°C, and also the absolute values of these parameters are very small, revealing a strong depression of structural development. However, the cross-over point from negative to positive in $R_H$ is observed on two glass-like carbons at the same HTT of 1800°C, almost the same as the coke (refer to Fig. 2.102).

## 2.5.3 Relations among structure parameters

The most exact parameter to evaluate the degree of graphitic structure development is the value $P_1$, which is the probability for two neighboring hexagonal carbon layers to have a graphitic AB stacking [299]. However, the determination of $P_1$ has not been employed so often, mainly because it is not easy to apply a number of samples. The value of $d_{002}$ has frequently been used as a parameter to characterize the degree of graphitization.

There are only a few works studied on the relation of $d_{002}$ to $P_1$ [300–302]. The relation between $d_{002}$ and $P_1$ derived from a number of experimental points on different carbon materials with a wide range of nanotextures is shown in Fig. 2.116 [302]. In previous works [299–301], the linear dependence of $d_{002}$ on $P_1$ was reported as the broken line in Fig. 2.116. However this relation has not been proved by measurements on a wide range of carbon materials [302], probably because $d_{002}$ for turbostratic stacking is not necessary to be a fixed value as 0.344 nm. A general trend, i.e., $d_{002}$ decreases with increasing $P_1$, is observed for all carbons, but the relation is not unique for all carbon materials, depending on their nanotexture. In

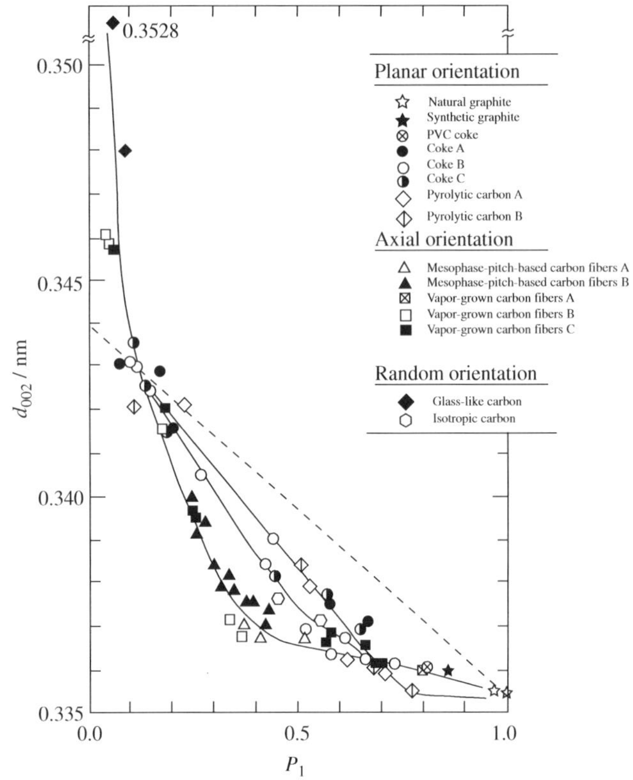


**FIGURE 2.116**

Relations between $d_{002}$ and $P_1$ [302].

the carbon materials with random orientation, the development of graphite structure is so difficult that $P_1$ cannot be more than 0.1 and $d_{002}$ is much larger than 0.345 nm. In the carbon materials with axial orientation (carbon fibers), $d_{002}$ decreases rather rapidly, but the decrease in $P_1$ is relatively slow, in comparison to those with planar orientation (cokes and pyrolytic carbons).

Based on the experimental results that turbostratic stacking occurs frequently with the spacing of around 0.344 nm and under the assumption that two kinds of spacing, 0.344 and 0.3354 nm, randomly co-exist in the crystallite, the conventional parameter $p$ was proposed [303] and used in some works. Later, the model was somewhat modified and a new relation between $d_{002}$ and the parameter $p$ was proposed [304]. Based on an idea that extra carbon atoms are located on carbon hexagonal layers, which is the reason why $d_{002}$ is larger than graphite (no extra carbon atoms), the parameter $g$ has been proposed [305,306], which can be calculated from the observed $d_{002}$. However, these parameters, $p$ and $g$, are not used

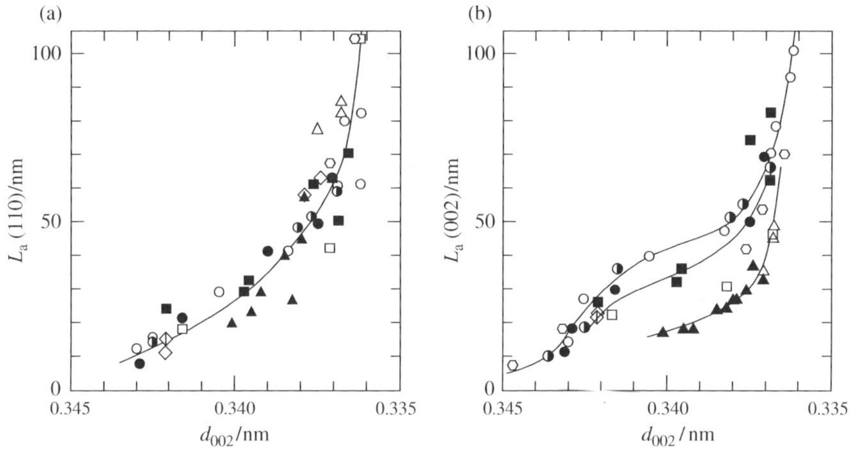


**FIGURE 2.117**

Relations of *La* and *Lc* to $d_{002}$ [302]. For the signs of carbon materials refer to Fig. 2.116.

now, mainly because some newly developed carbon materials, such as glass-like carbons, have much larger spacing than 0.344 nm.

The structure of carbon materials has to be characterized not only from $d_{002}$ but also *La* and *Lc*. In Fig. 2.117, *La*(110) and *Lc*(002) are plotted against $d_{002}$ for various carbon materials used in Fig. 2.116 [302]. The relation between *La* and $d_{002}$ looks unique for all carbon materials, irrespective of nanotexture of carbon materials (Fig. 2.117a). On the other hand, the relation between *Lc*(002) and $d_{002}$ depends strongly on nanotexture; the carbons with planar orientation show a plateau at around 0.34 nm, but the growth of *Lc*(002) is strongly depressed in the carbon fibers with axial orientation, particularly in mesophase-pitch-based carbon fibers, as shown in Fig. 2.117b.

As explained above on carbon materials with different nanotextures, the changes of physical properties of carbon materials with heat treatment are also dependent on their nanotextures. Therefore, there must be certain relations between the physical properties and X-ray parameters, $P_1$, $d_{002}$, *La* and *Lc*. In Fig. 2.118, $\rho$, $R_H$ and $(\Delta\rho/\rho_0)_{max}$ are plotted against $d_{002}$ on a pyrolytic carbon [260]. Since the pyrolytic carbon used has been prepared by a chemical vapor deposition process at rather high temperature such as 2200°C, $R_H$ passes by its maximum value and $(\Delta\rho/\rho_0)_{max}$ by its minimum value already. So, $\rho$ and $R_H$ decrease, and $(\Delta\rho/\rho_0)_{max}$ increases with the decrease in $d_{002}$. Below the $d_{002}$ values of 0.338 nm, $\rho$ starts to decrease, $R_H$ and $(\Delta\rho/\rho_0)_{max}$ change their signs. This suggests that the graphite structure becomes predominant in the pyrolytic carbons with $d_{002}$ of more than 0.338 nm.

In order to understand the correspondence of these galvanomagnetic properties on the development of graphitic structure in carbons, $R_H$ and $(\Delta\rho/\rho_0)_{max}$ are plotted against the degree of graphitization $P_1$ on cokes and pyrolytic carbons in

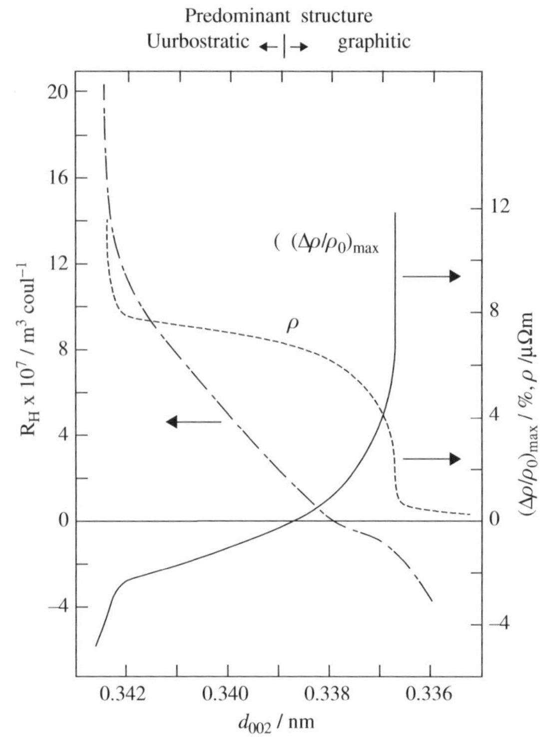


**FIGURE 2.118**

Dependences of $\rho$, $R_H$ and $(\Delta\rho/\rho_0)_{max}$ for a pyrolytic carbon.

*(Courtesy of Prof. Y. Kaburagi of Tokyo City Univ.)*

Fig. 2.119 [307]. Though the absolute values of these parameters are different on cokes and pyrolytic carbons, because of the difference in preferred orientation scheme, the dependence of these parameters on $P_1$ is common; $R_H$ decreases and $(\Delta\rho/\rho_0)_{max}$ increases with the increase in $P_1$ (qualitatively corresponds to the increase in HTT and the decrease in $d_{002}$), as shown in Fig. 2.119a. It has to be mentioned that $R_H$ and $(\Delta\rho/\rho_0)_{max}$ change their sign at the $P_1$ value of 0.5, i.e., the predominant structure changes from turbostratic stacking to a graphitic one. The dependences of $(\Delta\rho/\rho_0)_{max}$ on $La(110)$ are common for different pyrolytic carbons and cokes, as shown in Fig. 2.119b. The change in sign from negative to positive occurs at $La(110)$ of around 50 nm.

Relations between various structural parameters have been studied mostly by focusing on carbon materials having the same scheme of nanotexture in various literatures. Some of them are reproduced in Figs. 2.120. In Fig. 2.120a, the well-defined relation between $(\Delta\rho/\rho_0)_{cr}$ and $d_{002}$ for various cokes, including needle-like cokes, gilsonite coke, etc. is shown [252]. In Fig. 2.120b, thermal

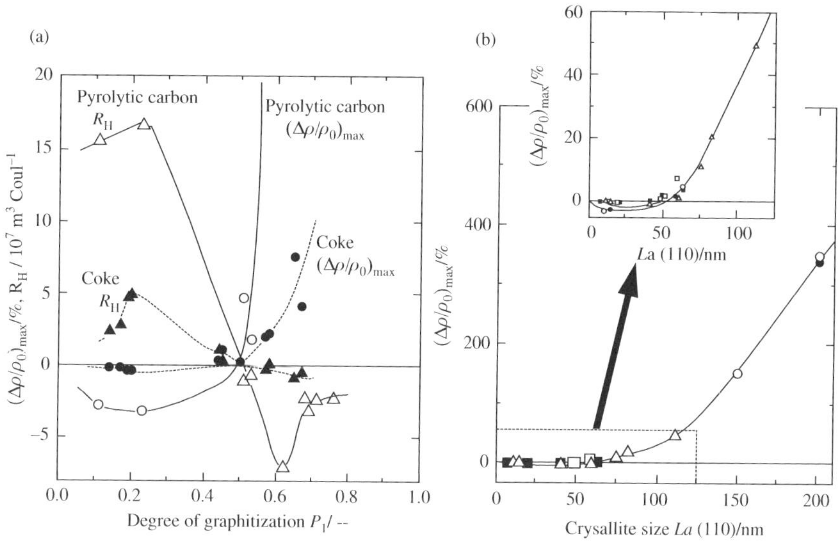


**FIGURE 2.119**

Dependences of $(\Delta\rho/\rho_0)_{max}$ and $R_H$ on $P_1$ and $La(110)$ [307].

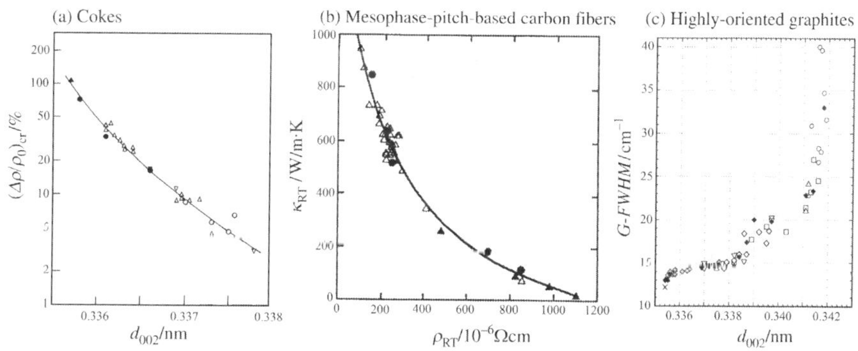


**FIGURE 2.120**

Relations between structure parameters for carbon materials: (a) $(\Delta\rho/\rho_0)_{cr}$ vs. $d_{002}$ for various cokes [6], (b) thermal conductivity $\kappa$ vs. $\rho$ at room temperature for various mesophase-pitch-based carbon fibers [308], (c) FWHM of G-band in Raman spectrum vs. $d_{002}$ for highly-oriented carbon films [309].

conductivity at room temperature $\kappa_{RT}$ is plotted against $\rho$ at room temperature ($\rho_{RT}$) for three mesophase-pitch-based carbon fibers prepared under different conditions [308]. Well-defined relation between $\kappa_{RT}$ and $\rho_{RT}$ along the fiber axis is observed. Mesophase-pitch-based carbon fibers having ribbon-like morphology and low $\rho_{RT}$ give very high $\kappa_{RT}$ as 1000 W/mK, which are interesting as one of

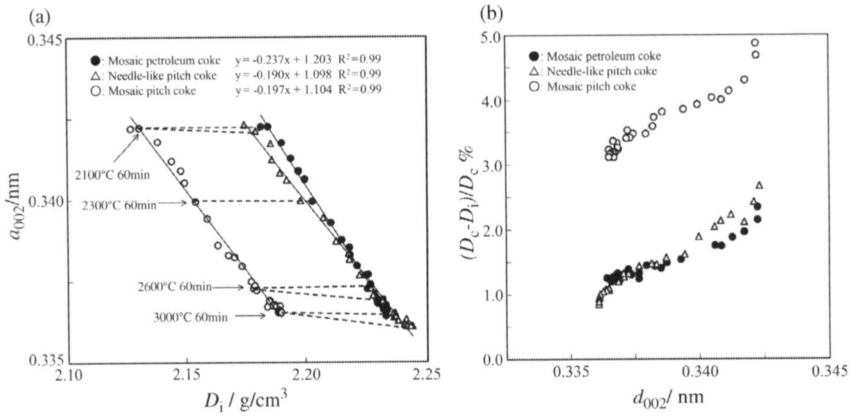


**FIGURE 2.121**

Dependence of $D_R$ (a) and closed porosity $(D_C\text{-}D_R)/D_C$ (b) on $d_{002}$ for cokes [311].

thermal management materials [310]. In Fig. 2.120c, the dependence of FWHM of G-band in Raman spectra on $d_{002}$ is shown for highly oriented carbon films derived from a polyimide at various heat-treatment conditions, together with HOPG [309]. A clear relation between *G-FWHM* and $d_{002}$ is observed.

For cokes, the linear relation between $d_{002}$ and apparent density measured by immersion into butanol (immersion density, $D_i$) is found in the range of HTT of 2100–3000°C, as shown in Fig. 2.121a [311]. The coke derived from pitch gives a different linear relation whether needle-like or mosaic texture. The difference in these relations is mainly due to the difference in the content of closed pores for butanol, which is estimated from $D_i$ and crystal density $D_c$ calculated from lattice parameters measured.

## 2.5.4 Graphitization process

In order to develop the graphite structure in carbon materials, high-temperature heat treatment is required, but its degree of development depends predominantly on the nanotexture of carbon materials, as explained above. Graphite structure is developed relatively easily in carbon materials with planar orientation by high-temperature treatment, but in those with random orientation the development of graphite structure is very much depressed even after high-temperature treatment. In Fig. 2.122, the dependences of $d_{002}$ and $La$ on heat treatment temperature are reproduced from Fig. 2.85.

The furnace black heat-treated up to 3400°C has values of $d_{002}$ and $La$ corresponding to the needle-like coke heat-treated to about 2300°C. The structure parameters for the 3400°C-treated thermal black correspond to those for the 2400°C-treated needle-like coke. In other words, structure development is depressed in the furnace and thermal blacks, in comparison with the needle-like

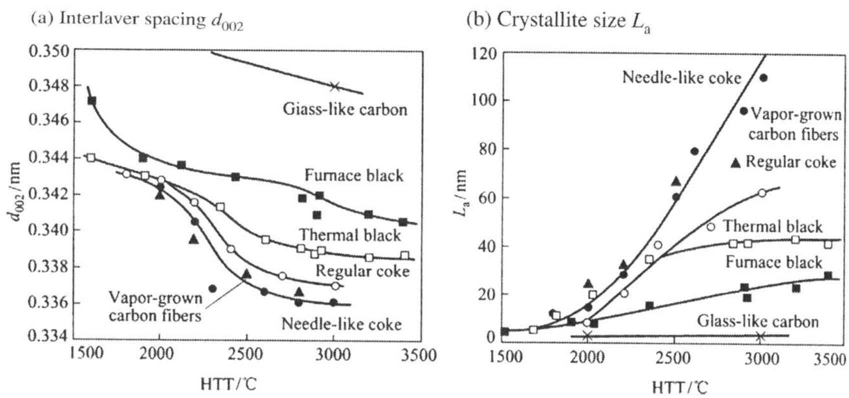


**FIGURE 2.122**

Dependences of $d_{002}$ and $La$ on HTT for different carbon materials.

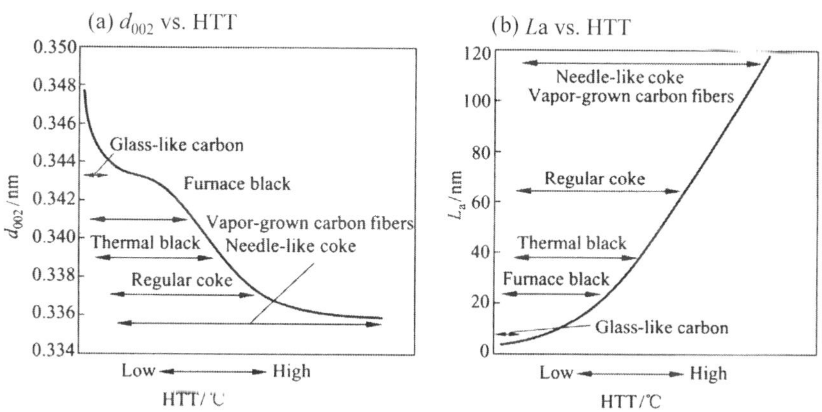


**FIGURE 2.123**

Master curves for the dependence of $d_{002}$ and $La$ on HTT.

coke. For the glass-like carbon, its structure parameters $d_{002}$ and $La$ after 3000°C treatment correspond to the needle-like coke heat-treated to a very low temperature, even $d_{002}$ value of the glass-like carbon not appearing in the needle-like coke. Based on this discussion, we may draw the master curves for these structure parameters. In Fig. 2.123, the master curves for the dependences of $d_{002}$ and $La$ on heat treatment temperature are shown. On these master curves, almost the whole range of the changes in these parameters is observed for the needle-like coke and also vapor-grown carbon fibers. On the dependence of $d_{002}$, the regular coke is very similar to the needle-like coke, but on $La$ its growth is suppressed in the regular coke. By two carbon blacks, however, the dependences up to the

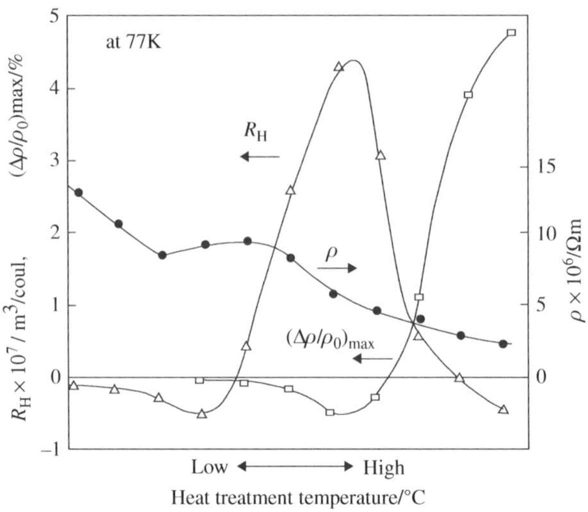


**FIGURE 2.124**

Master curves for the dependence of $\rho$, $(\Delta\rho/\rho_0)_{max}$ and $R_H$ on HTT.

*(Courtesy of Prof. Y. Hishiyama of Tokyo City Univ.)*

middle part of the master curves can be realized after high-temperature treatment above 3000°C. In the glass-like carbon, the structure development is very much suppressed, as explained in the previous section, and so only the very beginning of the master curves for the structure parameters can be achieved.

For galvanomagnetic properties of carbon materials, their master curves are shown in Fig. 2.124, $\rho$, $(\Delta\rho/\rho_0)_{max}$ and $R_H$ at 77 K. At low temperatures of heat treatment, $(\Delta\rho/\rho_0)_{max}$ and $R_H$ have very low absolute values and negative, and $\rho$ decreases. $R_H$ turns to increase, passing through a minimum, and change its sign to positive. At almost the same temperature, $(\Delta\rho/\rho_0)_{max}$ starts to decrease further. Positive $R_H$ increases rapidly and then turns to decrease, this turning point corresponding to the minimum point of $(\Delta\rho/\rho_0)_{max}$. With increasing temperature, $(\Delta\rho/\rho_0)_{max}$ increases rapidly to a very high value, as seen in highly oriented carbons (Section 2.5.3), and $R_H$ decreases and becomes negative. In highly graphitized carbon materials, two carriers, negative electrons and positive holes, coexist in almost equal concentration, but the former has a little larger mobility than the latter. This is the reason why the highly graphitized carbon materials have high $(\Delta\rho/\rho_0)_{max}$ and negative $R_H$.

For the cokes and vapor-grown carbon fibers mentioned in Fig. 2.123, almost the whole range of these properties can be realized by the heat treatment at different temperatures, being the same as X-ray parameters. For the glass-like carbon, however, only the beginning of the master curve is obtained, as shown in the Section 2.5.2a, where $R_H$ is positive but before its maximum, and $(\Delta\rho/\rho_0)_{max}$ is negative but before its minimum (Fig. 2.115), being the same as X-ray parameters (Fig. 2.122).

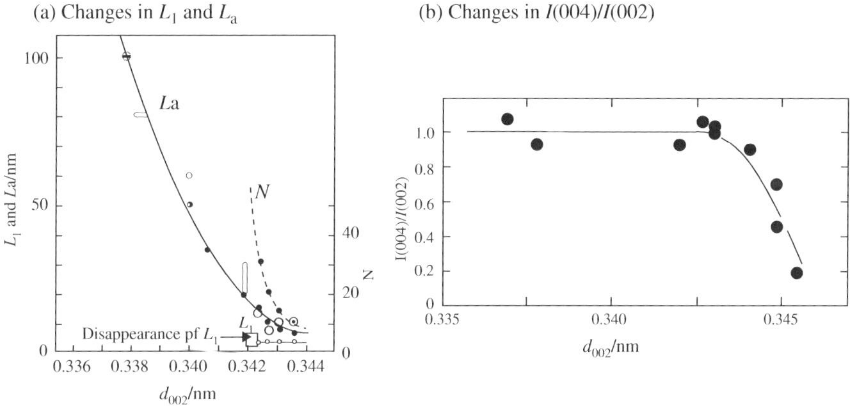


**FIGURE 2.125**

Crystallite sizes, *La*, $L_1$ and *N*, and intensity ratio $I(004)/I(002)$ as a function of $d_{002}$ [309,310].

Based on many experimental results on cokes, it has been proposed to divide the graphitization process into two, before and after reaching $d_{002}$ to around 0.342 nm [312]. The first process down to $d_{002}$ of around 0.342 nm may be called pre-graphitization and the one after 0.342 nm is the graphitization process, where a three-dimensional graphitic structure is gradually developed. The flaky particles of a coke have been shown through TEM analyses to consist of small domains of 2–3 carbon layers with about 1 nm size, which grow by heat treatment above 2000°C, i.e., after passing through $d_{002}$ of 0.342 nm [16,313,314]. In Fig. 2.125a, the size of small domains $L_1$, the size *La* of crystallites aggregated from these small domains, and the number of layers stacked in parallel *N* are plotted against average interlayer spacing $d_{002}$. After reaching $d_{002}$ of about 0.342 nm, crystallites grow rapidly with decreasing $d_{002}$, i.e., increasing heat treatment temperature. Almost the same conclusion has been derived from the measurement of the change in intensity ratio of 004 to 002 lines, $I(004)/I(002)$, with $d_{002}$ on a coke (Fig. 2.125b) [315]. Since the intensity ratio $I(004)/I(002)$ can be a measure of flatness of carbon layer planes, the layer planes are wrinkled in the cokes with $d_{002}$ of more than 0.342 nm, in other words, the process before reaching $d_{002}$ of about 0.342 nm being dewrinkling of layer planes.

Also, the results of structural strain measurements on cokes suggest a similar structural change in the pre-graphitization process [316–319]. In one carbon with random nanotexture, sugar char, the relation between structural strain along the a-axis $\delta_a^2$ and that along the c-axis $\delta_c^2$, which are measured through the detailed analysis of X-ray diffraction profiles, does not pass through the origin, where the main structure change occurs in the pre-graphitization process [316], as shown in Fig. 2.126. In contrast, the relation between $\delta_a^2$ and $\delta_c^2$ for different cokes in the second step of structure change (graphitization) passes through the origin,

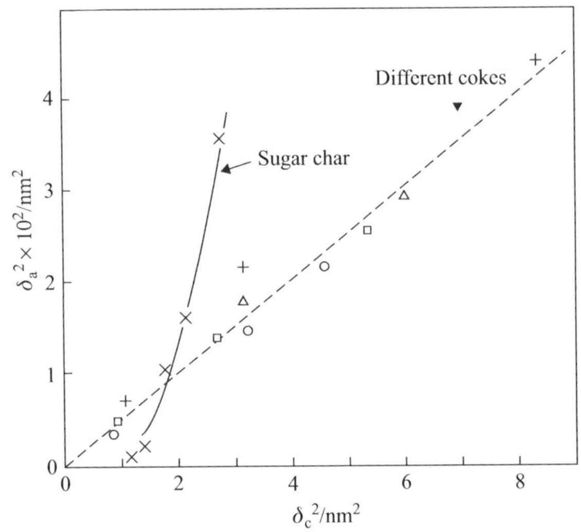


**FIGURE 2.126**

Relations between average structural strains along a and c axes, $\delta_a^2$ and $\delta_c^2$ [316].

suggesting that the strains measured are the *a*- and *c*-axes components of a strain due to the slight displacement of carbon atoms from their exact positions in hexagonal carbon layers. The result on structural strain along the *a*-axis suggests that the cokes heat-treated below 2000°C contain two kinds of strain, one remaining even in the graphitization process above 2000°C [317]. As shown in Fig. 2.99c, a large strain is contained in the cokes with $d_{002}$ larger than 0.342 nm, which decreases quickly with decreasing $d_{002}$ [271].

In conclusion, the graphitization process is common for all carbon materials, which may represent by the master curves for each structure parameter, as shown for $d_{002}$ and *La* in Fig. 2.123 and for $\rho$, $(\Delta\rho/\rho_0)_{max}$ and $R_H$, in Fig. 2.124. In some carbon materials, such as needle-like coke and pyrolytic carbons, the whole range of these master curves can be realized by the heat treatment at different temperatures up to 3000°C. In some carbons, such as glass-like carbons, however, only a first part of the master curves can be realized even by high-temperature treatment above 3000°C. Many carbon materials are located in between these two extreme cases. Some examples are shown on the master curves for $d_{002}$ and *La* by regular coke and carbon blacks in Fig. 2.123. It has to be mentioned that the range of the master curve, which can be realized by the heat treatment, depends also on the structure parameters; in a regular coke $d_{002}$ change is very similar to needle-like coke but *La* growth is suppressed, as shown in Fig. 2.123.

The graphitization proceeds in two steps, the first step (pre-graphitization) above $d_{002}$ of about 0.342 nm and the second step (graphitization) below 0.342 nm. The second graphitization step is accompanied by the decrease in $d_{002}$, the increase in $P_1$ and the growth of crystallite sizes. For the structural change in the first pre-graphitization step, a de-wrinkling model has been proposed. In glass-like carbons, even the pre-graphitization step seems to be difficult to be completed by 3000°C treatment.

The difference in graphitization behavior, in other words, what range of the master curves can be realized, is governed predominantly by the nanotexture of the carbon, as explained in previous sections. In order to realize the whole range of the master curves, the nanotexture of the carbon materials, except high degree of planar orientation, has to be destroyed, as explained in Section 2.6.

### 2.5.5 Graphitizing and non-graphitizing carbons

By high-temperature treatment at around 3000°C, graphite structure is markedly developed in some carbons, e.g., cokes, but not at all in other carbons, e.g., sugar chars and carbon blacks. These two different graphitization behaviors were reported on different carbon materials through XRD [303,318,319], and it was proposed to call them **graphitizing carbons** and **non-graphitizing carbons**, respectively. Later, the terms **soft carbons** and **hard carbons**, respectively, were used [320], which represented directly their appearance because most of the graphitizing carbons were soft and non-graphitizing carbons were hard. In International Committee for the Nomenclature and Terminology of Carbon Materials in IUPAC (International Union for Pure and Applied Chemistry), these two kinds of carbons were proposed to be called **graphitizable carbon** and **non-graphitizable carbons**, respectively [321]. However, it has to be mentioned that many carbon materials have an intermediate behavior at high temperatures, in other words, carbon materials cannot be clearly divided into these two classes, as demonstrated in Fig. 2.85 and discussed above.

The graphitization behavior of carbon materials looks somewhat different from what parameter is employed. Between needle-like and regular cokes, no pronounced difference is observed if $d_{002}$ and $P_1$ are employed (Fig. 2.85a and d), but the increases in $Lc$, $La$ and $(\Delta\rho/\rho)_{max}$ are much more marked in the former (Fig. 2.85b, c and e). If the needle-like coke is compared with vapor-grown carbon fiber, which can reach the highest degree of graphitization among carbon fibers from various precursors, no difference in changes of $d_{002}$, $P_1$ and $(\Delta\rho/\rho)_{max}$ is detected. However, crystallite growth measured by both $Lc$ and $La$ is much inferior in the latter, which is supposed to be due to the morphology of thin fibers (around 10 μm diameter). Two carbon blacks also show different behaviors in $d_{002}$, $Lc$ and $La$ with HTT; thermal black has average radius of primary particles of about 300 nm, but furnace black has about 20 nm.

It has to be pointed out that the words graphitizing and non-graphitizing carbons (either called soft and hard carbons or graphitizable and non-graphitizable carbons) are the prediction of graphitization behaviors, in other words, whether the carbon is easily or with difficulty graphitized at a high temperature. Although carbon materials cannot simply be classified into the two, they show two extremes. In Fig. 2.127, the models for these two extremes are shown [247].

The most important difference in structure between graphitizing and non-graphitizing carbons is the nanotexture; the former has the one with planar orientation, even though the degree of orientation is not so high, but the latter has random orientation nanotexture and contains a large amount of micropores. The crystallites in these two carbons are difficult to differentiate, which is a stack of one to three hexagonal carbon layers with very small size, about few nm, and they are locally oriented. With increasing temperature of heat treatment, crystallites grow in both layer size and stacking number. In non-graphitizing carbons, the growth of crystallites is very limited mostly within the size of their local orientation. Therefore, non-graphitizing carbons gave the carbon materials with very low graphitization degree, low apparent density and high porosity even after heat treatment at high temperatures. In graphitizing carbons, however, it is relatively easy because of planar orientation, even beyond the local orientation established

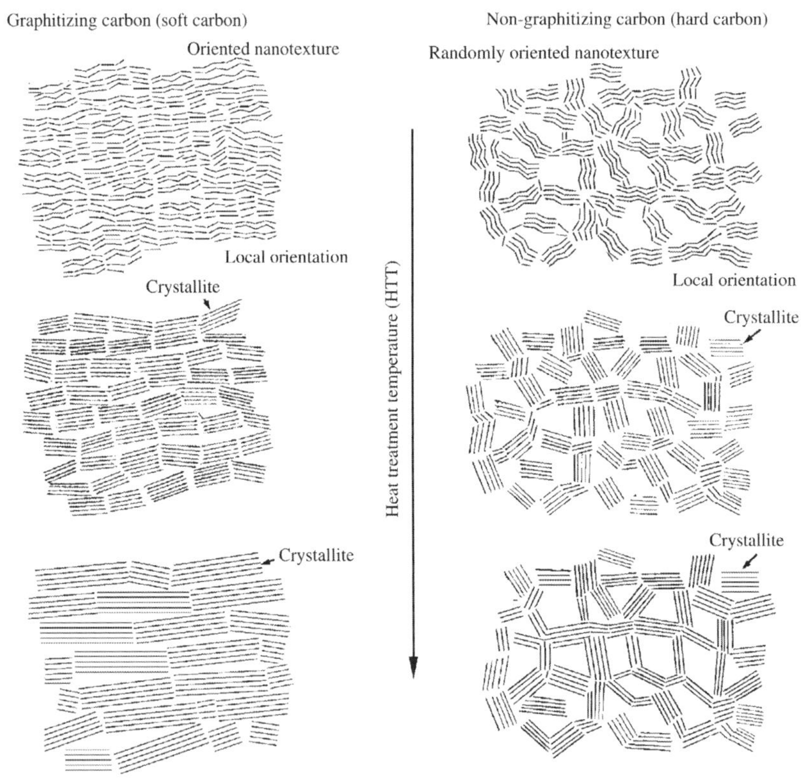


**FIGURE 2.127**

Structural models for graphitizing and non-graphitizing carbons with different HTTs [247].

during the carbonization process. Therefore, the carbon materials with a high degree of graphitization and with a high density were obtained from graphitizing carbons.

### 2.5.6 Heterogeneous graphitization (multiphase graphitization)

For most monolithic carbon materials, symmetrical 00*l* diffraction profiles of X-ray and their gradual shift to high diffraction angle side, i.e., gradual decrease in $d_{002}$, are observed, which is called **homogeneous graphitization**. For some carbon materials, however, diffraction profiles composed of more than two peaks are often observed and the relative intensity of the component peaks changes with increasing HTT, called **heterogeneous graphitization**. In Fig. 2.128a and b, changes in 004 diffraction profiles are shown in these two cases, respectively.

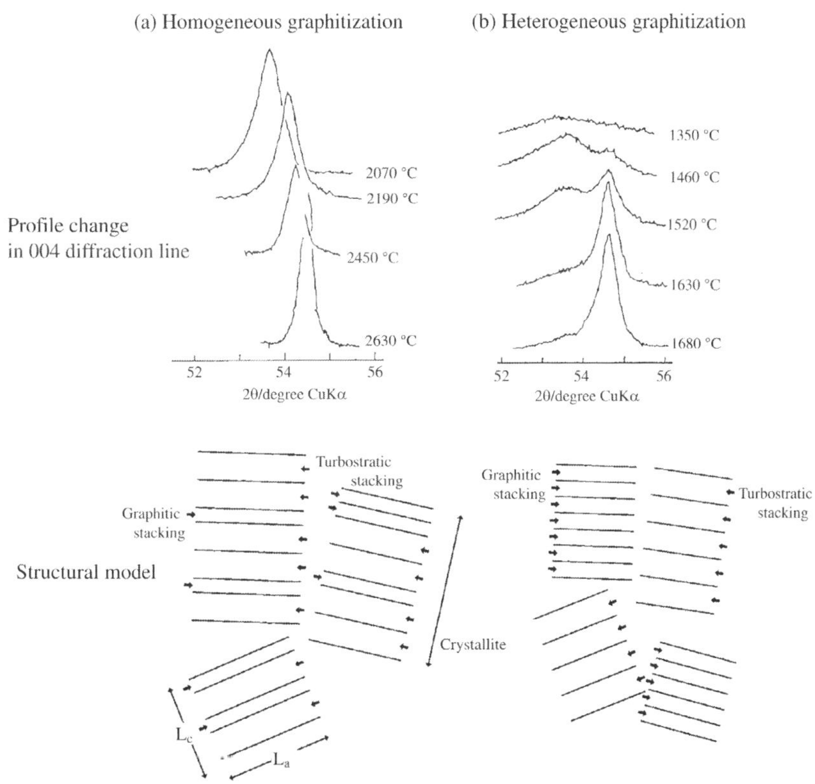


**FIGURE 2.128**

Profile changes in 00*l* diffraction and structural models for homogeneous and heterogeneous graphitization processes.

In the homogeneous graphitization process, graphitic stacking with smaller interlayer spacing increases at random in all crystallites with increasing HTT, as shown in the structural model. The graphitization processes discussed in the previous sections (2.5.1–2.5.8) are homogeneous ones. In the cases of heterogeneous graphitization process, on the other hand, the observed profile of 004 line consists of two component profiles at about 53.5 and 54.6° in 2θ (composite profiles), which corresponds to the interlayer spacing of about 0.343 and 0.336 nm, respectively. The former is reasonably supposed to be due to the component with turbostratic stacking and the latter to the component with graphitic stacking. 004 peak corresponding to the latter grows with sacrificing that to the latter, suggesting that the relative content of the component with graphite structure increases. Therefore, the structural model shown in Fig. 2.128b is considered and so this graphitization process has also been called **multi-phase graphitization** because the carbon consists of different phases.

The heterogeneous graphitization process has firstly been observed on the carbons prepared from poly(vinylidene chloride) and sugar [318], on the carbons from phenol and furfuryl alcohol resins [322,323] and also on a charcoal [324] by using X-ray diffraction. It has been studied in detail by TEM techniques [325,326].

The change in 002 profile of a charcoal with heat treatment temperature is shown in Fig. 2.129a [324]. Most of the observed composite profiles seem to

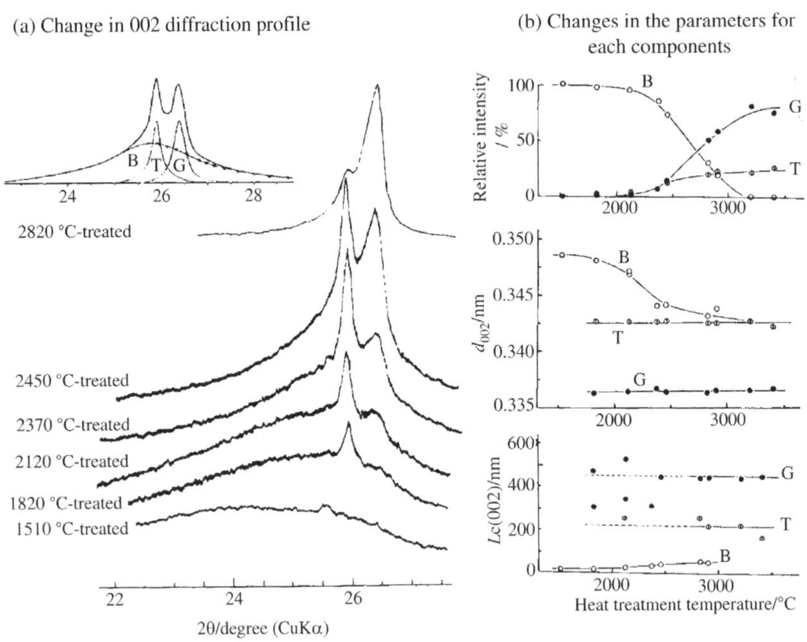


**FIGURE 2.129**

Heterogeneous graphitization process in a charcoal [315].

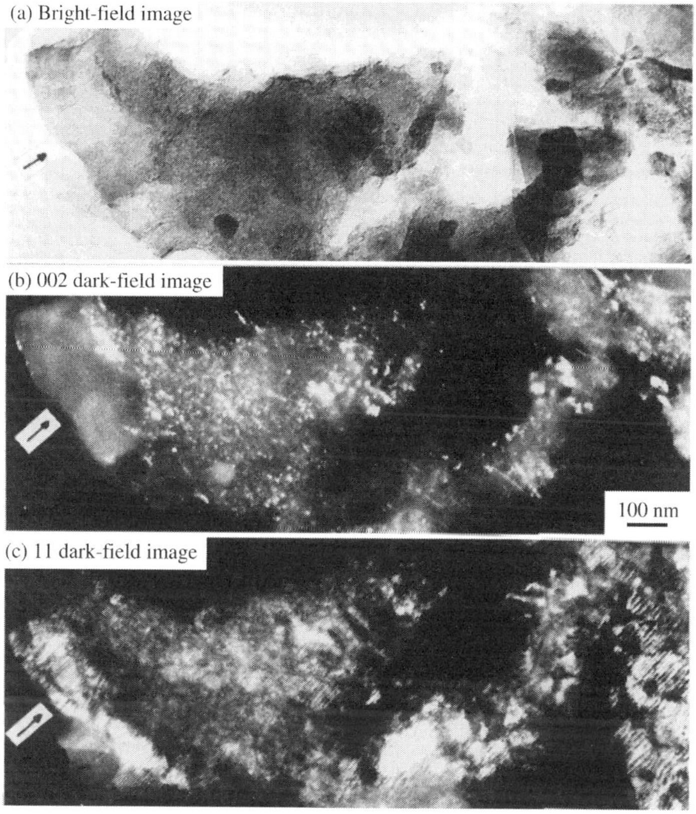


**FIGURE 2.130**

TEM images of the anthracite heat-treated at 2500°C [314].

consist of three components profiles, B, T and G, and are separated into each component profile under the assumption of a simple overlapping with each other, as shown in the figure. The relative intensity, $d_{002}$ and $Lc(002)$ of each component profile separated are plotted against HTT in Fig. 2.129b. Above 2000°C, the intensity of B component decreases and then disappears above 3000°C. Associated with the intensity decrease of the B component, the intensities for both T and G components increase and the G component becomes predominant. The T and G components have almost the constant $d_{002}$ as 0.343 and 0.336 nm and $Lc(002)$ as 20 and 45 nm, respectively. For the B component, however, $d_{002}$ decreases, approaching that for the T component, and $Lc(002)$ increases slightly. This change in X-ray diffraction profile with heat treatment suggests the consecutive transformations from B to G through T. The reason for this heterogeneous graphitization is not clear yet. Even after dissolving out of lignin and cellulose in the precursor woods, heterogeneous graphitization was observed [327].

The heterogeneous graphitization of anthracite, a high-rank coal, is due to the strain accumulated during coal formation (coalification) from a detailed study by transmission electron microscopy (TEM) [325]. Anthracite is strongly tectonized, i.e., submitted to shear stresses. Anthracite as-received had turbostratic structure, which is unable to be graphitized until a temperature range of 2000–2500°C. Above 2500°C, it transforms suddenly to a highly graphitized state [325]. The particles of the anthracite are porous, but their pores are flattened along their deposition beds. Shear stresses have thus permanently strained the material during coalification, giving preferred orientation of hexagonal carbon layers. Above 2500°C, plasticity is induced and so the preferential breakage of the less-developed pore walls occurs, resulting in the formation of crystalline lamellar domains with graphite structure, which is typically demonstrated at the part arrowed in Fig. 2.130.

During carbonization of the composites of carbon fiber and thermosetting resin, such as phenol resin and furfuryl alcohol condensates, to produce carbon/carbon composites, stress concentration is found to occur at the interface between filler fiber and matrix resin, which showed heterogeneous graphitization as discussed in detail in Section 2.6.

## 2.6 Acceleration of graphitization

### 2.6.1 Catalytic graphitization

The presence of foreign species, either atoms or compounds, in carbon precursors gives certain influences on the structure change at high temperatures, some of them accelerating the graphitization of carbon (catalytic graphitization), but some inhibiting the development of graphitic structure. The effect of foreign species on the structural change of carbon with temperature has been discussed and classified into three groups [328,329], which can be illustrated by the 002 diffraction profile, as shown in Fig. 2.131.

Some metals and metal compounds are known to give 002 profile after high-temperature treatment as Fig. 2.131a, which shows the formation of graphite separately from the matrix carbon, a sharp peak appearing at the position for graphite and the broad band being almost the same as that for the precursor without foreign material (the profile shown by a broken line). This means the partial acceleration of the graphitization of carbon matrix, which may be called **catalytic graphitization**. This effect is named as the ‘G effect’. With increasing content of foreign species, heat treatment temperature and residence time, the peak at graphite position grows, revealing the increase in the amount of graphite. Two mechanisms for the catalytic graphitization have been proposed, dissolution/precipitation of carbon atoms and formation/decomposition of metal carbide. On cooling process of molten metals which are almost saturated by dissolved carbon and are interstitial alloys containing a large amount of carbon, graphite can precipitate

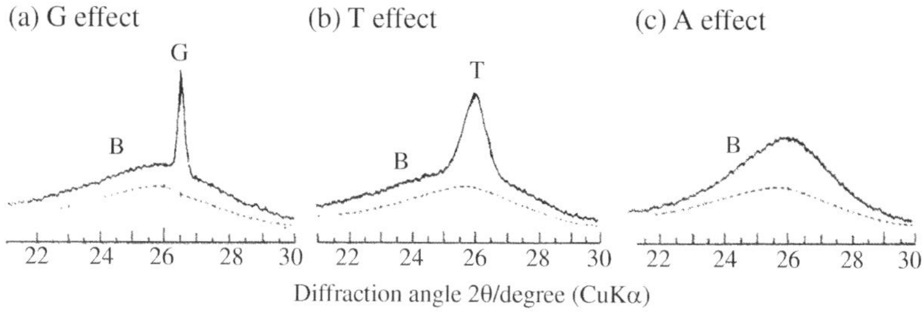


**FIGURE 2.131**

Effects of foreign species in carbon precursor after high temperature treatment observed on 002 diffraction profile. The broken line shows the profile obtained without any foreign species.

*(Courtesy of Prof. A. Oya of Gunma Univ.)*

because the solubility of carbon in metals decreases with decreasing temperature and free energy of graphite structure is lower than disordered carbon. Transition metals such as Fe and Ni are known to accelerate the graphitization through this mechanism [330,331]. In addition to the formation of graphite (G effect), the formation of the component, of which 002 diffraction line locates at around 26° in $2\theta$ (CuKα), as shown in Fig. 2.131b (the interlayer spacing $d_{002}$ of about 0.343 nm) has also been observed (T effect). In almost all cases showing heterogeneous graphitization, three components are formed, whether T component is detectable on X-ray diffraction profile or not [332]. When the starting carbons employed for catalytic graphitization already had well-developed texture, for example, a carbonized petroleum coke, the component B is not observed on X-ray diffraction profile and, as a consequence, only two components, T and G, are often observed.

The effect of these metallic catalysts depends also on the precursor carbons. The catalytic effects of a wide range of metals have been studied for two carbons prepared from 3,5-dimethyl-phenol-formaldehyde (3,5-DMPF) and phenol-formaldehyde (PF), the former has a graphitizing nature and the latter a non-graphitizing one [328,329]. On the former, the G effect is observed for most metals used. On the latter, the growth of T component is observed, giving a composite 002 profile, similar to Fig. 2.131b, and then the G component appears after the treatment at high temperatures. In these cases, the composite profile consisting of three component peaks, G, T and B peaks, is observed. In Fig. 2.132, the effects observed on various metals are summarized showing the results on two carbon precursors in the Periodic Table [329]. From the powder mixtures of carbon precursor such as poly(vinyl alcohol) with oxides of these metals, graphite is formed even at 900°C [333–335].

Synthesis of graphite single crystals has been performed using molten iron [336,337]. The formation of graphite through the formation of metal carbides and their decomposition is reported on Al, Be and Si [338,339]. Through the

| IIa | IIIb | IVb | Vb | VIb | VIIb | VIII | VIII | VIII | Ib | IIb | IIIa | IVa | Va |
|---|---|---|---|---|---|---|---|---|---|---|---|---|---|
| Be | | | | | | | | | | | B | C | N |
| | | | | | | | | | | | G<br>G | | |
| Mg | | | | | | | | | | | Al | Si | P |
| –<br>G | | | | | | | | | | | G<br>GT | –<br>G | |
| Ca | Sc | Ti | V | Cr | Mn | Fe | Co | Ni | Cu | Zn | Ga | Ge | As |
| G<br>G | | G<br>G, T | G<br>G, T | G<br>G, T | G<br>G, T | G<br>G, T | G<br>G, T | G<br>G, T | –<br>G | –<br>– | | –<br>G | |
| Sr | Y | Zr | Nb | Mo | Tc | Ru | Rh | Pd | Ag | Cd | In | Sn | Sb |
| | | | | G<br>G,T | | | | | | –<br>– | | –<br>– | –<br>– |
| Ba | La | Hf | Ta | W | Re | Os | Ir | Pt | Au | Hg | Tl | Pb | Bi |
| | | | | G<br>G,T | | | | | | | | –<br>– | –<br>– |

Upper position: Effect for 3,5-DMPF carbon with graphitizing nature
Lower position : Effect for PE carbon with non-graphitizing nature
G: Formation of graphitic component (G effect)
T : Formation of turbostratic component (T effect)

**FIGURE 2.132**

Catalytic effects of various metals for the carbons with graphitizing and non-graphitizing nature [328,329].

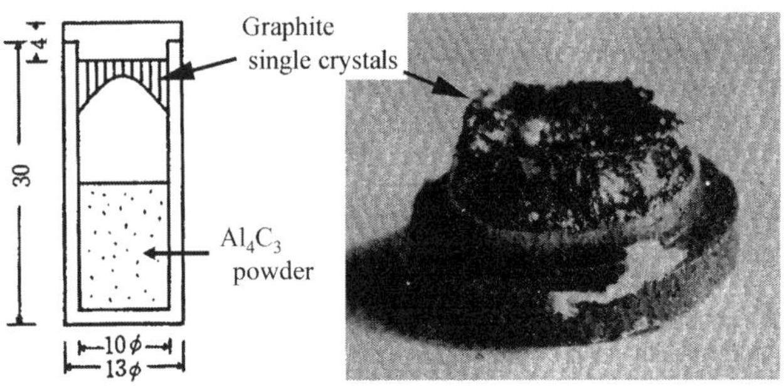


**FIGURE 2.133**

Formation of graphite single crystals through chemical transport from $Al_4C_3$ [340].

decomposition of aluminum carbide $Al_4C_3$ to $Al_2C_2$ vapor, carbon is transported to the graphite lib to deposit as small flakes of graphite single crystals, as shown in Fig. 2.133 [340].

When fine particles of metallic Ni were dispersed into carbon precursors, such as phenol resin, the resultant carbons were not able to be graphitized, with no evidence of graphitization, after the heat treatment at 300°C, the carbons being named TS carbon [341,342]. This TS carbon shows well-developed lattice fringes of 002 planes even though no graphitic stacking is formed, as shown in

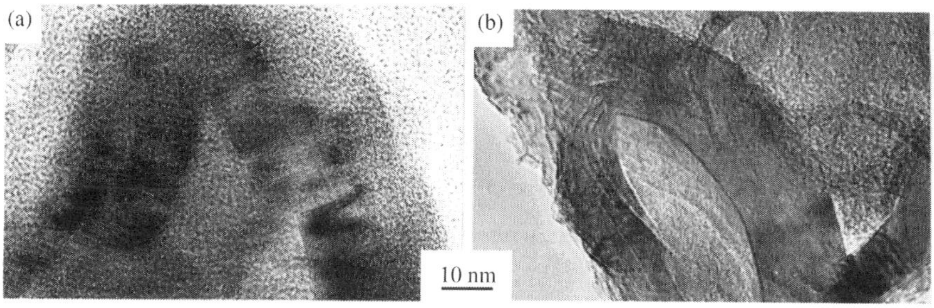


**FIGURE 2.134**

002 lattice fringe images of TS carbon [343].

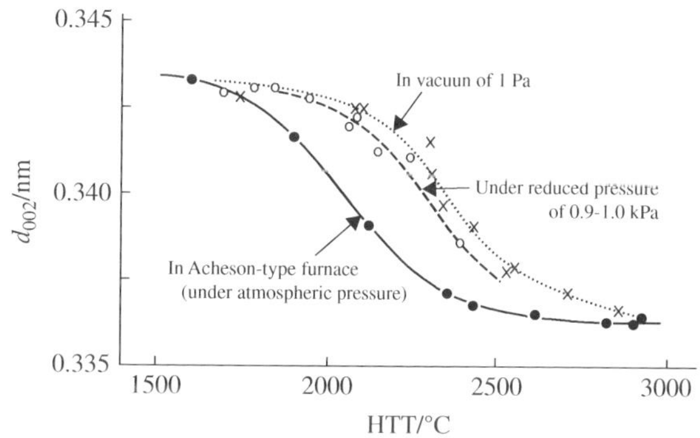


**FIGURE 2.135**

Acceleration effect of atmosphere on the graphitization of a coke [341].

Fig. 2.134 [343]. This carbon could not be graphitized even under 30 MPa pressure, though almost all carbons, even glass-like carbons, were graphitized under this pressure, as explained in Section 2.6.2. This effect of foreign atoms on carbon is represented by the X-ray diffraction profile of Fig. 2.131b ('T-effect').

Although the graphitization process with metallic catalysts proceeds heterogeneously, some catalysts accelerate the graphitization in the homogeneous process, as represented by the profile of 00*l* diffraction lines like Fig. 2.131c. Graphitization of a coke is accelerated under reduced pressure of atmosphere, as shown in Fig. 2.135 by plotting $d_{002}$ against heat treatment temperature (HTT) [273,344]. This result is explained by the preferential oxidation of C—C bonds bridging between two crystallites, which release the constraint preventing the coalescence of these crystallites. The apparent activation energy for this graphitization process is measured as 290—330 kJ/mole, though that in completely inert atmospheres is about 1130 kJ/mole.

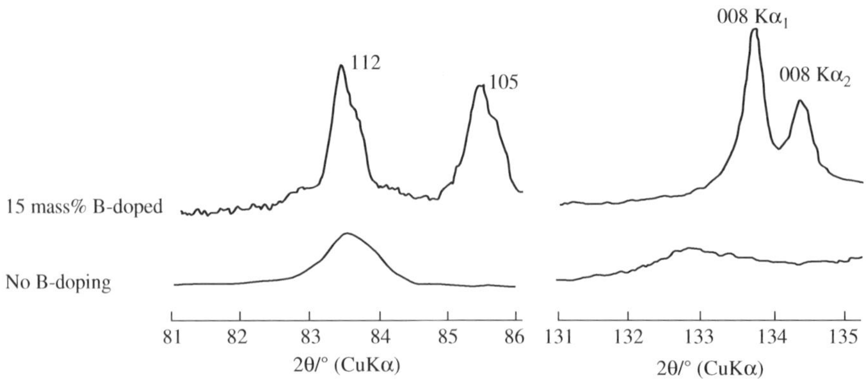


**FIGURE 2.136**

X-ray diffraction patterns of a coke heat-treated at 2800°C with or without $B_4C$ [344].

Boron is incorporated into hexagonal carbon layers by replacing carbon atoms [345]. Doping of B atoms into the hexagonal layers by replacing carbon atoms at high temperatures, the graphitization degree of matrix carbon is improved. In Fig. 2.136, X-ray diffraction profiles are compared on a coke heat-treated at 2800°C with and without boron carbide [346]. B-doped coke shows very distinct three-dimensional diffraction lines of 112 and even 105, and sharp 008 line with well-separated $K\alpha_1$ and $K\alpha_2$ peaks, though only broad lines are observed without B. In order to discuss catalytic graphitization of B, the measurement of $d_{002}$ is not enough because doping of B reduces $d_{002}$, and so the detection of three-dimensional diffraction lines, such as 112 and 105, is essential.

## 2.6.2 Stress graphitization

The graphitization process of carbon materials is greatly accelerated by heat treatment under pressures above 0.3 GPa [347–350]. This acceleration effect is expected because the graphitization reaction is associated with volume reduction and also because natural graphite with high crystallinity is supposed to be formed at a temperature of several hundred degrees Celsius and a pressure of several tenths GPa from geological investigations of surrounding minerals [351–353].

In Fig. 2.137, the changes in 004 diffraction profile with HTT are compared by using the powder of a typical graphitizing carbon, which is derived from poly (vinyl chloride) by the carbonization at 680°C, under atmospheric pressure and under the pressure of 0.5 GPa [354]. Under atmospheric pressure, the 004 profile shifts towards high-angle side, i.e. the decrease in $d_{002}$, and is sharpened by keeping its symmetrical profile with the increase in HTT (homogeneous graphitization), as shown in Fig. 2.137a. Under the pressure, however, the 004 profile heat-treated at around 1500°C appears to consist of two peaks at 54.5 and 53.5° in $2\theta$ (CuKα) (Fig. 2.137b). The structure component corresponding to the former

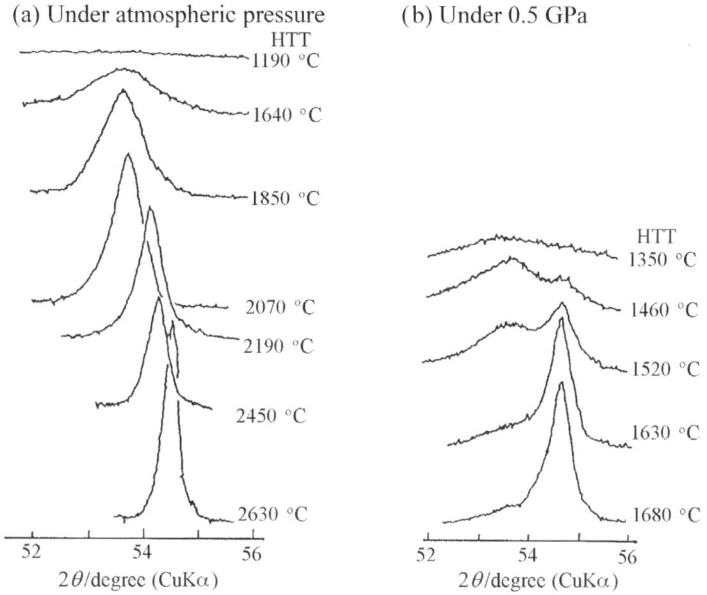


**FIGURE 2.137**

Changes of 004 diffraction profile of a coke with HTT under atmospheric pressure and 0.5 GPa [354].

peak is named graphitic component (G component) because it has a $d_{002}$ of 0.336 nm, almost the same value as graphite, but that to the latter peak has a large $d_{002}$ of about 0.343 nm, suggesting the turbostratic stacking, and is called turbostratic component (T component). Under transmission electron microscopy, two kinds of particles different in electron diffraction pattern are found, one having the diffraction spots with hexagonal symmetry corresponding to the G component, but the other having diffraction rings corresponding to the T component. With the increase in HTT under pressure, the peak for the G component grows with an expense of that for T component, suggesting that two different crystallites with graphitic and turbostratic structures coexist in the samples and the relative content of G component increases with increasing HTT and residence time under pressure. Therefore, graphitization under a pressure above 0.3 GPa proceeds through the heterogeneous process (heterogeneous graphitization).

When the starting carbon materials have random nanotexture, this heterogeneous graphitization under pressure occurs abruptly. In Fig. 2.138, change in 002 diffraction profile with HTT under 0.5 GPa is shown on a glass-like carbon [355]. Below 1450°C, only a broad peak at low-angle side corresponding to T component is observed. After 1500°C, however, the peak becomes sharp and located at a high-angle side, showing the change to G component.

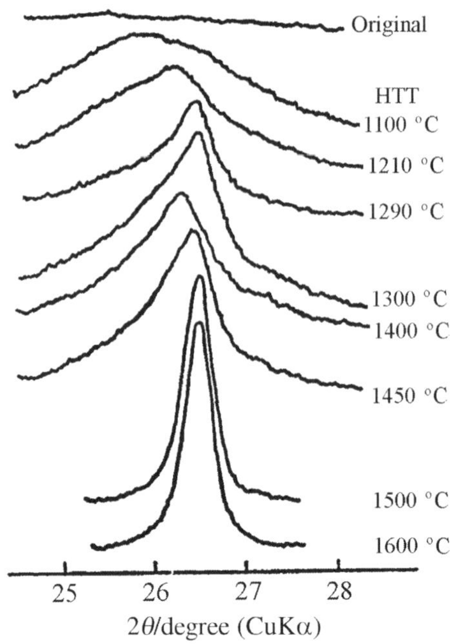


**FIGURE 2.138**

Change in 002 diffraction profile of glass-like carbon with HTT under 0.5 GPa [355].

By taking into consideration the fact that glass-like carbon used here is not graphitized even at a high temperature near the melting point (about 3400°C) [356], the result shown in Fig. 2.138 suggests that the characteristic nanotexture of glass-like carbon is destroyed during heat treatment under a high pressure and so it is graphitized, in other words, random nanotexture has to be destroyed in order to develop the graphite structure.

This heterogeneous process of graphitization is known to be due to the stress concentration at contact points between carbon particles. In order to release the concentrated stress, local orientation of hexagonal carbon layers is formed around the contact points, where the graphitic structure is developed above 1600°C. In Fig. 2.139, optical micrographs under polarized light are shown on the spheres of glass-like carbon heat-treated at different temperatures under 0.5 GPa [357]. Heat treatment at 1300°C under 0.5 GPa creates anisotropic areas at the boundary between adjacent spheres, which shows a small degree of alignment of these oriented areas along the compressing direction. At 1500°C, the anisotropic area appears to grow preferentially. Above 1600°C, the isotropic area completely disappears and the specimen is converted to the aggregate of small anisotropic areas.

Under the pressure of 0.5 GPa, graphitization of carbons, not only graphitizing but also non-graphitizing carbons, completes at about 1600°C, which is a markedly low temperature in comparison with the fact that graphitization starts

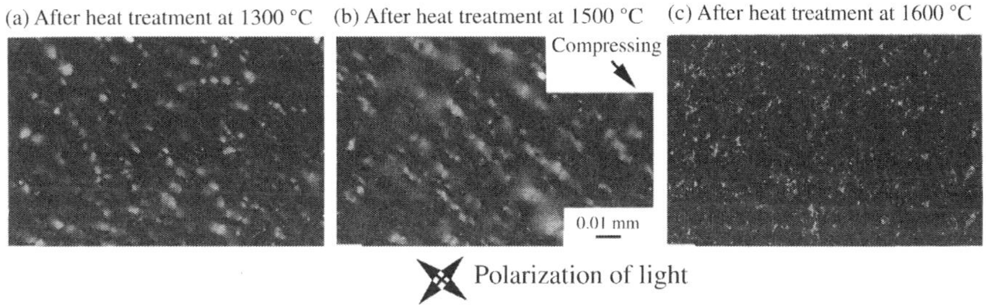


**FIGURE 2.139**

Polarized-light micrographs of carbon spheres heat-treated at different temperatures under 0.5 GPa [357].

around 2500°C in most graphitizing carbons and almost no graphitic structure is developed in non-graphitizing carbons even by heat treatment up to 3000°C under atmospheric pressure.

Graphitization can also be accelerated by stresses due to the shrinkage of organic precursors that are used to form the matrix in carbon/carbon composites [348]. This is possible if the shrinkage of the organic precursor is inhibited at the interface between the organic matrix and a carbon fiber [358–361]. For the precursors of the matrix, therefore, thermosetting resins, such as furfuryl alcohol condensates, are known to be effective. In the interface region, mechanical stress is produced and accumulated during the carbonization of matrix precursor, which is estimated to be a few tenths of a GPa. This causes the occurrence of graphitization, which appears to occur in the vicinity of stress accumulation [362,363]. In Fig. 2.140, a scanning electron microscope (SEM) image is shown at the boundary between matrix carbon and filler carbon fibers, which has been heat-treated up to 2800°C. The formation of oriented texture in the glass-like carbon matrix and the change to graphite structure are clearly observed.

This graphitization phenomenon has been extensively studied, scientifically and technologically, to develop the proper conditions for altering the thermal-mechanical properties of carbon/carbon composites because this type of carbon material has unique capabilities for use in advanced technology [364–372].

The temperature-pressure conditions for the formation of graphite crystals in nature were geologically estimated to be at a temperature of several hundreds of degrees Celsius and a pressure of several tenths in GPa [351–353]. In the laboratory, however, carbons are graphitized rapidly at temperatures above 1600°C and pressures above 0.3 GPa, as explained above. Therefore, the metamorphic rocks, in which the graphite crystals are embedded, may have an accelerating effect on the graphitization, in addition to the acceleration by pressure. Graphitization of carbons has experimentally been found to be accelerated in coexistence of different minerals under 0.3 GPa [348], where the sample carbon powders (coke prepared from poly(vinyl chloride) at 680°C) are sandwiched in between the tablets

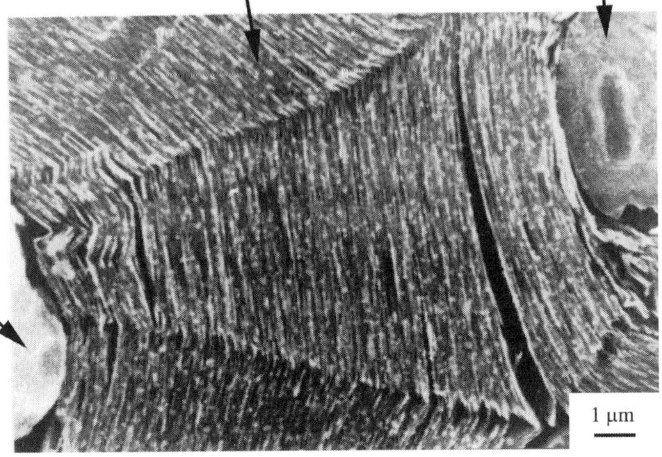


**FIGURE 2.140**

SEM image of the cross-section of carbon fiber filler and glass-like carbon matrix after the heat treatment at 2800°C.

*(Courtesy of Prof. E. Yasuda of Tokyo Inst. Tech.)*

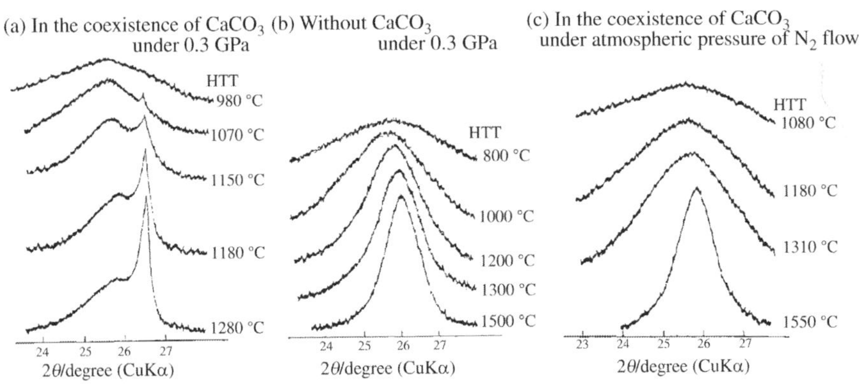


**FIGURE 2.141**

Changes in 002 diffraction profile of the coke with HTT [351].

of minerals. This graphitization process is the coupling of the stress graphitization with catalytic one. Different inorganic compounds have been used [373–378].

In Fig. 2.141, changes in 002 diffraction profile with HTT are compared with different conditions [373]. In the coexistence of $CaCO_3$ under 0.3 GPa, small peak appears at 26.6° in $2\theta$ (CuKα) at HTT of about 1000°C, which corresponds to the $d_{002}$ for graphitic structure, and grows with increasing HTT. Without

$CaCO_3$ under 0.3 GPa and also with $CaCO_3$ under atmospheric pressure, no formation of graphite structure is detected even at 1500°C. The same acceleration of graphitization is observed in the coexistence with natural limestone (main component of $CaCO_3$), in which good-quality graphite crystals are found in nature. Graphite formation at a temperature as low as 800°C has been found on heat treatment of carbon with $Ca(OH)_2$ [374].

Based on various experiments using various compounds, such as $CaCO_3$, CaO, $Ca(OH)_2$, $CaF_2$ $Al_2O_3$ $NaCO_3$, including natural minerals, the mechanism for accelerated graphitization has been discussed [378]. Two possible routes for graphitization of carbon are proposed, which can be expressed by the following chemical reactions using an oxide $M_xO_y$;

at low temperatures, $$M_xO_y + C(\text{turbostratic}) \rightarrow M_xO_{y-1} + CO \rightarrow M_xO_y + C(\text{graphite})$$

and

at high temperatures, $$mM_xO_y + (n + my)C(\text{turbostratic}) \rightarrow M_{mx}C_y + myCO \rightarrow mM_xO_y + (n + my)C(\text{graphite})$$

In the cases of calcium compounds and alumina, these two routes are possible at temperatures up to 1500°C under 0.3 GPa pressure. The reactivity of the calcium compounds is very important for these two reactions to proceed. The formation of reactive calcium oxide either by the decomposition of $CaCO_3$ and $Ca(OH)_2$ or by the reaction between the starting calcium fluoride and water vapor under high pressure and temperature is favorable for accelerating the graphitization [373,374,377]. Formation of calcium carbides is experimentally found at the interface above 1200°C, where the carbon specimen is contacted with a calcium oxide disk. When calcium carbonate is used, the decomposition product of carbide, i.e., $CO_2$, reacts also with the turbostratic carbon to form CO and then is disproportionated into $CO_2$ and solid carbon with graphitic structure, this reaction process being experimentally proved through the measurement of carbon isotope $C^{13}$ [378]. In the cases of silica and magnesia, only the first route seems to be possible under the present condition of heat treatment, because the formation of silicon carbide needs a temperature slightly higher than 1500°C.

The graphitization of anthracite occurs as low as 600°C in simple shear under pressures of 0.8–1.0 GPa [379]. There might also have been some accelerating graphitization effect of dunite, which is used to transfer the shear stress to anthracite, although the authors did not mention it.

These experimental results reveal that well-crystallized graphite can be formed in nature under mild conditions (below 1000°C and 0.3 GPa) with the catalytic action of some metal compounds, such as $CaCO_3$, during geological time period, which is an extremely long residence time in comparison with time applied in the laboratory. Graphitization under pressure described above gives certain information on the formation of well-crystalline to microcrystalline graphite in nature. Crystallinity of carbon in rocks is known to be a reliable indicator of metamorphic grade [380–382].

### 2.6.3 Graphitization of exfoliated carbon fibers

As discussed in Section 2.5.4, the development of graphitic structure in most carbon fibers is depressed, depending strongly on the nanotexture of carbon fibers [383,384]. In vapor-grown carbon fibers having annual ring texture in their cross-sections, graphitic structure develops markedly, but in isotropic-pitch-based and some PAN-based carbon fibers only a very low graphitization degree is observed even after heat treatment above 3000°C [279]. The behavior of mesophase-pitch-based carbon fibers is intermediate between these two, depending strongly on the nanotexture in their cross-sections. Dependence of graphitizability of carbon fibers on their structure and nanotexture has clearly been demonstrated through the measurements of magnetoresistance [279]. However, marked acceleration of graphitization has been found on exfoliated carbon fibers, even those prepared from PAN-based carbon fibers [385,386].

Exfoliated carbon fibers prepared from mesophase-pitch-based and PAN-based carbon fibers are shown in Fig. 2.142a and b [387]. On mesophase-pitch-based carbon fibers, single fiber was split into a number of thin filaments after exfoliation, as shown in Fig. 2.142a. In the case of PAN-based carbon fibers, however, small scale-like fragments are removed from the original fibers, and a number of cracks and fissures are formed preferentially along the original fiber axis, as shown in Fig. 2.142b. By high-temperature treatment, splitting into thin filaments for mesophase-pitch-based carbon fibers look to become clearer; just after exfoliation a memory of the original single fiber is recognized, but it seems to be less distinct after high-temperature treatment. However, no change in surface morphology in exfoliated PAN-based carbon fibers is observed after high-temperature treatment.

In Figs 2.143 and 2.144, 002 diffraction profiles and Raman spectra are compared for two kinds of carbon fibers heat-treated at 2800°C, before the exfoliation, just after exfoliation and after re-heat-treatment at high temperatures. Structural parameters, $d_{002}$, crystallite size along c-axis $Lc(002)$, relative

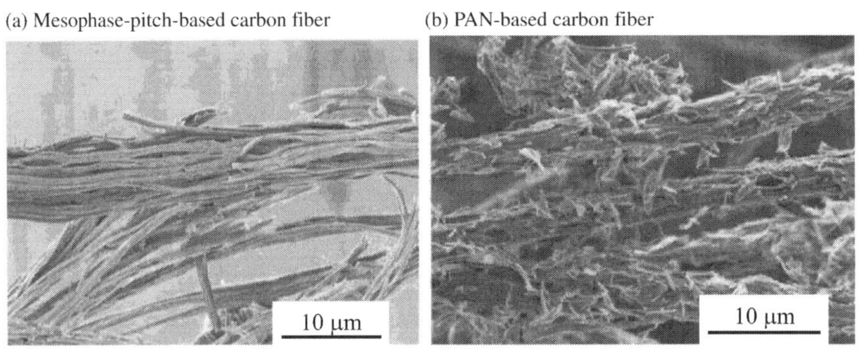


**FIGURE 2.142**

Exfoliated carbon fibers [387].

intensities of D band to G band, $I_D/I_G$, and of D′ band to G band, $I_{D'}/I_G$, in Raman spectra are listed in Table 2.10.

Pronounced acceleration of graphitization is observed, particularly on PAN-based carbon fibers. Pristine PAN-based carbon fibers have a low degree of graphitization, a high $d_{002}$ value of 0.343 nm and relatively small $Lc(002)$ value of 5.5 nm even though they have been heat-treated to 2800°C in advance.

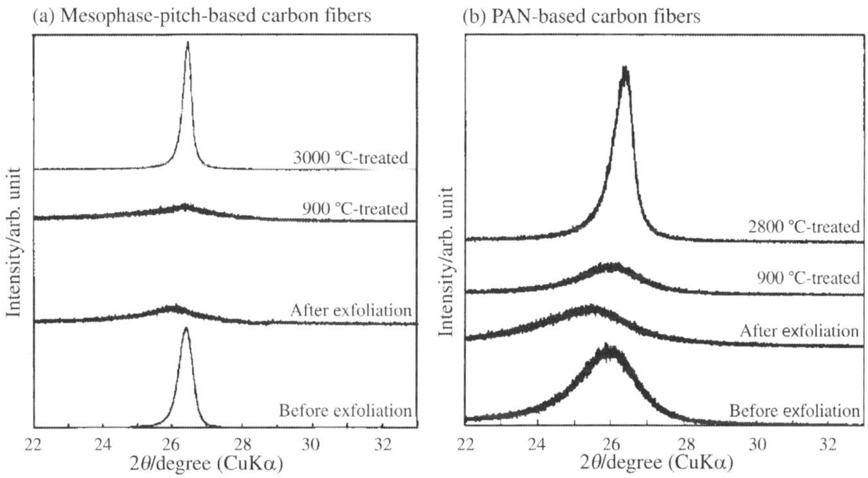


**FIGURE 2.143**

Changes in 002 diffraction profile with exfoliation and following high temperature treatment [386].

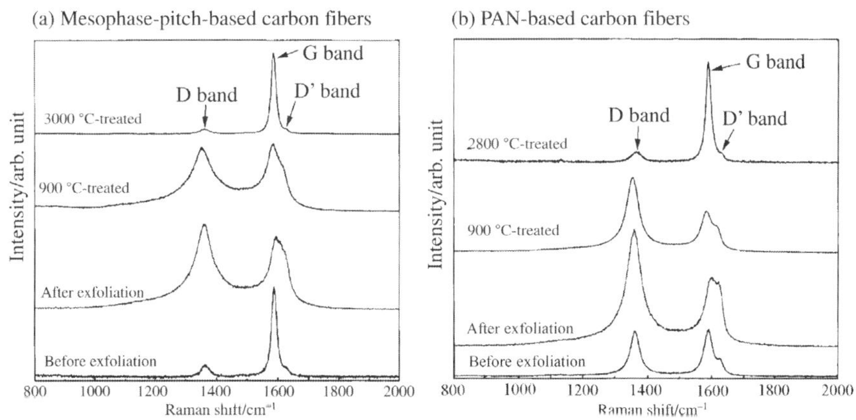


**FIGURE 2.144**

Changes in Raman spectrum with exfoliation and following high temperature treatment [386].

**Table 2.10** Structural Parameters Determined by X-ray Diffraction and Raman Spectroscopy [386]

| Carbon Fibers | Structure Parameter | Before Exfoliation | After Exfoliation | Re-heat-Treated at 900°C | Re-heat-Treated at 2800°C |
|---|---|---|---|---|---|
| Mesophase-pitch-based (2800°C-treated) | $d_{002}$ (nm) | 0.338 | 0.341 | 0.338 | 0.338 |
| | $Lc(002)$ (nm) | 28 | 9 | 14 | 31 |
| | $I_D/I_G$ | 0.3 | 1.2 | 0.9 | 0.05 |
| | $I_{D'}/I_G$ | 0.03 | 0.3 | 0.2 | 0.02 |
| PAN-based (2800°C-treated) | $d_{002}$ (nm) | 0.343 | 0.343 | 0.343 | 0.337 |
| | $Lc(002)$ (nm) | 6 | 5 | 6 | 16 |
| | $I_D/I_G$ | 1.0 | 1.7 | 1.9 | 0.09 |
| | $I_{D'}/I_G$ | 0.2 | 0.4 | 0.3 | 0.08 |

Pronounced decrease in $d_{002}$ to 0.337 nm and growth of $Lc(002)$ to 15.5 nm are observed after re-heat-treatment at 2800°C after exfoliation. Raman spectrum change leads to the same conclusion. Pristine PAN-based carbon fibers are not graphitized, a strong D band with almost the same intensity as G-band and a clear D′ band. Exfoliation introduced a large amount of structural defects, as being revealed by growth and broadening of D band. Re-heat-treatment at 900°C cannot anneal these structural defects. Re-heat-treatment at 2800°C, however, results in marked decreases in intensities of D and D′ bands, and in strong and sharp G band, which reveals a high degree of graphitization, much higher than pristine fibers.

These results reveal that the exfoliation through decomposition of intercalation compounds of carbon fibers releases the constraint, which is needed to keep fibrous morphology, associated with the introduction of various structural defects. The release of this constraint by exfoliation and following annealing by the re-heat-treatment at a high temperature such as 2800°C causes the acceleration of graphitization, in other words, improvement of structural perfection, of carbon fibers. These exfoliated carbon fibers give exceptionally high capacitance in electric double-layer capacitor with $H_2SO_4$ electrolyte [388–390] and also marked improvement in mechanical properties of the composites with organic polymers [391,392].

## 2.7 Pore development in carbon materials

### 2.7.1 Pores in carbon materials

All carbon materials, except highly oriented graphite, contain pores, because they are polycrystalline products of thermal decomposition of organic precursors, such

**Table 2.11** Classification of Pores in Solid Materials

| | |
|---|---|
| ***1) Based on the origin*** | |
| Intra-particle pores | Intrinsic intra-particle pores |
| | Extrinsic intra-particle pores |
| Inter-particle pores | Rigid inter-particle pores |
| | Flexible inter-particle pores |
| ***2) Based on the size*** | |
| Micropores <2 nm | Ultra-micropores <0.7 nm |
| | Super-micropores 0.7–2 nm |
| Mesopores 2-50 nm | |
| Macropores >50 nm | |
| ***3) Based on the state*** | |
| Open pores | |
| Closed pores (Latent pores) | |

as various resins and pitches. During their pyrolysis and carbonization, large amounts of decomposition gases are formed in a wide range of temperature. Since the gas evolution behavior from organic precursors is strongly dependent on both the precursor used and the heating condition, such as heating rate, pressure, etc., as pointed out in Section 2.3, carbon materials contain a large amount of pores consisting of a wide range of sizes and shapes. The pores in solids are classified as shown in Table 2.11.

Pores can be classified into two, intraparticle and interparticle pores, on the basis of their origin. The intraparticle pores are further classified into two, intrinsic and extrinsic. The former owes its origin to the crystal structure, of which a typical example is the pores in zeolite crystals. The graphite gallery between neighboring hexagonal carbon layers can be a slit-shaped pore with width of 0.3354 nm, which can accept various atoms, ions and even molecules, that is, intercalation. Therefore, graphite gallery is an intrinsic intraparticle pore. Into the gallery which is widened by the presence of some intercalates, such as Cs ions, hydrogen and some hydrocarbon molecules can be intercalated to form ternary intercalation compounds, these intercalated galleries being extrinsic intraparticle pores. In most activated carbons, a large amount of pores in various sizes of nanometer scale, micropores and mesopores, are formed because of random orientation of crystallites, which are rigid interparticle pores. In exfoliated graphite, which consists of worm-like particles, large pores in and among worm-like particles are formed, which are flexible interparticle and intraparticle pores, respectively, because they can be easily deformed by compression and by sorption of heavy oil.

The classification of pores based on pores sizes was proposed by IUPAC (International Union for Pure and Applied Chemistry). As illustrated in the first line in Fig. 2.145, pores are usually classified into three: macropores (>50 nm),

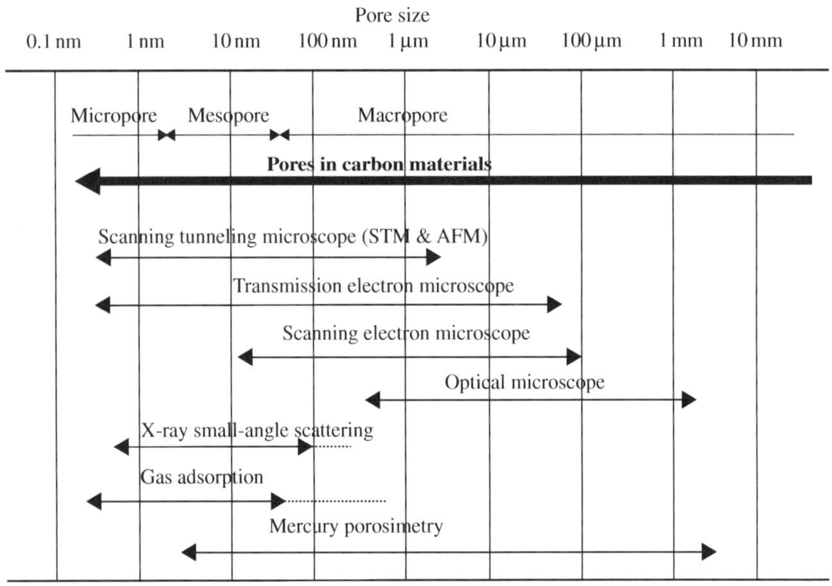


**FIGURE 2.145**

Pores in carbon materials and their identification techniques.

mesopores (2–50 nm) and micropores (<2 nm) [393]. Micropores are further divided into super-micropores with a size of 0.7–2 nm and ultra-micropores of less than 0.7 nm. The terms 'nanopores' and 'nanoporous carbons' are often used to show the presence of micropores and mesopores in the carbon.

Pores can also be classified on the basis of their state, either open or closed. In order to identify the pores by gas adsorption, which has frequently been used for activated carbons, they must be exposed to the adsorbate gas. If some pores are too small to accept gas molecules, however, they cannot be recognized as pores by the gas molecules, in other words, these pores are closed pores for the gas used. These pores are called latent pores including closed pores. Closed pores are not necessarily small. When carbon foam was prepared by the impregnation of polyimide into a poly(urethane) foam and the following carbonization, large macropores of few millimeter sizes were formed in the center of a block of foam, which gave an advantage that the foam could float on water [394].

Pores in carbon materials have been identified by different techniques depending mostly on their size. The techniques for the identification of pores are summarized in Fig. 2.145. Pores with nanometer sizes, i.e., micropores and mesopores, are identified by the analyses of gas adsorption isotherms, mostly of nitrogen gas at 77 K. The fundamental theories, instruments, measurement practices, analysis procedures and many results obtained so far by gas adsorption have been reviewed in different publications [395,396]. X-ray small-angle scattering has an

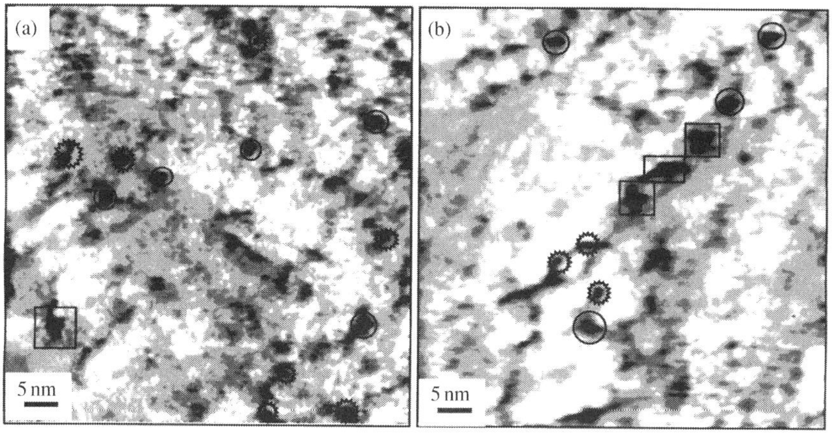


**FIGURE 2.146**

STM images of glass-like carbon oxidized in air at 400°C [403]. Square: mesopores, circle: super-micropores and star: ultra-micropores.

advantage to identify the latent pores, including closed pores, though gas adsorption can detect only open pores, which can accept gas molecules. For macropores, mercury porosimetry has been frequently applied [397]. Identification of intrinsic pores, interlayer space between hexagonal carbon layers in the case of carbon materials, is carried out by X-ray diffraction (XRD).

Direct observation of pores, extrinsic pores, on the surface of carbon materials has been reported by using microscopy techniques coupled with an image processing technique, scanning tunneling microscopy/atomic force microscopy (STM/AFM) [397,398] and transmission electron microscopy (TEM) [399] for micropores and mesopores, and scanning electron microscopy (SEM) and optical microscopy for macropores [400–402].

## 2.7.2 Identification of pores

### a. Scanning tunneling microscopy

In Fig. 2.146, scanning tunneling microscope (STM) images of the surface of a glass-like carbon oxidized in air are shown, where mesopores, super-micropores and ultra-micropores are indicated [403]. It has to be pointed out here that these STM observations give only information on the entrance of pores, but it is possible to show not only pores of a very small size but also the shape and fractal dimension along the pore surfaces. In order to get reliable information on pores, pore entrances, a large number of observations and statistical treatments are necessary.

An STM image on the surface of carbon spheres, which are prepared from phenol resin with heat treatment at 1000°C, is shown in Fig. 2.147a, and an

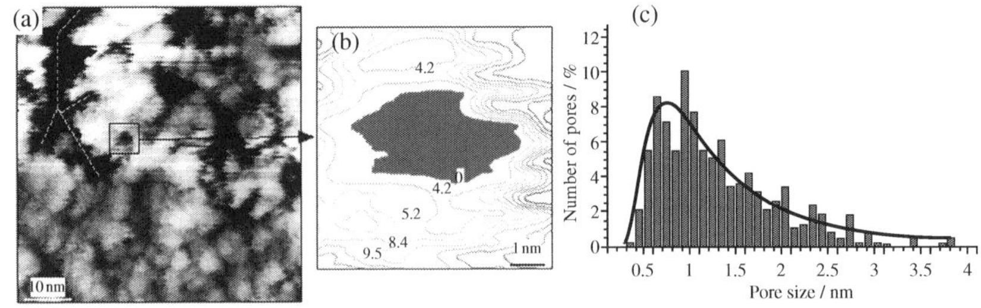


FIGURE 2.147

STM image of the surface of carbon sphere (a), contour map for a pore (b) and pore size distribution (c) [404].

**Table 2.12** Comparison of STM Image Analysis with Gas Adsorption [404]

| Sample | Pores with the size of 0.5–1.5 nm Analyzed by STM | | Gas Adsorption Analysis | |
|---|---|---|---|---|
| | Number in 1 μm$^2$ | Ratio to B | $S_{BET}$ (m$^2$/g) | Ratio to B |
| A | 75 | 0.04 | 2 | 0.06 |
| B | 1788 | 1.0 | 307 | 1.0 |
| C | 4078 | 2.3 | 993 | 3.2 |
| D | 6322 | 3.5 | 1193 | 3.9 |

example of a contour map around a pore is also shown in Fig. 2.147b [403,404]. For quantitative assessment of these pores, the application of the following two criteria has been proposed on measured contour map around each pore: (1) pore wall must be steep; its slope must be more than 1, and (2) the pore must be so deep that the tip of the microscope cannot reach the bottom. From observations on a large number of pores, a pore size distribution is determined, as shown in Fig. 2.147c. In Table 2.12, the results obtained from STM image analysis are compared with those from gas adsorption. The number of micropores with the size of 0.5 to 1.5 nm, per 1 μm$^2$ area of the surface measured on four different samples corresponds to the BET surface area ($S_{BET}$) determined by $N_2$ gas adsorption at 77 K. The results of STM characterization show quite good correspondence with those of gas adsorption analysis.

In the beginning of the oxidation reaction, atomic size pores are formed on the surface of carbon materials. In Fig. 2.148, the change in STM image of basal plane of natural graphite flake with increasing power of oxygen plasma is shown [405,406]. By a slight irradiation of oxygen plasma, pores are formed, which are clusters of atomic vacancies, as shown in the inserted high-magnification image of Fig. 2.148a. With increasing power of the plasma, both the number and size of

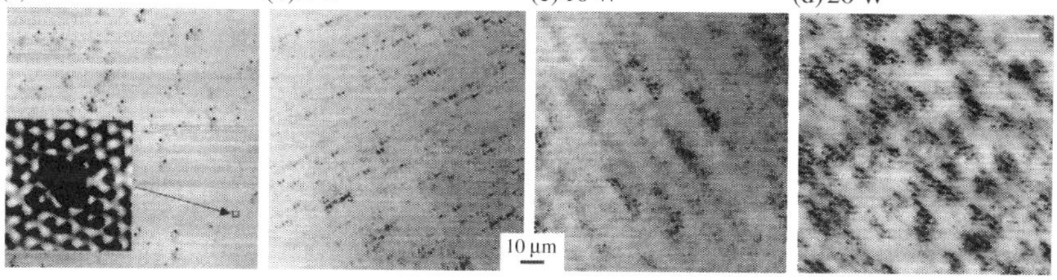


**FIGURE 2.148**

Change in STM image of basal plane of natural graphite flake with increasing power of oxygen plasma irradiated for 5 s at room temperature [405].

**Table 2.13** Pores Formed on the Basal Plane of Natural Graphite Flakes by Oxygen Plasma [405]

| | | | | |
|---|---|---|---|---|
| ***a) Effect of plasma power (irradiation time of 10 s, at room temperature)*** | | | | |
| Plasma power (W) | 1 | 5 | 10 | 20 |
| Concentration of pores ($\mu m^{-2}$) | 12 500 | 19 000 | 22 500 | 29 500 |
| Averaged area of pores ($nm^2$) | 0.5 | 0.6 | 0.9 | 1.4 |
| Fraction of shallow pores (%) | 99 | 99 | 97 | 94 |
| ***b) Effect of irradiation time (plasma power of 10 W, at room temperature)*** | | | | |
| Irradiation time (s) | 2 | 5 | 10 | |
| Concentration of pores ($\mu m^{-2}$) | 9000 | 12 500 | 22 000 | |
| Averaged area of pores ($nm^2$) | 0.4 | 0.5 | 0.7 | |
| Fraction of shallow pores (%) | 99 | 99 | 94 | |
| ***c) Effect of irradiation temperature (plasma power of 10 W, irradiation time of 10 s)*** | | | | |
| Irradiation temperature (°C) | 25 | 400 | | |
| Concentration of pores ($\mu m^{-2}$) | 22 500 | 12 500 | | |
| Averaged area of pores ($nm^2$) | 0.9 | 2.3 | | |
| Fraction of shallow pores (%) | 97 | 96 | | |
| ***d) Effect of annealing of irradiated samples*** | | | | |
| Annealed at 400°C for 1 h | Before annealing | After annealing | | |
| Concentration of pores ($\mu m^{-2}$) | 12 500 | 12 500 | | |

pores increase. The concentration of pores (number of pores per 1 $\mu m^2$ of the surface), their averaged size and the fraction of shallow pores with the depth of two atomic layers to total number of pores formed are calculated on the basis of the determinations of more than 6000 pores. The results are summarized in Table 2.13.

Pores formed by the slight irradiation of oxygen plasma are mostly shallow, a depth of less than two atomic layers. The increases in plasma power and irradiation time give the increases in both concentration and size of pores (Table 2.13a and b, respectively). Irradiation at 400°C results in the decrease in pore concentration but the increase in pore size, though the total area of pores formed is almost the same in comparison with the results at room temperature, suggesting that the pores formed are coagulated to form large pores (Table 2.13c). However, pores once formed are pretty stable for annealing at 400°C, with no changes in concentration of pores or in their sizes being observed (Table 2.13d).

### b. Transmission electron microscopy

In transmission electron micrographs taken on thin section of the carbon materials with sufficiently high magnification, pores look white because an electron beam passes through and pore walls look black because of scattering of the electron beam. The quantitative analysis of these micrographs with the aid of image processing gives information of pore size distribution and also smoothness of pore walls (fractal dimension). Detailed studies have been carried out mainly on activated carbon fibers [395,407,408].

In Fig. 2.149a, an electron micrograph of activated carbon fiber is shown. No characteristic distribution in orientation and size of pores is recognized. This bright-field image is transferred using two-dimensional Fourier function to the power spectrum (Fig. 2.149b), which shows the distribution of brightness. From this power spectrum, the curve showing the change in brightness with distance is obtained, which may be considered to reveal pore size distribution, as shown in Fig. 2.149c. The area under the power spectrum curve corresponds to the relative pore volume, as shown in the figure. On this power spectrum, low-frequency component, corresponding to the distance more than 15 nm, is erased to increase the sharpness in image, and then inversed Fourier transformation is applied to get the bimodal image (Fig. 2.149d). Usually good correspondence between original TEM image and bimodal image obtained by image processing is obtained on activated carbon fiber samples.

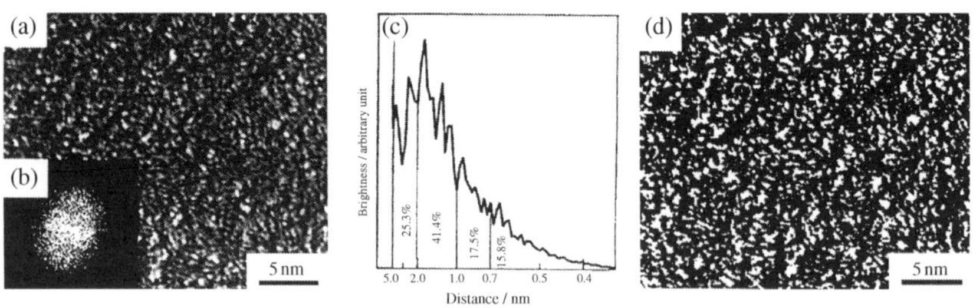


**FIGURE 2.149**

TEM image of activated carbon fiber (a), its power spectrum (b), its brightness *vs.* distance curve (c) and its bimodal image (d).

*(Courtesy of Prof. K. Oshida of Nagano Nat. Coll. Tech.)*

In Fig. 2.150, the power spectrum obtained from TEM image is shown with pore size distribution determined by gas adsorption for three activated carbon fibers with different $S_{BET}$. Since the power spectrum is expressed in reciprocal space, the distance in real space indicated on the abscissa increases to the left-hand side. Therefore, pore size distributions are plotted in the same manner. By taking into account that the magnification of TEM observation for this analysis does not give the information on the distance more than 5 nm, relatively good correspondence was obtained between the power spectrum from TEM observation and pore size distribution from gas adsorption. The distribution estimated from the TEM image is a little broader than that from gas adsorption, which is supposed to be due to the fact that three-dimensional averaging is performed in the former, but the minimum value of pore parameters is detected in the latter.

The fractal dimension of the pore wall can be calculated from the bimodal image. The values obtained for various activated carbon fibers are in the range of 1.66 to 1.75, which may correspond to three-dimensional fractal dimension of 2.65 to 2.75 [399]. Almost the same fractal dimensions are determined by the same procedure on viscose rayon-based activated carbon fibers [409].

## c. Gas adsorption

The pore structure of carbon materials, particularly of activated carbons, has usually been measured by using physical adsorption of various gases. From the convenience for the measurements, nitrogen adsorption has often been used at liquid nitrogen temperature 77 K. Many reviews and books have been published on pore structure determination by nitrogen gas adsorption focusing on carbon materials [395,396,410–412].

The isotherms of adsorption and desorption were classified from their shapes into 6 types by IUPAC, as shown in Fig. 2.151 [410]. Type I isotherm is typical

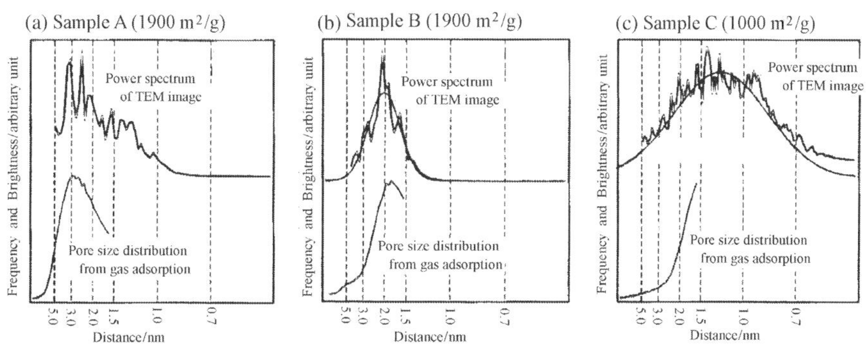


**FIGURE 2.150**

Power spectrum determined by TEM analysis and pore size distribution determined by gas adsorption for three activated carbon fibers.

*(Courtesy of Prof. K. Oshida of Nagano Nat. Coll. Tech.)*

for microporous solids, most activated carbons, where micropore filling occurs at relatively low partial gas pressure $P/P_0$. In most cases adsorption completes at $P/P_0$ below 0.5.

For the Type I isotherms measured by nitrogen at 77 K, the BET method has commonly been applied to have an information of surface area of the solid samples ($S_{BET}$). The adsorption data up to $P/P_0$ less than 0.3 are used for the analysis. However, it has been pointed out that the measured surface area might be 40% larger than the true area if the sample contains relatively large micropores, and might be too small if there are many ultramicropores. In order to evaluate the surface area and pore size distribution, various methods have been proposed, DR plot, BJH method, DH method, t plot, $\alpha_s$ plot, and DFT method [410], together with theoretical calculation based mainly on Grand Canonical Monte-Carlo method.

In Fig. 2.152a, isotherms for nitrogen gas adsorption at 77 K are compared for three activated carbon fibers [412,413]. For activated carbon fiber A5, the

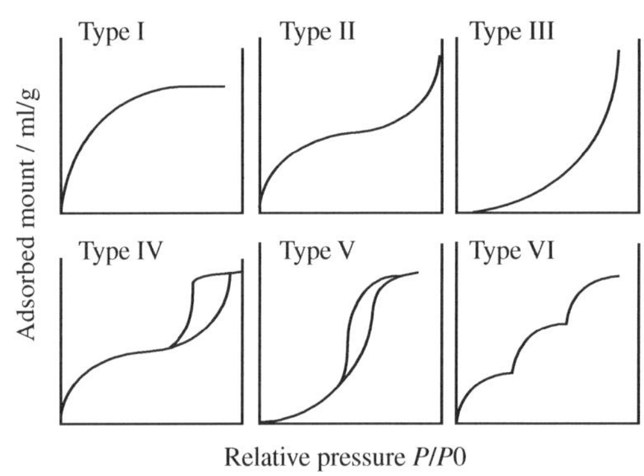


**FIGURE 2.151**

Classification of gas adsorption isotherms by IUPAC [410].

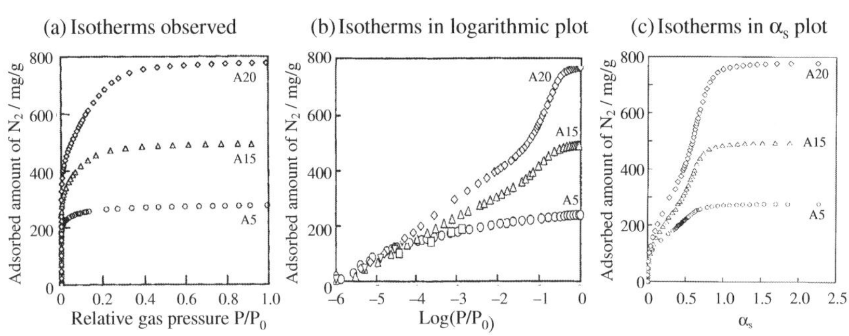


**FIGURE 2.152**

Isotherms of nitrogen adsorption at 77 K in different plots for three activated carbon fibers.

*(Courtesy of Prof. K. Kaneko of Chiba Univ.)*

adsorption of nitrogen gas reaches almost saturation at $P/P_0$ less than 0.05. For A20, however, a gradual increase in adsorption continues up to $P/P_0$ of 0.3, after an abrupt initial increase. This abrupt initial adsorption is due to filling of micropores with the size less than 1 nm. In order to know the information on micropores less than 1 nm, the detailed measurements of the amount of adsorption under the relative pressure less than $10^{-3}$ are required, in other words, the measurements in very low relative pressure range are necessary. In Fig. 2.152b, the adsorbed amount of nitrogen is plotted against the logarithm of $P/P_0$ for the same three activated carbon fibers. Adsorption for A5 at $P/P_0$ of $10^{-5}$ is slightly larger than for A20, suggesting that micropores smaller than 1 nm are formed in A5. To get quantitative evaluation of micropores, so-called $\alpha_s$ plot is often carried out, as shown in Fig. 2.152c.

The information, which can be obtained from $\alpha_s$ plot of isotherm, is summarized in Fig. 2.153. Adsorbed amount of nitrogen in the unit of mg/g is plotted against $\alpha_s$, which is nitrogen adsorption of the reference materials containing no micropores (usually that of a carbon black) at each relative pressure $P/P_0$. Often the $\alpha_s$ plot shows the deviations upward from the line passing through the origin, f-swing below $\alpha_s$ of 0.5 and c-swing over $\alpha_s$ of 0.5 to 1.0, as shown schematically in Fig. 2.153. The f-swing is mainly due to the strong adsorption into micropores and c-swing to quasi-micropore condensation.

External surface area $S_{ext}$ is calculated from the slope 1 in $\alpha_s$ plot and total surface area $S_{total}$ from the slope 2 by using the respective equations in Fig. 2.153, and microporous surface area $S_{micro}$ is obtained as a balance. The

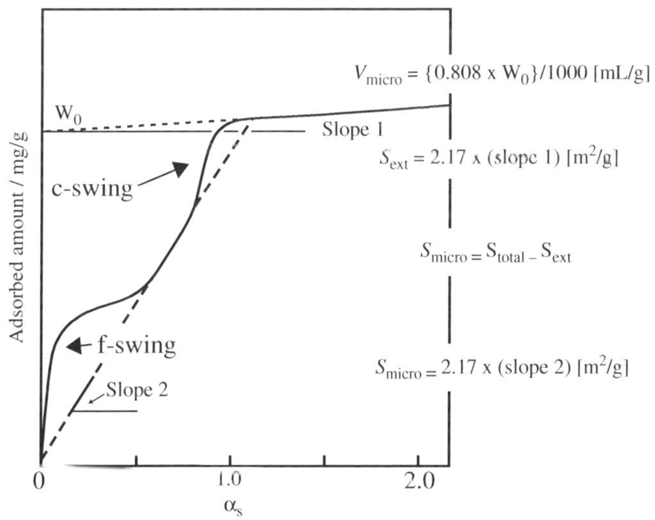


**FIGURE 2.153**

$\alpha_s$ plot of isotherm of nitrogen adsorption at 77 K.

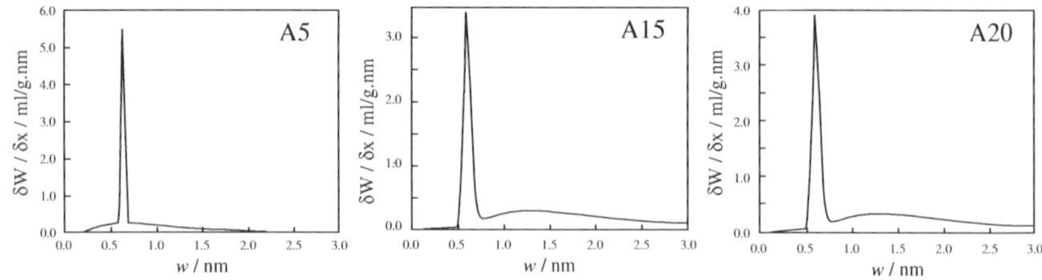


**FIGURE 2.154**

Pore size distribution determined by DFT method for three activated carbon fibers.

*(Courtesy of Prof. K. Kaneko of Chiba Univ.)*

values of surface area thus calculated are those subtracted from the effect of strong potential field in micropores and so-called SPE surface areas (SPE: subtracting pore effect). From the intercept $W_0$, micropore volume $V_{micro}$ can be obtained.

In order to determine the pore size distribution in micropore range, the density functional theory (DFT) has frequently been used. In Fig. 2.154, pore size distributions determined by DFT method are shown for three activated carbon fibers [414], of which adsorption isotherms are shown in Fig. 2.152. For A5, a very high concentration of micropores with a sharp distribution in the size at 0.6 nm is observed. In A20, size distribution of pores becomes a little broader. However, some problems for DFT method are still pointed out [414,415].

DFT analysis of the adsorption isotherm of $CO_2$ at 273 K is recommended for the determination of pore size distribution in a micropore range, and the adsorption of hydrogen at 77 K has been proposed to measure micropores less than 0.4 nm [415]. Pore size distribution curves measured by three different gases, $N_2$ and $H_2$ at 77 K and $CO_2$ at 273 K are compared on an activated carbon fiber with $S_{BET}$ of about 1000 $m^2/g$ in Fig. 2.155.

The fractal dimension of micropores has been determined from the adsorption isotherm measurements using adsorbate molecules with different sizes [416,417].

## d. Scanning electron microscopy

Image processing technique has successfully been applied to the micrographs of scanning electron microscope (SEM) for quantitative characterization of pore structure in exfoliated graphite [400,418]. Exfoliated graphite, which has low bulk density and is very fragile, has been characterized mostly by either its bulk density or exfoliation volume. By introducing techniques to prepare the fractured surface and image processing, quantitative characterization of pore structure, which consists of at least three kinds of pores, i.e., pores inside the work-like particles, crevice-like pores on the worm-like particles and pores formed by complicated entanglement of these fragile worm-like particles, became possible to obtain and a series of works has been performed [401,418–423]. In order to have

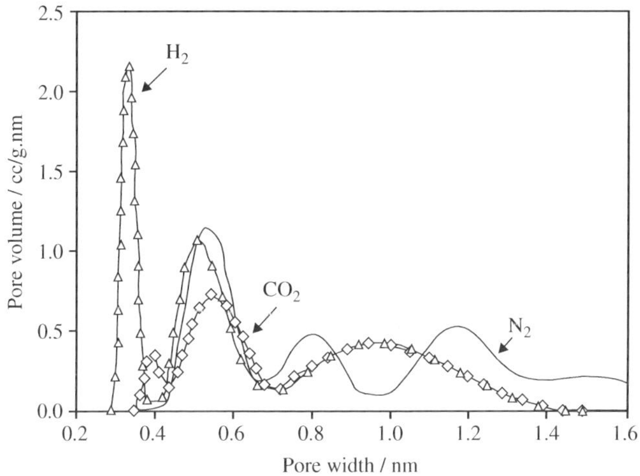


**FIGURE 2.155**

Pore size distributions of an activated carbon fiber measured by using three different gases [415].

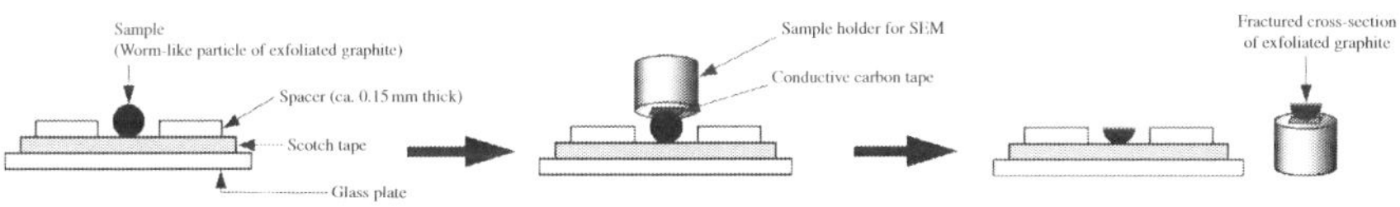


**FIGURE 2.156**

Procedure to prepare the fractured surface of a worm-like particle of exfoliated graphite [400].

the information on the pore structure inside the worm-like particles of exfoliated graphite, the technique to get the fractured surface of fragile worm-like particles had to be developed. In Fig. 2.156, the procedure to prepare the fractured surface of worm-like particle is shown [400] and a typical SEM image of the fractured surface in Fig. 2.157a. In Fig. 2.157b, one distribution of cross-sectional area of pores inside the particles determined from more than 7000 pores is shown.

One of the advantages of this technique is to be able to observe large-size pores. In Fig. 2.158, histograms showing distribution of cross-sectional area of pores inside the worm-like particles are compared on the exfoliated graphite prepared by microwave irradiation for 20 and 60 s at room temperature [423]. From the comparisons in histogram and accumulated frequency, and also averaged values, the differentiation between these two exfoliated graphites is difficult, as shown in Fig. 2.158. Comparing the histograms in large size range, however, the difference is clearly seen, as shown by the figures in Fig. 2.158.

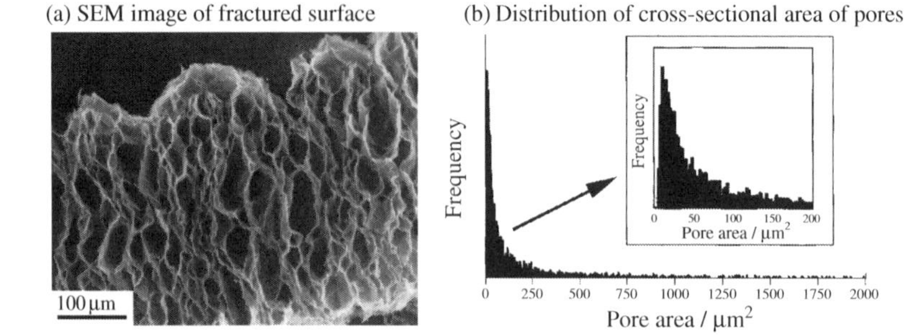


**FIGURE 2.157**

Appearance of pores inside the worm-like particle (a) and the distribution of cross-sectional area of the pores (b) [400].

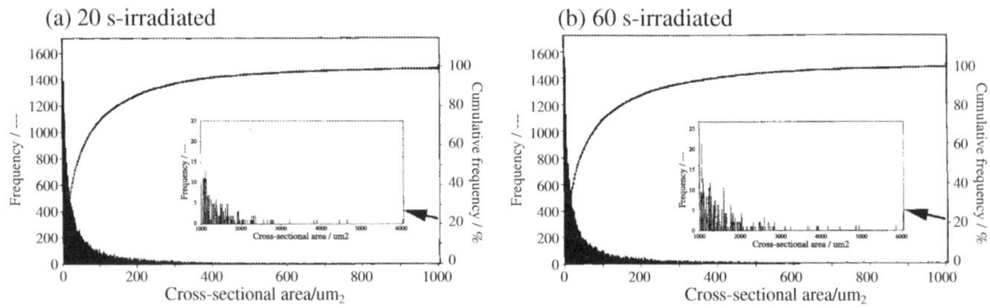


**FIGURE 2.158**

Distribution histograms and accumulated frequency curves of exfoliated graphite prepared by microwave irradiation at room temperature [423].

## e. Optical microscopy

For isotropic high-density graphite blocks, the application of image processing on optical micrographs of their cross-sections gives quantitative information on their macropores and makes the discussion on the dependences of various properties on pore structure possible [402].

In Fig. 2.159, optical micrographs are compared for six grades of commercially available isotropic high-density graphite blocks [402]. The size and number of macropores seen in these micrographs are reducing gradually from A to F. Applying an image processing technique to these micrographs, number of pores in a unit area (pore density), averaged cross-sectional area of pores, roundness of pore cross-section and fractal dimension along pore wall are able to be determined. The results are tabulated in Table 2.14, together with bulk density and porosity, which have been used as characterization parameters by measuring conventional techniques.

With increasing bulk density, the pore density increases drastically from D to F, but at the same time the cross-sectional area of the pore (i.e., pore size)

**FIGURE 2.159**

Optical micrographs of high-density isotropic graphite blocks listed in Table 2.14 [402].

**Table 2.14** Pore Parameters for High-Density Isotropic Graphite Blocks [402]

| Sample | Conventional Parameters | | Parameter Determined by Image Analysis | | | |
|---|---|---|---|---|---|---|
| | Bulk Density ($g/cm^3$) | Porosity (%) | Pore Density ($mm^{-2}$) | Average Cross-Sectional Area ($\mu m^2$) | Roundness of Pores | Fractal Dimension |
| A | 1.735 | 21.9 | 414 | 545 | 0.65 | 1.40 |
| B | 1.788 | 25.3 | 446 | 506 | 0.69 | 1.40 |
| C | 1.842 | 21.1 | 480 | 395 | 0.66 | 1.33 |
| D | 1.842 | 21.7 | 822 | 244 | 0.68 | 1.46 |
| E | 1.848 | 20.9 | 1275 | 155 | 0.71 | 1.47 |
| F | 1.802 | 12.0 | 2731 | 31 | 0.79 | 1.56 |

decreases drastically. However, the roundness of pores increases, in other words, the cross-section of pores becomes closer to a circle, but the fractal dimension increases, in other words, the walls of pores become more complicated. The correspondences of these pore parameters to mechanical properties, such as breaking strength, Young's modulus, etc. are discussed later.

## f. Liquid impregnation

For the characterization of macropores in solid materials, impregnation of mercury has often been applied (mercury porosimetry). However, it can apply only for rigid materials, like high-density graphite, but not for fragile materials, like exfoliated graphite. It has been proposed to use a new dilatometer for mercury porosimetry to apply it to exfoliated graphite [425], but there is still a limit on the largest size of the pores that can be measured.

By impregnating molten paraffin into a lump of exfoliated graphite, cutting thin slices after solidification, and then observing under optical microscope, the cross-sectional area of large spaces among the worm-like particles of exfoliated graphite are analyzed with the aid of an image-processing technique [401]. By this technique, the pore up to about 30 mm size can be measured, although mercury porosimetry using a new dilatometer can be measured only up to 600 μm. Averaged pore parameters measured on a commercially available exfoliated graphite are shown in Table 2.15 and distribution histograms of cross-sectional area and periphery length are shown in Fig. 2.160 [426]. Average pore size was reached as $0.49 \times 0.30\ mm^2$ ($490 \times 300\ \mu m^2$) with average cross-sectional area of about $0.14\ mm^2$ ($1.4 \times 10^5\ \mu m^2$). The most frequent pores, however, locate the cross-sectional area of around $0.010\ mm^2$.

**Table 2.15** Parameters for Spaces Among the Worm-Like Particles in Exfoliated Graphite [426]

| **Direction of slices*** | | **Perpendicular** | **Parallel** |
|---|---|---|---|
| **Exfoliation Volume ($m^3$/kg)** | | **0.089** | |
| Averaged pore parameters | Cross-sectional area ($mm^2$) | 0.137 | 0.138 |
| | Periphery length (mm) | 1.667 | 1.681 |
| | Major axis (mm) | 0.482 | 0.497 |
| | Minor axis (mm) | 0.292 | 0.306 |
| | Aspect ratio | 0.65 | 0.65 |
| | Fractal dimension | 0.77 | 0.77 |
| Number of pores measured | | 3285 | 3015 |

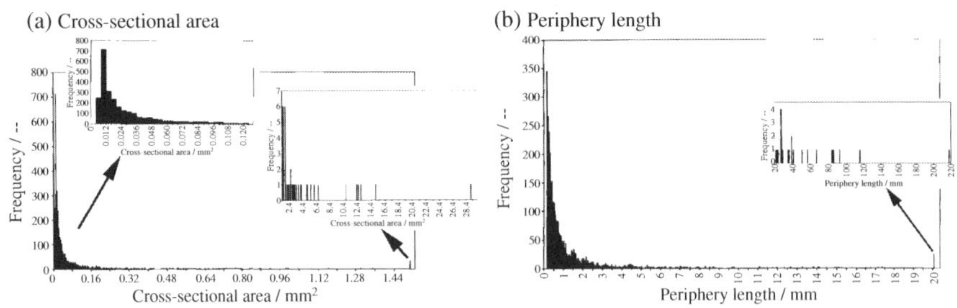


**FIGURE 2.160**

Distribution histograms of parameters for large spaces among worm-like particles of a commercial exfoliated graphite [426].

### 2.7.3 Pore development in carbon materials

Nano-sized pores are created in carbon materials and modified their structure mainly by oxidation under controlled conditions; the process involves the oxidative gasification of precursor carbons to CO and $CO_2$ and is called 'activation'. The products are called activated carbons, having played important roles since prehistoric times and now becoming even more important materials in various technological fields [80,427]. Different activation processes have been employed for carbon materials by using air, steam, $ZnCl_2$ and KOH as activating reagents in order to develop micropores, in particular KOH results in very high surface area as 3600 $m^2/g$ [424]. Mechanical treatment by compression and grinding, associated with activation, of single-wall carbon nanohorns has been reported to be effective to convert mesopores to micropores and to achieve high $CH_4$ storage capacity [428].

Various techniques to control pore structure in carbon materials without any activation process have also been proposed, mainly because of more exact control of pore structure and avoidance of sacrificial loss of carbon atoms; microporous carbons by template carbonization using zeolite [140,157] and metal-organic frameworks (MOFs) [168], mesoporous carbons by using mesoporous silicas [163,164], block copolymer surfactants [165,429] and MgO as template [143,430–432], by defluorination of poly(tetrafluoro ethylene) [238,433] and by carbonization of carbon aerogels [434,435]. Most of these techniques were introduced in Section 2.4 and discussed in detail in Section 3.6. Here, the developments of pores in carbon materials via different processes are discussed; activation of glass-like carbon spheres by air where pore formation starts preferentially on the surface of spheres, exfoliation of graphite flakes where flexible pores are formed without oxidation, and intrinsic pores formed in graphite intercalation compounds to create new functionalities.

#### a. Development of extrinsic nano-sized pores in glass-like carbons

In order to control the pore structure in carbon materials, the activation process has been employed by using different agents, $Zn(OH)_2$, KOH, $CO_2$, water vapor, etc. The pore development in glass-like carbon spheres by air oxidation was studied in detail [436,437], because activation surely starts at the surface of spheres and activation process is the simplest in the equipment, the mildest in thermal conditions, and also energy- and resource-saving in materials economy. The commercially available glass-like carbon spheres with a diameter of about 15 μm were oxidized at different temperatures between 355 and 430°C for various residence times in a flow of dry air with a rate of 50 mL/min (activation).

In Fig. 2.161a, the yield after the oxidation at different temperatures is plotted against residence time at each oxidation temperature in logarithmic scale [436]. Curves observed at different oxidation temperatures are similar to one another. Each experimental point on oxidation yield plotted against logarithm of residence time, log t, at different temperatures are superimposed on the curve for a

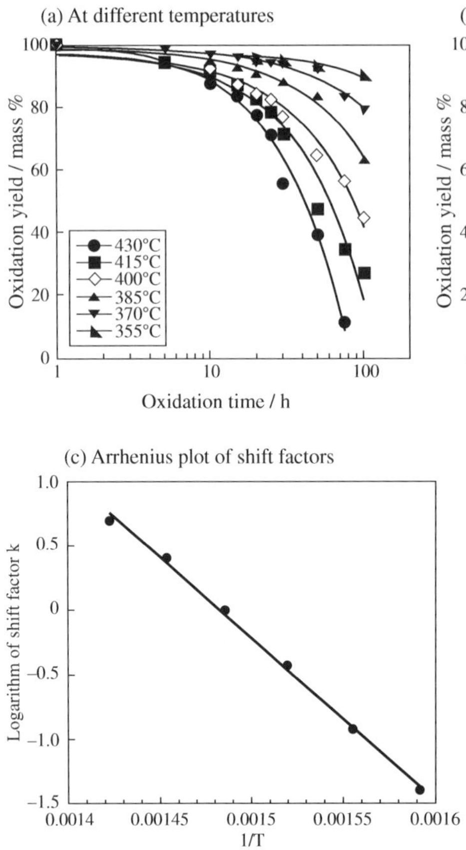


**FIGURE 2.161**

Oxidation yield against oxidation time (logarithmic scale) in dry air [436].

reference temperature 400°C by the translation along the log t axis. This superimposition of all experimental points gave a smooth curve, being called the master curve, as shown in Fig. 2.161b. Plot of shift factors against inverse of oxidation temperature can be approximated to be linear and gives apparent activation energy $\Delta E$ of about 150 kJ/mol from its slope (Fig. 2.161c) [436]. In wet air, however, $\Delta E$ of about 200 kJ/mol has been obtained [437]. The observed $\Delta E$ values are in between those for CO and $CO_2$ formation reactions (111 and 394 kJ/mol, respectively).

The same procedure as oxidation yield is applied to $S_{BET}$ measured and the master curve at 400°C oxidation is obtained (Fig. 2.162). Plot of shift factor against 1/T gave the same $\Delta E$ of about 150 kJ/mol. For $S_{total}$, $S_{ext}$, $S_{micro}$ and $V_{micro}$, which are determined from $\alpha_s$ plot, similar master curves were also obtained.

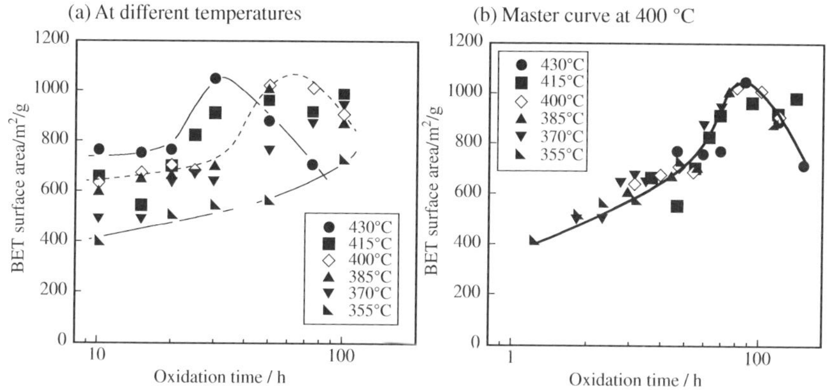


**FIGURE 2.162**

$S_{BET}$ against logarithmic oxidation time [436].

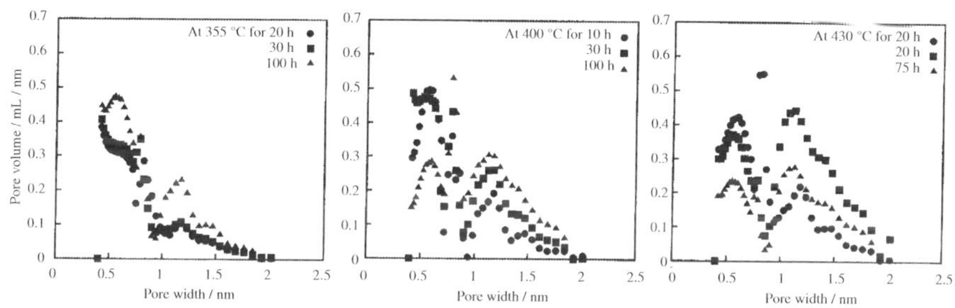


**FIGURE 2.163**

Pore size distribution in micropore range measured by DFT method [436].

In order to understand the development of pores, pore size distribution in micropore range is measured on some samples by DFT analysis, as shown in Fig. 2.163.

At 355°C, pores less than 0.4 nm seems to be principal after 20 and 30 h oxidation (corresponding to the oxidation at 400°C for 5 and 15 h), but after 100 h (corresponding to 25 h at 400°C) the pores with the sizes around 0.6 and 1.2 nm are predominant. At 400°C, the trends to increase the pore volume around 0.6 and 1.2 nm and to decrease the pores less than 0.4 nm are more markedly observed than at 355°C. At 430°C, the same trend can be recognized. After 75 h oxidation at 430°C (corresponding to 150 h at 400°C), however, pore volume of 0.6 and 1.2 nm decreases because pores are widened to more than 2 nm by oxidation.

Master curves for $V_{micro}$ determined by $\alpha_s$ plot, ultramicropore and supermicropore volumes by DFT method and mesopore volume $V_{meso}$ by BJH method are shown in Fig. 2.164, in order to make the comparison easier, together with some SEM images to show the appearance of a sphere. In the beginning of

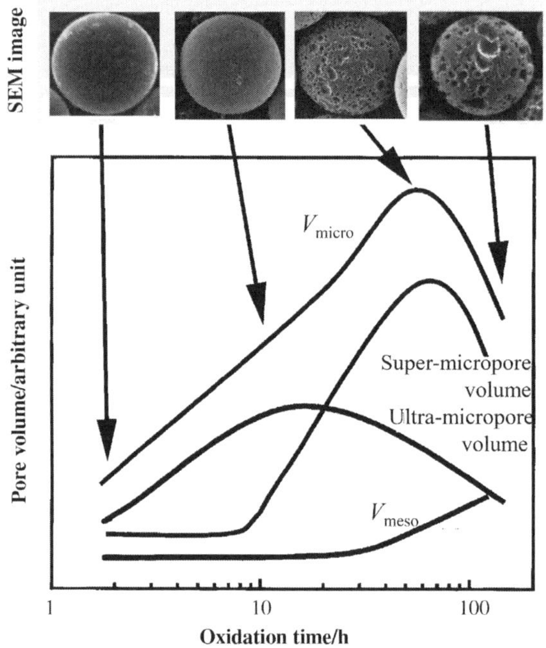


**FIGURE 2.164**

Change in pore structure for carbon spheres with oxidation time at 400°C [436].

oxidation, i.e., up to 10 h oxidation at 400°C, the main process is the formation of ultramicropores. Above 10 h up to about 60 h, the relative amount of ultramicropores decreases but super-micropores and also mesopores increase in number (volume) with increasing oxidation time. Above 65 h, micropores decreases rapidly but $V_{meso}$ increases slightly, which results in the decrease in surface areas measured by different methods. This change in pore volumes suggests the gradual enlargement of pore size from ultramicropore to macropore through super-micropore and mesopore.

SEM observation of sphere surface seems to agree with this pore development sequence, as shown in Fig. 2.164, and it gives some information on macropores, which could not be measured through the gas adsorption analyses. After a few hours oxidation, no change on the surface of the sphere is detected because micropores cannot be observed under SEM. After 20 h oxidation only a few pit-like holes are observed, where super-micropores and mesopores are developed. Around and after passing through the maximum of micropore volume, the surface of spheres becomes rough due to the formation of macropores on the surface.

For these air-oxidized carbon spheres, adsorption of methylene blue, diquat and phenol is determined after soaking for 20 h in their diluted aqueous solutions

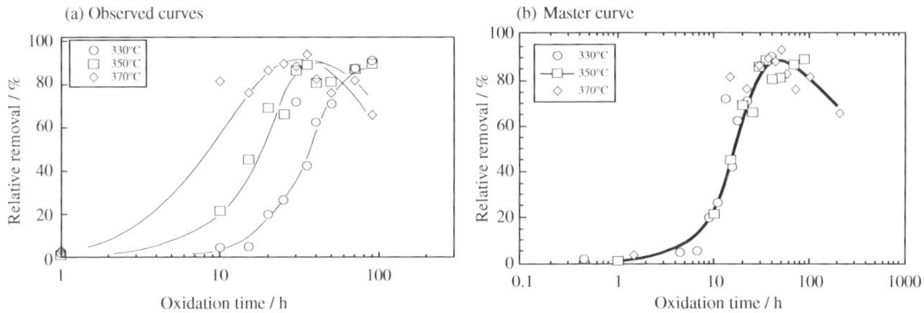


**FIGURE 2.165**

Relative removal of phenol from its $1 \times 10^{-3}$ mol/L solution against oxidation time of the adsorbent carbon spheres [438].

of $2.0 \times 10^{-2}$, $5.0 \times 10^{-5}$ and $1.0 \times 10^{-3}$ mol/L, respectively [438]. By shifting the experimental points on relative removal measured using the spheres oxidized at 330 and 370°C to those at 350°C along the axis of log t, a master curve was obtained for each adsorbate. The results on phenol are shown in Fig. 2.165. The master curves obtained reveal the removal behavior of phenol with oxidation of adsorbent carbon owing to the development in micropores in the carbon. The results show that it is also possible to convert oxidation temperature to oxidation time for adsorption by oxidized carbons and to get the master curve.

## b. Macropore development in exfoliated graphite

Exfoliated graphite is an important industrial raw material for the production of flexible graphite sheets, which are widely used as gaskets and seals [439–441]. Exfoliated graphite is usually prepared by rapid heating of residue compounds of natural graphite flakes with sulfuric acid to about 1000°C. It consists of fragile worm-like particles, which are formed by exfoliating preferentially along the normal to the basal plane of graphite flakes. It has attracted attention because of its very high sorption capacity of spilled heavy oils and easy recovery [442–445], and also of biological proteins and body fluids into its macropores [446,447].

In Fig. 2.166a and b, exfoliation volume and mass loss are plotted against exfoliation temperature on two runs of experiments on the same residue compound [418]. In Fig. 2.166a, exfoliation volume increases with increasing exfoliation temperature. Two kinks are clearly observed in the dependence of exfoliation volume on exfoliation temperature, around 650 and 800°C. Below 650°C and above 800°C, the increasing rate of exfoliation volume is smaller than that in an intermediate temperature range, i.e., 650–800°C. In Fig. 2.166b, mass loss occurs below 650°C and above 800°C, in intermediate temperature range only a slight increase in mass loss is observed. Exfoliation volume is a parameter for a lump of exfoliated graphite, including all kinds of pores. In most exfoliated graphite, pores detected by gas adsorption techniques are usually a negligibly small amount.

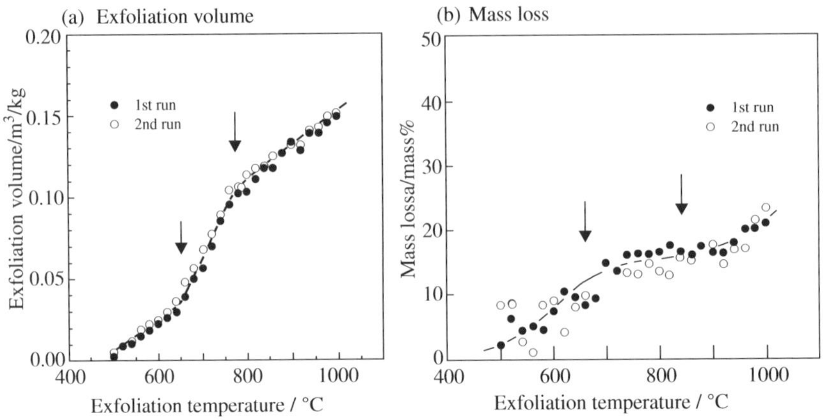


**FIGURE 2.166**

Changes in exfoliation volume and weight loss with exfoliation temperature [418].

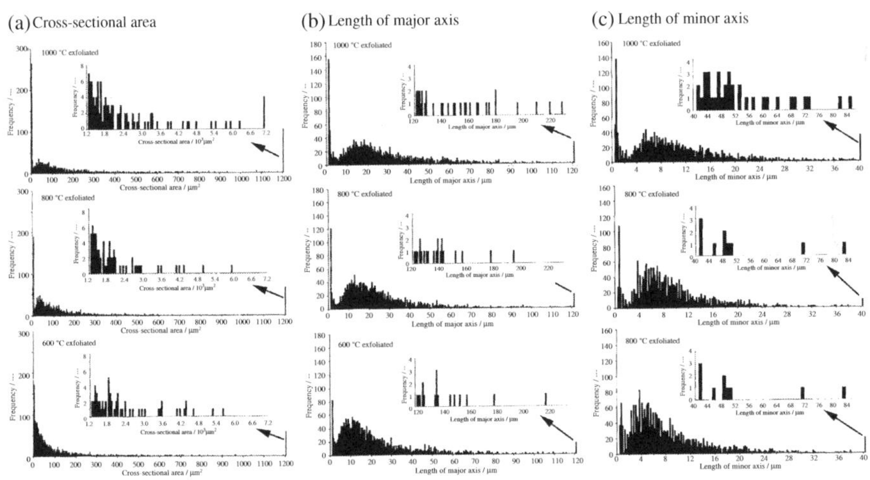


**FIGURE 2.167**

Effect of exfoliation temperature on the distributions of cross-sectional area, and lengths along the major and minor axes of the pores inside the worm-like particles [418].

As shown in Fig. 2.157, these worm-like particles contain a large amount of macropores, of which the cross-sections may be approximated by ellipses. In Fig. 2.167a–c, cross-sectional area, the length along the major axis and that along the minor axis of pores inside the worm-like particles are shown in histograms on the samples exfoliated at 600, 800 and 1000°C for 60 s [418–420]. Also the averaged values of pore parameters are listed in Table 2.16, together with the number of pores measured.

**Table 2.16** Parameters for Pores Inside the Worm-Like Particles of Exfoliated Graphite Prepared at 600, 800 and 1000°C [418]

| **Exfoliation Temperature** | | **600°C** | **800°C** | **1000°C** |
|---|---|---|---|---|
| **Exfoliation Volume ($m^3$/kg)** | | **0.025** | **0.114** | **0.152** |
| Averaged pore parameters | Cross-sectional area ($\mu m^2$) | 193 | 217 | 321 |
| | Major axis ($\mu$m) | 24.4 | 26.0 | 31.2 |
| | Minor axis ($\mu$m) | 8.8 | 9.7 | 11.2 |
| | Aspect ratio | 0.412 | 0.424 | 0.412 |
| | Fractal dimension | 1.09 | 1.10 | 1.09 |
| Number of pores measured | | 2583 | 2161 | 2059 |

Pore structures in the worm-like particles of 600°C- and 800°C-exfoliated samples are difficult to differentiate from averaged values of pore parameters, but their exfoliation volume increases markedly from 0.025 to 0.1 $m^3$/kg. On the other hand, exfoliated graphite prepared at 1000°C has slightly broadened histograms of three parameters and much larger averaged pore parameters than those prepared at low temperatures (Table 2.16). These differences in averaged values of parameters are due to the increase in frequency of large-size pores, as can be seen from the comparison between inserted figures in Fig. 2.170 [418].

Exfoliation of graphite proceeds in the three steps. In the first step below 650°C, exfoliation volume increases with temperature through exfoliation of each graphite flake to worm-like particles, in which ellipsoidal pores are developed. In the second step above 650°C, the main process in the second step may be the introduction of complicated entanglement of worm-like particles and result in the increase in exfoliation volume, because growth of pores inside the particles is not so marked but exfoliation volume increases more rapidly than the previous step. In this temperature range, large open spaces among the particles are expected to grow markedly, which is suggested from quite different exfoliation volume between 600°C- and 800°C-exfoliated graphite. In the third step above 800°C, the development of pores inside the worm-like particles is supposed to continue with temperature, because the increase in exfoliation temperature from 800 to 1000°C causes mainly widening of pores inside the particles, but relatively small increases in exfoliation volume.

Residence time at a high temperature for exfoliation has a certain effect on the pore structure of exfoliated graphite. With a long heating time, cross-sectional area and lengths along the major and minor axes of pores increased markedly.

Large spaces among the particles (intraparticle pores) have been measured by impregnating paraffin into a lump of exfoliated graphite, solidifying the lump of exfoliated graphite and then slicing into thin sections [140]. Fig. 2.168 shows distribution histograms of cross-sectional area and periphery length of the pores among the worm-like particles of exfoliated graphite prepared at 1000°C for 20

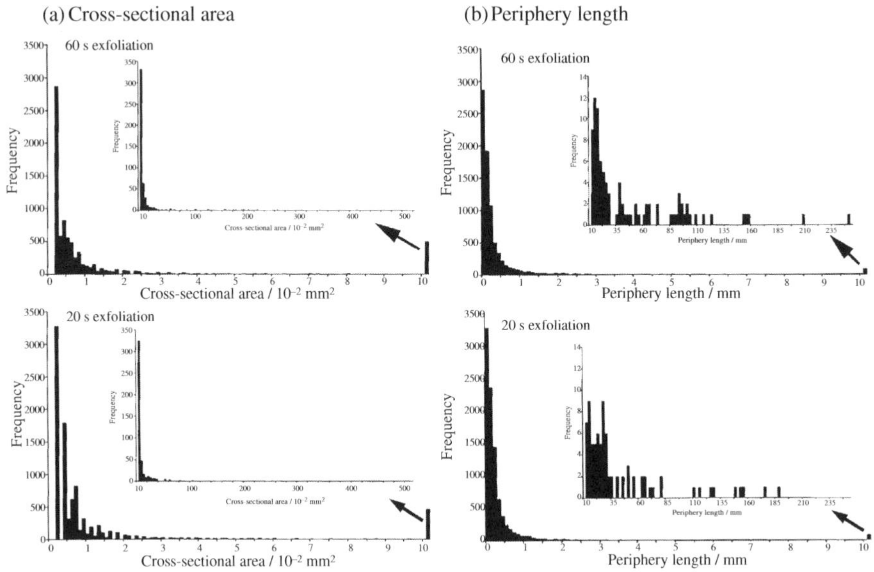


**FIGURE 2.168**

Histograms for cross-sectional area and periphery length of pores among the worm-like particles of exfoliated graphite prepared at 1000°C for the residence time of 20 and 60 s [418].

**Table 2.17** Average Values of Parameters of Pores Among the Worm-Like Particles for Exfoliated Graphite Prepared at 1000°C for the Residence Time of 20 and 60 s [418]

| Residence Time at 1000°C | | 20 s | 60 s |
|---|---|---|---|
| Average pore parameters | Area ($mm^2$) | 0.070 | 0.104 |
| | Periphery length (mm) | 0.647 | 0.847 |
| | Aspect ratio | 0.76 | 0.76 |
| | Fractal dimension | 1.72 | 1.74 |
| Number of pores used | | 9678 | 8194 |

and 60 s [418]. Averaged pore parameters are also tabulated in Table 2.17 [402]. By 60 s heating at 1000°C, pores among the particles become notably larger than by 20 s heating. On the distribution histogram of cross-sectional area, the effect of long-term heating at 1000°C is not apparent, but averaged cross-sectional area and periphery length are larger in 60 s heating.

The content of intercalates remaining in the residue compound has a marked influence on exfoliation behavior. The residue compounds are prepared by electrochemical intercalation of sulfuric acid [422]. In Fig. 2.169, volatiles content in the starting residue compounds and exfoliation volume of resultant graphite exfoliated at 1000°C are plotted against electricity consumed during electrolysis to prepare intercalation compounds. With increasing electricity consumption, volatiles content, i.e., content of intercalates, and exfoliation volume increase roughly proportionally [422].

On five residue compounds with different intercalate contents, the parameters of pores inside the worm-like particles are summarized in Table 2.18, together

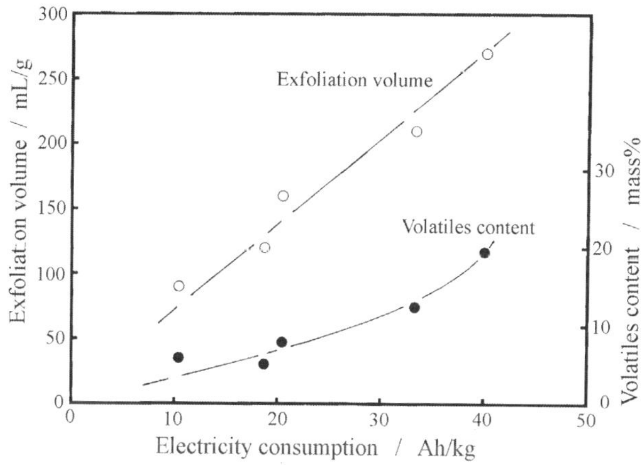


**FIGURE 2.169**

Changes of volatiles content of residue compounds and exfoliation volume of resultant exfoliated graphite with electricity consumed during electrolysis to prepare the original intercalation compounds [422].

**Table 2.18** Pore Parameters of Exfoliated Graphite Prepared from Residue Compounds with Different Intercalate Contents [421]

| Sample Code | | No. 1 | No. 2 | No. 3 | No. 4 | No. 5 |
|---|---|---|---|---|---|---|
| Electricity consumption (A·h/kg) | | 10.38 | 18.67 | 20.41 | 33.33 | 40.00 |
| Content of volatiles at 1000°C (mass%) | | 5.9 | 5.1 | 7.9 | 12.4 | 19.5 |
| Exfoliation volume (mL/g) | | 90 | 120 | 160 | 210 | 270 |
| Averaged pore parameters | Area ($\mu m^2$) | 95.3 | 102.2 | 110.4 | 92.8 | 113.3 |
| | Major axis ($\mu m$) | 16.0 | 18.3 | 17.4 | 16.1 | 17.2 |
| | Minor axis ($\mu m$) | 6.04 | 5.78 | 6.33 | 6.35 | 6.87 |
| | Aspect ratio | 0.43 | 0.38 | 0.40 | 0.44 | 0.45 |
| | Fractal dimension | 1.13 | 1.16 | 1.11 | 1.12 | 1.10 |
| Number of pores | | 6476 | 6647 | 6346 | 6541 | 6811 |

with the electricity quantity consumed for the preparation of the intercalation compounds, content of volatiles and exfoliation volume after the exfoliation at 1000°C [421]. Even though exfoliation volume of the lumps of exfoliated graphite prepared changes with increasing electricity consumption, the parameters of pores inside the worm-like particles depend slightly on electricity consumption and volatiles content in the residue compounds.

From these analyses, the content of intercalates in the starting residue compounds does not give pronounced influence on the pore structure inside the worm-like particles, but governs exfoliation volume, in other words, bulk density.

Four samples of EG prepared from residue compounds with different contents of intercalates, consequently having different exfoliation volumes, have been prepared through electrochemical intercalation with different electricity consumptions and subsequent exfoliation at 1000°C for 10 s. Large spaces among the particles are measured by impregnating paraffin into a lump of exfoliated graphite, solidifying the lump of exfoliated graphite and then slicing into thin sections [402]. Volumes occupied by worm-like particles $V_W$ and those of large spaces among particles $V_S$ are plotted against exfoliation volume in Fig. 2.170. Both $V_W$ and $V_S$ increase linearly with increasing exfoliation volume [401]. The relative proportion $V_S$ is in a narrow range of 73–77% for samples shown in Fig. 2.170, irrespective of exfoliation volume.

Development of pores during the exfoliation process, which has been studied through image analysis techniques, is summarized in review papers [426,448].

Graphite residue compounds are successfully exfoliated under microwave irradiation [449–451]. The microwave exfoliation can be performed in a short time without any furnace to heat, but we have to be careful to apply this technique because it might give inhomogeneity, some flakes exfoliating to worm-like particles but some flakes no exfoliation. The exfoliated particles are easily separated from un-exfoliated particles by a simple sieving [451].

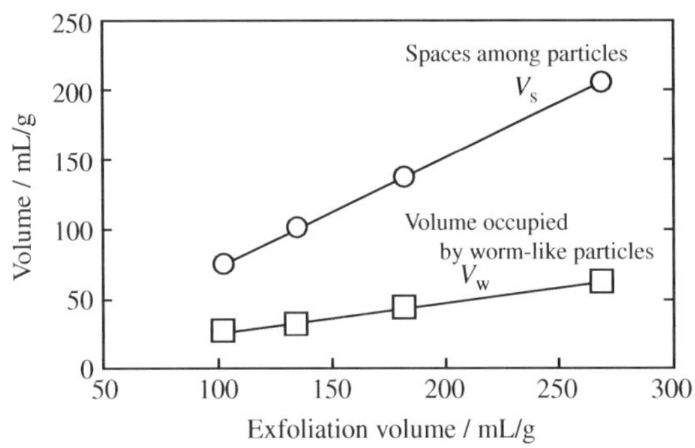


**FIGURE 2.170**

Changes in $V_W$ and $V_S$ with exfoliation volume [401].

### c. Development of extrinsic pores in graphite intercalation compounds

Intercalation of various species into graphite gallery can be considered to be the chemical adsorption of various species into intrinsic pores, which are formed between neighboring graphite layers and are two-dimensional slits. These intercalation phenomena are one of the characteristics of carbon materials in graphite family. This is discussed in Section 2.8 in detail.

In these intercalation compounds of various carbon materials, new but extrinsic pores are formed. In graphite intercalation compounds (GICs) of alkali metals with stage-2 structure, nanospace is formed surrounded by alkali metal ions and graphite layer planes [452,453]. As an example, arrangement of $Cs^+$ ions in graphite gallery is shown in Fig. 2.171. Since the interlayer spacing intercalated by $Cs^+$ is widened to 0.601 nm, the carbon layer plane of graphite is considered to have a thickness of 0.335 nm and $Cs^+$ intercalated are apart by 0.311 nm from each other, the effective space which the third component can be inserted, is the size of about $0.311 \times 0.266$ nm. Changing the alkali metal ions to $K^+$ (ion radius of 0.133 nm) and $Rb^+$ (ion radius of 0.148 nm) from $Cs^+$ (ion radius of 0.169 nm), the size of the space for the third component changes a little. These calculated data are summarized in Table 2.19.

Into this nanospace in alkali metal-GICs, different gases are adsorbed, as summarized in Table 2.19. Hydrogen molecule $H_2$, as well as deutrium $D_2$, is adsorbed into these three GICs. However, adsorption of nitrogen, argon and methane shows a certain dependence on GICs. By taking into account the fact that the nanospace in stage-2 alkali metal-GICs consists of the gate, which has the width $w$ determined by two alkali metal ions and the thickness $t$ determined by graphite

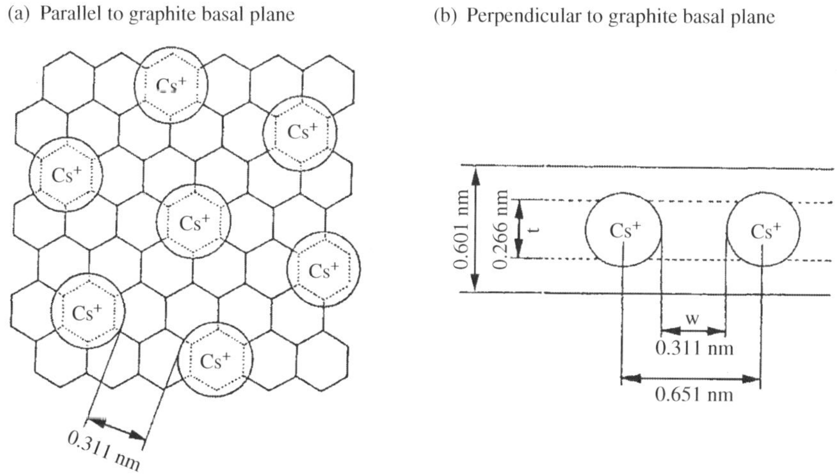


**FIGURE 2.171**

Alignment of $Cs^+$ in stage-2 Cs-GIC.

*(Courtesy of Prof. N. Akuzawa of Tokyo Nat. Coll. Tech.)*

**Table 2.19** Nanospaces Formed in Graphite Intercalation Compounds with Stage-2 Structure of Alkali Metals and Gases Adsorbed at Low Temperatures of 196–63 K [452,453a, 453b]

| Stage-2 GIC | Interlayer Spacing of Intercalated Space $d_1$ (nm) | Nanospace in GIC $w \times t$* ($nm^2$) | Gas Adsorbed (Gas Molecules/alkali Metal) | | | | |
|---|---|---|---|---|---|---|---|
| | | | $H_2$ | Ne | $N_2$ | Ar | $CH_4$ |
| $KC_{24}$ | 0.539 | 0.385 × 0.204 | 2.1 | 0.0 | (0.7) | 0.0 | 0.0 |
| $RbC_{24}$ | 0.570 | 0.355 × 0.235 | 2.05 | 0.0 | 1.0 | 1.2 | 0.9 |
| $CsC_{24}$ | 0.601 | 0.311 × 0.266 | 2.00 | 0.0 | 1.3 | 1.4 | 1.2 |

*Width × thickness, refer to Fig. 2.171b.*

layers, $w$ being 0.311 nm and $t$ 0.266 nm in $CsC_{24}$ as shown in Fig. 2.171, the adsorption of $N_2$, Ar and $CH_4$ molecules seems to be predominantly restricted by $t$ of the nanospace in GICs. On the other hand, almost the same amount of hydrogen molecules is adsorbed into nanospaces in these three GICs. For $CsC_{24}$ compounds prepared from petroleum cokes heat-treated at different temperatures from 1300–2800°C, adsorption isotherms of $H_2$, $D_2$, $N_2$ and Ar gases are expressed by Langmuir equation, suggesting monolayer adsorption [453b]. The saturated amount of adsorbed molecules depended on heat treatment temperature, i.e., on crystallite size $Lc$(002) of host cokes, and became constant above $Lc$(002) of 40–60 nm.

For stage-1 GIC of potassium $KC_8$, hydrogen is adsorbed even at room temperature to a composition of $KC_8H_{2/3}$, but this product has the stage-2 structure where the intercalates formed the hydrogen layer sandwiched by potassium layers [454]. When deuterium molecules $D_2$ were adsorbed into $KC_8$ together with hydrogen molecules $H_2$, the molecules HD were found in the desorbed gas. These results may suggest a certain interaction of hydrogen molecules with the alkali metals intercalated in the graphite gallery. For hydrogen storage into alkali metal-GICs, a pronounced effect of host carbons has been reported [455].

Adsorption of hydrocarbon gas molecules into stage-2 Cs-GICs ($CsC_{24}$) has been extensively studied [456–458]. All saturated hydrocarbon molecules, such as $CH_4$, $C_3H_8$ n-$C_6H_{14}$, etc., are adsorbed into the nanospace in $CsC_{24}$ and desorbed reversibly. However, unsaturated hydrocarbon molecules, such as $C_2H_4$ and $C_2H_2$, are adsorbed irreversibly, probably because of oligomerization of hydrocarbon molecules in the nanospace of GICs [458] (refer to Section 3.7.5).

Into graphite oxide, various organic molecules can be intercalated to form ternary compounds. In the ternary compounds, such as surfactant-intercalated graphite oxide, new spaces to adsorb organic compounds are created [459,460]. In Fig. 2.172, the change in XRD pattern with adsorption of various aromatic molecules into $C_{14}$TMA-intercalated graphite oxide ($C_{14}$TMA: $C_{14}H_{29}(CH_3)_3N^+$) is shown [459]. All aromatic molecules, except anthracene, are adsorbed and

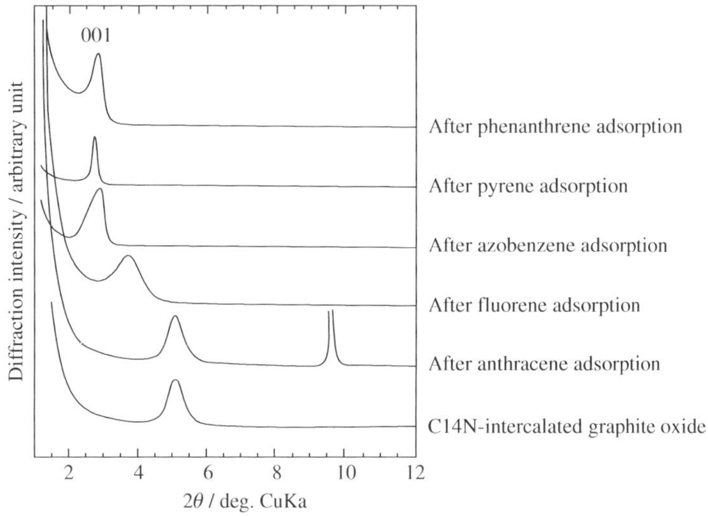


**FIGURE 2.172**

X-ray diffraction patterns of $C_{14}$N-intercalated graphite oxide after adsorption of various aromatic molecules.

*(Courtesy by Prof. Y. Matsuo of Hyougo Pref. Unv.)*

make identity period *Ic* larger, from 1.72 nm for the pristine $C_{14}$TMA-intercalated graphite oxide to 3.09 nm after adsorption of phenanthrene. This increase in *Ic* is explained by the changes in orientation of alkyl changes in $C_{14}$TMA cations to carbon layer. Into $C_{18}$TMA-intercalated graphite oxide ($C_{18}$TMA: $C_{18}H_{27}(CH_3)_3N^+$), adsorption of polyaniline has been reported [460].

# 2.8 Introduction of foreign species

## 2.8.1 Possibility to introduce foreign species into carbon materials

Carbonization of organic precursors (Sections 2.3 and 2.4) can be said to be the process to exclude the foreign atoms to get pure carbon, and this process is very important for the establishment of the carbon network (structure and nanotexture). In contrast, foreign species, not only atoms but also ions and molecules, can be deliberately introduced into carbon networks. As explained previously (Section 2.1), each carbon family has its characteristic ways to accept foreign species into its structure. In Fig. 2.173, the ways each family have for the introduction of foreign species is introduced. Diamond can accept foreign atoms by substitution for carbon atoms, but there is no other way to accept foreign atoms

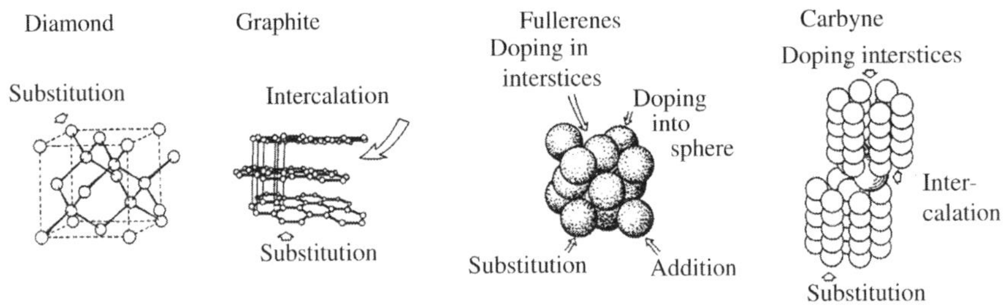


**FIGURE 2.173**

Possibility to introduce foreign atoms into carbon material of each family.

because of its rigid three-dimensional network. In the graphite family, intercalation into interlayer space (graphite gallery) is very characteristic and also substitution for carbon atoms in hexagonal carbon layers can occur to modify the properties. The fullerene family has different possibilities, such as the substitution for carbon atoms constructing fullerene cages, accepting foreign atoms not only into the interstices formed by closest packing of fullerene cages, but also inside each cage, and chemical bonds of organic radicals on the surface of the cluster. Although not studied in detail, there are many possibilities in the case of the carbyne family, intercalation between the layers composed of linear carbyne chains, accepting among the chains and substitution for carbon atoms in the chains.

These possibilities to accept foreign species may be classified into two groups, substitution and doping. The substitution for carbon atoms in carbon network can occur in all families, but it happens with limited atoms, boron and nitrogen. Doping may be further divided into two classes, intercalation into graphite gallery and introduction of foreign species into different interstices of carbon networks. Foreign species, which can be doped into carbon materials, spread in a wide range of molecules and ions, as well as atoms.

In this chapter, therefore, intercalation into graphite gallery was firstly explained, because it is one of the characteristics of the graphite family, and then substitution and doping are discussed.

## 2.8.2 Intercalation

### a. Characteristics of intercalation compounds

It is well known that graphite can accept various atoms, ions and even molecules between its interlayer space (gallery) of hexagonal layer planes of carbon atoms. The phenomenon of such an insertion of foreign species is called **intercalation** and the species, which can be inserted into graphite gallery, are usually called **intercalates**, the products of this intercalation reaction being **graphite intercalation compounds**. The properties of host graphite are strongly modified by the intercalation; for example, the intercalation compound of potassium up to a composition of $KC_8$ (stage-1 structure) has a golden color and that to $KC_{24}$ (stage-2) blue, while the compound with fluorine $(CF)_n$ is white. However, even carbon materials, in which the graphite structure is not developed and crystallite sizes along both *a*- and *c*-axes are small, can be intercalated by some intercalates, such as potassium, but not by sulfuric acid. In order to cover all of these compounds, from those of highly crystallized graphite to those of low crystallinity carbons, therefore, the word graphite intercalation compounds (GICs) is not appropriate and so the term **intercalation compounds** (ICs) is used [461]. These ICs have many possibilities for novel applications [462–465], including the remarkable development of lithium ion rechargeable batteries. Hereafter, 'ICs' is mostly used to express a whole group of compounds, including GICs, but GICs is used if the host is clearly known to be graphite.

A wide range of carbon materials in the graphite family have been used as the host materials for various ICs by aiming to develop their applications, which are explained below and also in Section 3.8.

One of the characteristics of ICs is the charge transfer between intercalate and the carbon layer, which is the main cause of different functions of ICs. This charge transfer can occur both ways, from the carbon layer to intercalate and vice versa, because carbon can have two kinds of carriers, negative electrons and positive holes, in its band structure, particularly graphite having two carriers in almost equal concentrations. This is the main reason why intercalates for carbon are so numerous. Some representative intercalates are tabulated in Table 2.20, where the intercalates are classified on the basis of the direction of charge transfer between intercalate and carbon layer; donor-type intercalates which give electrons to the carbon layer and become positive ion in the gallery of carbon layers, and acceptor-type ones which accept electrons from the carbon layer and become

**Table 2.20** Representative Intercalates for ICs

| Bond Nature | Electronic State of Intercalate | Examples of Intercalates |
|---|---|---|
| Ionic | Donor type | Li, Na, K, Rb, Cs |
| | | Ca, Sr, Ba |
| | | Mn, Fe, Ni, Co, Zn, Mo |
| | | Sm, Eu, Yb |
| | | K-Hg, Rb-Hg |
| | | K-$NH_3$, Ca-$NH_3$, Eu-$NH_3$, Be-$NH_3$ |
| | | K-H, K-D |
| | | K-THF, K-$C_6H_6$, K-DMSO |
| | Acceptor type | F, Br, ICl, IBr, $IF_5$ |
| | | $FeCl_2$, $FeCl_3$, $NiCl_2$, $AlCl_3$, $SbCl_5$ |
| | | $AsF_5$, $SbF_5$, $NbF_5$, $XeF_5$ |
| | | $CrO_3$, $MoO_3$ |
| | | $HNO_3$, $H_2SO_4$, $HClO_4$, $H_3PO_4$ |
| Covalent | – | F, O(OH) |

negative ion in the gallery. This charge transfer between intercalates and carbon does not occur at 100%, in other words, the ionization ratio of intercalates in the carbon gallery is not always 100%. In general, donor-type intercalates, such as alkali metals, have a high ionization ratio in ICs, close to 100%, but acceptor-type ones have a rather low ionization ratio of 10–30%. In acceptor-type ICs, therefore, there are neutral species, $Br_2$ molecules intercalating together with $Br^-$ in ICs with bromine and both $H_2SO_4$ and $HSO_4^-$ in ICs with sulfuric acid. In these donor and acceptor types of ICs, the carbon layers are kept as flat as in the host carbons.

Another characteristic of ICs is the stage structure, which can be in a wide range from 1 to more than 10. The stage number n defines how many carbon layers are located between two intercalate layers, as shown in Fig. 2.174a and b. In a stage-1 structure, every gallery of carbon is filled with intercalates, the carbon and intercalate layers being stacked alternatively along the normal to these layers, and therefore one carbon layer between every two intercalate layers. There are two carbon layers between the intercalate layers in stage-2, three carbon layers in stage-3, and so on. This stacking regularity between carbon and intercalate layers (staging) has been observed for various intercalates and their stage number n has been reported to reach 10 and more. This wide range of staging leads to a wide range of composition of ICs and, as a consequence, a wide range of functions.

In other layered compounds, such as clays and transition metal sulfides, the bimodal intercalation phenomenon, i.e., donor- and acceptor-type, has never been observed; for example, clays can accept only positive ions such as sodium. Also, a wide range of staging, seen on ICs of carbons, has never been observed in other intercalation compounds of clays, metal sulfides, etc.

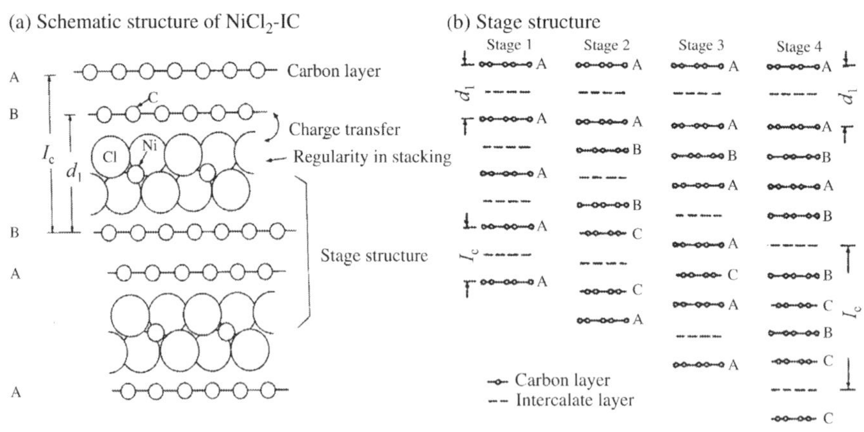


**FIGURE 2.174**

Structural characteristics of ICs.

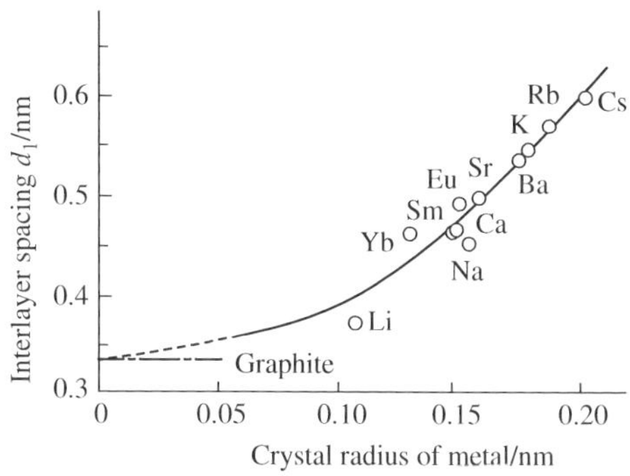


**FIGURE 2.175**

Dependence of $d_I$ on crystal radius of some metallic atoms.

In the case of graphite, it has to be pointed out that the stacking sequence between two intercalated graphite layers changes to AA (or BB) in most cases, although pristine graphite has an AB stacking sequence and the graphite layers that are not intercalated keep an ABA stacking, as shown in Fig. 2.174a and b. The interlayer spacing between these non-intercalated graphite layers is believed to keep that of host graphite, i.e., 0.3354 nm, although the spacing between intercalated graphite layers $d_I$ (refer Fig. 2.174a) depends primarily on the size of intercalate. In Fig. 2.175, the spacing between intercalated graphite layers $d_I$ by different metal ions is plotted against their sizes (crystal radius), which exhibits a strong dependence of interlayer spacing $d_I$ on the size of intercalates. In the case of carbon materials which have turbostratic stacking and larger spacing than 0.3354 nm for graphite, however, there have been no studies on the stacking regularity between two carbon layers and a sandwiched intercalate layer, and also on the spacing for a non-intercalated gallery, whether it retains the original value larger than 0.3354 nm or not.

This stage structure of ICs can be determined from their X-ray diffraction (XRD) patterns. In Fig. 2.176, XRD patterns of GICs with $MoCl_5$ having stage-1, -2 and -4 structures, where only a series of $00l$ diffraction lines is detected, because of the strong orientation of layer planes parallel to the holder surface for an X-ray diffractometer. The index $l$ is determined from the observed values of spacing, $d_{obs}$, on each diffraction line and the following equation:

$$I_c = d_I + 0.3354 \times (n - 1) = l \times d_{obs}, \tag{2.7}$$

to give a unique value of identity period $I_c$ (refer to Fig. 2.174b) for each compound.

By using a common $d_I$ value of 0.922 nm, stage-1, -2 and -4 structures for three $MoCl_5$-GICs in Fig. 2.176 are decided, which have $I_c$ values of 0.922, 1.257 and 1.927 nm, respectively. The $I_c$ value thus determined is only the distance of repeating periodicity of the layers of carbon and intercalate along their normal,

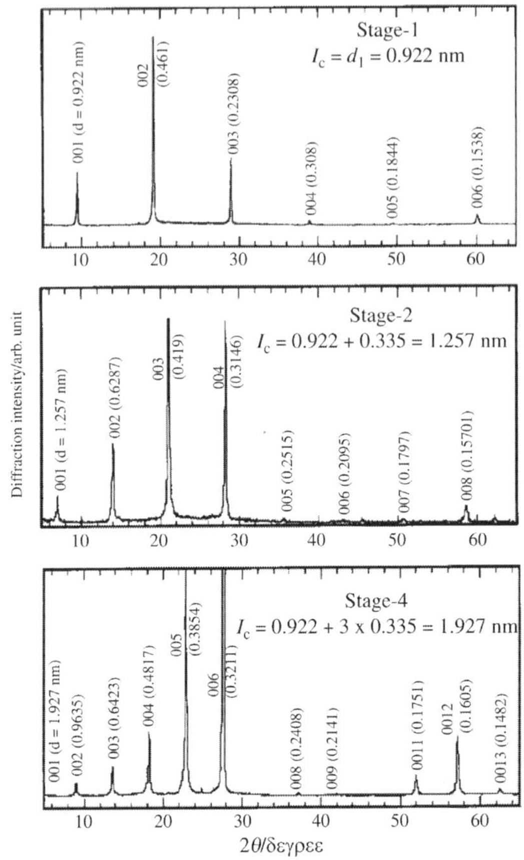


**FIGURE 2.176**

XRD patterns for stage-1, -2 and -4 GICs with $MoCl_5$.

not necessarily the lattice constant of the crystal, this being the reason why $I_c$ has been called identity period. In stage-1 K-GIC, the layer of potassium intercalates is known to stack with the mutual relation of $\alpha\beta\gamma\delta$ and so the crystallographic lattice constant must be four times the $I_c$ value determined from the XRD pattern.

In ICs with most transition metal chlorides, the intercalate layers are composed of octahedra of chloride ions with metal ions at the center which are shearing edges, and so the interlayer spacing for intercalated gallery $d_1$ does not depend on metal species. In these ICs, three layers of atoms, chloride-metal-chloride, are formed in the graphite gallery, as can be seen from Fig. 2.174a on $NiCl_2$-IC.

It was pointed out by many authors that a simple model for stage structure, as in Fig. 2.174b, cannot be explained by the stage transition phenomena during either formation or decomposition of various ICs. For instance, the transition

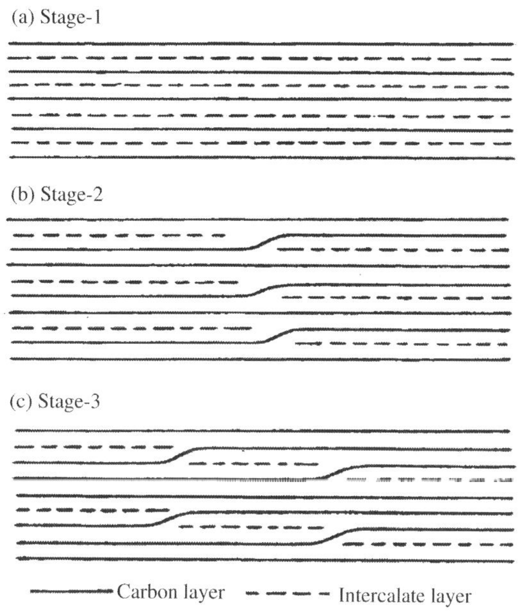


**FIGURE 2.177**

Daumas-Herold model for stage structure of GICs [466].

from stage-3 to -2 during formation requires the movement of intercalates already presented in the gallery to the neighboring non-intercalated gallery. A model, shown in Fig. 2.177, has been proposed [466], where intercalates are located in every gallery of graphite but make domains with a stage structure. In this model, the stage transition can be explained only by the diffusion of intercalates in the same gallery and so has been accepted by many researchers. However, clear direct evidence for this model has not yet been presented. It gives difficulty in interpreting the formation reactions of bi-intercalation structures (refer to Fig. 2.179); in the reaction of stage-2 $FeCl_3$-GICs with $AlCl_3$, for example, $AlCl_3$ has to pass by the intercalate layer of $FeCl_3$ in order to reach the non-intercalated region to form the stage-1 IC with bi-intercalation structure.

Under scanning tunneling microscope, the domain intercalated can be differentiated from non-intercalated domain, in the former all carbon atoms can be recognized, i.e., hexagonal pattern corresponding to carbon atoms are seen, but in the latter only a half of carbon atoms are shown up, i.e., triangle pattern, are seen (refer to Section 2.2.1 and Figs 2.10a and 2.13). The STM observation of $MoCl_5$-GIC shows the presence of two kinds of domains, intercalated and non-intercalated, and also clear-cut boundaries between these two domains, as shown in Fig. 2.178a [467]. At the boundary between these two domains, the alignment of carbon atoms is seen not to be disturbed on the STM image. Corrugation at some boundaries for large intercalated domains is observed, as shown in Fig. 2.178b.

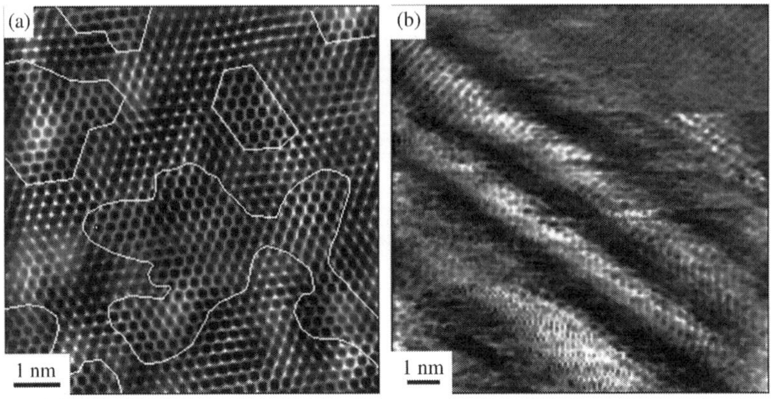


**FIGURE 2.178**

STM images of GIC with $MoCl_5$ [467].

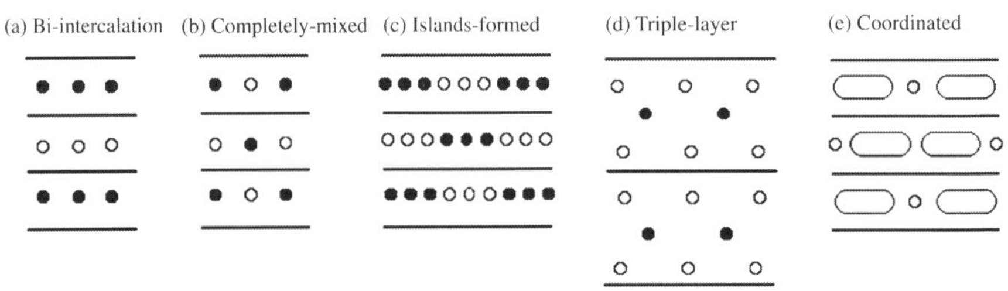


**FIGURE 2.179**

Schematic alignment of two kinds of intercalates in graphite gallery.

ICs with two different intercalates were synthesized, which were called ternary ICs, i.e. two different intercalates and carbon. In these ternary ICs, various alignments of intercalates are formed in the carbon gallery, bi-intercalation, completely mixed, islands-formed, triple-layer and coordinated types, as shown schematically in Fig. 2.179.

Ternary ICs with bi-intercalation structure have been synthesized by different couples of two intercalates. When stage-2 $FeCl_3$-GIC reacts with $AlCl_3$, $AlCl_3$ molecules intercalate preferentially into the gallery where no $FeCl_3$ is present, and so bi-intercalation compound was obtained, which has the periodicity of graphite–$FeCl_3$–graphite–$AlCl_3$–graphite [468,469]. When stage-1 $FeCl_3$-GIC, where all graphite galleries are occupied by $FeCl_3$, is reacted with $AlCl_3$, however, $AlCl_3$ molecules replace partly $FeCl_3$ in the gallery of graphite and resulted in completely mixed ternary compounds. An example of a ternary GIC with a triple-layer structure is $KC_8H_{2/3}$, as shown schematically in Fig. 2.180a, which is synthesized by the reaction between stage-1 K-GIC with hydrogen gas and has a

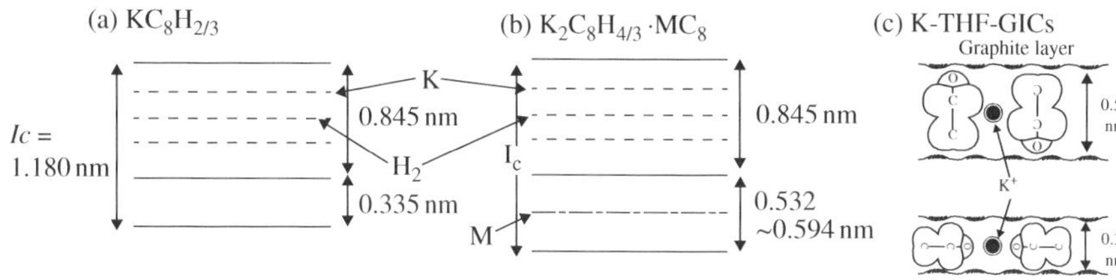


**FIGURE 2.180**

Structure models for ternary GICs with triple-layer type structure [467,468].

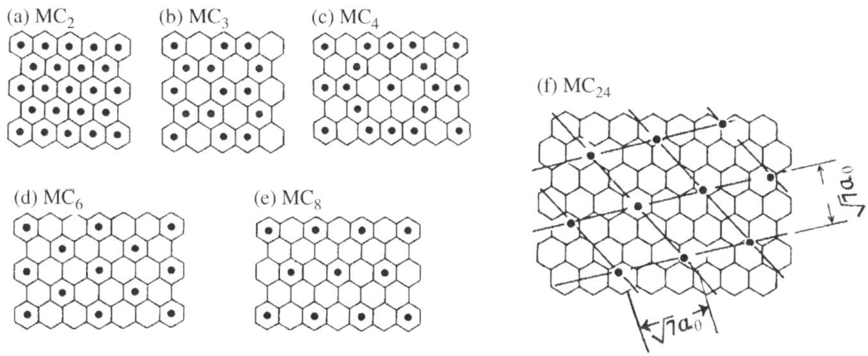


**FIGURE 2.181**

Alignment of alkali metal ions relative to the hexagons in carbon layer.

stage-2 structure [467]. In this compound, K forms double layers in the gallery of graphite and hydrogen is located in between these two K layers. On this $KC_8H_{2/3}$ compound, additional alkali metals intercalate to form a bi-intercalation structure (Fig. 2.180b) [470]. The ternary ICs with coordinated structure can be found in the compound of alkali metal with tetrahydrofurane (THF) [471,472], where various alignments and orientation of THF molecules relative to alkali metal ions and graphite layers have been observed, as shown in Fig. 2.180c.

A variety of alignment of intercalates relative to the hexagons in carbon layer has been determined in binary ICs. Alkali metals are located at registered positions to carbon hexagons in the different ways shown in Fig. 2.181, which correspond to the compositions of $MC_2$, $MC_3$, $MC_4$, $MC_6$ and $MC_8$ if a stage-1 structure is assumed. $MC_8$ is realized in ICs with alkali metals of K, Rb and Cs in the stage-1 structure, but with Li the stage-1 compound has the composition and alignment of $MC_6$. In these $MC_8$ and $MC_6$, all metal ions are located just above the center of the hexagon of the carbon layer (registered) and the distance to nearest-neighbor ions is a multiple of the lattice constant of graphite $a_o$ (commensurate), $2a_o$ ($2 \times 2$ superstructure) and $\sqrt{3}a_o$ ($\sqrt{3} \times \sqrt{3}$ superstructure), respectively. In ICs of heavy alkali metals with a stage structure greater than 2, the metal ions do not have a simple relation with carbon hexagons. At a temperature between room temperature and 125 K, K layers in the gallery have a random alignment and, between 125 and 95 K, their alignment is discommensurate with the hexagonal lattice of the carbon layers. Below 95 K, commensurate structure with a regularity of $\sqrt{7} \times \sqrt{7}$ R19.1° is formed, as shown on stage-2 K-GIC of the composition $KC_{24}$ in Fig. 2.181f.

The ICs with alkali metals having higher compositions of $MC_2$, $MC_3$ and $MC_4$ have been synthesized under high pressures [473]. A detailed study on $LiC_2$ formed under high pressure shows that some Li–Li bonds in the intercalate layers are smaller than those in metal, but the charge transfer to carbon layers is comparable to that in $LiC_6$, suggesting a certain degree of covalent bonding between Li atoms in their layer.

In addition to two types of intercalates, donor- and acceptor-type, fluorine and oxygen can make covalent bonds with carbon atoms in hexagonal carbon layers (Table 2.20). The structure and chemical composition of the compounds with fluorine are well understood [474], in which the graphite layers are no longer flat but the structures corresponding to stage-1 and -2 are synthesized with chemical compositions of CF and $C_2F$ (often written as $(CF)_n$ and $(C_2F)_n$ in the references), of which structural models are shown in Fig. 2.182. In $(C_2F)_n$, half of the carbon atoms form a covalent bond with fluorine and the other half are also covalently bonded with carbon atoms in neighboring layers.

Covalent compound, graphite oxide, has been synthesized in aqueous solution of sulfuric acid with strong oxidants, such as nitric acid and $KMnO_4$, and called graphite oxide [475,476]. Graphite oxides contain oxygen and hydroxide radicals in various amounts, which make covalent bonds with carbon and, as a consequence, a layer of carbon atoms is not flat. Structure models have been proposed by different authors [477,478], some of them being shown in Fig. 2.183.

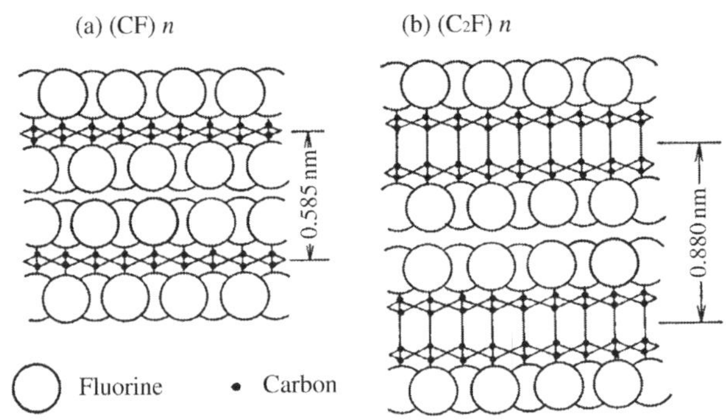


**FIGURE 2.182**

Structure models of $(CF)_n$ and $(C_2F)_n$.

*(Courtesy of Prof. Touhara of Shinshu Univ.)*

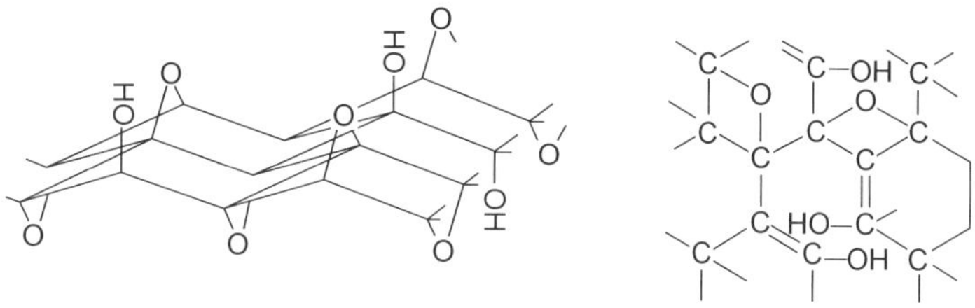

FIGURE 2.183

Structure models of graphite oxide.

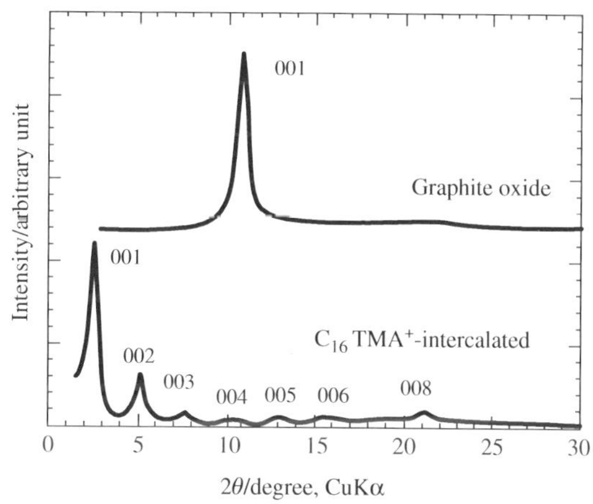


FIGURE 2.184

X-ray diffraction patterns of graphite oxide before and after the intercalation of $C_{16}TMA^+$.

*(Courtesy of Prof. Y. Matsuo of Hyougo Pref. Univ.)*

Into this graphite oxide, various organic compounds can be intercalated to form ternary compounds [479–481]. Organic compounds cannot be intercalated into graphite by themselves, and either the presence or co-intercalation of metal ions, such as K, is essential to intercalate organic molecules into the graphite gallery. Into graphite oxide, however, organic molecules, such as various tri-methylammonium surfactant molecules, are possible to be intercalated by making ionic bonding with oxygen in graphite oxide [479]. By the intercalation of organic molecules, interlayer spacing between carbon layers is widened, depending on the size of molecules intercalated (Fig. 2.184). When lauryl-trimethylammonium cation ($C_{12}TMA^+$, ($C_{12}N_{25}(CH_3)_3N^+$)) and tetradecyl-trimethylammonium cation ($C_{14}TMA^+$, ($C_{14}H_{29}(CH_3)_3N^+$)) are intercalated into graphite oxide, the compounds with various identity periods are obtained, due to the orientation of long aliphatic chain in trimethylammonium cations, as illustrated in Fig. 2.185. For the graphite oxide with much longer chain molecules, like $C_{16}TMA$

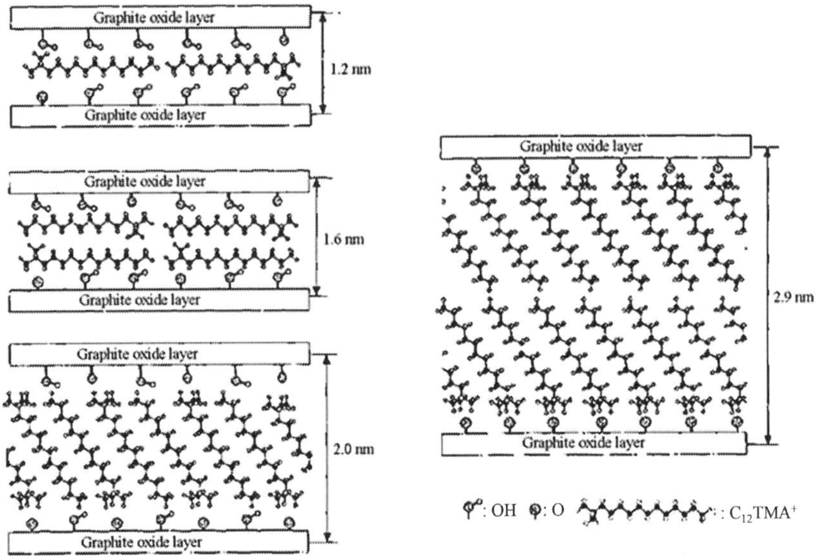


**FIGURE 2.185**

Structure models for graphite oxide co-intercalated by $C_{12}TMA^+$.

*(Courtesy of Prof. Y. Matsuo of Hyougo Pref. Univ.)*

($C_{16}H_{33}(CH_3)_3N^+$, hexadecyl-trimethylammonium cation), however, only the structure with the longest identity period is obtained, where $C_{16}TMA^+$ molecules are supposed to orientate almost perpendicular to the graphite oxide layer [479].

## b. Synthesis of intercalation compounds

Various methods have been employed to synthesize ICs. The methods for the synthesis of ICs are summarized in Table 2.21, where they are classified based firstly on whether the host carbon is separated from or directly in contact with the intercalate source, and secondly on the state of intercalates during reactions.

The two-bulb method has often been used to synthesize various ICs with minor modifications, where the host carbon and intercalates are separated in respective bulbs, as shown in Fig. 2.186a. The intercalation reaction, therefore, proceeds due to the reaction between solid carbon and intercalate vapor, the pressure of which is controlled by the temperature of the intercalate source $T_i$. The temperature of host carbon $T_h$ is usually kept a little higher than $T_i$, in order to avoid the deposition of intercalate vapor on the surface of ICs formed. The stage structure can be controlled by the vapor pressure of intercalate, practically the temperature difference $T_h - T_i$ Fig. 2.186b shows an isotherm for potassium intercalation by plotting the composition of the compound against the temperature difference $T_h - T_i$ [482]. This method has some advantages, i.e., the ability to obtain a homogeneous stage structure and to introduce fewer defects in the resultant ICs

**Table 2.21** Synthesis Methods for Intercalation Compounds of Carbons

| Mutual Position Between Host and Intercalate | State of Intercalate | Motive Force for Reaction | Examples of Intercalate |
|---|---|---|---|
| Separated (two-bulb method) | Vapor | Thermal | K, $FeCl_3$ |
| Contacted (mixing method) | Solid powder | Thermal | $FeCl_3$, $MoCl_5$ |
| | Liquid | Thermal | $Br_2$ |
| | Alloy | Thermal | Li-Na |
| | Molten salt | Thermal | $CuCl_2$-KCl |
| | Aqueous solution | Chemical | $H_2SO_4$, $HNO_3$ |
| | | Electrochemical | $H_2SO_4$ |
| | Non-aqueous solution | Chemical | K-THF, $FeCl_3$-DME |
| | | Electrochemical | Li, Li-DMSO |
| | | Light assisted | $FeCl_3$-$CHCl_3$ |

*DMSO: dimethylsulfoxide, THF: tetrahydrofuran, DME: dimethylethane.*

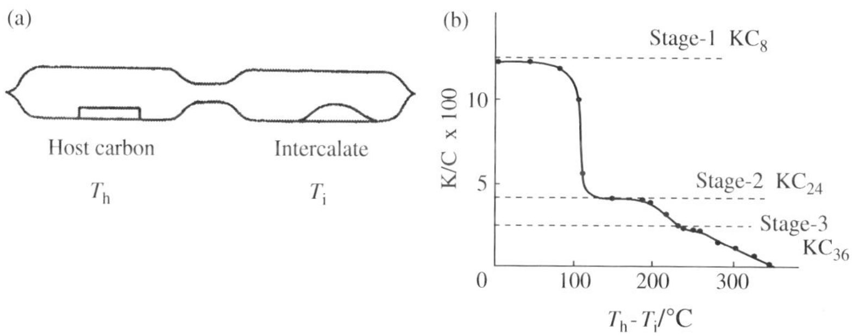


**FIGURE 2.186**

Two-bulb method for the synthesis of ICs. (a) Set-up of host carbon and intercalate, and (b) an isotherm for potassium intercalation [482].

because of the slow reaction rate. However, some disadvantages are also pointed out, e.g., the difficulty in synthesizing a large amount of the compound and a long time needed for the reaction.

The methods for the synthesis of ICs where host and intercalate are in direct contact (mixing method) are classified on the basis of the state of intercalates, solid and liquid (Table 2.21). The first synthesis of Li-GIC with stage 1 structure, $LiC_6$, has been achieved from a mixture of metallic lithium with host graphite under pressure [483].

The intercalation from liquid state of intercalates can be achieved from their own melts, their molten salts with other species which do not intercalate into graphite, or their solutions, either aqueous or non-aqueous. Intercalation

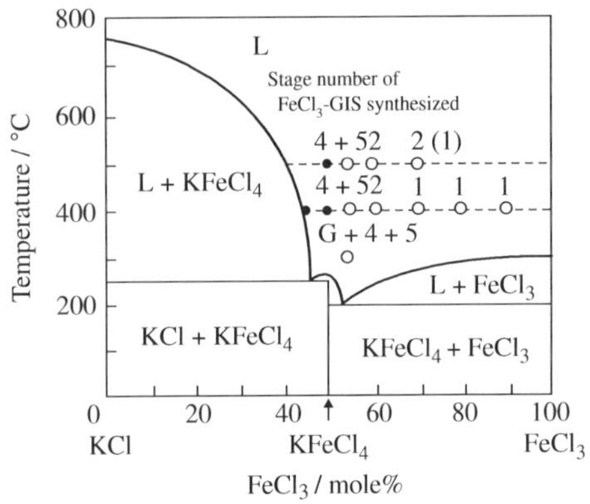


**FIGURE 2.187**

Stage number of GICs synthesized in molten salts of $FeCl_3$-KCl system with different compositions in relation to the phase diagram of the system [485].

compounds with bromine can be synthesized easily by direct dipping the host carbon into bromine, which is in liquid phase at room temperature. Using Li–Na molten alloy, Li-GICs have been obtained at a temperature lower than the formation temperature of lithium carbide [484]. Various transition metal chlorides are intercalated at relatively low temperatures by using their molten salts, e.g., $FeCl_3$–KCl system. In Fig. 2.187, the stage number of GICs formed is shown in the phase diagram of the $FeCl_3$–KCl system, where KCl lowers the melting point of the system and changes the chemical activity of intercalate $FeCl_3$, although it cannot be intercalated into the graphite gallery [485,486].

In Fig. 2.188, three methods to synthesize $CuCl_2$-GICs, two-bulb, molten salt and powder mixing methods, are compared [487]. Since the reaction temperature of 380°C is lower than the melting point of $CuCl_2$, its vapor pressure is so low that the two-bulb method cannot produce GICs. In the molten salt of $CuCl_2$–KCl, intercalation reaction proceeds rapidly at the beginning, and stage-3 structure is formed within 1 h. However, only stage-2 structure is obtained even after a long reaction time, probably because of insufficient chemical activity of $CuCl_2$ in the molten system to reach stage-1 structure. The direct mixing method leads to the formation of stage-1 structure, although it takes rather a long time because of the low vapor pressure.

In order to intercalate $NiCl_2$ into graphite by mixing method, the co-existence of chlorine gas is essential [488]. In the case of $MoCl_5$, however, it is sufficient that a mixture of reagent-grade $MoCl_5$ powder with host graphite, either powder or film, is sealed in an ampoule under vacuum and heated to a temperature between 150 and 300°C, because chlorine gas is generated by the disproportionation reaction of $MoCl_5$ to $MoCl_4$ and chlorine. Through this process, only a mixture of stage-1 and -2 structures of GICs is obtained, not pure stage-1. The

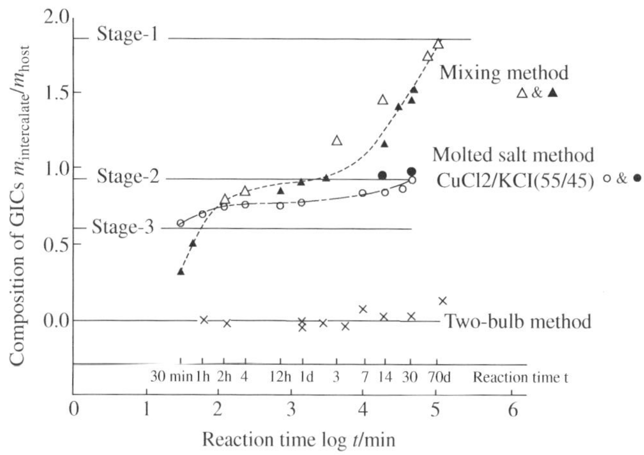


**FIGURE 2.188**

Comparison among two-bulb, powder mixing and molten salt methods for the synthesis of $CuCl_2$-GICs at 380°C [487].

addition of either $MoOCl_3$ or $MoO_3$ into $MoCl_5$ results in the formation of the single phase of stage-1 structure [489], probably due to the formation of a large amount of reactive chlorine gas from the reaction between $MoCl_5$ and either $MoOCl_3$ or $MoO_3$ [490].

A typical example of intercalation in an aqueous solution is the formation of $H_2SO_4$-GICs. The intercalation of $H_2SO_4$ proceeds by the addition of some oxidants, $HNO_3$, $KMnO_4$, etc., into concentrated sulfuric acid (chemical oxidation process). This process is understood to be comparable to electrochemical oxidation through the measurement of potential changes during these two processes, as shown in Fig. 2.189a [493]. The onset potential for each plateau, which corresponds to each stage structure, is the same, irrespective of the oxidation processes, whether chemical oxidation by $HNO_3$, that by $KMnO_3$ or electrochemical oxidation. However, they are known from electrochemical studies to depend strongly on the concentration of sulfuric acid. In the case of chemical oxidation, the maximum potential depends strongly on the oxidant used and also the concentration of sulfuric acid. In Fig. 2.189b, the values of onset potentials for each stage structure (solid lines) and the maximum potential values, which can be achieved either by $KMnO_4$ or $HNO_3$ (broken lines), are plotted against concentration of sulfuric acid. From these relations, it has been understood that in sulfuric acid of 14 mol/dm$^3$, for example, only the compounds with a stage number higher than 3 can be synthesized by $HNO_3$ oxidation because the maximum saturated potential of $HNO_3$ is slightly lower than the onset potential for stage-2, but by $KMnO_4$ oxidation stage-2 structure is possible to be synthesized because the maximum saturated potential of $KMnO_4$ is above the onset potential for stage-2 [491,492].

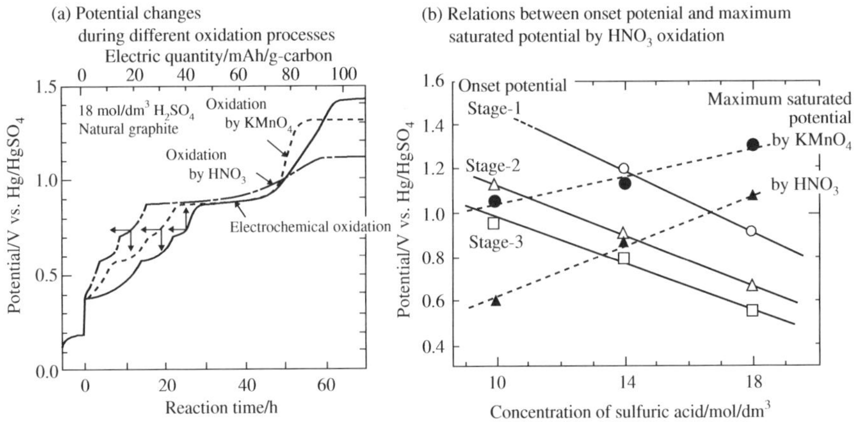


**FIGURE 2.189**

Formation of $H_2SO_4$-GICs in sulfuric acid [491,492].

Intercalation of alkali metal in organic solution has been found [471] and studied extensively, reactions of which can be expressed by the following equations:

$$M + E + A \rightarrow M^+[E]_m + A^-, \tag{2.8}$$

$$C + A^- \rightarrow C^- + A, \tag{2.9}$$

$$C^- + M^+[E]_m \rightarrow C^-\{M^+[E]_m\}, \tag{2.10}$$

where M is alkali metals (Li, Na, K, etc.), E is ethers (tetrahydrofuran, 2-methylhydrofuran, diethoxyethane, etc.), A is an aromatic compound (anthracene, phenanthrane, etc.) and C is host carbon. M dissolves into E by the addition of A through electron transfer from M to A, resulting in the formation of a complex of M ions coordinated by the molecules of solvent E (Eq. 2.8), because C has a slightly higher electron affinity than the A used, and so charges on A in the solution are transferred gradually to solid C in the solution (Eq. 2.9). Then, the complexes $M^+[E]_m$ can be intercalated into C by electrostatic attraction between these two (Eq. 2.10). ICs thus formed are usually ternary compounds because intercalates are the complexes of M ions coordinated by E molecules. However, high-stage binary compounds with alkali metals are formed in some ether solutions, such as 2-dimethylhydrofuran (2-MeTHF) [472,494,495], depending not only on the solvents used but also the host carbon materials [496]. Although direct contact of Na vapor with graphite gives only a high-stage structure, a low stage, even stage-1, is possible to be obtained easily in some organic solution, such as tetrahydrofurane (THF) [497]. Such a behavior of Na ions has been explained by 'misfitting' of the Na lattice in the gallery with carbon hexagons [464].

The intercalation of acceptor-type transition metal chlorides in organic solutions is also possible. In chloroform, two types of $FeCl_3$–$CHCl_3$-GICs with different structures in the intercalate layers have been formed [498]. In dimethylethane (DME), $FeCl_3$ intercalation depends strongly on the graphitization degree of host carbons [499].

Electrochemical intercalation of Li into various carbon hosts, i.e., the formation of binary compounds with Li in organic solutions such as ethylene carbonate, has attracted attention as the fundamental electrode reaction in lithium-ion rechargeable batteries and numerous studies have been performed [500] (Section 3.9.1).

The intercalation reaction of fluorine to graphite gives two different compounds, covalent and ionic ones, depending on the reaction temperature [501,502]. At a reaction temperature of 350–600°C, it gives covalent compounds, such as $(CF)_n$, with puckered carbon layers (Fig. 2.182). At low temperatures such as 150–250°C in the presence of some metal fluorides such as LiF and $CuF_2$, however, fluorine is intercalated into the gallery of flat carbon layers by forming ionic interactions, where no covalent bond between carbon and fluorine is formed but electron transfer from carbon to fluorine occurs. The nature of the bond between carbon and fluorine can be controlled by selecting conditions for synthesis reactions, which gives new possibility to create novel functions [503].

### 2.8.3 Substitution

#### a. Substitution for carbon atoms

Boron B has been known to be able to substitute for carbon atoms in all families of carbon materials, diamond, fullerenes and also graphite. It is a neighbor of carbon C in the Periodic Table, of which the number of electrons is smaller by one. In the case of the graphite family, therefore, the substitution of B for C is expected to result in a reduction in the number of carrier electrons, in other words, B acts as electron acceptor. Another neighbor in the Periodic Table, nitrogen N, has recently been known to substitute for C in its hexagonal layers. The substituted N is expected to act as electron donor because it has one electron more than C, although a theoretical calculation predicts that excess electrons are localized around N in practice. In Fig. 2.190, an illustration of the change in band structure is shown for the substitution of either B or N and both into graphite.

#### b. Substitution of B

B can be introduced into each carbon family by substituting for C, in other words, by making solid solution. The substitution of B into hexagonal carbon layers in graphite family has been extensively studied.

The substitution of B into graphite has been reported already in 1967 and the maximum substitution, solid solubility, of B has been determined to be 2.35 at% at the temperature of 2350°C [345]. The phase diagram for the C-B system is shown in Fig. 2.191. Boundaries between solid solution and liquid, and the segregation of $B_4C$ from solid solution are difficult to be determined accurately, mainly because it is difficult to confirm the exact equilibrium condition at such

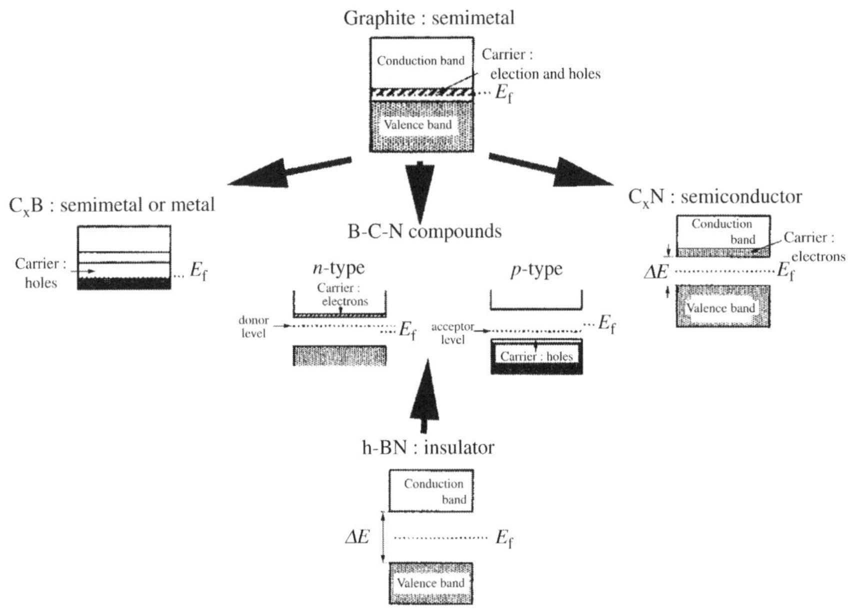


**FIGURE 2.190**

Illustration of the changes in band structure with the substitution of B and N in graphite.

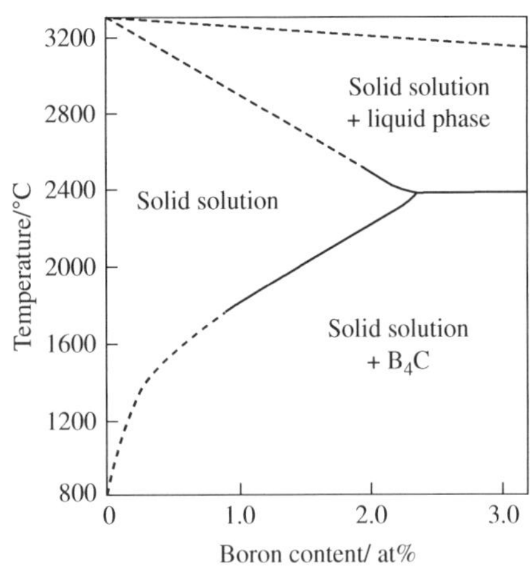


**FIGURE 2.191**

Phase diagram for C-B system [345].

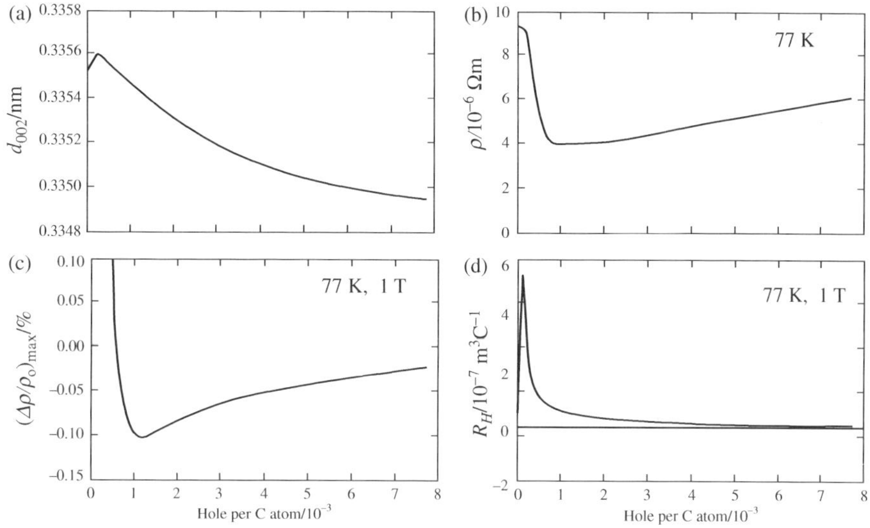


**FIGURE 2.192**

Changes in $d_{002}$, $\rho$, $(\Delta\rho/\rho_0)_{max}$ and $R_H$ with B doping.

*(Courtesy of Prof. Y. Hishiyama of Tokyo City Univ.)*

high temperatures. This phase diagram, however, suggests that the substitution of B for C is a very limited amount.

In Fig. 2.192, the changes in $d_{002}$, $\rho$, $(\Delta\rho/\rho_0)_{max}$ and $R_H$ with number of positive holes per C (concentration of holes), which is supposed to be proportional to the content of B, are shown on a graphite sheet [504]. B doping is performed by the heat treatment of the flexible graphite sheets with 100 μm thickness in the powder mixture of $B_4C$ and petroleum coke. The amount of B substituted into the graphite layers can be controlled by the ratio of $B_4C$ to petroleum coke in the powder mixture. With increasing B doped, $d_{002}$ increases slightly, but then decreases gradually to 0.3349 nm, which is smaller than the value for graphite, 0.3354 nm. In contrast to the decrease in $d_{002}$, the length along the $a$-axis $a_0$ increases from 0.2461 nm for the pristine graphite sheet to 0.4682 nm. $\rho$ does not change by a small amount of doping, but decreases rapidly with further doping and then turns to gradual increase after passing through a minimum. Corresponding to the change in $\rho$, $(\Delta\rho/\rho_0)_{max}$ decreases very quickly and changes its sign to negative. By the small amount of B doping, $R_H$ changes its sign from negative to positive and then decreases rapidly. With increasing B doping, $(\Delta\rho/\rho_0)_{max}$ increases slightly but keeps its sign negative, and $R_H$ keeps a small positive value.

The changes in galvanomagnetic properties with B doping are explained by the changes in carriers, from a two-carriers system for the pristine graphite to a one-carrier (positive hole) system with lowering the Fermi level because B works

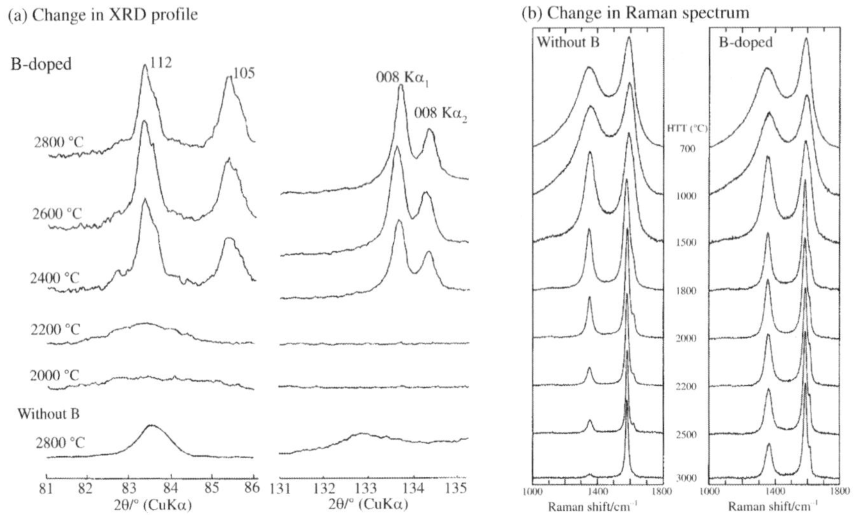


**FIGURE 2.193**

Changes in XRD pattern and Raman spectrum with HTT for the B-doped coke (a) [505] and mesophase-pitch-based carbon fibers (b) [506].

as an acceptor [504]. The transition from a two- to a one-carrier system seems to occur at a very low concentration of B, i.e., about $0.3 \times 10^{-3}$ B per C.

With B doping, 00*l* X-ray diffraction lines shift to high angle side, because of the decrease in $d_{002}$, and become sharp, suggesting that the crystallinity is improved. In Fig. 2.193a, the change in XRD pattern with HTT is shown for a coke doped by B of about 2 at% (close to maximum solubility, 2.35 at%) [505]. The 008 diffraction line shows clear separation into $K\alpha_1$ and $K\alpha_2$ lines by the heat treatment above 2400°C, even though a broad and weak 008 line is detected after 2800°C treatment without doping of B. The B-doped coke shows clearly three-dimensional diffraction lines as 112 and 105, as shown in Fig. 2.193a. On Raman spectrum, in contrast, the growth of D band, i.e., the increase in structural disorder (refer to Section 2.5.1), is observed, as shown for mesophase-pitch-based carbon fibers in Fig. 2.193b [506]. The same results on B doping have been reported on a highly oriented graphite sheet [504]. The substitution of foreign atoms for C gives certain deformation of the carbon hexagonal layers. In the case of B, which is the neighbor in the Periodic Table and has a little larger size than C (covalent bond radius of 0.086 nm for B and 0.077 nm for C), the deformation of layer plane is too little to detect by X-ray diffraction, but is large enough for Raman spectroscopy. In other words, the presence of B in hexagonal carbon layer planes cannot be detected by X-ray diffraction averaged over layer planes, their flatness and stacking order being kept. By Raman spectroscopy which can detect lattice vibration of each atom, however, foreign atom B gives disorder.

By substituting B, the lattice parameter $c_0$ decreases, i.e., decrease in $d_{002}$, and increase in $a_0$. These changes are explained as follows [507]. The length for the C–B bond is supposed to be a little longer than the C–C bond length (0.148 nm and 0.1421 nm, respectively) and the substitution of one B forms three C–B bonds. Therefore, average atomic distance in the layer plane increases, in other words, lattice parameter $a_0$ increases. On the other hand, the interaction between $\pi$-electron clouds becomes slightly weaker, because of the decrease in the concentration of $\pi$-electrons by B substitution, which may increase $d_{002}$, but it might be counterbalanced by the decrease in $d_{002}$ due to Poisson contraction, which is induced by the expansion of $a_0$. The dependences of average lattice parameters $c_0$ and $a_0$ on the concentration of substituted B $x_B$ have been derived as follows [507];

$$c_0 = 0.6708 - 0.0017x_B \text{and } a_0 = 0.2461 + 0.0003x_B. \tag{2.11}$$

B-doping into carbon materials changes their electronic properties. The effect of boron addition into anode carbons for lithium-ion rechargeable batteries has been discussed in different papers [508]. In many papers, however, the state of Bs in carbon materials, i.e., how much is substituted and how much being formed $B_4C$, is not carefully analyzed. The effects of boron addition on the performance of carbon anode are discussed in Section 3.9.1 and the carbon composites with $B_4C$ are discussed in Section 3.7.4c.

### c. Substitution of N

The substitution of N for C is very difficult to prove, one of the reasons being a very small amount of substitution, less than 1 at%. Sudden release of nitrogen gas from carbon films prepared from polyimides at a high temperature above 2200°C has been observed [509], which destroys the highly oriented nanotexture of the resultant carbon films [108]. The nitrogen released is supposed to have been substituted in carbon layers.

The chemical state of N in the carbon materials has been analyzed by X-ray photoelectron spectroscopy (XPS) and predicted theoretically by using semi-empirical molecular orbital calculations. In Fig. 2.194, the change in N1s spectrum of XPS with HTT is shown for the carbon film prepared from the polyimide Kapton [268]. The peak at 400.5 eV, which is attributed to N in imide ring, gradually changes to the peak at 398–399 eV by heat treatment up to 1000°C, which seems to be secondary N and is found in pyridine molecule. Above 1000°C the peak changed rapidly to that at 401 eV, which was attributed to tertiary N as seen in pyrrole molecule, and stays up to 2200°C treatment. After 2300°C, no N is detected. The N1s peak at 410 eV corresponds to the chemical state to have the bonds with three carbon atoms, which seems to substitute for C in hexagonal layer. Therefore, the structure of N-substituted layer plane has theoretically been simulated. The results of a simulation are shown in Fig. 2.195 for single and triple carbon layers. Since lone-paired electrons are located in anti-bonding $\pi$ orbital $\pi^*$ in C–N bond, N tend to come out from the plane [510], as shown in Fig. 2.195a.

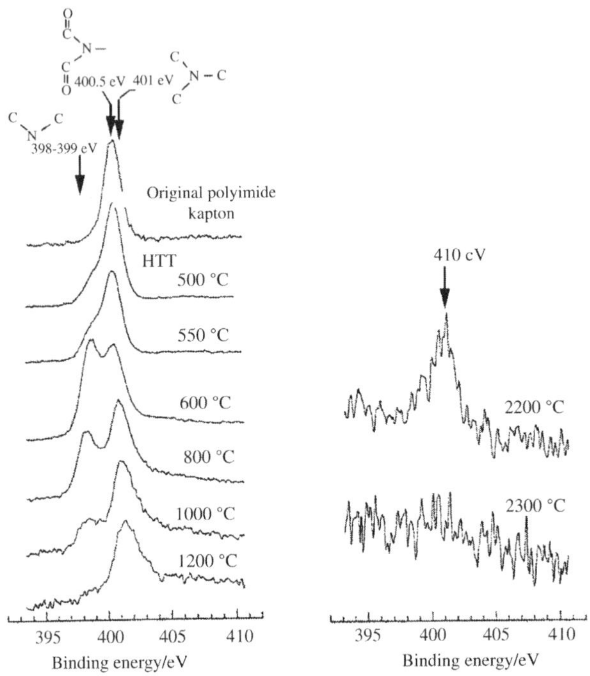


**FIGURE 2.194**

Change in N1s XPS spectrum with HTT [268].

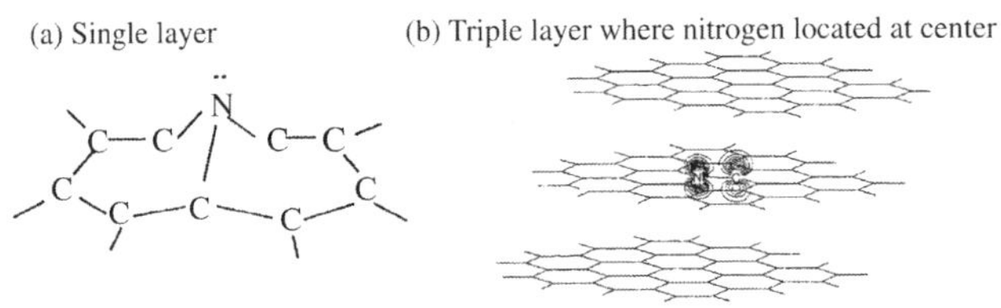


**FIGURE 2.195**

Simulation of hexagonal carbon layers substituted by N [510].

If this N is sandwiched between two carbon layers, being predicted to be the most stable position for N, the deformation of layer is not detected, but the excess electron associated with nitrogen is mostly localized near N, as shown in Fig. 2.195b. If the interlayer spacing makes changeable on the course of simulation, localized electron concentration near N tends to increase, suggesting that the structural strain due to substitution of N is pretty strong.

Local atomic and electronic structures of N-doped graphite, which is prepared by nitrogen ion bombardment followed by thermal annealing, are studied by

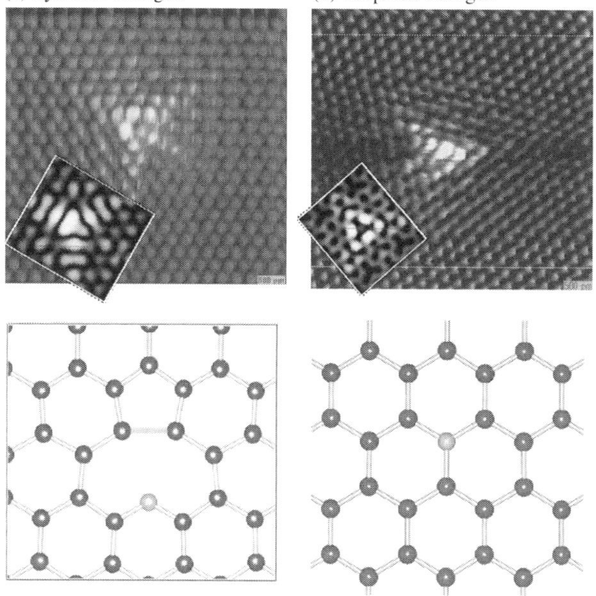


**FIGURE 2.196**

AFM images and geometry calculated by DFT of (a) pyridinic and (b) graphitic nitrogen [511].

atomic force microscopy (AFM) [511]. Two types of N species, pyridinic and graphitic N, are identified. AFM image and geometry calculated by DFT for two Ns are shown in Fig. 2.196.

In all carbon films prepared from the molecules containing nitrogen, such as poly(phenylene-1,3,4-oxadiazole) (POD) and various polyimides, N remains as the same chemical state [510]. The carbonization and graphitization behavior of the films of these polymers are discussed in Section 3.2.4.

A relatively large amount of N-doping is performed by a chemical vapor deposition from acetonitrile in the presence of nickel and cobalt metal powder as catalyst [512]. At 900–1000°C, fine powder with a composition of about $CN_{0.07}$ ($C_{14-16}N$) is obtained. In Fig. 2.197, XRD pattern and Raman spectrum are compared between the carbons obtained with and without nickel catalyst. Improvement in crystallinity of N-doped carbon by using nickel catalyst is observed from both XRD and Raman spectroscopy.

### 2.8.4 Doping

#### a. Doping of boron and nitrogen

Doping of B and N at the same time seems to be interesting from two points of view. At first the compound BN has the same structures as carbon, either layer

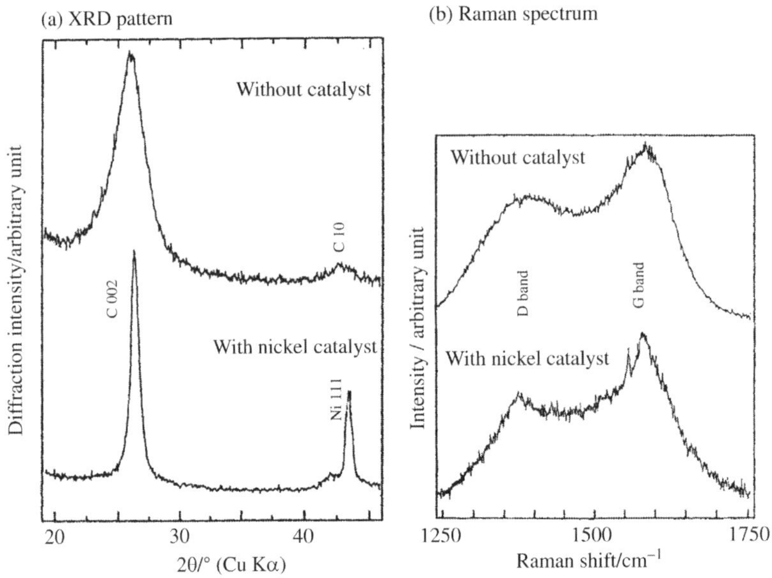


**FIGURE 2.197**

Effect of nickel catalyst on the crystallinity of N-doped carbon [512].

structure as graphite or three-dimensional zincblende and wurtzite type structures as diamond (refer to Section 2.1.3). Secondly, substituted B is expected to act as an electron acceptor and N as an electron donor, although it is not the case when only nitrogen is substituted, as discussed in the previous section. In order to synthesize the carbon materials doped by B and N, various methods have been employed; CVD, thermal decomposition, etc. [513,514]. As explained before, the substitutional doping of either B or N is possible only in a very limited amount. Consequently, a large amount of doping of these atoms into carbon materials seems to occur under non-equilibrium state.

In Fig. 2.198, the range of composition is shown to get the compound consisting of single crystalline phase under CVD method from acetylene, $BCl_3$ and $NH_3$ gases [515]. The compound is confirmed to contain B, C and N, and is stable up to 1700°C, its crystallinity being improved with increasing HTT. Above 2000°C, it releases $N_2$ gas to give a mixture of graphite and $B_4C$. Two types of structure are observed, which are supposed to consist of a basic structural unit of $C_5B_2N$ with various structural defects, as shown in Fig. 2.199a and b.

From the reaction of poly(acrylonitrile) solid with $BCl_3$ gas at a low temperature, the compound with the chemical composition of $BC_3N$ has been synthesized [515,516]. Since its XRD pattern is very similar to low-temperature-treated carbons, consisting of broad 002 and 10 diffraction peaks, the structure is supposed to be composed of hexagonal layers, as shown in Fig. 2.199c.

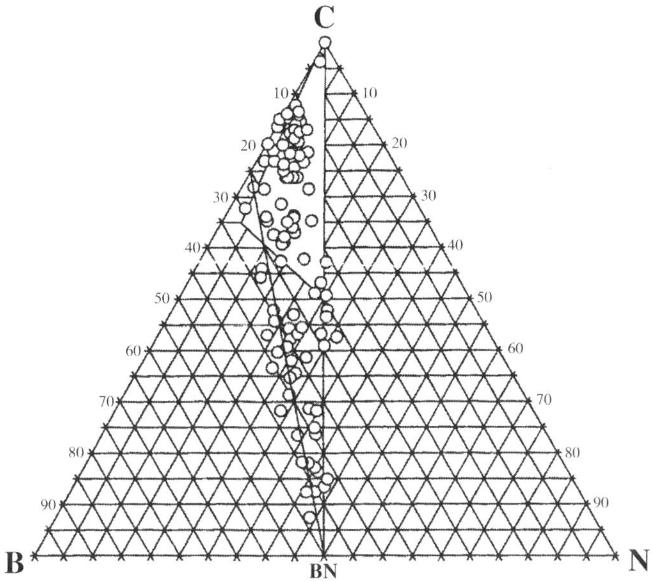


**FIGURE 2.198**

Formation range in composition of the compound in B-C-N system by CVD method.

*(Courtesy of Dr. A. Marchand.)*

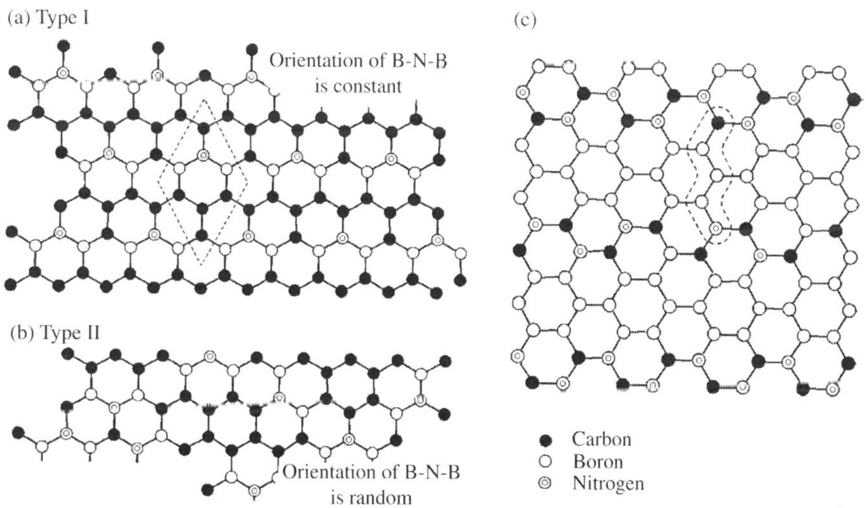


**FIGURE 2.199**

Structure models of carbons co-doped by B and N [515,516].

By substituting stylene for a part of polyacrylonitrile, the compound with a composition $BC_7N$ is also synthesized [516]. When aromatic polyimides, which contain N in their molecules, together with organic boron compounds or polyimides containing Bs in their side chains are heated at a high temperature, the B—N bonds are preferentially formed [517].

### b. Doping of metallic atoms into fullerenes

Doping of metal atoms and ions into fullerenes has been carried out by two ways [518], doping into the interstices of close-packed fullerene spheres and that into the inside of fullerene cages, as illustrated in Fig. 2.173.

All alkali metals and different combinations of alkali metals are able to be doped into the interstices, tetrahedral and octahedral sites, of cubic closest packing of $C_{60}$, as shown in Fig. 2.200a. This phenomenon has frequently been called 'intercalation', though it does not agree with the customarily used terminology of intercalation. Doping of alkali metal ions gives electrical superconductivity to fullerene crystals [520]. The critical temperature for superconductivity $T_c$ is closely related to the lattice constant $a_0$ of the $C_{60}$ crystals doped by alkali metal ions [519], as shown in Fig. 2.200b. The dependence of $T_c$ on $a_0$ shows a tendency to

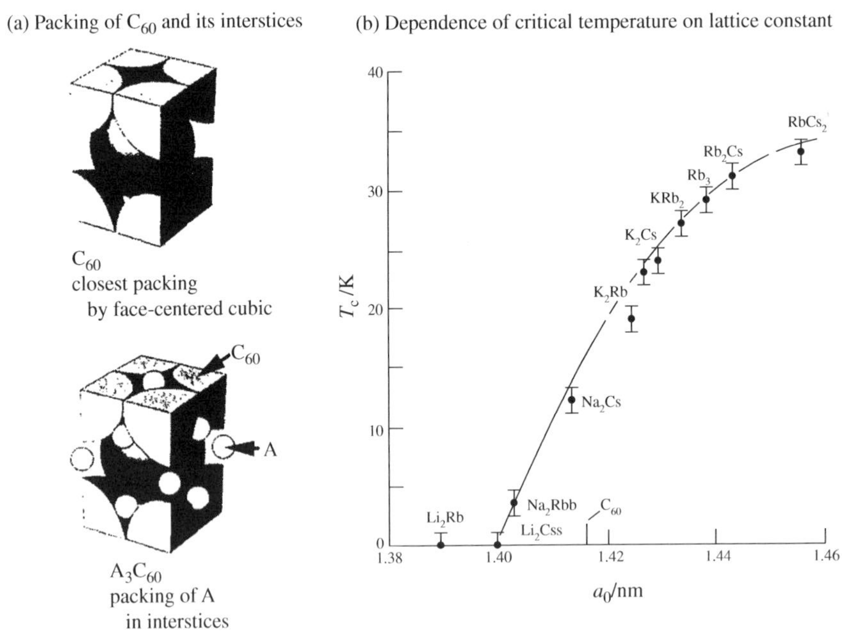


**FIGURE 2.200**

Doping of alkali metals into the interstices of cubic closest packing of $C_{60}$ and the dependence of $T_c$ on $a_0$ [519].

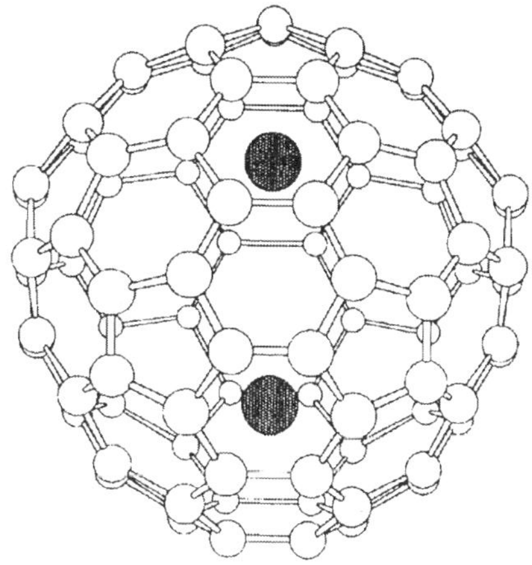

FIGURE 2.201

Endohedral metallofullerene $Sc_2@C_{84}$ [522].

be saturated. Doping of alkali metal ions is possible to be the composition of $A_6C_{60}$, but these compounds do not show superconducting phenomena [521].

Some metals are also known to be encapsulated into fullerene cages, which are called endohedral metallofullerenes. They are expressed, for example, by $Sc_3@C_{84}$, two atoms of Sc being encaged (encapsulated) in $C_{84}$ fullerene, the structure of which is shown in Fig. 2.201 [522].

Different fullerenes encaged different metals, mostly rare earth metals, have been synthesized through the vaporization of carbon atoms with metals by laser ablation, arc discharge, direct heating of the mixtures of graphite and metal oxides. The endohedral metallofullerenes synthesized and isolated so far are listed in Table 2.22.

## 2.8.5 Dispersion of fine metal particles

Dispersion of fine metal particles, especially transition metals, has attracted attention for the applications as catalysts [523]. It has been done by different means, such as deposition of metal salts from their solution and following heat treatment, implantation of metals, etc.

Carbon films containing fine particles of transition metals have successfully been prepared through the carbonization of polyimide films, in which either some complexes or salts of transition metals are dispersed [524–528]. Poly(amic acid) solutions, in which an acetylacetonate ion(III) complex are mixed with the content of 0.6 at% Fe, are cast into thin films, followed by imidization and carbonization at different temperatures [524,525]. Most iron particles are metallic,

**Table 2.22** Endohedral metallofullerenes synthesized and isolated [517]

| Single Metal Atom | Two Metal Atoms | Three Metal Atoms |
|---|---|---|
| $La@C_{82}$ | $La_2@C_{80}$ | $Sc_3@C_{82}$ |
| $Y@C_{82}$ | $Y_2@C_{82}$ | $Ti_3@C_{80}$ |
| $Sc@C_{82}$ | $Y_2@C_{84}$ | |
| $Sc@C_{84}$ | $Y_2@C_{90}$ | |
| $Ce@C_{82}$ | $Sc_2@C_{74}$ | |
| $Gd@C_{82}$ | $Sc_2@C_{76}$ | |
| $Pr@C_{82}$ | $Sc_2@C_{82}$ | |
| $Ca@C_{72}$ | $Sc_2@C_{84}$ | |
| $Ca@C_{82}$ | $Gd_2@C_{90}$ | |
| $Ca@C_{84}$ | $Er_2@C_{82}$ | |
| | $Ti_2@C_{80}$ | |

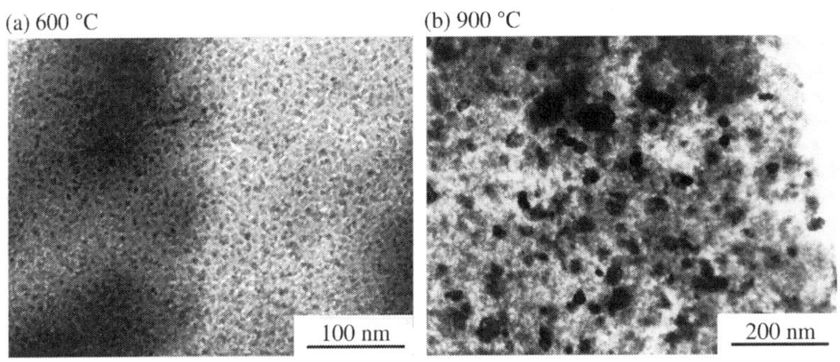


**FIGURE 2.202**

TEM images for Fe-dispersed carbon films heat-treated up to different temperatures.

*(Courtesy of Prof. Y. Kaburagi of Tokyo City Univ.)*

i.e., α-Fe, but small amounts of γ-Fe, $Fe_3O_4$ and $Fe_3C$ are detected. TEM images of a carbon film are shown in Fig. 2.202. Fine particles of Fe are dispersed homogeneously in the films. The size of these particles increases from about 4 nm after the heat-treatment at 600°C to about 13 nm after 700°C and to about 25 nm after 900°C.

In Table 2.23, X-ray parameters on the carbon films prepared from Fe-dispersed polyimide films are summarized. Carbonization of polyimide films seems to be accelerated, because 002 diffraction line of carbon is detected distinctly on that containing Fe, but not detected on that without Fe up to 800°C. The size of Fe particles increases with increasing temperature, being consistent with TEM observation in Fig. 2.202.

The advantage of this process is to be able to prepare the film, in which metal particles are dispersed homogeneously in small size. The detailed analysis of magnetic properties of these Fe-dispersed carbon films are performed [526,527]. For the carbon films heat-treated at different temperatures, mass magnetization is plotted as a function of magnetic field applied parallel to carbon film at different temperatures of measurement in Fig. 2.202a and b [527]. Magnitudes of the magnetization of Fe-dispersed carbon films are independent of the direction of magnetic field relative to the film surface, indicating random orientation of the crystallographic axes of iron particles [528]. The carbon films heat-treated up to 800°C are superparamagnetic, while they become ferromagnetic above 900°C. In Fig. 2.203c, hysteresis loops due to ferromagnetism are shown on the Fe-dispersed carbon films prepared at 1000°C. However, all films have two

**Table 2.23** X-ray Parameters of Fe-Dispersed Carbon Films [526]

| **Heat Treatment Temperature (°C)** | **Fe-dispersed Carbon Films** | | | **Carbon Films without Fe** | |
|---|---|---|---|---|---|
| | **$d_{002}$ of C** | **Lc(002) of C** | **L(110) of α-Fe** | **$d_{002}$ of C** | **Lc(002) of C** |
| 700 | 0.3444 | 3.6 | – | – | – |
| 750 | 0.3441 | 4.0 | – | – | – |
| 800 | 0.3432 | 4.5 | 16 | – | – |
| 900 | 0.3436 | 4.1 | 18 | 0.3505 | 1.2 |
| 1000 | 0.3429 | 4.5 | 23 | 0.3516 | 1.2 |
| 1200 | 0.3432 | 6.0 | 25 | 0.3525 | 1.2 |

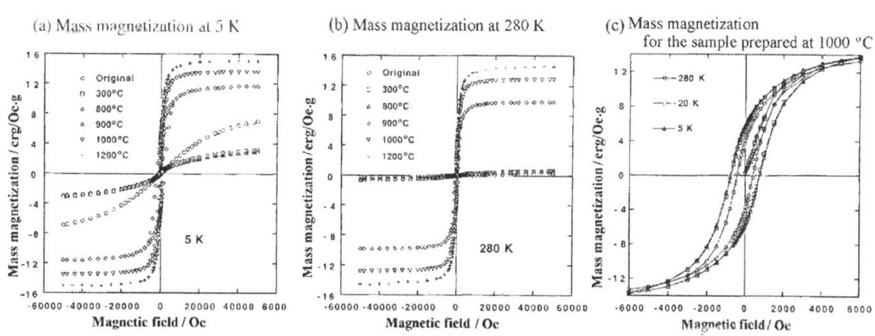


**FIGURE 2.203**

Mass magnetization against strength of magnetic field applied parallel to the surface of carbon film at different temperatures.

*(Courtesy of Prof. Y. Kaburagi of Tokyo City Univ.)*

magnetic components, superparamagnetic and ferromagnetic, with increasing HTT the latter becomes predominant.

Dispersion of transition metal particles in a graphite matrix is also tried through the reduction of graphite intercalation compounds with metal chlorides such as $FeCl_3$ [529–531]. However, a difficulty to get homogeneity in particle size of metals has to be pointed out. Dispersion of fine particles of Fe metal on graphite surface is successfully carried out by the heat treatment of exfoliated graphite sorbed tris(acetylacetonate)Fe(III) at 5f00–1200°C [532].

Metal particles of Ce and Zr are dispersed into carbon aerogels through sol-gel polymerization of the mixture of resorcinol, formaldehyde Ce-nitrate and Zr-nitrate [533,534]. After activation in $CO_2$, highly microporous carbon materials are obtained, as summarized in Table 2.24. Dispersion of metal particles attributes to the increase in microporosity.

Dispersion of transition metals into carbon aerogels has also been carried out by adding acetates of Fe, Co and Ni prepared from resorcinol with formaldehyde [535]. Cu-dispersed organic aerogels are prepared by an ion-exchange method and then they are successfully converted to Cu-dispersed carbon aerogels, which contain 9–10 mass% Cu with particle sizes from 10–50 nm and have a high pore volume of 2.2 mL/g in the range of 2–20 nm [536,537].

Fine metallic Pd particles of less than 10 nm in size are successfully loaded on the surface of activated carbons from acidic aqueous solutions of $PdCl_2$ at room temperature [538]. The amount of Pd loaded is possible to be changed by controlling the concentration of $PdCl_2$ in the solution. Even the smallest loading of 3.5 mass% Pd, however, can give 100% conversion of NO gas at a temperature above 350°C. This loading process has some advantages; a high efficiency of Pd catalysts for NO conversion and no necessity to be heat-treated at high temperatures in order to get Pd metal, in addition to the merit of no necessity of reducing gas. Maximum loading of total amount of Pd is estimated to be 11 mass% to the substrate-activated carbons. Low amount of loading of Pd, lower than 6 mass%, is recommended in order to get high dispersion of fine Pd particles on the surface of activated carbon, and to achieve high efficiency of NO conversion.

**Table 2.24** Pore Parameters for Ce, Zr-Dispersed Carbon Aerogels

| Activation Condition in $CO_2$ | Total Surface are ($m^2/g$) | External Surface Area ($m^2/g$) | Micropore Volume (mL/g) |
|---|---|---|---|
| No activation | 500 | 11 | 0.17 |
| 900°C for 1 h | 1150 | 40 | 0.40 |
| 900°C for 3 h | 1750 | 67 | 0.78 |
| 900°C for 5 h | 2240 | 600 | 1.29 |

*(Courtesy of Prof. K. Kaneko of Shinshu Univ.)*

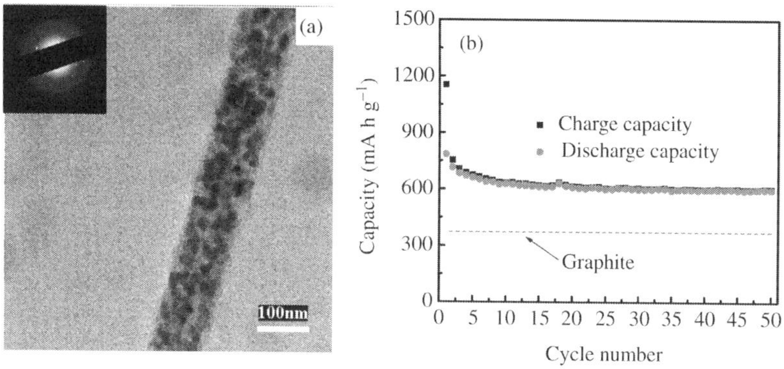


**FIGURE 2.204**

Carbon nanofibers dispersed $MnO_x$ nanoparticles: (a) TEM image, (b) cycle performance in lithium-ion rechargeable battery [541].

Various transition metal nanoparticles have been dispersed in carbon nanofibers via electrospinning, in order to improve the performance of electrochemical capacitors and lithium-ion rechargeable batteries, and also dispersing platinum to carbon nanofiber webs has been carried out for fuel cell applications [193]. By the addition of Ru acetylacetonate into polyacrylonitrile (PAN) solution, metallic Ru particles of 2–15 nm size are dispersed in carbon nanofibers. Nanofibers dispersed by 7.31 mass% Ru show the capacitance of 391 F/g in 6M KOH aqueous solution, although the nanofibers without Ru gave 140 F/g [539]. Porous carbon nanofibers dispersed $MnO_x$ and Cu nanoparticles are prepared by electrospinning of PAN nanofibers containing different amounts of either $Mn(CH_3COO)_2$, or Cu $(CH_3COO)_2$, followed by stabilization and carbonization [540,541]. As shown in Fig. 2.204, carbon nanofibers prepared from 50 mass% $Mn(CH_3COO)_2$/PAN solution, having the diameter of about 100 nm, give high capacity, higher than graphite, and steady performance for cycling with the current density of 50 mA/g. By electrospinning of lignin/ethanol solution containing 0.2 and 0.4 wt% Pt acetylacetonate, followed by stabilization and carbonization at 600–1000°C, microporous Pt-dispersed carbon nanofibers are obtained [542].

## References

[1] J.D. Bernal, Proc. Roy. Soc. A 106 (1924) 749.
[2] O. Hassel, H. Mark, Z Phys. 25 (1924) 317.
[3] H. Lipson, A.R. Stokes, Nature 149 (1942) 328.
[4] H. Lipson, A.R. Stokes, Proc. Roy. Soc. A 181 (1942) 101.
[5] J.B. Nelson, D.P. Riley, Proc. Phys. Soc. 57 (1945) 477.
[6] B.E. Warren, J. Chem. Phys. 2 (1934) 551.
[7] M. Endo, K. Oshida, K. Kobori, et al., J. Mater. Res. 10 (1995) 1461.

[8] G.E. Bacon, Acta Cryst. 3 (1950) 137.
[9] H.P. Boehm, U. Hofman, Z Anorg. Chem. 278 (1955) 58.
[10] M. Inagaki, H. Mugishima, K. Hosokawa, TANSO (1973) 76 (in Japanese).
[11] H.P. Boehm, R.W. Conghlin, Carbon 2 (1964) 1.
[12] I.L. Spain, Chem. Phys. Carbon 16 (1980) 119.
[13] S. Mrozowski, Carbon 9 (1971) 97.
[14] M. Inagaki, TANSO (122) (1985) 114 (in Japanese).
[15] M. Inagaki, New Carbons -Control Struct. Funct., Elsevier, 2000.
[16] A. Oberlin, G. Terriere, J. Microsc. 14 (1972) 1.
[17] M. Inagaki, K. Kuroda, M. Sakai, High Temp-High Press. 13 (1981) 207.
[18] J.D. Brooks, G.H. Taylor, Chem. Phys. Carbon 4 (1968) 243.
[19] D. Augie, M. Oberlin, A. Oberlin, P. Hyvernay, Carbon 18 (1980) 337.
[20] J.B. Donnet, A. Voet, Carbon Blacks, Marcel Dekker 34 (1976).
[21] Y. Hishiyama, A. Yoshida, M. Inagaki, Carbon 20 (1982) 79.
[22] R.E. Franklin, Proc. Roy. Soc. London A 209 (1951) 196.
[23] G.M. Johnson, K. Kawamura, Nature 231 (1971) 175.
[24] M. Shiraishi, Kaitei Tansozairyou Nyuumon 29 (1984) (in Japanese).
[25] J.B. Donnet, A. Voet (Eds.), Carbon Black, Phys., Chem. and Elastomer Reinforcement, 1976.
[26] J. Lahaye, G. Prado, Chem. Phys. Carbon 14 (1978) 167.
[27] S. Maeno, TANSO 222 (2006) 140.
[28] J.C Bokros, Chem. Phys. Carbon 5 (1969) 1.
[29] J.C. Bokros, L.D. LaGrange, F.J. Shoen, Chem. Phys. Carbon 9 (1972) 103.
[30] R.J. Diefendorf, J. Chim. Phys. 57 (1960) 815.
[31] S. Yajima, T. Satow, T. Hirai, J. Nucl. Mater. 17 (1965) 127.
[32] S. Kimura, E. Yasuda, N. Takase, S. Kasuya, High Temp-High Press. 13 (1981) 193.
[33] E. Yasuda, S. Kimura, Y. Tanabe, Sci New Appl Carbon Fibers, Toyohashi Univ, Tech (1984) 173.
[34] R.L. Bard, H.R. Baxman, J.P. Berting, et al., Carbon 6 (1968) 603.
[35] M. Endo, T. Koyama, Y. Hishiyama, Jpn. J. Appl. Phys. 13 (1974) 1933.
[36] M. Endo, T. Koyama, Y. Hishiyama, Jpn. J. Appl. Phys. 15 (1976) 2073.
[37] M. Endo, T. Koyama, Y. Hishiyama, Carbon 14 (1976) 133.
[38] T. Koyama, M. Endo, Ouyou Buturi 42 (1973) 690 (in Japanese).
[39] A. Oberlin, M. Endo, T. Koyama, J. Cryst. Growth 32 (1976) 335.
[40] H. Katsuki, K. Matsunaga, M. Egashira, et al., Carbon 9 (1981) 148.
[41] G.G. Tibbetts, J. Cryst. Growth 73 (1985) 431.
[42] T. Kato, K. Kusakabe, S. Morooka, J. Mat. Sci. Lett. 11 (1992) 674.
[43] M. Ishioka, T. Okada, K. Matsubara, et al., Carbon 30 (1992) 859.
[44] M. Ishioka, Y. Hishiyama, M. Inagaki, TANSO (169) (1995) 218 (in Japanese).
[45] M. Endo, M. Shikata, Ouyou Butsuri 54 (1985) 507 (in Japanese).
[46] M. Ishioka, Y. Hishyama, M. Inagaki, TANSO (169) (1995) 218 (in Japanese).
[47] H.P. Boehm, Carbon 11 (1973) 583.
[48] T. Baird, J.R. Fryer, B. Grant, Carbon 12 (1974) 591.
[49] R.T.K. Baker, P.S. Harris, Chem. Phys. Carbon 14 (1978) 83.
[50] M. Audier, M. Coulon, A. Oberlin, Carbon 18 (1980) 73.
[51] M. Audier, A. Oberlin, M. Coulon, J. Cryst. Growth 55 (1981) 549.
[52] S. Motojima, S. Ueno, T. Hattori, et al., J. Cryst. Growth 96 (1989) 383.
[53] N.M. Rodriguez, J. Mater. Res. 8 (1993) 3233.

[54] Y. Soneda, M. Makino, Carbon 38 (2000) 478.
[55] S. Motojima, I. Hasegawa, S. Kagiya, et al., Carbon 33 (1995) 1167.
[56] M. Endo, Y.A. Kim, T. Hayashi, et al., Carbon 41 (2003) 1941–1947.
[57] S. Iijima, Nature 354 (1991) 56.
[58] S. Iijima, T. Ichihashi, Y. Ando, Nature 358 (1992) 220.
[59] S. Iijima, T. Ichihashi, Nature 363 (1993) 603.
[60] D.S. Bethune, C.H. Kiang, M.S. deVries, et al., Nature 363 (1993) 605.
[61] T.W. Ebbessen, P.M. Ajayan, Nature 358 (1992) 220.
[62] R. Bacon, J. Appl. Phys. 31 (1960) 283.
[63] Z.W. Pan, S.S. Xie, B.H. Chang, et al., Nature 394 (1998) 631.
[64] S.S. Fan, M.G. Chapline, N.R. Franklin, et al., Science 283 (1999) 512.
[65] Y. Murakami, S. Chiashi, Y. Miyauchi, et al., Chem. Phys. Lett. 385 (2004) 298.
[66] K. Hata, D.N. Futaba, K. Mizuno, et al., Science 306 (2004) 1362.
[67] D.N. Futaba, K. Hata, T. Namai, et al., J. Phys. Chem. B 110 (2006) 8035.
[68] E.W. Kroto, J.R. Hearth, S.C.O. O'Brien, et al., Nature 318 (1985) 162.
[69] R.F. Curl, R.E. Smalley, Sci. Am. (1991) 32.
[70] W. Kraetschmer, K. Fostiropoulos, D.R. Huffman, Chem. Phys. Lett. 170 (1990) 167.
[71] D.M. Cox, R.D. Sherwood, P. Tindall, et al., ACS Symp. series (481) (1992) 117.
[72] W. Kraetschmer, D. Lowell, K. Lamb, et al., Nature 347 (1990) 354.
[73] R.F. Curl, R.E. Smalley, Sci. Am. 265 (1991) 54.
[74] T. Wakabayashi, K. Kikuchi, H. Shiromaru, et al., Physik D 26 (1993) S258.
[75] C. Fabre, A. Rassat, C R. Acad. Sci. Paris 308 (1988) 1223.
[76] M. Inagaki, Y.A. Kim, M. Endo, J. Mater. Chem. 21 (2011) 3280.
[77] L. Gao, W. Ren, H. Xu, et al., Nat. Commun. (2012). DOI:10.1038/ncomms1702.
[78] Y. Sanada, M. Suzuki, K. Fujimoto, Kasseitann (Activated Carbons) (1992) (in Japanese).
[79] J.W. Patrick (Ed.), Porosity in Carbons: Characterization and Applications, Edward Arnold, London (1995).
[80] F. Rodriguez-Reinoso, Introduction to Carbon Technol. (1997) 35.
[81] H. Marsh, D.S. Yan, T.M. O'Grady, et al., Carbon 22 (1984) 603–611.
[82] T. Mays, Carbon Mater. Adv. Technol. (1999) 95.
[83] M. Inagaki, T. Nishikawa, K. Sakuratani, et al., Carbon 42 (2004) 890.
[84] T. Noda, M. Inagaki, S. Yamada, J. Non-Cryst. Solids 1 (1969) 285.
[85] E. Fitzer, W. Schaffer, S. Yamada, Carbon 7 (1969) 643.
[86] H. Sato, S. Yamada, Symposium on Carbon, Tokyo, Extended Abstracts (1964) IX–3.
[87] K. Nishikawa, K. Fukuyama, T. Nishizawa, Jpn. J. Appl. Phys. 37 (1998) 6486.
[88] K. Fukuyama, T. Nishizawa, K. Nishikawa, Carbon 39 (2001) 1863.
[89] M. Inagaki, T. Nishikawa, K. Oshida, et al., Ads. Sci. Technol. 24 (2006) 55.
[90] J. Kakinoki, Acta Cryst. 18 (1965) 578.
[91] K. Furukawa, Nihon Kessyou Gakkaishi 6 (1964) 101 (in Japanese).
[92] G.M. Jenkins, K. Kawamura, Polym. Carbons (1976) 68.
[93] T. Noda, M. Inagaki, Bull. Chem. Soc. Jpn. 37 (1964) 1534.
[94] A. Oberlin, G. Terriere, J.L. Boulmier, TANSO 80 (1975) 29.
[95] A. Oberlin, G. Terriere, J.L. Boulmier, TANSO 83 (1975) 153.
[96] Y. Takahashi, E.F. Westrum, J. Chem. Thermodyn. 2 (1970) 847.
[97] R.R. Saxena, R.H. Bragg, Carbon 12 (1974) 210.
[98] M. Shiraishi, Tanso Zairyou Nyuumon (1984) 29 (in Japanese).

[99] O.P. Bahl, Carbon Fibers (1998) 1.
[100] T. Takahagi, I. Shimada, M. Fukuhara, et al., Symposium on Carbon, Tokyo, Extended Abstracts (1982) 321.
[101] A. Oberlin, Carbon 22 (1984) 521.
[102] M. Guigon, A. Oberlin, G. Desarmot, Fibre Sci. Technol. 20 (1984) 55.
[103] M. Guigon, A. Oberlin, G. Desarmot, Fibre Sci. Technol. 20 (1984) 177.
[104] H. Ogawa, K. Saito, Carbon 33 (1995) 783.
[105] K. Sato, F. Sato, N. Tomioka, TANSO (1993) (in Japanese).
[106] M. Inagaki, Y. Hishiyama, T. Takeichi, et al., Chem. Phys. Carbon 26 (1999) 246.
[107] M. Inagaki, S. Harada, T. Sato, et al., Carbon 27 (1989) 253.
[108] Y. Hishiyama, A. Yoshida, Y. Kaburagi, et al., Carbon 30 (1992) 517.
[109] M. Inagaki, M. Sato, T. Takeichi, et al., Carbon 30 (1992) 903.
[110] M. Murakami, H. Yasujima, Y. Yumoto, et al., Solid State Commun. 45 (1983) 1085.
[111] T. Onishi, I. Murase, T. Noguchi, et al., Synth. Met. 14 (1987) 207.
[112] J. Yamashita, M. Shioya, H. Hatori, et al., TANSO (2010) 196–199.
[113] M. Inagaki, T. Ishida, K. Yabe, et al., TANSO (1992) 244.
[114] M. Kakuta, TANSO (105) (1981) 60 (in Japanese).
[115] S. Higuchi, R. Otsuka, M. Shiraishi, J. Mat. Sci. 19 (1984) 270.
[116] Y. Yamada, K. Imamura, TANSO (103) (1980) 158 (in Japanese).
[117] H. Honda, TANSO (113) (1983) 66 (in Japanese).
[118] A. Oberlin, S. Bonnamy, P.G. Rouxeht, Chem. Phys. Carbon 26 (1998) 1.
[119] D. Augie, M. Oberlin, A. Oberlin, et al., Carbon 18 (1980) 337.
[120] Y. Yamada, K. Imamura, H. Kakiyama, H. Honda, S. Oi, K. Fukuda, Carbon 12 (1974) 307.
[121] H. Honda, Y. Yamada, S. Oi, et al., TANSO (1973) 3 (in Japanese).
[122] Y. Yamada, H. Honda, S. Oi, TANSO (73) (1973) 51 (in Japanese).
[123] C.A. Kovac, I.C. Lewis, 13th Conference on Carbon, Extended Abstrats (1977) 199.
[124a] K. Imamura, M. Nakamizo, H. Honda, 4th Annual Meeting of Carbon Society of Japan, (1977) 4 (in Japanese).
[124b] J.L. White, R.J. Price, Carbon 12 (1974) 321.
[125] Y. Sanada, Nennryo Kyokaishi 57 (1978) 117 (in Japanese).
[126] J.E. Zimmer, Symposium on Carbon, Toyohashi, Extended Abstracts (1982) 131.
[127] J.E. Zimmer, J.L. White, Adv. Liq. Cryst. 5 (1982) 157.
[128] I. Mochida, Y. Kourai, TANSO (116) (1984) 35.
[129] L.S. Singer, I.C. Lewis, 11th Conference on Carbon, Extended Abstracts, (1973) 27.
[130] Y. Sanada, T. Furuta, H. Kimura, et al., Carbon 10 (1972) 644.
[131] Y. Yamada, K. Imamura, H. Honda, et al., TANSO (75) (1973) 173 (in Japanese).
[132] Y. Hishiyama, Y. Kaburagi, M. Inagaki, et al., Carbon 13 (1975) 540.
[133] I. Ogawa, M. Sakai, M. Inagaki, J. Mater. Sci. 20 (1985) 17.
[134] I. Ogawa, M. Sakai, K. Kobayashi, et al., J. Mater. Sci. 20 (1985) 414.
[135] I. Ogawa, M. Sakai, M. Inagaki, Yougyo Kyoukaishi 91 (1983) 456 (in Japanese).
[136] T. Yokono, T. Ohara, S. Iyama, et al., Nennryou Kyokashi 63 (1984) 239 (in Japanese).
[137] Y. Yamada, H. Honda, M. Tanaka, et al., Sekiyu Gakkaishi 18 (1975) 765 (in Japanese).
[138] M. Sakai, M. Yoshihara, M. Inagaki, Carbon 19 (1981) 83.
[139] T. Fujimoto, K. Shiraishi, M. Hasegawa, International Conference on Chemistry and Technology on Molten Salts, Kyoto (1983).

[140] T. Kyotani, Carbon 38 (2000) 269.
[141] B. Sakintunas, Y. Yueruem, Ind. Eng. Chem. Res. 44 (2005) 2893.
[142] A. Stein, Z. Wang, M.A. Fierke, Adv. Mater. 21 (2009) 265.
[143] M. Inagaki, H. Orikasa, T. Morishita, RSC Adv. 1 (2011) 1620.
[144] H. Nishihara, T. Kyotani, Adv. Mater 24 (2012) 4473.
[145] T. Kyotani, N. Sonobe, A. Tomita, Nature 331 (1988) 331.
[146] N. Sonobe, T. Kyotani, A. Tomita, Carbon 26 (1988) 573.
[147] N. Sonobe, T. Kyotani, Y. Hishiyama, et al., J. Phys. Chem. 92 (1988) 7029.
[148] T. Kyotani, L. Tsai, A. Tomita, Chem. Mater. 7 (1995) 1427.
[149] T. Kyotani, L. Tsai, A. Tomita, Chem. Mater. 8 (1996) 2109.
[150] T. Kyotani, L. Tsai, A. Tomita, J. Chem. Soc. Chem. Commun. (1997) 701.
[151] T. Kyotani, B.K. Pradhan, A. Tomita, Bull. Chem. Soc. Jpn. 72 (1999) 1957.
[152] B.K. Pradhan, T. Kyotani, A. Tomita, J. Chem. Soc. Chem. Commun. (1999) 1317.
[153] B.K. Pradhan, T. Toba, T. Kyotani, et al., Chem. Mater. 10 (1998) 2510.
[154] Y. Hattori, Y. Watanabe, S. Kawasaki, et al., Carbon 37 (1999) 1033.
[155] T. Kyotani, S. Nakazaki, W.H. Xu, et al., Carbon 39 (2001) 771.
[156] H. Konno, S. Sato, H. Habazaki, et al., Carbon 42 (2004) 2756.
[157] T. Kyotani, T. Nagai, S. Inoue, et al., Chem. Mater. 9 (1997) 609.
[158] Z. Ma, T. Kyotani, A. Tomita, Carbon 40 (2002) 2367.
[159] T. Kyotani, A. Tomita, J. Jpn. Petrol. Inst. 45 (2002) 261.
[160] R. Ryoo, S.H. Joo, S. Jun, J. Phys. Chem. B 103 (1999) 7743.
[161] J. Lee, S. Yoon, S.M. Oh, et al., Adv. Mater. 12 (2000) 359.
[162] S.B. Yoon, J.Y. Kim, J.-S. Yu, Chem. Commun. (2001) 559.
[163] R. Ryoo, S.H. Joo, M. Kruk, et al., Adv. Mater. 13 (2001) 677.
[164] J. Lee, S. Han, T. Hyeon, J. Mater. Chem. 14 (2004) 478.
[165] C. Liang, K. Hong, G.A. Guiochon, et al., Angew. Chem. Int. Ed. 43 (2004) 5785.
[166] S. Tanaka, N. Nishiyama, Y. Egashira, et al., Chem. Commun. (2005) 2125.
[167] H. Kosonen, S. Valkama, A. Nykaenen, et al., Adv. Mater. 18 (2006) 201.
[168] B. Liu, H. Shioyama, T. Akita, et al., J. Am. Chem. Soc. 130 (2008) 5390.
[169] B. Liu, H. Shioyama, H. Jiang, et al., Carbon 48 (2010) 456.
[170] J. Hu, H. Wang, Q. Gao, et al., Carbon 48 (2010) 3599.
[171] H.-L. Jiang, B. Liu, Y.-Q. Lan, et al., J. Am. Chem. Soc. 133 (2011) 11854.
[172] T. Morishita, T. Tsumura, M. Toyoda, et al., Carbon 48 (2010) 2690.
[173] T. Morishita, L. Wang, T. Tsumura, et al., TANSO (242) (2010) 60 (in Japanese).
[174] J. Wang, Electrochim. Acta 26 (1981) 1721.
[175] A.G. Chakovskoi, C.E. Hunt, G. Forceberg, et al., J. Vac. Sci. Tech. B 21 (2003) 571.
[176] M. Inagaki, T. Morishita, A. Kuno, et al., Carbon 42 (2004) 497.
[177] J. Ozaki, N. Endo, W. Ohizumi, et al., Carbon 35 (1997) 1031.
[178] T. Takeichi, M. Zuo, A. Ito, High Perform Polym. 11 (1999) 1.
[179] T. Takeichi, Y. Yamazaki, A. Ito, M. Zuo, J. Photopolym. Sci. Technol. 12 (1999) 203.
[180] T. Takeichi, Y. Yamazaki, M. Zuo, et al., Carbon 39 (2001) 257.
[181] T. Takeichi, Y. Yamazaki, T. Fukui, et al., TANSO 2000 (2000) 388.
[182] S. Koyama, H. Haniu, S. Tanaka, et al., Mol. Cryst. Liq. Cryst. 388 (2002) 587.
[183] D. Hulicova, F. Sato, K. Okabe, M. Koishi, A. Oya, Carbon 39 (2001) 1438.
[184] D. Hulicova, K. Hosoi, S. Kuroda, H. Abe, A. Oya, Adv. Mater. 14 (2002) 452.
[185] A. Oya, N. Kasahara, Carbon 38 (2000) 1141.

[186] N. Patel, K. Okabe, A. Oya, Carbon 40 (2002) 315.
[187] D. Li, Y.N. Xia, Adv. Mater. 16 (2004) 1151.
[188] T. Subbiah, G.S. Bhat, R.W. Tock, et al., J. Appl. Polym. Sci. 96 (2005) 557.
[189] A. Greiner, J.H. Wendorff, Angew. Chem. Int. Ed. 46 (2007) 5670.
[190] D.H. Reneker, A.L. Yarin, E. Zussman, et al., Adv. Appl. Mechan. 41 (2007) 43.
[191] D.H. Reneker, A.L. Yarin, Polymer (Guildf). 49 (2008) 2387.
[192] Z. Dong, S.J. Kennedy, Y. Wu, J. Power Sources 196 (2011) 4886.
[193] M. Inagaki, Y. Yang, F. Kang, Adv. Mater. 24 (2012) 2547.
[194] M. Wang, Z.-H. Huang, Y. Bai, et al., Curr. Org. Chem. 17 (2013) 1434.
[195] E. Zussman, X. Chen, W. Ding, et al., Carbon 43 (2005) 2175.
[196] Z. Zhou, C. Lai, L. Zhang, et al., Polymer (Guildf). 50 (2009) 2999.
[197] E. Zussman, A.L. Yarin, A.V. Bazilevsky, et al., Adv. Mater. 18 (2006) 348.
[198] C. Kim, Y. Jeong, B. Ngoc, et al., Small. 3 (2007) 91.
[199] K.S. Yang, D.D. Edie, D.Y. Lim, et al., Carbon 41 (2003) 2039.
[200] C. Kim, Y.J. Cho, W.Y. Yun, et al., Solid State Commun. 142 (2007) 20.
[201] N.T. Xuyen, E.J. Ra, H.-Z. Geng, et al., J. Phys. Chem. B 111 (2007) 11350.
[202] K. Suzuki, H. Matsumoto, M. Minagawa, et al., Polym. J. 39 (2007) 1128.
[203] M.-X. Wang, Z.-H. Huang, F. Kang, et al., Mater. Lett. 65 (2011) 1875.
[204] L. Wang, Z.-H. Huang, M. Yue, et al., Chem. Eng. J. 218 (2013) 232.
[205] M. Inagaki, M. Ishihara, S. Naka, High Temp-High Press. 8 (1976) 270.
[206] M. Inagaki, K. Kuroda, M. Sakai, E. Yasuda, S. Kimura, Carbon 22 (1984) 335.
[207] H. Marsh, F. Dachille, J. Melvin, P.L. Walker, Carbon 9 (1971) 159.
[208] S. Hirano, F. Dachille, P.L. Walker, High Temp. High Press. 5 (1973) 207.
[209] S. Hirano, M. Ozawa, S. Naka, J. Mater. Sci. 16 (1981) 1989.
[210] M. Inagaki, K. Kuroda, M. Sakai, High Temp. High Press. 13 (1981) 207.
[211] M. Inagaki, K. Kuroda, M. Sakai, Carbon 21 (1983) 231.
[212] M. Inagaki, K. Kuroda, N. Inoue, M. Sakai, Carbon 22 (1984) 617.
[213] M. Inagaki, M. Washiyama, M. Sakai, Carbon 26 (1988) 169.
[214] M. Washiyama, M. Sakai, M. Inagaki, Carbon 26 (1988) 303.
[215] N. Urata, M. Sakai, M. Inagaki, Annu. Meeting Carbon Soc. Jpn. (1983) 26.
[216] M. Inagaki, M. Sakai, TANSO (134) (1988) 295 (in Japanese).
[217] K.C. Park, H. Tomiyasu, S. Morimoto, et al., Carbon 46 (2008) 1804.
[218] Q. Wang, H. Li, L. Chen, et al., Carbon 39 (2001) 2211.
[219] X. Sun, Y. Li, Angew. Chem. Int. Ed. 43 (2004) 597.
[220] M. Zheng, J. Cao, X. Chang, et al., Mater. Lett. 60 (2006) 2991.
[221] M.-M. Titirici, M. Antonietti, N. Baccile, Green. Chem. 10 (2008) 1204.
[222] Y. Shin, L.-Q. Wang, I.-T. Bae, et al., J. Phys. Chem. C 112 (2008) 14236.
[223] M. Sevilla, A.B. Fuertes, Chem. Eur. J. 15 (2009) 4195.
[224] B. Hu, K. Wang, L. Wu, et al., Adv. Mater. 22 (2010) 813.
[225] E. Yasuda, H. Kajiura, Y. Tanabe, TANSO (1995) 286 (in Japanese).
[226] H. Kajiura, Y. Tanabe, M. Kodama, et al., TANSO (174) (1996) 201 (in Japanese).
[227] H. Kajiura, Y. Tanabe, E. Yasuda, Carbon 35 (1997) 169.
[228] N. Miyajima, T. Akutsu, T. Ikoma, et al., Carbon 38 (2000) 1831.
[229] N. Miyajima, T. Akutsu, O. Ito, et al., Carbon 39 (2001) 647.
[230] N. Miyajima, S. Dohi, T. Akatsu, et al., Carbon 40 (2002) 1533.
[231] H. Kajiura, Y. Yanabe, E. Yasuda, et al., J. Mater. Res. 13 (1998) 302.
[232] J.P. Amistead, T.M. Keller, S.B. Sastri, Carbon 32 (1994) 345.
[233] M. Itoh, K. Inoue, K. Iwata, et al., Macromolecules 30 (1997) 694.

[234] M. Kijima, H. Tanimoto, H. Shirakawa, Synth. Met. 119 (2001) 353.
[235] M. Kijima, H. Tanimoto, H. Shirakawa, et al., Carbon 39 (2001) 297.
[236] H. Hatori, Y. Yamada, M. Shiraishi, et al., Carbon 30 (1992) 305.
[237] Y. Hishiyama, A. Yoshida, Y. Kaburagi, et al., Carbon 30 (1992) 333.
[238] J. Jansta, F.P. Dousek, V. Patzelova, Carbon 13 (1975) 377.
[239] F.P. Dousek, J. Jansta, J. Baldrian, Carbon 18 (1980) 13.
[240] L. Kavan, Chem. Phys. Carbon 23 (1991) 69.
[241] T.T. Liang, Y. Yamada, N. Yoshizawa, et al., Chem. Mater. 13 (2001) 2933.
[242] S. Shiraishi, D. Hiruma, Y. Onuma, et al., TANSO (2000) 395 (in Japanese).
[243] M. Shiraishi, H. Kurihara, H. Tsubota, et al., Electrochem. Solid-State Lett. 4 (2001) A5.
[244] O. Tanaike, H. Hatori, Y. Yamada, et al., Carbon 41 (2003) 1759.
[245] Y. Yamada, O. Tanaike, S. Shiraishi, TANSO (215) (2004) 285 (in Japanese).
[246] O. Tanaike, H. Hatori, Y. Yamada, Conference on Carbon, Oviedo, Spain., Extended Abstracts (2003).
[247] M. Inagaki, New Carbons, Elsevier (2000) 14.
[248] B.E. Warren, Phys. Rev. 59 (1941) 693.
[249] N. Iwashita, M. Inagaki, Y. Hishiyama, Carbon 35 (1997) 1073.
[250] N. Iwashita, R.A. Park, H. Fujimoto, et al., Carbon 42 (2004) 701.
[251] S. Ergun, Carbon 12 (1976) 139.
[252] Y. Hishiyama, Y. Kaburagi, M. Inagaki, Chem. Phys. Carbon 23 (1991) 1.
[253] A. Oberlin, Chem. Phys. Carbon 22 (1989) 1.
[254] A. Yoshida, Y. Hishiyama, TANSO (130) (1987) 110 (in Japanese).
[255] A. Yoshida, Y. Hishiyama, TANSO (137) (1989) 93 (in Japanese).
[256] M. Nakamizo, TANSO (90) (1977) 105 (in Japanese).
[257] Y. Kawashima, G. Katagiri, Phys. Rev. B 52 (1995) 10053.
[258] A. Yoshida, Y. Kaburagi, Y. Hishiyama, TANSO (234) (2008) 227.
[259] D.B. Fischbach, Chem. Phys. Carbon 7 (1971) 1.
[260] Y. Kaburagi, R.H. Bragg, Y. Hishiyama, Phil. Mag. B 63 (1991) 417.
[261] C.A. Klein, M.G. Holland, Phys. Rev. A. 136 (1964) 576.
[262] Y. Hishiyama, S. Yasuda, A. Yoshida, et al., J. Mater. Sci. 23 (1988) 3272.
[263] C. Bourgerette, A. Oberlin, M. Inagaki, J. Mater. Res. 7 (1992) 1158.
[264] M. Inagaki, T. Takeichi, Y. Hishiyama, et al., Chem. Phys. Carbon 26 (2000) 246.
[265] Y. Kaburagi, A. Yoshida, Y. Hishiyama, et al., TANSO (1995) 19.
[266] Y. Hishiyama, A. Yoshida, Y. Kaburagi, TANSO (2012) 176.
[267] Y. Hishiyama, K. Igarashi, I. Kanaoka, et al., Carbon 35 (1997) 657.
[268] H. Konno, T. Nakahashi, M. Inagaki, Carbon 35 (1997) 669.
[269] M. Inagaki, H. Tachikawa, T. Nakahashi, et al., Carbon 36 (1998) 1021.
[270] Y. Kaburagi, Y. Hishiyama, Carbon 33 (1995) 773.
[271] M. Inagaki, Y. Murase, T. Noda, TANSO (54) (1968) 80.
[272] M. Inagaki, TANSO (71) (1972) 124.
[273] T. Noda, M. Inagaki, T. Sekiya, Carbon 3 (1965) 175.
[274] T. Noda, M. Inagaki, Nature 196 (1962) 772.
[275] M. Inagaki, Y. Murase, T. Noda, Yogyou Kyoukaishi 76 (1968) 184 (in Japanese).
[276] M. Inagaki, Y Hishiyama, New Carbon Mater. (1994) 24 (in Japanese).
[277] Y. Hishiyama, A. Ono, M. Hashimoto, Jpn. J. Appl. Phys. 10 (1971) 416.
[278] M. Inagaki, Tanso Zairyou Kougaku (1985) 95 (in Japanese).
[279] Y. Hishiyama, Y. Kaburagi, A. Yoshida, Sci. New Appl. Carbon Fibers (1984) 21.
[280] A. Hamwi, H. Alvergnat, S. Bonnamy, et al., Carbon 35 (1997) 723.

[281] H.T. Pinnick, J. Chem. Phys. 20 (1952) 756.
[282] H. Akamatsu, Bull. Chem. Soc. Jpn. 29 (1956) 574.
[283] T. Noda, M. Inagaki, N. Fujisawa, Kogyou Kagaku Zasshi 64 (1961) 1518.
[284] M. Inagaki, T. Noda, Bull, Chem. Soc. Jpn. 35 (1962) 1652.
[285] M. Inagaki, S. Hayashi, K. Kamiya, et al., High Temp-High Press. 3 (1971) 355.
[286] M. Inagaki, M. Ishihara, S. Naka, Carbon 14 (1976) 88.
[287] Y. Yamada, K. Imamura, H. Kakiyama, et al., Carbon 12 (1974) 307.
[288] H. Honda, Mol. Cryst. Liq. Cryst. 94 (1983) 97.
[289] Y. Yamada, K. Kobayashi, H. Honda, et al., TANSO (86) (1976) 101 (in Japanese).
[290] M. Inagaki, Y. Tamai, S. Naka, et al., Carbon 12 (1974) 639.
[291] M. Inagaki, Y. Tamai, S. Naka, TANSO (75) (1973) 118 (in Japanese).
[292] M. Inagaki, TANSO (129) (1987) 68 (in Japanese).
[293] M. Inagaki, A.P. Meyer, Chem. Phys. Carbon 26 (1998) 149.
[294] T. Noda, M. Inagaki, Bull. Chem. Soc. Jpn. 1964 (1709) 37.
[295] A. Yoshida, Y. Kaburagi, Y. Hishiyama, Carbon 29 (1991) 1107.
[296] S. de Fonton, A. Oberlin, M. Inagaki, J. Mater. Sci. 15 (1980) 909.
[297] T. Yamaguchi, Carbon 1 (1963) 535.
[298] D.F. Baker, R.H. Bragg, J. Non-Cryst. Solids 58 (1983) 57.
[299] S.R. Houska, B.E. Warren, J. Appl. Phys. 25 (1956) 1503.
[300] J.C. Bowman, Proceedings of the 1st and 2nd Conference on Carbon (1956) 59.
[301] T. Noda, M. Iwatsuki, M. Inagaki, TANSO (47) (1966) 14 (in Japanese).
[302] N. Iwashita, M. Inagaki, Carbon 31 (1993) 1107.
[303] R.A. Franklin, Acta Cryst. 4 (1951) 253.
[304] G.E. Bacon, Acta Cryst. 4 (1951) 558.
[305] J. Maire, J. Mering, Les Carbones 1 (1965) 129.
[306] M. Inagaki, K. Kamiya, K. Taoka, TANSO (48) (1964) 24 (in Japanese).
[307] N. Iwashita, Y. Hishiyama, M. Inagaki, Carbon 35 (1977) 1073.
[308] J.G. Lavin, D.R. Boyington, J. Lahijani, et al., Carbon 31 (1993) 1001.
[309] A. Yoshida, Y. Kaburagi, Y. Hishiyama, Carbon 44 (2006) 2333.
[310] M. Inagaki, Y. Kaburagi, Y. Hishiyama, Adv. Eng. Mater. DOI: 10.1002/adem.201300418.
[311] K. Taguchi, M. Okada, H. Morota, et al., TANSO (254) (2012) 165 (in Japanese).
[312] M. Inagaki, A. Oberlin, T. Noda, TANSO (81) (1975) 68.
[313] H. Kuroda, Bull. Chem. Soc. Jpn. 32 (1959) 728.
[314] A. Oberlin, G. Terriere, Carbon 13 (1975) 367.
[315] M. Inagaki, TANSO (1972) 124 (in Japanese).
[316] C. Schiller, J. Mering, P. Cornaullt, et al., Carbon 5 (1967) 385 & 507.
[317] A. Bouraoui, J. Mering, Carbon 1 (1964) 465.
[318] R.E. Franklin, Acta Cryst. 3 (1950) 107.
[319] R.E. Franklin, Proc. Roy. Soc. Lond. A 209 (1951) 196.
[320] S. Mrozowski, Proc. Conf. Carbon (1956) 31.
[321] International Committee for Characterization and Terminology of Carbon, Carbon 20 (1982) 445.
[322] K. Kobayashi, S. Sugawara, S. Toyoda, et al., Carbon 6 (1965) 359.
[323] H. Honda, K. Kobayashi, S. Sugawara, Carbon 6 (1968) 517.
[324] M. Inagaki, K. Kamiya, TANSO (66) (1971) 76 (in Japanese).
[325] A. Oberlin, F. Rousseaux, J. Appl. Cryst. 1 (1968) 218.
[326] A. Oberlin, F. Rousseaux, J. Appl. Cryst. 3 (1970) 105.

[327] S. Otani, A. Oya, TANSO (64) (1971) 10 (in Japanese).
[328] A. Oya, S. Otani, Carbon 17 (1979) 131.
[329] A. Oya, TANSO (102) (1980) 118 (in Japanese).
[330] E. Fitzer, B. Kegel, Carbon 6 (1968) 433.
[331] W. Weisweiler, N. Subramanian, B. Terwiesch, Carbon 9 (1971) 755.
[332] A. Oberlin, J.P. Rouchy, C. R. Acad. Soc. Paris 268 (1969) 660.
[333] M. Inagaki, K. Fujita, Y. Takeuchi, et al., Carbon 39 (2001) 921.
[334] M. Inagaki, T. Imase, H. Iwata, et al., TANSO (201) (2002) 12 (in Japanese).
[335] H. Konno, K. Fujita, H. Habazaki, et al., TANSO (203) (2002) 113.
[336] S.B. Austerman, S.M. Nyron, J.W. Wagner, Carbon 5 (1967) 548.
[337] T. Noda, Y. Sumiyoshi, N. Ito, Carbon 6 (1968) 813.
[338] T. Ishii, Denki Kagaku 35 (1967) 688 (in Japanese).
[339] L.M. Foster, G. Long, H.C. Stumpt, Am. Mineralogists 43 (1958) 285.
[340] M. Inagaki, K. Taoka, T. Noda, Carbon 9 (1971) 94.
[341] S. Otani, A. Oya, M. Nishina, TANSO (76) (1974) 2.
[342] A. Oya, M. Mochizuki, S. Otani, et al., Carbon 17 (1979) 71.
[343] M. Inagaki, A. Oberlin, TANSO (97) (1979) 66.
[344] T. Noda, M. Inagaki, Nature 196 (1963) 772.
[345] C.E. Lowell, J. Am. Ceram. Soc. 50 (1967) 142.
[346] T. Sogabe, K. Nakajima, M. Inagaki, J. Mater. Sci. 31 (1996) 6469.
[347] M. Inagaki, TANSO (129) (1987) 68.
[348] M. Inagaki, R.A. Meyer, Chem. Phys. Carbon 26 (1999) 149.
[349] T. Noda, H. Kato, Carbon 3 (1965) 289.
[350] T. Noda, K. Kamiya, M. Inagaki, Bull. Chem. Soc. Jpn. 41 (1968) 485.
[351] P.J. Wyllie, O.F. Tuttle, Nature 183 (1959) 770.
[352] P.J. Wyllie, O.F. Tuttle, Am. Mineralogists 44 (1959) 453.
[353] P.R. Buseck, B.J. Huang, Geochim. Cosmochim. Acta 1985 (2003) 49.
[354] M. Inagaki, K. Kamiya, T. Noda, Kogyo Kagaku Zasshi 71 (1968) 652 (in Japanese).
[355] K. Kamiya, M. Mizutani, T. Noda, et al., Bull. Chem. Soc. Jpn. 41 (1968) 2169.
[356] T. Noda, M. Inagaki, Bull. Chem. Soc. Jpn. 37 (1964) 1709.
[357] M. Inagaki, K. Horii, S. Naka, Carbon 13 (1975) 97.
[358] Y. Hishiyama, M. Inagaki, S. Kimura, et al., Carbon 12 (1974) 249.
[359] S. Kimura, E. Yasuda, Zairyo Kagaku 20 (1983) 36.
[360] S. Kimura, E. Yasuda, H. Tanaka, et al., Yogyo Kyokai Shi 83 (1975) 122 (in Japanese).
[361] H. Tanaka, Y. Kaburagi, S. Kimura, J. Mat. Sci. 13 (1978) 2555.
[362] K. Kamiya, M. Inagaki, Carbon 11 (1973) 429.
[363] S. Kimura, Y. Tanabe, N. Takase, et al., Nihon Kogaku Kaishi 1981 (1981) 1474.
[364] R.A. Meyer, S.R. Gyetvay, Amer. Ceram. Soc. Symp. Series 303 (1986) 380.
[365] R.L. Hales, E.M. Woodruff, Proc. 4th Carbon Conf. 1 (1959) 456.
[366] R.D. Reiswig, L.S. Levinson, T.D. Baker, Carbon 5 (1967) 603.
[367] R.A. Meyer, S.R. Gyetvay, A.B. Chase, Proc. 17th Carbon Conf. (1985) 505.
[368] R.A. Meyer, J.E. Zimmer, M.C. Almond, Aerospace Report; ATR-74(7408)-2 (1984).
[369] R.A. Meyer, Proceedings of Carbon Conference, Baden-Baden (1986).
[370] L.A. Feldman, S.R. Gyetvay, Aerospace Report; TOR-0086(6728-020)-1 (1986).
[371] R. Pleger, W. Braue, R.A. Meyer, Proc. 20th Carbon Conf. (1991) 399.
[372] L.H. Peebles Jr, R.A. Meyer, J. Jortner, SAMPE Tech. Series 20 (1988) 109.
[373] T. Noda, M. Inagaki, S. Hirano, et al., Kogyo Kagaku Zasshi 72 (1969) 643.

[374] T. Noda, M. Inagaki, S. Hirano, et al., Bull. Chem. Soc. Jpn. 1969 (1738) 42.
[375] S. Hirano, H. Saito, M. Inagaki, Bull. Chem. Soc. Jpn. 43 (1970) 2599.
[376] S. Hirano, M. Inagaki, H. Saito, Bull. Chem. Soc. Jpn. 43 (1970) 2624.
[377] S. Hirano, M. Inagaki, H. Saito, Carbon 17 (1979) 395.
[378] S. Hirano, Ph.D.Thesis, Nagoya Univ., 1970.
[379] R.M. Bustin, J.-N. Rouzaud, J.V. Ross, Carbon 33 (1995) 679.
[380] P.E. Buseck, B.-J. Huang, Geochim. Cosmochim. Acta 49 (1985) 2003.
[381] O. Beyssac, B. Goffe, C. Chopin, et al., J. Metamorphic. Geol. 20 (2002) 859.
[382] F.J. Luque, L. Ortega, J.F. Barrenechea, et al., Geology 37 (2009) 275–278.
[383] A. Oberlin, S. Bonnamy, K. Lafdi, Carbon Fibers (1998) 85.
[384] M. Inagaki, New Carbons -Control Struct. Funct. (2000) 82.
[385] M. Toyoda, Y. Kaburagi, A. Yoshida, et al., Carbon 40 (2002) 628.
[386] M. Toyoda, Y. Kaburagi, A. Yoshida, et al., Carbon 42 (2004) 2567.
[387] M. Toyoda, A. Shimizu, H. Iwata, et al., Carbon 39 (2001) 1697.
[388] Y. Soneda, M. Toyoda, K. Hashiya, et al., Carbon 41 (2003) 2680.
[389] Y. Soneda, M. Toyoda, Y. Tani, et al., J. Phys. Chem. Solids 65 (2004) 219.
[390] M. Toyoda, Y. Tani, Y. Soneda, Carbon 42 (2004) 2833.
[391] M. Toyoda, R. Kohara, T. Tsumura, et al., Mater. Sci. Eng. B 161 (2009) 202.
[392] M. Toyoda, Y. Kohara, S. Yoshinaga, et al., TANSO (238) (2009) 92 (in Japanese).
[393] K.S. Sing, D.H. Everett, R.A.W. Haul, et al., Pure. Appl. Chem. 57 (1985) 603.
[394] M. Inagaki, T. Morishita, A. Konno, et al., Carbon 42 (2004) 497.
[395] F. Rodriguez-Reinoso, A. Linares-Solano, Chem. Phys. Carbon 21 (1995) 1.
[396] Deleted in Review.
[397] J.B. Donnet, E. Papier, W. Wang, et al., Carbon 32 (1994) 183.
[398] C. Daulan, A. Derre, S. Flandrois, et al., J. Phys. France 5 (1995) 1111.
[399] K. Oshida, K. Kogiso, K. Matsubayashi, et al., J. Mater. Res. 10 (1995) 2507.
[400] M. Inagaki, T. Suwa, Carbon 39 (2001) 915.
[401] Y.P. Zheng, H.N. Wang, F.Y. Kang, et al., Carbon 42 (2004) 2603.
[402] K. Oshida, N. Ekinaga, M. Endo, et al., TANSO (173) (1996) 142 (in Japanese).
[403] V. Vignal, A.W. Morawski, H. Konno, et al., J. Mater. Res. 14 (1999) 1102.
[404] M. Inagaki, V. Vignal, H. Konno, et al., J. Mater. Res. 14 (1999) 3152.
[405] E. Bourelle, M. Inagaki, Y. Kaburagi, et al., Synth. Met. 125 (2002) 239.
[406] E. Bourelle, H. Konno, M. Inagaki, Carbon 37 (1999) 2041.
[407] M. Endo, K. Oshida, K. Takeuchi, et al., Dennki Jouhou Tsuushin Gakkai Ronbunnshi 77 (1994) 139.
[408] M. Endo, T. Furuta, F. Minoura, et al., Supramol. Sci. 5 (1998) 261.
[409] Z.H. Huang, F. Kang, W.L. Huang, et al., J. Colloid Interface Sci. (2002) 249.
[410] S.J. Gregg, K.S.W. Sing, Adsorption, Surface Area and Porosity, Academic Press, London, (1982).
[411] Y. Hanzawa, K. Kaneko, in: E. Yasuda, et al. (Eds.), Carbon Alloys, Elsevier (2003) 319.
[412] B. McEnaney, J.T. Mays, F. Rodriguez-Reinoso, Carbon 36 (1998) 1.
[413] T. Ohba, T. Suzuki, K. Kaneko, Chem. Phys. Lett. 326 (2000) 158.
[414] M.E. Merraoui, M. Aoshima, K. Kaneko, Langmuir 16 (2000) 4300.
[415] J. Jagiello, M. Thommes, Carbon 42 (2004) 1227.
[416] D. Avnir, D. Farin, P. Pfeifer, J. Chem. Phys. 79 (1983) 3566.
[417] N. Setoyama, K. Kaneko, Zairyo Kagaku 32 (1995) 213 (in Japanese).
[418] M. Inagaki, N. Saji, M. Toyoda, et al., TANSO (215) (2004) 258.

[419] M. Inagaki, R. Tashiro, T. Suwa, Res. Rep. Aichi Inst. Tech. (37) (2002) 53.
[420] M. Inagaki, R. Tashiro, Y. Washino, et al., J. Phys. Chem. Solids 65 (2004) 133.
[421] M. Inagaki, R. Tashiro, M. Toyoda, et al., J. Ceram. Soc. Jpn. 112 (2004) S1513.
[422] F. Kang, Y. Leng, T.Y. Zhang, et al., Carbon 40 (2002) 1575.
[423] M. Inagaki, H. Kobayashi, B. Tryba, TANSO (215) (2004) 249.
[424] H. Marsh, D.S. Yan, T.M. O'Grady, et al., Carbon 22 (1984) 603.
[425] Y. Nishi, N. Iwashita, Y. Sawada, et al., TANSO (201) (2002) 31 (in Japanese).
[426] M. Inagaki, M. Toyoda, F. Kang, et al., New Carbon Mater. 18 (2003) 241.
[427] T.D. Burchell, Carbon Mater. Adv. Technol. (1999).
[428] E. Bekyarova, K. Murata, M. Yudasaka, et al., J. Phys. Chem. 107 (2003) 4681.
[429] S. Tanaka, N. Nishiyama, Y. Egashira, et al., Chem. Commun. 2005 (2005) 2125.
[430] T. Morishita, Y. Soneda, T. Tsumura, et al., Carbon 44 (2006) 2360.
[431] T. Morishita, T. Tsumura, M. Toyoda, et al., Carbon 49 (2010) 2690.
[432] W.F. Zhang, Z.-H. Huang, C.J. Zhou, et al., J. Mater. Chem. 22 (2012) 7158.
[433] S. Shiraiashi, H. Kurihara, H. Tsuboi, et al., Electrochem. Solid State Lett. 4 (2001) A5.
[434] R.W. Pekala, C.T. Alviso, F.M. Kong, J. Non-Cryst. Solids 145 (1992) 90.
[435] Y. Hanzawa, K. Kaneko, Langmuir 5 (1996) 7.
[436] T. Nishikawa, M. Inagaki, Adsorption Sci. Tech. 23 (2005) 827.
[437] M. Inagaki, T. Suwa, Mol. Cryst. Liq. Cryst. 386 (2002) 197.
[438] M. Inagaki, M. Sakanishi, Adsorption Sci. Tech. 21 (2003) 587.
[439] D.D.L. Chung, J. Mater. Sci. 22 (1987) 4190.
[440] G. Furdin, Fuel 77 (1998) 475.
[441] M. Inagaki, M. Toyoda, F. Kang, Chem. Phys. Carbon 29 (2004) 1.
[442] M. Toyoda, J. Aizawa, M. Inagaki, Desalination 115 (1998) 199.
[443] M. Toyoda, M. Inagaki, Carbon 38 (2000) 199.
[444] M. Inagaki, M. Toyoda, N. Iwashita, et al., Carbon Sci. Korea 2 (2001) 1.
[445] M. Inagaki, M. Toyoda, N. Iwashita, et al., TANSO (201) (2002) 16.
[446] W.E. Shen, S.Z. Wen, N.Z. Cao, et al., Carbon 37 (1999) 351.
[447] F. Kang, Y.-P. Zheng, H. Zhao, et al., New Carbon Mater. 18 (2003) 161.
[448] M. Inagaki, N. Saji, Y.P. Zheng, et al., TANSO (215) (2004) 258.
[449] O.Y. Kwon, S.W. Choi, K.W. Park, et al., J. Ind. Eng. Chem. 9 (2003) 743.
[450] M. Inagaki, N. Kobayashi, B. Tryba, TANSO (215) (2004) 249.
[451] B. Tryba, A.W. Morawski, M. Inagaki, Carbon 43 (2005) 2417.
[452] Y. Takahashi, TANSO (160) (1993) 301 (in Japanese).
[453a] Y. Takahashi, N. Akuzawa, Zairyo Kagaku 32 (1995) 226 (in Japanese).
[453b] N. Akuzawa, T. Tajima, Y. Soneda, et al., TANSO (214) (2004) 179 (in Japanese).
[454] D. Saeher, A. Herold, Bull. Soc. Chim. Fr. 1965 (1965) 3130.
[455] P. Lagrange, A. Herold, C.R. Acad. Sci. Paris C 281 (1975) 381.
[456] M. Goldmann, H. Pilliere, F. Beguin, Synth. Met. 34 (1989) 59.
[457] Y. Takahashi, K. Oi, T. Terai, N. Akuzawa, Carbon 29 (1991) 283.
[458] H. Pilliere, Y. Takahashi, T. Yoneoka, T. Otosaka, N. Akuzawa, Synth. Met. 59 (1993) 191.
[459] Y. Matsuo, K. Hatase, Y. Sugie, Chem. Lett. 1999 (1999) 1109.
[460] S. Higashika, K. Kimura, Y. Matsuo, Y. Sugie, TANSO 1999 (189) (1999) 171 (in Japanese).
[461] M. Inagaki, New Carbons, Elsevier (2000) 146.
[462] M. Inagaki, Chem. Phys. Intercalation (1987) 105.

[463] M. Inagaki, J. Mater. Res. 4 (1989) 1560.
[464] M.S. Dresselhaus, G. Deresselhaus, Adv. Phys. 30 (1981) 139.
[465] T. Enoki, M. Suzuki, M. Endo, Graphite Intercalation Comp. Appl. (2003) 440.
[466] N. Daumas, A. Herold, C. R. Acad. Sci. Paris 268 (1969) C-373.
[467] V. Vignal, H. Konno, M. Inagaki, et al., J. Mater. Res. 14 (1999) 270.
[468] M. Inagaki, M. Ohira, Carbon 31 (1993) 777.
[469] G. Furdin, P. Lagrange, A. Herold, C. R. Acad. Sci. Paris 283 (1976) C-563.
[470] P. Lagrange, A. Metrot, A. Herold, C. R. Acad. Sci. Paris 278 (1974) C-701.
[471] M. Nomine, L. Bonnetain, J. Chim. Phys. 66 (1969) 1731.
[472] O. Tanaike, M. Inagaki, Synth. Met. 90 (1997) 69.
[473] V.A. Nalimova, V.V. Avdeev, K.N. Semenenko, Mater. Sci. Forum. 91–93 (1992) 11.
[474] H. Touhara, K. Kakuno, N. Watanabe, TANSO (117) (1984) 98 (in Japanese).
[475] B.C. Brodie, Ann. Chem. 114 (1860) 6.
[476] T. Nakajima, Graphite Intercalation Comp. (1990) 195 (in Japanese).
[477] U. Hofman, Naturewissenshaften 18 (1939) 229.
[478] G. Ruess, Monatsh. Chem. 76 (1947) 381.
[479] Y. Matsuo, T. Niwa, Y. Sugie, Carbon 37 (1999) 897.
[480] Y. Matsuo, K. Hatase, Y. Sugie, Chem. Lett. (1999) 1109.
[481] S. Higashika, K. Kimura, Y. Matsuo, et al., TANSO (189) (1999) 171 (in Japanese).
[482] A. Herold, Intercalation Layered Mater. (1979) 329.
[483] D. Guerard, A. Herold, Carbon 13 (1975) 337.
[484] S. Basu, C. Zeller, P.J. Flander, et al., J. Mater. Sci. Eng. 38 (1979) 275.
[485] Z.D. Wang, M. Inagaki, J. Mater. Chem. 2 (1992) 629.
[486] M. Inagaki, Z.D. Wang, TANSO (153) (1992) 184.
[487] Z.D. Wang, M. Inagaki, TANSO (145) (1990) 243.
[488] S. Flandrois, J.M. Masson, J.C. Rouillon, et al., Synth. Met. 3 (1981) 1.
[489] J. Mittal, M. Inagaki, Synth. Met. 95 (1998) 21.
[490] J. Mittal, M. Inagaki, Solid State Ionics 121 (1999) 183.
[491] N. Iwashita, M. Inagaki, Synth. Met. 34 (1989) 139.
[492] M. Inagaki, N. Iwashita, H. Konno, Carbon 28 (1990) 49.
[493] N. Iwashita, M. Inagaki, Noppon Kagaku Kaishi (1992) 1414.
[494] Y. Mizutani, E. Ihara, T. Abe, et al., J. Phys. Chem. Solids 57 (1996) 799.
[495] Y. Mizutani, T. Abe, K. Ikeda, et al., Carbon 35 (1997) 61.
[496] O. Tanaike, M. Inagaki, Synth. Met. 96 (1998) 109.
[497] M. Inagaki, TANSO (170) (1995) 298 (in Japanese).
[498] Y. Soneda, M. Inagaki, Z Anorg. Allg. Chem. 610 (1992) 157.
[499] O. Tanaike, Ph D Thesis, Hokkaido Univ, 1998.
[500] J.R. Dahn, A.K. Sleigh, H. Shi, et al., Lithium Batteries (1994) 1.
[501] T. Nakajima, M. Kawaguchi, N. Watanabe, Z Naturforsch 36 (1981) 1419.
[502] T. Nakajima, M. Kawaguchi, N. Watanabe, Carbon 20 (1982) 287.
[503] H. Touhara, F. Okino, Carbon 38 (2000) 241.
[504] Y. Hishiyama, H. Irumano, Y. Kaburagi, Phys. Rev. B 63 (2001) 245406.
[505] T. Sogabe, K. Nakajima, M. Inagaki, J. Mater. Sci. 31 (1996) 75.
[506] M. Endo, C. Kim, T. Karaki, et al., Phys. Rev. B 58 (1998) 8991.
[507] Y. Hishiyama, M. Inagaki, Carbon 39 (2001) 150.
[508] G. Pistoia, (ed.) Lithium Batteries, Elsevier (1994).

[509] H. Hatori, Y. Yamada, M. Shiraishi, Carbon 32 (1994) 359.
[510] M. Inagaki, H. Tachikawa, T. Nakahasji, et al., Carbon 36 (1998) 1021.
[511] T. Kondo, S. Casolo, T. Suzuki, et al., Phys. Rev. B 86 (2012) 035436.
[512] T. Nakajima, M. Koh, Carbon 36 (1997) 203.
[513] S. Marinkovic, Chem. Phys. Carbon 19 (1984) 1.
[514] M. Kawaguchi, N. Bartlett, Fluorine-Carbon and Fluoride-Carbon Mater. (1995) 187.
[515] F. Sugnac, F. Teyssandier, A. Marchand, J. Am. Chem. Soc. 75 (1992) 161.
[516] M. Kawaguchi, Adv. Mater. 9 (1997) 615.
[517] H. Konno, H. Oka, K. Shiba, et al., Carbon 37 (1999) 887.
[518] H. Shinohara, Y. Saito, Chem. Phys. Fullerenes (1997) (in Japanese).
[519] K. Tanigaki, J. Phys. Chem. Solids 54 (1993) 1645.
[520] A.F. Hebard, M.J. Rosseinsky, R.C. Haddon, et al., Nature 350 (1991) 600.
[521] O. Zhou, J.E. Fischer, N. Coustel, et al., Nature 351 (1991) 462.
[522] E. Yamamoto, M. Tansho, T. Tomiyama, et al., J. Am. Chem. Soc. 118 (1996) 2293.
[523] L.R. Radovic, F. Rodriguez-Reinoso, Chem. Phy. Carbon 25 (1997) 243.
[524] H. Hatori, S. Hishiki, T. Kobayashi, et al., TANSO (189) (1999) 165 (in Japanese).
[525] H. Oka, M. Inagaki, Y. Kaburagi, et al., Solid State Ionics 121 (1999) 157.
[526] Y. Kaburagi, T. Toriyama, A. Yoshida, et al., J. Mater. Res. 16 (2001) 352.
[527] Y. Kaburagi, Y. Hishiyama, H. Oka, et al., Carbon 39 (2001) 593.
[528] Y. Kaburagi, H. Hatori, A. Yoshida, et al., Synth. Met. 125 (2002) 171.
[529] A. Messaoudi, M. Inagaki, F. Beguin, J. Mater. Chem. 1 (1991) 735.
[530] M. Ohira, A. Messaoudi, M. Inagaki, et al., Carbon 29 (1991) 1233.
[531] H. Shioyama, H. Sakakibara, N. Iwashita, et al., TANSO (156) (1993) 37 (in Japanese).
[532] H. Konno, Y. Takahashi, H. Habazaki, TANSO (214) (2004) 191 (in Japanese).
[533] E. Bekyarova, K. Kaneko, Langmuir 15 (1999) 7119.
[534] E. Bekyarova, K. Kaneko, Adv. Mater. 12 (2000) 1625.
[535] F.J. Maldonado, C. Moreno-Castilla, J. Rivera-Utrilla, et al., Langmuir 16 (2000) 4367.
[536] N. Yoshizawa, M.S. Dresselhaus, R. Fu, et al., Mol. Cryst. Liq Cryst. 388 (2002) 489.
[537] T.F. Baumann, G.A. Fox, J.H. Satcher Jr., et al., Langmuir 18 (2002) 7073.
[538] T. Ohkuni, Y. Sahashi, A. Satsuma, et al., TANSO (217) (2005) 95.
[539] Y.W. Ju, G.R. Choi, H.R. Jung, et al., J. Electrochem. Soc. 154 (2007) A192.
[540] L. Ji, A.J. Medford, X. Zhang, J. Mater. Chem. 19 (2009) 5593.
[541] L. Ji, Z. Lin, R. Zhou, et al., Electrochim. Acta 55 (2010) 1605.
[542] R. Ruiz-Rosas, J. Bedia, M. Lallave, et al., Carbon 48 (2010) 696.

CHAPTER

# Engineering and Applications of Carbon Materials

# 3

## 3.1 Polycrystalline graphite blocks

### 3.1.1 Production

#### a. Graphite blocks

A generalized flow chart for the production of polycrystalline graphite blocks is shown in Fig. 3.1. For the filler, petroleum and coal tar pitch cokes are usually used, and natural graphite, carbon blacks and also recycled graphite particles are sometimes used. Particle size distribution of the fillers has to be controlled. For the binder, petroleum and coal tar pitches are used in most cases, because they have relatively high carbon yield as about 60 mass% and also because they give the carbon similar to the filler cokes after carbonization. Thermosetting resins, such as phenol and epoxy resins, are sometimes employed. The filler and binder, of which mixing ratio has also to be controlled in accordance with the requirements from the applications, are mixed at a temperature higher than the softening point of the binder. The mixtures thus prepared, which are usually called carbon pastes, are formed after warming up at a temperature around 150°C either by extrusion, molding or cold isostatic pressing (CIP). The formed blocks are sent to carbonization process at a temperature of 700–1000°C (sometimes called calcination) and then to graphitization at a high temperature above 2500°C. The carbon blocks heat-treated at a high temperature are often called 'polycrystalline graphite', although the particles of fillers are not always converted to graphitic structure, as explained in Section 2.5.

The forming process is an important process for the fabrication of polycrystalline graphite, because it governs the preferred orientation of crystallites. Three forming processes, which are commonly used in industries, i.e., extrusion, molding and CIP, are schematically illustrated in Fig. 3.2. The extrusion process is applied for the pastes with thermoplastic binders, such as pitches, and gives the preferred orientation of either flaky or needle-like filler particles along with the direction of extrusion, as schematically shown in Fig. 3.2a. Electrodes for metal processing with large diameter, various jigs with different sizes and also leads for automatic pencil with a diameter as thin as 0.3 mm are fabricated by this forming process. In the molding process, the filler particles are statistically aligned perpendicular to the compressing direction (Fig. 3.2b). Carbon brushes for electric

Materials Science and Engineering of Carbon: Fundamentals. DOI: http://dx.doi.org/10.1016/B978-0-12-800858-4.00003-6

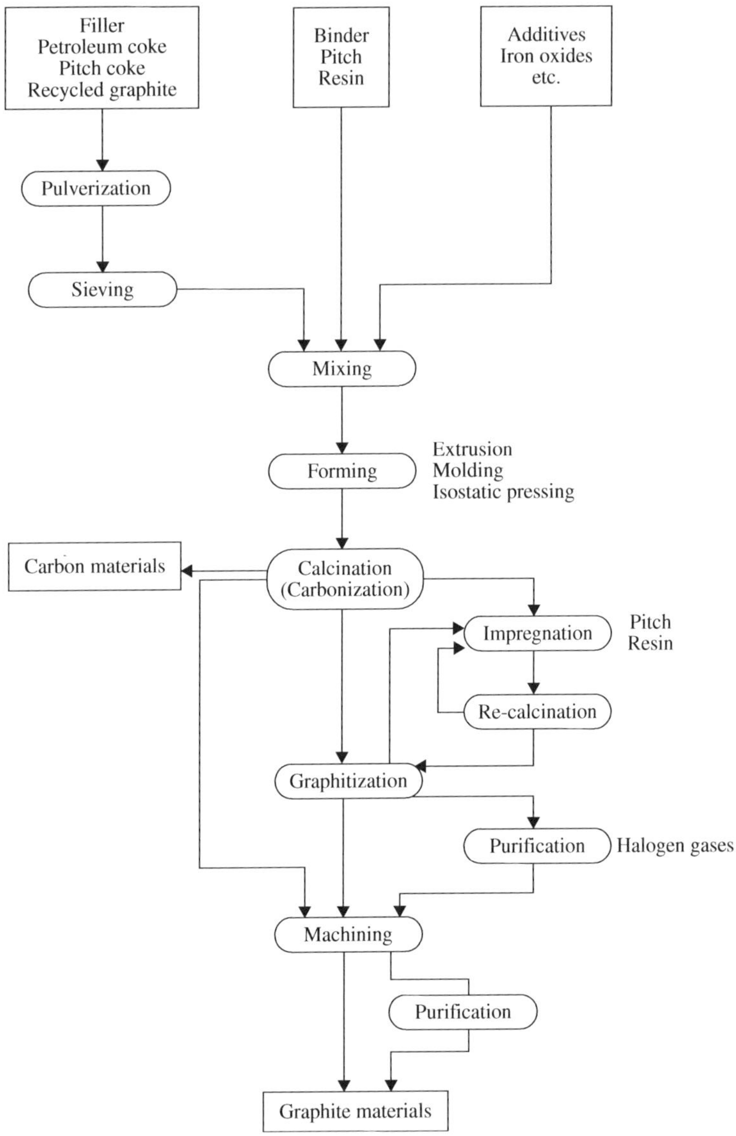


**FIGURE 3.1**

Production process of polycrystalline graphite blocks.

motors and electric contacts are fabricated by molding the carbon pastes. By CIP, compressive force is applied hydrostatically (compressed equally from all the directions) and so the orientation of filler particles is random, giving the blocks isotropic nature (Fig. 3.2c). Most high-density isotropic graphite blocks are produced through this process.

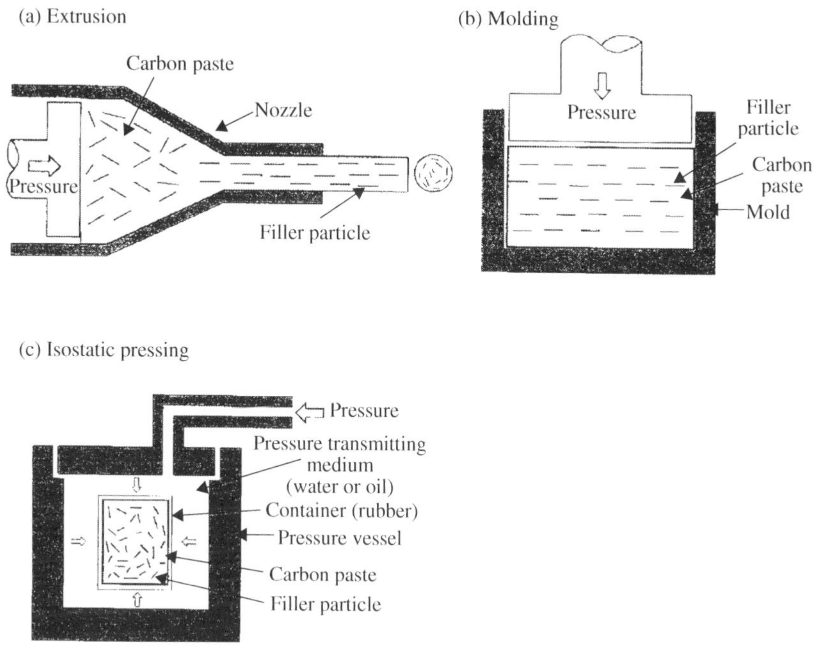


**FIGURE 3.2**

Forming processes commonly used for graphite blocks.

Just after the carbonization, three-dimensional graphite structure is not developed yet, except the case where natural graphite or graphitized recycled materials are used, and the products obtained after carbonization are carbon materials and some of them are sent to machining for applications. But most carbon materials thus prepared are heat-treated at a high temperature as 2600–3000°C in so-called Acheson-type furnace, in which carbon materials are packed with the mixture of coke and sand, and get high temperature by the contact resistance of packing mixture. The industrial process heating the carbonized rods by direct passing of the electricity (lengthwise graphitization furnace, LWG) has also been developed. For the process using conventional Acheson furnace, about one month is required to complete the graphitization treatment, but only a little more than one week for the newly developed LWG process. For some of these graphitized blocks, impregnation of molten pitches is applied in order to densify. To get high density, the cycle of impregnation, carbonization and graphitization processes is repeated, as shown in Fig. 3.1. If necessary, purification at high temperatures in a flow of halogen gases is employed either before or after machining.

## b. High-density isotropic graphite blocks

The production procedure for high-density isotropic graphite is shown as a block diagram in Fig. 3.3, which is classified into two on the basis of raw materials

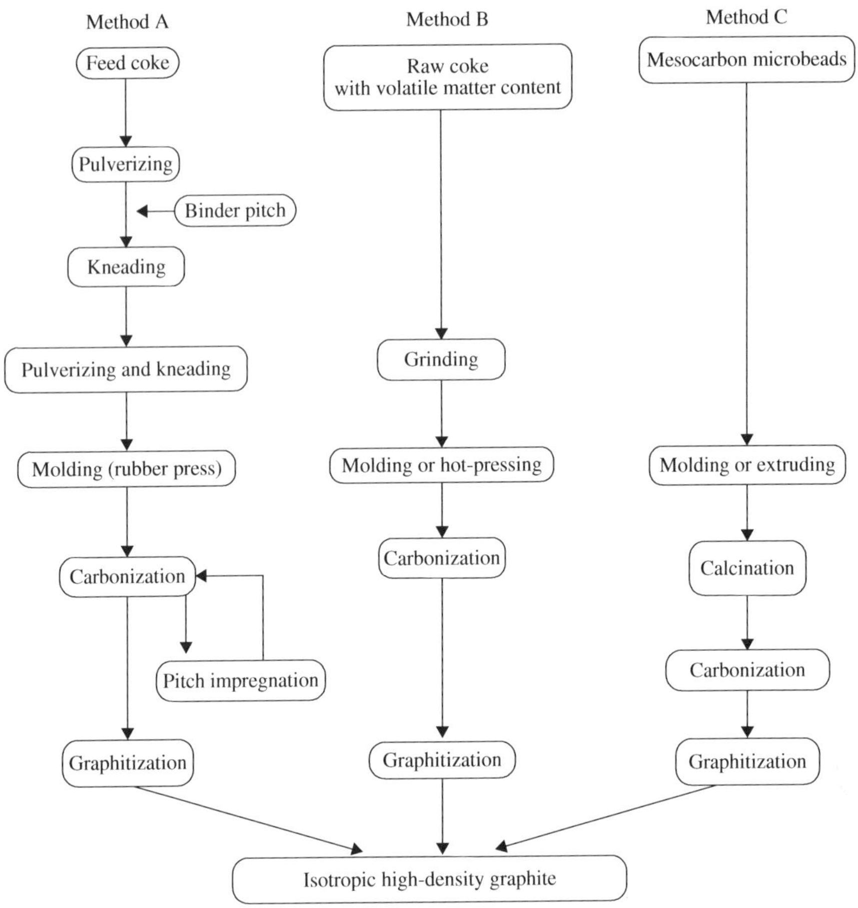


**FIGURE 3.3**

Production processes for high-density isotropic graphite.

used and whether the binder is used or not. The fundamental procedure for method A is the same as the conventional one, as shown in Fig. 3.1, using the carbon pastes of filler coke particles with binder pitch followed by forming, carbonization and graphitization. Key techniques to produce high-density isotropic graphite are the use of fine particles of filler cokes, less than 5 μm, and CIP (often called rubber pressing because rubber container is used) for forming.

In Fig. 3.2c, a principle of CIP is illustrated. By applying this forming process, various advantages in production are obtained; fine particles of filler cokes become possible to be used, high homogeneity in the products are attained because of good conductance of pressure, near-shaped products are possible to be obtained, high density of the final products can be reached by using a relatively large amount of binder pitch, etc. However, very accurate control during carbonization and graphitization processes is required in order to avoid deformation and cracking of the blocks, such as selection of raw materials (filler coke and binder pitch), pressurizing program during CIP, temperature control during carbonization, etc.

The detailed conditions for the production of high-density isotropic graphite blocks are not published because they are highly confidential issues for the industries. Some characteristics of commercially available high-density isotropic graphite prepared by CIP forming process are tabulated in Table 3.1. On these graphite blocks the difference in various properties along different directions of the blocks is less than 3%. To select the binder giving high carbon yield after carbonization is also important for densification of the final products. Carbon yield from pitch depends on pitch itself and on the preparation conditions as a binder, but also on its mixing ratio with filler cokes. If filler content becomes larger than a certain amount, the binder pitch forms a thin film on the surface of the filler particles because of its good wettability, and its flow at high temperatures during carbonization is supposed to be disturbed. In order to get high carbon yield, there have been some trials to use additives for binder pitch, but it is not accepted as an industrial process because of lowering of the purity of the final products, high purity of these graphite blocks being one of the important characteristics for some applications.

These have been reported to carbonize the carbon pastes under pressure [1]. In Table 3.2, the characteristics of two graphite blocks are compared, which are prepared from the formed carbon paste with the bulk density of 1.67 $g/cm^3$ by the carbonization in argon under 5 MPa pressure and under atmospheric pressure. By pressure carbonization up to 600°C, bulk density, bending strength, Young's modulus and electrical conductivity of the block showed a marked increase.

In method B in Fig. 3.3, raw coke powder is used as the fillers and can be sintered without any binder, because it contains a large amount of volatile matters (about 8.5 mass%) [2]. An advantage of this method B is self-sintering of raw coke, in other words, no binder is necessary, but severe control of the carbonization process is required. This method is now used to produce the carbon/metal carbide composites in industry, which is described in Section 3.7.4c.

The use of spherical particles of filler has been proposed to be an alternative production process (method C in Fig. 3.3) for high-density isotropic graphite. Mesophase spheres, optically anisotropic spheres formed at the very beginning of pyrolysis of pitch which are separated from isotropic matrix pitch (mesocarbon microbeads, MCMB) (Section 2.3.4b), are formed into a block and then carbonized. Some properties of three commercially available mesocarbon microbeads are compared in Table 3.3; their particle size and ash content are different in the products.

The method C to produce high-density isotropic graphites has some advantages; isotropy in bulk is obtained even by using molding because the raw materials are spheres, though forming by CIP is in practice recommended in order

**Table 3.1** Characteristics of Commercial Products of High-Density Isotropic Graphite Formed by CIP Forming Process

| Company | Trade Name | Bulk Density ($g/cm^3$) | Electrical Resistivity ($\mu\Omega m$) | Bending Strength (MPa) | Young's Modulus (GPa) | Shore Hardness | Thermal Expansion Coefficient (/K) | Thermal Conductivity (W/m.K) |
|---|---|---|---|---|---|---|---|---|
| Toyo | IG-11 | 1.77 | 11,0 | 39.2 | 9.8 | 51 | $4.5\times10^{-6}$ | 116 |
| Tanso | IG-43 | 1.82 | 9.2 | 53.9 | 10.8 | 55 | $4.8\times10^{-6}$ | 139 |
| | ISO-88 | 1.90 | 15.5 | 93.1 | 12.7 | 90 | $6.5\times10^{-6}$ | 70 |
| | ISEM-3 | 1.85 | 10.0 | 49.0 | 11.8 | 60 | $5.6\times10^{-6}$ | 128 |
| Tokai | G250 | 1.70 | 12.0 | 39.2 | 9.8 | 46 | $3.5\times10^{-6}$ | 108 |
| Carbon | G348 | 1.92 | 10.0 | 63.7 | 12.3 | 68 | $4.2\times10^{-6}$ | 128 |
| | G535 | 1.82 | 17.0 | 63.7 | 10.8 | 72 | $4.2\times10^{-6}$ | 81 |
| Hitachi | PD-320 | 1.78 | 11.0 | 47 | | 47 | $5.2\times10^{-6}$ | |
| Kasei | PD-610 | 1.88 | 11.0 | 55 | | 65 | $6.0\times10^{-6}$ | |
| Nippon | IGS-603 | 1.80 | 12.5 | 44 | | 57 | $4.6\times10^{-6}$ | 116 |
| Techno- | IGS-895 | 1.93 | 15.0 | 74 | | 87 | $6.2\times10^{-6}$ | 90 |
| Carbon | EGS-763 | 1.81 | 12.0 | 55 | | 53 | $4.5\times10^{-6}$ | 130 |

**Table 3.2** Properties of the Graphite Blocks Carbonized Under Different Pressures [1]

| Carbonization Conditions | Under Pressure of 5 MPa at 600°C | Under Atmospheric Pressure at 800°C |
|---|---|---|
| Bulk density ($g/cm^3$) | 1.686 | 1.471 |
| Electrical resistivity ($\times 10^{-6}$ Ωcm) | 79 | 106 |
| Bending strength (MPa) | 26.6 | 14.6 |
| Young's modulus (MPa) | 111 | 69 |
| Shore hardness | 41 | 32 |
| Thermal expansion coefficient (75 – 375°C, $\times 10^{-6}$/°C) | 2.39 | 2.40 |

**Table 3.3** Characteristics of Commercially Available Mesocarbon Microbeads

| Company | Kawasaki Steel | NKK | Osaka Gas |
|---|---|---|---|
| Toluene-insoluble (mass%) | 97.0–99.5 | – | 94–98 |
| Quinoline-insoluble (mass%) | 92.0–96.0 | 95.9–97.0 | 80–93 |
| Fixed carbon (mass%) | – | 85.0–89.0 | 87–90 |
| Volatile matter (mass%) | 0.5–0.9 | – | 10–13 |
| Ash content (ppm) | 400 | Trace | 100–300 |
| Particle size (μm) | 13–17 | 5–50 (av. 10) | $D_{10}$ : 2–3, $D_{50}$ : 5–7, $D_{90}$ : 9–15 |

to avoid a possible inhomogeneity in the formed blocks, and no binder pitch is necessary because spheres still contain a relatively large amount of volatile matters which can act as binder during carbonization. However, some disadvantages are also pointed out; low yield of these spheres from pitch, difficulty in their separation from matrix pitch, both of which lead to an increase in the cost of production.

## 3.1.2 Applications

### a. Metal processing

Large amounts of artificial graphite blocks have been used as electrodes for metal processing. In Fig. 3.4, graphite electrodes for electric arc furnaces are shown. A scheme of the electric arc furnace is illustrated in Fig. 3.5. Three graphite electrodes (poles) are perpendicularly placed in the furnace, and the electric arc between these poles and ferrous scraps gives high temperature to melt the scraps. Graphite electrodes must have high electrical conductivity and good refractory properties, such as low thermal expansion and high thermal shock resistance. In

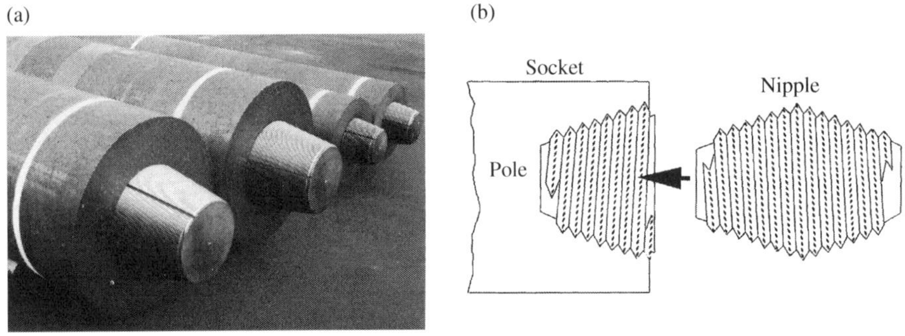


FIGURE 3.4

Graphite electrode for steel refining.

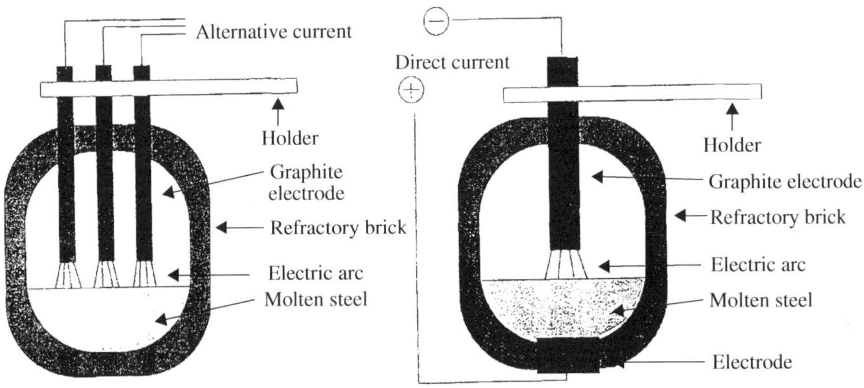


FIGURE 3.5

Schemes of electric-arc furnaces using graphite electrodes.

order to have these properties, a highly crystallized graphite structure is required and therefore high-temperature treatment, usually above 2500°C, is essential. During the use of these electrodes in the furnace, they are gradually eroded by dissolving into molten steel and also by oxidation. Therefore, the graphite poles have to be lowered into their position in regular intervals and also new poles have to be connected by using nipples, as shown in Fig. 3.4b, which are made from the same polycrystalline graphite. A photograph of a practical arc furnace for steel refining in operation is shown in Fig. 3.6.

The worldwide production of steel is slowly increasing, but its production using electric arc furnaces increases rapidly. By the improvement of operation conditions of the furnace using high electrical power, however, the consumption of the graphite electrodes in order to produce one ton of steel decreases gradually, about 7.5 kg in 1975, but about 5 kg in 1990.

**FIGURE 3.6**

Electric arc furnace in operation.

For the operation under high power, the graphite electrodes are required to be improved in their properties, particularly reduction of the thermal expansion and resulting improvement in thermal shock resistance. For this purpose, so-called needle-like cokes are necessary to be used. Some characteristics of industrial products of graphite electrodes for ultrahigh power operation of electric arc furnace are summarized in Table 3.4.

Aluminum metal is produced by electrolysis of molten alumina (aluminum oxide) using some fluorides as flux (Fig. 3.7). Polycrystalline graphites, therefore, are used as both anode and cathode. The anodes are similar in properties and also in fabrication to those used in electric arc furnaces described above, and the cathodes are mostly made of carbon materials heat-treated at relatively low temperatures (calcined carbon blocks). For cathodes, carbon blocks based on natural anthracite have been often used, but coke-based carbons have also been applied in recent years.

Graphite materials are used as molds, dies, guides, run-out tables, furnace linings, boats and crucibles for the processing of not only ferrous metals and their alloys, such as gray and ductile irons, but also non-ferrous metals and alloys, such as copper, nickel, brass, bronze, zinc, noble metals, etc. For most of these purposes, high-density isotropic graphite blocks are selected, because of their high thermal conductivity, high mechanical strength and high density.

**Table 3.4** Characteristics of Commercially Available Graphite Electrodes for Ultrahigh Power Operation of Electric Arc Furnace

| Company | Ash Content (mass %) | True Density ($g/cm^3$) | Bulk Density ($g/cm^3$) | Porosity (%) | Bending Strength (MPa) | Young's Modulus (GPa) | Electrical Resistivity ($\mu\Omega.m$) | Thermal Expansion Coefficient ($\times 10^{-6}/°C$) |
|---|---|---|---|---|---|---|---|---|
| Tokai Carbon | <0.2 | 2.20–2.23 | 1.68–1.74 | 21–26 | 10.0–15.0 | 9–14 | 4.5–6.5 | 0.5–1.0 |
| SEC | <0.2 | 2.22–2.25 | 1.65–1.80 | 21–27 | 9.8–14.7 | 8.8–12.7 | 4.5–6.0 | 0.8–1.5 |
| Nippon Carbon | <0.2 | 2.20–2.23 | 1.65–1.75 | 20–25 | 11.0–12.0 | 8.3–11.8 | 4.3–6.0 | 1.1–1.6 |

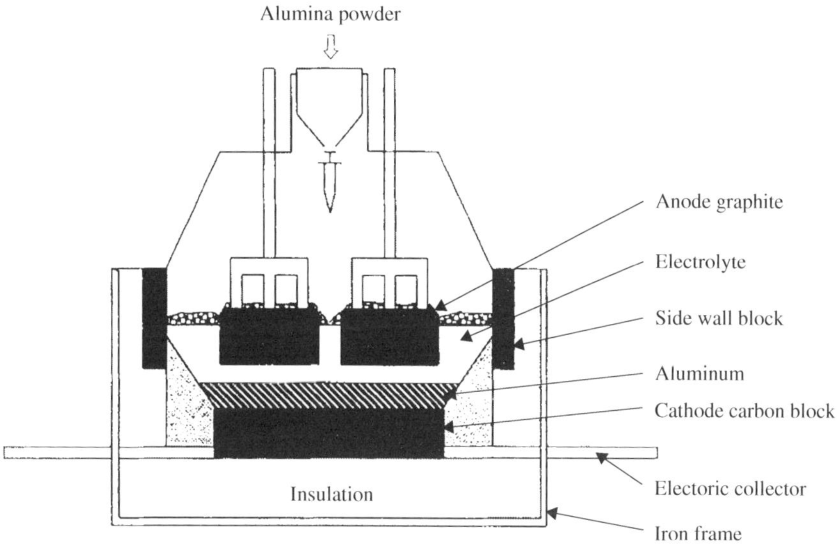


**FIGURE 3.7**

Construction of the furnace for aluminum metal production.

### b. Semiconductor production and electrical discharge machining

Recent development in semiconductor technology led to the development of various microelectronic devices. Polycrystalline graphite blocks have given and are giving an important contribution to the production of semiconductor crystals in various respects.

For the synthesis of single crystals of silicon, germanium, and I-V and II-VI semiconductors by either ribbon or Czochralski crystal-pulling techniques, graphite heater and crucible (Fig. 3.8a) are essential. For these applications, high purity of graphite blocks is required and so purification at high temperatures by using halogen gases has to be applied to reduce the impurity content to less than 100 ppm. High-density isotropic graphite blocks are mostly employed because of their accurate and easy machining, high strength, and isotropic nature in electrical resistivity. In addition, there are so many applications, such as boats and assemblies for liquid epitaxy, susceptors and wafer trays for different CVD processes, shields, electrodes and ion sources for ion implantation, etc. (Fig. 3.8b).

High-density isotropic graphite blocks have firstly been produced in an industrial scale for the applications of hot-pressing molds, nozzles for continuous casting of metals, and electrodes for electrical discharge machining (EDM). In Fig. 3.9, some electrodes for EDM are shown.

### c. Electrical and electronic devices

Electrical applications of polycrystalline graphite blocks have been well established and some of them are well specified in industrial standards. Brushes for

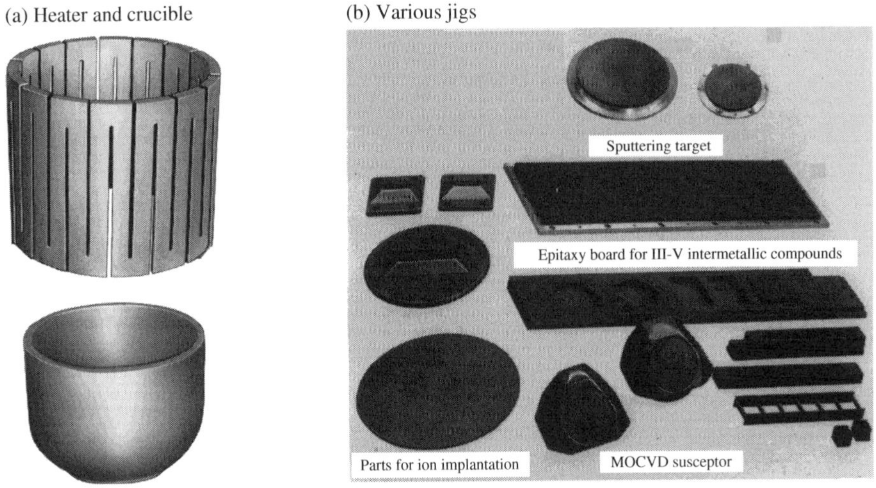


**FIGURE 3.8**

Jigs for semiconductor crystals.

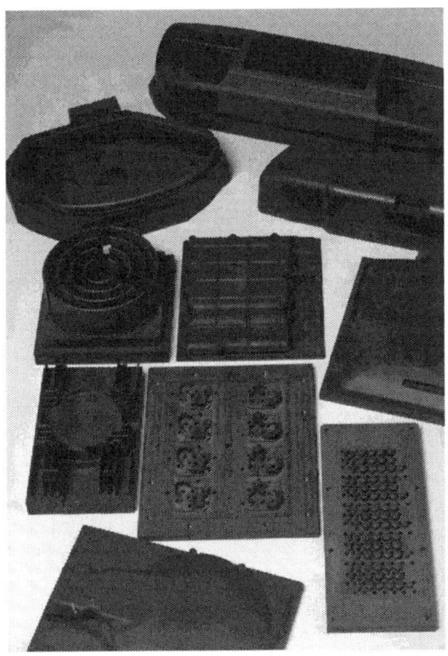

**FIGURE 3.9**

Electrodes for electrical discharge machining.

electric motors have been used since 1890 and their usage is increasing now mostly due to automation, for example, windows of automobiles and various daily necessaries, such as refrigerators, electric ovens, videos, computers, players, etc. Some brushes are shown in Fig. 3.10. The brushes are widely dispersed in size and shape, even as small as a few millimeters being produced. To produce brushes, different raw materials, such as carbon blacks (soots), cokes and natural graphite, as fillers and either pitches or phenol resin as binder are employed and calcined to about 1000°C in inert atmosphere. Some of them are heated up to 3000°C to be graphitized. Also some metals, such as copper and silver, are mixed, in order to control the electrical resistivity and lubricity.

Graphite plates are used as current-collecting shoes for electric trains, though their usage decreases because of high-speed driving. Graphite materials having high electrical conductivity and lubricity do not form insulation oxides and also do not melt, and so they are used as electrical contacts.

Various carbon materials, including polycrystalline graphite rods, have been used in various primary cells, manganese, manganese-alkali, air-zinc, and lithium batteries (Section 3.9.2).

### d. Nuclear applications

Polycrystalline graphite is one of the best materials for nuclear-fission applications because of its high moderating efficiency and low absorption cross-section

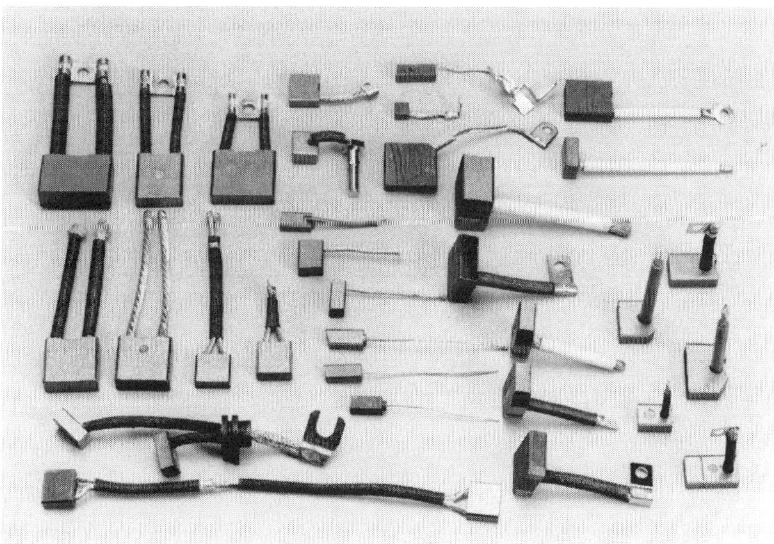

**FIGURE 3.10**

Carbon brushes.

*(Courtesy of Fuji Carbon Manufacturing Co., Ltd.)*

for neutron, in addition to their high mechanical strength, chemical stability, high machinability and light weight. It has been used as the building block of many fission reactors, since the first one at Chicago, USA. It is also used in high-temperature gas-cooled reactors. In Fig. 3.11, the construction is shown for a test reactor of this type in Japan, which is now producing 950°C water steam to be used for hydrogen production. In this reactor, graphite materials are used as permanent reflectors at the outer part, replaceable reflector at the inner part, fuel-element blocks at the center, and coating of small fuel particles (Fig. 2.29). For the nuclear application, graphite materials have to have high purity, high strength and isotropic nature. In Table 3.5, specification on graphite materials used for Japanese test reactor is summarized.

Graphite materials, such as high-density isotropic graphite and carbon fiber/carbon composite, are also used in nuclear-fusion reactors as interior liners, movable limiters and specialized fixtures. In Fig. 3.12, the inner structure of Japanese test reactor JT-60 is shown. Low atomic number of graphite materials is the most important factor in reducing the interference with plasma, but many problems for practical use of graphite materials, such as physical and chemical sputtering resistance, thermal shock resistance, recycling of hydrogen, etc., have to be solved.

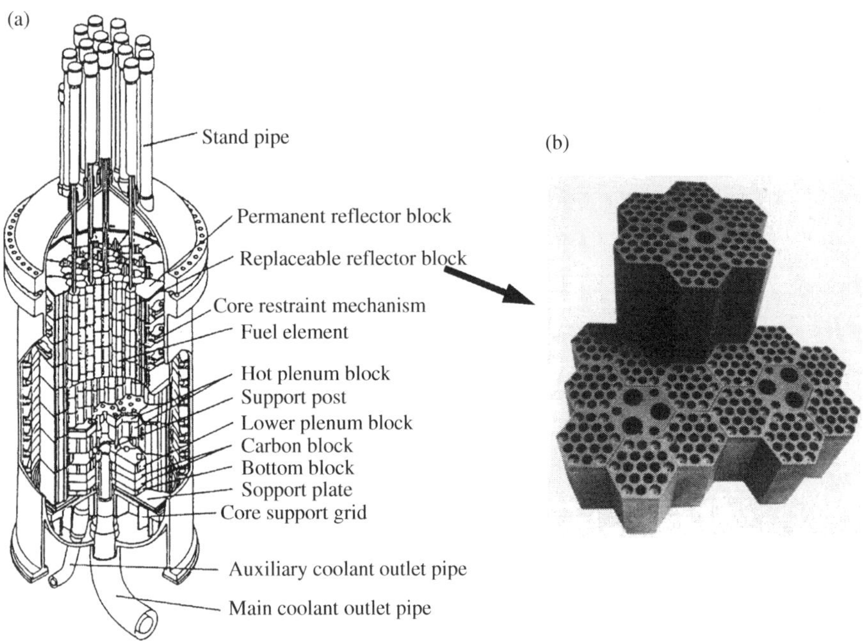


**FIGURE 3.11**

Construction of Japanese high-temperature gas-cooled reactor (a) and replaceable reflector blocks of high-density isotropic graphite (b).

*(Courtesy of Japan Atomic Energy Research Institute.)*

**Table 3.5** Specification of Graphite Materials for Japanese High-Temperature Gas-Cooled Test Nuclear Reactors

| | Replaceable Reflectors & Fuel Element Blocks | Permanent Reflectors |
|---|---|---|
| Bulk density ($g/cm^3$) | 1.78 | 1.73 |
| Tensile strength (MPa) | 25.8 | 8.3[a] |
| Compressive strength (MPa) | 78.4 | 32.1[a] |
| Thermal expansion coefficient ($\times 10^{-6}$/°C) | 4.06 | 2.34[a] 2.87[b] |
| Thermal conductivity (cal/cm·s.°C) | 0.19 | 0.18[a] |
| Ash content (ppm) | <100 | <7000 |
| Grain size of filler coke (μm) | 20 | <800 |

[a]*Radial direction*
[b]*axial direction.*
*(Courtesy of Japan Atomic Energy Research Institute.)*

**FIGURE 3.12**

Inner structure of Japanese test fusion reactor JT-60.

*(Courtesy of Japan Atomic Energy Research Institute.)*

#### e. Mechanical applications

Fine-grain high-density graphite blocks are extensively used for bearing and seals, particularly in high-temperature circumstances. This application is mostly due to the following characteristics of graphite materials; lubricating and frictional properties, high thermal conductivity, and high compressive strength at high temperatures. However, their practical usage is somewhat limited, because they are easily oxidized and also attacked by some chemicals, such as *aqua regia*, perchloric acid, fumic sulfuric acid, chromic oxide, etc. Seal rings for gas-turbine engines, chemical pumps for corrosive fluid transfer, water pumps of diesel engine, compressors for air conditioning, and water pumps of home washing machine and dishwasher can be mentioned as commercial applications.

### 3.1.3 Filler cokes and binder pitches

#### a. Filler cokes

Cokes, which are used as fillers for polycrystalline graphite blocks, are manufactured from pitches of petroleum and coal tar origins. In a pitch, all constituent molecules are mobile above the softening point of the pitch and the pitch has a sufficiently low viscosity to flow along movement of the bubbles of decomposition gases (thermoplastic nature). At high temperatures, however, the plasticity of the decomposed pitches is lost because of polymerization and condensation reactions, and therefore out-gases leave pores behind. When the pitch has enough plasticity to flow at low temperatures or it is made to flow by additional forces, such as stirring, needle-like cokes are produced, which are important raw materials for various carbon materials, such as graphite electrodes and high-density isotropic graphites. An example of such a flow pattern and elongated pores in a needle-like coke particle is shown in Fig. 3.13. These flow patterns are known to be due to the preferred orientation of polycondensed aromatic molecules, which leads to the preferred orientation of hexagonal carbon layers after carbonization and graphitization.

The preferred orientation of crystallite in the particle of cokes is characterized by measuring transverse magnetoresistance $\Delta\rho/\rho_0$ (Section 2.5.1). In Fig. 3.14, the dependences of $\Delta\rho/\rho_0$ on the angles from the direction max (refer Fig. 2.84b) are shown for a needle-like coke and a gilsonite coke, the former having the nanotexture of planar orientation and the latter point orientation (Section 2.5.3) [3]. On the needle-like coke, $\Delta\rho/\rho_0$ changes markedly with both rotation angles $\theta$ and $\varphi$. On the gilsonite coke, on the other hand, the changes in $\Delta\rho/\rho_0$ are very small (the scales of ordinates for the gilsonite coke are very small in comparison with those for the needle-like coke). If anisotropy ratios $r_{TL}$ and $r_T$ were employed, the needle-like coke heat-treated up to 3000°C has the values $r_{TL}$ of 0.12 and $r_T$ of 0.31, though the gilsonite coke heat-treated at the same temperature has $r_{TL}$ of 0.86 and $r_T$ of 0.93.

Maximum transverse magnetoresistance $(\Delta\rho/\rho_0)_{max}$, a measure for the degree of graphitization, is quite different for these two cokes; for the needle-like coke

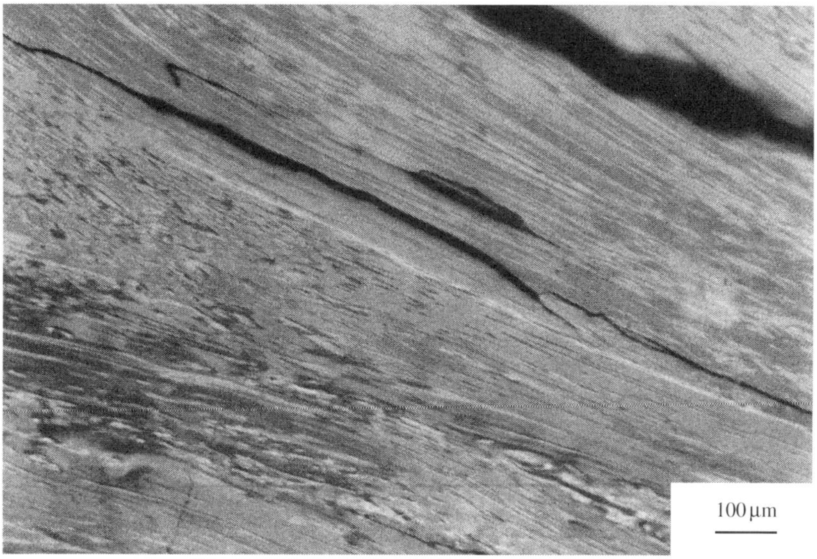


**FIGURE 3.13**

Polarized-light micrograph of a needle-like coke.

close to 50 but for the gilsonite coke only 3 after the heat treatment at 3000°C, revealing that the needle-like coke is highly graphitizable.

The elongated pores observed in Fig. 3.13 are formed during the cooling process of carbonization of the precursor pitch to coke, and kept after the graphitization of the formed blocks. These elongated pores can absorb large thermal expansion along the c-axis of each crystallite, which are oriented along the wall of these pores, and are known to absorb thermal shock very effectively. High thermal shock resistance is one of the important characteristics of graphite electrodes for steel refining, as mentioned above.

Structural change in coke particles has also been studied by TEM [4]. In Figs. 3.15 and 3.16, changes in 002 lattice fringes and 11 dark-field images, are shown on the coke prepared from anthracene. The structural units in the coke heat-treated up to 1000°C are carbon layers with the size of about 1 nm, 2–3 layers being stacked in parallel (Fig. 3.15a). In these coke particles, however, the layers are highly oriented with planar orientation scheme. After 1300°C treatment, the orientation of the layers is improved little (Fig. 3.15b) and after 1800°C the size of layers becomes larger (roughly 3 nm) (Fig. 3.15c). Above 2000°C, layers grow quickly with the increase in heat treatment temperature (HTT), as shown in Fig. 3.15d. In this stage of structure change, however, the stacking regularity is not improved much, still turbostratic stacking. Layer size in the coke heat-treated above 2000°C is evaluated from 11 dark-field images (Fig. 3.16). After 2600°C treatment, the regions showing regularly spaced Moire fringes, i.e., layers with regular stacking, become larger, about 300–400 nm.

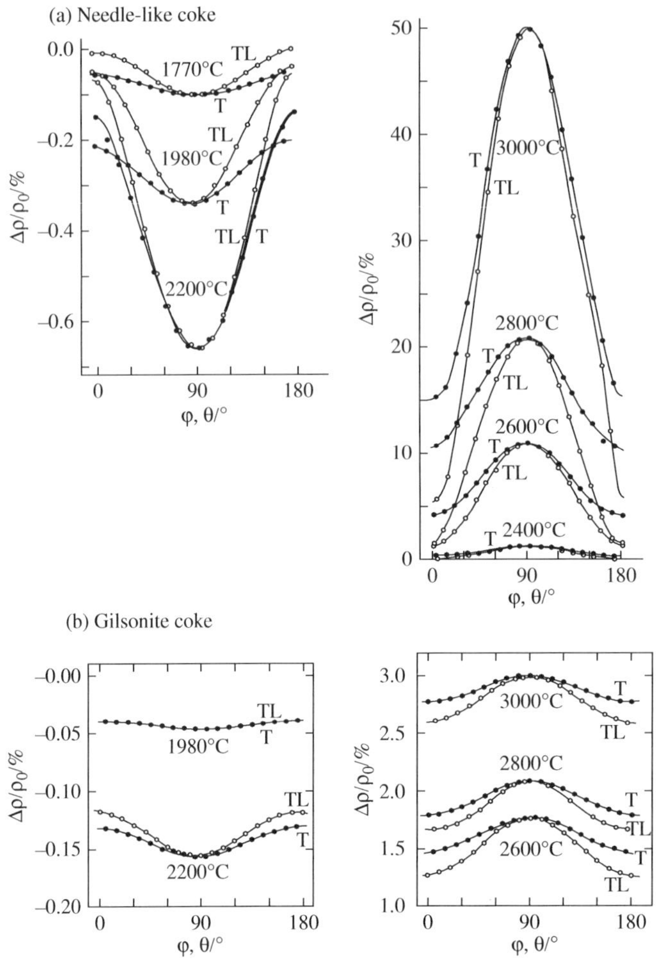


**FIGURE 3.14**

Dependences of $\Delta\rho/\rho_0$ on magnetic field direction for needle-like and gisonite cokes.

*(Courtesy of Prof. Y. Hishiyama of Tokyo City Univ.)*

## b. Binder pitches

Binder pitches have a very important role in the forming process to produce various polycrystalline graphite blocks. Their characterization has usually been performed by the measurements of softening point and viscosity. The estimation of chemical structure of molecules consisting of pitches is also tried by using various analytical techniques [5]. All pitches, prepared either from petroleum oils or coals, consist of various organic compounds and are viscoelastic. Viscoelastic characterization of pitches has been carried out, coupling with their fractionation (Section 2.3.4e) [6].

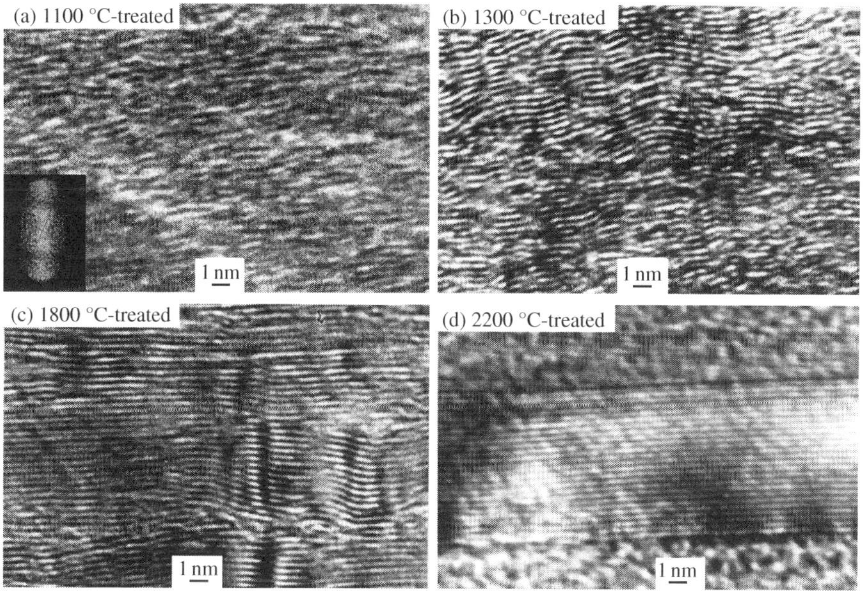


**FIGURE 3.15**

002 lattice fringe images of the cokes prepared from anthracene at different temperatures.

*(Courtesy of Mm. A. Oberlin.)*

For the fractionation of pitches, different procedures have been proposed; fractionation of benzene-soluble part using carbon tetrachloride, n-heptane, etc. [7], fractionation of benzene-insoluble part using the mixing solvent of benzene and n-hexane, which is shown in Fig. 2.61 [8], fractionation of toluene-insoluble part using the mixed solvent of pentane and pyridine [9]. In Table 3.6, number-averaged molecular weight $Mn$, carbon/hydrogen ratio C/H and softening point are tabulated on a pitch and the fractions of its toluene-insoluble but pyridine-soluble part fractionated by using the mixed solvents of pentane and pyridine [9]. The original pitch shows low values of $Mn$, C/H and softening point because the toluene-soluble part is included. The toluene-insoluble part is included from fraction 5 to fraction 1 so that its values of $Mn$, C/H and softening point are intermediate among the values for these five fractions.

In organic molecules, the intrinsic viscosity $[\eta]$ is related to their molecular weight $M$ by the following formula:

$$[\eta] = KM^{a}, \tag{3.1}$$

where $\alpha$ depends on the shape and size of the molecules. In Fig. 3.17, $[\eta]$ in diluted chloroform solution is plotted against $Mn$ for different fractions of a petroleum pitch at different temperatures [8,10,11]. In Fig. 3.17, the value $\alpha$ is not dependent on temperature, and to be about 0.12 in the low-molecular-weight region and about

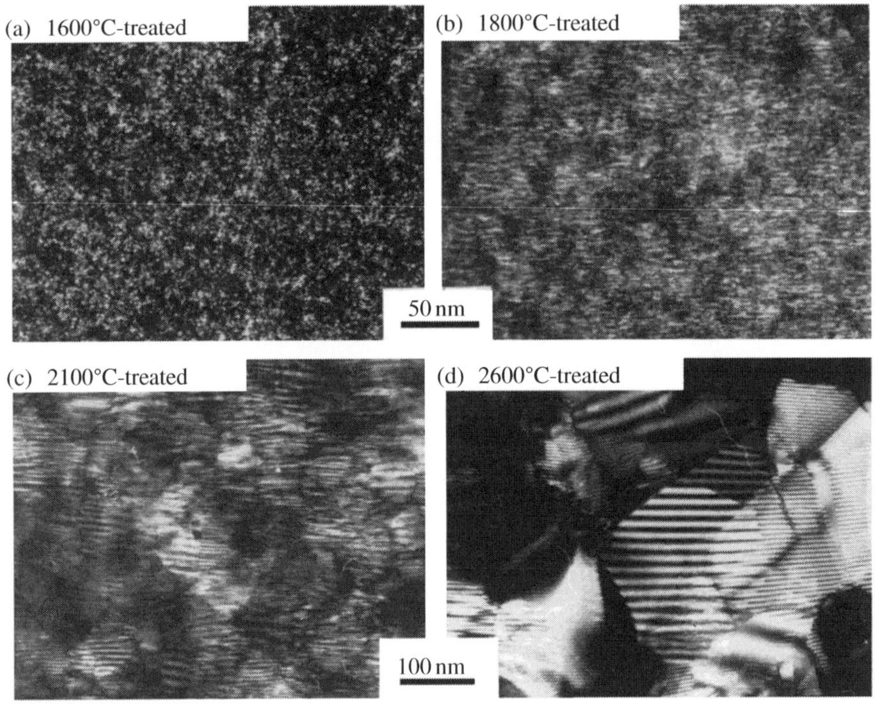


**FIGURE 3.16**

11 dark field images of the cokes prepared from anthracene at different temperatures.

*(Courtesy of Mm. A. Oberlin.)*

**Table 3.6** The *M*n, Carbon/Hydrogen Ratio C/H and Softening Point of a Pitch and the Fractions from its Toluene-Insoluble But Pyridine Soluble Part [9]

| Fraction | *Mn* | C/H | Softening Point (°C) |
|---|---|---|---|
| Original pitch | 430 | 1.45 | 100 ~ −20 |
| Toluene-insoluble part | 1140 | 1.75 | 295 |
| Fraction 5 | 890 | 1.57 | 250 |
| Fraction 4 | 1120 | 1.65 | 290 |
| Fraction 3 | 1200 | 1.72 | 305 |
| Fraction 2 | 1250 | 1.72 | 320 |
| Fraction 1 | 1420 | 1.76 | 335 |
| Pyridine insoluble | 1560 | 1.78 | 355 |

0.21 in the high-molecular-weight region, suggesting that pitch molecules have three-dimensional shape, near spherical rather than plate-like, because α-value is 0.0 for spherical molecules and 0.5 for plate-like molecules. With increasing molecular weight, the shape of molecules is supposed to be more oblate ellipsoidal and their sizes are about 1–2 nm in the equatorial diameter [11].

Since pitches are viscoelastic materials, they have to be characterized by the steady state creep compliance $J_e^0$ and the steady state viscosity $\eta$, the former being a measure of elastic energy stored and the latter of energy dissipation during steady state flow of the pitch.

Creep behavior of pitches has been studied by measuring the torsional creep at temperatures of 30–90°C for different pitches and their fractions [12,13]. The curves of creep compliance $J(t)$ measured at different temperatures are shifted along the axis of log $t$ to overlap the curve at the reference temperature, so that the master curve of $J(t)$ at the reference temperature can be obtained. The curves of $J(t)$ at different temperatures are shown for a fractionated pitch in Fig. 3.15a and the master curves of $J(t)$ at 60.0°C are shown for fractionated and blended pitches in Fig. 3.18b [13].

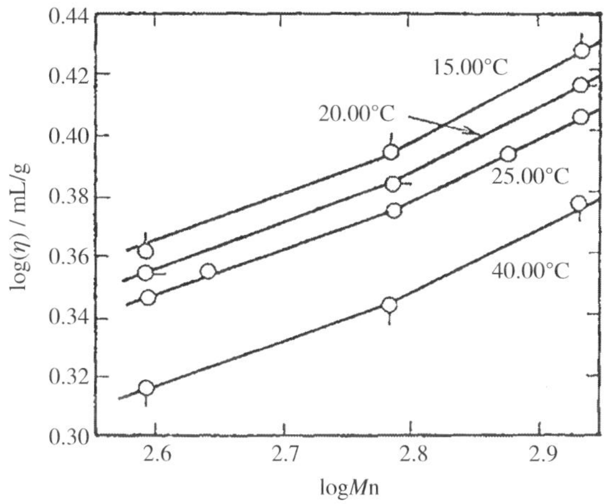


**FIGURE 3.17**

Dependences of [$\eta$] in diluted chloroform solutions on $Mn$ of different fractions of a petroleum pitch.

*(Courtesy of Prof. M. Sakai of Toyohashi Univ. Tech.)*

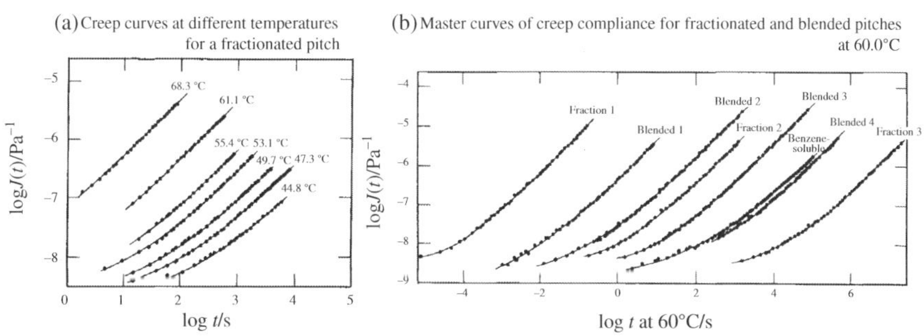


**FIGURE 3.18**

Creep curves for fractionated and blended pitches.

*(Courtesy of Prof. M. Sakai of Toyohashi Univ. Tech.)*

From creep compliance $J(t)$, the steady state viscosity $\eta$ and the steady state creep compliance $J_e^0$ can be obtained by the extrapolation of $J(t)$ according to the following equations:

$$1/\eta = \lim(dJ(t)/\mathrm{d}t), \tag{3.2}$$

$$J_e^0 = \lim\,[J(t) - t/\eta]. \tag{3.3}$$

The temperature dependences of these two viscoelastic parameters, $\eta$ and $J_e^0$, obtained for the fractionated and blended pitches used in Fig. 3.18 are shown in Fig. 3.19 [13].

Fig. 3.19a shows that $\eta$ depends strongly on the constituent molecules of pitch; the fraction 1, which is composed of small molecules (*M*n of 380) can flow at much lower temperature and has much lower *T*g than the fraction 3 (*M*n of 590), and the pitch blended with the fraction 1 to fraction 3 in mass ratio of 7/3 (blended 1) flows in a little higher temperature range than the fraction 1 and the blended 4 in 1.5/8.5 mass ratio flows in a much higher temperature range than the fraction 1 and in a little lower temperature range than the fraction 3. The temperature dependence of $[\eta]$ obtained in Fig. 3.19a can be expressed by the following equation:

$$\ln(\eta/\eta_g) = (\Delta H_\eta/R)(1/T - 1/T_g). \tag{3.4}$$

Viscosity at the glass transition temperature $T_g$ is almost the same for various pitches, about $10^{12}$ Pa s, and activation energy for viscous flow is also constant of about 270–290 kJ/mol.

The $J_e^0$ increases linearly with increasing temperature and so the dependence can be written in the following equation:

$$J_e^0 = J_g + A(T - T_g), \tag{3.5}$$

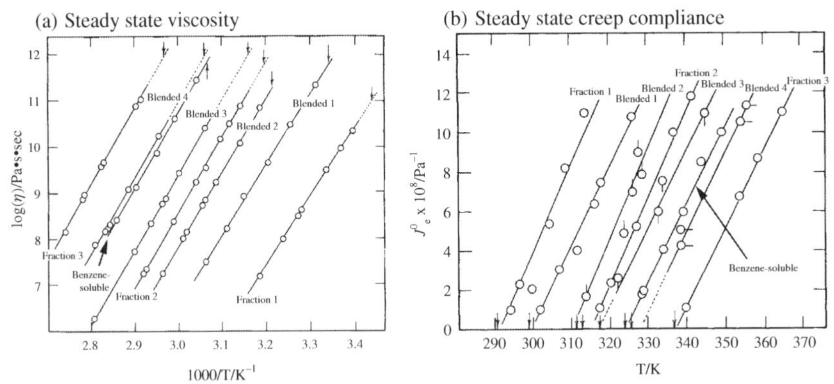


**FIGURE 3.19**

Temperature dependences of $\eta$ and $J_e^0$ for the fractionated and blended pitches. Small arrow for each pitch shows the glass transition point

*(Courtesy of Prof. M. Sakai of Toyohashi Univ. Tech.)*

where $J_g$ is the compliance at $T_g$. As shown in Fig. 3.19b, $J_g$ is very close to zero. The same studies on $\eta$ of different pitches have been reported [12].

It is well known in the field of polymers that viscoelastic properties abruptly change at the glass transition temperature. The glass transition temperature $T_g$ is measured for different pitches and discussed as a function of number-averaged molecular weight $M$n [14]. In Fig. 3.20a, $T_g$ is plotted against $M$n [14]. The relation is linear and expressed by the following equation:

$$T_g = T_{g0} - B/Mn, \tag{3.6}$$

where $T_{g0}$ and B are constants, being determined from Fig. 3.20a to be 423 K and $5.1 \times 10^4$ K, respectively. Through $T_g$, $\eta$ of pitches can be related to $M$n, as shown in Fig. 3.20b. If attention is paid that both $\eta$ and $M$n are plotted in logarithmic scale and the slope of its approximated linear relation is 4.2, the molecular weight dependence of $\eta$ of pitches is very strong, almost one order of magnitude stronger than most molten polymers.

Creep behavior of the carbon paste of needle-like coke with molten coal tar pitch has also been studied by using simple compressive creep apparatus [15,16]. The master curves of creep compliance are shown for the paste with different coke contents in Fig. 3.21a and the dependence of the steady state compliance $J_e^0$ on coke content is shown in Fig. 3.21b.

As shown in Fig. 3.21a, the master curves do not show a plateau for the paste with a coke content $W_{coke}$ of less than 40 mass%, suggesting that the pastes are viscous. For $W_{coke}$ of more than 40 mass%, a plateau is observed on the master curves, which is more flat and more long term, in other words, the paste becomes more elastic, with increasing $W_{coke}$. As shown in Fig. 3.21b, $J_e^0$ decreases rapidly and

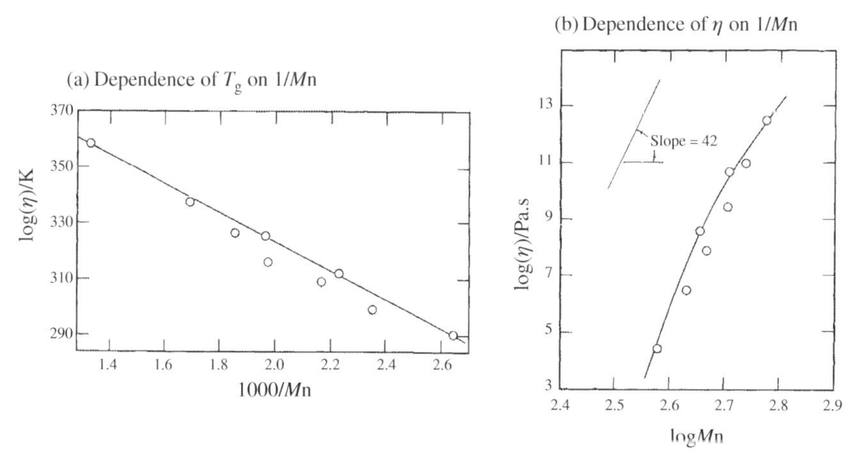


**FIGURE 3.20**

Relations among $T_g$, $M$n and [$\eta$].

*(Courtesy of Prof. M. Sakai of Toyohashi Univ. Tech.)*

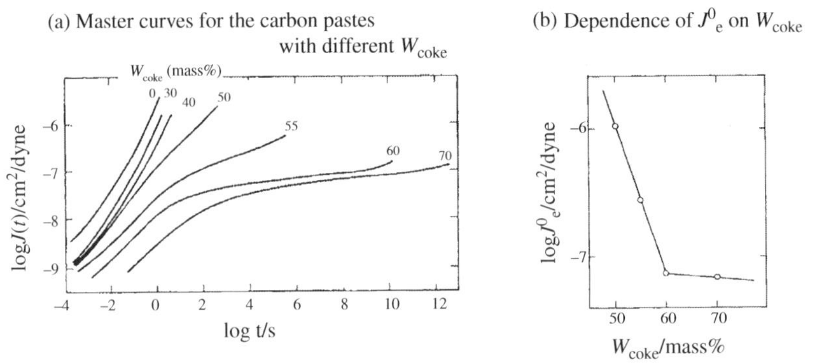


**FIGURE 3.21**

Creep compliance of the carbon pastes with different coke contents.

*(Courtesy of Prof. M. Sakai of Toyohashi Univ. Tech.)*

tends to be saturated with increasing $W_{coke}$, which is due to the interaction between coke particles. For these viscoelastic pastes, very high viscosity is observed, but the activation energy for viscous flow is the same as the pitch itself, about 145 kJ/mol. The size of filler coke influences also on creep behavior, i.e., viscoelastic properties [16]. The addition of thermal black, of which average diameter is about 0.3 μm, makes the viscosity of the paste increase markedly, by the addition of only 14 mass% into a pitch the viscosity becoming 3 times larger [17].

## 3.1.4 Properties

Most properties of polycrystalline graphite blocks are very sensitive for their structure and texture, as many ceramics do. In addition, anisotropy of basic structure units (crystallites) and also that of coke particles lead to make their preferred orientation important for understanding the properties.

Mechanical properties of graphite blocks depend strongly on their texture. In Fig. 3.22, optical micrographs of the cross-section of the pole and nipple of graphite electrode are shown. From these observations, the coke particles for the pole are much larger and more porous than those for the nipple. This is due to the requirements for these two parts of graphite electrode, the latter having to have higher mechanical properties. In Table 3.7, some properties are compared between pole and nipple of graphite electrode; the nipple has higher bulk density and bending strength, but lower electrical resistivity and thermal expansion coefficient than the pole.

Not only size but also shape of pores was pointed out to give certain influence on mechanical properties. Optical micrographs of the cross-sections of high-density isotropic graphite blocks with different bulk densities are shown in Fig. 2.160. From these micrographs, different pore parameters, pore density, average cross-sectional area, roughness of pore surface and fractal dimension, have been determined with the aid of image processing, as shown in Table 3.8, and the

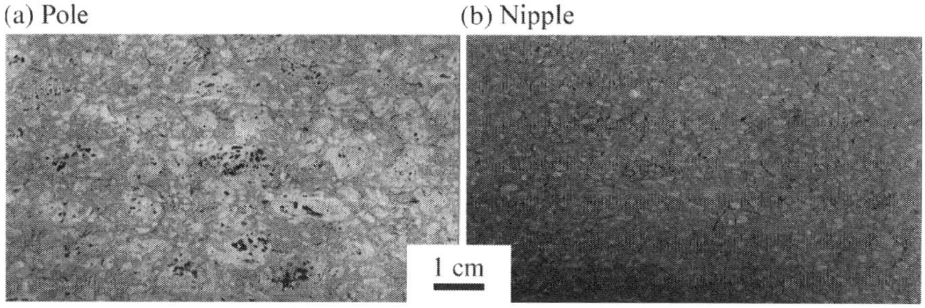


**FIGURE 3.22**

Optical textures of the pole and nipple of graphite electrode.

**Table 3.7** Comparison in Properties Between Pole and Nipple of Graphite Electrode

| | Bulk Density ($g/cm^3$) | Bending Strength (MPa) | Young's Modulus (GPa) | Electrical Resistivity ($\mu\Omega \cdot m$) | Thermal Expansion Coefficient ($\times 10^{-6}/°C$) |
|---|---|---|---|---|---|
| Pole | 1.68–1.74 | 10–15 | 9–14 | 4.5–6.5 | 0.5–1.0 |
| Nipple | 1.76–1.86 | 25–36 | 20–28 | 2.8–3.6 | 0.3–0.5 |

*(Courtesy of Tokai Carbon Co., Ltd.)*

**Table 3.8** Pore Parameters Measured by Image Analysis on High-Density Isotropic Graphite Blocks [18]

| Sample | Bulk Density ($g/cm^3$) | Porosity (%) | Pore Density ($mm^{-2}$) | Average Cross-Sectional Area ($\mu m^2$) | Roundness of Pores | Fractal Dimension |
|---|---|---|---|---|---|---|
| A | 1.735 | 21.9 | 414 | 545 | 0.65 | 1.40 |
| B | 1.788 | 25.3 | 446 | 506 | 0.69 | 1.40 |
| C | 1.842 | 21.1 | 480 | 395 | 0.66 | 1.33 |
| D | 1.842 | 21.7 | 822 | 244 | 0.68 | 1.46 |
| E | 1.848 | 20.9 | 1275 | 155 | 0.71 | 1.47 |
| F | 1.802 | 12.0 | 2731 | 31 | 0.79 | 1.56 |

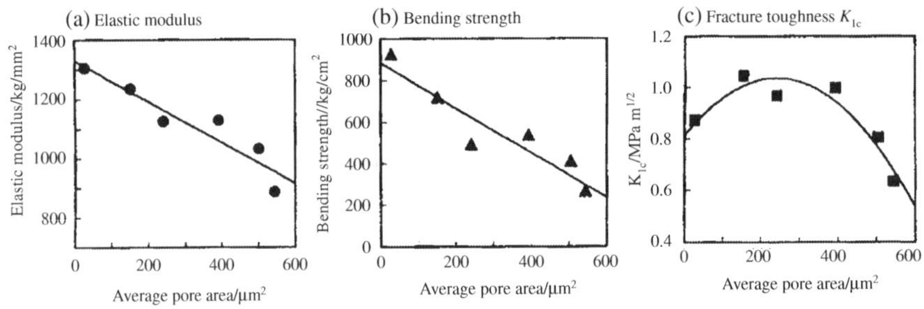


**FIGURE 3.23**

Relations between averaged pore area and mechanical properties [18].

relations between these pore parameters and different mechanical properties are discussed [18].

With increasing bulk density, pore density tends to increase, average cross-sectional area of pores decreases and roundness of pores slightly increases, i.e., pores become circular, but the fractal dimension increases slightly, i.e., pore walls become complicated.

Pore parameters of the polycrystalline graphite blocks, which are produced from fine particles of a coke using a pitch binder, are closely related to mechanical properties. As shown in Fig. 3.23, elastic modulus and bending strength decrease with increasing average pore area, but fracture toughness shows a maximum at an area of around 200–300 $\mu m^2$.

The dependence of Young's modulus parallel to the extrusion on HTT for carbon blocks fabricated from coke particles is shown in Fig. 3.24 [19]. The modulus shows a maximum at HTT of about 1000°C, and then decreases up to 2000°C, the value at 2200°C being about one half of the maximum value. The decrease of modulus can be explained by the increase in dislocation density and the formation of micropores by heat treatment. Above 2200°C, the modulus increases a little and then turns to decrease slightly, mainly due to the growth of graphite structure.

Breaking strength of polycrystalline graphite blocks increases with increasing temperature in both tensile and compressive modes [20–27]. In Fig. 3.25, temperature dependence of tensile strength on temperature is shown on graphite blocks with different mixing ratios of the filler coke with three different particle sizes [22]. The tensile strength at 2500°C is roughly 2 times larger than that at room temperature. Since mechanical strength is texture-sensitive, the production conditions strongly influence on the breaking strengths, as shown on the composition of filler coke particles in Fig. 3.25. Above 2500°C, strength decreased rapidly because of creep (plastic deformation). Under compression, rod-shape specimen deforms to drum-shape due to creep at a temperature above 2500°C [24].

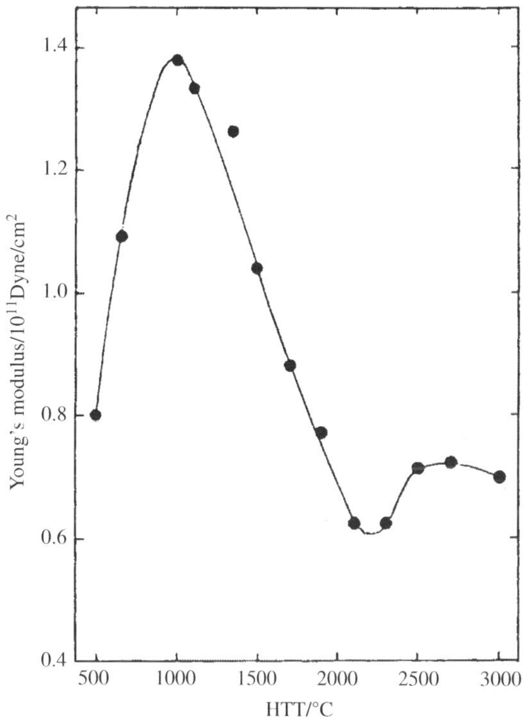


**FIGURE 3.24**

Dependence of Young's modulus along the extrusion axis on HTT for coke-based carbon blocks [19].

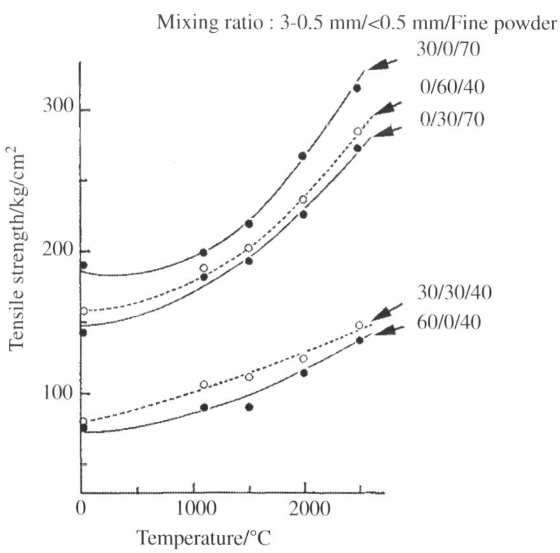


**FIGURE 3.25**

Temperature dependences of tensile strength of graphite blocks [22].

Young's modulus of commercially available reactor-grade polycrystalline graphite blocks increases slightly with increasing temperature, its maximum occurring at a temperature from 1800 to 2000°C, as shown in Fig. 3.26a [27].

Based on the development in non-linear fracture mechanics, the resistance to fracture of the materials is characterized by fracture toughness $K_{Ic}$. In Fig. 3.26b, temperature dependence of $K_{Ic}$ determined using a disk-shaped specimen is shown for different nuclear-grade polycrystalline graphite blocks [27,28]. Fracture toughness increases with increasing temperature, as breaking strengths do.

Non-linear fracture behavior of polycrystalline graphite materials has been analyzed by applying the non-linear energy principle for elastic-plastic materials [29]. The relations between load and loadpoint displacement (P-u diagram) are measured by repeating loading–unloading cycles on various carbon materials. Some examples of the P-u diagram are shown in Fig. 3.27. By unloading on the way of crack propagation, the loadpoint usually does not return to the origin, which is considered to be due to a plastic displacement. By subtracting these plastic displacements from the observed loadpoint displacement, the P-u diagram due to elastic displacement could be estimated (broken curves in Fig. 3.27). The areas under these two P-u diagrams give the energies for fracture, total fracture energy (work of fracture, $\gamma_{WOF}$) from the observed P-u diagram (solid line in the figure) and elastically stored energy $\gamma_e$ from the broken diagram. The balance between $\gamma_{WOF}$ and $\gamma_e$ gives the plastic energy dissipation $\gamma_p$. In Table 3.9, fracture energies for four polycrystalline graphite blocks, including three blocks in Fig. 3.27, are summarized [30–32].

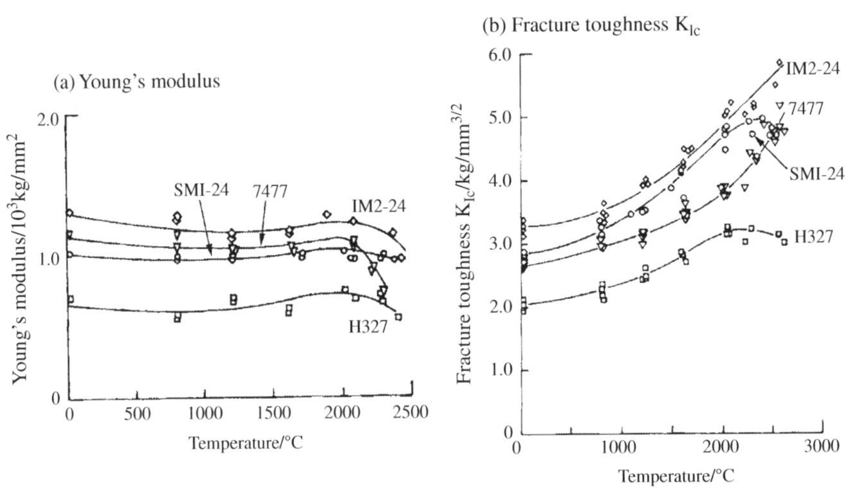


**FIGURE 3.26**

Temperature dependences of Young's modulus and $K_{Ic}$ of reactor-grade graphite blocks.

*(Courtesy of Prof. S. Sato of Meijo Univ.)*

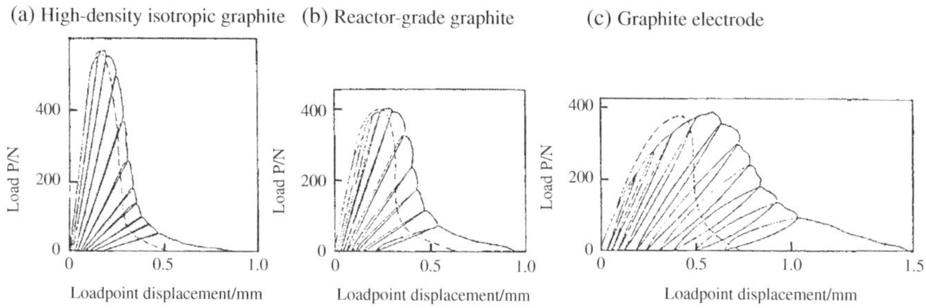


**FIGURE 3.27**

The P-u diagrams for different graphite blocks.

*(Courtesy of Prof. M. Sakai of Toyohashi Univ. Tech.)*

**Table 3.9** Fracture Energies for Polycrystalline Graphite Blocks

| Sample | Production | Bulk Density (g/cm³) | Breaking Strength (MPa) | Young's Modulus (GPa) | Fracture Energy (J/m²) | | | |
|---|---|---|---|---|---|---|---|---|
| | | | | | $\gamma_{WOF}$ | $\gamma_e$ | $\gamma_p$ | $\gamma_p/\gamma_{WOF}$ |
| Grade for arc-discharge machining | High-density, isotropic | 1.83 | 106 | 16.5 | 92 | 73 | 19 | 0.21 |
| Nuclear-grade | Needle coke, CIP | 1.75 | 35.0 | 9.83 | 103 | 66 | 37 | 0.36 |
| | Gilsonite coke | 1.78 | 27.6 | 9.07 | 120 | 75 | 45 | 0.37 |
| Electrode-grade | Coarse coke, extrusion | 1.68 | 8.72 | 6.47 | 191 | 92 | 99 | 0.52 |

*(Courtesy of Prof. M. Sakai of Toyohashi Univ. Tech.)*

From Fig. 3.27 and Table 3.9, high-density isotropic graphite has a small proportion of plastic energy dissipation (i.e., small plastic deformation) and a large breaking strength, but a small $\gamma_{WOF}$. On the other hand, electrode-grade graphite with low bulk density and low bending strength has a high $\gamma_{WOF}$, which is mainly due to a large proportion of plastic energy dissipation $\gamma_p$.

The detailed investigations on crack propagation behavior based on the energy principles were carried out on different graphite materials, including carbon-fiber-reinforced graphite blocks [30–36].

Non-linear fracture behavior and apparent plastic deformation of polycrystalline graphite blocks is due to different reasons, plastic deformation at the boundary between crystallites, deformation of pores and formation of microcracks. Formation of microcracks is exhibited by SEM to occur in the places far from the notch tip, where the load is concentrated [34].

A single crystal of graphite is known to have a strong anisotropy in thermal expansion coefficient $\alpha$. According to direct measurements of $\alpha$ on highly oriented pyrolytic graphite (HOPG), $\alpha$ along the basal plane is extremely small negative (i.e., shrinkage) of the order of $10^{-6}\,K^{-1}$ below 600 K, showing a minimum at about 210 K [37,38]. After sign reversal it increases gradually with the increase in temperature. The $\alpha$ perpendicular to the basal plane at room temperature is quite large, $26.3 \times 10^{-6}\,K^{-1}$ and its temperature dependence is positive in the whole temperature range, increasing gradually with increasing temperature.

On the graphite blocks which are the aggregates of anisotropic crystallites, thermal expansion coefficient $\alpha$ is positive in the whole temperature range above room temperature, the value being not so high as that perpendicular to the basal plane in HOPG, as shown in Fig. 3.28 on some blocks produced from coke particles and heat-treated at high temperatures [39]. For the blocks, not only the value of $\alpha$ but also its anisotropy depend strongly on the preparation conditions, particularly forming process; the carbon formed by isostatic pressing (B in Fig. 3.28) is almost isotropic, but those by either extrusion or compression (A and C) show a pronounced anisotropy, which is mainly due to preferred orientation of crystallites along the pressing direction. By the addition of $B_4C$ with hot-pressing (D), graphitization, i.e., crystallite growth, is accelerated and resulted in enhanced anisotropy [40].

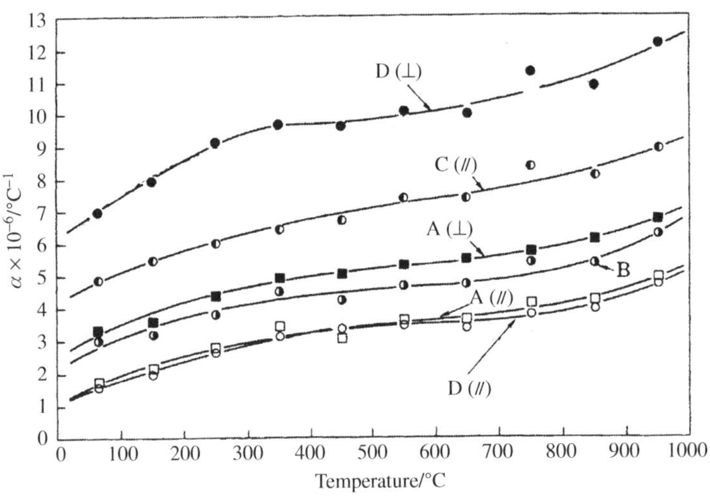


**FIGURE 3.28**

Dependences of $\alpha$ on various graphite blocks (courtesy of Prof. K. Kobayashi). A: Extruded and heat-treated at 2800°C. B: CIP-formed and heat-treated at 2800°C. C: Molded and heat-treated at 2800°C. D: Hot-pressed at 2200°C under 40 MPa with 5 mass% $B_4C$.

The temperature dependence of $\alpha$ for the polycrystalline blocks has been explained by the relaxation of anisotropic large expansion perpendicular to the layer planes in each crystallite by microcracks and micropores in the bulk, of which the formation is shown for the particle of needle-like coke in Fig. 3.13. The $\alpha$ observed on the blocks $\alpha_{obs}$ can be approximated by the following simple equation;

$$a_{obs} = Aa_c + (1 - A)a_a,$$

where $\alpha_a$ and $\alpha_c$ are thermal expansion coefficients along and perpendicular to the basal plane of single crystal of graphite, respectively, and A is an experimental parameter which depends on development of structure and preferred orientation of crystallites in the block [41].

Thermal conductivity shows also strong anisotropy. The thermal conductivity $\kappa$ of natural graphite parallel to the basal plane gives a maximum around 80 K, the value being $2.8 \times 10^3$ W/m · K [42]. At temperatures above 80 K, $\kappa$ decreases rapidly with increasing temperature. The anisotropy ratio of $\kappa$ along the basal plane to that perpendicular to the basal plane has been estimated to be not smaller than 200 for HOPG.

Experimental results on thermal conductivity $\kappa$ are reported on commercially available and laboratory-prepared carbons [42,43]. The dependences of $\kappa$ on temperature along a direction parallel to the extrusion axis is shown for the extruded blocks heat-treated at different temperatures in Fig. 3.29 [44]. The value of $\kappa$ at a constant temperature increases with increasing HTT, more markedly in the blocks

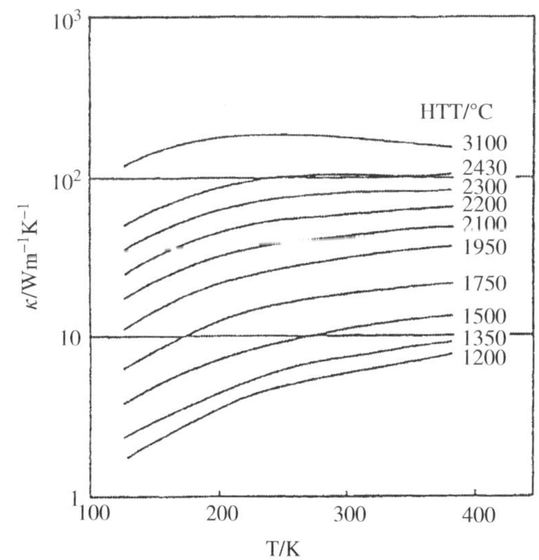


**FIGURE 3.29**

Temperature dependences of $\kappa$ along the extrusion direction of carbon blocks heat-treated at different temperatures [44].

heat-treated at low temperatures. However, the values of $\kappa$ for these carbons are much smaller than that perpendicular to the basal plane of natural graphite, even after heat treatment at 3100°C. Theoretical considerations on thermal conductivity on carbon materials heat-treated at different temperatures have been presented [44,45].

For the blocks which have been extruded a coke particles with a binder pitch and heat-treated between 1700 and 3000°C after being baked at 1300°C, experimental results of the absolute thermoelectric power *S* along the extrusion axis are shown in Fig. 3.30 [46]. It has been confirmed by the measurements of X-ray diffraction and $(\Delta\rho/\rho_0)_{max}$ that the layer order is turbostratic for the specimens heat-treated below 1900°C and graphitic structure is developed gradually with the increase in HTT above 2100°C. The experimental result on a kish graphite flake [47] is included in the figure for comparison.

The blocks with turbostratic structure, i.e., the blocks heat-treated below 1900°C, show a positive thermoelectric power *S* which increases almost linearly with temperature, indicating a hole-type conduction (Fig. 3.30a). The slope of the *S* versus temperature relation increases with the increase in HTT, in other words, the Fermi level rises with increasing HTT. The blocks heat-treated above 2100°C show a negative peak in the range between 20 and 35 K, as observed in the kish graphite (Fig. 3.30b). This negative peak is attributed to a phonon drag effect caused by the coupling between carriers and long wavelength in-plane phonons [48]. The width of the negative peak for the carbons is broader than that for the kish graphite. This behavior was explained by Rayleigh scattering due to

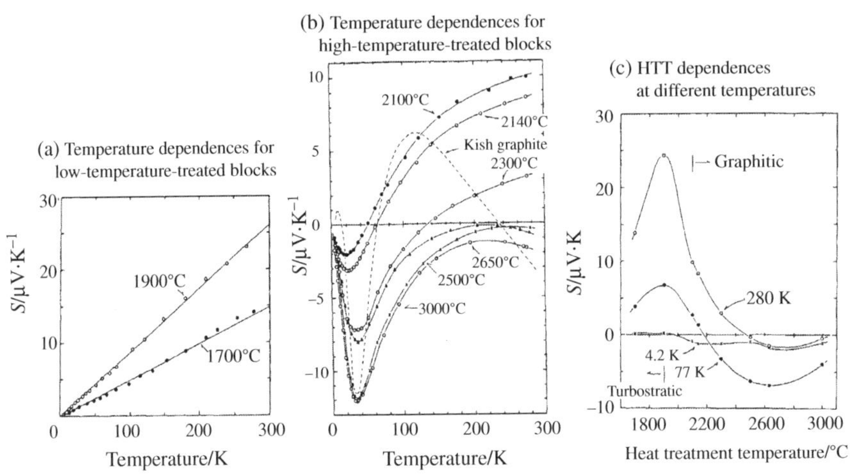


**FIGURE 3.30**

The *S* for the extruded coke-based carbon blocks heat-treated at different temperatures (courtesy of Prof. Y. Hishiyama of Tokyo City Univ.). For comparison, the data for kish graphite flake is shown.

crystalline defects. The thermoelectric power $S$ measured at 4.2, 77, and 280 K are plotted as a function of HTT in Fig. 3.30c [46].

Thermal and mechanical properties have been measured as functions of ambient temperature and also HTT on the blocks, which are fabricated from different cokes with different particle sizes by using binder pitches. These properties are mostly texture-sensitive, so that they depend strongly on the conditions of fabrication, particularly on forming process, and only a little on the structure development.

One of the important applications of high-density isotropic graphite is the electrode for electric discharge machining (EDM) (Fig. 3.9). However, published data on the performance for EDM electrode are quite rare. In Fig. 3.31, the effect of particle size of the coke on EDM performance is shown [49]. Average particle size of the filler coke governs wearing of the electrode, surface roughness of machined metal and machining speed. In order to attain small wear ratio and smooth surface of machined metal, small particle size of coke is required (Fig. 3.31a and b). To perform machining speedy, however, it is desired to use large particle size (Fig. 3.31c). Anisotropy in wearing of the electrode increases markedly with increasing anisotropy ratio in thermal expansion (Fig. 3.31d).

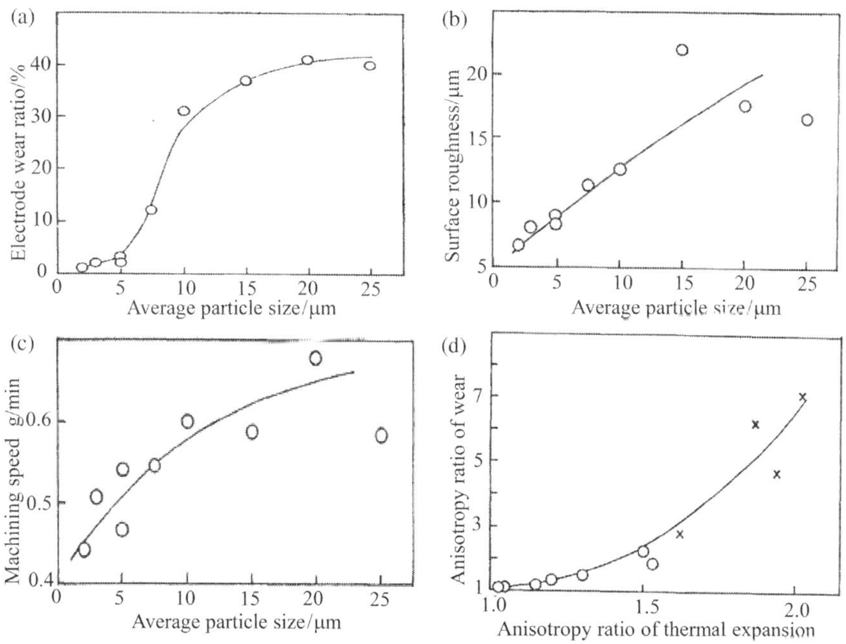


**FIGURE 3.31**

Dependences of EDM performances on the production parameters of high-density isotropic graphite [49].

## 3.2 Highly oriented graphite

### 3.2.1 Highly oriented graphite

The blocks or particles, which are composed of large-sized layers stacked in large numbers with graphitic regularity, are called highly oriented graphite. The extreme case is a single crystal of graphite. In practice, however, single crystals with large size are difficult to be obtained, and it is almost impossible to get those of more than a few $mm^2$. Since graphite has typical layered structure and so is very easy to be cleaved along the layer, it is impossible to get single crystals with certain thickness. There are only two possibilities to find single crystals of graphite, in natural graphite ores and so-called kish graphite. The resources of high-quality natural graphite are very limited: Sri Lanka, Madagascar and China. Even in these ores, the possibility to find single crystals of a certain size is only occasional. Kish graphite is formed by the precipitation of supersaturated carbon from molten iron, and cannot be in a large size. However, some of them have been found to have very high crystallinity, which can be called single crystal.

The alternatives for single crystal graphite are the synthetic graphites of which layers are highly oriented, such as highly oriented pyrolytic graphite (HOPG) and graphite films derived from some organic polymer precursors, such as polyimides. These two materials are not single crystals, but have an extremely high degree of orientation of *c*-axis of small graphite crystals (crystallites) along the normal to the flakes and films. In these flakes and films, however, *a*-axes of graphite crystallites are randomly oriented.

Large-sized graphite sheets have been produced from natural graphite flakes, which are in small sizes and irregular shapes, via their intercalation and following exfoliation. They are usually called flexible graphite sheets and are industrially produced in large amounts. In these sheets, graphite layer planes are preferentially oriented along the sheet surface.

Therefore, five kinds of highly oriented graphite are available, natural graphite, kish graphite, pyrolytic graphite, graphite films derived from some polymers, and flexible graphite sheets.

### 3.2.2 Natural graphite

Natural graphite is usually recovered as a powder from the ore through the following process; milling, purification by floating, chemical purification with either NaOH or mixed solution of HCl/HF. Natural graphite powder contains 85–94 mass% C after the purification by floating and more than 99 mass% C after chemical purification. Some natural graphites consist of flaky particles having highly crystalline structure and highly oriented nanotexture (flaky graphite), as shown as scanning electron micrograph (SEM) and transmission electron micrograph (TEM) with electron diffraction pattern in Fig. 3.32a. But some of them consist of small crystals, as shown its SEM and TEM in Fig. 3.32b, and used to be called amorphous graphite, but they are now clarified to be microcrystalline [50]. These

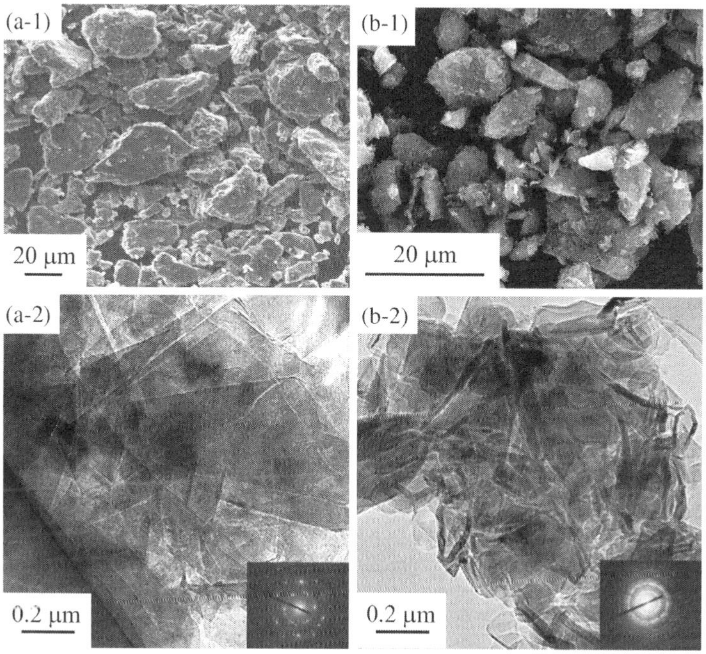


**FIGURE 3.32**

SEM and TEM images of natural graphite powder. (a) Flaky graphite and (b) microcrystalline graphite.

two kinds of natural graphite can be differentiated by X-ray diffraction (XRD) technique, as shown in Fig. 3.33. Flaky graphite shows sharp 100 and 101 XRD peaks and some of them contain metastable rhombohedral modification due to shear stress during milling process (Section 2.2.1), but microcrystalline graphite is characterized by broad 101 peak.

Natural occurrence of these natural graphite crystals was geologically estimated from the facts that minerals coexisted to be very mild, at a temperature lower than 1000°C and a pressure of about 0.5 GPa [51,52]. Under atmospheric pressure without catalysts, however, graphitization of carbons requires a temperature higher than 2500°C. By using some metals, such as iron and nickel, as catalysts, graphitization of carbons is accelerated, but still it needs a temperature above 1000°C [53,54]. Pressure accelerates graphitization, but still high temperature as 1600°C is required to graphitize carbon materials [55–58]. Therefore, the metamorphic rocks, in which natural graphite is embedded, may have an accelerating (catalytic) effect on the graphite formation (graphitization), in addition to the acceleration by pressure. Experimentally graphitization of carbons is proved to be accelerated in coexistence of different minerals, such as $CaCO_3$, CaO, Ca $(OH)_2$, $CaF_2$ $Al_2O_3$ $NaCO_3$ under the pressure of 0.3 GPa [56,59–62]. From these

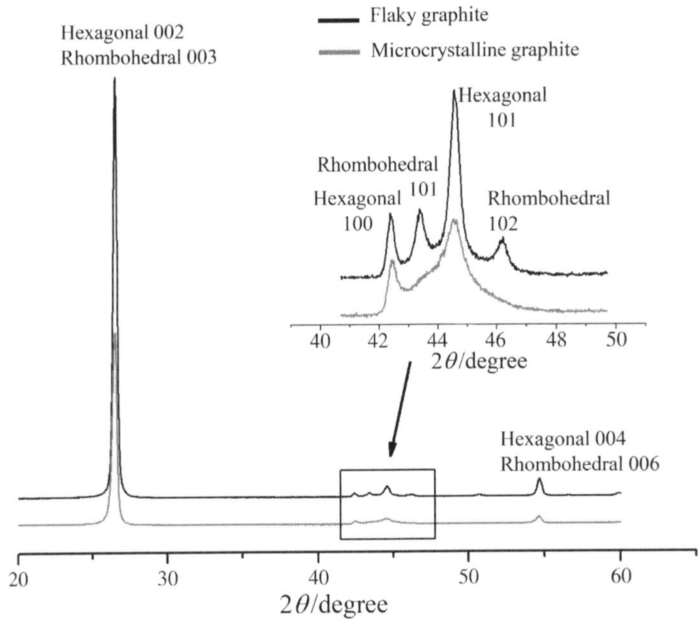


**FIGURE 3.33**

XRD patterns of flaky and microcrystalline natural graphite [50].

experimental results, it is concluded that graphite can be formed under mild conditions with the catalytic action of some metallic species, such as calcium. If we could realize extremely long residence time under these mild conditions, it seems to be possible to reproduce graphite crystals in the laboratory that are similar to those formed geologically in nature.

Natural graphite is a promising anode material for lithium ion rechargeable batteries due to its high reversible capacity, high charge/discharge performance and low cost. In Fig. 3.34, the charge/discharge performance for the first cycle is shown for flaky graphite and microcrystalline graphite. Both of them exhibit high discharge capacity with marked potential plateau at low voltage. However, flaky graphite suffers from poor cycle stability because of the anisotropic volume expansion/shrinkage during charge/discharge process. On the other hand, microcrystalline graphite shows much better cycle stability than flaky graphite, but it suffered from lower coulombic efficiency [50].

Carbon coating has been proved to be effective to improve battery performance, especially for anode graphite [63–66]. However, the cycle stability of carbon-coated flaky graphite is still not enough for practical application. To resolve this problem, a so-called 'mild-expansion modification' has been introduced before carbon coating [63]. First, spherical particles agglomerated flaky natural graphite are dipped into $H_2SO_4$ to be partially converted to graphite intercalation compounds,

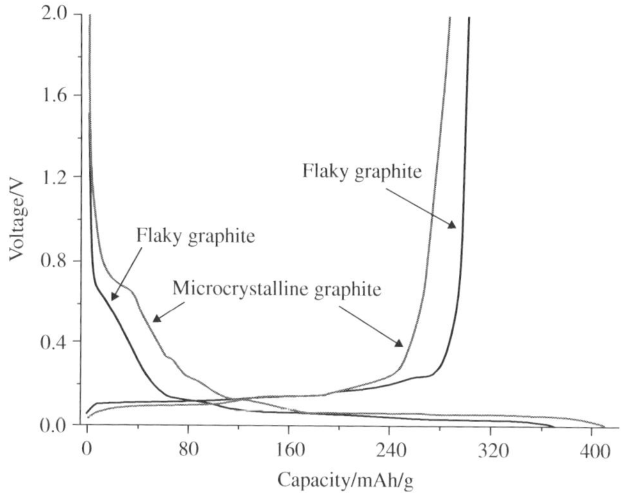


**FIGURE 3.34**

The first charge/discharge cycle of flaky and microcrystalline graphites [50].

followed by heating at 360°C to decompose intercalation compounds. SEM images of original spherical particles and after mild-expansion modification are shown in Fig. 3.35. During the intercalation/de-intercalation of $H_2SO_4$, the interlayer distance between graphite layers is slightly increased, because of the introduction of defects, such as stacking fault, and pores. The larger interlayer distance and porous structure act as buffer for the volume change of flaky graphite particles, and the porous structure can offer more places to store the Li ions according to the microporous storage mechanism, which greatly improves the reversible capacity. As shown in Fig. 3.36, the spheres composed of flaky natural graphite modified by the 'mild-expansion' treatment accommodate a capacity of 378 mAh/g and keep 100% capacity up to 100 cycles after carbon coating.

Microcrystalline graphite has not been used as anode material in commercial cells, mainly because of low efficiency at the first cycle as 70%. However, control of particle size and its distribution via powder processing improves cycle efficiency after carbon coating. The first-cycle efficiency after powder processing and carbon coating increased to 89.9% from 86.2%. Microcrystalline graphite coated by carbon twice shows a capacity of 340 mAh/g and very good stability, which enables microcrystalline graphite to become a very promising commercial anode material.

Microcrystalline graphite is also a suitable filler material for isotropic graphite [67a]. Graphite artifacts can be prepared with microcrystalline graphite via isostatic pressing, as shown in Fig. 3.37. The manufacture process is the same as traditional isostatic graphite which is made from coke, and the final density of more than 1.7 g/cm$^3$ is achieved by the repetition of pitch impregnation and following baking. By adjusting the particle size distribution of microcrystalline graphite, the

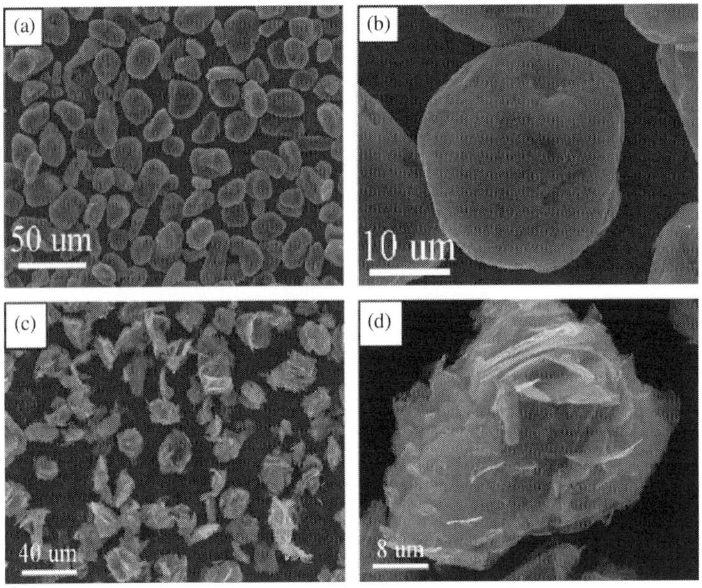


FIGURE 3.35

SEM images of spherical agglomeration of flaky natural graphite (a and b) and after "mild-expansion" treatment (c and d).

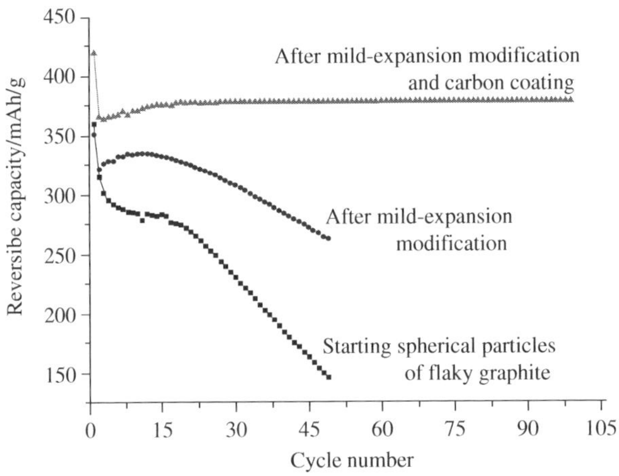


FIGURE 3.36

Cycle performance of spherical particles of flaky natural graphite, after mild-expansion modification and carbon coating [63].

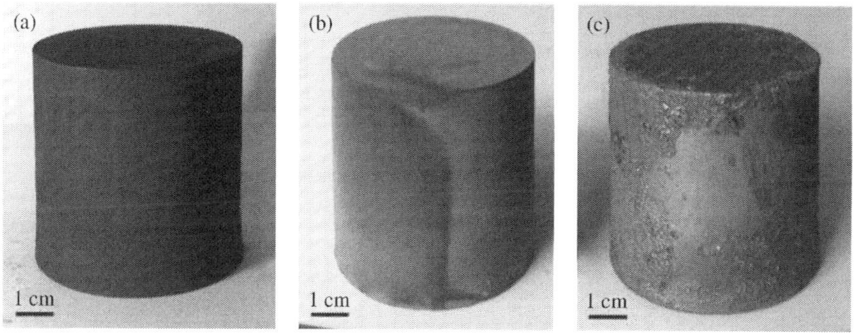


**FIGURE 3.37**

Photographs of (a) green body, (b) after first baking, (c) after pitch impregnation and graphitization [67a].

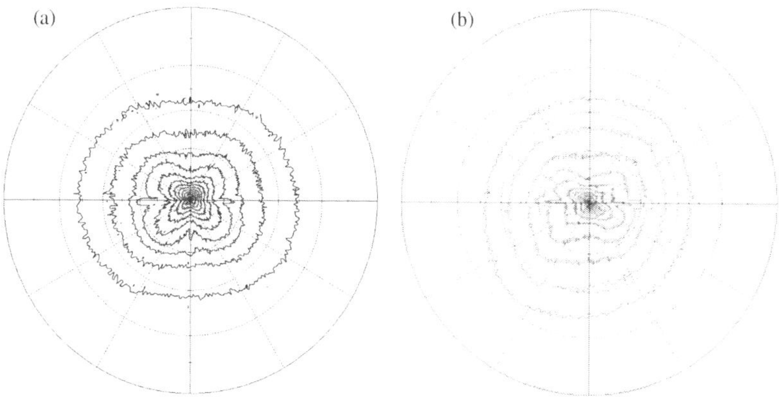


**FIGURE 3.38**

002 pole figures of (a) artifact with microcrystalline graphite filler after baking and (b) IG-110.

flexure strength after graphitization ranges from 20–45 MPa. There are several advantages for microcrystalline fillers [67a]. One of the advantages is that the product has very low preferred orientation, in other words, isotropic graphite or near isotropic graphite is obtained easily. Fig. 3.38 compares the 002 pole figures for the artifact prepared from microcrystalline graphite and IG-110, nuclear grade graphite with an isotropy ratio of 1.05. The pole figure represents the orientation distribution of $<002>$ crystal direction, for an ideal isotropic situation concentric circles have to be observed. The pole figure for the artifact of microcrystalline graphite shows no obvious preferred orientation, and is very similar to that of IG-110, indicating an isotropic nature. Another advantage for microcrystalline filler lies in the high thermal diffusivity or conductivity of green bodies. The thermal diffusivity is a measure of how quickly a body can change its temperature, in other words, the substance with high thermal diffusivity is able to adjust their

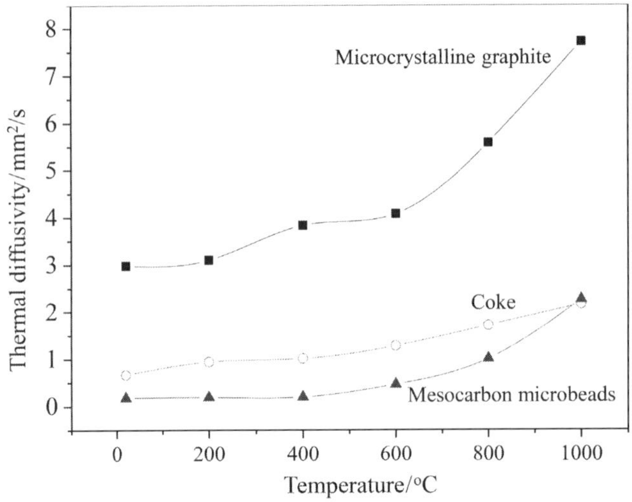


**FIGURE 3.39**

Temperature dependence of thermal diffusivity of the green bodies.

temperature to that of its surroundings rapidly. The body prepared from microcrystalline graphite has a high graphitization degree, which makes thermal diffusivity of the green body high, much higher than that of coke and mesocarbon microbeads, during baking, as shown in Fig. 3.39. The higher thermal diffusivity gives rise to a more homogeneous temperature distribution, which can avoid cracking of green bodies, and in turn allow a faster heating rate.

## 3.2.3 Kish graphite

Into metal iron carbon can be dissolved, the amount depends strongly on the temperature. A part of carbons dissolved into molten iron at high temperatures is incorporated into the crystal lattice of iron after solidification and forms alloys, this giving us different steels, but another part of dissolved carbon precipitates as a separated phase from iron, in most cases as graphite. These graphite flakes precipitated from molten iron in the course of cooling from a high temperature are called kish graphite [67b–67d]. In the course of the production of iron and alloys, rather large amounts of kish graphite are obtained as one of the byproducts, but all of them do not have very high crystallinity, because it depends strongly on the precipitation condition. When they are produced at a very high temperature, as high as iron evaporates, the kish graphite flakes were found to have single crystal nature. The flakes of kish graphite usually have irregular shapes, as shown in Fig. 3.40, and are very thin.

In these kish graphite flakes formed either during the cooling of molten iron or by evaporation of iron, a relatively large amount of iron remains as impurity.

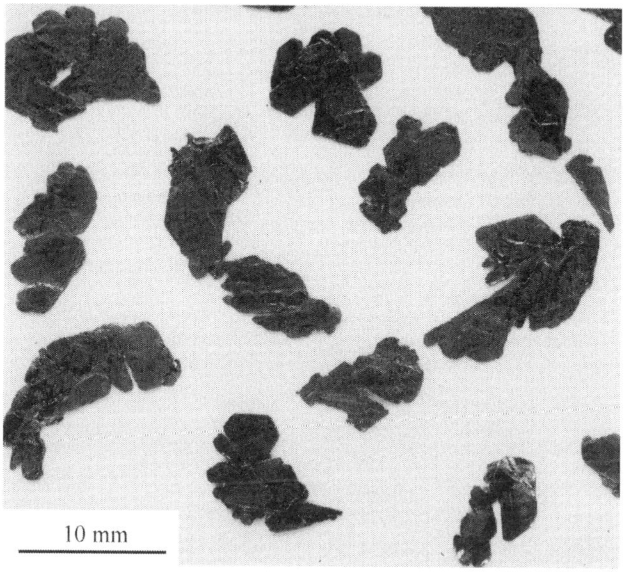


**FIGURE 3.40**

Kish graphite flakes.

*(Courtesy of Prof. A. Yoshida of Tokyo City Univ.)*

In order to have highly crystallized graphite flakes which can be regarded as single crystals, therefore, purification is essential, for example, heat treatment in a flow of halogen gas at a high temperature.

High crystallinity of the particles of these kish graphite flakes has been shown from the observations by scanning electron microscopy (SEM) on the edge surface of the flake and by electron channeling techniques on the surface [68,69]. Their edge surfaces are constructed from regular stacking of layers, as shown in Fig. 3.41a and b by SEM with a low accelcration voltage of 3 kV. The surfaces of these flakes give a well-organized electron channeling pattern, as shown in Fig. 3.42, indicating that it is a single crystal. In an electron channeling contrast image, no contrast was observed and its brightness changed in every 60° rotation of the flake, which is another proof of the single crystal.

From the examination of the crystallinity of a large number of flakes of kish graphite by the measurement of galvanomagnetic properties, each flake is found to have a wide range of structural perfection. Table 3.10 shows resistivity ratio $\rho_{300K}/\rho_{4.2K}$, i.e., the ratio of electrical resistivity at room temperature (300 K) to that at liquid helium temperature (4.2 K), maximum transverse magnetoresistance $(\Delta\rho/\rho_0)_{max}$ measured at liquid nitrogen temperature (77 K) under a magnetic field perpendicular to the flakes with a strength of 1 T, and anisotropy ratios $r_{TL}$ and $r_T$ (Section 2.5.1). The larger values of $\rho_{300k}/\rho_{4.2K}$ and $(\Delta\rho/\rho_0)_{max}$ reveal the higher perfection in crystal structure [71,72].

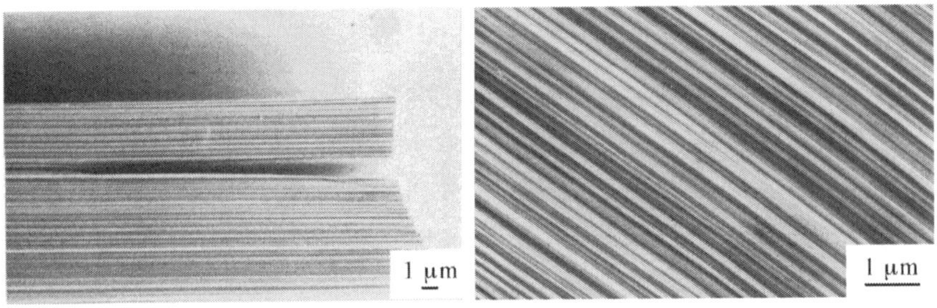


**FIGURE 3.41**

SEM images of edge surface of a flake of kish graphite.

*(Courtesy of Prof. Y. Hishiyama of Tokyo City Univ.)*

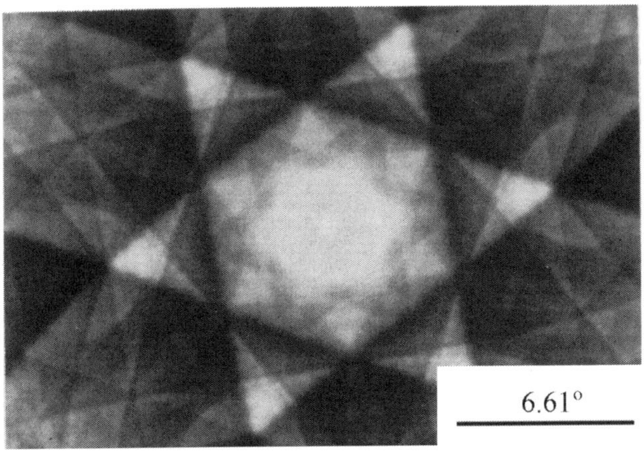


**FIGURE 3.42**

Electron channeling pattern of a flake of kish graphite.

*(Courtesy of Prof. A. Yoshida of Tokyo City Univ.)*

From Table 3.10, the flakes of kish graphite have a wide range of crystallinity from 50 to 4 in $\rho_{300K}/\rho_{4.2K}$ and 3800 to 700% in $(\Delta\rho/\rho_0)_{max}$, even though they have been prepared through high-temperature evaporation of iron. In kish graphite flakes formed through the precipitation from molten iron, the range of structural perfection of the flake is expected to be extended toward much lower values of $\rho_{300K}/\rho_{4.2K}$ and $(\Delta\rho/\rho_0)_{max}$. The flakes which have high $\rho_{300K}/\rho_{4.2K}$ and $(\Delta\rho/\rho_0)_{max}$ tend to have low anisotropy ratios $r_{TL}$ and $r_T$, revealing that the flakes are composed of the crystallites taking planar orientation scheme in a high degree.

On the kish graphite flakes with high crystallinity, Shubnikov-de-Haas oscillation in magnetoresistance, which is characteristic for high crystallinity, is observed at as low a temperature as 4.2 K [73]. In Fig. 3.43, the dependences of $\Delta\rho/\rho_0$ on magnetic field strength $B$ are shown for some kish graphite flakes with

**Table 3.10** Galvanomagnetic Characteristics for Different Flakes of Kish Graphite [70]

| Sample | $\rho_{300K}/\rho_{4.2K}$ | $(\Delta\rho/\rho_0)_{max}$ (%) | $r_{TL}$ | $r_T$ |
|---|---|---|---|---|
| KG12 | 47.6 | 3880 | 0.004 | 0.005 |
| KG13 | 35.7 | 3680 | 0.007 | 0.010 |
| KG3 | 35.6 | 3460 | 0.007 | 0.009 |
| KG18 | 34.5 | 3510 | 0.003 | 0.007 |
| KG100 | 30.2 | 3450 | 0.007 | 0.009 |
| KG36 | 29.6 | 3250 | 0.006 | 0.011 |
| KG6 | 28.8 | 2900 | 0.007 | 0.010 |
| KG17 | 19.2 | 2580 | 0.007 | 0.010 |
| KG127 | 12.7 | 2200 | 0.017 | 0.011 |
| KG2 | 7.96 | 1380 | 0.004 | 0.008 |
| KG20 | 6.62 | 1260 | 0.012 | 0.014 |
| KG19 | 5.95 | 1000 | 0.015 | 0.026 |
| KG835 | 4.71 | 781 | 0.005 | 0.007 |

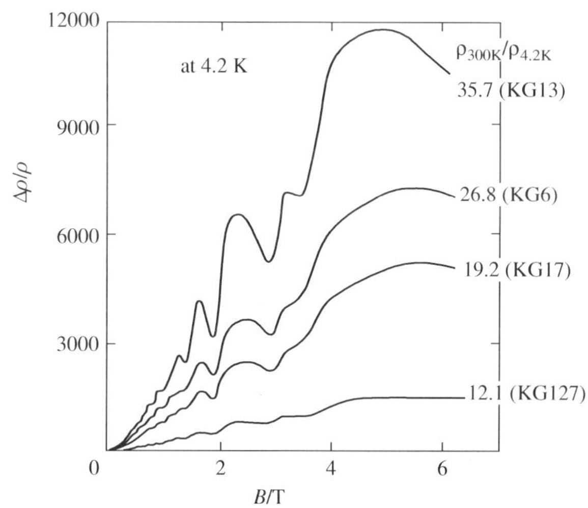


**FIGURE 3.43**

Dependences of $\Delta\rho/\rho_0$ on $B$ for kish graphite flakes with different $\rho_{300K}/\rho_{4.2K}$.

*(Courtesy of Prof. Y. Hishiyama of Tokyo City Univ.)*

different values of $\rho_{300K}/\rho_{4.2K}$. The flake with the higher $\rho_{300K}/\rho_{4.2K}$ shows Shubunikov-de-Haas oscillation more markedly.

Absolute thermoelectric power $S$ is also very sensitive to the crystal perfection. In Fig. 3.44, the changes of $S$ with temperature are compared with different flakes of kish graphite and HOPG having relatively high values of $\rho_{300K}/\rho_{4.2K}$ [74,75]. These experimental results are interpreted as follows: there are three anomalies in $S$ due to a phonon-drug, i.e., a sharp minimum around 35 K, small but sharp peak around 10 K and a shallow minimum around 3 K, which are overlapped on the usual dependence of $S$ on temperature due to scattering. These phonon-drug phenomena depend strongly on the crystallinity of graphite. For example, the minimum at 35 K, which is known to be due to the interaction between the phonons of lattice vibration within the graphite layer and carriers, depends strongly on the crystallinity, becoming deep with the decrease in $\rho_{300K}/\rho_{4.2K}$, i.e., with the decrease in crystal perfection.

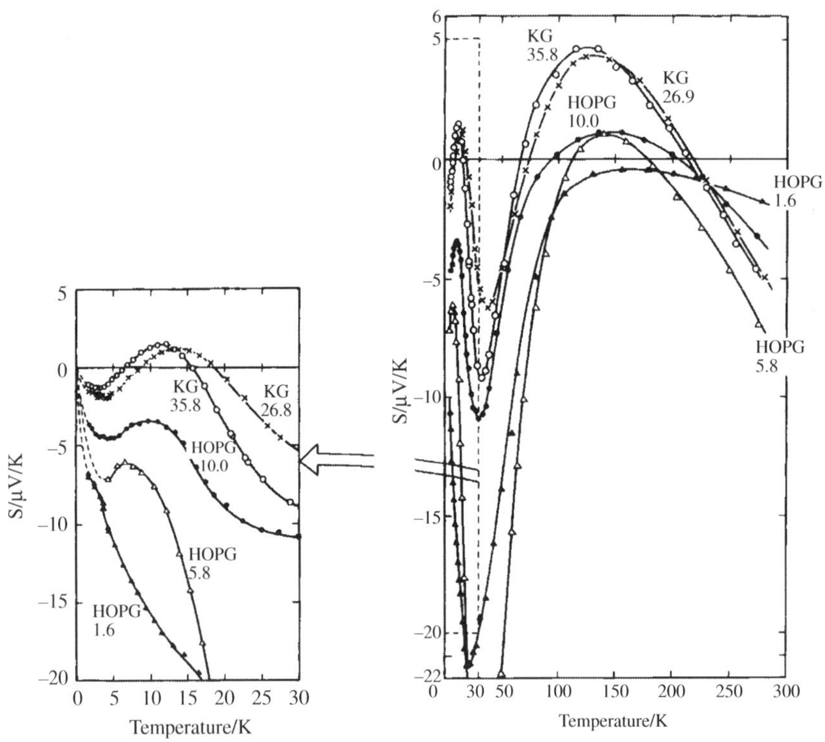


**FIGURE 3.44**

Temperature dependences of $S$ for kish graphite (KG) and HOPG. The value of $\rho_{300K}/\rho_{4.2K}$ is shown for each sample.

*(Courtesy of Prof. Y. Hishiyama of Tokyo City Univ.)*

The precipitation of graphite crystals from molten iron has been controlled to synthesize the graphite single crystals [76–78]. Iron powder is melted in a graphite crucible and saturated by carbon at 1800–1900°C and then cooled down below 1500°C with a rate of 5°C/min or less [77]. A single crystal of graphite has been formed on the surface of molten iron, of which the maximum size is 30 mm across and 60 μm thick. The addition of a small amount of silicon, less than 1 mass%, in the iron melt is favorable. Graphite crystals, including flaky and dendritic, have been synthesized via precipitation from molten metals and metal borides, and also via transportation of decomposition gases from metal carbides [79–84].

### 3.2.4 Highly oriented pyrolytic graphite (HOPG)

Hydrocarbon gases, such as methane and propane, give a deposition of carbon at high temperatures on a substrate, which has been called pyrolytic carbon and used for carbon coating of nuclear fuel particles in a reactor (Section 2.3.2). It has extensively been studied to control the structure and texture of pyrolytic carbons formed by changing the deposition conditions, such as precursor hydrocarbon, its concentration and flow rate, temperature of deposition and also geometry of the furnace [85]. When the graphite substrate is used as a heater by passing electric current directly and as a substrate for deposition, pyrolytic carbon deposited close to the substrate, in other words, deposited in the beginning of the process, is graphitized more than the part near the surface [86]. This inside part is reasonably supposed to experience rather high compressive stress in addition to the temperature higher than the outside, because of a large temperature gradient across the thickness of the pyrolytic carbon deposited. This seems to give an idea to produce highly oriented pyrolytic graphite (HOPG) [87,88].

Fig. 3.45 shows a typical procedure to produce HOPG. The first step of the preparation is hot-pressing at 2800–3000°C, which is the process to destroy the growth cone texture and to improve preferred orientation, evaluated by mosaic

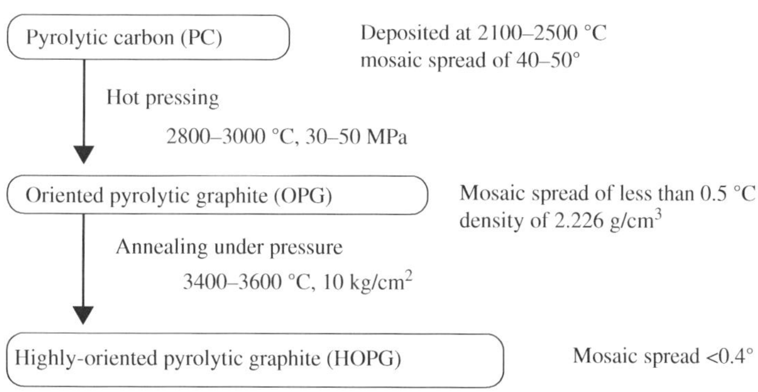


**FIGURE 3.45**

Procedure for the production of HOPG.

spread *MS*. A similar *MS* value is obtained by either elongation or twisting of pyrolytic carbons at high temperatures, as shown in Table 3.11. After this hot-pressing, the plate is easily cleaved and its surface gives a mirror reflection, but its physical properties are usually far from those observed on a single crystal of graphite. The second step of the preparation, annealing under pressure at a high temperature above 3400°C, is essential to improve the physical properties. At such a high temperature, graphite has some ductility and so the compressive stress is not necessary to be very high. In order to attain high crystallinity and a high degree of preferred orientation of graphite layers, certain structure and texture are required for the starting pyrolytic carbons, singularly nucleated growth cone texture being preferable.

As described above, the structure of pyrolytic carbons and, in consequence, their properties depend primarily on the deposition condition. In Table 3.11, $\rho_{300K}/\rho_{4.2K}$, $(\Delta\rho/\rho_0)_{max}$, $r_{TL}$ and $r_T$ of so-called HOPG prepared under compression at the temperature shown in the sample code are tabulated. The crystallinity of HOPG depends strongly on the preparation conditions, particularly the temperature of the final heat treatment. However, even HOPG heated up to 3600°C, near the evaporation temperature of graphite, cannot overcome kish graphite in crystallinity measured by $\rho_{300K}/\rho_{4.2K}$ and $(\Delta\rho/\rho_0)_{max}$ (Table 3.10). Twisting and elongation at a high temperature is an effective way to improve the orientation of crystallites and so the crystallinity of pyrolytic carbons. However it has to be pointed out that the temperature is primarily important.

In Fig. 3.46, a representative electron channeling pattern and its contrast image are shown on one of HOPG [69]. This channeling pattern is a little distorted, in

**Table 3.11** Galvanomagnetic Parameters for HOPG [70]

| Sample | $\rho_{300K}/\rho_{4.2K}$ | $(\Delta\rho/\rho_0)_{max}$ (%) | $r_{TL}$ | $r_T$ |
|---|---|---|---|---|
| HOPG 3600-1 | 4.50 | 1210 | 0.016 | 0.013 |
| HOPG 3600-2 | 4.03 | 1110 | 0.005 | 0.006 |
| HOPG 3600-3 | 2.83 | 812 | 0.008 | 0.009 |
| HOPG 3200-2 | 1.60 | 356 | 0.013 | 0.015 |
| HOPG 3300 | 1.35 | 416 | 0.009 | 0.008 |
| HOPG 3100T-1 | 1.17 | 284 | 0.007 | 0.012 |
| HOPG 3200-1 | 1.14 | 304 | 0.019 | 0.019 |
| HOPG 3100T-2 | 1.13 | 296 | 0.011 | 0.014 |
| HOPG 3100E | 1.06 | 254 | 0.017 | 0.028 |
| HOPG 2760E-1 | 0.798 | 118 | 0.003 | 0.040 |
| HOPG 2760E-2 | 0.709 | 101 | 0.004 | 0.042 |
| HOPG 2800T-1 | 0.546 | 16.2 | 0.024 | 0.044 |
| HOPG 2800T-2 | 0.502 | 8.92 | 0.029 | 0.048 |

*T: under twisting, E: under elongation.*

contrast to that observed on a kish graphite flake (Fig. 3.40), indicating the aggregation of crystallites with slightly different orientations. The detailed analysis of electron channeling contrast image with different orientations shows clearly that HOPG is composed from crystalline domains; domains show different orientation of *a*-axis with each other in the flake, although all *c*-axes of almost all domains are exactly perpendicular to the flake surface. The SEM on the cross-section, which is obtained by breaking the HOPG block at liquid nitrogen temperature applying a shock, shows a regular stacking of the basal planes of graphite, as shown in Fig. 3.47 [42]. In this SEM image, a few streaks running perpendicular to the stacked layers are observed, which are supposed to come from the growth cone texture in the starting pyrolytic carbons and to cause the distortion in orientation.

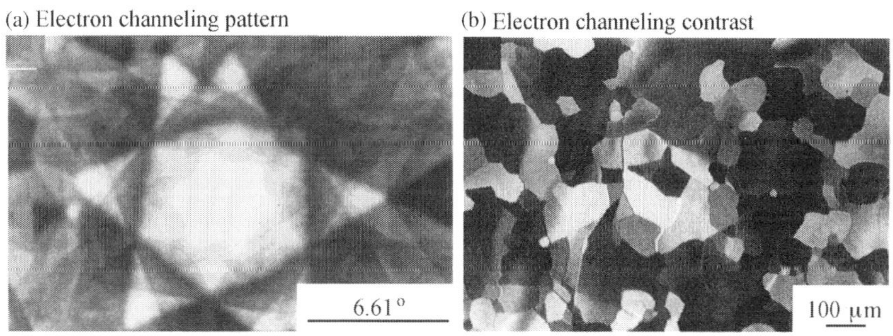


**FIGURE 3.46**

Electron channeling pattern and its contrast image of HOPG.

*(Courtesy of Prof. A. Yoshida of Tokyo City Univ.)*

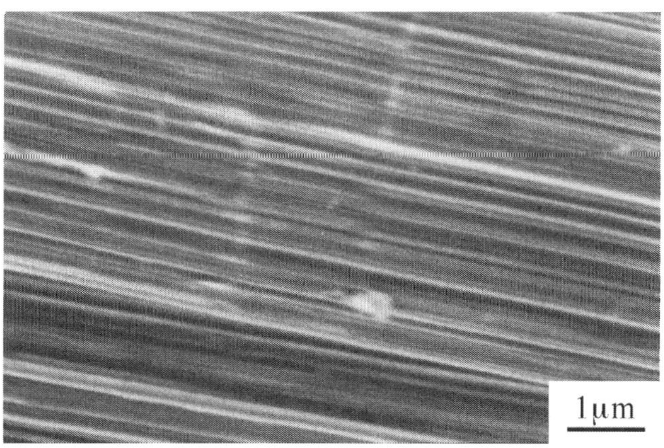


**FIGURE 3.47**

SEM image of the edge surface of HOPG.

*(Courtesy of Prof. A. Yoshida of Tokyo City Univ.)*

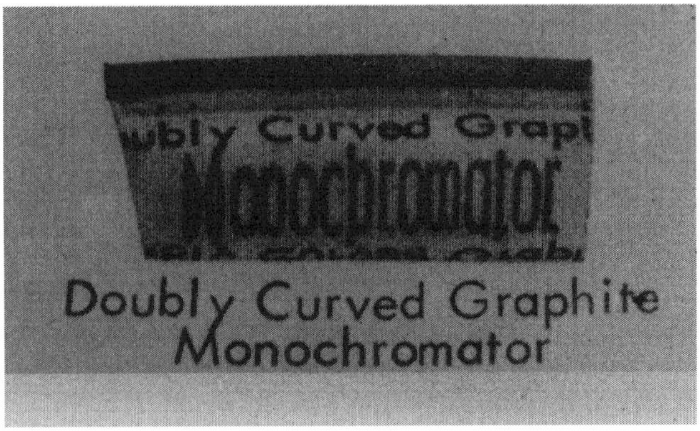


**FIGURE 3.48**

Curved monochrometer produced from HOPG [87].

As explained above, HOPG is inferior in crystallinity to kish graphite, but has an advantage to be able to have a large size. HOPG has been used for various fundamental scientific researches, instead of single crystals. It developed application as a monochrometer for X-ray and neutron diffractometry [88]. For monochrometer applications, the primary demand is a high degree of orientation of graphite layers, in practice of which mosaic spread must be less than 0.2°. In the course of the preparation of HOPG, the compression of the pyrolytic carbon blocks can be applied to give a curved surface. An example of a doubly curved monochrometer prepared from HOPG is shown in Fig. 3.48.

## 3.2.5 Graphite films derived from polyimide films

### a. Preparation

Polyimides have been developed as thermoresistant polymers, and have been used in different fields, especially in the field of electronics. From some commercially available polyimide films, graphite films with rather high crystallinity are possible to be prepared through a simple heat treatment under atmospheric pressure [89]. Because of their practical and promising applications, polyimide films with different molecular structures have been synthesized, which give a wide variety of structure in the carbon films after heat treatment at high temperatures, from highly crystalline graphite to amorphous glass-like carbon films [90,91]. This is a typical case where the molecular structure of organic precursors and the texture of their polymer films govern the structure and texture of the resultant carbon films, i.e., crystallinity and preferred orientation of hexagonal layers.

In Fig. 3.49, the imide molecules which have been studied on carbonization and graphitization are summarized by aligning the imide part in the

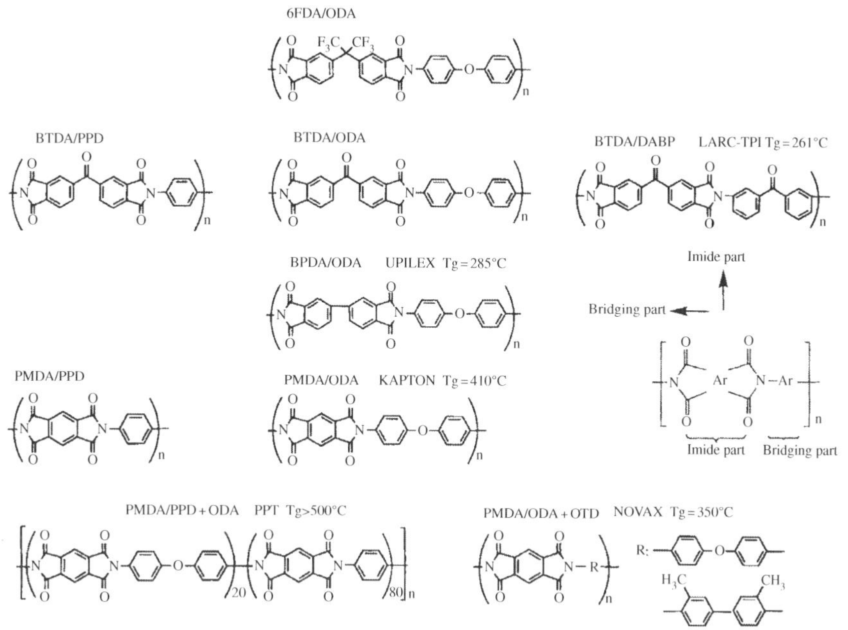


**FIGURE 3.49**

Molecular structure and code of the polyimides used for carbonization studies [90].

perpendicular direction and the bridging part in the horizontal [90]. For commercially available polyimide films, the trade names are also cited on the corresponding structures of imide molecules. It has experimentally been shown that, for instance, the commercially available polyimide film 'Kapton' is constituted from the imide molecules of PMDA/ODA, but has a quite different behavior during carbonization and graphitization to the film prepared in the laboratory from the same molecules of PMDA/ODA (Section 2.3.3d). Hereafter, therefore, the trade names like Kapton are used when the commercial films are employed as precursor film, and the structure names, like PMDA/ODA, when the laboratory-made films are used.

Most polyimide films homogeneously shrink during carbonization and give carbon films without any cracks, no cracks being observed even under scanning electron microscopy. Such a homogeneous shrinkage is demonstrated by the crane made from a polyimide film by a paper-folding technique (Fig. 3.50a), which can be carbonized and graphitized (Fig. 3.50b and c, respectively) without any deformation.

On Kapton film with 25 μm thickness, the changes in weight and shrinkage along the film with heat treatment temperature (HTT) are shown in Fig. 3.51a and the change in the composition of decomposition gas during carbonization is shown in Fig. 3.51b [89,92].

**FIGURE 3.50**

Crane made of a polyimide film and after carbonization and graphitization.

*(Courtesy of Dr. A. Hatori of AIST, Japan)*

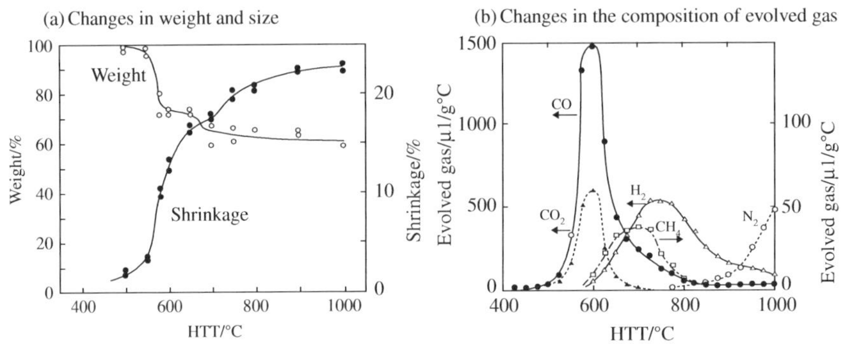


**FIGURE 3.51**

Changes in weight and thickness (a) and those in the composition of decomposition gases (b) of a Kapton film with carbonization temperature [89].

The results in Fig. 3.51 show that the carbonization of the polyimide Kapton proceeds in 2 steps, the first step in rather narrow HTT range of 500–600°C showing an abrupt weight decrease associated with an evolution of a large amount of CO and with a pronounced shrinkage along the film, and the second step with small weight loss, evolution of small amounts of methane, hydrogen and nitrogen, and little shrinkage over a temperature range from 800°C to more than 1000°C. The first step of decomposition is seen to be mainly due to a breakage at carbonyl groups in the imide part. The etheric oxygen is supposed to be released at the end of the first step, from the comparison of the result on the molecule without etheric oxygen (for example, PMDA/PPD). The release of nitrogen in the second step of carbonization continues up to high temperatures above 2000°C, leaving a large amount of pores in the film if it is heated continuously up to 2400°C. Similar profile of the evolution of decomposition gases from other polyimide films has been reported.

The structural change in the second step of carbonization reflects markedly on properties of the film [90]. In Fig. 3.52, changes of electrical conductivity at room temperature $\sigma_{RT}$ along the film with HTT are plotted on different laboratory-made polyimide films [92]. All films show a pronounced increase in $\sigma_{RT}$, more than one order of magnitude, between 700–800°C, although only small weight loss and shrinkage are observed (Fig. 3.51). Further carbonization up to 1100°C gives another increase in $\sigma_{RT}$, roughly one more order of magnitude. Higher $\sigma_{RT}$ of PMDA/PPD film (Fig. 3.49) than other films is explained from higher orientation of hexagonal carbon layers in the films, which is attributed to higher orientation of the planar imide molecules. A little lower $\sigma_{RT}$ of 6FDA/ODA film is also expected from its steric structure of precursor molecule (Fig. 3.49). The facts that methane and hydrogen are released and $\sigma_{RT}$ increases in the second step suggest that the carbonization process occurs mainly in the second step, and the first step may be pyrolysis of polyimide molecules.

In Fig. 3.53a, a schematic illustration is shown for the nanotexture change in the cross-section of the film with HTT [93a]. At low HTTs as 1000°C, crystallites are very small but preferentially oriented along the film surface. Above 2000°C, these crystallites started to coalesce with each other to grow and consequently pores are formed, which are flattened along the film surface, as illustrated in Fig. 3.53b. Above 2550°C, the pores collapse so that largely extended flat and perfect layers are suddenly observed (Fig. 3.53a), which are stacked in parallel but still have graphitic regularity only partially. Above 2700°C, a sudden progress of improvement in stacking regularity (graphitization) is observed, which is also shown by X-ray diffraction and galvanomagnetic property measurements, as explained later.

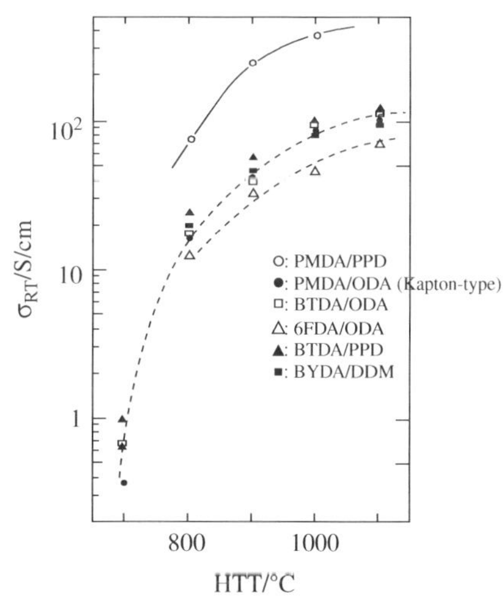


**FIGURE 3.52**

Changes in $\sigma_{RT}$ along the film with HTT for various polyimide films [80].

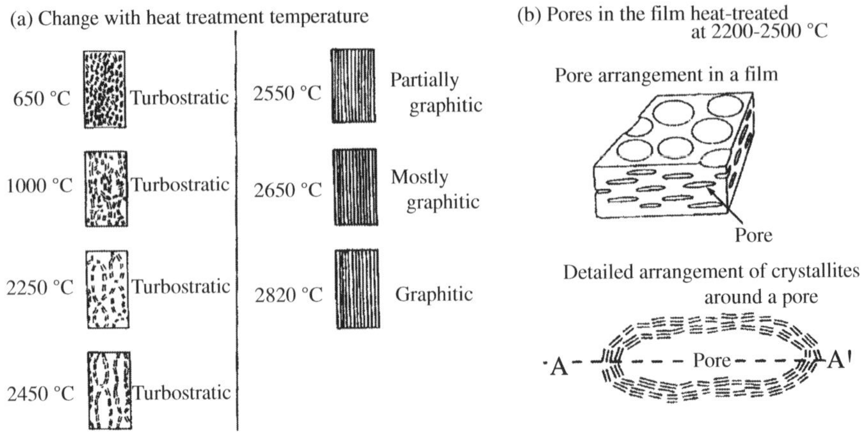


**FIGURE 3.53**

Scheme of structure change in carbon film derived from Kapton with HTT.

*(Courtesy of Mm. A. Oberlin.)*

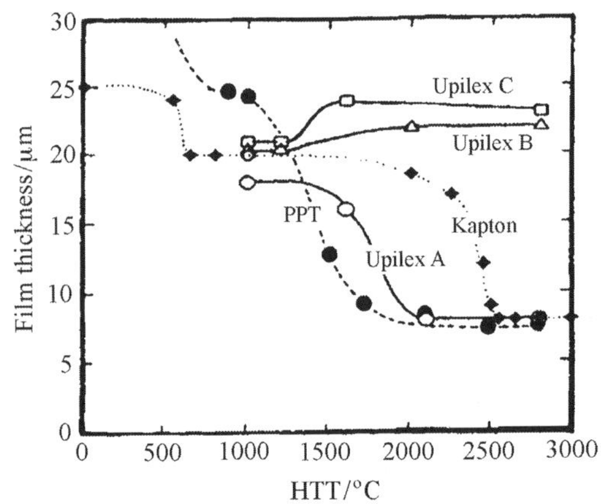


**FIGURE 3.54**

Changes in film thickness of various polyimides with HTT.

*(Courtesy of Mm. A. Oberlin.)*

The thickness of film decreases from initially an 25 μm with the increase in HTT, as shown in Fig. 3.54 on different polyimide films [93a,b]. For Kapton, the first abrupt decrease at 550°C in Fig. 3.52 corresponds to a massive release of oxygen as CO and $CO_2$ (Figs. 3.51b) in the first step of carbonization. However, the release of $CH_4$, $H_2$ and also $N_2$ in the second step of carbonization with a wide range of temperature from 800–1300°C does not give any thickness change, only a little shrinkage along the film (Fig. 3.51a). Although no appreciable change in film thickness occurs, a considerable improvement occurs in the stacking order of the carbon layers in this HTT range, as shown schematically in Fig. 3.53. Above 2000°C, particularly above 2300°C, the second thinning process is observed. This process corresponds to the large increase in anisotropy and the increasing contrast in the 002 lattice fringes of the flattened pores observed in TEM. It corresponds to a sudden breakage of all the defective areas near the edge of flattened pores, insuring complete collapsing of flattened pores and, as a consequence, lateral coalescence of the aromatic layer stacks, which yields maximum compactness and a complete annealing of the layer distortions. The partial graphitization thus suddenly introduced does not give any further changes in thickness.

Other polyimide film PPT reaches the same thickness as Kapton after high-temperature treatment and a high degree of graphitization, although the behaviors are a little different from Kapton. On Upilex films, quite different changes in thickness of carbonized films, film by film and also place by place in one film; some shrink as other polyimide films and reach rather high graphitization degree (Upilex A in Fig. 3.54), and others show even expansion above 1500°C and give glass-like carbon films (Upilex B and C in Fig. 3.54).

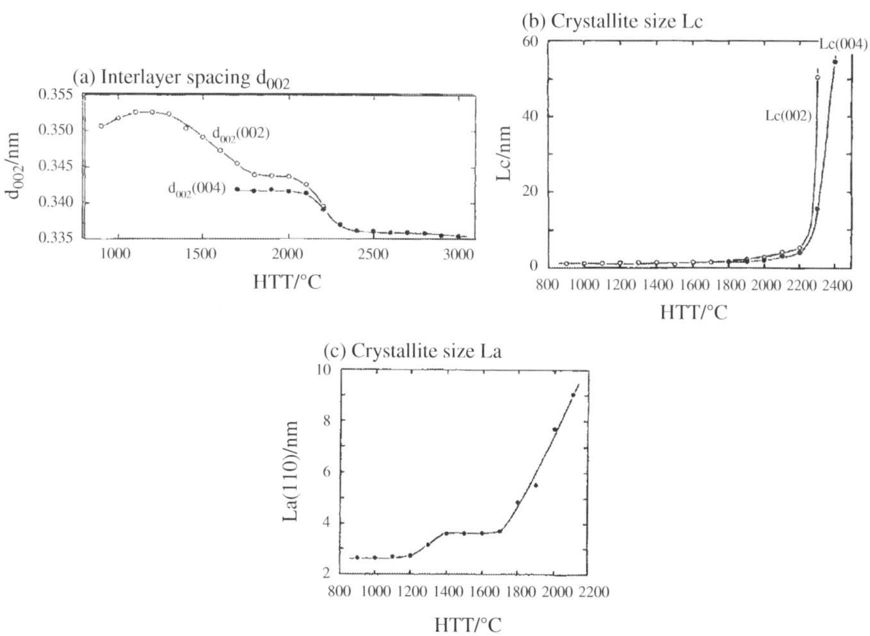


**FIGURE 3.55**

Changes in structure parameters with HTT for the carbon film derived from Kapton.

*(Courtesy of Prof. Y. Hishiyama of Tokyo City Univ.)*

In Fig. 3.55, the values of interlayer spacing $d_{002}$ and crystallite thickness along c-axis $Lc$ and also crystallite size along $a$-axis $La$ are plotted against HTT on the carbon prepared from Kapton film with 25 μm thickness by the carbonization up to 900°C [94]. The parameters $d_{002}$ and $Lc$ are determined from both 002 and 004 diffraction lines and $La$ from 110 line. The galvanomagnetic properties, $\rho$, $\rho_{300K}/\rho_{77K}$, Hall coefficient $R_H$ and $(\Delta\rho/\rho_0)_{max}$, are also shown on the same films in Fig. 3.56 [3].

Only 002 diffraction line is observed for the films heat-treated up to 1500°C. In the HTT range between 1700–2200°C, the $d_{002}$ values measured from 002 and 004 lines are different, suggesting the presence of defects in the layers, as shown in the discussion on graphitization process (Section 2.5.4), and above 2300°C two values are the same. Above 2200°C, even 006 line is observed clearly and above 2700°C both 004 and 006 lines show splitting due to $K\alpha_1$ and $K\alpha_2$ radiation, which reveal a marked improvement in crystallinity in the film, in accordance with marked increase in crystallite sizes both in $Lc$ and $La$ (Fig. 3.55b and c). Above 2200°C, the $Lc$ values determined from 002 and 004 lines, $Lc$(002) and $Lc$(004), are different, suggesting the presence of some stacking disorder (Section 2.5.4).

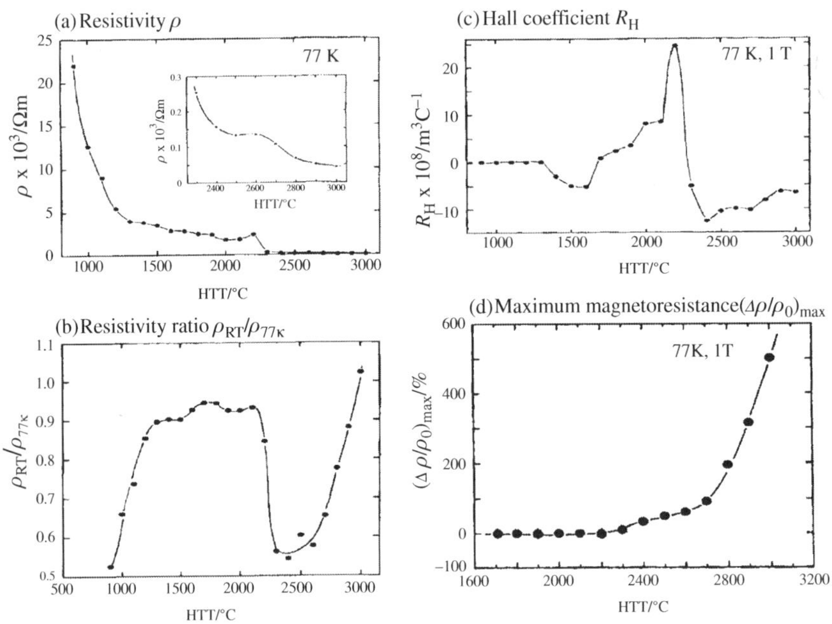


**FIGURE 3.56**

Changes in galvanomagnetic properties with HTT for the carbon film derived from Kapton.

*(Courtesy of Prof. Y. Hishiyama of Tokyo City Univ.)*

As shown in Fig. 3.56a, $\rho$ decreases rapidly at first and then gradually with increasing HTT, but in the HTT range between 2200 and 2300°C it drops discontinuously and then decreases again gradually. From the HTT dependence of $\rho_{300K}/\rho_{77K}$ in Fig. 3.56b, three HTT ranges can be divided; 900–1200, 1300–2100 and above 2300°C. In the first range of HTT semiconductor-like conduction is observed, and $\rho_{300K}/\rho_{77K}$ is almost constant and weakly temperature-dependent in the second range. Above 2300°C (the third step), the electrical conduction becomes semiconductive again which agrees with the development of graphitic structure. The $\rho_{300K}/\rho_{77K}$ higher than 1.0 for the film heat-treated at 3000°C suggests good crystallinity. Its discontinuous change in the HTT range of 2100–2300°C is due to the abrupt transformation of the structure from turbostratic to graphitic, which agrees with the changes in X-ray diffraction parameters (Fig. 3.55).

The $R_H$ at 77 K and 1 T gives a maximum at the HTT of 2200°C (Fig. 3.56c). From the detailed measurements of its magnetic field dependence, $R_H$ values on the films heat-treated below 2100°C are independent of magnetic field strength, being negative in the range of 1400–1600°C and positive in 1700–2100°C. Above 2300°C, $R_H$ drops down to negative value again and then gradually increases with increasing HTT, its dependence on magnetic field being weak but becoming similar to those for well-crystallized HOPG, which suggests the coexistence of two carriers, electrons and holes.

The $(\Delta\rho/\rho_0)_{max}$ at 77 K and 1 T is not detected at HTT below 1500°C, negative up to 2200°C, then changes to positive and increases very quickly (Fig. 3.56d). The magnetic field dependence of $(\Delta\rho/\rho_0)_{max}$ as a function of HTT shows a general trend common with graphitizing carbons (Fig. 2.102) [50]. Below HTT of 2200°C, it is negative and increases its absolute value with the increase in magnetic field. At 2200°C, $(\Delta\rho/\rho_0)_{max}$ shows a trend of saturation at high field, suggesting the coexistence of a large amount of turbostratic structure with a small amount of graphitic one. Above 2300°C it becomes positive and increases markedly with magnetic field. It reaches about 500% in the film heat-treated at 3000°C under a magnetic field of 1 T. The anisotropy ratios $r_T$ and $r_{TL}$ are as low as 0.05, showing highly oriented texture in planar orientation scheme for the films heat-treated at high temperatures. These changes in galvanomagnetic properties around 2200–2300°C correspond well to the abrupt changes in X-ray parameters shown in Fig. 3.55.

These abrupt changes in structural parameters, $d_{002}$, $Lc$, $La$ and lattice fringes, and in galvanomagnetic properties, $\rho$, $\rho_{300K}/\rho_{77K}$, $R_H$ and $(\Delta\rho/\rho_0)_{max}$, around 2200–2300°C are related to the structural change from turbostratic to graphitic, associated with the final departure of nitrogen atoms which are supposed to be substitutionally located in hexagonal carbon layers [95,96]. Abrupt departure of nitrogen gas has been observed in the course of heating from 2000 to 2700°C, as shown in Fig. 3.57 [97]. From the calculation by using semi-empirical molecular orbital method, extra-electron-spins accompanied by nitrogen seem to be localized around the C–N bond, in other words, nitrogen acting as an acceptor [92] (Section 2.8.3c).

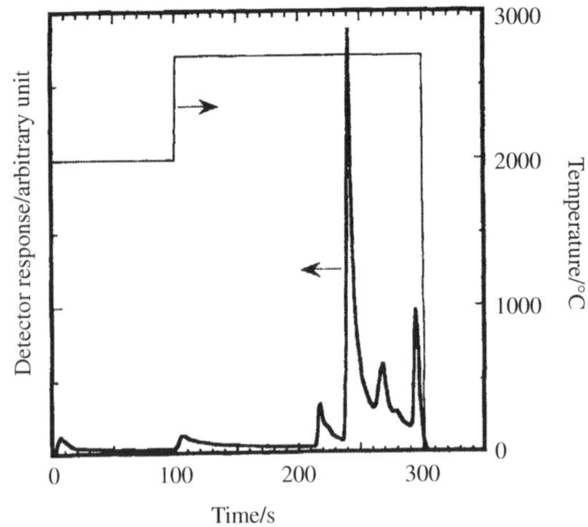


**FIGURE 3.57**

Evolution profile of nitrogen gas from Kapton film (125 μm thick) [97].

From the studies on different polyimide films, the following three fundamental conditions have been concluded for obtaining the well-crystallized graphite films [90,98];

1. flatness of starting imide molecules,
2. high degree of orientation of these flat molecules along the film,
3. less disturbance in this orientation during carbonization and graphitization due to the out-gas of non-carbon atoms.

The factor **(1)** concerns the molecular structure of polyimide used as a precursor. As shown in Fig. 3.49, the polyimides can have a wide variety of molecular structures, but not all do give graphite with high quality. From Kapton of which molecules are known to be flat [99], graphite film with high crystallinity can be obtained by selecting the appropriate conditions of film preparation and heat treatment at high temperatures. The molecule of PMDA/PPD is completely flat and expected to give highly graphitized film. In practice, however, the PMDA/PPD film is so brittle that it is broken easily only by touching with the tip of a knife, and as a consequence there is no application as plastic films. By using a small amount of tetramine, enough flexibility can be obtained, which is named as PPT film [100] and is confirmed to give highly graphitized films [93a,101,102]. A thin film of pure PMDA/PPD is graphitized at temperatures as low as 2100°C [103]. On the other hand, the film of Larc-TPI, which has a steric arrangement of constituent atoms in a molecule (Fig. 3.49), does give a film of glass-like carbon even after the heat treatment at high temperatures as 3000°C [104].

The factor **(2)** can only be controlled by the conditions of film preparation, particularly by those at imidization and selection of film thickness. Even started with the molecules of PMDA/ODA (constituent molecule of Kapton), a high degree of graphitization in the film has not been obtained if an appropriate procedure for the film preparation is not employed [92,105].

The factor **(3)** is related with the conditions during carbonization and graphitization. The heating rate in carbonization process has to be selected in the relation with the glass transition temperature $T_g$ of the film used, because the orientation of molecules may be modified [106a]. During graphitization, the parallel alignment of hexagonal carbon layers is disturbed by the release of a small amount of nitrogen remains, for Kapton film being recommended to keep at 2200°C before going to a higher temperature [106b].

## b. Properties

The size and *a*-axis orientation of the graphite layers aligned along the film surface has been characterized by using electron channeling contrast techniques. In Fig. 3.58, the electron channeling contrast images are shown on the graphite films prepared from three polyimide precursors [102]. The boundaries between the grains with different contrasts are defined more clearly in the films derived from Kapton and PPT, almost comparable to HOPG, than that from Novax, suggesting better crystallinity. This qualitative comparison in crystallinity among films coincides with the quantitative evaluation of crystallinity by using galvanomagnetic

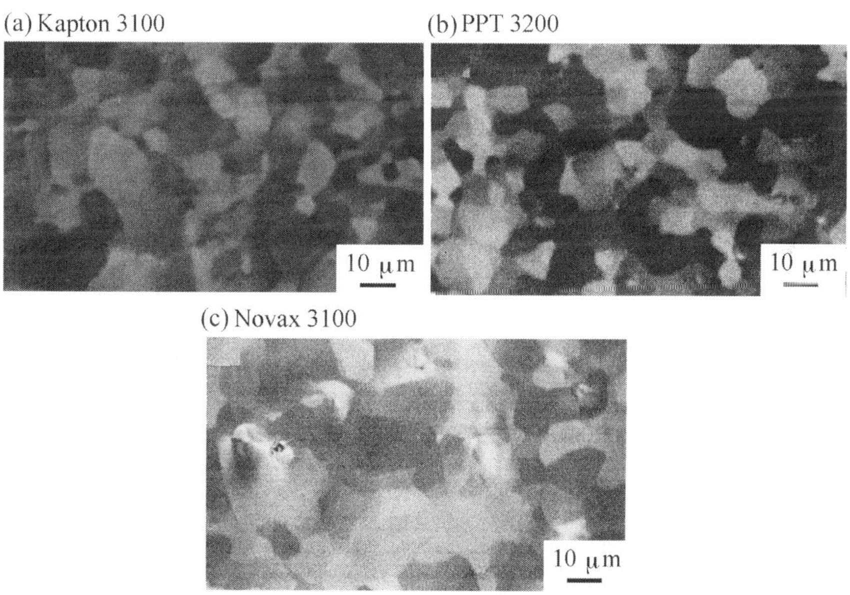


**FIGURE 3.58**

Electron channeling contrast images of polyimide films heat-treated at high temperatures.

*(Courtesy of Prof. A. Yoshida of Tokyo City Univ.)*

**Table 3.12** Characteristics of Highly Crystalline Graphite Films Prepared from Polyimide Films in Comparison With Pyrolytic Carbon Heat-Treated at 3200°C and HOPG

| Precursor | HTT (°C) | MS (°) | $\rho_{300K}/\rho_{4.2K}$ | $(\Delta\rho/\rho_0)_{max}$ at 77 K | $(\Delta\rho/\rho_0)_{max}$ at 4.2 K | anisotropy Ratio *r* | λ (μm) | *D* (μm) |
|---|---|---|---|---|---|---|---|---|
| Kapton | 3100 | 6.7 | 3.32 | 1254 | 7180 | 0.0113 | 3.5 | 8 |
| Novax | 3100 | 6.9 | 2.67 | 872 | – | 0.0173 | – | 8 |
| PPT | 3200 | 5.7 | 3.45 | 1206 | 5791 | 0.0170 | 2.6 | 8 |
| Pyrolytic carbon | 3200 | 8.6 | 1.60 | 338 | 1575 | 0.0186 | 1.5 | 5 |
| HOPG | 3600 | 0.9 | 5.50 | 1394 | 11252 | 0.0051 | 5.4 | 60 |

*(Courtesy of Prof. Y. Hishiyama of Tokyo City Univ.)*

properties, as discussed below. The averaged grain size $D$, which corresponds to the average size of graphite layers in the films, is measured from these channeling contrast images. The results on the present five films are listed in Table 3.12, in order to compare to the mean free path determined from magnetoresistance.

In Fig. 3.59a and b, $(\Delta\rho/\rho_0)_{max}$, at 77 and 4.2 K is plotted against the magnetic field $B$ [102]. The field dependences of $(\Delta\rho/\rho_0)_{max}$ on $B$ at 77 K are approximated by $B^2$, as reported on well-crystallized graphites. The values of $(\Delta\rho/\rho_0)_{max}$ for the films from Kapton and PPT are close to that for HOPG (Table 3.12). Under high magnetic fields and at 14.2 K, $(\Delta\rho/\rho_0)_{max}$ reveals the Shubunikov-de-Haas oscillation, as shown in Fig. 3.59b.

From the measurement of $R_H$, high crystallinity of the films from Kapton and PPT, comparable to HOPG, has also been concluded [102]. Its field dependences up to 1.0 T at 77 K and up to 6.5 T at 4.2 K are shown in Figs. 3.60a and b, respectively. In Fig. 3.60b, the Shubnikov-de-Haas oscillation in $R_H$ is observed, as in $(\Delta\rho/\rho_0)_{max}$ in Fig. 3.59b. Because the oscillation in $R_H$ is more pronounced than in $(\Delta\rho/\rho_0)_{max}$, Landau levels for majority carriers can be assigned, as indicated by 3e and 3h (majority electrons and holes with $n = 3$, respectively), for example, in Fig. 3.60b.

In Table 3.12, $(\Delta\rho/\rho_0)_{max}$ at 77 and 4.2 K under 1 T, $r$ which is an average of $r_{TL}$ and $r_T$, $\rho_{300K}/\rho_{4.2K}$, and mean free path for carriers λ calculated from the value of $(\Delta\rho/\rho_0)_{max}$ at 4.2 K and 0.1 T are listed on five films, together with the mosaic spread $MS$ measured by X-ray diffraction and averaged grain size $D$ determined from electron channeling contrast image [102].

The preferred orientation of graphite basal planes along the film surface is characterized by either $MS$ or $r$, both showing that HOPG is superior to the other four films, but the films derived from polyimides are better than the film of PG3200 which has been heated up to 3200°C without pressure. The crystallinity in the films from Kapton and PPT, evaluated by either $\rho_{300K}/\rho_{4.2K}$, $(\Delta\rho/\rho_0)_{max}$ at 77 and 4.2 K, λ, and $D$, is comparable to HOPG and much better than PG3200.

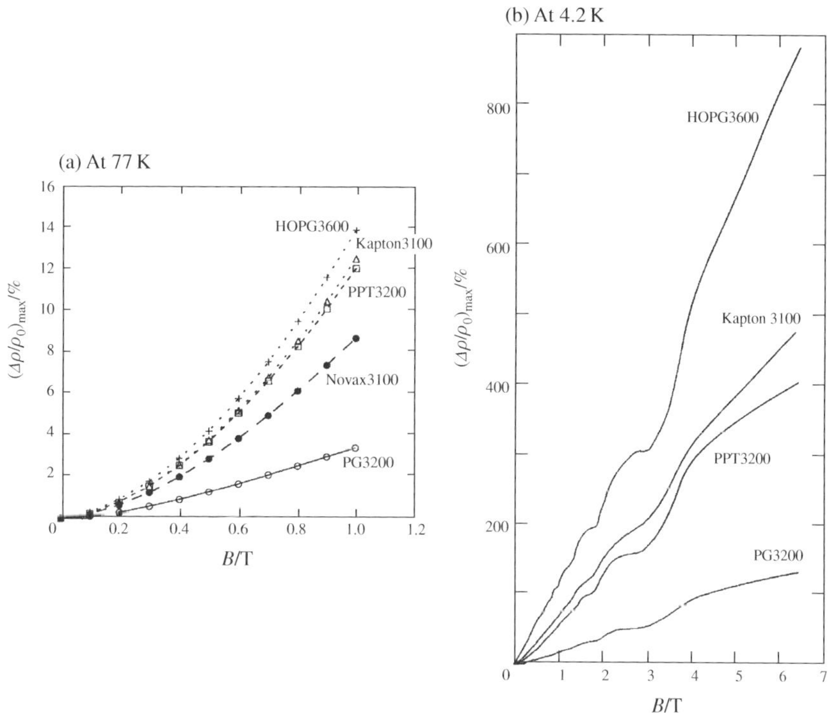


**FIGURE 3.59**

Dependences of $(\Delta\rho/\rho_0)_{max}$ on $B$ for highly-oriented graphite films from different precursors.

*(Courtesy of Prof. Y. Kaburagi of Tokyo City Univ.)*

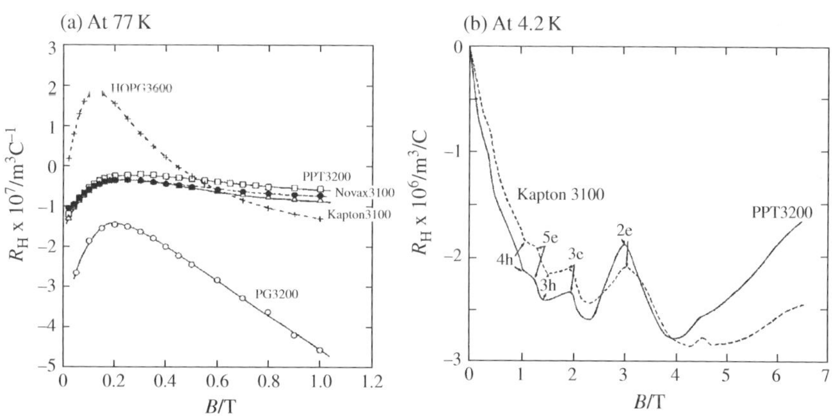


**FIGURE 3.60**

Dependences of $R_H$ on $B$ for highly-oriented graphite films from different precursors.

*(Courtesy of Prof. Y. Kaburagi of Tokyo City Univ.)*

These results exhibit the fact that, by a simple heat treatment, i.e., without pressure, at high temperatures, excellent crystallinity can be obtained from some polyimide films.

Highly oriented graphite blocks are prepared by compressing a pile of Kapton films with 25 μm thickness under 10–15 kg/cm$^2$ during carbonization and then hot-pressing at 100–300 kg/cm$^2$ up to a temperature 2800–3000°C, which are commercialized as an alternative of HOPG for the use of monochrometers of X-ray and neutron beams [107,108].

Highly crystallized graphite films were prepared from different organic precursor films, poly(p-phenylene vinylene) PPV [109], poly(p-phenylene-1,3,4-oxadiazole) PPD [110–112], and benzimidazobenzophenanthroline ladder polymer BBL [113], in addition to aromatic polyimide films [114,115]. On aromatic polyimides, detailed studies have been carried out by many reserch groups, as reviewed [91]. Highly crystalline and highly oriented graphite films have been prepared at high temperatures above 3100°C under simple mechanical constraint, and are shown to have high thermal conductivity, close to 2000 W/m.K, comparable to HOPG and single crystal of graphite [116,117].

## 3.2.6 Flexible graphite sheets

### a. Preparation

Since large flakes of natural graphite are not commonly available, the development of a technique to prepare graphite sheets has promoted the applications of graphite. Graphite sheets provide the characteristic advantages, such as flexibility, compactability, resilience and easy forming into various shapes, in addition to the intrinsic properties of graphite, i.e., lubricity, chemical and thermal stability, high electrical and thermal conductivities, etc. Based on these properties, they have been widely used in modern technology [118–120]. In order to satisfy the severe demands from recent modern technology, various researches are still conducted actively.

Fig. 3.61 shows the manufacturing process for flexible graphite sheets. Natural graphite flakes are chemically or electrochemically intercalated to form graphite intercalation compounds (GICs), usually sulfuric acid–GICs being employed

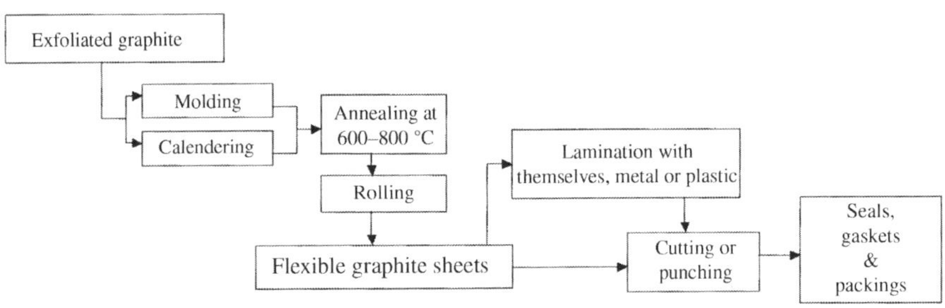


**FIGURE 3.61**

Procedure for the production of flexible graphite sheet from exfoliated graphite.

because of a simple process and low cost for their production. After rinsing and drying, residue compounds, in which regular stage structure is lost but still some of sulfuric acid derivatives remain in graphite gallery, are obtained. These residue compounds are then rapidly heated to 900–1200°C, in which intercalates remaining in graphite gallery are decomposed to gaseous products, and exfoliated graphite is yielded. This exfoliated graphite is either molded or rolled into a sheet without any adhesives or binders. These are called flexible graphite sheets.

During this forming process, several factors are very critical in determining the properties, particularly mechanical properties, of the final sheets. Besides exfoliation and forming processes, the properties of the raw residue compounds, i.e., their contents of residual intercalates, moisture and ash, as well as particle size of the starting natural graphite flakes, are very important.

During intercalation with sulfuric acid and exfoliation at high temperatures, most of the mineral impurities, such as silica and iron oxides, are removed, which have come from the ores of original natural graphite. Therefore, the resultant graphite sheets have pretty high purity, but still contain some minerals, which are referred to as an ash, and a small amount of sulfur, which comes from intercalate (sulfuric acid). Exfoliation temperature has a pronounced influence on the exfoliation volume and also on residual sulfur content. Usually, for industrial manufacturing, the temperature is kept in the range of 900–1200°C in order to ensure complete decomposition of intercalated sulfuric acid and to minimize the residual sulfuric oxides. If the temperature is too high, the consumed energy increases and graphite is partially oxidized. If the temperature is too low, exfoliation does not occur sufficiently and sulfur remains in relatively large amount. The duration of exfoliation also influences the final sulfur content, too short a time resulting in a higher sulfur content. Tensile properties are also affected by prior exposure to high temperature and residence times [121]. Exfoliation process of the residue compounds at high temperatures is discussed in Section 2.7.3b and also the formation of macropores in exfoliated graphite in Section 3.6.3.

Molding or rolling assists mechanical interlocking among worm-like particles of exfoliated graphite [122] and the rolling process determines the thickness, density, as well as preferred orientation of graphite layer planes of the finished sheets. Annealing before rolling of the sheet is usually carried out at 600–800°C for industrial manufacturing. This annealing process is effective to reduce the final sulfur content and to give homogeneity in surface quality and texture after the subsequent rolling process. The fabricated flexible graphite sheet is stocked as a roll, as shown in Fig. 3.62.

Flexible graphite sheets can be laminated with themselves, and also with various metals and plastics to form gaskets for a variety of applications. Lamination of graphite sheets is aimed to increase the thickness of a gasket or packing ring, each sheet being bonded with an adhesive. Lamination of a graphite sheet with a metal or plastic sheet is carried out in order to improve its handling performance and also mechanical strength. However, their thermal, mechanical and chemical performances as a gasket are somewhat restricted [123]. The metal interlayer may

**FIGURE 3.62**

Flexible graphite sheet rolled for store.

be foil, screen or perforated metal, the former two being adhesively bonded to the graphite sheet and the latter is mechanically bonded.

## b. Properties

The Material Safety Data Sheet on flexible graphite sheets claims that it does not contain any hazardous ingredients, is not a fire or explosion hazard, is not reactive, and requires no special protection or precaution. Therefore, it may be a good replacement for asbestos gaskets, particularly in the places where fire safety is demanded. Graphite sheet is also superior to conventional elastomeric bonded gaskets, because it is more thermally stable and chemically inert with considerably less creep relaxation [124].

The physical properties of a flexible graphite sheet are listed in Table 3.13. The bulk density of a flexible graphite sheet for industrial uses is ca. $1.1 \times 10^3$ kg/m$^3$, which is only one-half of the theoretical density of graphite ($2.25 \times 10^3$ kg/m$^3$). This makes the sheets compressible, which is required to produce an effective sealing in gasket applications. The bulk density can be controlled from 0.8 to $1.4 \times 10^3$ kg/m$^3$ by the degree of compaction during its forming processes, particularly rolling [123]. The density of the sheet also affects its other properties, with increasing density its recovery, tensile strength, thermal conductivity, Young's modulus, and hardness increasing, but its compressibility, gas permeability, oxidation rate, flexibility, and electrical conductance perpendicular to the sheet decreasing [123].

The typical tensile strength of flexible graphite sheets varies from 4.8 to 6.9 MPa along the sheet. It depends on flake size of starting natural graphite, ash content and exfoliation volume of exfoliated graphite, as well as bulk density of the sheet, and increases approximately linearly with bulk density, as shown in Fig. 3.63a [123]. Almost the same relationship has been reported [124]. At a given density, tensile strength decreases with increasing ash content. The slope of tensile strength $\sigma$ versus

**Table 3.13** Room Temperature Properties of Flexible Graphite Sheets

| Property | Value | Remarks |
|---|---|---|
| Bulk density | $1.1 \times 10^3$ kg/m$^3$ | 0.8 to $1.4 \times 10^{-3}$ kg/m$^3$ |
| Thickness | $0.15–1.0 \times 10^{-3}$ m | |
| Tensile strength along the sheet | 4.4 MPa | 4.0 to 6.9 MPa |
| Compressive strength | 240 MPa | |
| Creep relaxation | < 5% | 100°C × 22 h |
| Compressibility | 40% | |
| Recovery | 20% | |
| Young's compressive modulus | 1.4 GPa | |
| Sealability | <0.5 mL/h | |
| Static friction coefficient | 0.052 (at 0.07 MPa) | Against steel |
| Working temperature | −240 to 3000°C | Oxidation occurs above ca. 455°C in air |
| Thermal conductivity at 21°C | 140 W/m.K | |
| Along the sheet surface | 5 W/m.K | |
| Perpendicular to the sheet | | |
| Electrical resistivity | $8 \times 10^{-6}$ ohm.m | |
| Along the sheet surface | $15 \times 10^{-6}$ ohm.m | |
| Perpendicular to the sheet under 690 kPa | | |
| Coefficient of thermal expansion | $-0.4 \times 10^{-6}$ m/m°C | linear |
| | $27 \times 10^{-6}$ m/m°C | from 21 to 1094°C |
| Along the sheet surface | | from 21 to 2206°C |
| Perpendicular to the sheet | | |
| Specific heat at 24°C | 711 J/kg.K | |
| Oxidation rate in air | Mass loss | |
| at 500°C | 0.03 to 0.75 g/m$^2$.h | |
| at 700°C | 12 to 130 g/m$^2$.h | |
| Carbon content | 95.0 to 99.5 mass% | 99.8 mass% for nuclear |
| Ash content | <0.5 to 5.0 mass% | |
| Sulfur content | <500 to 1000 ppm | <450 ppm for nuclear |
| Leachable chlorides and fluorides | <50 to 100 ppm | <20 ppm for nuclear |

bulk density $\rho$, $d\sigma/d\rho$, is a measure of the ability of the graphite sheet to be strengthened by densification. Fig. 3.63b shows that the value of $d\sigma/d\rho$ is inversely related to ash content [123]. Therefore, ash content has to be low in order to achieve high strength by densification of graphite sheets. Tensile strain to failure is also reduced by the presence of impurities. The flake size, which is often controlled by screen aperture size, affects the exfoliation volume and, as a consequence, the tensile strength of the resultant graphite sheet. Fig. 3.64a shows a linear relationship

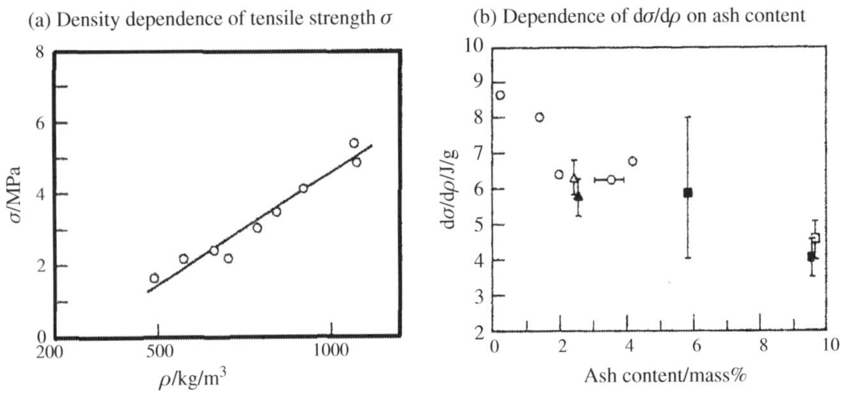


**FIGURE 3.63**

Linear relation between $\sigma$ and $\rho$ of flexible graphite sheets, and dependence of $d\sigma/d\rho$ on ash content in starting graphite [123].

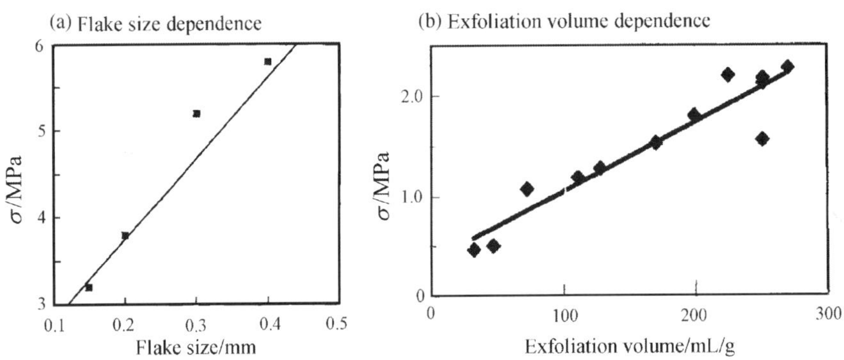


**FIGURE 3.64**

Dependences of $\sigma$ of flexible graphite sheet on the size of starting natural graphite flakes and exfoliation volume of exfoliated graphite [125].

between tensile strength and flake size [124]. As shown in Fig. 3.64b, the tensile strength of sheets also depends on exfoliation volume of graphite used. The increase in screen aperture size results in a tensile strength increase through a relationship between exfoliation volume and flake size [125].

Flexible graphite sheets can also be strengthened by the addition of inorganic salts, which had to be uniformly distributed in graphite sheets. When boric acid or its salt is dispersed into exfoliated graphite, the tensile strength of the molded sheet increases by 50–80%, although it becomes somewhat brittle [125].

High purity of the sheet is required for nuclear and other applications, where corrosion of metals is of critical concern. For standard industrial use, maximum sulfur level of ca. 1000 ppm meets the specifications. For nuclear power plants, however, sulfur content has to be controlled below 500 ppm. Some of the sheets contain very small amounts of halides (chlorides and fluorides), which catalyze corrosion, and their levels have to be kept also very low in critical applications. Leachable halides in aqueous solution accelerate pitting corrosion by permeating stainless steel's chromium oxide film and eventually cause its complete physical breakdown. A maximum value for the total amount of leachable chlorides and fluorides is about 100 ppm in most industrial grades of the sheets. For nuclear power plants, this value must be controlled below 50 ppm, even less than 20 ppm.

The properties related to the performance of flexible graphite sheet as a gasket are the following: sealability, compressibility, recovery and creep relaxation. Density of the sheets is known to affect these properties; a sheet with the lower density has the higher compressibility but lower resilience. Flexible graphite sheets with a bulk density of $1.1 \times 10^3$ kg/m$^3$ present usually 40–50% compressibility and 15–25% resilience, resilience being virtually unaffected by service conditions, showing that it is an effective, long-term sealing material. It possesses very low creep relaxation and is thermally stable.

### c. Applications

Flexible graphite sheets are formed to different shapes and sizes by punching. Some of the industrial products, including tapes, O-shaped rings, etc., are shown in Fig. 3.65. Most of these products are used as seals, packings and gaskets. Fig. 3.66 shows schematically how many seals and gaskets made of flexible graphite sheets are used for an engine in an automobile.

Flexible graphite sheets can be used over the widest temperature range among all the sealing materials, from −240°C to 3000°C, if the atmosphere is maintained inert. In a reducing or vacuum environment, graphite sheet stiffens slightly between 1100 and 2000°C, but remains usable. Above 3000°C, it begins to sublime. In air or oxidizing atmosphere, pure graphite sheet begins to be oxidized at

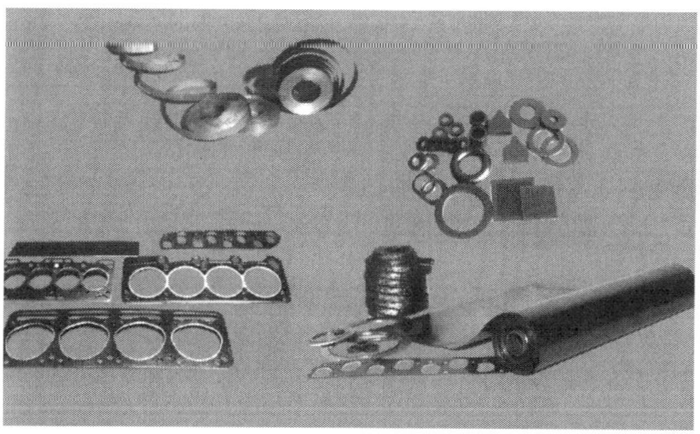

**FIGURE 3.65**

Seals, packing, tapes and sheet of flexible graphite.

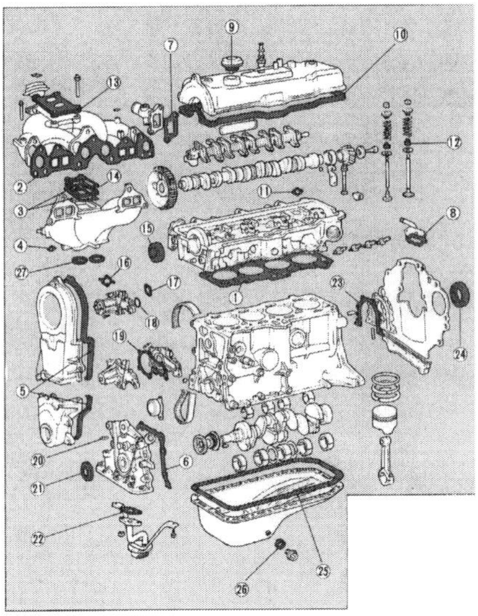

Gaskets
1. Cylinder head
2. Mainfold to cylinder head
3. Intake to exhaust mainfold
4. Oxygen senser
5. Timing belt cover
6. Oil pump
7. Water outlet
8. Cylinder head rear plate
10. Cylinder head cover
11. Fuel pump
16. Water inlet
17. Heater pipe
19. Water pump
22. Oil strainer flange
23. Retainer rear oil seal
25. Oil pan
26. Oil pan drain plug
27. Exhaust pipe

Seals
9. Oil filler cap seal
12. Valve stream oil seal
13. Carbureter insulator
15 &.21. front oil seal
24. Rear oil seal.

**FIGURE 3.66**

Seals and gaskets made of flexible graphite sheet in an engine for automobile.

*(Courtesy of Nippon Gasket Co., Ltd.)*

455°C, but it can be used at a much higher temperature, about 800°C, for gasket and packing materials, because in those applications only thin edges are exposed to air. In reinforced laminates the temperature usable is governed by the adhesives and inserted materials.

Flexible graphite sheets exhibit an outstanding fluid-sealing function and can be made into a variety of shapes to fit virtually any fluid-sealing application. Their chemical resistance and thermal stability make them an effective material where fire-safe sealant is required. They are safe in regard to people's health and so an ideal replacement for gaskets based on asbestos. They are also superior to conventional elastomeric bonded gaskets, because they are more thermally stable and chemically inert with considerably less creep relaxation. They are also preferable to other non-asbestos sheet gaskets, such as aramids, glass fiber and mica, because they have lower thermal stability, higher creep and often resulting in poor performance under load.

Resilience of flexible graphite sheets may develop new applications for damping of mechanical loads [126,127]. Damping of hazardous loads due to accidental loading, for example, wind, ocean waves or earthquakes, is required.

Flexible graphite sheets have a minimum thermal conductivity of 5 W/m K perpendicular to the sheet surface but approximately 140 W/m K along the sheet. This anisotropy made them thermal insulators perpendicular to the sheet and conductors along the sheet. Graphite, including highly crystallized graphite sheets and flexible graphite sheets, has been reviewed as an excellent thermal management material [128]. One of the advantages of these sheets is their emissivity of 0.5 at high temperatures, which means that 50% of the heat radiated upon the surface of the sheet is reflected back to the hot zone.

Electromagnetic interference shielding, which can be done by the reflection and/or absorption of electromagnetic radiation by a material, is increasingly needed in our daily life; high-frequency radiation emanating from cellular phone interferes with many electronics (e.g., computers). Shielding of these radiations is required with high-efficiency, using light-weight and thin layers. Flexible graphite sheet is one such candidates [129,130].

Remarkable development in microelectronics opens new application fields for carbon materials, such as membrane switches, variable resistors and rechargeable batteries. In Fig. 3.67, some of the membrane switches, which consist of fine graphite powders printed by using a binder onto polyester films, and a schematic illustration of their construction are shown. These switches are very thin, as thin as 10–15 μm, and very light. By pushing the key in a keyboard of a computer or the switch in the control panel of various instruments, two membranes contact and the signals are sent. Their usage is going to increase, accompanied by the development of microelectronic devices. Variable resistors, which are produced by screen printing of carbon pastes of a mixture of natural graphite, carbon black and other carbon materials with some thermosetting resin on substrate of phenol resin or alumina, are also widely used in electronic devices.

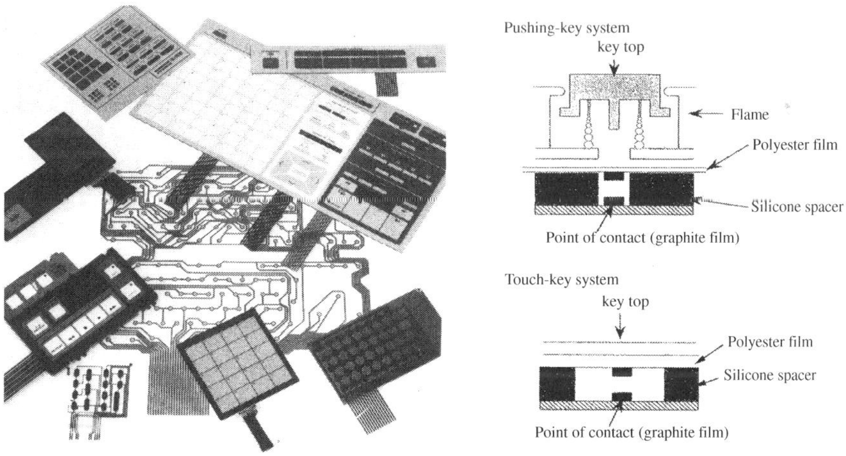


**FIGURE 3.67**

Membrane switches for various electronic devices.

*(Courtesy of Nihon Kokuen Co., Ltd.)*

# 3.3 Non-graphitizing and glass-like carbons

## 3.3.1 Structural characteristics

Most carbons produced from thermosetting resins are non-graphitizing, i.e., no marked development of graphite structure after the heat treatment at high temperatures as 3000°C, even near the melting point of carbon, about 3400°C, as described in Section 2.5.4. The non-graphitizing nature of these carbons is due to their nanotexture, random orientation of carbon layers. Since the fundamental structural units are anisotropic hexagonal carbon layers, their random orientation results in the formation of minute pores, showing a sharp contrast to highly oriented graphites with a high degree of planar orientation and dense nanotexture. In Fig. 3.68, change in 002 lattice fringe image of transmission electron microscopy (TEM) with heat treatment temperature (HTT) is shown on a typical non-graphitizing carbon obtained from poly(vinylidene chloride). Below 2000°C, the fringes, i.e., hexagonal carbon layers, are oriented randomly, though they seem to

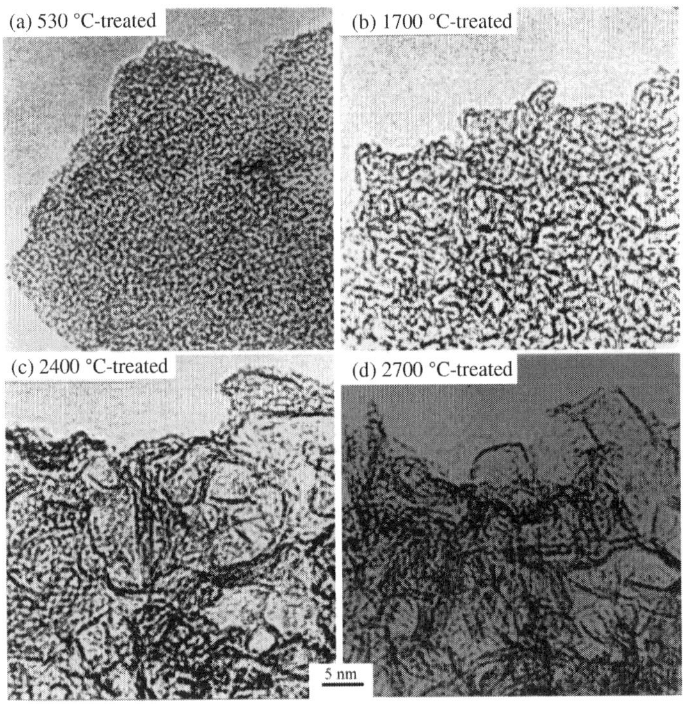


**FIGURE 3.68**

002 lattice fringe images of the carbon obtained by the heat treatment of poly(vinylidene chloride) at different temperatures.

*(Courtesy of Mme. A. Oberlin.)*

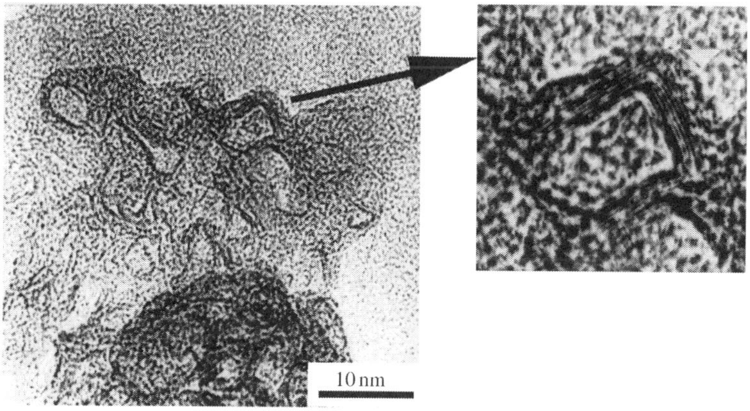


**FIGURE 3.69**

Lattice fringe image for a carbon obtained from sugar by the heat treatment at a high temperature.

*(Courtesy of Mme. A. Oberlin.)*

grow a little larger with increasing HTT. Above 2000°C, they grow much larger to be able to recognize their stacking. Above 2400°C, the stacks of carbon layers (called basic structural units or crystallites) seem to form shells. Fig. 3.69 shows clearly a shell surrounded by stacked carbon layers. For different non-graphitizing carbons, very similar structural characteristics and their change with heat treatment have been observed.

For non-graphitizing or hard carbons, 002 lattice fringe image of Fig. 3.70a is usually obtained after the heat treatment at high temperatures as 2500°C [131–133], which is very similar to those shown in Figs 3.68 and 3.69. Through detailed studies using high-resolution TEM techniques, such as selected area electron diffraction (Fig. 3.70b) and 110 dark-field image (Fig. 3.70c), a structure model, as shown in Fig. 3.71a, has been proposed [134]. The regions surrounded by 002 layers in lattice fringe image (Fig. 3.70a) are lightened in 110 dark-field image (Fig. 3.70c), revealing that carbon layers are perpendicular to the incident electron beams, in other words, are laying down along the micrograph. Therefore, the postulation of pores of which walls consist of carbon layers seems to be more reasonable than to assume the presence of the ribbons of stacked carbon layers proposed before [135]. Fig. 3.71b shows another scheme of the pores in non-graphitizing carbons after high-temperature treatment, which are surrounded by carbon layers [136].

Different glass-like carbons have been developed [137], some of which are now industrially produced and widely used. They are characterized by conchoidal fracture surface and gas impermeability, in addition to their nanotexture of random orientation, which is the same as the characteristics for non-graphitizing carbons explained above. In the present book, glass-like carbons are defined to be

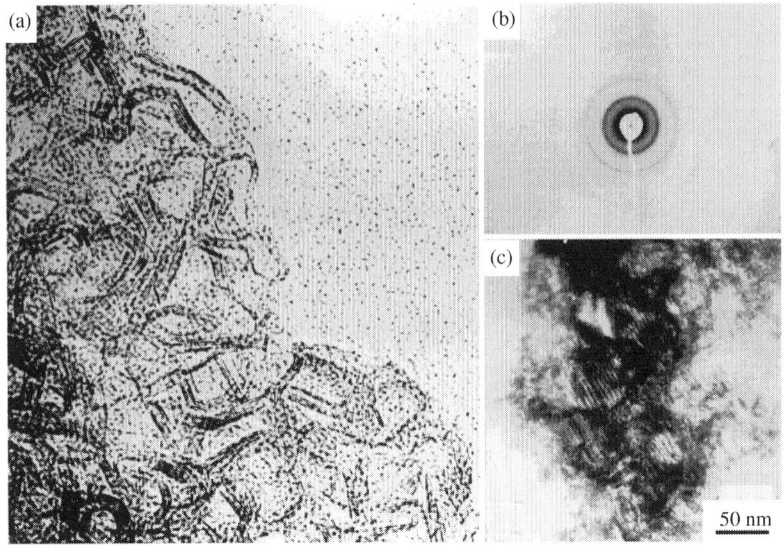


**FIGURE 3.70**

TEM images on 2500°C-treated sugar coke.

*(Courtesy of Mme. A. Oberlin.)*

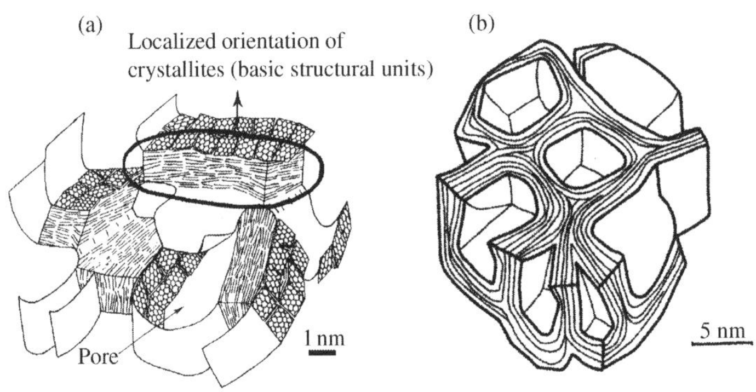


**FIGURE 3.71**

Structural models for the non-graphitizing carbon after the heat treatment at high temperatures.

*(Courtesy of Mme. A. Oberlin [a] and Prof. M. Shiraishi of Tokai Univ. [b].)*

the carbon blocks with controlled or designed shape, not particles with irregular shapes or powders, with random orientation of small basic structural units, which has been called amorphous structure. In glass-like carbons, most of the pores are closed, this being the reason why they are gas-impermeable even though low bulk density. In order to give these characteristics for carbon materials, exact control of the production process, such as the selection of an appropriate precursor and the heating rate, is required.

In Fig. 3.72, SEM micrographs are shown on the fractured surface of the glass-like carbons heat-treated at different temperatures [138], which show granular texture, agreeing with the structural model shown in Fig. 3.71. The size of constituent grains is measured from these micrographs and listed in Table 3.14, together with XRD and magnetoresistance parameters (Section 2.5.1) measured on commercially available glass-like carbons. The average grain size $D$ increases slightly with the increase in HTT, associated with a small increase in crystallite sizes, $Lc(002)$ and $La(110)$, and a slight decrease in $d_{002}$. On two samples (GC-10 and AGC), both of which are supposed to be treated at a low temperature around 1000°C, no $\Delta\rho/\rho_0$ is measured even at a high magnetic field strength such as

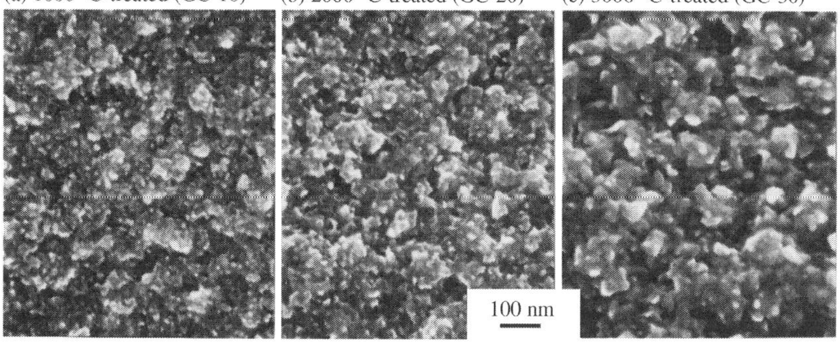


**FIGURE 3.72**

SEM images of the fractured surface of the glass-like carbon heat-treated at different temperatures.

*(Courtesy of Prof. A. Yoshida of Tokyo City Univ.)*

**Table 3.14** Structure Parameters of the Commercially Available Glass-Like Carbons

| Grade | Heat Treatment Temperature (°C) | Grain Size $D$ (nm) | $d_{002}$ (nm) | $Lc$ (002) (nm) | $La$ (110) (nm) | $(\Delta\rho/\rho_0)_{cr}$ (%) | $r_T$ | $r_{TL}$ |
|---|---|---|---|---|---|---|---|---|
| GC-10 | 1000 | 7.0 | 0.3468 | 1.9 | 2.5 | – | – | – |
| GC-20 | 2000 | 10.0 | 0.3442 | 3.3 | 3.1 | −0.0846 | 0.771 | 0.887 |
| GC-30 | 3000 | 13.1 | 0.3436 | 3.6 | 3.5 | −0.1821 | 0.962 | 0.860 |
| AGC | 1000 | 6.0 | 0.3470 | 1.9 | 2.0 | – | – | – |
| UDAC | 3000 | 9.4 | 0.3443 | 3.7 | 3.0 | 0.0696 | 1.00 | 1.00 |

*(Courtesy of Prof. A. Yoshida of Tokyo City Univ.)*

6.5 T. Anisotropy ratios $r_T$ and $r_{TL}$ are close to 1, indicating a random nanotexture.

From TEM observation on non-graphitizing carbon heat-treated above 2500°C, pore diameter of 3–6 nm and pore wall thickness of 2–4 nm are measured and so grain size observed on fractured surface by SEM seems to be about 7–10 nm, which agrees well with SEM observations on the glass-like carbons shown in Table 3.14. The pore wall thickness determined from TEM observation roughly corresponds to $Lc(002)$ value determined by XRD on the glass-like carbons. Therefore, bulk density is calculated by assuming the grain with the size $D$ and wall thickness of $Lc(002)/2$ and by using observed $d_{002}$ and $a_0$, which showed a rough agreement with the measured bulk density [138]. Therefore, the structure model consisting of closed pores shown for non-graphitizing carbons in Fig. 3.71b and the model of crumpled sheets shown in Fig. 3.71a are exactly the same, just a little different visualization of the structure. The ribbon model for non-graphitizing carbons [135] seems not to be effective to interpret the structure, nanotexture and properties of glass-like carbons, particularly being difficult to explain the presence of a large amount of closed pores and gas impermeability.

On glass-like carbons, there is a thin surface layer with better crystallinity than the inside, particularly on the ones heat-treated at high temperatures. This phenomenon has markedly been observed on thin films derived from thin cellulose film and a polyimide film Larc-TPI (Fig. 3.43), and studied in detail by measuring different properties [139,140]. SEM images of Fig. 3.73 on the cross-section of the film show that these well-graphitized and well-oriented regions are

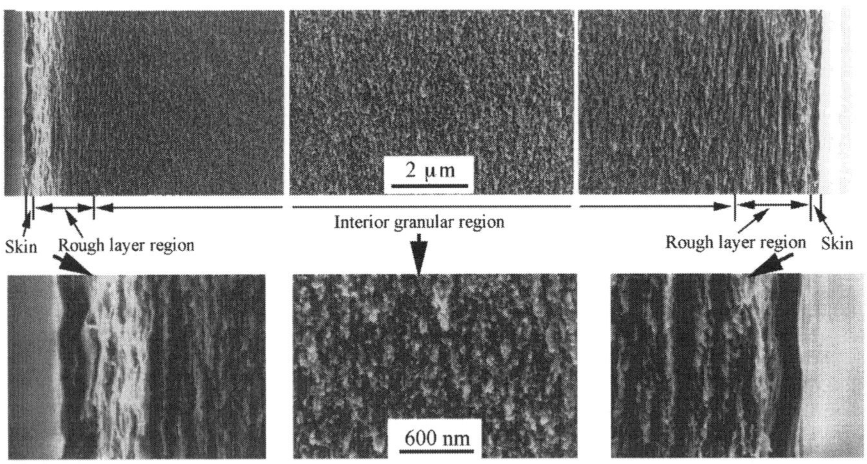


**FIGURE 3.73**

SEM images of the cross-section of the polyimide film Upilex after the heat treatment at 3000°C.

*(Courtesy of Prof. Y. Hishiyama of Tokyo City Univ.)*

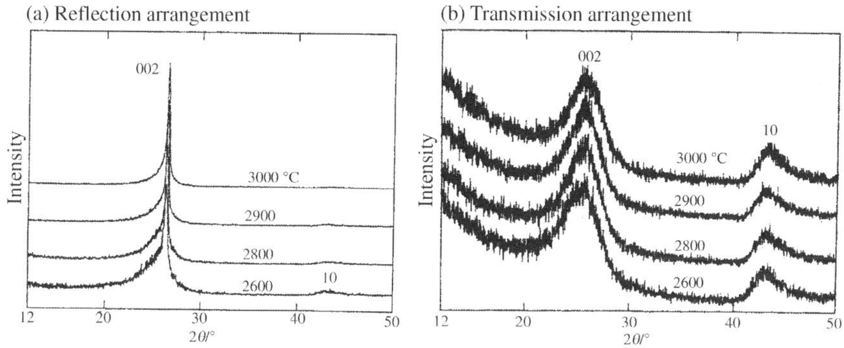


**FIGURE 3.74**

X-ray diffraction patterns of the glass-like carbon films heat-treated at high temperatures by two diffraction arrangements.

*(Courtesy of Prof. Y. Hishiyama of Tokyo city Univ.)*

formed on the surface of the film. On the films heat-treated at high temperatures, XRD patterns measured by reflection and transmission arrangements of the film to incident X-ray beam are shown in Fig. 3.74. 002 profile measured by reflection mode shows a sharp peak overlapped on a broad one, but that by transmission mode only a broad peak, suggesting the formation of well-crystallized regions by orienting along the film surface.

Graphitization on the surface after the treatment at high temperatures above 3000°C has been observed for non-graphitizing bacteria-cellulose-derived film and nanofiber [141,142] and phenol-based carbon nanofibers [143]. For phenol-based carbon nanofibers after activation show more marked graphitization than those without activation, of which the mechanism is supposed to be evaporation of carbon species at high temperatures and following deposition [143].

## 3.3.2 Properties

Various properties of non-graphitizing carbons have been measured mostly on the glass-like carbons, because they are obtained in a size enough to measure the properties, as plates and films. Due to the non-graphitizing nature of glass-like carbons, their electromagnetic and mechanical properties are quite different from highly oriented graphites described in Section 3.2 and isotropic high-density graphites in Section 3.1.

Electrical resistivity $\rho$ of a glass-like carbon measured at 20 and 300 K is plotted against HTT in Fig. 3.75a [144]. With increasing HTT, $\rho$ decreases at first, passes through a shallow minimum and then increases gradually. The HTT dependences of Hall coefficient $R_H$ and maximum transverse magnetoresistance $(\Delta\rho/\rho_0)_{max}$ are shown for two glass-like carbons with different thicknesses at different temperatures in a magnetic field of 0.65 T in Fig. 3.75b and c, respectively

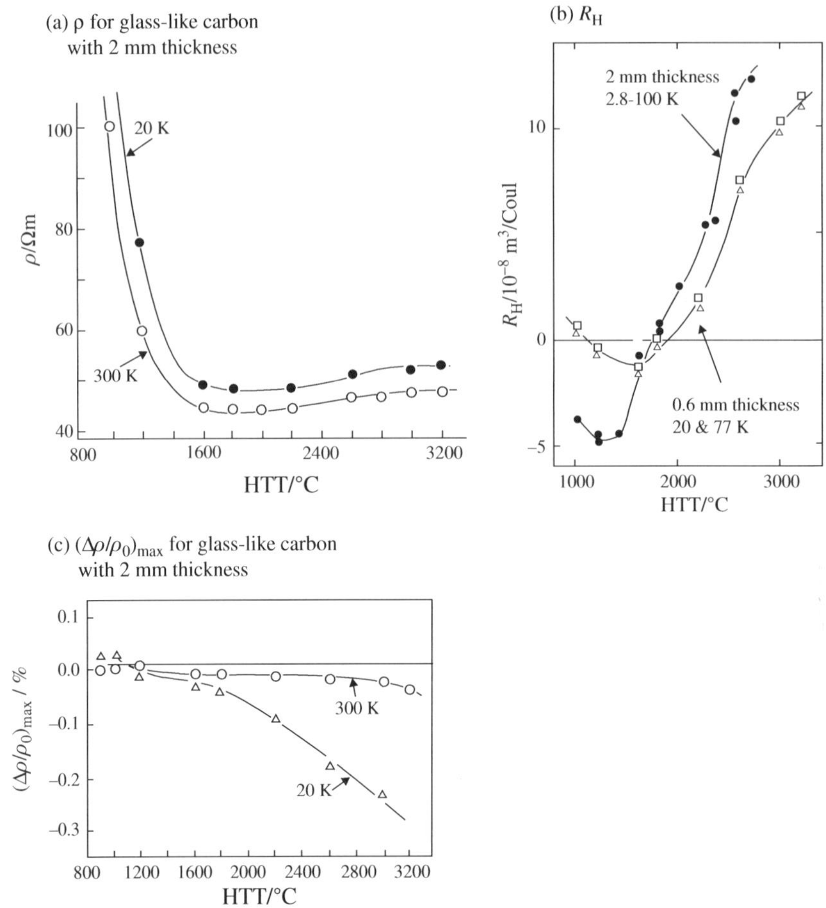


**FIGURE 3.75**

Dependences of galvanomagnetic properties of the glass-like carbons on HTT.

*(Courtesy of Dr. T. Yamaguchi of Tokai Carbon Co., Ltd.)*

[145,146]. In comparison with graphitizing carbons, such as cokes (Fig. 2.101), the changes in these parameters are markedly retarded. Referring to the general scheme for the changes in these parameters (Fig. 2.123), glass-like carbons are located just in the beginning of the changes; $R_H$ and $(\Delta\rho/\rho_0)_{max}$ do not reach a maximum and a minimum, respectively. Even after the heat treatment at 3200°C, glass-like carbons exhibit positive $R_H$ and negative $(\Delta\rho/\rho_0)_{max}$, which are characteristic for non-graphitizing carbons. $R_H$ for glass-like carbons is independent of magnetic field and also of temperature of measurement in a range of 2.8 to 300 K. However, it shows a strong dependence on HTT and a change from negative to positive values around 1600–1700°C, as shown in Fig. 3.75b.

From non-graphitizing carbons heat-treated at low temperatures, we could understand how the parameters for structure and properties change in the very beginning of structure development in carbon materials.

The dependence of thermoelectric power $S$ on temperature of measurement $T$ has been determined on the glass-like carbons with different HTTs [139,147], as shown in Fig. 3.76. The dependence of $S$ on $T$ can empirically be written as:

$$S = aT + bT^{1/2} + S_B, \tag{3.7}$$

where the first term is a strong scattering due to metallic component, the second is attributed to variable-range hopping and the third is a peaked component. The signs of the constants a and b give those of majority carriers for metallic and hopping conductions, respectively. According to the Equation (3.7), the plots of $S/T^{1/2}$ against $T^{1/2}$ are shown for three glass-like carbons heat-treated at different temperatures in Fig. 3.80a [139,147], and the dependences of two constants a and b in the equation (3.7) on HTT are shown in Fig. 3.77b [139]. The majority carriers for metallic conduction in glass-like carbons estimated from the constant a in Equation (3.7) change from negative electrons to positive holes around 1500°C, which shows a good correspondence to the change in the sign of $R_H$ in Fig. 3.75b. The sign of majority carriers for variable-range hopping, estimated from the constant b, changes at the same temperature from positive to negative, but its contribution to the observed value of $S$ becomes small with the increase in HTT.

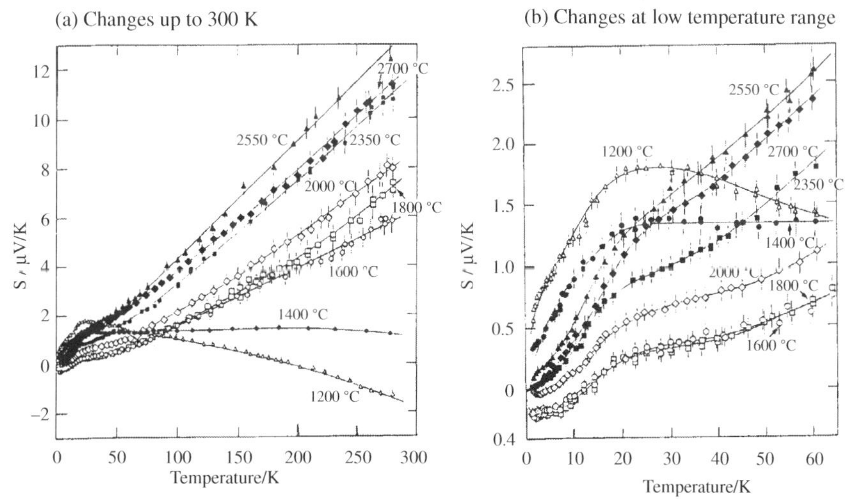


**FIGURE 3.76**

Temperature dependences of $S$ for glass-like carbons heat-treated at different temperatures.

*(Courtesy of Prof. Y. Kaburagi of Tokyo City Univ.)*

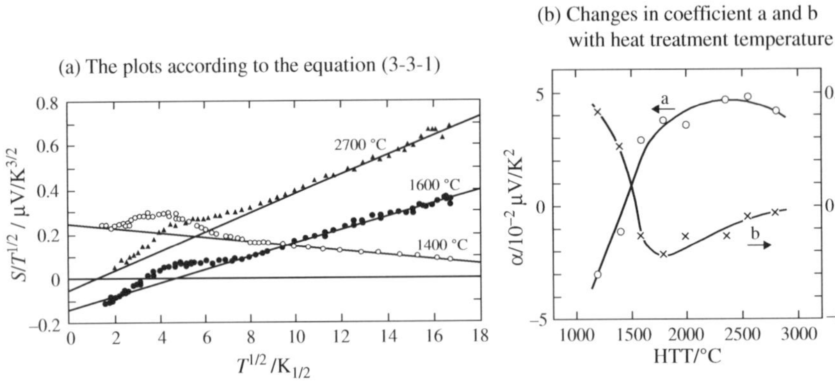


**FIGURE 3.77**

Temperature dependences of the parameters related to $S$ for glass-like carbons.

*(Courtesy of Prof. Kaburagi of Tokyo City Univ.)*

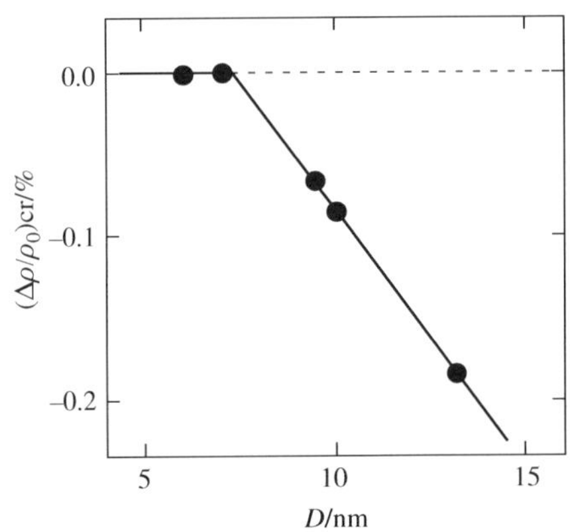


**FIGURE 3.78**

Relation between $(\Delta\rho/\rho_0)_{cr}$ and $D$ for glass-like carbons shown in Table 3.14.

*(Courtesy of Prof. A. Yoshida of Tokyo City Univ.)*

In Fig. 3.78, $(\Delta\rho/\rho_0)_{cr}$ of the glass-like carbons shown in Table 3.14 is plotted against average grain size $D$ determined from SEM micrograph. This result suggests a strong dependence of galvanomagnetic properties on structure.

In Fig. 3.79a and b, $\rho$ and $(\Delta\rho/\rho_0)_{cr}$ at 77 K under 1 T are plotted against HTT for the two glass-like carbon films. By the comparison of Figs. 3.82a and b with Figs. 3.75a and c, respectively, the behaviors of $\rho$ and $(\Delta\rho/\rho_0)_{cr}$ on these films are very similar to other glass-like carbons up to HTT of 2600°C. Above this temperature, however, their HTT dependences are quite different; $\rho$ decreases gradually with the increase in HTT and $(\Delta\rho/\rho_0)_{cr}$ increases abruptly to become

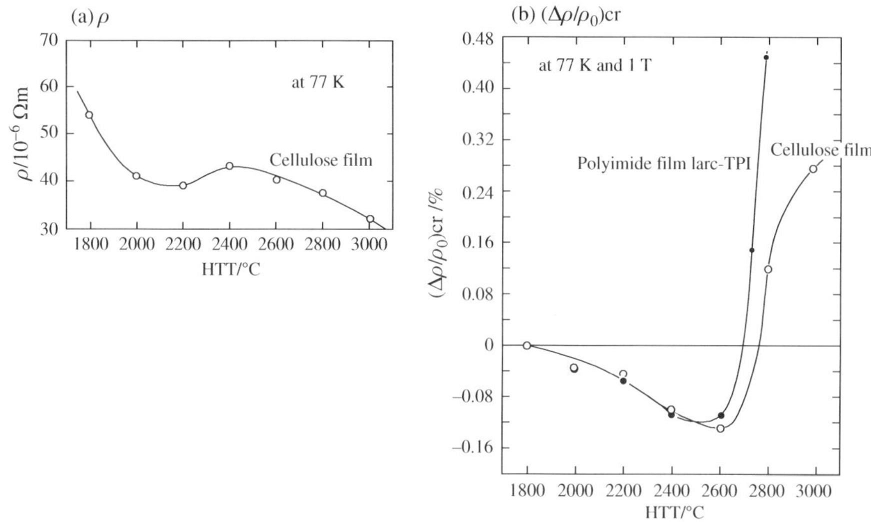


**FIGURE 3.79**

Dependences of $\rho$ and $(\Delta\rho/\rho_0)_{cr}$ on HTT for glass-like carbon films.

*(Courtesy of Prof. Y. Hishiyama of Tokyo City Univ.)*

positive, instead of gradual increase of $\rho$ and a continuous decrease in negative $(\Delta\rho/\rho_0)_{cr}$ in other glass-like carbons (Fig. 3.75). This is proved to be due to the formation of thin graphitized layers on the surface of films from XRD and SEM studies, as shown in Figs. 3.73 and 3.74.

Among carbon fibers, isotropic-pitch-based and PAN-based carbon fibers have also non-graphitizing nature: their changes in $(\Delta\rho/\rho_0)_{cr}$ with HTT are compared in Fig. 2.105 with those for vapor-grown and mesophase-pitch-based carbon fibers, which have graphitizing nature, and their nanotexture and various properties are discussed in detail in Section 3.4.

Most carbon blacks also show non-graphitizing behavior with heat treatment, mainly because of their point orientation nanotexture, as explained in Section 2.5.5. The size of particles is shown to govern the growth of crystallites (Fig. 2.112), where the non-graphitizing carbons prepared from thermosetting resins are located at the position extrapolated the relation to the size of closed shells observed in these carbons.

## 3.3.3 Glass-like carbons

In Table 3.15, the glass-like carbon is compared qualitatively with high-density isotropic graphite and quartz glass in various properties. In thermal resistance, electrical conductivity and thermal expansion, glass like carbon is very similar to isotropic high-density graphite, because these properties are basic for all carbon

**Table 3.15** Qualitative Comparison of Glass-Like Carbon with High-Density Isotropic Graphite and Quartz Glass

| | Isotropic High-Density Graphite | Glass-Like Carbon | Quartz Glass |
|---|---|---|---|
| Thermal resistance | Excellent | | Good |
| Electrical conductivity | High | | None |
| Thermal expansion | Relatively high | | Low |
| Thermal conductivity | High | Low | |
| Particle generation | Formed | None | |
| Gas permeability | Permeable | Non permeable | |
| Fractured surface | Granular | Conchoidal | |

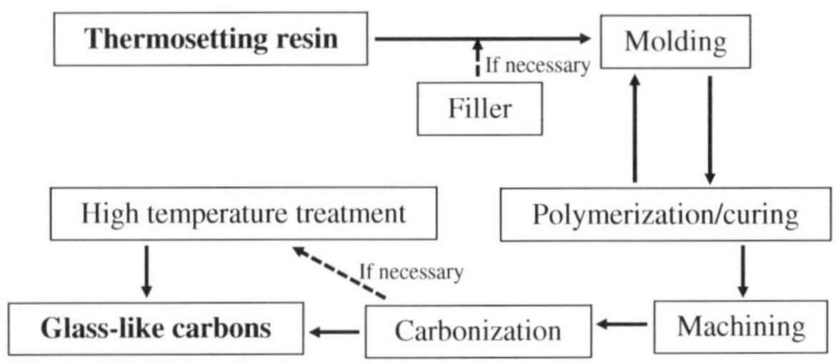


**FIGURE 3.80**

Scheme for the production of glass-like carbons.

materials. In thermal conductivity, gas permeability, particle generation by rubbing and morphology of fractured surface, however, glass-like carbon is very similar to quartz glass.

Glass-like carbon is produced by pyrolysis of thermosetting resins, such as phenol-formaldehyde, poly(furfuryl alcohol), etc. [148], and also of cellulose [149,150]. Glass-like carbon is a prototype of non-graphitizing carbon which possesses really glass-like nature, such as high hardness, brittle conchoidal fracture and gas impermeability [137]. The apparent density of commercially available glass-like carbons ranges from 1.46–1.50 g/cm$^3$, irrespective of heat treatment temperature, which implies the existence of thermally stable closed pores in the matrix.

Production of glass-like carbon is a very slow process, mostly limited by the rate at which decomposition gases are evolved out of the precursor blocks [150]. The scheme of the production of glass-like carbons is shown in Fig. 3.80, the procedure being almost the same as that of granular non-graphitizing carbons, except very slow heating rate during heat treatment, carbonization and high-temperature treatment. In order to produce the glass-like carbons with designed shapes and accurate size, however, more exact control of the conditions in each process, particularly carbonization process, are essential.

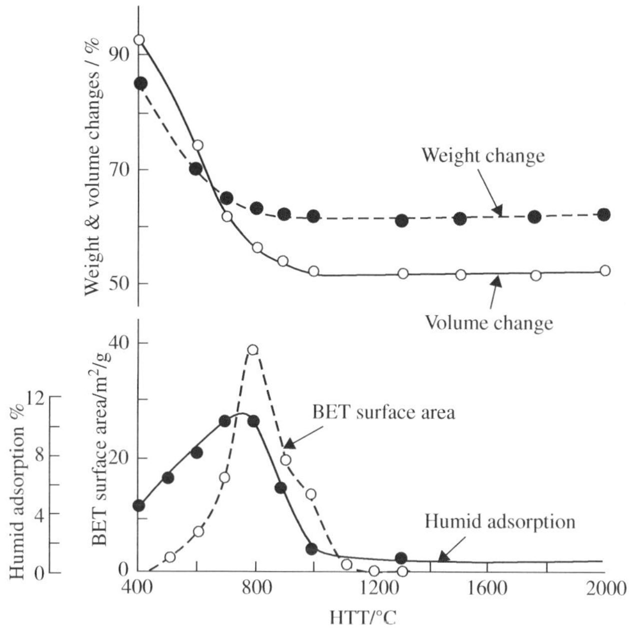


**FIGURE 3.81**

Changes in weight, volume, BET surface area and water adsorption of the glass-like carbon prepared from poly(furfuryl alcohol) with HTT [17].

In Fig. 3.81, changes in weight and volume with HTT are shown on a poly (furfuryl alcohol) condensate, together with the changes in BET surface area and adsorptivity of water vapor [151]. BET surface area and water adsorption increase with the decrease in weight due to the pyrolysis of the precursor, revealing the formation of many open pores. By a very slow heating, slower than the volume decrease occurring at a little higher than 700°C, both BET surface area and water adsorption decrease to negligibly small values. This result shows that a slow heating process to make open pores formed during pyrolysis closed is essential for the production of glass-like carbons.

Glass-like carbons are very hard and brittle, and so their machining is difficult after their production, so that they have to have a shape near the final product before carbonization. Therefore, their forming process has to be controlled exactly by taking account of rather large shrinkage during pyrolysis and carbonization.

In Table 3.16, some commercially available glass-like carbons are tabulated with their properties. All of them have very low gas permeability as $10^{-12}$ to $10^{-9}$ $cm^3/s$ and also very low open porosity, even though their bulk density is rather low, 1.4–1.5 $g/cm^3$, indicating the presence of a large amount of closed pores. Low gas impermeability is kept even after the heat treatment at 3000°C.

**Table 3.16** Characteristics of Commercially Available Glass-Like Carbons

| Company | Tokai Carbon | | | Kao |
|---|---|---|---|---|
| Trade name | GC-10 | GC-20 | GC-30 | S-100 |
| Heat treatment temperature (°C) | 1000 | 2000 | 3000 | 1200 |
| Bulk density ($cm^3/g$) | 1.47–1.51 | 1.46–1.50 | 1.43–1.47 | 1.45 |
| Gas permeability ($cm^2/s$) | $10^{-11}$–$10^{-12}$ | $10^{-11}$–$10^{-12}$ | $10^{-7}$–$10^{-9}$ | – |
| Bending strength (MPa) | 88–98 | 98–118 | 49–59 | 98 |
| Young's modulus (GPa) | 29–32 | 29–32 | 22–25 | 20 |
| Shore hardness | 100–120 | 100–110 | 70–80 | 120 |
| Electrical resistivity (μΩm) | 45–65 | 40–45 | 35–40 | 45 |
| Thermal conductivity (W/m·K) | 3.76–4.6 | 8-36–9.19 | 15.0–17.6 | 3.34–3.76 |
| Thermal expansion coefficient ($\times 10^{-6}$/°C) | 2.0–2.2 | 2.0–2.2 | 2.0–2.2 | 3.5 |

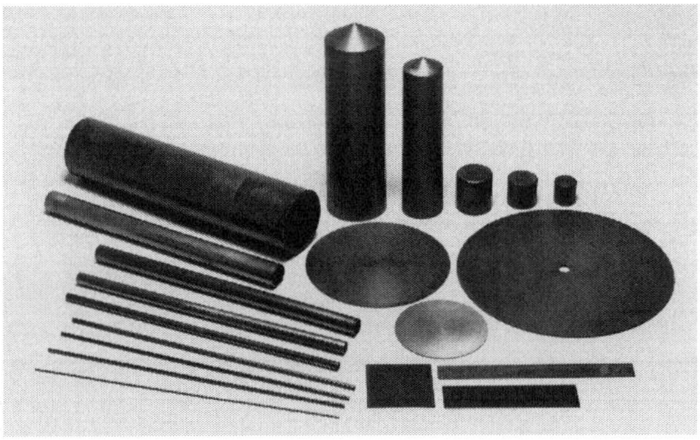

**FIGURE 3.82**

Products of glass-like carbon.

*(Courtesy of Tokai Carbon Co., Ltd.)*

In Fig. 3.82, some industrial products of glass-like carbon are shown. Thin plates, pipes, rods and crucibles are produced and used in various fields of industries.

## 3.4 Carbon fibers

### 3.4.1 Classification of fibrous carbons

A variety of carbon materials with fiber-like morphology, fibrous carbons, have been prepared and attracted attention of scientists and engineers in different

**Table 3.17** Classification of Fibrous Carbon Materials

| Diameter of Fibers | Short Carbon Layers | Long Carbon Layers |
|---|---|---|
| Micrometer-size | Carbon fibers | Graphite whiskers |
| | PAN-based carbon fibers | |
| | Pitch-based carbon fibers | |
| | Vapor-grown carbon fibers → | Graphite fibers |
| | Carbon microcoils | |
| Nanometer-size | Carbon nanofibers | |
| | Tubular-type → | Carbon nanotubes |
| | Herringborne-type | Single-wall |
| | Platelet-type | Multi-walled |
| | Crooked | |

fields. Table 3.17 is a proposal to classify those fibrous carbons whether the 002 lattice fringes of carbon layers are short or long, and apparent diameter of fibrous carbons, whether their diameters are in micrometer or nanometer size.

By arc-discharging between graphite electrodes, carbon nanotubes, single-wall carbon nanotube (SWCNT) and multi-walled carbon nanotubes (MWCNT), were obtained, of which carbon layers are long and oriented exactly parallel to the tube axis, mainly because of their formation at a high temperature. Through CVD processes using metal catalysts, carbon nanofibers with various nanotextures were also produced. Many tubular type nanofibers can be converted to MWCNTs by high-temperature treatment.

Carbon fibers have developed since 1960, which was one of new carbons (refer to Section 1.2), from different precursors, either organic resins, like poly (acrylonitrile) (PAN) and various pitches, or hydrocarbon gases by CVD process. The carbon fibers produced from benzene vapor in hydrogen atmosphere are named vapor-grown carbon fibers, of which characteristics are high graphitizability by high temperature treatment. In the vapor-grown carbon fibers heat-treated at a high temperature such as 3000°C, carbon layers grow markedly and so graphite structure is developed. Therefore, these vapor-grown carbon fibers heat-treated at high temperatures can be called graphite fibers. Fibrous carbons with a high degree of graphitization are obtained under arc discharging and called graphite whiskers. In PAN-based carbon fibers, on the other hand, graphite structure is difficult to develop by high-temperature treatment. Through CVD processes using different catalysts and precursor gases, various crooked nanofibers are formed and carbon materials with coil morphology with homogeneous pitches, carbon microcoils, are synthesized with high reproducibility.

In this chapter, carbon fibers are explained by classifying on the basis of precursor used. Carbon nanotubes and nanofibers are going to be explained as one of nanocarbons in Section 3.5.

## 3.4.2 Characteristics of carbon fibers

Carbon fibers, which have been produced on an industrial scale, are classified by the precursors used, as shown in Table 3.18 together with some characteristics of each fiber. The PAN-, isotropic-pitch- and mesophase-pitch-, cellulose- and phenol-based carbon fibers are produced by melt spinning of each precursor, followed by stabilization and carbonization up to about 1300°C. In contrast to these carbon fibers derived from organic precursor fibers, vapor-grown carbon fibers are prepared by thermal decomposition of gas of organic precursors, such as benzene vapor, by using catalyst of fine metal particles, where stabilization is not needed. The carbon fibers produced by spinning of organic precursors can be obtained as continuous ones, but vapor-grown carbon fibers are only short fibers.

Carbon fibers are also classified according to their mechanical performance, tensile strength and modulus, as shown in Fig. 3.83a. Carbon fibers, which have relatively low tensile strength and modulus, around 1000 MPa and 100 GPa are classified into the general purpose grade (GP-grade). Isotropic-pitch-based, cellulose-based, phenol-based and some of PAN-based carbon fibers belong to this grade and have the application to use its low weight and bulkiness, for example, thermal insulation for high-temperature furnaces. Recently a novel application of these GP-grade carbon fibers to cement reinforcement has been developed (refer to Section 3.7.4). Carbon fibers having higher strength and modulus than GP-grade are called high performance grade (HP-grade), which are further classified into high strength type (HT-type) and high modulus type (HM-type). Most PAN-based carbon fibers produced in industry are HT-type, and mesophase-pitch-based

**Table 3.18** Classification of Carbon Fibers Based on their Precursors and their Characteristics

| Precursor | Carbon Fibers | Characteristics |
|---|---|---|
| Polyacrylonitrile (PAN) | PAN-based carbon fibers | Different grades and types |
| Isotropic pitch | Isotropic-pitch-based carbon fibers | General purpose grade<br>Random cross-sectional nanotexture |
| Mesophase pitch | Mesophase-pitch-based carbon fibers | High modulus types<br>Various cross-sectional nanotextures |
| Cellulose | Cellulose-based carbon fibers | General purpose grade<br>Random nanotexture |
| Phenol | Phenol-based carbon fibers | General purpose grade<br>Random nanotexture |
| Hydrocarbon gases | Vapor-grown carbon fibers | High graphitizability<br>Annual-ring texture in cross-section |

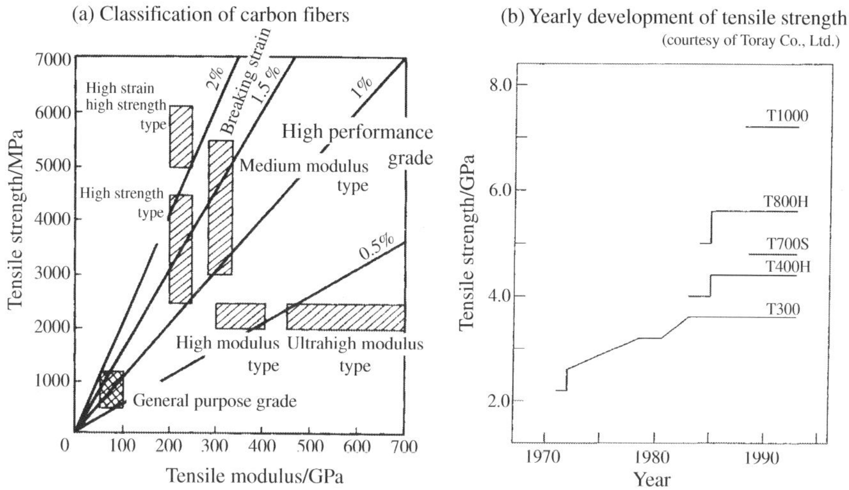


**FIGURE 3.83**

Mechanical properties of carbon fibers.

carbon fibers, which are produced through spinning of anisotropic mesophase pitches, are HM-type. In the production of PAN-based carbon fibers, a great effort has been paid to get high modulus, and some of them could have relatively high modulus. On the contrary, mesophase-pitch-based carbon fibers have relatively high tensile modulus and are desired to improve their strength.

The industrial effort on the production of carbon fibers in various respects resulted in a great improvement in mechanical properties and also in their reproducibility. Fig. 3.83b shows the changes in tensile strength on commercially available PAN-based carbon fibers produced in a company, tensile strength reaching to more than 7 GPa, about double that before 1985.

These mechanical properties of carbon fibers depend strongly on the precursor, as well as heat treatment conditions, mainly temperature, as other properties do. For most commercially available carbon fibers, however, these conditions have never been declared in detail. Therefore, the understanding of mechanical properties as a function of heat treatment conditions is limited to the data obtained in laboratories.

In Table 3.19, representative values of some physical properties are compared on four kinds of carbon fibers, after carbonization at about 1300°C and heat treatment as a high temperature as 2800°C. By the heat treatment to a high temperature above 2800°C, bulk density increases, and tensile strength, tensile modulus and electrical resistivity decrease in each carbon fiber. However, the precursor predominantly governs these properties, isotopic-pitch-based carbon fibers having the lowest values of all properties and vapor-grown carbon fibers the highest ones. Though vapor-grown carbon fibers heat-treated at temperatures around

**Table 3.19** Characteristics of Four Kinds of Carbon Fibers

| | Heat Treatment | Diameter ($\mu m$) | Density ($g/cm^3$) | Tensile Strength (MPa) | Tensile Modulus (GPa) | Electrical Resistivity ($10^{-3}\ \Omega cm$) | CTE ($10^{-6}/°C$) |
|---|---|---|---|---|---|---|---|
| Isotropic-pitch-based | C | 14.5 | 1.65 | 720 | 32 | 15 | 1.7 |
| | G | 14.5 | 1.57 | 600 | 30 | 5 | 1.5 |
| PAN-based | C | 7.0 | 1.74 | 3300 | 230 | 1.6 | −0.7 |
| | G | 6.5 | 1.81 | 2500 | 300 | 0.8 | −1.2 |
| Mesophase-pitch-based | C | 10 | 1.98 | 3500 | 200 | 13 | −0.5 |
| | G | 10 | 2.18 | 2100 | 520 | 0.5 | −1.2 |
| Vapor-grown | C | 5–8 | 1.80 | 3000 | 300 | 1 | – |
| | G | 5–8 | 2.20 | 5000–8000 | 500 | 0.05 | – |

*C: carbonized at about 1300°C; G: treated at a high temperature as 2800°C; CTE: thermal expansion coefficient.*

3200°C have similar values of physical properties to the graphite whisker, their applications are limited because of short fibers. Mesophase-pitch-based carbon fibers have relatively high tensile modulus and PAN-based ones have high tensile strength.

Industrial applications of carbon fibers are summarized in Table 3.20. The main applications of carbon fibers are in various composites not only with plastics, but also with ceramics, including concrete, and with carbon. These composites are described in Section 3.7. GP-grade carbon fibers, either isotropic-pitch-based or PAN-based, have successfully been used as heat-insulating materials for high-temperature furnaces, because of their low weight and bulkiness. Also carbon fiber braids have widely been used for sealing. By activation of GP-grade carbon fibers, high surface area and high proportion of micropores are obtained (activated carbon fibers), which created various new applications (Section 3.6.1a). Strands of HP-grade carbon fibers are recently used for reinforcement of piers of highways and also pillars in buildings (Section 3.7.4). A novel and important application of carbon fibers for environmental remediation has been developed, which is explained in Section 3.10.3.

**Table 3.20** Applications of Carbon Fibers

| Materials | | Applications | Carbon Fibers | Related Fields of Technology |
|---|---|---|---|---|
| Fibers | | Insulation materials | GP-grade | Electronics, automobile, aircraft, nuclear energy |
| | | Sealing materials | GP- and HP-grades | Chemicals, petrochemicals |
| | | Adsorption of bacteria | HP-grade | Environmental |
| | | Microporous materials | activated | Capacitors, adsorbents |
| | | Reinforcing materials | HP-grade | Construction, bridge |
| Composites | with resin | Functional materials | GP- and HP-grades | Electronics, mechanics, automobiles, aircraft, sports, etc. |
| | with carbon | Structural materials | HP-grade | Sports, medical, space, aircraft, automobiles, electrodes, etc. |
| | with metal | Ablation & frictional materials | GP- and HP-grades | Space, military, aircraft, automobiles, railways, etc. |
| | with ceramics | Cell electrodes | GP- and HP-grades | Automobiles |
| | with concrete | Construction materials | GP- and HP-grades | Buildings, housing, bridges, etc. |

Change in structure of carbon fibers depends strongly on the precursors, as is described in the following sections. For four kinds of carbon fibers, magnetoresistance parameters (Section 2.5.1) are summarized in Table 3.21 as a function of heat treatment temperature. Isotropic-pitch-based carbon fibers show negative values of crystallite magnetoresistance $(\Delta\rho/\rho_0)_{cr}$ and anisotropy ratio $r_{TL}$ of around 0.5, though $r_T$ is close to 1.00 even after the heat treatment up to 3000°C, indicating turbostratic structure with less degree of axial orientation. PAN-based carbon fibers also have negative values of $(\Delta\rho/\rho_0)_{cr}$ even after 3000°C-treatment, but axial orientation is formed, values of $r_{TL}$ being small but $r_T$ close to 1.00. In mesophase-pitch-based and vapor-grown carbon fibers, a transition from negative to positive value of $(\Delta\rho/\rho_0)_{cr}$ is observed with increasing heat treatment temperature (HTT); the former requiring much higher HTT in order to attain positive value than the latter. For vapor-grown carbon fibers, $r_{TL}$ is a very small value even after treatment at around 2000°C, showing a high degree of axial orientation nanotexture. A slight decrease in $r_T$ from 1.00 is observed after high-temperature treatment above 2600°C, mainly due to the polygonization of the cross-section of fibers, as shown in Fig. 3.107.

A general scheme of the change in $(\Delta\rho/\rho_0)_{cr}$ for these four carbon fibers is discussed in Section 2.5.4, by showing the dependence of $(\Delta\rho/\rho_0)_{cr}$ on HTT

**Table 3.21** Magnetoresistance Parameters for Carbon Fibers Heat-Treated at Different Temperatures

| Carbon Fibers | HTT (°C) | $(\Delta\rho/\rho_0)_{cr}$ (%) | $r_{TL}$ | $r_T$ |
|---|---|---|---|---|
| Isotropic-pitch-based | 3000 | −0.230 | 0.615 | 0.988 |
| | 2800 | −0.152 | 0.512 | 0.992 |
| | 2600 | −0.075 | 0.679 | 0.982 |
| PAN-based | 3000 | −0.547 | 0.063 | 0.977 |
| | 2900 | −0.954 | 0.049 | 0.994 |
| | 2800 | −0.618 | 0.056 | 0.964 |
| Mesophase-pitch-based | 3000 | 1.61 | 0.012 | 0.417 |
| | 2800 | −0.763 | 0.079 | 0.929 |
| | 2500 | −0.469 | 0.050 | 0.858 |
| | 2000 | −0.214 | 0.119 | 1.000 |
| Vapor-grown | 3000 | 153.0 | 0.152 | 0.854 |
| | 2800 | 138.7 | 0.047 | 0.883 |
| | 2600 | 38.6 | 0.026 | 0.843 |
| | 2400 | 11.7 | 0.069 | 0.952 |
| | 2200 | 0.963 | 0.000 | 0.915 |
| | 2000 | −1.20 | 0.087 | 0.913 |
| | 1800 | −0.275 | 0.000 | 0.897 |

*(Courtesy of Prof. Y. Hishiyama of Tokyo City Univ.)*

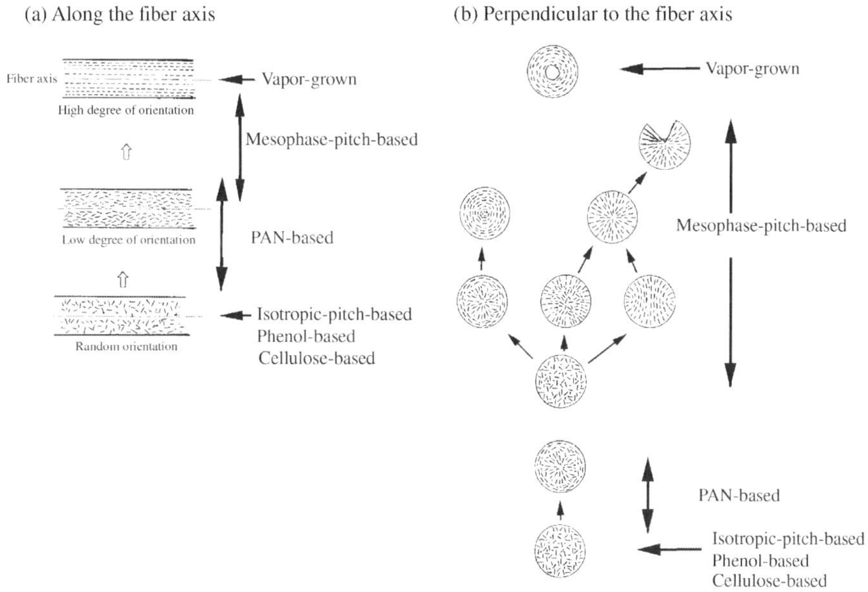


**FIGURE 3.84**

Scheme of nanotextures in various carbon fibers.

reduced by referring to that for vapor-grown carbon fibers, including the data listed in Table 3.21 [152].

Carbon fibers have a wide range of nanotextures in their cross-sections both along and perpendicular to the fiber axis, as shown schematically in Fig. 3.84. The carbon fibers prepared from isotropic pitch, phenol resin and cellulose have random nanotexture in both cross-sections along and perpendicular to their fiber axis. In PAN-based carbon fibers, the degree of orientation of carbon layers is pretty low, but oriented along the fiber axis. By stretching of the fibers during their carbonization and graphitization, the orientation of layers along the fiber axis was improved markedly, as explained in the following section. Nanotexture in mesophase-pitch-based carbon fibers is spread in a wide range, mostly depending on precursor pitches and their spinning conditions, in both cross-sections along and perpendicular to the fiber axis. In their cross-section perpendicular to the axis, various nanotextures from almost random to either concentric or radial orientation are produced. Vapor-grown carbon fibers have a high degree of concentric axial orientation with a hollow tube at their center, by high-temperature treatment becoming almost complete orientation in both along and perpendicular directions.

Detailed scientific aspects on carbon fibers, particularly on PAN-based ones, were reviewed in Reference [153].

### 3.4.3 PAN-based carbon fibers

The production process is shown as a flow chart in Fig. 3.85. It consists of spinning of precursor PAN, stabilization of spun fibers and heat treatment of stabilized fibers to high temperatures. The key technology for the production of PAN-based carbon fibers is the design of precursor and the selection of its stabilization condition. Because of the confidential aspects for industries, however, the practice of these key technologies has never been published.

For the production of carbon fibers, not only high spinnability of the precursor PAN but also high rate of stabilization of the spun PAN fibers and high mechanical performance of the resultant carbon fibers have to be attained. For high spinnability of PAN, various additives have been tried. Sodium acrylate co-monomer is effective to reduce the stabilization time and also to have high strength of carbon fibers [154,155]. Structural changes in PAN during stabilization and carbonization processes have been studied by various authors [152,156] and are discussed in Section 2.3.3c.

During stabilization and carbonization, the stress is accumulated mainly due to hindrance of thermal motion of PAN molecules, shrinkage caused by oxidation reactions during stabilization, and also shrinkage by pyrolysis and carbonization accompanied by the departure of hydrocarbon gases. A scheme of the change in stress during heating up to 600°C with a rate of 4.5°C/min in either air or nitrogen atmosphere is shown in Fig. 3.86 [155].

The change in stress with temperature is divided into four regions, as indicated in Fig. 3.86. In the region I, stress increases up to a maximum (region I-1) and

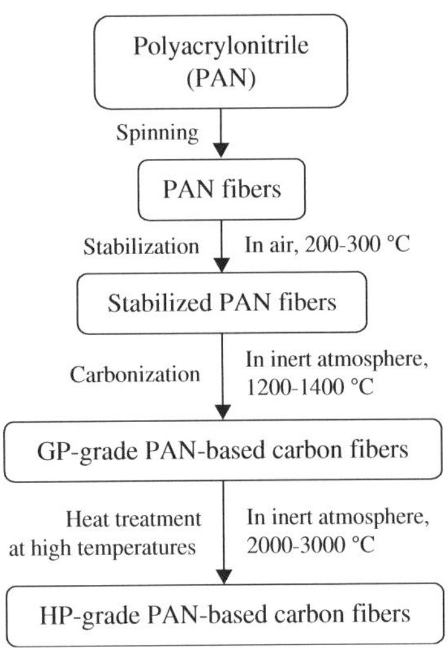


**FIGURE 3.85**

Production process of PAN-based carbon fibers.

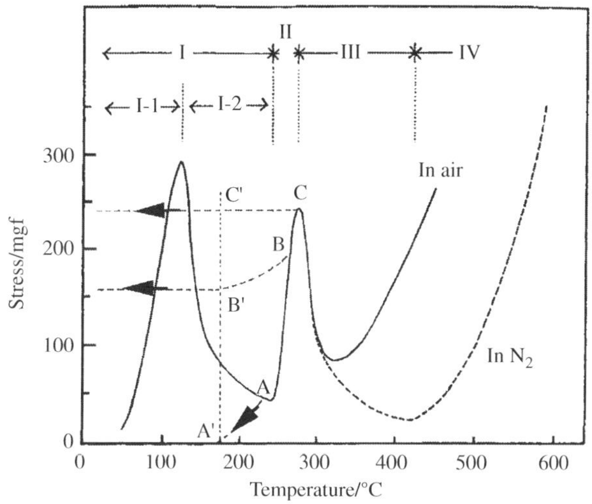


**FIGURE 3.86**

Stress behavior on PAN fibers during stabilization and carbonization [155].

then relaxes down to a minimum at around 240°C (region I-2), which is supposed to be due to the motion of PAN molecules at high temperatures. If the fibers are cooled from this region, the shrinkage stress always became zero. For example, cooling down from the point A in Fig. 3.86, a little before reaching the minimum, results in zero stress below 180°C. The region II, abrupt increase of stress and reaching a maximum again, is due to the change in chemical structure of the fibers, such as network formation by oxygen, polymerization of nitril side group, etc. The cooling from this region leaves certain stress in the fibers, which becomes constant below 180°C, as shown by the line from B to B′ or from C to C′ in the figure.

In Fig. 3.87, the changes of shrinkage stress with time at different temperatures in air and at 242°C in an atmosphere with different oxygen contents are shown. From the results in Fig. 3.87, the scheme of stress change is concluded to be exactly the same as that with a constant rate heating, but a pronounced dependence on oxidation conditions, temperature and oxygen content, is observed in this second region, though it is not clearly seen in Fig. 3.86 because of so rapid increase with temperature. Based on these experimental results, a parameter, stabilization index (SI), which is a ratio of the stress at 180°C in cooling (for example, the point B' in Fig. 3.86) to that at the temperature where cooling starts (the point B), has been proposed to evaluate the degree of modification in chemical structure in fibers by oxidation (stabilization) [155]. This parameter SI has a good correspondence to the content of oxygen in the fibers after stabilization (stabilized fibers) and also to the tensile strength of resultant carbon fibers, as shown in Fig. 3.88.

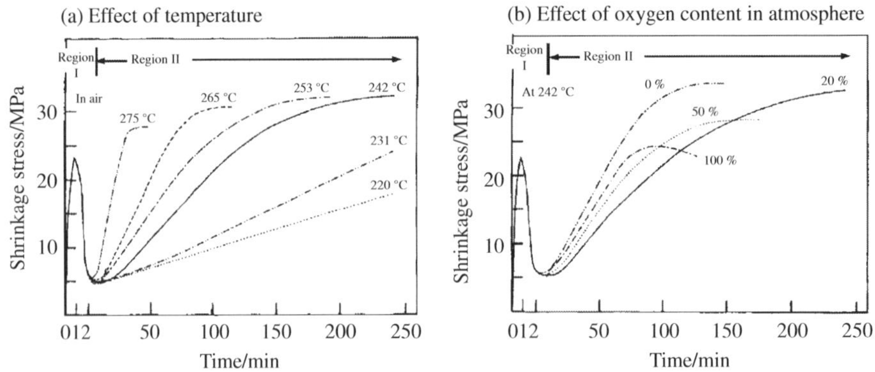


**FIGURE 3.87**

Shrinkage stress on the PAN fibers with time [155].

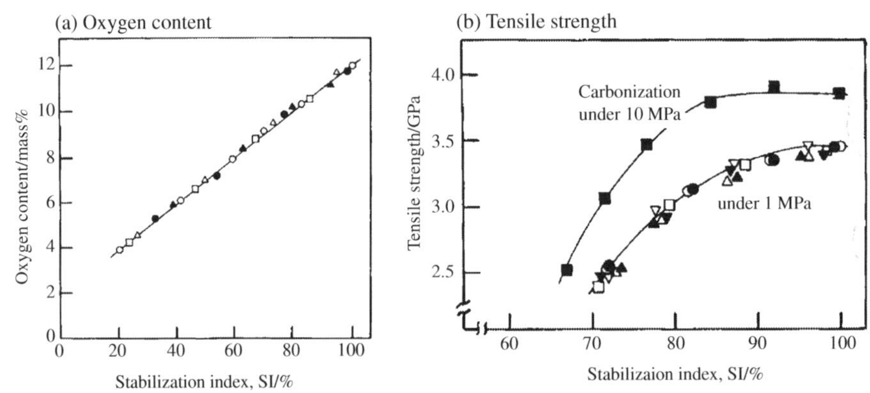


**FIGURE 3.88**

Dependences of oxygen content after stabilization and tensile strength after carbonization on SI for PAN fibers [155].

In the region III in Fig. 3.86, the fibers show different behaviors in stress depending on atmosphere; in air stress decreases quickly and reaches a minimum at a low temperature, but in nitrogen it decreases to a lower minimum at a higher temperature. This region corresponds to further pyrolysis and plasticization of polymers formed in previous regions. The region IV with gradual increase in stress is due to the formation and growth of carbon hexagonal layers.

Because of plasticization during stabilization and carbonization, stretching of fibers during these processes is effective to improve the performance of carbon fibers [157], which is caused by the improvement in preferred orientation of carbon layers along the fiber axis. Fig. 3.89 shows how the improvement in orientation is effective to get high modulus by plotting modulus of fibers against

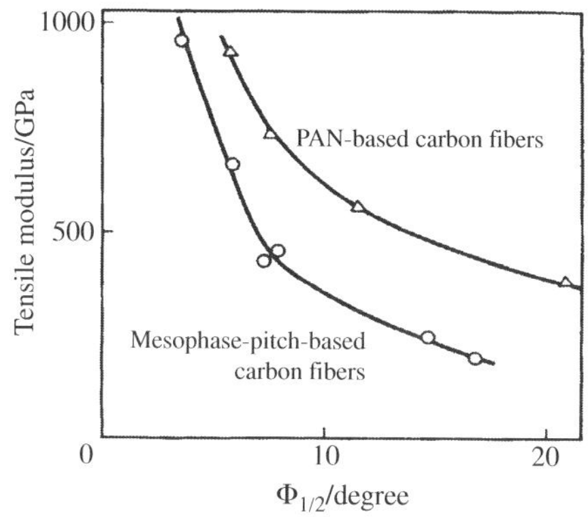


**FIGURE 3.89**

Dependence of tensile modulus of carbon fibers on $\Phi_{1/2}$ [158].

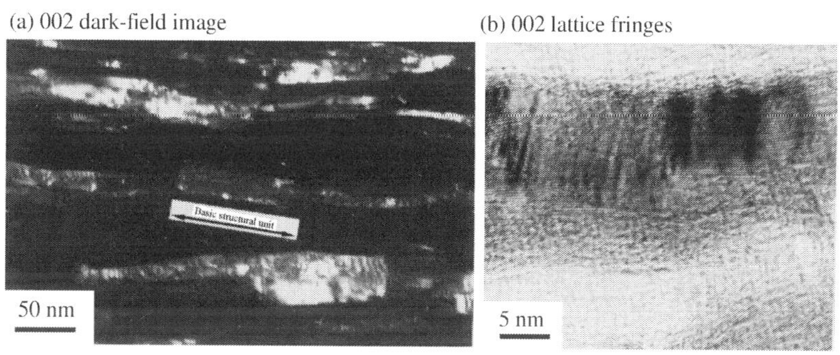


**FIGURE 3.90**

TEM images of HM-type of PAN-based carbon fibers [159].

orientation degree (width at half maximum of orientation function) $\Phi_{1/2}$ [158]. From the analysis of stress in fibers shown in Fig. 3.86, stabilization in keeping the fiber length constant, where the fibers are under tension in the region I-2 and under shrinkage in the region II, and carbonization under stretching in the temperature range of the region III and IV are effective to get high mechanical strength with shortened processing time for the production [155].

Using high-resolution transmission electron microscopy (TEM) techniques, the structure of commercial PAN-based carbon fibers of HM- and HT-types has been studied in detail [159]. Representative 002 dark field and 002 lattice fringe images of these two types of fibers are shown in Figs. 3.90 and 3.91, respectively. It is clearly seen that there is a pronounced difference in size of carbon layer stacks (crystallites) between HM and HT types; in the former crystallites are much thicker and longer than those in the latter. In addition, the curvature of the

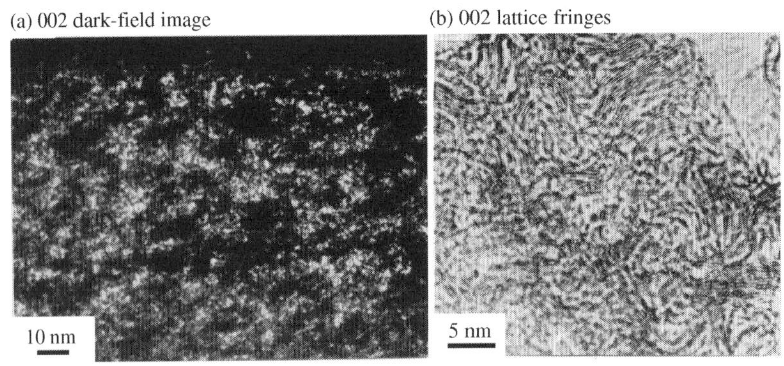


**FIGURE 3.91**

TEM images of HT-type of PAN-based carbon fibers [159].

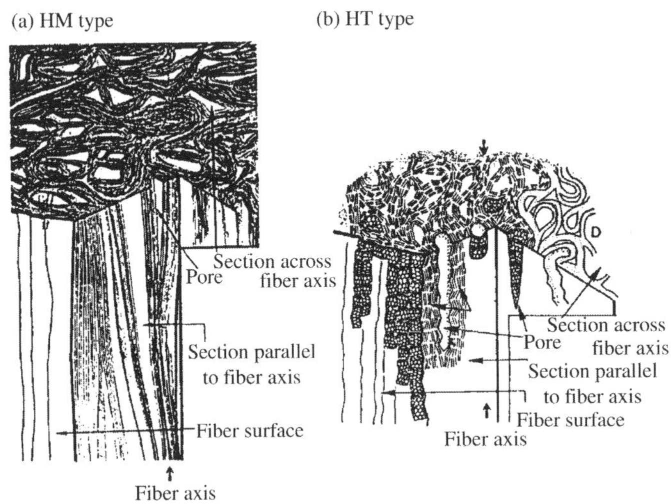


**FIGURE 3.92**

Structure models for HM- and HT-type PAN-based carbon fibers.

*(Courtesy of Mme. A. Oberlin.)*

fringes is more pronounced in the former. This difference is caused because of thermal history between these two types of fibers; the former is heat-treated up to a high temperature possibly under stretching, but the latter to a relatively low temperature. However, the exact reason is not known because detailed preparation conditions are not published on these commercial fibers.

From the detailed analysis of the cross-sections of fibers, the structure models for HM- and HT-types of fibers have been constructed [159,160]. The proposed models are reproduced in Fig. 3.92. In HM-type, stacked carbon layers are

**Table 3.22** Effect of Stretching on Magnetoresistance Parameters of PAN-Based Carbon Fibers Prepared Under Stretching

| Stretching Conditions | HTT (°C) | $(\Delta\rho/\rho_0)_{cr}$ (%) | $r_{TL}$ | $r_T$ |
|---|---|---|---|---|
| Without stretching | 3000 | −0.547 | 0.063 | 0.977 |
| | 2900 | −0.954 | 0.049 | 0.994 |
| | 2800 | −0.618 | 0.056 | 0.864 |
| 40 mg/fiber during stabilization and carbonization | 3000 | 26.18 | 0.027 | 0.753 |
| | 2830 | 6.37 | 0.022 | 0.403 |
| 40 mg/fiber during 2700°C-treatment of PAN-S | 3000 | 26.20 | 0.110 | 0.200 |
| | 2830 | 9.19 | 0.011 | 0.303 |
| | 2700 | 2.10 | 0.045 | 0.545 |
| 80 mg/fiber during 2700°C-treatment of PAN-S | 3000 | 21.18 | 0.225 | 0.665 |
| | 2830 | 9.23 | 0.023 | 0.945 |
| | 2700 | 4.74 | 0.000 | 0.789 |
| 120 mg/fiber during 2700°C-treatment of PAN-S | 3000 | 22.99 | 0.083 | 0.874 |
| | 2830 | 4.73 | 0.091 | 0.209 |
| | 2700 | 1.54 | 0.000 | 0.540 |

*(Courtesy of Prof. Y. Hishiyama of Tokyo City Univ.)*

tangled with each other in the central part of the fiber, but the layers are straight and stacked parallel in a wide range in the periphery (Fig. 3.92a). In HT-type, on the other hand, carbon layer stacks are much smaller and thinner, and no marked difference between central and peripheral parts (Fig. 3.92b).

Stretching of the fibers during carbonization and high-temperature treatment gives a pronounced effect on structure and nanotexture of the resultant carbon fibers. The effect of stretching is exhibited through the measurements of $(\Delta\rho/\rho_0)_{cr}$, as shown in Table 3.22 [1]. PAN-based carbon fibers show negative $(\Delta\rho/\rho_0)_{cr}$ even after heat treatment to 3000°C, but $(\Delta\rho/\rho_0)_{cr}$ becomes positive even after 2700°C treatment on the fibers stabilized and carbonized under a load of 40 mg/filament. This marked increase in $(\Delta\rho/\rho_0)_{cr}$ is due to the development of graphitic structure in the fibers [161]. The $r_{TL}$ becomes small value by stretching, suggesting that the orientation scheme of carbon layers in the fiber changes from axial to planar orientation by stretching during heat treatment processes. This expectation has been proved by SEM observation in Fig. 3.93 [152].

### 3.4.4 Pitch-based carbon fibers

Pitch-based carbon fibers have to be divided into two classes, isotropic-pitch-based and anisotropic mesophase-pitch-based. These two carbon fibers are different not only in precursor but also in their structure, nanotexture and properties. The

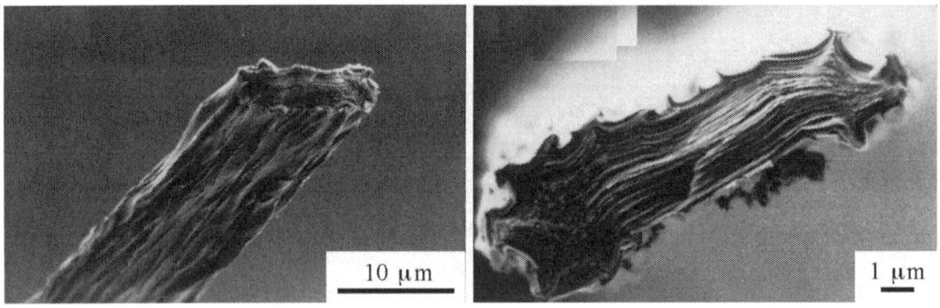


**FIGURE 3.93**

SEM images of PAN-based carbon fibers stretched during 2700°C-treatment and then heat-treated at 3000°C.

*(Courtesy of Prof. Y. Hishiyama of Tokyo City Univ.)*

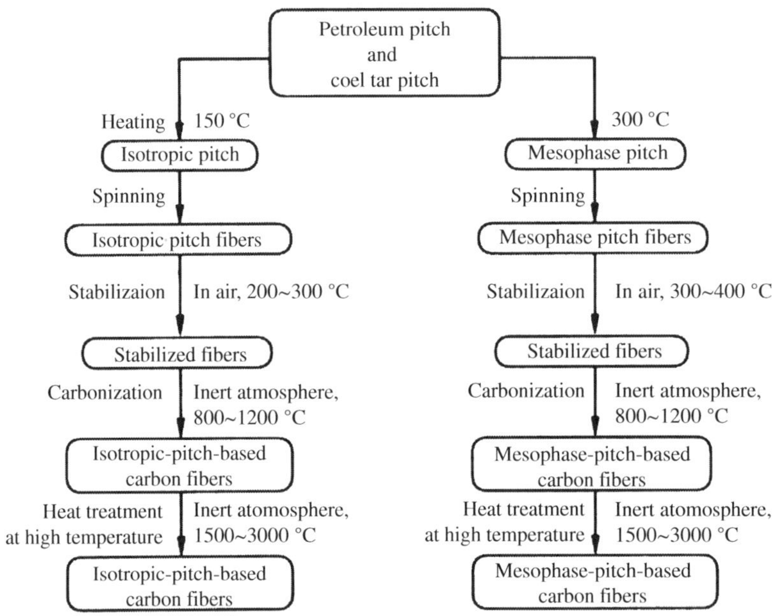


**FIGURE 3.94**

Production processes for pitch-based carbon fibers.

production processes for these two fibers are shown in Fig. 3.94. Fundamental processes for these carbon fibers are spinning, stabilization and then carbonization, as for PAN-based ones (Fig. 3.85).

In Table 3.23, some examples of pitches used for carbon fiber production are listed, showing softening point, toluene insoluble fraction, percentage of optically anisotropic regions and chemical analysis data [162]. For the production of

**Table 3.23** Pitches for Carbon Fiber Production

| Pitches | | Isotropic | | | Anisotropic | | | |
|---|---|---|---|---|---|---|---|---|
| Origin of pitch | | Coal | Coal | Petroleum | Coal | Coal | Synthetic | Synthetic |
| Softening point (°C) | | 267 | 250 | 250 | 308 | 299 | 257 | 265 |
| Toluene insoluble (mass%) | | 74.6 | 44.1 | 58 | 86 | 95 | 43 | 74 |
| Anisotropy ratio (%) | | 0 | 0 | -- | 100 | 100 | 100 | 100 |
| Elemental composition | C | 94.3 | 93.4 | 93.9 | 94.2 | 95.2 | 94.4 | 95.2 |
| | H | 3.89 | 4.44 | 4.65 | 3.92 | 3.82 | 5.91 | 5.02 |
| | N | 1.01 | 0.85 | 0.2 | 1.05 | 0.89 | <0.1 | 0.1 |
| | S | 0.32 | 0.29 | 0.7 | 0.28 | 0.15 | <0.01 | 0.16 |
| H/C ratio | | 0.5 | 0.57 | 0.59 | 0.50 | 0.48 | 0.75 | 0.63 |

*(Courtesy of Dr. K. Sato of Nippon Steel Tech. Co., Ltd.)*

isotropic-pitch-based carbon fibers, not so many problems have been pointed out, only a requirement of the pitch with homogeneity and appropriate fluidity. In contrast, the preparation of optically anisotropic pitches (mesophase pitches) is the most important process for the production of mesophase-pitch-based carbon fibers. Mesophase and its formation in pitches are described in Section 2.3.4. The factors influencing on carbon fiber production have been pointed out to be the content of mesophase (percentage of optically anisotropic regions, anisotropy ratio, as listed in Table 3.23), softening point and degree of hydrogenation of the pitch. In Fig. 3.95, the relations among softening point and mesophase content are shown before and after hydrogenation treatment. The softening point of a pitch depends strongly on mesophase content. Even if mesophase content is the same, however, softening point decreases markedly by hydrogenation treatment. In order to be able to spin the fibers at low temperatures, where thermal stability of the pitch is assured, hydrogenation is an essential process. Various processes for that purpose have been proposed; direct addition using hydrogen gas, employing different solvents with a hydrogen-donation ability, etc. After hydrogenation, the pitches are heated to complete the condensation reactions and to remove volatile matters, mesophase content of the pitch becoming close to 100%.

Viscosity of pitch increases by the formation of mesophase, which is one important factor for the spinning of fibers. For an isotropic pitch, the viscosity decreases markedly with increasing temperature and, after reaching a minimum, it increases abruptly, mainly due to its change to a solid (coke), as shown in Fig. 3.96. The pitch containing a small amount of mesophase shows much higher viscosity, and the decrease in viscosity due to temperature increase shifts to the high-temperature side. For the pitch consisting of almost 100% mesophase (mesophase pitch), its viscosity becomes much higher and solidified at a lower temperature. Mesophase pitches show an abrupt increase and decrease in viscosity at a

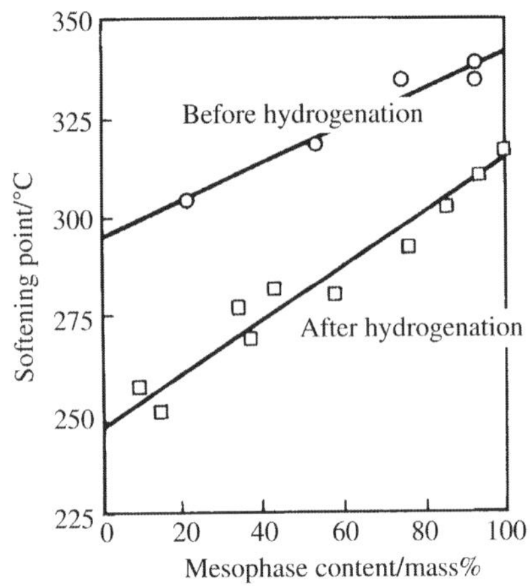


**FIGURE 3.95**

Relations between the softening point and mesophase content in the pitch before and after hydrogenation.

*(Courtesy of Dr. K. Sato of Nippon Steel Tech. Co.)*

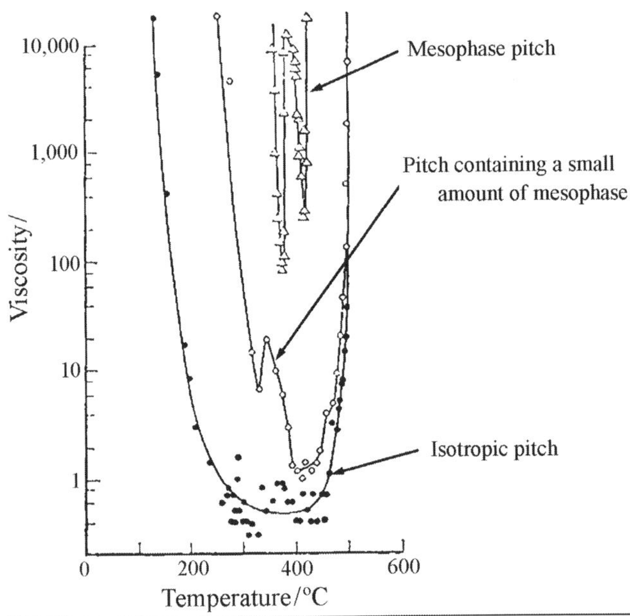


**FIGURE 3.96**

Temperature dependences of viscosity of three kinds of pitches [163].

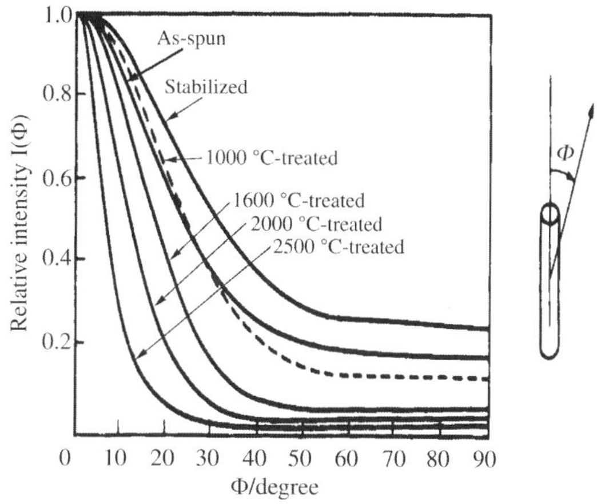


**FIGURE 3.97**

Orientation functions measured at different steps of the production of mesophase-pitch-based carbon fibers [166].

narrow temperature range, as shown in Fig. 3.96 [163]. Viscoelastic properties of mesophase pitches have been studied, revealing that they depended strongly not only on the content of mesophase but also on the precursors [164,165].

One of the advantages of mesophase-pitch-based carbon fibers is that it is possible to get a high degree of orientation of carbon hexagonal layers along the fiber axis during spinning and so not necessarily to stretch the fibers during stabilization and carbonization. This fact reveals the importance of spinning process for getting high performance in the resultant carbon fibers. Fig. 3.97 shows the change in orientation function determined by X-ray diffraction during each step of processing of carbon fiber production [166]. Just after spinning an already relatively high degree of orientation is obtained. It is slightly disturbed by stabilization, but improved by heat treatment at high temperatures, the improvement becoming pronounced with the increase in HTT.

It is also an advantage of mesophase-pitch-based carbon fibers to be able to realize various nanotextures in their cross-sections perpendicular to the fiber axis, as schematically shown in Fig. 3.84. By the control of viscosity of the pitch, the nanotexture from radial alignment of carbon hexagonal layers with wedge to onion-like one (concentric around fiber axis) through either random or doubly textured with concentric and radial alignments can be obtained. In Fig. 3.98, SEM micrographs are shown on the cross-section of the fibers, which have been produced from exactly the same pitch by changing the spinning condition [167]. This nanotexture in cross-section has a great influence on the structural change with high-temperature treatment and the properties of resultant carbon fibers.

In Fig. 3.99, the nanotexture of cross-section is shown by using SEM on the fibers, the same as Fig. 3.132, after heat treatment at 3000°C [167]. Radial and

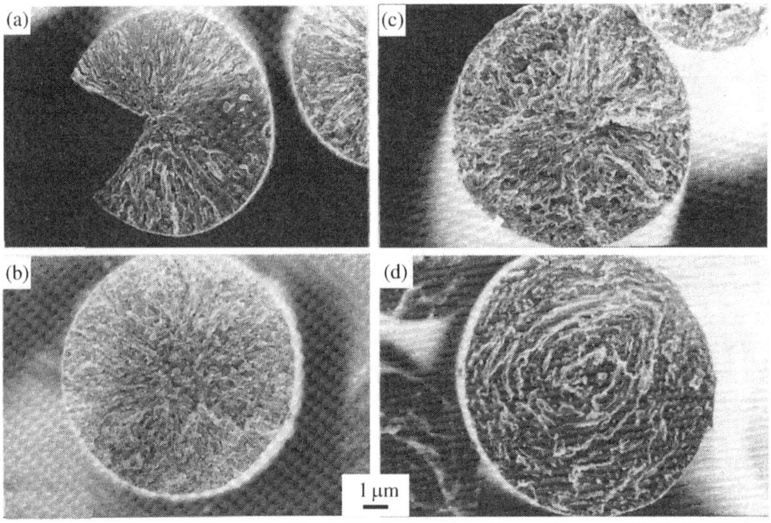


**FIGURE 3.98**

SEM micrographs of the cross-section of mesophase-pitch-based carbon fibers after the carbonization at 1200°C [167].

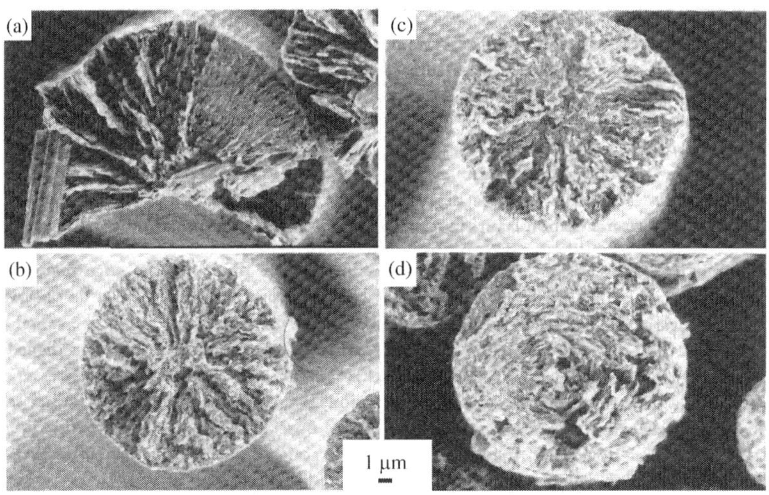


**FIGURE 3.99**

SEM images of the cross-section of mesophase-pitch-based carbon fibers after the heat treatment at 3000°C [167].

concentric textures are seen more clearly, radial texture with wedge showing distinct straightness of the layers. On the fiber with radial texture, the wedge shows bigger opening after high-temperature treatment (Fig. 3.99a) than in carbonized one (Fig. 3.98a). This difference in nanotexture reflects on the development of graphite structure with 3000°C treatment, various parameters as a measure of degree of graphitization are summarized in Table 3.24. The fibers having the radial texture with wedge (straight radial) have particularly high values of *La*, *Lc* and $(\Delta\rho/\rho_0)_{cr}$, in comparison to the fibers without wedge, i.e., zig-zag radial, corrugate radial and concentric. On these carbon fibers, mechanical and thermal properties are not so much different, but Young's modulus increases markedly in the fiber of radial texture with wedge after high-temperature treatment [168]. Young's modulus of the fiber, which is calculated on the basis of a simple model by taking account of orientation function of crystallites along the fiber axis, shows a good correspondence to the measured one, indicating that modulus of carbon fibers is predominantly governed by the preferred orientation of crystallites and also their sizes.

From high-resolution TEM observations, the nanotexture in precursor pitch composed from isotropic and anisotropic parts has been shown to be retained into the fibers after spinning, carbonization and graphitization [167]. In 1200°C-carbonized fibers, two regions with different nanotexture are observed, small layers being oriented in one region and being random in another region, reasonably supposed to come from anisotropic and isotropic parts, respectively, in the precursor pitch. By heat treatment at high temperatures, the layers grow in the oriented region, but the growth of layers is so restricted in the random region that it becomes porous. The preparation of precursor pitch and its spinning are decisive unit operations to control the structure and properties of mesophase-pitch-based carbon fibers.

One of the characteristics of mesophase-pitch-based carbon fibers is their high thermal conductivity, the value close to 1000 W/m.K at room temperature being reported. For a number of fibers, a close relation between thermal conductivity and electrical resistivity was found, as shown in Fig. 3.100 [169].

### 3.4.5 Vapor-grown carbon fibers

Vapor-grown carbon fibers [170] are characterized by the presence of minute iron particles at the tip of the fibers associated with hollow thin-tube at the center of the fibers, as shown by the TEM image in Fig. 3.101. The preparation of vapor-grown carbon fibers was performed by two methods; seeding catalyst and floating catalyst methods, of which the principles are schematically illustrated in Fig. 3.102a and b, respectively [171]. In most cases reported, high-purity hydrogen gas was used to reduce and activate the catalyst metal particles, iron particles in most cases, to synthesize carbon fibers with high yield. The impurity of inert gases, such as nitrogen and argon, and oxidative gases, such as oxygen and water vapor, was found to contaminate the catalyst particles and markedly decrease the yield of fibers.

**Table 3.24** Structure Parameters of Mesophase-Pitch-Based Carbon Fibers After the Heat Treatment at 3000°C

| Carbon Fibers | Cross-Sectional Texture | $d_{002}$ (nm) | $Lc(002)$ (nm) | $La(110)$ (nm) | $P_1$ | $P_{ABA}$ | $P_{ABC}$ | $(\Delta\rho/\rho_0)_{cr}$ (%) | $r_T$ | $r_{TL}$ |
|---|---|---|---|---|---|---|---|---|---|---|
| A | Straight radial | 0.3367 | 49 | 81 | 0.41 | 0.25 | 0.04 | 12.447 | 0.681 | 0.01 |
| B | Zig-zag radial | 0.3385 | 24 | 40 | 0.30 | 0.10 | 0.03 | 0.372 | 0.297 | – |
| C | Corrugate radial | 0.3380 | 27 | 45 | 0.32 | 0.12 | 0.01 | 0.849 | 0.597 | – |
| D | Concentric | 0.3379 | 27 | 56 | 0.35 | 0.13 | 0.04 | 0.352 | 0.480 | – |

*(Refer to Fig. 3.103.)*

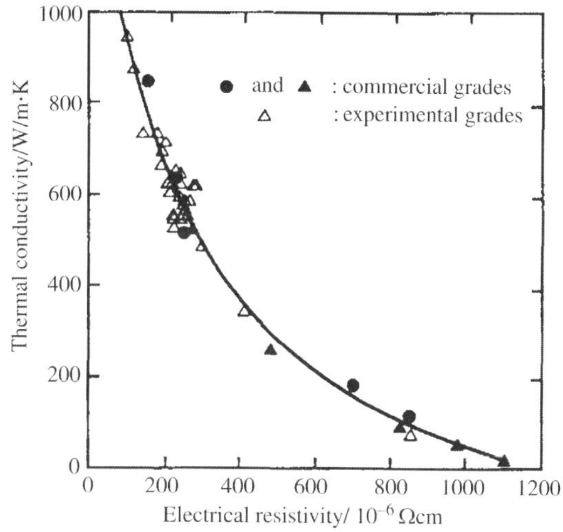


**FIGURE 3.100**

Relation between thermal conductivity and electrical resistivity at room temperature on various mesophase-pitch-based carbon fibers [169].

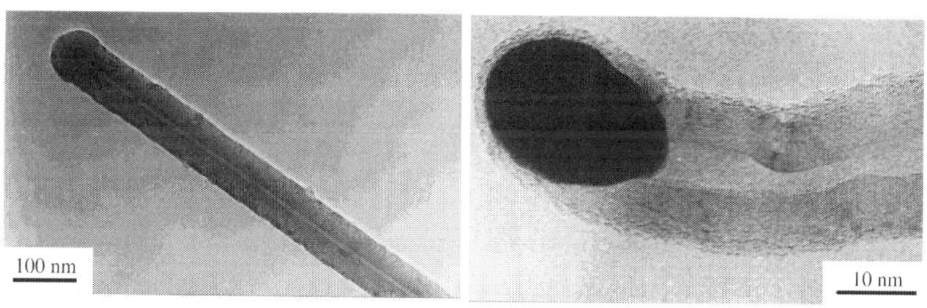


**FIGURE 3.101**

TEM images of the tip of a vapor-grown carbon fiber.

*(Courtesy of Prof. M. Endo of Shinshu Univ.)*

In the seeding catalyst method, catalyst particles are seeded on a substrate, such as graphite or ceramics, by either painting very fine metal particles or spraying alcohol solution of metal nitrate, followed by its decomposition at high temperatures. By the development of these seeding techniques, the synthesis of vapor-grown carbon fibers becomes more reproducible with higher yield. The metal particles with the size less than 20 nm are useful to grow fibers, the finer particles giving the higher yield and the faster growth in fiber length, growth rate of fiber length being roughly 1 mm/min in the elongation step and that of fiber diameter being roughly 5–10 μm/h in the thickening step (Section 2.3.3c). By this seeding catalyst method, the fibers with few tens micrometers in diameter

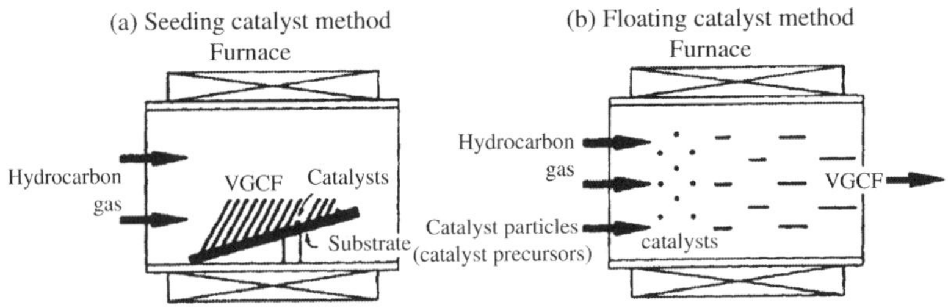


**FIGURE 3.102**

Schema of the systems for the production of vapor-grown carbon fibers (VGCFs).

*(Courtesy of Dr. M. Ishioka of Nippon Steel Co., Ltd.)*

and a few tens of centimeters in length are obtained. The following disadvantages, however, are pointed out; inhomogeneity in radius and length of fibers, relatively low fiber yield, and discontinuity in processes for the synthesis, i.e., seeding of catalyst, reduction and purification of catalyst metal, growth of fibers and recovery of fibers.

The floating catalyst method is developed in order to improve the fiber yield. In this method, two ways to feed catalyst particles in the reaction zone are employed, direct introduction of fine metal particles and introduction of organometallic compounds, which form fine metal particles by thermal decomposition in reaction zone. By using ferrocene in benzene-hydrogen system, very high yield of carbon fibers with the diameter of 0.1–1.5 μm and the length of about 1 mm has been reported [172]. There is an optimum amount of ferrocene to achieve a maximum yield of vapor-grown carbon fibers [173]. The vapor-grown carbon fibers prepared by floating catalyst method are usually thin but homogeneous in diameter and also short in length, in comparison with those produced by the seeding catalyst method. By the control of gas flow pattern in the furnace and temperature of inlet gas, straight fibers have been synthesized, though most of the fibers synthesized by the floating catalyst method are crooked [174]. The floating catalyst method has the advantages in continuous operation of the synthesis, high fiber yield and, as a consequence, low cost.

The use of liquid pulse injection technique with floating catalyst method has been proposed [175], and long fibers with high yield, higher than by transferring catalyst precursor with carrier gas, are obtained [176–180]. The concept of the use of liquid pulse injection technique is schematically shown in Fig. 3.103 [175]. Catalyst precursor solution, e.g., ferrocene dissolved in benzene, is injected into reactor through a nozzle, separately from carbon precursor and carrier gas. The injected precursor hits the hot wall of the reactor and is simultaneously decomposed to ultrafine catalyst particles, which meet carbon precursor to form vapor-grown carbon fibers. By using this technique, vapor-grown carbon fibers are successfully synthesized from various carbon precursor, benzene, toluene, xylene and their mixtures [176] and methane, ethane, propane, i-butane, n-pentane, n-hexane and

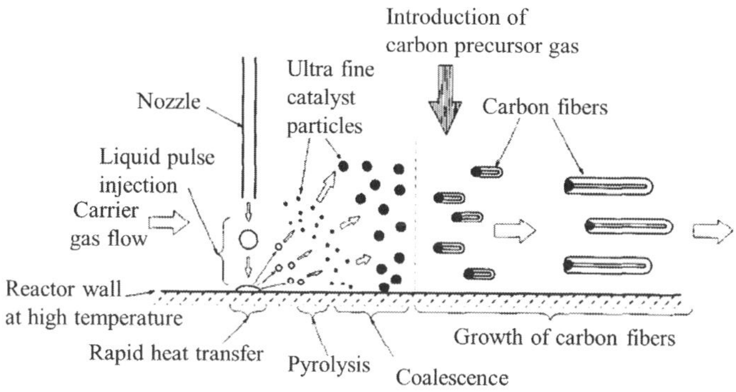


**FIGURE 3.103**

Scheme of the use of liquid pulse injection technique with floating catalyst method [24].

cyclohexane [177] with ferrocene as catalyst precursor. The yield of fibers increases with the increase in the number of catalyst particles in the range of 20–30 nm [179], and the flow rate of carrier gas influences on fiber yield [180].

Instead of high-purity hydrogen gas, Linz-Donawitz converter gas (LDG), which is an industrial byproduct and consists mainly of CO, has been shown to be successfully used as a carrier gas for benzene to synthesize fibers by floating catalyst method [181]. The elimination of oxygen and water vapor from LDG (67%CO, 1.2%$H_2$, 16%$CO_2$, 15%$N_2$, 0.2%$O_2$ and 0.6%$H_2O$) is essential to obtain fibers with a high yield; about 50 mass% yield of the fibers with the diameter of about 3 μm and the length of about 3 mm. Using gas mixtures, the effect of composition of this carrier gas, such as contents of CO and $CO_2$, is studied in detail [182,183]. When a mixture of ferrocene, thiofene and cobaltacetylacetonate as a precursor of catalyst particles and a mixed gas of 77%CO, 19%$CO_2$ and 4%$H_2$ were used, a high fiber yield as 70 mass% has been obtained.

Characteristics in structure and nanotexture of vapor-grown carbon fibers are summarized into three points; concentric alignment of small carbon hexagonal layers along fiber axis (Fig. 3.104d), the presence of catalyst particle at the tip of fibers (Fig. 3.101) and the formation of a hollow tube at the center of the fibers (Fig. 3.104a). This central tube is now recognized to be the carbon nanotube, although it is not called so at the time of finding the presence of this tube [184]. Even a single-wall carbon nanotube has been observed in the beginning of the growth of vapor-grown carbon fibers, as shown in Fig. 3.105. The diameter of the hollow tube is very similar to that of the catalyst particles located at the tip of the fiber, but always a little smaller. The outer part of the central tube is a piling-up of small carbon layers with the size of about 1 nm and stacking 2–3 layers in parallel (Fig. 3.104d) [184]. The structure of the outer part is very similar to that of pyrolytic carbons deposited on a substrate through thermal pyrolysis of hydrocarbon gases (Section 2.3.2b). These structural characteristics do not depend on the synthesis method, hydrocarbon gas and also catalyst metal used.

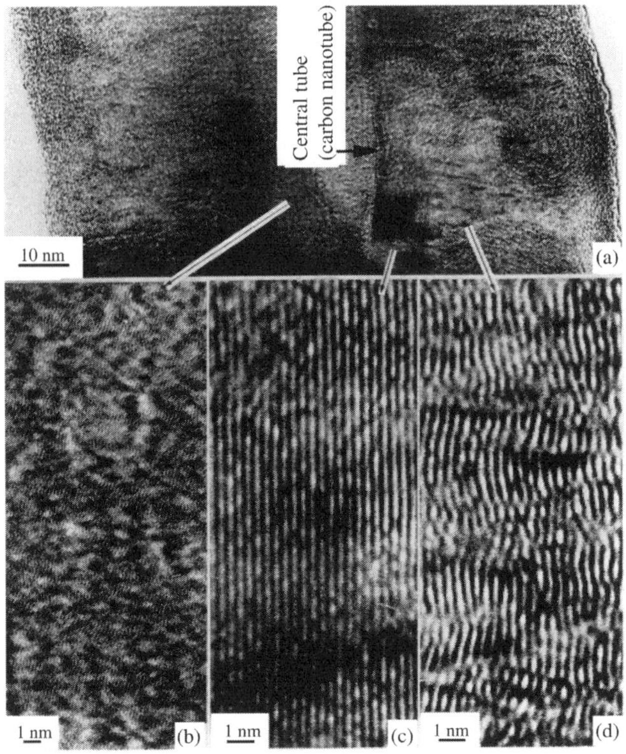


**FIGURE 3.104**

TEM images of vapor-grown carbon fibers.

*(Courtesy of Prof. M. Endo of Shinshu Univ.)*

Mainly based on the structural studies, the growth model for vapor-grown carbon fibers has been proposed for two production methods, as shown in Fig. 3.106. In the first step, metal oxide particles, which are either formed from a catalyst precursor or supplied directly, are reduced to metal by hydrogen gas. In this step, the metal particles formed are in liquid phase and either stay on the substrate or are floating. Carbonaceous materials, which are formed by pyrolysis of hydrocarbon gas supplied, get contact with these liquidous metal particles and diffuse along the surface of metal particles, and at the same time pyrolysis proceeds by some catalytic effect of metal to form hexagonal carbon layers. In the seeding catalyst method, the carbon layers thus formed are deposited at the boundary between substrate and catalyst particle. As a consequence of continuous supply of carbonaceous materials (carbon supply in Fig. 3.106), catalyst particles are raised up from the substrate and thin tubes (carbon nanotubes) are formed. The formation of these carbon nanotubes seems to be terminated if the surface of a metal particle is covered by deposited carbon or if it is contaminated by impurity such

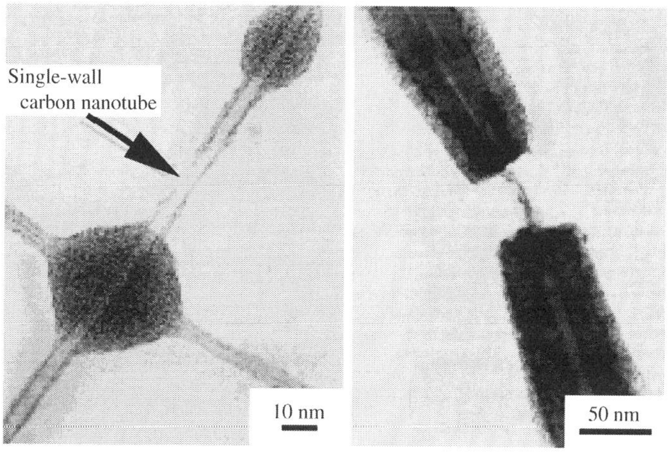


**FIGURE 3.105**

Formation of a single-wall carbon nanotube in the beginning of the growth of vapor-grown carbon fibers.

*(Courtesy of Prof. M. Endo of Shinshu Univ.)*

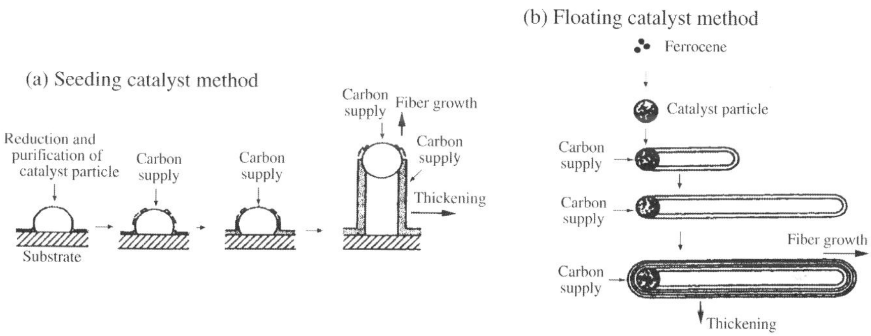


**FIGURE 3.106**

Models for the growth of vapor-grown carbon fibers.

*(Courtesy of Prof. M. Endo of Shinshu Univ.)*

as oxygen. In the floating catalyst method, the carbon layers formed on the surface of catalyst particles are deposited at the boundary between carbon already accumulated and catalyst particle. Therefore, it results also in fibrous texture, in other words, the formation of carbon nanotubes. Onto the surface of these carbon nanotubes, pyrolytic carbon is deposited in the next step and the fibers grow in the radial direction (thickening). Therefore, the growth of these fibers occurs in three steps, i.e., purification of catalytic metal particles and nucleation of carbon

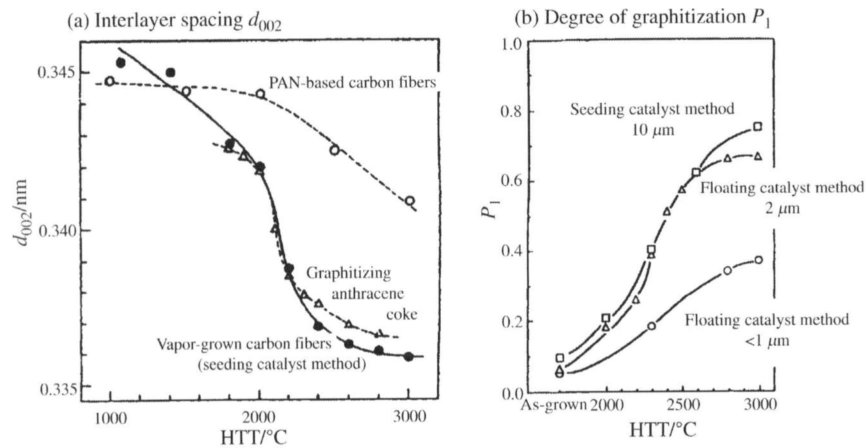


**FIGURE 3.107**

Changes of $d_{002}$ and $P_1$ of vapor-grown carbon fibers with HTT.

*(Courtesy of Prof. M. Endo of Shinshu Univ. [a] and Dr. N. Iwashita of AIST, Japan [b].)*

deposition, growth along fiber length (elongation step) and then growth in the radius direction (thickening step). It was experimentally shown that the second step occurs in a short period but the third step takes time (refer α) [170,185].

Another feature of this fiber is high graphitizability, marked development of graphite structure with heat treatment. In Fig. 3.107a, $d_{002}$ is plotted against HTT on vapor-grown carbon fibers prepared from benzene by seeding catalyst method, in comparison with PAN-based carbon fibers and an anthracene coke which is a typical graphitizing carbon [186]. Very high graphitizability of the vapor-grown carbon fibers is clearly seen, even better than anthracene coke at high temperatures and a marked difference from PAN-based carbon fibers. The graphitizability is also known to depend on the diameter of the fibers. In Fig. 3.107b, the changes in $P_1$ with HTT are shown on the vapor-grown carbon fibers with different diameters [187]. The fibers having the diameter less than 1 μm show a rather low degree of graphitization. With the heat treatment, morphology of the fibers also changes; after high-temperature treatment above 3000°C, the fibers are found to be polygonized owing to the growth of graphite layers, as shown in Fig. 3.108 [188].

For the vapor-grown carbon fibers heat-treated at different temperatures, Fig. 3.109 shows the dependences of $(\Delta\rho/\rho_0)_{max}$ on $B$ at 77 K and Fig. 3.110 shows the changes of $(\Delta\rho/\rho_0)$ with TL- and T-rotation (refer to Section 2.5.1) under 1 T at 77 K [152]. With increasing HTT, $(\Delta\rho/\rho_0)_{max}$ increases its absolute value with negative sign, and above 2200°C changes its sign to positive and increases markedly. By plotting $(\Delta\rho/\rho_0)_{cr}$ against HTT for different carbon fibers in Fig. 2.105, a marked progress in graphitization in vapor-grown carbon fibers has been discussed before (Section 2.5.4). A pronounced increase in $\Delta\rho/\rho_0$ with TL-rotation but almost no change along T-rotation (Fig. 3.110) show that

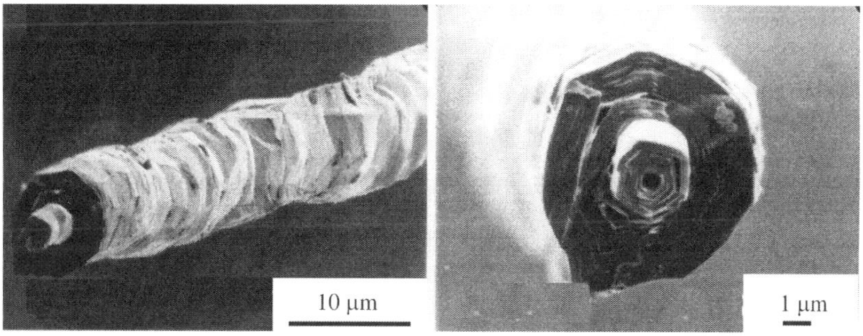


**FIGURE 3.108**

SEM images of vapor-grown carbon fibers heat-treated at 3000°C.

*(Courtesy of Prof. A. Yoshida of Tokyo City Univ.)*

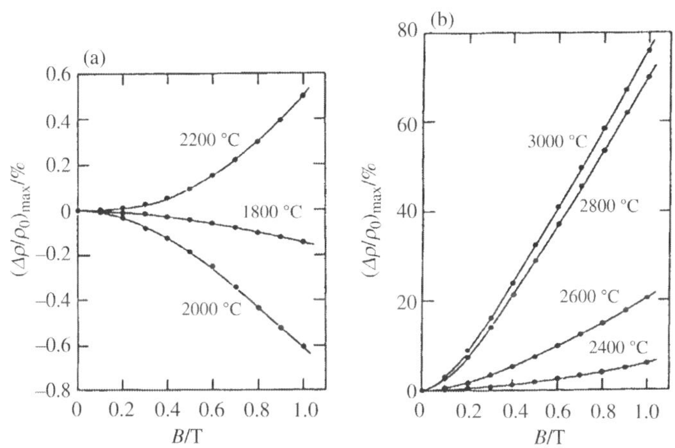


**FIGURE 3.109**

Dependences of $(\Delta\rho/\rho_0)_{max}$ on $B$ at 77 K for vapor-grown carbon fibers heat-treated at different temperatures.

*(Courtesy of Prof. Y. Hishiyama of Tokyo City Univ.)*

preferred orientation of crystallites along the fiber axis becomes marked with increasing HTT, but its axial orientation scheme is kept the whole range of temperature. Above 2800°C, however, the change of $\Delta\rho/\rho_0$ with T-rotation becomes a little irregular, which is reasonably supposed to come from polygonization of this fiber, as shown in Fig. 3.108.

Galvanomagnetic properties depend also on diameter of fibers, as well as HTT and residence time. Fig. 3.111 shows the dependence of $(\Delta\rho/\rho_0)_{max}$ on diameter of vapor-grown carbon fibers at different temperatures and residence times [189].

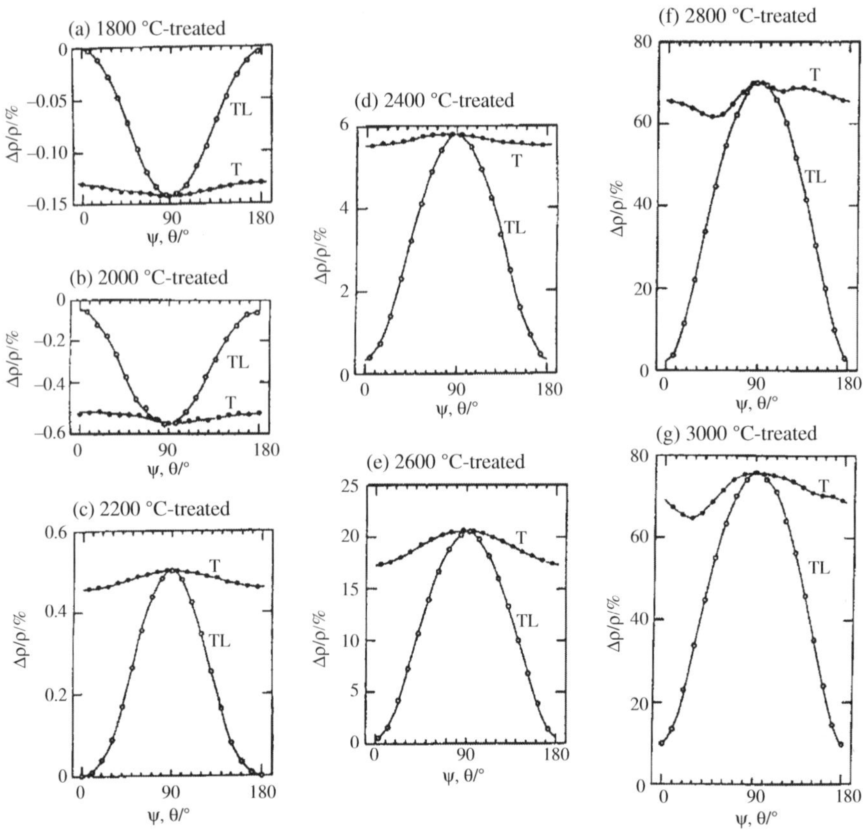


**FIGURE 3.110**

Changes in $\Delta\rho/\rho_0$ with two modes of rotation, TL- and T-rotations, for vapor-grown carbon fibers heat-treated at different temperatures.

*(Courtesy of Prof. Y. Hishiyama of Tokyo City Univ.)*

The larger diameter gives the higher $(\Delta\rho/\rho_0)_{max}$, i.e., the more marked development of graphite structure, being consistent with the results on $P_1$ (Fig. 3.107b).

Even by high-temperature treatment above 2800°C, the central hollow tube of vapor-grown carbon fibers (carbon nanotube), which is formed in the beginning of growth process of the fiber (elongation step), is supposed not to change in structure, although the outer part of the fiber, which is deposited in the thickening step and consists of small carbon layers, changes to straight and long carbon layers. Fig. 3.112a shows the 002 lattice fringe image of the vapor-grown carbon fibers heat-treated at 3000°C, revealing straightened carbon layers. In this graphitized fiber, the central thin tube is easily pulled out with breaking the fiber, as shown in Fig. 3.112b.

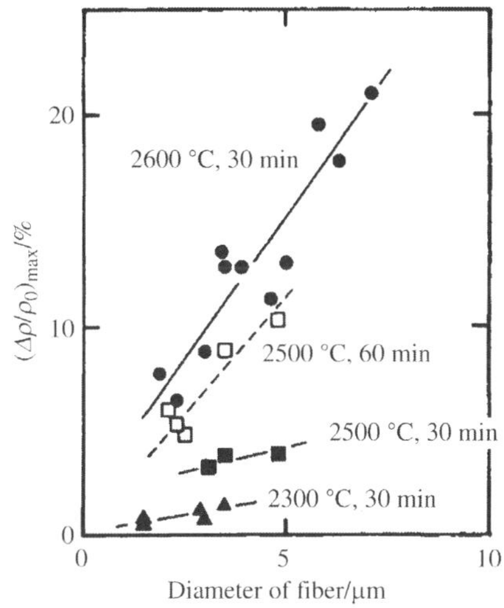


**FIGURE 3.111**

Dependences of $(\Delta\rho/\rho_0)_{max}$ on the diameter of the vapor-grown carbon fibers at different HTTs and residence times.

*(Courtesy of Dr. M, Ishioka of Nippon Steel Co. Ltd.)*

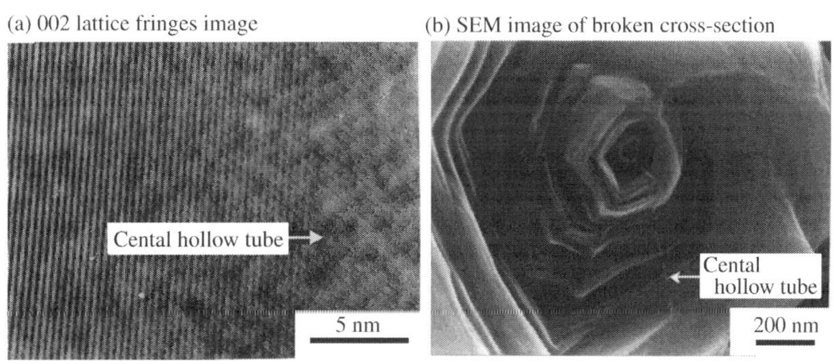


**FIGURE 3.112**

002 lattice fringe image (a) and SEM image after breaking (b) for the vapor-grown carbon fibers heat-treated at 3000°C.

*(Courtesy of Prof. M. Endo of Shinshu Univ.)*

## 3.4.6 Glass-like carbon fibers

Carbon fibers with glass-like nanotexture have been prepared from organic precursors, such as phenol resin and cellulose, in addition to isotropic-pitch-based carbon fibers. Because most of them show multiphase graphitization at high temperatures, their graphitization behavior has been attracting attention.

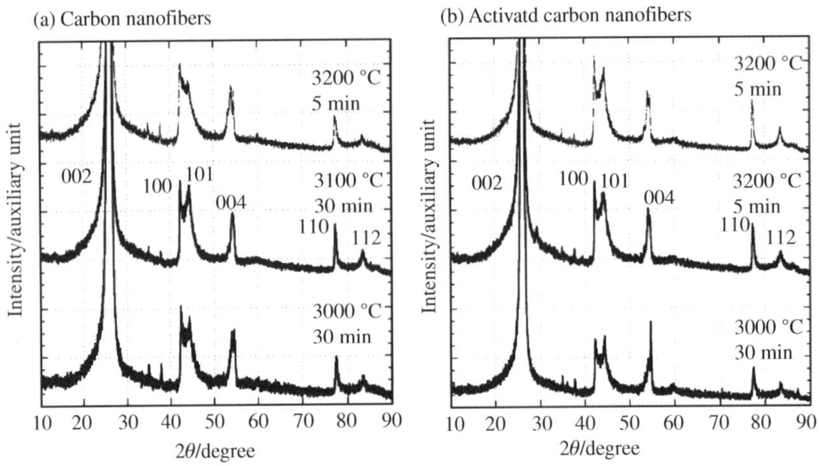


**FIGURE 3.113**

XRD patterns for phenol-based carbon fibers with and without activation after the heat treatment at high temperatures [143].

Carbon fibers were prepared via melt-spinning of polyethylene with dispersed novolac-type phenol-formaldehyde resin particles (polymer blend method in Section 2.4.2), followed by carbonization at 1000°C [190]. The carbon fibers showed multiphase graphitization by the heat treatment at 3000°C, consisting of three phases, amorphous, turbostratic and graphitic phases [191,192]. More pronounced graphitization is observed on activated carbon nanofibers derived from phenol resin with activation than those without activation, as shown in Fig. 3.113 [143].

Changes in structure and nanotexture with heat treatment up to 3200°C have been studied on carbon fibers prepared from commercially available cellulose fibers [142]. The formation of graphitic skin on the surface of fibers is confirmed by TEM, XRD and Raman spectra. Carbon nanofibers with diameters of 30–60 nm, which are prepared from bacteria cellulose, show marked multiphase graphitization under the heat treatment up to 3100°C [193]. Graphitization behavior depends on the preparation condition, either dispersed in ethanol, water or no dispersion, as shown in Fig. 3.114. Carbon nanofibers prepared after dispersion in ethanol show marked graphitization after 3100°C treatment, although those derived from the same cellulose without dispersion show turbostratic structure even after 3100°C treatment. Intermediate behavior is observed on carbon nanofibers prepared via dispersion in water. The detailed graphitization behavior of these bacteria-cellulose-based carbon nanofibers was studied by the heat treatment at temperatures between 2400 and 3200°C [194].

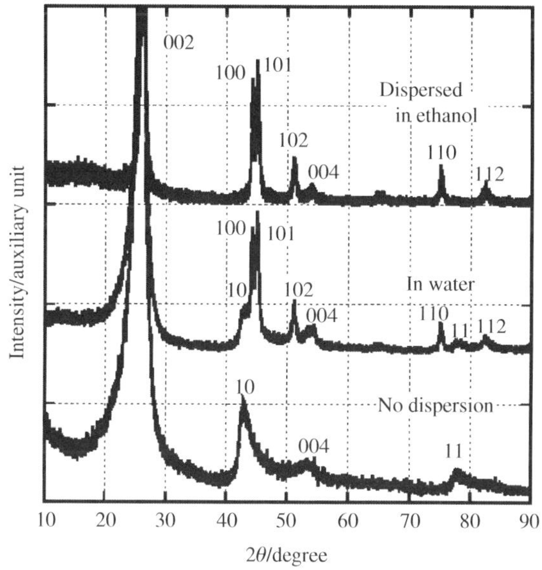


**FIGURE 3.114**

XRD patterns of cellulose-based carbon nanofibers heat-treated at 3100°C [193].

Multiphase graphitization has been observed on thin carbon films derived from different precursors [104,141].

### 3.4.7 Carbon microcoils

Helical carbon microcoils with various morphologies shown in Fig. 3.115 are reproducibly prepared by the catalytic pyrolysis of acetylene with a small amount of additives, such as sulfur or phosphor, by using metal catalysts, such as Ni [195–199]. Most of them are the coils with different diameter and different pitches. Even rolls of fibrous carbon with relatively flat cross-section are observed, as shown in Fig. 3.115f.

Most of these microcoils show pyramidal fracture surfaces, as can be seen on the microcoils heat-treated at a high temperature. The nanotexture of these microcoils is concluded to have a herring-bone type, mainly based on TEM analysis of the fractured tips of the microcoils, as shown in Fig. 3.116. After heat treatment at 3000°C, the microcoils showed only G-band of Raman spectrum, no D-band being detected, and gave $d_{002}$ of about 0.335 nm.

The applications of these carbon microcoils are extensively studied for various fields, such as electromagnetic shielding, tactic sensor, cosmetics, cancer therapy, etc. [200].

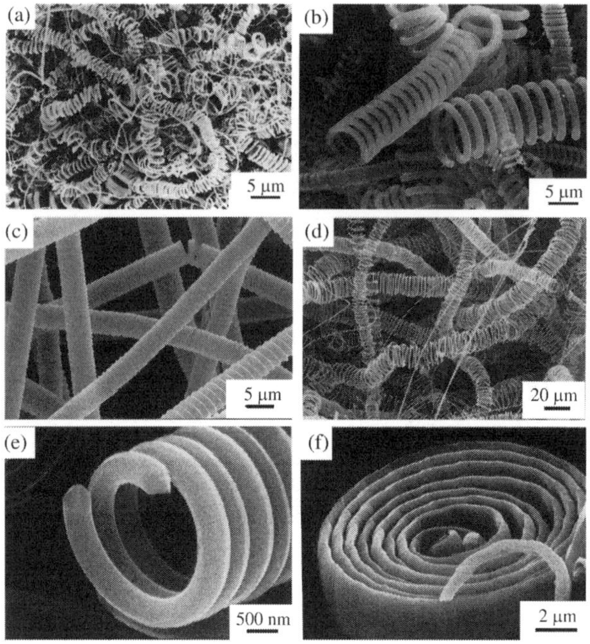


**FIGURE 3.115**

Carbon microcoils with various morphologies.

*(Courtesy of Prof. S. Motojima of Gifu Univ.)*

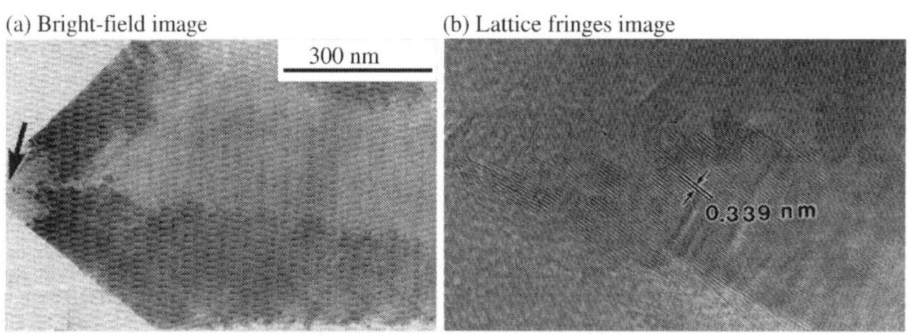


**FIGURE 3.116**

TEM image of a tip of carbon microcoil heat-treated at 3000°C.

*(Courtesy of Prof. S. Motojima of Gifu Univ.)*

# 3.5 Nanocarbons

## 3.5.1 Carbon nanotubes and nanofibers

### a. Nanotubes and nanofibers

Carbon nanotubes (CNTs), of which the diameter is in a wide range from 1 to 50 nm and the wall consists of different number of carbon layers, have been

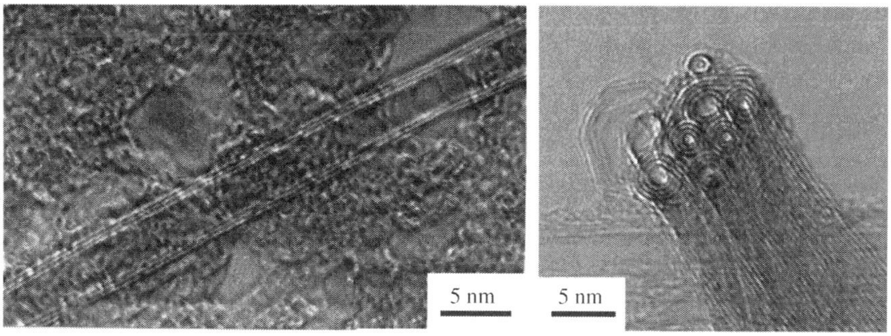


**FIGURE 3.117**

TEM image of CNTs.

*(Courtesy of Prof. M. Endo of Shinshu Univ.)*

found in the carbon deposits on the graphite anode during arc discharging in He atmosphere [201]. The temperature of the graphite electrode was estimated to be as high as 2500°C. CNTs consisting of a single wall are found later [202,203]. Most of them are closed at the end [204]. In Fig. 3.117, transmission electron microscopy (TEM) images of CNT and a bundle consisting of few nanotubes are shown.

Under very similar conditions of arc discharge, fibrous carbons have been obtained in 1960, of which the electron diffraction pattern is very similar to those observed on CNTs found later and a structure model consisting of a scroll of hexagonal carbon layers has been proposed [205]. They are called graphite whisker. Similar CNTs, even single-walled nanotubes, are formed at the beginning of the formation of vapor-grown carbon fibers, as explained in Section 3.4.5 [184].

With the increasing interest in nanotechnology, CNTs have attracted the attention of scientists and engineers and the necessity of more efficient synthesis methods has been pointed out, which promotes the related researches in the world. Many people worked on the synthesis of carbon nanotubes through chemical vapor deposition (CVD) using fine particles of various metals as catalysts and announced that they could get CNTs. However, most of the products are not CNTs actually, as shown in Fig. 3.118a, of which wall consists of small carbon layers even though they orient preferentially along the tube axis. By heat treatment at a high temperature such as 2800°C, carbon layers become long, as shown in Fig. 3.118b, which is similar to CNT synthesized by arc discharge technique. Therefore, the name 'carbon nanofibers' (CNFs) is now often used for these fibrous carbons consisting of small carbon layers. CNT, CNF and carbon fiber are described in Section 3.4.1 in relation to carbon fibers and are here schematically compared in Fig. 3.119. The diameters of the tubes and fibers are different. CNTs and CNFs are tubular, but most carbon fibers, except vapor-grown ones, are not tubular and can have various nanotextures in their cross-sections, as described in Section 3.4. CNTs can be synthesized either with or without metallic catalysts. In order to synthesize CNFs, however, metallic catalysts are essential.

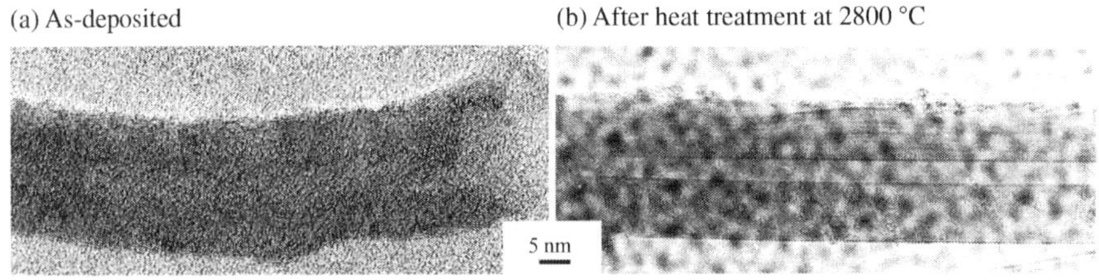


**FIGURE 3.118**

CNFs prepared from acetylene gas using Co catalyst at 900°C.

*(Courtesy of Dr. S. Bonnamy of CNRS, Orleans.)*

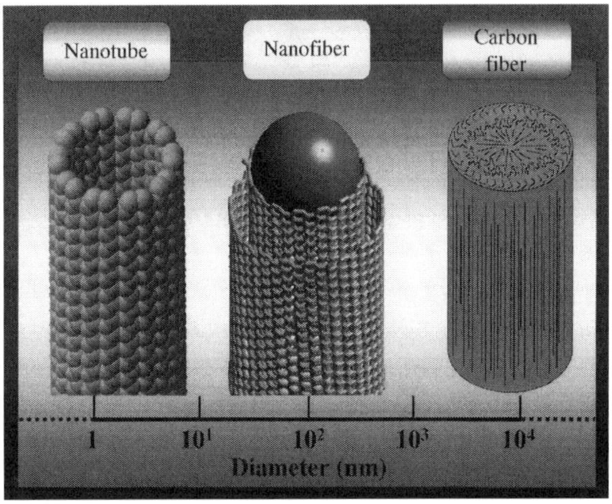


**FIGURE 3.119**

Schematic illustration of the difference among carbon nanotube, nanofiber and fiber.

*(Courtesy of Prof. M. Endo of Shinshu Univ.)*

In the preparation of CNTs by arc discharge, other forms of carbons, such as carbon blacks and pyrolytic carbons are formed together. Therefore, the purification process, that is, separation of CNTs from other forms of carbon, is unavoidable in order to have CNTs in high purity. Also, the control of diameter and length of CNTs is difficult in arc discharge process. In order to solve these problems, the preparations of CNTs and CNFs through various CVD processes using metallic catalysts have been carried out, though the exclusion of metal catalyst is unavoidable on some CNTs and CNFs. For CNFs, high-temperature treatment is necessary to obtain long carbon layers with high degree of orientation along the tube axis. Different ideas to prepare CNFs, which do not contain any metal particles, have been proposed [206].

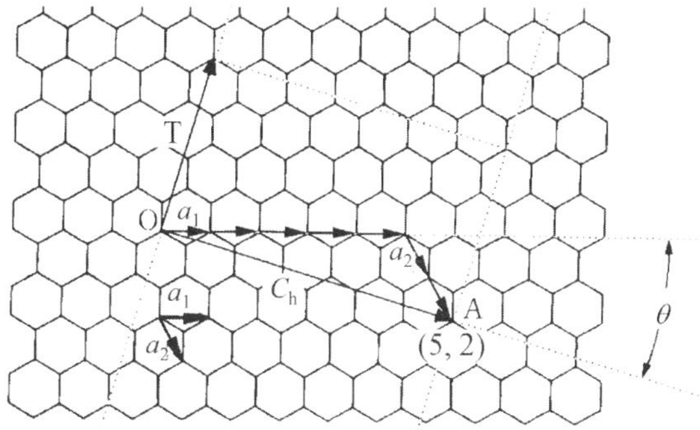


**FIGURE 3.120**

Definition of chiral vector $C_h$ on the graphene sheet, which can be converted to a SWCNT by rolling.

## b. Carbon nanotubes

Single-wall carbon nanotubes (SWCNTs) are characterized by diameter $d_t$ of tubes and chiral angle $\theta$, which is determined by the chiral vector $C_h$. It is defined by the vector OA in Fig. 3.120, connecting the points A and O, which are overlapped with each other when the sheet of carbon hexagon (graphene sheet) makes a tube. The vector OA is usually indicated by the numbers of unit translation vector $a_1$ and $a_2$ in hexagons, as shown by the equation

$$C_h = na_1 + ma_2, \tag{3.8}$$

and expressed by (n, m). In the case of Fig. 3.120, the vector OA is expressed by (5, 2).

By using this chiral vector (n, m), the diameter of the tube $d_t$ and chiral angle $\theta$ are expressed as follows;

$$d_t = \sqrt{3d_{C-C}}\sqrt{n^2 + nm + m^2}/\pi \tag{3.9}$$

and

$$\theta = \tan^{-1}[3m/(2n + m)] \quad \theta \leqq \pi/6^{\circ}, \tag{3.10}$$

where $d_{C\text{-}C}$ stands for the distance between neighboring carbon atoms and equals to 0.142 nm.

The structure of SWCNTs is often classified into three, as shown in Fig. 3.121. Fig. 3.121a–c are called arm-chair, chiral and zig-zag type. Arm-chair type nanotube has n = m (i.e., $\theta = \pi/6^{\circ}$) and zig-zag type has m = 0 (i.e., $\theta = 0^{\circ}$). Chiral type structure can have various n and m values (n≠m≠0), i.e., various chi ral angle $\theta$ values between 0 and $\pi/6$.

Electronic properties are governed by the chiral vector $C_h$: the tubes with a value of (n-m) of the multiples of 3 are metallic and those with other (n-m) values

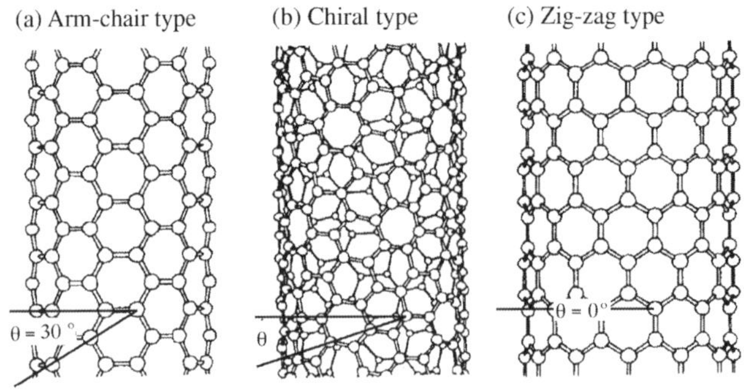


**FIGURE 3.121**

Structures of SWCNTs.

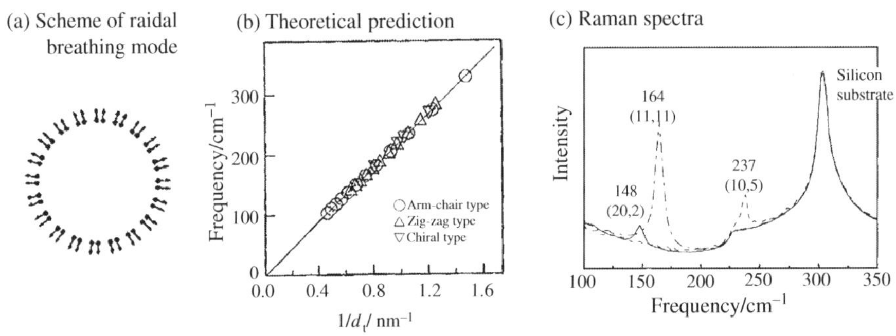


**FIGURE 3.122**

Breathing mode of Raman scattering for SWCNT.

*(Courtesy of Prof. R. Saito of Tohoku Univ.)*

are semiconductive [207]. The band structure and various electronic properties have been theoretically and experimentally studied in detail [207,208].

The detailed structure of a SWCNT is analyzed from resonant-Raman spectroscopy [209]. The diameter of the tube is evaluated from wave number of vibration mode of $A_{1g}$ (radial breathing mode, RBM), which is measured as resonant-Raman scattering at around 150–300 $cm^{-1}$, the frequency of RBN $\omega_{RBM}$ being proportional to the inverse of diameter $d_t$ of SWCNTs by the equation:

$$\omega_{RBM} = 248/d_t. \tag{3.11}$$

In Fig. 3.122, a schematic illustration of RBM, theoretically calculated dependence of its frequency $\omega_{RBM}$ on $1/d_t$ and some examples of Raman spectra are shown. The tube with the peak at 164 $cm^{-1}$ has the chiral vector of (11,11) and that at 237 $cm^{-1}$ the chiral vector of (10,5).

Vertically aligned SWCNTs, as shown in Fig. 3.123, have been synthesized by CVD process of alcohol using metal catalysts, the length of nanotubes reaching up to 5 μm [210,211].

The synthesis of multi-walled carbon nanotubes (MWCNTs) has been carried out through various processes [212–214]. The carbon nanotubes obtained after high-temperature treatment of CNFs with tubular type nanotexture, which could be prepared by various methods, are mostly MWCNTs. These carbon nanotubes usually form bundles, as shown in Fig. 3.124. By controlling preparation

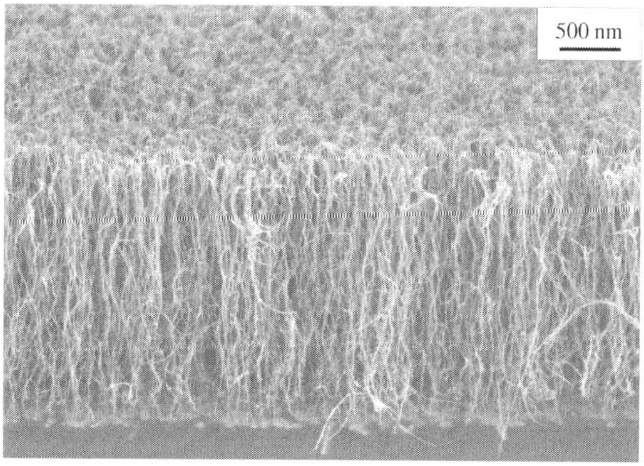


**FIGURE 3.123**

SWCNTs grown preferentially perpendicular to the substrate.

*(Courtesy of Prof. S. Maruyama of Univ. Tokyo.)*

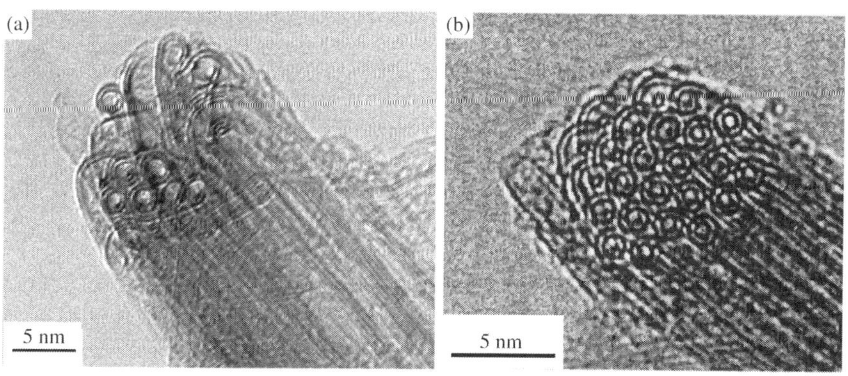


**FIGURE 3.124**

Bundles of MWCNTs.

*(Courtesy of Prof. Y. A. Kim of Shinshu Univ.)*

conditions, double-walled carbon nanotubes (DWCNTs) could be prepared [215], as shown Fig. 3.124b.

CNTs are known to have excellent electrical, thermal and mechanical properties [216–221]. However, it is very difficult to show these intrinsic properties of individual CNTs on the aggregates of CNTs. In order to address this point, various problems have to be solved, efficient method for the preparation of well-characterized CNTs, purification of the prepared CNTs, establishment of characterization methods, etc. Many of these problems are not yet solved completely and considerable efforts to solve these problems have been devoted. As-prepared CNTs are usually accompanied by various impurities, metal catalysts, carbonaceous impurities with amorphous and fullerene-like structures, and CNT structure variations, such as different wall thickness (single-, double- and multi-walled), capped and opened, various diameters and lengths of tubes. For the purification of these CNTs, various processes, chemical oxidation, physical separation and a combination of chemical and physical processes, have been proposed. A comprehensive review on the purification of CNTs has been published [222].

In SWCNTs, metallic and semiconducting nanotubes are usually mixed. From as-grown CNTs, semiconducting SWCNTs are extracted without detectable traces of metallic SWCNTs and impurities by using polyfluorene as an extracting agent in toluene under ultracentrifugation [223]. Continuous separation of metallic and semiconducting SWCNTs was performed by passing SWCNT/sodium sulfate dispersion through a column containing agarose gel beads [224], semiconducting ones being trapped by the beads and metallic ones passed through the column.

Most CNTs are bundled by strong van der Waals force, which is often troublesome for some applications: for example, bundled SWCNTs give a smaller surface area than the theoretical one. Debundling of SWCNTs is performed through intercalation of lithium with solvent molecules between the tubes in dimethyl sulfoxide [225]. This process was confirmed not to give any change in the quality of SWCNTs by TEM observation and Raman spectroscopy. In order to realize the excellent properties of CNTs, it has been tried to make CNTs into various forms, such as fibers, mats, pellets, etc. [218,225–231], but the exhibition of the intrinsic properties is still difficult.

Vertically aligned SWCNTs have been grown by CVD of ethylene on various metal catalysts in either Ar or He with $H_2$, the gas containing a small but controlled amount of water vapor [230]. Dense and vertically aligned SWCNTs with millimeter-length can be grown in 10 min by using thin catalyst films of $Al_2O_3$ and Fe sputtered on Si wafer. The process was named water-assisted CVD and the product was called 'SWNT-forest', an example of the product being shown in Fig. 3.125. Balancing the relative ratios of ethylene and water is crucial for production. The length of SWCNTs grows to 2.5 mm in 10 min. The as-grown SWNT-forest has very low content of metal catalyst, SWCNT/catalyst weight ratio being more than $5 \times 10^4$ %, and is free from either amorphous carbon or MWCNTs, which has been confirmed from the observation of RBM of Raman spectrum. The SWNT-forest obtained can be easily removed from the catalyst

**FIGURE 3.125**

SWCNT forest prepared through the water-assisted CVD method [230].

and substrate with a razor blade and the substrate recovered could be re-used for the growth of SWNT-forest. By applying a water-assisted CVD process, the formation of the SWNT forest consisting of more than 95% SWCNT is possible to be formed on conducting metal foils of a wide composition range of Ni–Cr–Fe alloys [232]. On Ni, Fe and Fe/Cr alloy, MWCNTs are formed together with SWCNTs.

SWNT-forest thus prepared is easily densified through liquid-induced collapse process after being removed from the substrate, which is called 'SWNT-solid' [231]. Liquid-induced collapsing introduces a 4.5-fold decrease in the two lateral dimensions but no detectable change in the height, resulting in an increase in bulk density from 0.029 $g/cm^3$ for an as-grown SWNT-forest to about 0.57 $g/cm^3$ for a collapsed SWNT-solid. The SWNT-solids are successfully applied to the electrodes of supercapacitors, giving high capacitance [231,233,234]. Via controlled oxidation, SWNT-forest is converted to a material with high surface area of 2240 $m^2/g$, where 85% of carbon atoms constitute a surface [235].

SWCNTs have been synthesized without using any metal catalyst through CVD of $CH_4$ at 900°C on $SiO_2$ film deposited on Si or Si/$SiO_2$ substrate [236]. The patterned growth of SWCNTs is shown to be possible by scratching the Si/$SiO_2$ substrate.

DWCNTs having high purity have been fabricated through the catalytic CVD of $CH_4$/Ar gas mixture at 875°C [237–239]. The catalyst consisted of Mo on $Al_2O_3$ for enhancing the active carbon species is placed at the end of the furnace and Fe on MgO for nanotube formation at the center of the furnace. The yield of DWCNTs is more than 95% after purification. From these high-purity DWCNTs, paper-like thin sheets (so-called 'buckypaper') are successfully prepared through their stable suspension.

Large amount of CNTs, up to 100 tons/year, has been used as one of the fillers in the anode in commercial lithium-ion rechargeable batteries (LIBs), in which the resilience and electrical properties of CNTs are believed to play an important role (Section 3.9.1). CNTs are also successfully used for the reinforcement of the cathode of LIBs [240–244]. Lithium-containing oxides in the cathode are

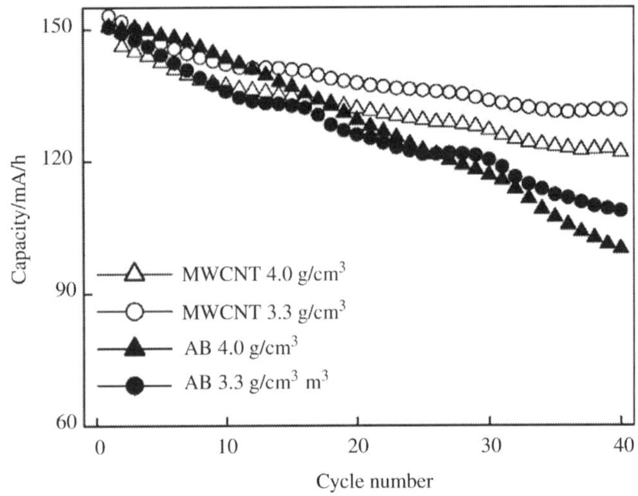


**FIGURE 3.126**

Cyclic performance of $LiCoO_2$ cathodes added different amounts of acetylene black (AB) and CNTs.

*(Courtesy of Prof. M. Endo of Shinshu Univ.)*

electrically insulating and so the addition of a small amount of carbon black was commonly employed to get electrical conductivity for the cathode. In Fig. 3.126, cyclic performance is compared for the $LiCoO_2$ cathode formed by using different amounts of acetylene black (AB) and MWCNTs [244]. The addition of MWCNTs into $LiCoO_2$ is effective for improvement in cyclic performance, as well as for giving higher electrical conductivity, greater structural integrity, higher thermal conductivity, higher density and a shortened electrolyte absorption time.

Hydrogen storage in CNTs is revisited and reported to be less than 1.7 mass% under the pressure up to 12 MPa at room temperature [245a]. The results suggest that a hydrogen storage capacity of CNTs is far below the DOE benchmark.

### c. Carbon nanofibers

By the catalytic decomposition of different precursor gases, not only hydrocarbons but also carbon monoxide, using fine particles of different metals as catalysts, various fibrous carbons have been formed [245b,246], which are called carbon nanofibers (CNFs) to be differentiated from both CNTs and carbon fibers. Their sizes along and across the fiber axis, morphology and nanotexture depend strongly on the decomposition conditions, temperature, catalyst metal, precursor gas, atmosphere, etc. [246–249].

Nanotexture of CNFs is usually classified into three, platelet, herring-bone and tubular types, as shown schematically in Fig. 3.127. In Fig. 3.128, an example of the nanofibers with platelet type nanotexture is shown, prepared from a mixture

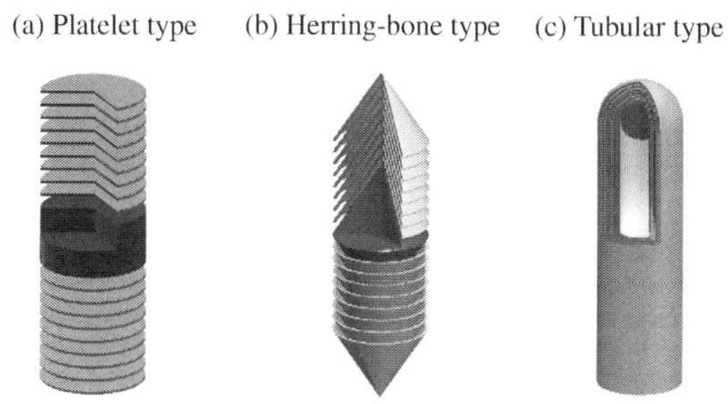


**FIGURE 3.127**

Nanotexture of CNFs.

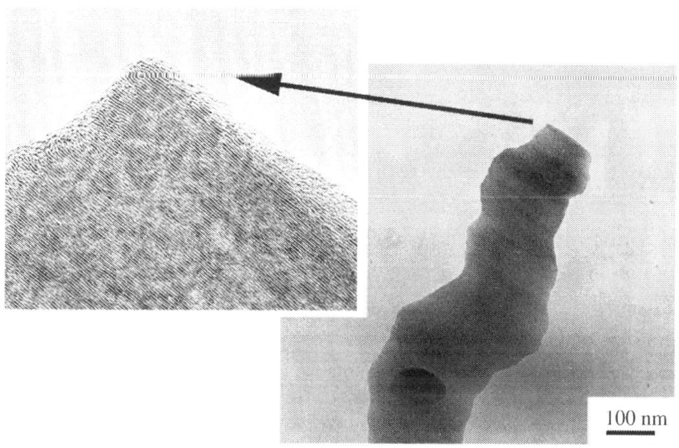


**FIGURE 3.128**

CNFs with platelet type nanotexture obtained by CVD method of CO gas.

*(Courtesy of Dr. Y. Soneda of AIST, Tsukuba.)*

of 80% CO and 20% $H_2$ at 550°C [250]. By the decomposition of CO gas under different conditions herring-bone type CNFs have also been obtained, as shown in Fig. 3.129 [248]. An example of tubular-type CNFs is shown in Fig. 3.118. The impregnation of carbonaceous products from poly(vinyl alcohol) into the channels of anodic aluminum oxide (AAO) films (template) at around 1000°C has given carbon nanorodes, having platelet nanotexture [251].

Long CNFs have been prepared by the floating catalyst method by using ferrocene or iron pentacarbonyl as a catalyst precursor, hydrogen sulfide as a co-catalyst, and natural gas as a carbon precursor, which have nanotexture a little different from herring-bone type, as shown in Fig. 3.130, and are named cup-stacked CNFs (CupNFs) [252–254]. The diameter of CupNFs is in the range of

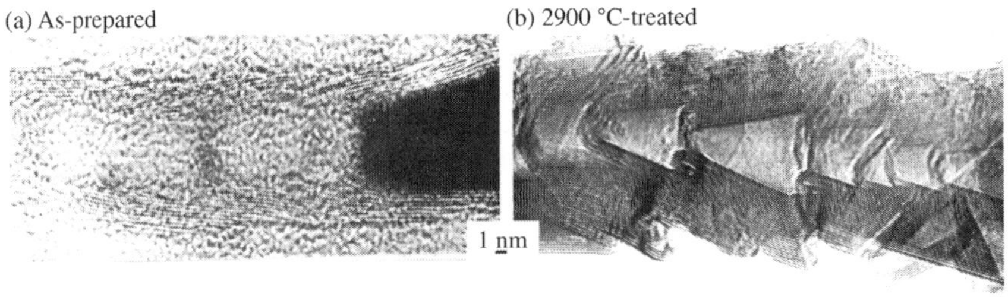


**FIGURE 3.129**

TEM images of CNF with herring-bone type nanotexture synthesized from CO at 400°C and heat-treated at 2900°C.

*(Courtesy of Mme A. Oberlin.)*

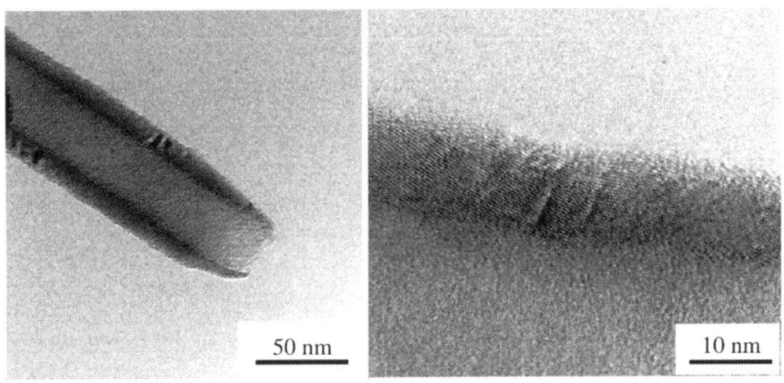


**FIGURE 3.130**

TEM images of CupNF.

*(Courtesy of Prof. M. Endo of Shinshu Univ.)*

50 to 150 nm and the wall consists of 30–35 truncated cups with a spacing between fringes of 0.34 nm.

Some CNFs having tubular-type nanotexture, which are able to convert to CNTs by the heat treatment at high temperatures, are successfully prepared by the techniques different from either arc-discharge or CVD, such as template carbonization technique [255,256], polymer blend technique coupled with spinning [257,258], under a steep thermal gradient in an alcohol [259,260], and the decomposition of SiC single crystals under reduced pressure [261–263]. The preparation of CNFs via template carbonization is described in Section 2.4.1 and that in polymer blend in Section 2.4.2.

CNFs have also been synthesized in alcohols, such as ethanol and methanol [259,264]. The substrate of Si(100), of which the surface is coated by Fe metal film with a thickness of 2.4–3.0 nm by magnetron sputtering, is electrically heated up to 500–1000°C in high-purity alcohol. CNFs grow in a high density on

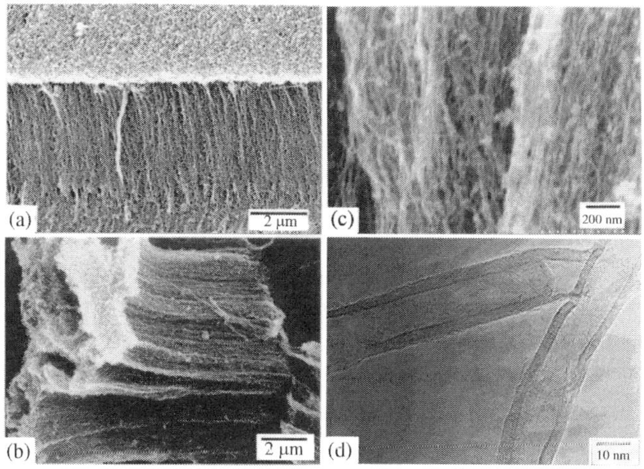


**FIGURE 3.131**

SEM and TEM images of aligned CNFs produced in methanol; (a-c) SEM and (d) TEM images.

*(Courtesy of Dr. T. Ando of CREST, Japan.)*

the surface of the substrate, as shown in Fig. 3.131, their growing direction being perpendicular to the surface of the substrate, in other words, along the thermal gradient. Nanofibers shown in Fig. 3.131 are synthesized on a 2.4 nm thick Fe film at a substrate temperature of 930°C in methanol, which are aligning in a density of more than $1 \times 10^{11}$ fibers per 1 $cm^2$ over the whole surface of the substrate. Nanofibers thus synthesized consist of carbon layers parallel to the tube axis and their tips were capped. The diameter of fibers is rather homogeneous in the range of 13–26 nm, 20 nm being the most frequent. Their length increases with increasing time and reached about 20 μm after 2 h.

Formation of CNFs has been found on the surface of SiC particles by heating to 1700°C under a reduced pressure ($1 \times 10^{-4}$ Torr), nanofibers aligning perpendicularly on the surface [261]. By using commercial α-SiC single crystal wafer with (0001) crystal plane, CNFs aligned perpendicular to the surface of (0001) plane of SiC have been obtained [262,263]. A TEM image of self-aligned nanofibers on SiC is shown in Fig. 3.132. CNFs thus formed were multi-walled nature, consisting of 2–5 layers and having a diameter of 2–5 nm and a length of about 0.3 μm, all of which were capped at their tips. Graphene layers composing nanofibers are well oriented along the tube axis and spaced with each other by 0.344 nm. The formation of self-aligned CNF film over a large area, as 12 mm in diameter (5 inch disk), can be one of the advantages of this process.

Heating SiC single crystal wafers, which contained $N_2$ dopant with a concentration of $\sim 10^{18}$ $cm^3$, to 1500°C at a rate of 1°C/min under vacuum of $10^{-2}$ Pa results in the formation of CNFs with the diameter of 3 nm on $(000\bar{1})$ plane [260]. On $(11\bar{2}0)$ planes of SiC, CNFs formed do not show alignment

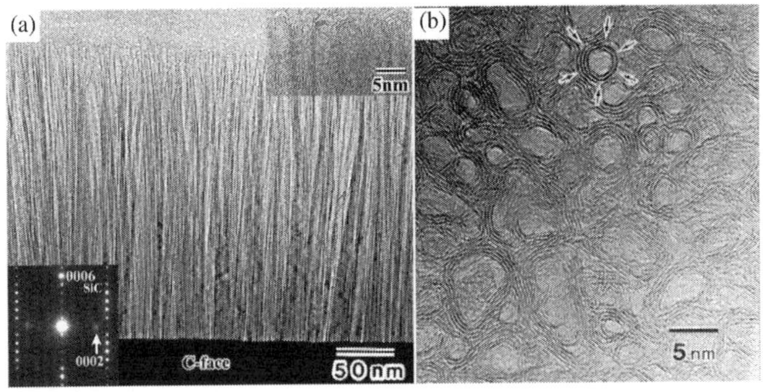


**FIGURE 3.132**

TEM image of well-aligned CNFs on SiC substrate.

*(Courtesy of Dr. M. Kusunoki of Japan Fine Ceramic Center.)*

perpendicular to the surface. On $(1\bar{1}00)$ plane, however, self-aligned CNFs are obtained to form film with 138 nm thickness, but their chirality was not constant.

These techniques, template, polymer blend, steep thermal gradient in alcohol and SiC decomposition, have certain advantages; pure CNFs, almost no other forms of carbon such as carbon blacks, are obtained, diameter and length of CNFs are pretty homogeneous and all of them are aligned in one direction (self-aligned). However, some disadvantages have also been pointed out; some techniques are costly and some are difficult to expand the process to a large scale.

Electron field emission performance from CNTs and CNFs has extensively been studied [265,266]. CNFs prepared by the template method using AAO film have various advantages, mainly due to the fact that CNFs aligned in one direction in electrically insulating alumina substrate. A top view of the electrode shown in Fig. 3.133a demonstrates clearly that all nanofibers are aligned vertically and their edges are on the surface of the electrode. Its electron emission characteristics are revealed by the relation between emitted electron density and electric field applied. The threshold electric field for electron emission is as low as 0.4 V/μm, as shown in Fig. 3.133b. Field emission performance was also studied on CNF films prepared by the decomposition of SiC wafer. CNF film after removal of caps shows a good performance, revealing threshold voltage of about 1.5 V/μm, a steep rise of total current and luminance of $10^5$ cd/m$^2$. Application of CNTs and CNFs to the probe for ultra-fine scanning has also been studied [267,268].

## 3.5.2 Fullerenes

Fullerenes can be classified into geodesic polyarenes, of which structures are ‘closed’. Subunits of fullerenes, that lack one or more of the rings but remain curved, are classified as ‘open’ geodesic polyarenes, and their chemical syntheses

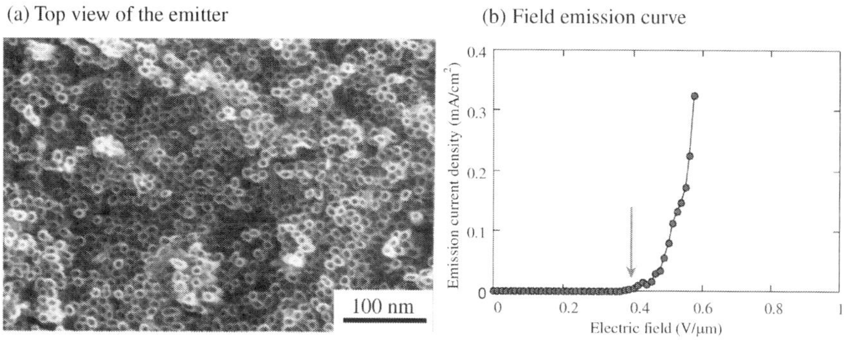


**FIGURE 3.133**

Top view of the electrode of CNFs prepared through template method and its electron emission characteristics.

*(Courtesy of Prof. T. Kyotani of Tohoku Univ.)*

have been studied [269]. In Fig. 3.134, geodesic polyarenes that have been prepared by flash vacuum pyrolysis are shown. The first chemical synthesis of $C_{60}$ has been done by zipping up a conventional polyarene via flush vacuum pyrolysis at 1100°C [270,271]. However, $C_{70}$ and higher fullerenes have not yet succeeded in synthesizing.

Structural modification of fullerenes is interesting and important to develop their applications. The fullerene family has a wide possibility for modification, as pointed out in Chapter 2.8, i.e., the substitution for carbon atoms constructing fullerene cages, accepting foreign atoms not only into the interstices formed by closest packing of fullerene cages, but also inside of each cage, and chemical bonds of organic radicals on the surface of the cluster. Establishment of chemical processes for these modifications of fullerenes, in addition to the chemical synthesis of fullerenes themselves, is a big challenge and so various proposals have been presented [271–275], some novel chemical routes having been developed to open fullerene cages, as described below.

Substitution of carbon atoms in fullerene cages has been studied to synthesize heterofullerenes [276]. Heterofullerenes are important role to synthesize fullerenes modified their surface by organic radicals. A series of hydroazaheterofullerene derivatives $RC_{59}N$ with different side chains R, shown in Fig. 3.135, have been synthesized and their properties have been compared with those of $C_{60}$ [277].

Encapsulation of atoms or molecules inside fullerene cages has been pointed out to give rise to novel molecules and new applications (endhedral fullerenes) [274]. In particular, endohedral metallofullerenes are very important for their potential application as new types of superconductors, organic ferromagnets, nonlinear optical materials, functional molecular devices, magnetic resonance imaging agents, and biological tracing agents, etc. [278]. Some endohedral metallofullerenes are shown in Fig. 3.136. A variety of gas molecules, such as

**FIGURE 3.134**

Geodesic polyarenes synthesized [269].

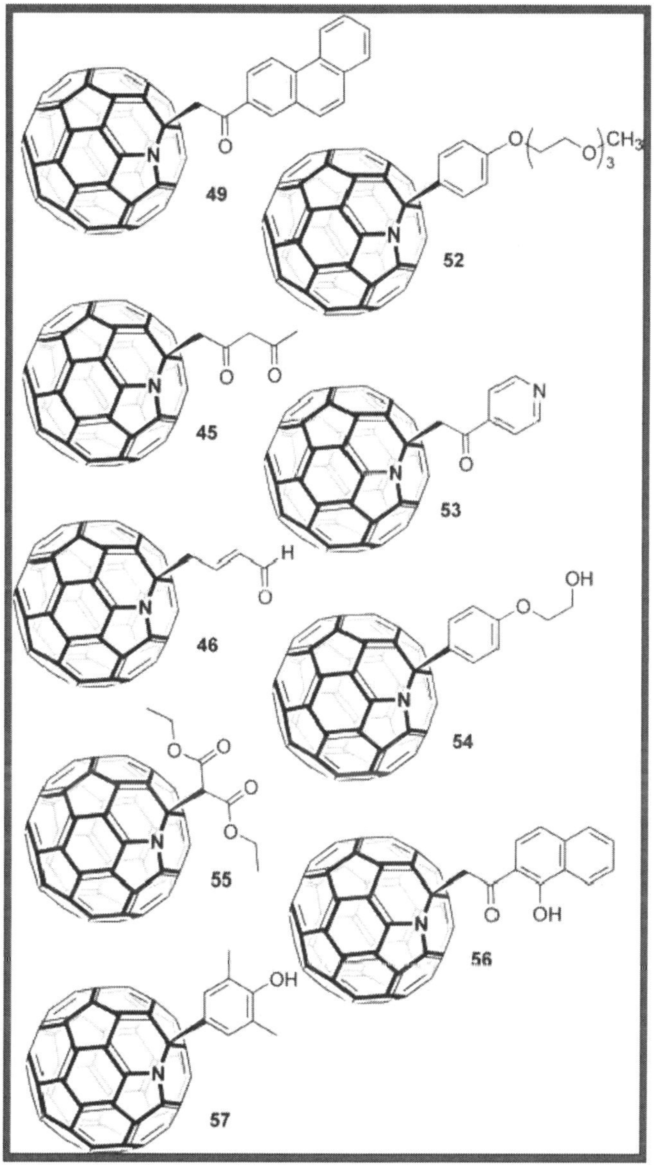


**FIGURE 3.135**

Hydroazaheterofullerene derivatives $RC_{59}N$ synthesized [276].

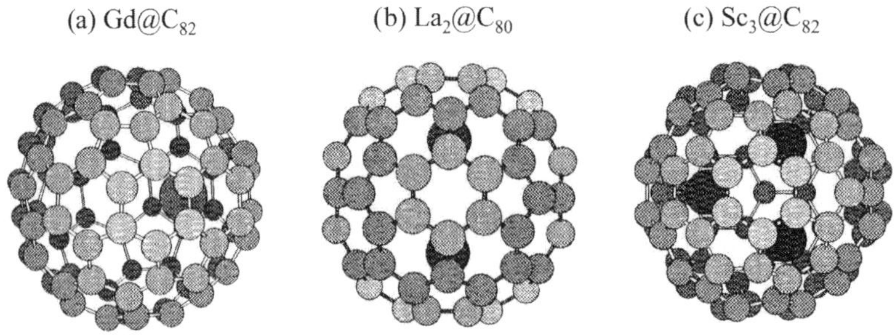


**FIGURE 3.136**

Endohedral metallofullerenes.

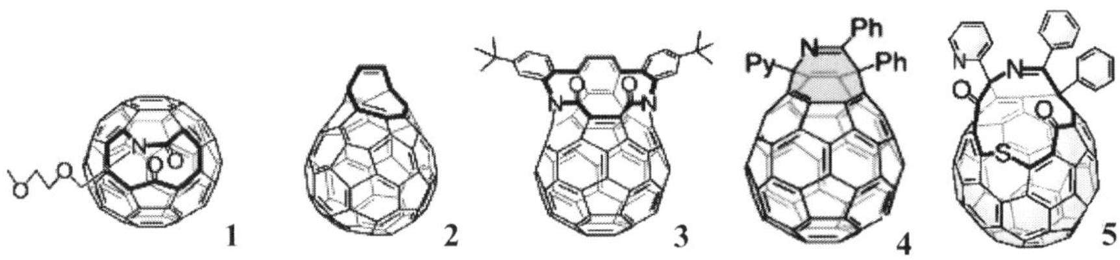


**FIGURE 3.137**

Open-cage fullerenes synthesized [282].

noble gases, and metals, such as transition and rear earth metals, have been inserted into fullerene cages.

Fullerenes synthesized by arc discharging in He atmosphere contain He in about 1 ppm [279]. Insertion of different noble gases into fullerene cage has been carried out by heating the fullerene in the presence of gas at 650°C under a high pressure of 303 MPa [280]. However, the fraction of recoverable fullerene incorporated noble gas was quite low: 0.1% for He, Ne, Ar, and Kr and about 0.03% for Xe. By adding KCN to the reaction mixture, the incorporation fraction was enhanced to 2% [281].

A chemical process for the opening of a large orifice in fullerene shell has been developed, as some open-cage $C_{60}$ synthesized shown in Fig. 3.137, compound **1** having an 11-membered-ring, compound **2** an 8-membered-ring, compound **3** a 14-membered-ring and compound **4** an 8-membered-ring orifice. Compound **4** is synthesized by the reaction of diazidobutadiene with $C_{60}$ at 55°C [282] and also by the liquid-phase reaction of 3-(2-pyridyl)-5,6-diphenyl-1,2,4-triazine in *o*-dichlorobenzene (ODCB) at 180°C [283,284]. Into the compound **3**, $^3He$ isotope gas is inserted under high pressure of about 48 MPa at about 300°C and $^3He$-inserted open $C_{60}$ is heated at 130°C to close the shell, i.e., to obtain endohedral $^3He@C_{60}$ [285]. Roughly half of He-inserted open $C_{60}$ can be converted to endohedral $^3He@C_{60}$. By using one of the derivatives of the compound

**4** (compound **5**), 100% encapsulation of $H_2$ can be performed [283], although only 5% yield from compound **3** [282]. Endohedral $H_2@C_{60}$ is produced from $H_2$@compound **4** under laser irradiation, without any deformation of $C_{60}$ cage, which can be recovered via chemical treatment with a yield of 61% [286].

As described in Section 2.8.4c, doping of alkali metals into the interstices of fullerene crystal is effective to increase the critical temperature of electrical superconductivity $T_c$, which is closely related to the lattice constant $a_0$ of doped $C_{60}$ crystal [287], as shown in Fig. 2.196b. Doping of K into the interstices of the crystal of endohedral $Ar@C_{60}$, $K_3Ar@C_{60}$, shows superconductivity, though $T_c$ is lower than $K_3@C_{60}$ (17.8 K for $K_3Ar@C_{60}$ and 19.2 K for $K_3@C_{60}$) [288].

Polyhydroxylated gadolinium metallofullerenols $Gd@C_{82}(OH)_n$ (Gd-metallofullerenol), where Gd atoms are inserted into the $C_{82}$ cage and -OH radicals are attached on the carbon atoms of $C_{82}$, has high performance as a contrast agent for magnetic resonance imaging (MRI) [289,290]. Since the metallofullerenes have to be modified to the water-soluble forms for biomedical applications, $Gd@C_{82}$ is converted to Gd-fullerenol by treating in tetrabutylammonium hydroxide, the resultant fullerenol containing about 40 hydroxyl [289]. In Fig. 3.138, MRI signal of $Gd@C_{82}(OH)_n$ is compared with those of Gd-chelate complex, Gd-DTPA (diethylenetriaminepentaacetic acid), and water for different concentration of Gd. $Gd@C_{82}(OH)_n$ gives much stronger signal than conventional contrast agent Gd-DTPA, even with a low concentration as low as 0.05 mM/L [289]. Water-soluble multi-hydroxyl lanthanoid (La, Ce, Gd, Dy, and Er) endohedral metallofullerenols ($M@C_{82}(OH)_n$) have been compared on the performance of MRI contrast agents [290]. The performance of these metallofullerenols is significantly higher than that of the corresponding lanthanoid-DTPA chelate complexes. Among these metalofullerenols, $Gd@C_{82}(OH)_n$ has the highest performance.

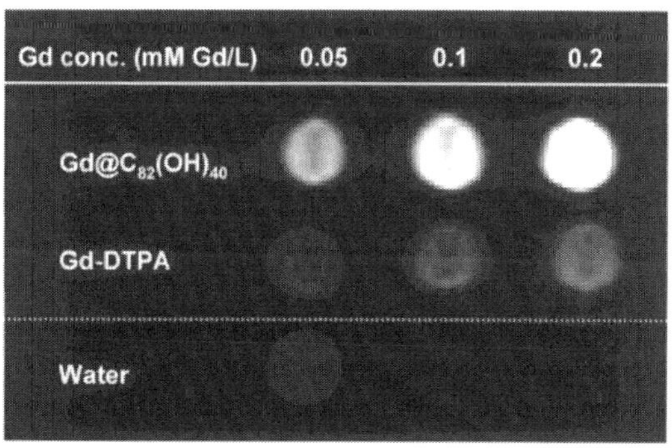


**FIGURE 3.138**

MRI signals obtained by $Gd@C_{82}(OH)_n$, Gd-DTPA chelate complex and water with different Gd concentrations [289].

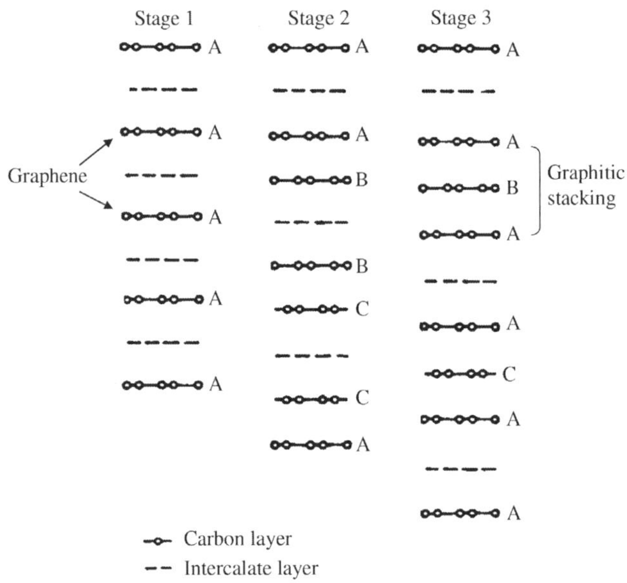


**FIGURE 3.139**

Occurrence of graphene layers in graphite intercalation compound.

### 3.5.3 Graphene and its derivatives

The term 'graphene' was first proposed in 1986 as a single two-dimensional carbon sheet, which occurs in a graphite intercalation compound [291]. In the first stage structure, two-dimensional carbon layer is neighbored with the intercalate layers, as schematically shown in Fig. 3.139. In the structures higher than second stage, however, more than 2 carbon layers stacked in parallel with the same regularity as graphite. Therefore, the carbon layer exists in single and is independent from other carbon layers, as in the first stage intercalation compounds, is proposed to be called 'graphene', which comes from the suffix '-ene' for polycondensed aromatic hydrocarbons, such as naphthalene, anthracene, etc., and the prefix 'graph-' for graphite.

If the strictly defined graphene, i.e., an isolated planar sheet of carbon hexagons consisting of $sp^2$ hybridized bonding with $\pi$-electron cloud, can be obtained, it is a two-dimensional crystal. In 2010, the Nobel Prize in physics was given to Profs A. K. Geim and K. S. Novoselov, who worked on graphene, which accelerated research activity related on graphene, as reviewed from different viewpoints [292–295].

Although there have been numerous trials to synthesize graphene, successes have been obtained only under very limited conditions, such as under electron microscope, under the dispersion in a liquid media, etc. Most of the synthesis of graphene has been carried out through 4 routes, (1) peeling of graphite crystals, (2) exfoliation of graphite through its intercalation compounds, (3) chemical vapor deposition of a hydrocarbon gas, and (4) processing via organic chemistry.

Scientific and engineering interests for graphene introduced the derivatives of graphene, such as hydrogenated graphene (graphane), graphene floride (fluorographene), and graphene oxide.

## a. Graphene

***i. Preparation by peeling*** The peeling (cleavage, mechanical exfoliation) of graphite crystals has been done by aiming to synthesize a two-dimensional crystal and so various properties of thin flakes obtained have been measured [296–299]. By peeling of a kish graphite flake with double-sided adhesive tape repeatedly, thin transparent flakes with various thicknesses of 18–108 nm are obtained and electrical resistivity $\rho$, Hall coefficient $R_H$ and transverse magnetoresistance $\Delta\rho/\rho$ are measured at a temperature between 4.2 and 300 K [198,297]. Analysis of the results on highly crystalline thin flakes applying the simple two band model and Sugihara's theory for lattice vibration shows that the overlap energy between conduction and valence bands decreases and the relaxation rate due to lattice defects increases with decreasing flake thickness. Thin flakes with a thickness of 10–100 nm and lateral size of about 2 μm have been obtained through the micromechanical cleavage technique [298,299] from the micropillars (about $2 \times 2 \times 5\ \mu m^3$) of a highly oriented pyrolytic graphite (HOPG) formed by a patterning technique using oxygen plasma [300]. Electric-field-dependent conductance measurement on these flakes shows a marked modulation as a function of gate voltage, the more markedly on the thinner flakes [298], as shown in Fig. 3.140. Single-layer graphene could be found on the surface rubbed on a fresh surface of graphite [301].

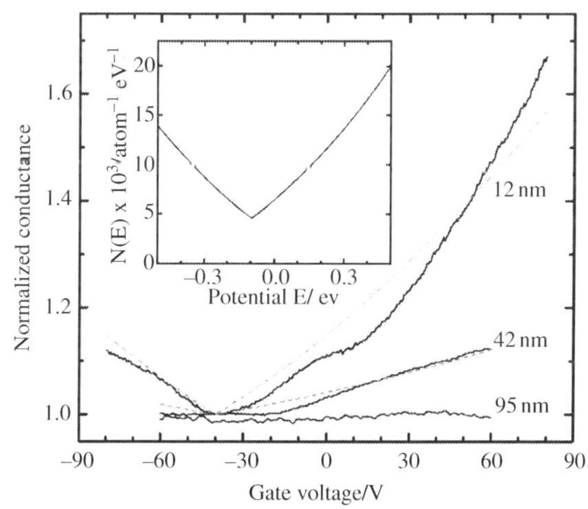


**FIGURE 3.140**

Electric-field-dependent conductance on thin flakes with different thicknesses prepared by peeling [298].

Through micromechanical cleavage of graphite, single-layer graphene together with multilayer flakes are obtained as a suspension in a liquid [302]. The flakes are attached to a micrometer-sized metallic scaffold to identify the single-layer graphene under TEM. The edges of the flakes obtained are often folded and the number of layers is easily determined through 002 lattice fringe image at this folding region.

***ii. Preparation through exfoliation*** Trials to prepare graphene through graphite oxide (GO) have been reported in a number of papers. In most trials, GO is synthesized through Hummers method [303], consisting of the oxidation of graphite in concentrated $H_2SO_4$ with $NaNO_3$ and $KMnO_4$, the exclusion of excess $KMnO_4$ by reducing to water-soluble $MnSO_4$ with $H_2O_2$, and then washing by methanol, although some modifications in the process are introduced for the preparation of the thin flakes. In some trials, Brodie method [304] has been employed, where the oxidation of graphite is carried out in fuming $HNO_3$ with $KClO_3$. To synthesize GO, electrochemical oxidation of graphite has been done on natural graphite [305]. The GO synthesized can have a wide range of chemical compositions, such as $C_8O_{3.5-4.3}H_{2.5-2.9}$ [306], $C_8O_{3.78-5.05}H_{2.9-4.4}$ [307], $C_8O_{2.54}H_{3.91}$ and $C_8O_{4.61}H_{6.70}$ [308], depending strongly on the starting graphite and the detailed conditions for the synthesis, and composing from different oxygen-containing functional groups, such as hydroxyl, carbonyl, ether, epoxy and peroxy groups, in different amounts. An example of thin flakes prepared from GO through exfoliation and reduction is shown in Fig. 3.141.

In order to obtain a thin sheet, the exfoliation of GO at a high temperature around 1000°C is often applied for a short time. During this high-temperature exfoliation, a

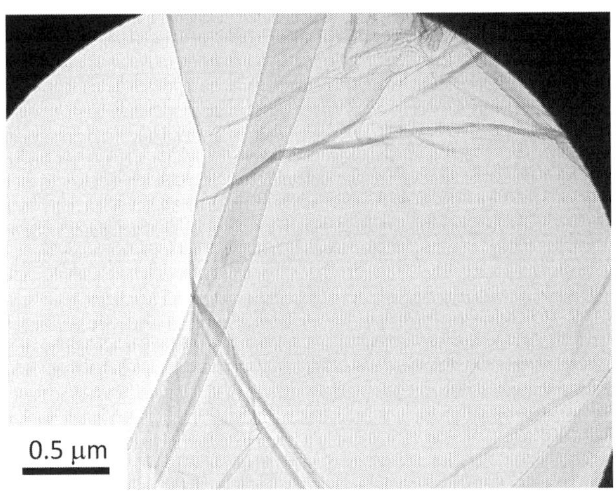


**FIGURE 3.141**

Ultrathin flakes prepared from GO through exfoliation.

*(Courtesy of Prof. M. Toyoda of Oita University, Japan.)*

part of functional groups on the surface of GO are removed and the thin flakes tended to coalesce. A strong sonication is exerted on the GO particles suspended in a liquid media and is also effective for the separation of layers. In addition, the reduction process with different methods is usually included in order to produce high-purity and electric conductive flakes. Heat treatment in a flow of mixed gas of Ar and $H_2$ has been applied [309]; single-layer graphene with the thickness of 0.37 nm is obtained after the reduction at 500°C for 2 days. The treatment of GO flakes by phenyl-isocianate is effective for having colloidal suspension of GO flakes with the thickness of ~1 nm [310,311], because the treatment gives hydrophobic chemical groups on the GO surface to keep the flakes separated [312]. These isocianate-treated GO flakes are reduced by $N_2H_4$, the resultant flakes giving relatively high electrical conductivity after making composite with styrene; 1 vol% mixing of the flakes gives ~0.1 S/m and 2.5 vol% mixing ~1 S/m [313]. Annealing at a high temperature, in addition to reduction, is needed to have pure graphene, because of large contents of oxygen-containing functional groups and also structural defects. In GO, carbon atoms take $sp^3$ hybridized bonding with neighboring carbon atoms and oxygen atom. Therefore, some structure imperfections are expected to remain even after removal of all oxygen-containing groups by reduction, because all C C bonds have to change to $sp^2$ bonding. Usually, GO shows a strong D band, often stronger than G band, in Raman spectrum, revealing the presence of a large amount of defects, and remains even after an exfoliation at a high temperature such as 1000°C.

The size and crystallinity of the starting graphite have a marked influence on the thickness of the flakes [314]. Among various starting graphites, an artificial graphite gave a sharp distribution of thickness of less than 2 nm, suggesting that about 80% of the flakes obtained from an artificial graphite are single-layer graphene. From HOPG having large lateral size and high crystallinity of crystallites, however, a broad distribution of flake thickness, consisting of mostly 4–10 layers and even more than 10 layers, is obtained.

Conductive and transparent films with a thickness of 14–86 nm have been prepared from GO by exfoliation under sonication and reduction with $N_2H_4$ in a water dispersion, followed by making films on a quartz surface and annealing up to 1100°C [315]. Photographs of the sheets are shown in Fig. 3.142. The optical transparency of the film with 14 nm thickness is more than 80% in the wavelength range of 1100 and 3000 nm, and its electrical conductivity is over 200 S/cm.

Thermal exfoliation at low temperature (200°C) under high vacuum (<1 Pa) is effective for obtaining thin flakes of graphene. The resultant graphenes gave much higher capacitance than that of the high-temperature exfoliated ones in supercapacitors [316,317a], and also were efficient conductive additive in the electrodes of Li-ion batteries [317b,c].

The exfoliation of graphite has been performed by exposing the first stage of graphite intercalation compound with potassium (K-GIC, $KC_8$) prepared from HOPG to isoprene vapor and heat-treated above 500°C [318]. Isoprene molecules

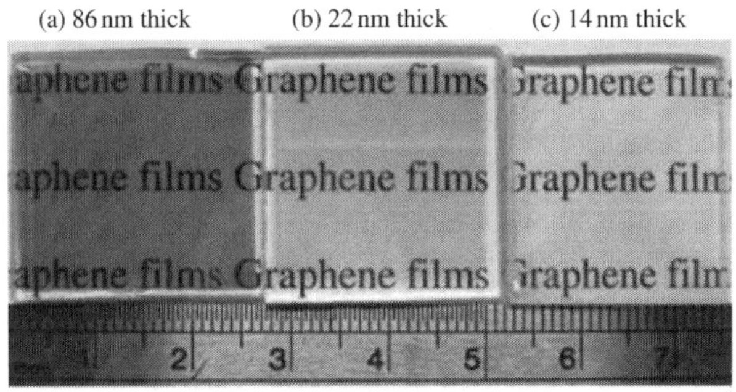


**FIGURE 3.142**

Thin sheets prepared from GO through exfoliation, reduction and annealing at 1100°C [315].

are supposed to be intercalated into graphite gallery with potassium and polymerized in the gallery, resulting in exfoliation of host graphite. Most thin flakes obtained are wavy and rolled up, especially at the edge. A highly exothermic reaction of $KC_8$ with ethanol also causes exfoliation of graphite, resulting in the formation of nanoscrolls composed from graphene [319].

On HOPG thin film which is partially intercalated by K, local electronic structure at the atomic level has been examined focusing on the surface, which neighbors the K-intercalated region (non-intercalated region), with scanning tunneling microscopy (STM) and scanning tunneling spectroscopy (STS), and discussed on electronic structure of single-layer graphene [320].

***iii. Preparation by chemical vapor deposition*** Deposition of graphite thin layers onto (111) plane of Pt crystal has been carried out in a scanning tunneling microscope (STM) by exposing to ethylene gas of very low pressure at 300 K and then annealed at different temperatures [321]. Small deposits are formed with the size of 2–3 nm after annealing to 900 K, and grow at the expense of the smaller deposits with annealing to higher temperatures. Detailed analysis of STM images observed on these deposits proves the formation of single-layer graphene. Thin carbon layers are obtained on a Ni(110) surface by reacting with CO at a temperature of 600 K and a pressure of $6.5 \times 10^{-3}$ Pa [322]. A strong interaction between deposited carbon atoms with Ni substrate is concluded. Two mechanisms for the growth of a thin carbon sheet have to be taken into account; chemical vapor deposition from hydrocarbon gas on the substrate Ni and carbon precipitation at a low temperature from the substrate Ni, because Ni can absorb a large amount of carbon into its interstitial sites at a high temperature [323].

Mono-layer graphene with a large area, containing few-layer regions of less than 5%, is grown on Cu foil by CVD of methane at a temperature up to 1000°C [324]. An efficient roll-to-roll transfer process of graphene formed on a flexible Cu foil has been developed, which consists of three steps: (1) adhesion of a polymer support, such as PMMA, to the graphene/Cu, (2) etching of the Cu foil in a

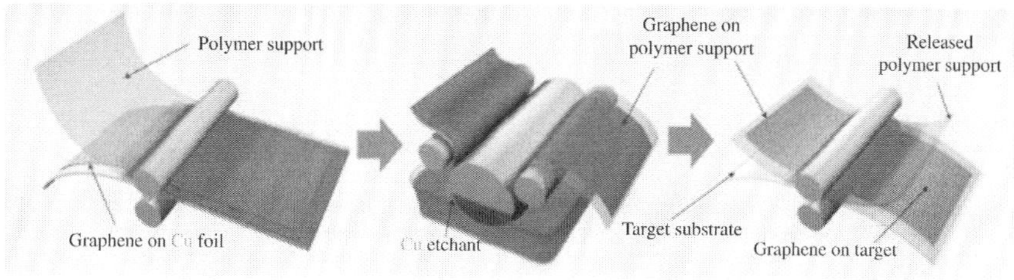


**FIGURE 3.143**

Roll-to-roll process for the preparation of a large-area graphene sheet via CVD [325].

$(NH_4)_2S_2O_8$ aqueous solution, and (3) transfer the graphene film from the polymer support onto a target substrate, as illustrated in Fig. 3.143 [325]. This process is applicable to prepare a large-sized graphene sheet and multi-layered graphene films can be obtained by repeating the process on the same substrate.

Single-layer graphene has been synthesized on TiC(111) surface by the reaction of ethylene gas with the substrate at 1100°C [326]. The strength of C–C bond in single-layer graphene is weakened because of the orbital hybridization between the graphene layer and the substrate TiC(111) surface. The sheets composed of few layers are grown on the (0001) surface of 6H-SiC by its thermal decomposition [327] and their electronic transport properties have been determined [328]. The thin sheet obtained is evidenced by low-energy electron diffraction (LEED) and STM measurements to grow epitaxially on the substrate and by Auger spectroscopy to consist of a few layers. The properties of these sheets reveal the existence of electron gas having marked anisotropy, high mobility and two-dimensional localization.

***iv. Synthesis via organic chemistry*** Organic synthesis route to graphene has been investigated, although different problems to obtain a large-sized graphene due to the limited solubility and the side reactions are recognized. In Fig. 3.144, one of the synthesis strategies toward graphene nanoribbons is outlined [329]. Polymer **5** was successfully synthesized from polymer **1** via Suzuki-Miyaura coupling process. Polymer **5** in the figure is obtained as a black solid with a yield of 65% [330]. Drop-casting of its THF solution onto silica surface gives nanoscroll with a diameter of about 100 nm and a length of up to 5 μm [331]. Stacking of the molecules of these polymers can lead to nanotubes with a platelet nanotexture and interlayer spacing of 0.34 nm [332].

## b. Graphane

Fully hydrogenated derivative of graphene with a stoichiometric formula unit CH, is named graphane, which can be considered as an extended two-dimensional hydrocarbon [333]. Graphane was theoretically predicted in the investigation on hydrogen chemisorption of graphene [334]. It was first synthesized in 2009 by exposing graphene to cold hydrogen plasma [335]. It consists of $sp^3$ C–C

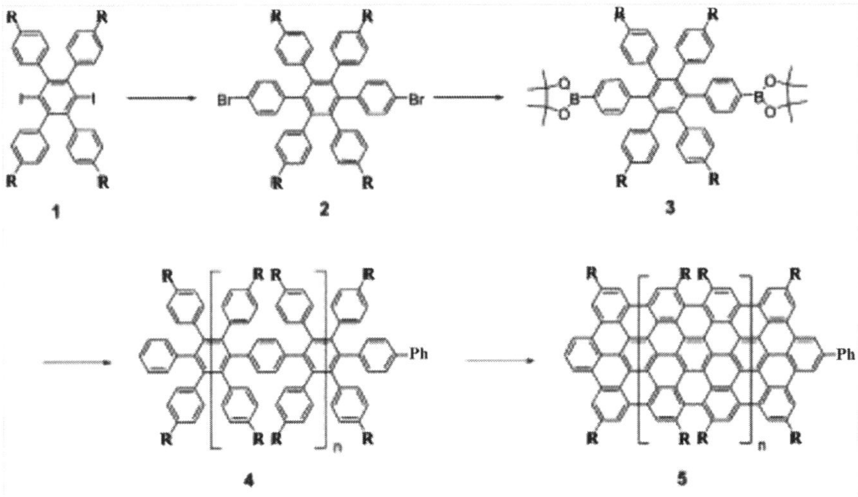


**FIGURE 3.144**

Strategy to synthesize graphene by organic chemistry route [329].

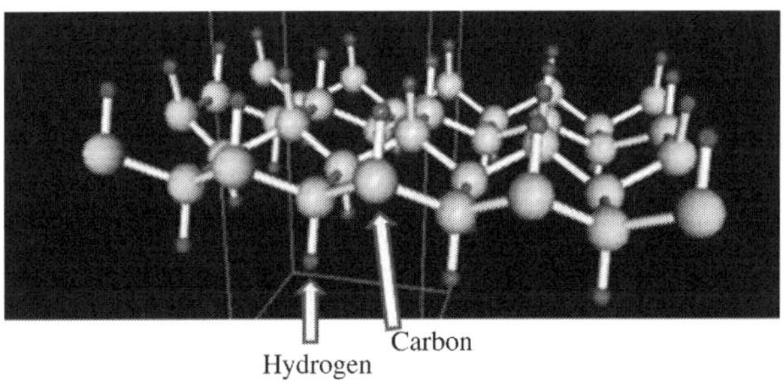


**FIGURE 3.145**

Structure scheme of graphane.

bondings, as opposed to graphene's $sp^2$ bondings, as a consequence, carbon atom layer is puckered, as shown in Fig. 3.145. It is also a two-dimensional analog of cubic diamond. It has been predicted to be a stable structure with a binding energy comparable to other hydrocarbons such as benzene, cyclohexane, and polyethylene [333].

Graphane and graphenes hydrogenated in various extents are formed by electrolytic hydrogenation of graphene [336] and exfoliation of graphite oxide under high hydrogen pressure [337,338]. In Fig. 3.146, Raman spectrum of graphane (fully hydrogenated graphene) is compared with that of the pristine graphene [336]. Marked development of D-band, broadening of G-band and 2D-band, and appearance of a band (denoted as E) are characteristic for graphane, the

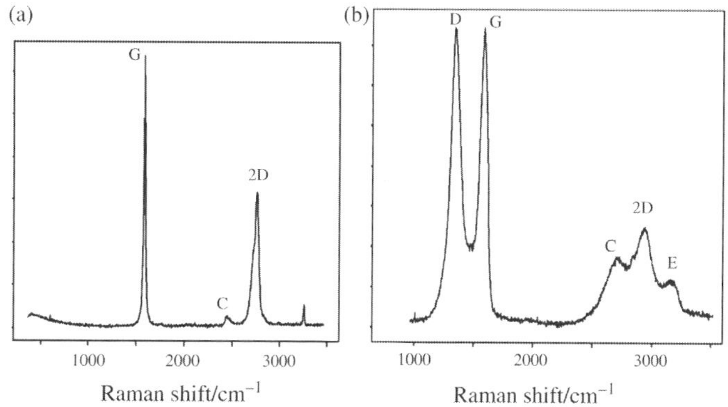


**FIGURE 3.146**

Raman spectra of the pristine graphene and graphane [336].

developments of D-band and E-band being correlated to hydrogen content. Hydrogen atoms on one side of graphane can be unloaded by applying an external electric field, where the hydrogen atoms on the other side are kept and the unpaired electrons in the unsaturated C sites give rise to magnetic moments [339,340]. Hydrogen treatment at higher temperatures (550°C) results in unzipping of some SWNTs into graphene nanoribbons, which are partially hydrogenated [341].

The electronic properties of graphene sheets and nanoribbons hydrogenated in different degrees, which are performed under radio-frequency hydrogen plasma (13.56 MHz, 300 W) in ultrahigh-vacuum, have been studied [342]. A rapid decrease in mobility is observed with increasing exposure to the hydrogen plasma, suggesting that graphane (fully hydrogenated graphene) is insulating, although graphene is highly conductive.

The discovery of graphane may open the floodgates for other graphene-based chemical derivatives, such as nanoribbons with metallic graphene at one end and insulating graphane at the other and, graphane substituted foreign atoms, etc.

### c. Fluorographene

Fluorinated graphene, fluorographene, monolayer of graphite fluoride (Section 2.8.1 and Fig. 2.182) and with similar configuration to graphane, is another important structural derivative of graphene. Structure of fully fluorinated graphene is schematically shown in Fig. 3.147a. It has been synthesized by reacting graphene with either $XeF_2$ [343–345] or $CF_4$ plasma at room temperature [346], and by mechanical and chemical exfoliation of graphite fluoride [347,348]. Fluorographene cannot be suspended in ethanol but the pristine graphene can be suspended by sonication for 30 s, as shown in Fig. 3.147b [343]. New methods for its synthesis in large-scale process with a low cost and less toxicity have also been proposed [349,350].

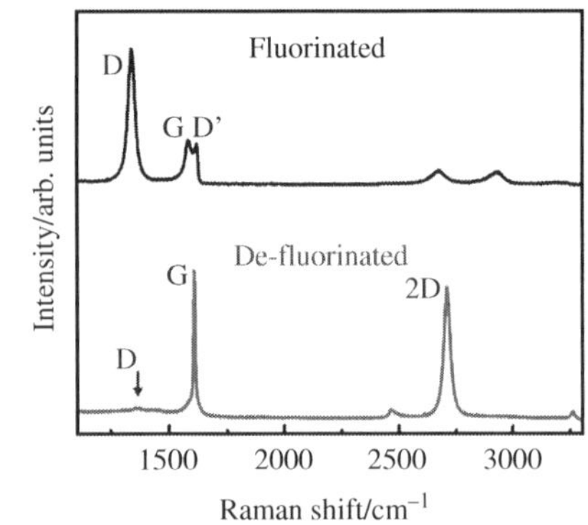


**FIGURE 3.147**

Scheme of structure (a), and dispersed state of fluorographene and graphene in ethanol [343].

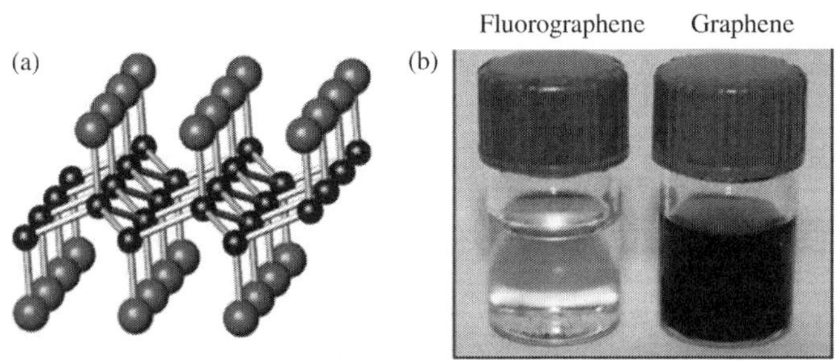


**FIGURE 3.148**

Raman spectra of partially fluorinated and de-fluorinated graphene [346].

Fluorographene has a band gap of 3.8 eV, close to that calculated for fluorinated single layer graphene, and shows luminesces broadly in the UV and visible light regions. Its optical properties resemble diamond, with excitonic and direct optical absorption/emission features. Fluorographene is a high-quality insulator (resistivity of higher than $10^{12}$ Ω) with an optical gap of 3 eV, exhibits a Young's modulus of 100 N/m, and is stable up to 400°C even in air, similar to Teflon [344]. In Fig. 3.148, Raman spectra are shown for partially fluorinated graphene and de-fluorinated graphene at 365°C in $Ar + H_2$ gas, showing marked development of D-band and broadening of 2D-band [346].

Theoretical predictions of the properties of fluorographene in relation to its applications have been published [351–354]. Fluorographene exhibits strong insulating properties and low friction, because of the low interlayer interaction induced by the repulsive electrostatic forces between F [354]. The same low friction can be expected for graphane, but fluorographene is predicted to show much lower friction than graphane. Doping of fluorographene with either metals [351] or boron and nitrogen [353], and hybrid with graphene [352] have been studied to learn its properties during functionalization. The p-type doping is suggested by density functional theory to be difficult to achieve in fluorographene.

#### d. Graphene oxide

Graphene oxide is a single-layer sheet of graphite oxide, which was proposed one and a half century ago. In recent years, graphene oxide has attracted great concern mainly because it is a potential starting material for the mass production of graphene. Chemistry of graphene oxide has been reviewed [355].

Graphene oxide can be a semiconductor or insulator, depending on the degree of oxidation, and their electronic and optical properties can be tuned in large scope. The controllable optical and electronic properties enable graphene oxide to be used in many fields. The major concern about graphene oxide is mainly focused on its chemical structure [355–357], electronic properties [358,359], reduction reaction [360,361] and chemical functionalization [362,363].

The structure of graphene oxide is still unclear due to its complicated non-stoichiometric nature. There are different kinds of oxygen species and bondings to carbon in graphene layer, such as epoxy, hydroxyl, carbonyl, carboxylic groups (Fig. 3.149).

The electronic properties of graphene oxide mainly depend on the oxidation level and chemical composition; it can be tailored by removal or addition of certain oxygen groups to adjust the proportion of $sp^2$ and $sp^3$ carbon [364,365]. Graphene oxide is easily converted into graphene, as explained above. The application fields of graphene oxide are mainly focused on sensor [366] and drug delivery [367]. With a large amount of functional groups, graphene oxide can react with many chemical groups and can be easily modified to improve its functionalities and create new functionalities.

### 3.5.4 Graphyne and graphdiyne

Instead of a flat single carbon layer consisting of hexagons (i.e., graphene), as occurred in nature as graphite, a flat single atomic layer of carbon is possible by connecting carbon hexagons by linear carbon chains. Graphyne structure, in which carbon hexagons are bonded by linear acetylenic chain, was predicted in 1987 [368], one year later than the proposal of the name 'graphene'. Later, graphdiyne which consist of two acetylenic chains and other structures was proposed. Many theoretical calculations have been reported, mostly focusing on stability of two structures, namely graphyne and graphdiyne [369–374]. The structure of these two is shown in Fig. 3.150, by comparing with graphene. The name 'graphyne' comes from its chemical structure, one-third of the carbon–carbon bonds in graphene being replaced by acetylenic linkages, and 'graphdiyne' from the presence of two scetylenic (di-acetylenic) linkages. A comprehensive review on graphyne and graphdiyne is presented [375].

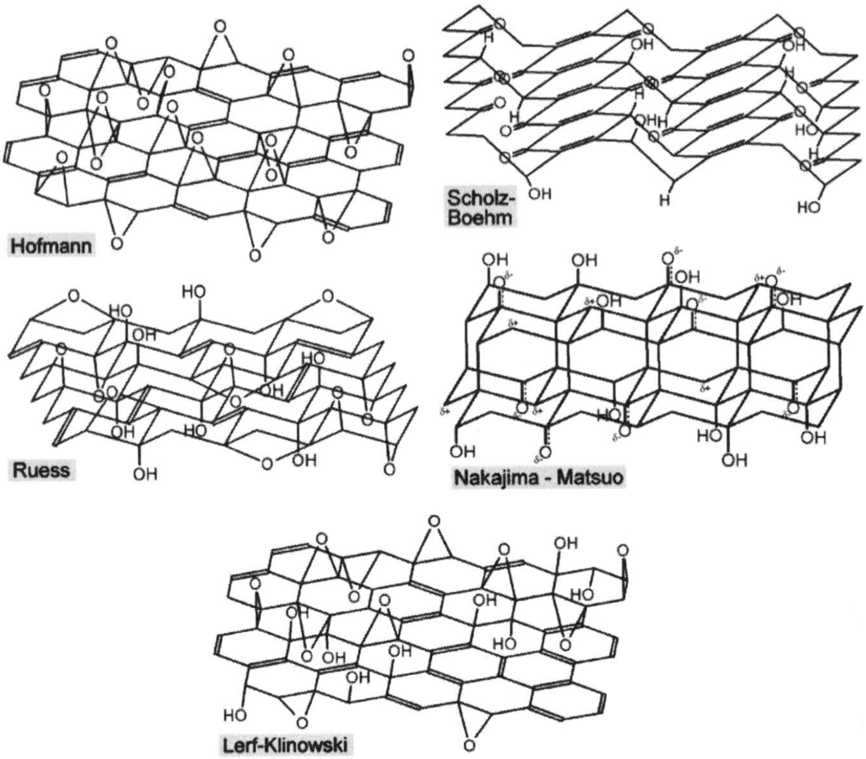


**FIGURE 3.149**

Proposed structure of graphene oxide [356].

The linear carbon chain between carbon hexagons is shown to be more stable to be composed of acetylenic linkages (–C≡C–) than cumulative linkages (=C=C=) [369]. The binding energies are 7.95 eV/atom for graphyne and 7.78 eV/atom for graphdiyne, and both are semiconductive [369]. The band gap for graphdiyne sheet is predicted theoretically to be 0.46 eV and its in-plane intrinsic electron mobility can reach the order of $10^5$ $cm^2/Vs$ at room temperature, while the hole mobility is about an order of magnitude lower [370].

Two approaches to synthesize graphyne and graphdiyne have been proposed, trimerization of cyclocarbon and polymerization of hexaethynylbenzene [371], and their substructures have been synthesized [376–380]. A large area ($\sim 3.6$ $cm^2$) of graphdiyne film has been successfully prepared via cross-linking reaction using hexaethynylbenzene on a copper surface, the resultant film exhibiting semiconducting property with electrical conductivity of $2.516 \times 10^{-4}$ S/m. [381]. The same process is applied to synthesize nanowires of graphydine using ZnO nanorod arrays on a silicon slice as a substrate, which has a conductivity of

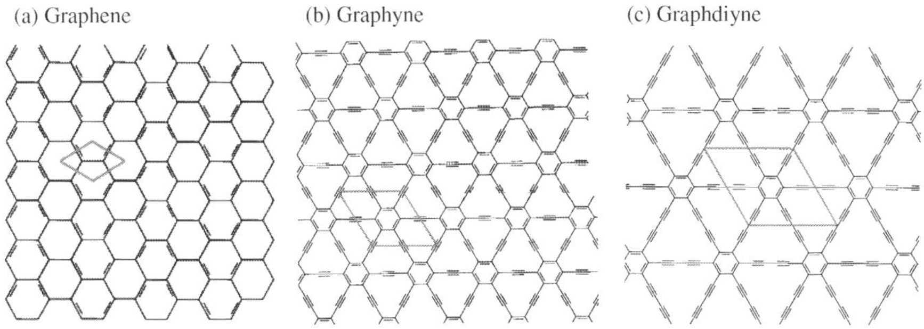


**FIGURE 3.150**

Structure of graphene (a), graphyne (b) and graphdiyne (c). The parallelogram drawn with a red line represents a unit cell.

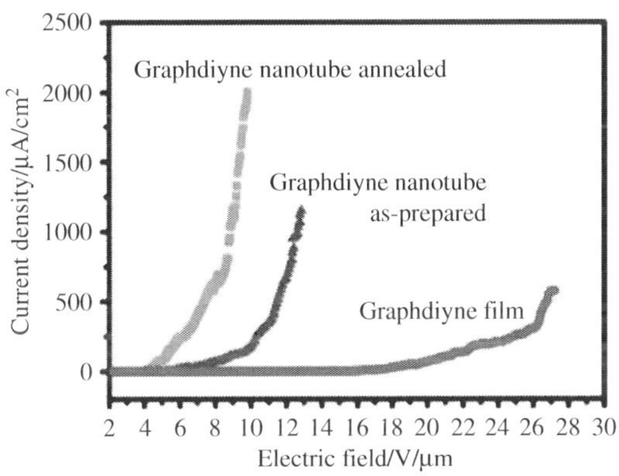


**FIGURE 3.151**

Field emission performance of graphdiyne [383].

$1.9 \times 10^3$ S/m and a mobility of $7.1 \times 10^2$ $cm^2/Vs$ at room temperature [382]. Graphdiyne is synthesized as nanotube array by using AAO film template and Cu catalyst, the nanotubes having a diameter of about 200 nm, a length of about 40 μm and a wall thickness of about 40 nm [383]. By annealing at 650°C, graphdiyne nanotubes are shrunk to the wall thickness of 15 nm, without noticeable change in the length. Annealed graphdiyne nanotube array shows high performance in field emission, as shown by comparing to nanotube before annealing and film in Fig. 3.151. Loading of photocatalytic $TiO_2$ nanoparticles on graphdiyne nanosheet is shown to be effective to improve photoactivity, a little better than that on graphene [384,385].

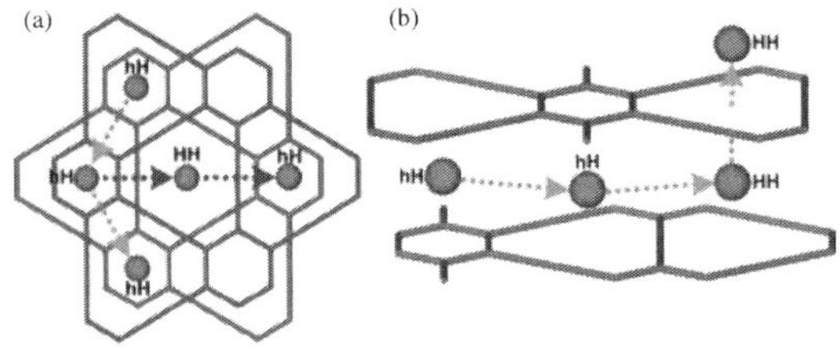


FIGURE 3.152

Diffusion pathway of Li in bulk graphyne [183]. (a) In-plane and (b) out-plane modes.

Theoretical considerations focusing on the functions related to the possible applications of graphyne and graphdiyne have been published. In contrast to graphite where Li diffusion is confined in the interlayer space (in-plane diffusion), graphyne and graphdiyne can give a possibility for both in-plane and out-plane diffusion of Li ions, as illustrated on graphyne in Fig. 3.152, owing to the unique atomic arrangement and electronic structures. Diffusion barriers calculated for two diffusion pathways are moderate, as 0.35–0.57 eV and the maximum Li intercalation density can be $LiC_4$ in graphyne [386], and $LiC_3$ in graphdiyne [387], exceeding the upper limit of $LiC_6$ in graphite. The first-principles density functional calculations predicts that Ca-decorated graphyne can have high hydrogen-storage capacity as 6–7 mass% at 25°C under 3 MPa [388–390], although Ca atoms tends to form clusters on the surface of graphene and fullerene. Graphdiyne can serve as a membrane for $H_2$ purification from $CH_4$ and CO [391].

### 3.5.5 Single-wall carbon nanohorns

Single-wall carbon nanohorns (SWNHs) have been produced by laser ablation of graphite substrate (purity of 99.99%) without any catalyst [392,393]. Ablation in Ar atmosphere of 0.1 MPa gives spherical assemblies of SWNHs with about 80 nm diameter, like dahlia flower, as shown in Fig. 3.153. The tube diameter of each SWNH is in the range of 2–4 nm. On the other hand, SWNH assemblies produced in He have the shape of a bud of the flower; the primary SWNHs in the bud-form assemblies are shorter than those in the dahlia-flower type. The bud-SWNH has different porosity from the dahlia-flower type one [394]. Yield of these SWNHs is very high, up to 95%, much higher than carbon nanotubes.

Liquids with surface tension less than 100–200 mN/m can wet SWNH surface, and so the liquid-bridged compressed bundles of SWNH particles can be prepared. Ethanol and glycerol as the liquid binders have been used for an efficient compression. SWNHs are dispersed in the liquid under sonication,

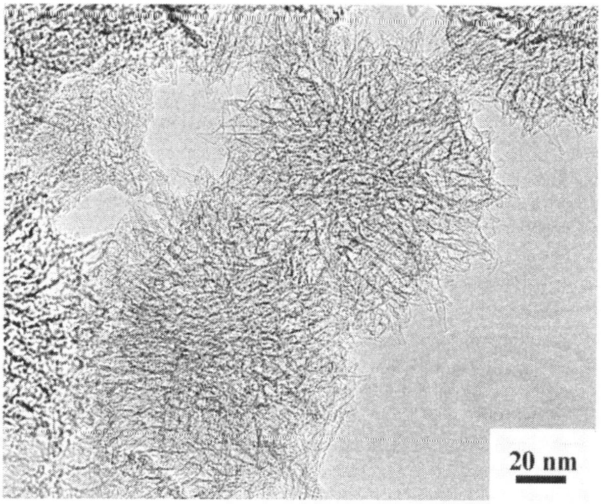


**FIGURE 3.153**

TEM images of single-walled carbon nanohorns.

*(Courtesy of Prof. K. Kaneko of Chiba Univ.)*

dried, and then compressed under a pressure of 10–50 MPa. The $N_2$ adsorption isotherm of compressed SWNHs at 77 K has a considerably high uptake at a low relative pressure and a clear adsorption hysteresis. Accordingly the oxidation and compression of SWNHs induce a pronounced increase in micropores and mesopores [395,396].

Compressed tablets prepared from these SWNHs have very high capacity for methane gas storage [397]. After chemical treatment in order to open the pores and following elimination of oxygen functionalities on the surface, SWNHs are subjected to the repetition of compression at 50 MPa and crushing in order to increase the bulk density. Through 9 times repetition of this mechanical cycle, SWNHs are strongly compressed to a bulk density of 0.97 g/cm$^3$ and their pore structure is a little modified; the original SWNH having surface area of 1030 m$^2$/g and micropore volume of 0.55 cm$^3$/g, and the compressed one having 1097 m$^2$/g and 0.55 cm$^3$/g, respectively. On this compressed SWNHs, methane adsorption uptake at 303 K reaches 160 cm$^3$/cm$^3$ under 3.5 MPa, higher than the requirement from U.S. Department of Energy, i.e., 150 cm$^3$/cm$^3$ (Section 3.9.4).

SWNH assemblies are oxidized in oxygen, donating nano-sized windows on the wall, of which the size and concentration are controlled by the oxidation temperature [398]. The molecular sieving effect has been proved by the adsorption capacity of oxidized SWNHs for probe molecules with various diameters: He, Ar, $N_2$, $CH_4$, $SF_6$, and $C_{60}$, which have diameters of 0.26, 0.335, 0343, 0.372, 0.525, and 0.92 nm, respectively. A spherical aggregate of about 0.1 μm in diameter can support various nanoparticles after oxidation [399]. Fig. 3.154 shows an aggregate

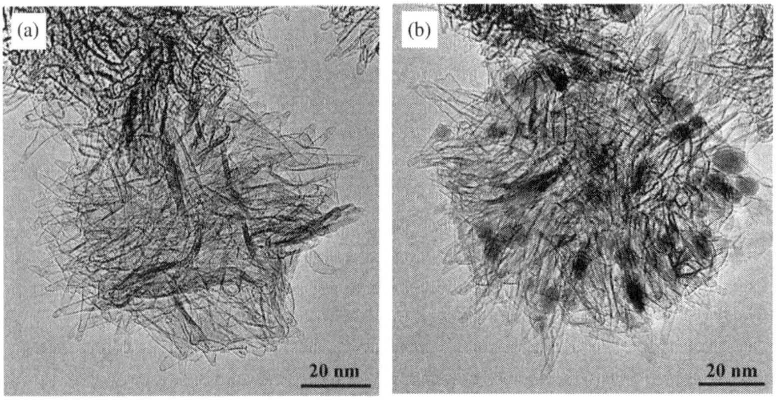


FIGURE 3.154

TEM images of an aggregate of SWNHs [399]. (a) After oxidized and (b) after loading of $Fe_2O_3$ nanoparticles.

of SWNHs loaded $Fe_2O_3$ nanoparticles after oxidation to open windows. In vitro experiments have demonstrated that SWNHs are also potential drug carriers.

## 3.5.6 Helical carbon films

Helical carbon nanofibrils have been synthesized from iodine-doped polyacetylene by carbonization at 800°C [400]. Iodine-doping is effective to keep the helical structure of the precursor polyacetylene. TEM images and scheme of helical structure after the heat treatment at 2600°C are shown in Fig. 3.155. After 2600°C treatment, helical nanofibril shows sharp 002 diffraction profile of XRD and sharp G- and D-bands in Raman spectrum.

Thin carbon films consisting of helical carbon fibrils have been prepared from helical polyacetylene film, in which bundles of nanofibrils with diameters less than 100 nm are twisted and concentrically curled by carbonization at 800°C. By doping of iodine with I/C atomic ratio of 0.26–0.30 before carbonization, characteristic morphology of polyacetylene film, nanofibril and concentrically curling of nanofibril bundles, are reserved after carbonization, as shown in Fig. 3.156 [401]. The yield of carbon films after carbonization at 800°C is 71–90 mass% of the weight of the pristine polyacetylene film before iodine doping, and the resultant films have the hydrogen content of about 1.4 mass% and bulk density of 0.59 $g/cm^3$. After heat treatment at 2600°C, the films keep their hierarchical morphology and bulk density becomes about 0.39 $g/cm^3$. The carbon film prepared without iodine-doping preserves neither its fibrous structure nor its spiral morphology after the carbonization at 800°C.

Helical structure of the fibrils and spiral structure in the film can be controlled in the pristine polyacetylene films and kept until carbon films after high

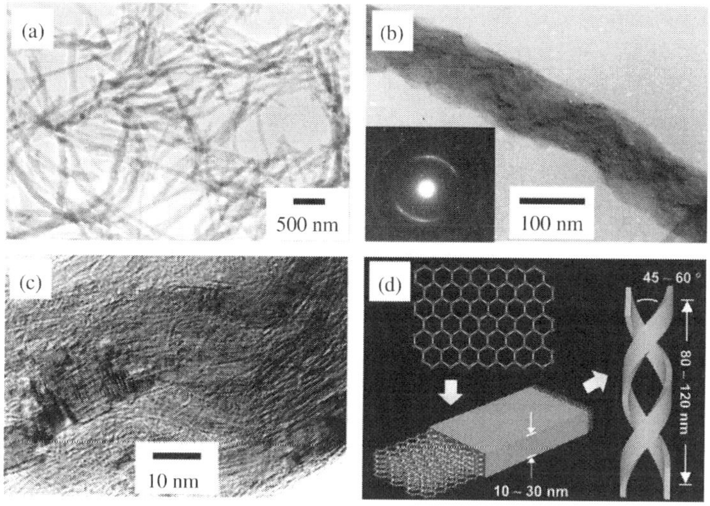


**FIGURE 3.155**

TEM images of helical carbon nanofibril prepared at 800°C after the heat treatment at 2600°C [400]. (a) TEM image of dispersed nanofibrils, (b) TEM image of a single nanofibril with electron diffraction pattern (inset), (c) high-resolution TEM image of a nanofibril, and (d) illustration of helical structure of nanofibril.

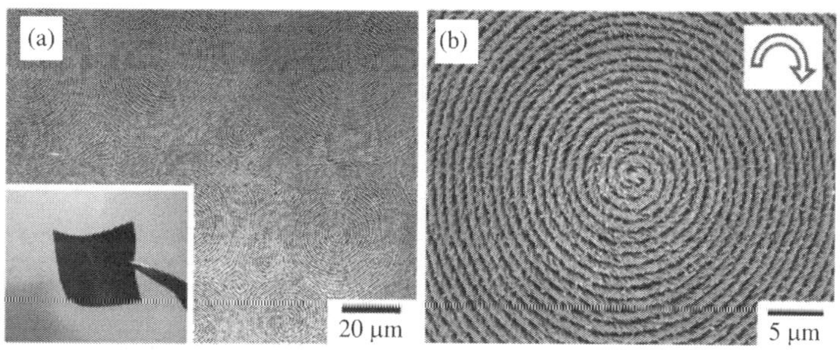


**FIGURE 3.156**

SEM images of helical carbon film [401]. (a) Helical carbon film showing multidomain spiral morphology and flexible freestanding carbon film (inset), and (b) right-handed spiral morphology.

temperature treatment up to 2600°C [401]. Left- and right-handed spiral structures of the carbon fibrils are shown in Fig. 3.157. The films, of which helical fibrils are aligned either vertically or horizontally, are also prepared.

The preparation and properties of pristine polyacetylene films and carbon films consisting of helical fibrils are reviewed in detail [402].

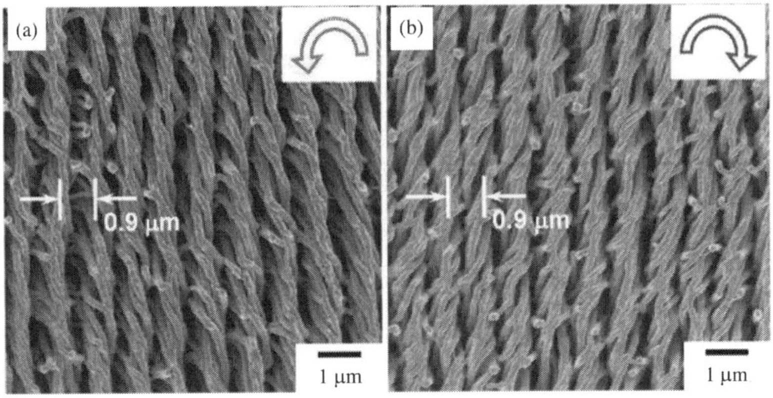


**FIGURE 3.157**

Helical structure of carbon fibrils [401]. (a) Left- and (b) right-handed helical structures.

## 3.6 Porous carbons

### 3.6.1 Activated carbons

The history of activated carbons goes back to the prehistoric era, when charcoals are known to be used for the purification of water and as a medicine to stop diarrhea. Granular activated carbons are now prepared from different precursors and used in a wide range of industries. Their preparation, structure and applications have been reviewed in different books [403–406]. In Table 3.25, some properties of different adsorbents, activated carbon, silica gel, alumina gel and zeolite, are compared with each other. High surface area and light weight are the advantages of activated carbon. Activated carbons with much higher surface area, as high as 3000 $m^2/g$, are now commercialized. Usually activated carbons have a wide range of pore size distribution from micropore to macropore, which shows a marked contrast to a definite pore size of zeolite.

Different pore sizes in carbon materials are required in their applications. Therefore, pore-size distribution in carbon materials has to be controlled during their preparation by the selection of precursor, process and condition of carbonization, and also those of activation. Rather a broad distribution in pore size and shape is usually obtained in carbon materials. In order to compete in the adsorption performance with porous inorganic materials, such as silica gels and zeolites, and to use the advantages of carbon materials, such as high chemical stability, high-temperature resistance and low weight, the control of pore size is essential. From the applications for modern technology fields, not only high surface area and a large pore volume but also the sharp pore-size distribution at a definite size and the surface nature of the pore walls are strongly required to be controlled. In order to control the pore structure in carbon materials, studies on the selection of

**Table 3.25** Properties of Adsorbents Produced in Industries

| | Activated Carbon | | Silica Gel | Alumina gel | Zeolite |
|---|---|---|---|---|---|
| | Granular | Powder | | | |
| Bulk density (g/cm$^3$) | 0.6–1.0 | | 0.8–1.3 | 0.9–1.9 | 0.9–1.3 |
| Porosity (--) | 0.33–0.45 | 0.45–0.75 | 0.4–0.45 | 0.4–0.45 | 0.32–0.4 |
| Pore volume (cm$^3$/g) | 0.5–1.1 | 0.5–1.4 | 0.3–0.8 | 0.3–0.8 | 0.4–0.6 |
| BET surface area (m$^2$/g) | 700–1500 | 700–1600 | 200–600 | 150–350 | 400–750 |

**Table 3.26** Comparison Between ACFs and Granular Activated Carbons

| | ACFs | Granular Activated Carbons |
|---|---|---|
| Size | 10–20 μm in diameter | 1–3 mm |
| BET surface area (m$^2$/g) | 700–2500 | 900–1200 |
| Physical surface area (m$^2$/g) | 0.2–2.0 | ~0.001 |
| Average size of pores (nm) | <40 | From micropores to macropores |

precursors and preparation conditions have been extensively carried out and certain successes have been achieved [403–408].

Pore sizes and their distributions in activated carbons have to be fit with the requirements from the applications. Relatively small pores are needed for gas adsorption and large pores for liquid adsorption, and very narrow pore-size distribution is required for molecular sieving applications, as shown in Fig. 2.43. Macropores in carbon materials for sorption of viscous heavy oils are effective. Porous carbons can respond to these widely ranged requirements from the applications, which is one of the advantages of carbon materials even though pore sizes distribute in certain range in most porous carbons. Novel techniques to control pore structure in carbon materials without activation process may solve this problem, as explained in the following.

## a. Activated carbon fibers

Fibrous activated carbons, activated carbon fibers (ACFs), have been prepared and developed a new field of applications. ACFs have a number of advantages over granular activated carbons. The principal merit to prepare in fibrous morphology is its pore structure and a large physical surface area. Their pore structure is schematically shown in Fig. 2.42, by comparing with that of granular activated carbon, and the differences between ACFs and granular activated carbons are listed in Table 3.26.

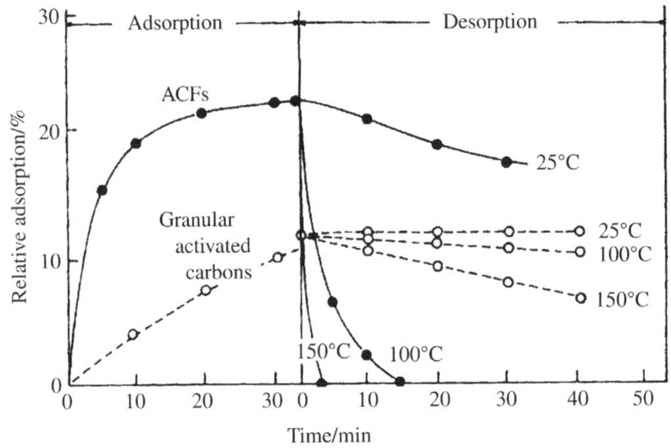


**FIGURE 3.158**

Adsorption and desorption performance of ACFs and granular activated carbons for toluene vapor.

*(Courtesy of Dr. Shimada.)*

Granular activated carbons have different sizes of pores from micropores to macropores, whereas ACFs have mostly micropores on their surfaces. In granular activated carbons, adsorbates have to reach micropores by passing through macropores and mesopores, whereas in ACFs they can directly reach most micropores because micropores are exposed directly to adsorbates. Therefore, the adsorption rate, as well as the amount of adsorption, of gases into ACFs is much higher than those into granular activated carbons. In Fig. 3.158, the adsorption and desorption behaviors of toluene vapor are compared on these two carbons. The amount of adsorption of toluene molecules is much higher and desorption proceeds much faster in ACFs than granular activated carbons. Desorption of gas was markedly accelerated by raising the temperature on ACFs.

In Table 3.27, characteristics of commercially available ACFs prepared from different precursors are compared. A very high surface area determined by Brunauer-Emmett-Teller method (BET surface area, $S_{BET}$) up to 2500 $m^2/g$ and a high micropore volume up to 1.6 mL/g can be obtained in isotropic-pitch-based carbon fibers. For the preparation of these carbon fibers with a very high surface area, such as 2500 $m^2/g$, the precursors, which give the carbon with poor crystallinity, are recommended; mesophase-pitch-based carbon fibers did not give a high surface area, whereas isotropic-pitch-based carbon fibers did [409].

Another advantage of ACFs is the ability to prepare woven clothes and non-woven mats, which developed new applications in small purification systems for city water and also as a deodorant in refrigerators in private houses, for example. Effective elimination of $SO_x$ from exhausted gases by oxidation to either sulfuric acid or sulfates is possible on the surface of ACFs [410,411]. In order to give the

**Table 3.27** Characteristics of Commercially Available Activated Carbon Fibers

| Precursor | Isotropic Pitch | PAN | Phenol | Cellulose |
|---|---|---|---|---|
| Fiber diameter (μm) | 10–18 | 7–15 | 9–11 | 15–19 |
| Physical surface area ($m^2/g$) | 0.2–0.6 | 0.9–2.0 | 1.0–1.2 | 0.2–0.7 |
| $S_{BET}$ ($m^2/g$) | 700–2500 | 500–1500 | 900–2500 | 500–1500 |
| Micropore volume (mL/g) | 0.3–1.6 | – | 0.22–1.2 | – |
| Tensile strength (MPa) | 100–200 | 200–370 | 300–400 | 60–100 |
| Elastic modulus (GPa) | 2–12 | 70–80 | 10–15 | 10–20 |
| Tensile strain (%) | 1.0–2.8 | ~2.0 | 2.7–2.8 | – |
| Adsorption of benzene (%) | 22–68 | 17–50 | 22–90 | 30–58 |
| Iodine adsorption (mg/g) | 900–2200 | – | 950–2400 | – |

fibers an antibacterial function and to increase their deodorant function, some trials on supporting minute particles of different metals, such as Ag, Cu and Mn, are reported [412,413].

## b. Molecular sieving carbons

Molecular sieving carbons (MSCs) have a smaller pore size with a sharper distribution in a range of micropores, as shown in Fig. 2.40, in comparison with other activated carbons for gas and liquid phase adsorbates. They have been used for adsorbing and eliminating unnecessary species with a very low concentration, ethylene gas adsorption to keep fruits and vegetables fresh, filtering of hazardous gases in power plants, etc.

An important application of these MSCs is developed in gas separation systems. The adsorption rate of gas molecules, such as nitrogen, oxygen, hydrogen and ethylene, depends strongly on the pore size of MSC; the adsorption rate of a gas becomes the slower for MSC with the smaller pore size, and the temperature also governs the rate of adsorption of a gas because of the activated diffusion of adsorbate molecules in micropores, the higher the temperature the faster the adsorption. Gas separation is performed by controlling (swinging) either temperature or pressure of adsorbate gas, temperature swing adsorption (TSA) and pressure swing adsorption (PSA).

In Figure 3.159a, time dependence of relative adsorption for oxygen and nitrogen is shown for a column of MSCs. Adsorption of oxygen into the MSC completes within 5 min, but nitrogen is adsorbed very slowly, less than 10% of equilibrium adsorption even after 15 min. From the column of MSC, therefore, nitrogen-rich gas comes out on the adsorption process, and oxygen-rich gas is obtained on the desorption process. By using more than two columns of MSCs and the repetition of these adsorption/desorption processes, nitrogen gas is

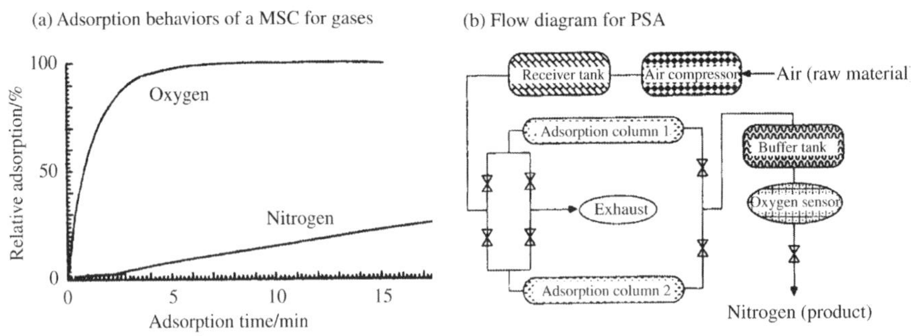


**FIGURE 3.159**

PSA method for the separation of nitrogen gas from air using MSCs.

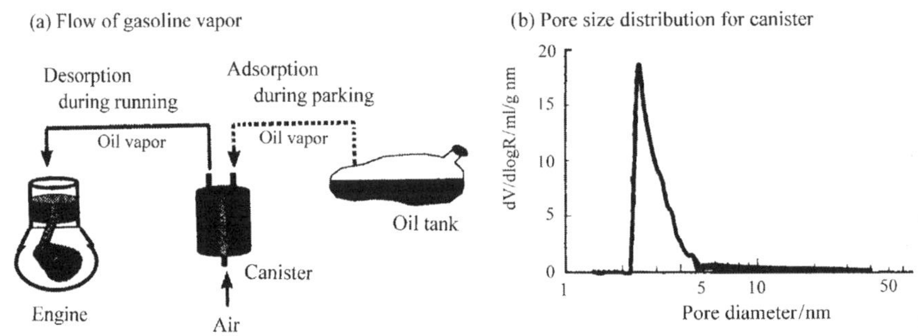


**FIGURE 3.160**

Application of activated carbon for car canister.

isolated from oxygen. A flow diagram of PSA for the separation of nitrogen gas from air is shown in Fig. 3.159b. This swing adsorption method, practically PSA, for gas separation has the advantages of low energy cost, room temperature operation, compact equipment, etc.

### c. Mesoporous carbons for car canisters

Activated carbon has been used for adsorbing gasoline vapor in the tank of a car during parking and for using it during running. The flow of gasoline vapor in a car is illustrated in Fig. 3.160a. Even during parking, gasoline vaporizes, particularly in hot weather. Some statistics show that running time of a car is usually very short in comparison with parking time. The gasoline vapor formed during parking has to be collected in a canister consisting of activated carbons, in order to save gasoline and to avoid contamination of the air. The adsorbed gasoline is desorbed during running by passing air through the canister and transferred to the engine.

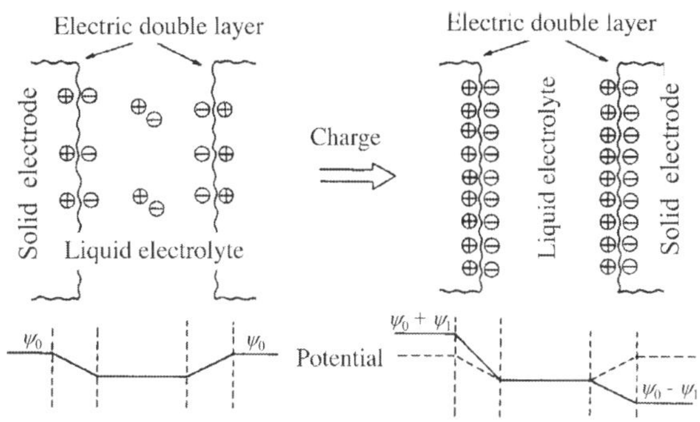


**FIGURE 3.161**

Fundamental concept for the electric double-layer capacitor.

For this purpose, porous carbons have to have rather large pores, as shown in Fig. 3.160b, compared with activated carbons for general purposes, because gasoline vapors are mostly adsorbed as liquid due to their capillary condensation and their surface tension in the pores.

### d. Porous carbons for electric double-layer capacitors

By using porous carbon materials with a very high surface area in both positive and negative electrodes, a large amount of electrical charge can be stored owing to the formation of electric double-layers on the surface of electrodes (electric double-layer capacitors, EDLCs) [414,415]. The fundamental concept of this capacitor is the formation of electric double-layers on the surface of electrodes, as illustrated in Fig. 3.161. The total amount of electric charges aligned in double layers on both electrodes increases by the application of potential difference and it is easily understood to depend on the area of this interface, i.e. the surface area of solid electrodes. The large surface area of activated carbons, as well as their electrical conductance, is one of the merits for the electrode material of EDLCs. Particularly by using ACF cloths prepared from phenol resin and organic electrolytes such as propylenecarbonate with tetraethyl-ammonium perchlorate, small capacitors with a coin shape are realized, with a high capacitance, high working voltage and high reliability. In Fig. 3.162a, capacitance per gram of carbon electrode in a capacitor is plotted against specific surface area of the ACFs used [415]. At room temperature, a linear relation is observed, revealing that a high surface area is desired for high capacitance.

The discharge behavior of a capacitor using ACFs is shown in Fig. 3.162b; an extremely long time discharge is possible if the discharge current is small enough, smaller than 10 μA, but the discharge voltage decreases quickly with time if the current is larger than 100 μA. This discharge behavior of the capacitors with high capacitance is not affected by ambient temperature in a relatively wide range, from 25 to +85°C. Even though the discharge current is limited to a small value,

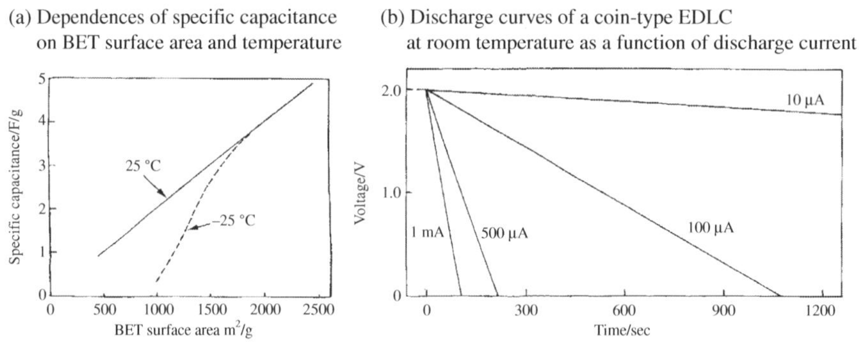


**FIGURE 3.162**

Performance of electric double-layer capacitor using ACFs.

*(Courtesy of Dr. A. Nishino)*

certain advantages of EDLCs over secondary batteries, such as Ni—Cd and lithium-ion rechargeable batteries, can be pointed out: the possibility of semi-permanent repetition of charge—discharge cycles, very short charging time and very simple circuit for charging.

Based on these characteristics, EDLCs are firstly used together with small batteries for back-up source of memory in computers. In particular, its use together with solar batteries expands markedly the application possibility of solar batteries due to the increase in the stability of the discharge voltage and the life. The applications of EDLCs to electric vehicles have now been extensively studied. To store the energy, which is dissipated during breaking of heavy tracks and buses, EDLCs are successfully used [416]. EDLCs are discussed in more details by paying special attention to pore structure of electrode carbons in Section 3.9.2.

## 3.6.2 Novel techniques to control pore structure

Various novel techniques have been developed to control the pore structure of carbon materials without any activation process. Since the pore structure is governed by nanotexture of carbon materials, its control has to be done during carbonization.

### a. Microporous carbons using zeolite template

Microporous carbons with surface area higher than 2000 $m^2/g$ have been prepared through the carbonization of a carbon precursor in nanochannels of zeolite [417—419]. The preparation procedure is the so-called template carbonization technique (Section 2.4.1), using three-dimensional channels of zeolites, of which size and shape are strictly defined by their crystal structures. Introduction of acrylonitrile into the channels of zeolite has been carried out through its vapor, followed by $\gamma$-irradiation for polymerization and by carbonization at 700°C. By impregnation of furfuryl alcohol in liquid phase and chemical vapor deposition

**FIGURE 3.163**

High resolution TEM images with selected-area diffraction patterns of zeolite Y and carbon prepared from furfuryl alcohol using zeolite Y as a template.

*(Courtesy of Prof. T. Kyotani of Tohoku Univ.)*

(CVD) from propylene gas, the similar microporous carbons are also prepared. The template zeolite is dissolved out by HF and HCl solutions. High-resolution transmission electron microscopy (TEM) image shows that regular alignment of super cages with the size of 1.4 nm in template zeolite is inherited in the resultant carbon as the periodicity of about 1.3 nm, as shown in Fig. 3.163. Most carbons give a diffraction peak at an angle of around 6° in $2\theta$ of CuK$\alpha$ X-rays, as the template zeolite shows, which corresponds to the periodicity of about 1.3–1.4 nm due to superstructure.

The carbons thus prepared are highly microporous, even though any activation is not included in the preparation process. In Table 3.28, some of the data on pore structure parameters are listed. The carbon prepared by carbonization of poly (furfuryl alcohol) with the following CVD of propylene at 700°C and then heat-treated at 900°C shows a special pore structure; exceptionally high $S_{BET}$ of 3600 $m^2/g$ and a large amount of micropores ($V_{micro}$) of about 1.5 mL/g, but relatively low mesopore volume ($V_{meso}$). The very high surface area indicates the presence of the curved carbon surfaces, which has been predicted by Grand Canonical Monte Carlo (GCMC) simulation [420].

### b. Porous carbons by controlling carbonization process

Carbon films with molecular sieving performance have successfully been developed through detailed studies on carbonization of polyimide films [105,421–423]. In Fig. 3.164, adsorbed amounts of gas molecules with different sizes are shown on the carbon films heat-treated at different temperatures, ethane $C_2H_6$ with the size of 0.40 nm permeating in a large amount through the film heat-treated at 700°C, but a very small amount through the one at 1000°C. This result suggests a strong dependence of pore sizes in carbon films on carbonization

**Table 3.28** Pore Structure Parameters of the Carbons Prepared by Zeolite Templates

| Templating Conditions | | | Pore Structure Parameters of Templated Carbon | | |
|---|---|---|---|---|---|
| **Carbon Precursor** | **Template Zeolite** | **Carbonization Conditions** | **$S_{BET}$ ($m^2/g$)** | **$V_{micro}$ ($cm^3/g$)** | **$V_{meso}$ ($cm^3/g$)** |
| PAN | NaY | 700°C, 3 h | 580 | 0.28 | 0.10 |
| PFA | NaY | 700°C, 3 h | 590 | 0.28 | 0.15 |
| | | 700°C 12 h | 1660 | 0.66 | 0.79 |
| | | 700°C, 18 h | 2260 | 1.11 | 0.76 |
| Propylene | USY | 800°C, 12 h | 2060 | 0.82 | 0.75 |
| | | 800°C, 15 h | 2200 | 0.88 | 0.83 |
| | | 800°C, 18 h | 1790 | 0.72 | 0.62 |
| PFA + Propylene | Y | 700°C, 4 h | 2170 | 0.9 | 0.4 |
| | | 700°C, 4 h +900°C 3 h | 3600 | 1.5 | 0.0 |

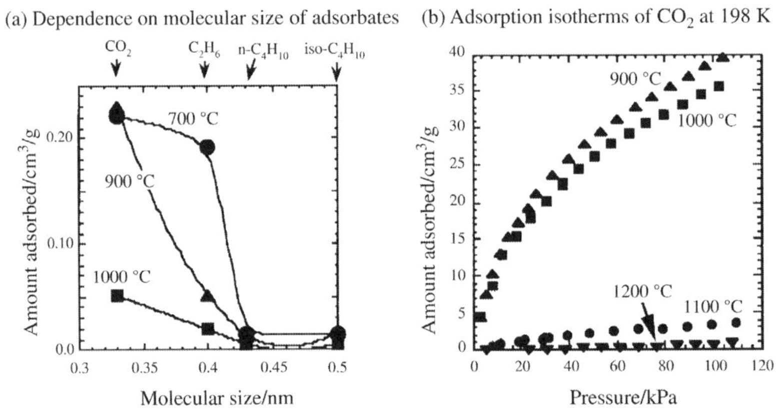


**FIGURE 3.164**

Adsorption behavior of polyimide carbon films with different heat treatment temperatures.

*(Courtesy of Dr. H. Hatori of AIST, Japan)*

temperature, which is supposed to be due to shrinkage during carbonization. A marked development of micropores below 0.4–0.5 nm size is observed on polyimide-derived carbon films. On these carbon films, adsorption of $CO_2$ molecules is measured (Fig. 3.164b): the films prepared at a temperature below 1000°C can adsorb a large amount of $CO_2$, but those above 1100°C no adsorption of $CO_2$, showing that pores become smaller than the size of $CO_2$ molecules, 0.33 nm above 1100°C.

**Table 3.29** Gas Permeability of Hydrogen and the Permselectivity for Hydrogen Against CO and $CO_2$ at 323 K

| Heat Treatment Conditions | $H_2$ Permeability $P_{H2}$ (mol/msPa) | Selectivity | |
|---|---|---|---|
| | | $P_{H2}/P_{CO}$ | $P_{H2}/P_{CO2}$ |
| at 1000°C for 20 min | $2.40 \times 10^{-15}$ | 5900 | 161 |
| at 1100°C for 20 min | $1.06 \times 10^{-16}$ | – | 343 |

On carbon films prepared from commercially available polyimide films with the thickness of 0.1 mm, selective permeation of hydrogen gas is found [423], permeability for $H_2$ molecules and selectivity against CO and $CO_2$ being listed in Table 3.29. The value of selectivity $P_{H2}/P_{CO}$ of 5900 observed on the carbon film heat-treated at 1000°C indicates that the permeability of $H_2$ gas is 5900 times larger than that of CO gas, in other words, the CO content of 1% in the $H_2$ gas can reduce to 2 ppm after passing through the film. This permselectivity of these carbon films may develop the application in fuel cells for vehicles, since CO in $H_2$ gas supplied from an on-board reformer has to be removed through molecular sieving film.

Carbon films prepared from different precursors, including polyimides, give high permselectivity for different gas mixtures, $H_2/N_2$, $He/N_2$, $CO_2/N_2$, $O_2/N_2$, $H_2/N_2$, $CO_2/CH_4$, $C_2H_4/C_2H_6$, etc. [105,423–433]; permselectivity of carbon film reaches 4700 for $H_2/N_2$, 2800 for $He/N_2$, 122 for $CO_2/N_2$, and 36 for $O_2/N_2$ at 308 K [430]. On a film of polyimide, permeability for $O_2$ increases by carbonization up to 535°C and then permselectivity increases markedly with increasing carbonization temperature up to 800°C, though permeability decreases slightly [433]. The separation properties of carbon films for $O_2/N_2$ are much better than most polymer films, well above the so-called 'upper-bound trade-off line' in the relation between permselectivity and permeability.

Porous carbons having hierarchical pore structure (consisting of macro-, meso- and micropores) are prepared by carbonization of mono-disperse nanospheres of styrene-divinylbenzene (ST-DVB) copolymers, which are cross-linked by intra- and inter-sphere —CO—bonds [434,435]. The formation of hierarchical porous carbon is schematically illustrated in Fig. 3.165.

## c. Mesoporous carbons through defluorination of PTFE

Mesoporous carbon with high surface area has been prepared through defluorination of poly(tetrafluoroethylene) (PTFE) using lithium amalgamate [436,437], the reaction being also used to synthesize one-dimensional carbyne structure [438]. PTFE film with 100 μm thickness is pressed with Li metal foil with 200 μm thickness under 4 MPa in Ar atmosphere to be defluorinated, followed by washing with methanol to remove excess Li metal, heat-treating at 700°C and then

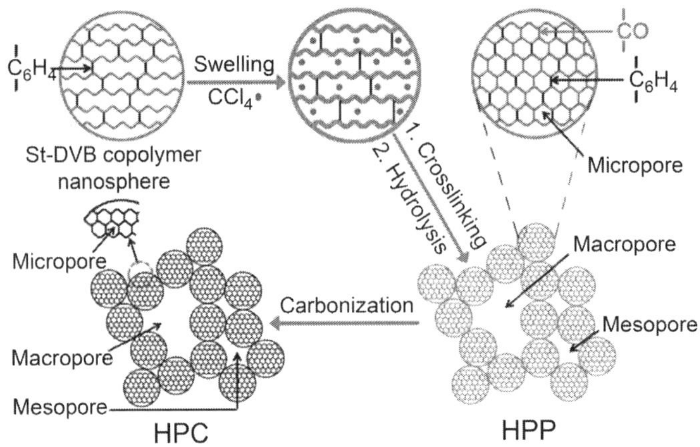


**FIGURE 3.165**

Schematic illustration of the formation of hierarchical porous carbon from St-DVB.

washing with dilute HCl to eliminate LiF [439]. The carbon thus prepared from PTFE has a large content of mesopores [439], and give high EDLC capacitance [439–444], as 200 F/g in 1 mol/L $H_2SO_4$ electrolyte.

Defluorination of PTFE is also possible through heating a mixture of PTFE powders with alkali metals, Na, K and Rb, in vacuum at 200°C in a closed vessel [440–443]. Nitrogen adsorption isotherm and pore size distribution of resultant carbon depend strongly on whether γ-rays are irradiated and also on alkali metal used, as shown in Fig. 3.166a and b, respectively. Defluorination of PTFE with Na metal gives mesopore-rich carbon and very high $S_{BET}$ as 2225 $m^2/g$ [442]. $S_{BET}$ of these carbons prepared using Na increases with heat treatment at a high temperature to 1000°C, probably due to gasification of carbon with surface oxygen functional groups. Heat treatment of these carbons was found to be desirable for giving high EDLC capacitance, a high capacitance of 240 F/g being observed on a 800°C-treated one. Deflurination of PTFE using Na metal not only gives mesoporous carbons with high $S_{BET}$ but also makes the preparation process simple, Na metals being much cheaper and easier to handle than other alkali metals. γ-irradiation of PTFE prior to its defluorination helps the formation of mesopores in resultant carbons [440,444], as shown in Fig. 3.166a.

Representative values of $S_{BET}$ and EDLC capacitance in 1 mol/L $H_2SO_4$ obtained on the carbons derived from PTFE are summarized in Table 3.30, together with defluorination reagent and additional treatment.

The idea to dehalogenize by using alkali metal to form mesopores has been applied to carboxymethylcellulose sodium (CMC-Na); CMC-Na is treated with iodine vapor at 115°C and carbonized at 600°C [445]. Microporous carbon with $S_{BET}$ of 1070 $m^2/g$ is obtained, nanoparticles of NaI formed during iodination working as a template for micropores.

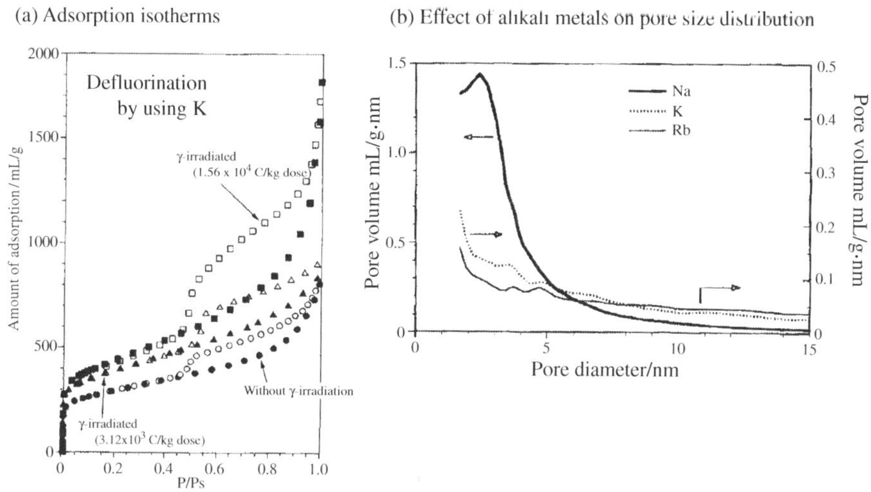


**FIGURE 3.166**

Pore structure of carbons derived from PTFE using different alkali metals at 200°C.

*(Courtesy of Prof. Y. Yamada of Fukui Univ.)*

**Table 3.30** $S_{BET}$ and EDLC Capacitance of Carbons Derived from PTFE with Different Defluorination Reagents

| Defluorination Reagent | Additional Treatment | $S_{BET}$ ($m^2/g$) | EDLC Capacitance (F/g) |
|---|---|---|---|
| Li metal | None | 1045 | 200 |
| K metal | None | 999 | 229 |
| K metal | γ-ray irradiation | 1516 | 237 |
| Na metal | 800°C-heated | 2764 | 240 |
| | 1000°C-heated | 2860 | 200 |

## d. Carbide-derived microporous carbons

Microporous carbons have been prepared by chlorination of metal carbides [446–450], pore size being controllable by metal carbide and its chlorination temperature. In Fig. 3.167, changes in average pore size, $S_{BET}$ and capacitance of EDLC with 1 mol/L $H_2SO_4$ with chlorination temperature are shown [449]. Pore size and $S_{BET}$ depend strongly on starting carbide and also chlorination temperature, $Tl_2AlC$-derived carbon gives a maximum $S_{BET}$ at 1000°C, and $B_4C$-derived one at 800°C. Capacitance changes with chlorination temperature almost in parallel to $S_{BET}$. Pore size distribution is compared for microporous carbons derived from different carbides in Fig. 3.168 [448], abscissa (pore size) being in logarithmic scale.

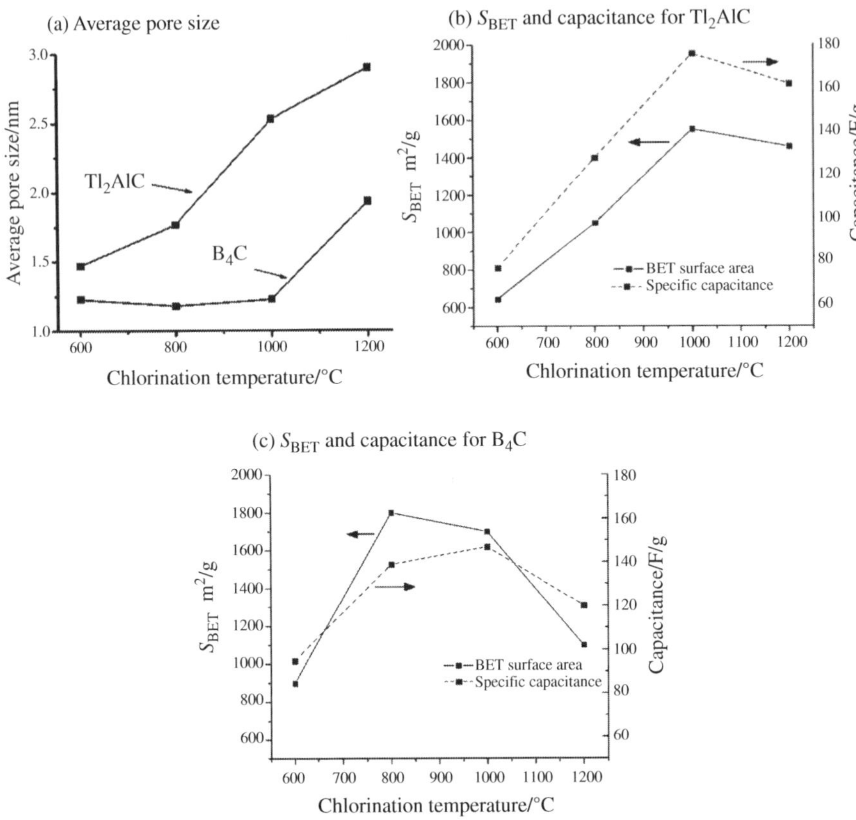


**FIGURE 3.167**

Changes in pore size, surface area and capacitance with chlorination temperature [449].

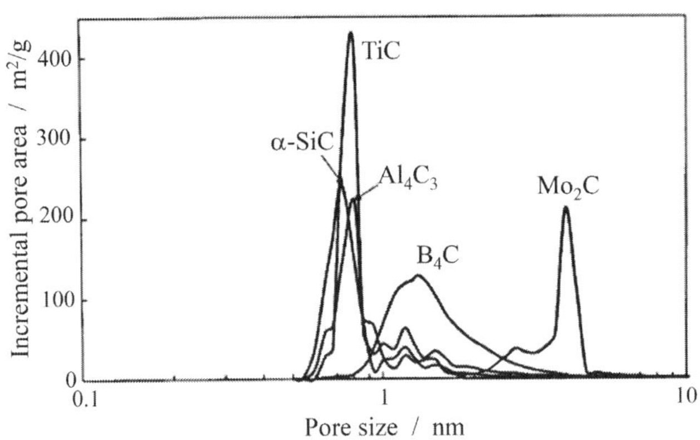


**FIGURE 3.168**

Pore size distribution of microporous carbons derived from different carbide [448].

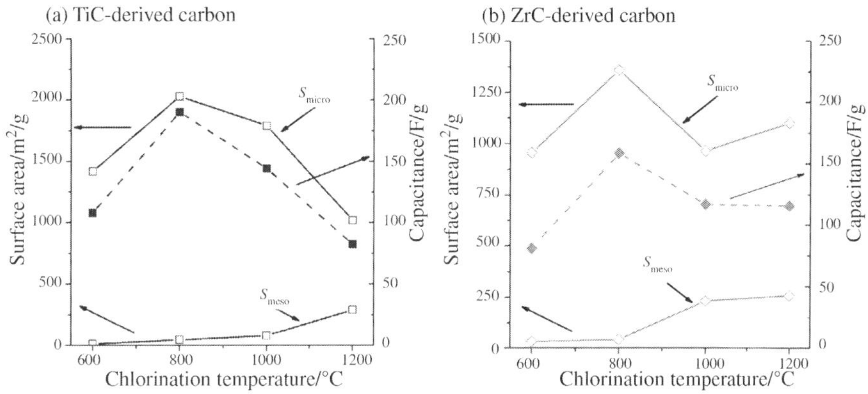


**FIGURE 3.169**

Capacitance determined from CV curve with 5 mV/s scan rate in 1 mol/L $H_2SO_4$ aqueous electrolyte and surface areas, $S_{micro}$ and $S_{meso}$, as a function of chlorination temperature [455].

Carbide-derived carbons have been tested as electrode materials in electric double-layer capacitors (EDLCs) [449,451–459]. In Fig. 3.169, changes in EDLC capacitance and surface area, $S_{micro}$ and $S_{meso}$, with chlorination temperature are shown for TiC- and ZrC-derived carbons [445]. Capacitance changes in parallel with $S_{micro}$ in both carbide-derived carbons. Via electrospinning of THF solution of polycarbomethylsilane and following pyrolysis at 800–900°C, fibrous SiC is prepared, which can be converted to porous carbon fibers through chlorination [451]. The resultant carbon fibers have very high $S_{BET}$ and pore volume; the carbon prepared by pyrolysis at 800 and 900°C followed by chlorination at 850°C gives $S_{BET}$ of 2700 and 3100 $m^2/g$ and pore volume of 1.03 and 1.66 $cm^3/g$, respectively. EDLC characteristics for the carbide-derived nanoporous carbon prepared from TiC, $\alpha$-SiC, $Mo_2C$, $Al_4C_3$, and $B_4C$ have been studied in various non-aqueous electrolyte solutions, 1 mol/L various tetraalkyl ammonium tetrafluoroborate, as $(C_2H_5)_3CH_3NBF_4$, $C_2H_5(CH_3)_3NBF_4$, $(C_2H_5)_3C_3H_7NBF_4$, and $(C_2H_5)_3C_4H_9NBF_4$, with various solvent, as acetonitrile, $\gamma$-butyrolactone, and propylene carbonate [456].

Carbide-derived microporous carbons have been studied for hydrogen storage [454–457]. They have high hydrogen uptake at 77 K, higher than MOF-5, SWCNT and MWCNT [460,461]. Annealing of as-prepared carbide-derived carbon is effective to increase the $H_2$ uptake. Chlorination of ordered mesoporous SiC prepared by templating by using triblock surfactant P123 gives nanoporous carbon consisting of cubic ordered mesopores with microporous carbon wall, which has high methane uptake and good EDLC performance [462].

### e. Carbon aerogels

Mesoporous carbon aerogels are prepared from the pyrolysis of organic aerogels of resorcinol and formaldehyde [463–467]. In Fig. 3.170a, a TEM image of a carbon aerogel with an apparent density of 0.4 g/cm$^3$ is shown. Primary carbon particles have the size of about 4–9 nm and form a network. A typical adsorption isotherm for carbon aerogels is shown in Fig. 3.170b, which belongs to type IV and has a clear hysteresis [468]. Pore structure parameters calculated through $\alpha_s$ plot are listed on three carbon aerogels prepared from resorcinol and formaldehyde at 1000°C in Table 3.31 [468]. These carbon aerogels contain predominantly mesopores, which are formed in a three-dimensional network of interconnected minute carbon particles and only a small amount of micropores are formed in primary carbon particles.

Carbon aerogels can be activated to increase $V_{micro}$ by carbon dioxide at 900°C [469]. The activation increases both micropores and mesopores: $V_{micro}$ of

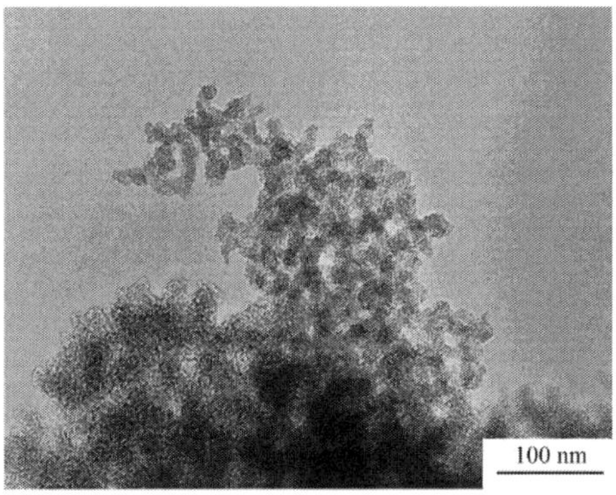


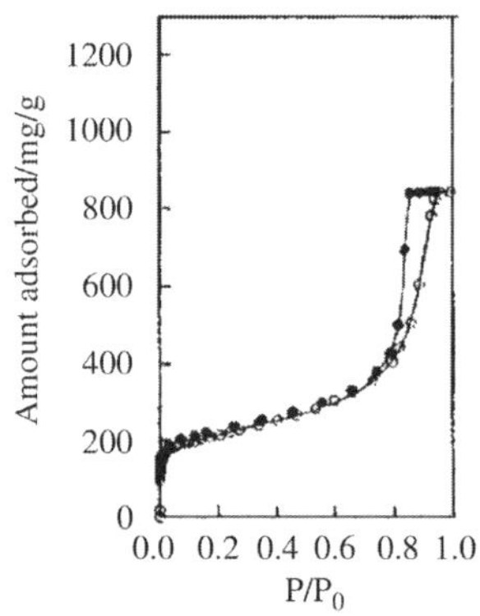


**FIGURE 3.170**

TEM image and adsorption/desorption isotherm at 77 K for a carbon aerogel with an apparent density of 0.4 g/cm$^3$.

*(Courtesy of Prof. K. Kaneko of Chiba Univ.)*

**Table 3.31** Pore Structure Parameters of Carbon Aerogels

| Heat Treatment Temperature (°C) | Apparent Density (g/cm$^3$) | $S_{total}$ (m$^2$/g) | $V_{total}$ (cm$^3$/g) | $S_{micro}$ (m$^2$/g) | $V_{micro}$ (cm$^3$/g) | $S_{meso}$ (m$^2$/g) | $V_{meso}$ (cm$^3$/g) |
|---|---|---|---|---|---|---|---|
| 1000 | 0.58 | 653 | 1.04 | 309 | 0.11 | 344 | 0.93 |
| | 0.43 | 577 | 1.51 | 222 | 0.08 | 355 | 1.43 |
| | 0.30 | 424 | 0.82 | 100 | 0.04 | 324 | 0.78 |

0.68 $cm^3/g$ and $S_{micro}$ of 1750 $m^2/g$, $V_{meso}$ of 2.04 $cm^3/g$ and $S_{meso}$ of 510 $m^2/g$. The detailed studies on adsorption of nitrogen at 77 K and of water vapor at 303 K on activated carbon aerogels, whose surface functional groups are nil, show clearly that the amount of adsorbed water corresponds mainly to $V_{micro}$, not to $V_{meso}$ [470–473]. Addition of Ce and Zr into carbon aerogels results in microporous carbons [474,475].

Changes in pore parameters of carbon aerogels with the heat treatment at high temperatures are shown in Table 3.32 [476]. With increasing HTT, both total surface area and volume decrease, which are mainly due to the decrease in micropores. As a consequence, heat treatment of the carbon aerogel above 2000°C gives a carbon containing only mesopores, mesoporous carbons.

Pore structure of carbon aerogels is governed by that of precursor organic aerogels, which is controlled by the mole ratios of resorcinol to formaldehyde (R/F), to water (R/W) and to basic catalyst of sodium carbonate (R/C) [477–479]. Aqueous gels synthesized are dried under supercritical condition with carbon dioxide. In Fig. 3.171, pore-size distribution in mesopore region is shown

**Table 3.32** Effect of Heat Treatment on Pore Structure Parameters of Carbon Aerogels [476]

| HTT (°C) | $S_{total}$ ($m^2/g$) | $V_{total}$ ($cm^3/g$) | $S_{micro}$ ($m^2/g$) | $V_{micro}$ ($cm^3/g$) | $S_{meso}$ ($m^2/g$) | $V_{meso}$ ($cm^3/g$) |
|---|---|---|---|---|---|---|
| 1000 | 850 | 1.63 | 416 | 0.12 | 361 | 1.51 |
| 1600 | 493 | 1.54 | 1.5 | 0.06 | 333 | 1.48 |
| 2000 | 456 | 1.44 | 0 | 0 | 396 | 1.44 |
| 2400 | 425 | 1.11 | 0 | 0 | 383 | 1.11 |
| 2800 | 325 | 0.76 | 0 | 0 | 285 | 0.76 |

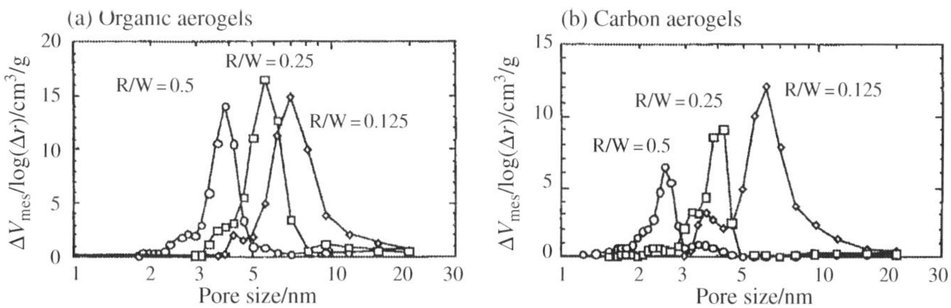


**FIGURE 3.171**

Pore-size distributions of organic and carbon aerogels prepared with different R/W ratios. The factors R/F and R/C are constant at 0.5 and 75, respectively.

*(Courtesy of Prof. K. Kaneko of Chiba Univ.)*

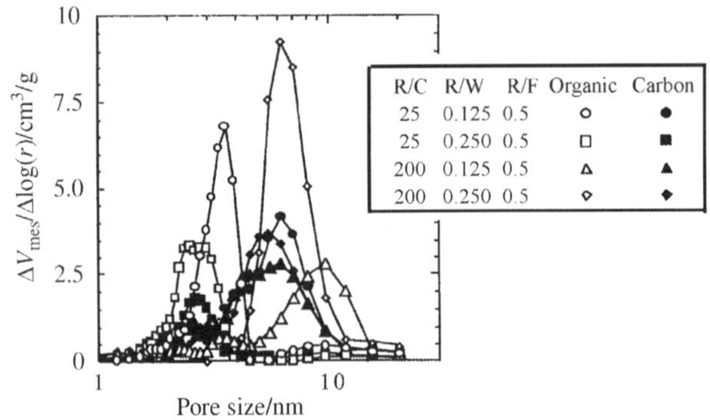


**FIGURE 3.172**

Pore-size distributions of the carbons prepared from cryogels with different R/C, R/W and R/F by the carbonization at 1000°C.

*(Courtesy of Prof. H. Tamon of Kyoto Univ.)*

for both original organic gels and resultant carbon gels as a function of R/W, the other controlling factors R/F and R/C being the constant as 0.5 and 75, respectively. Pore-size distribution of organic aerogels is rather sharp and its maximum decreases with increasing R/W ratio (Fig. 3.171a). By carbonization of these organic aerogels, pore-size distribution shifts to smaller size, mainly due to the shrinkage of gels during thermal decomposition (Fig. 3.171b).

Instead of supercritical drying of aqueous gels, freeze drying method has also been applied [480]. On the gels prepared through freeze drying (cryogels), much smaller shrinkage in pore size during carbonization is observed, as shown in Fig. 3.172. Conditions for the preparation of resorcinol-formaldehyde gels through sol-gel condensation and those for freeze drying are studied in detail in order to control mesoporosity in resultant carbon materials [481–483]. Resorcinol-formaldehyde gels are prepared from its hydrogels by either freeze drying, microwave drying or hot air drying, and converted to carbon gels by heating at 1000°C. To obtain mesoporous carbon, the first two drying methods are effective [484].

Mesopores of carbon aerogels have the possibility for reaction field and thereby carbon aerogels can be applied as the template for production of new materials. Carbon aerogels have been used as a template for the preparation of highly crystalline zeolite (ZSM-5 and Y) with uniform mesoporous channels [485,486].

### f. Mesoporous carbons via template carbonization

Carbons containing a large amount of mesopores with homogeneous size have been prepared by using various templates, mesoporous silicas [487,488], block-copolymers surfactants [489–494] and MgO [495,496].

Ordered mesoporous structures of silicas, which are formed by templating self-assembly of surfactants and named as MCM-48, MCM-41 and SBA-15, etc.,

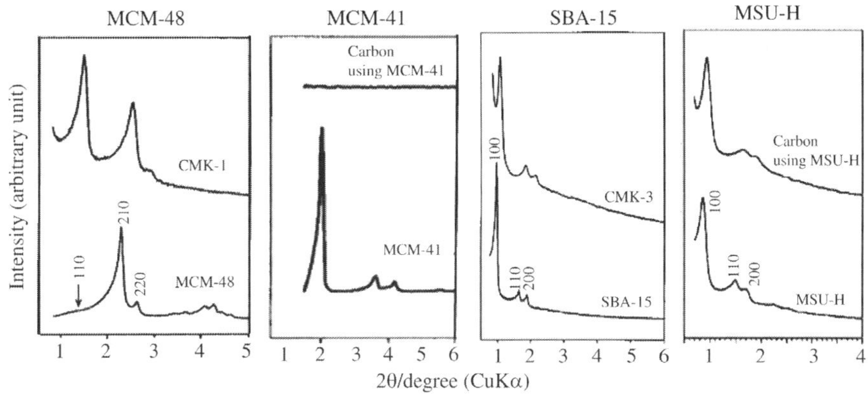


**FIGURE 3.173**

XRD pattern at low diffraction angle region for the template silica and the resultant carbon.

are successfully inherited by impregnation of carbon precursor, followed by carbonization and removal of the templates mostly with HF, to form ordered mesoporous carbons. In Fig. 3.173, XRD pattern at low diffraction angle region is compared for different silica templates and the carbons synthesized by using them as template to show correspondence in pore structure symmetry. Phenol and formaldehyde impregnated under a reduced pressure were polymerized in the mesopores of silica template MCM-48, of which the surface has to be implanted by Al in advance to generate strong acidic sites for the acceleration of polymerization [497]. Into the same template, sucrose is impregnated twice in the presence of sulfuric acid [498]. The composites of MCM-48 with carbon precursors thus prepared are carbonized and then etched with HF to remove the template framework. The resultant carbon is not an exact negative replica of the template, but has a periodically ordered structure, as shown in Fig. 3.168a. When SBA-15 and MSU-H are used as templates, the resultant carbons are exactly inherited in the pore symmetry in the templates. In contrary, MCM-41 template produces the carbon with disordered mesopores. Carbons with large mesopores (>20 nm) interconnected through uniform windows are also prepared through similar procedures [499]. Carbon with ordered mesopores in the same symmetry as the template SBA-15 has been obtained by impregnation of sucrose aqueous solution with sulfuric acid [500], of which a TEM image is shown in Fig. 3.174.

Mesoporous carbons with ordered pore structure were successfully prepared by using some surfactants as templates, applying the solvent evaporation-induced self-assembly (EISA) method to the mixture of a carbon precursor and a surfactant. The key to this process is the direct use of the self-assembly of block copolymer surfactants as template, although the same surfactants are used for the preparation of mesoporous silicas. A mixture of commercially available triblock copolymer, poly(ethylene oxide)-*b*-poly(propylene oxide)-*b*-poly(ethylene oxide)

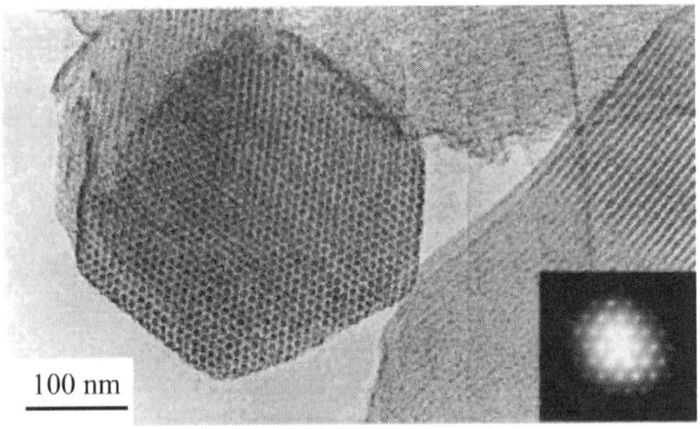


**FIGURE 3.174**

Lattice fringe image and diffraction pattern (inset) for the carbon with ordered mesopores prepared using SBA-15 as a template [500].

(PEO106-PPO70- PEO106, Pluronic F127), resorcinol, triethyl orthoacetate and formaldehyde in a water/ethanol/HCl mixed solvent is spin-coated on a silicon substrate and then carbonized up to 800°C to obtain mesoporous carbon films [489]. The periodic texture of resorcinol-formaldehyde/triethyl orthoacetate (carbon precursors) and F127 is established in the original organic films due to EISA with a spacing of 9.2 nm, and it is retained even after carbonization at 800°C to give mesoporous carbon. Adsorption/desorption isotherms for starting polymer and resultant carbon are shown in Fig. 3.175: both belong to type IV with a pronounced hysteresis and give sharp pore-size distribution in mesopore region [490]. By using another triblock copolymer Pluronic P123 (PEO20-PPO70-PEO20) with resorcinol/phloroglucinol-formaldehyde in aqueous solution, carbons with bicontinuous mesopores are obtained [491]. Different pore structures, 2D hexagonal, 3D bicontinuous, body-centered cubic and lamellar, are obtained in carbon materials by simply adjusting the ratio of phenol to template surfactant [492–494].

By using MgO nanoparticles as template, mesoporous carbons have been prepared by a simple process; mixing MgO precursor with carbon precursor in powder, carbonized at 900°C and then washing by diluted acidic aqueous solution [501–510]. MgO has been selected as a template for the following reasons; it is easily dissolved into non-corrosive diluted acid, it has a high melting point at 2800°C, and stable even in the coexistence of carbon precursors at the carbonization temperature. Various organic compounds, such as poly(vinyl alcohol) PVA, poly(ethylene terephthalate), phenol resins, polyimides and pitches, can be used as carbon precursor to prepare mesoporous carbons. From nitrogen-containing organic compounds, such as poly(vinyl pyrrolidone) nitrogen-doped mesoporous carbons are prepared [511]. Porous carbon nanofibers are prepared from polyacrylonitrile mixed with $MgCl_2$ by electrospinning, followed by stabilization, carbonization and dissolution of MgO [512]. For MgO precursor, magnesium compounds

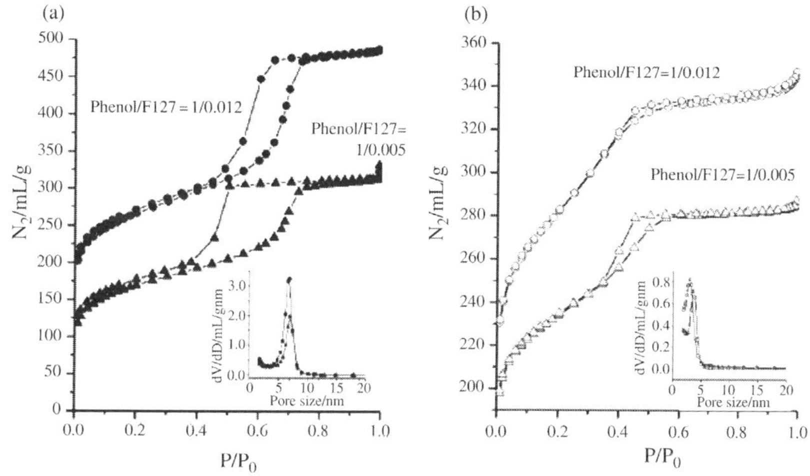


**FIGURE 3.175**

$N_2$ adsorption/desorption isotherms for the mesoporous polymers and the resultant mesoporous carbons with pore-size distribution (inset) [490]. (a) Polymers and (b) carbons prepared at 900°C.

which give MgO by pyrolysis can be used. Mg acetate, Mg citrate and Mg gluconate have been successfully used [503–506] giving homogeneous but different sizes of MgO nanoparticles, consequently different sizes of mesopores in the resultant carbons, as shown in Fig. 3.176a. When Mg-acetate is used, mixing method also influences on the pore-size distribution of the resultant carbon. Reagent-grade MgO with about 100 nm size is also used by mixing with different carbon precursors [501,502]. Mg-citrate and -gluconate work as carbon precursor as well as MgO precursor' as shown in Fig. 3.176b, the carbon having a large amount of micropores with some mesopores is obtained from Mg-citrate (i.e., MgO/PVA = 100 in Fig. 3.176), although small MgO/PVA ratios give mesopore-rich carbons [505]. Using needle-like nano-sized $Mg(OH)_2$ as MgO precursor, carbons having tunnel-like mesopores from different carbon precursors [513–515]. Natural ore of magnesium hydroxy-carbonate (magnesite) [509,510] and $MgCl_2$ [512] have also been used.

Colloidal silicas have also been used as templates to synthesize mesoporous carbons. A mixture of colloidal silica stabilized by cetyltrimethylammonium bromide with resorcinol/formaldehyde resulted in mesoporous carbon with $S_{BET}$ of 1512 $m^2/g$ and $V_{meso}$ of 3.6 $cm^3/g$, consisting mainly of mesopores with the size of about 12 nm [516]. The mixture of a mesophase pitch with colloidal silica in ethanol is kept at a temperature of slightly higher than the softening point of the pitch and then carbonized at 900°C to result in mesoporous carbon [517]. Silica-templated carbon films are prepared by spin-coating of an acidic aqueous solution

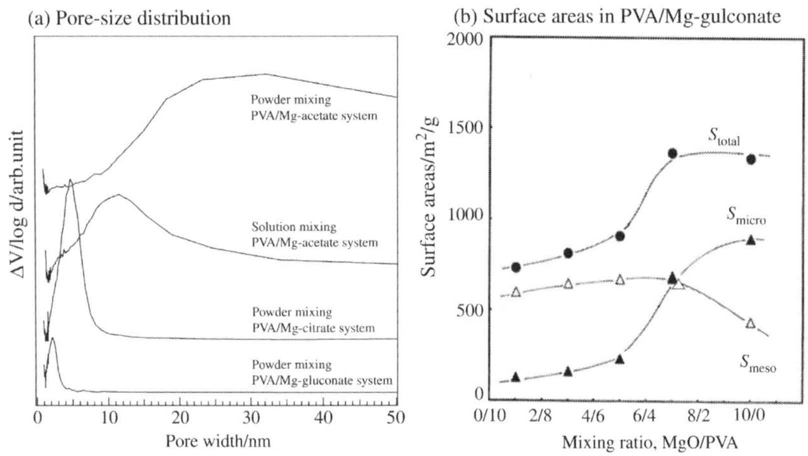


**FIGURE 3.176**

Pore-size distributions of Mg-templated carbons (a) and changes in surface areas with mixing ratio of MgO/PVA in Mg-gluconate system. $S_{total}$: total surface area, $S_{micro}$: microporous surface area and $S_{meso}$: mesoporous surface area.

of sucrose with tetraethyl orthosilicate onto a silicon wafers, followed by carbonization at 400°C and dissolution of silica template [518]. The resultant carbon film has $S_{BET}$ of 2600 m$^2$/g and $V_{total}$ of 1.4 cm$^3$/g, consisting mainly of mesopores with the size of about 2.4 nm. Using sintered silica spheres having diameters of 150–300 nm and a phenolic resin, carbon consisting of 3D periodical alignment of mesopores is prepared [519]. Gels of sucrose with sulfuric acid, tetraethyl orthosilicate (TEOS) and HF mixed in aqueous solution gives nanoporous carbon after carbonization and dissolution of silica [520,521]. Morphology and size of pores can be controlled by HF content, wormhole-like mesopores being obtained by HF/TEOS molar ratio less than 1/7 [520]. From THF solution of a petroleum pitch with TEOS and HF, bimodal mesoporous carbons are obtained [522]. From the mixtures of polystyrene (PS) latex and colloidal silica, macroporous carbons with mesopores are prepared, macropores being controllable by PS latex and mesopores by silica colloid [523]. By adding $SiO_2$ nanoparticles to the solution of triblock copolymer F127 and resolcinol/formaldehyde, ordered mesoporous carbon with an interconnected channel structure was synthesized [524]. Grafting of polymers on the surface of silica nanoparticles is effective to create nanopores in the resultant carbons, grafting of PS [525] and poly(styrene-co-acrylonitrile) [526] being applied. Formation of nanoporous carbon from PS-grafted $SiO_2$ is schematically shown in Fig. 3.177.

The results on the preparation and structure of mesoporous carbons using various templates have been reviewed in detail [527,528].

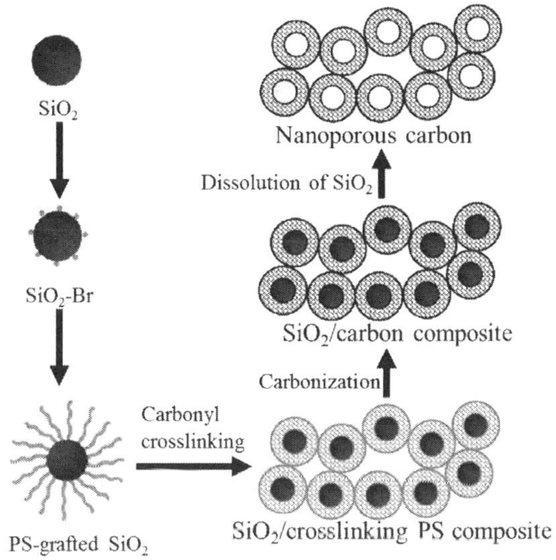


**FIGURE 3.177**

Scheme of the formation of nanoporous carbon from PS-grafted $SiO_2$ nanoparticles.

### 3.6.3 Carbon foams (macroporous carbons)

Through the synthesis of poly(urethane-imide) films and their carbonization, carbon films have been obtained, of which macropore structure is controlled by changing the molecular structure of polyurethane prepolymer [529–532]. Poly (urethane-imide) films are prepared by blending poly(amic acid), consisting of pyromellitic dianhydride (PMDA) and 4,4'-oxydianiline (ODA), with phenol-terminated polyurethane prepolymers. The films show phase-separation of the matrix PI and small islands of PU by heating up to 200°C. By heat treatment up to 400°C, PU component is pyrolyzed to gases and porous PI film is obtained, which is able to convert easily to macroporous carbon film by carbonization. Pore sizes in the PI and carbon films are controlled by the blending ratio of PI to PU and also by the molecular structure of PU. In Fig. 3.178, size distribution of macropores in the carbon films is shown as a function of PU content. With increasing blending ratio of PU, average size of macropores increases and pore-size distribution becomes broad. The film prepared using 10% PU shows very narrow distribution of pore size at around 1 μm. The carbon films prepared were suitable as a medium for culturing biological cells, macropores in the range of 0.6 to 3.0 μm sizes being shown to be preferable [533].

Carbon foams have been prepared from pitches through either blowing or pressure release of molten pitches followed by stabilization in air [534]. Since the foams with high thermal conductivity may give a potential reinforcing material

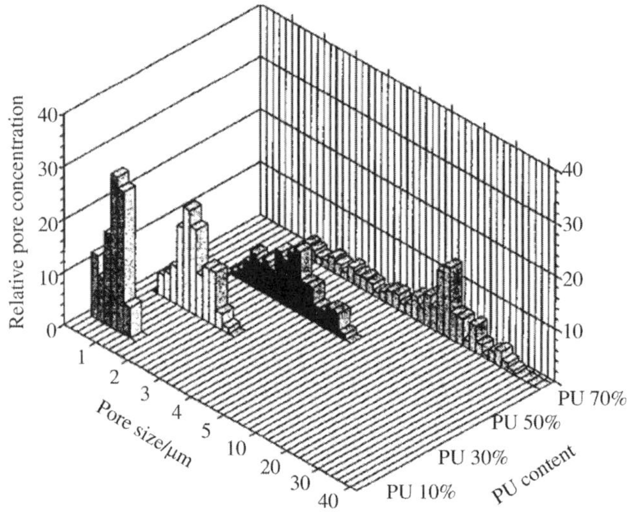


**FIGURE 3.178**

Change in pore-size distribution of macroporous carbon films with PU content [531].

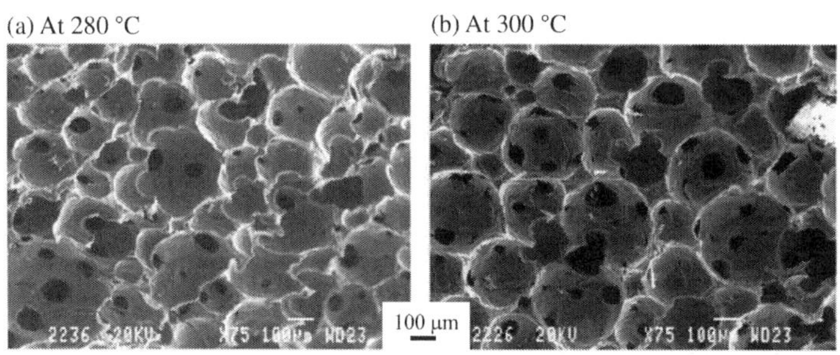


**FIGURE 3.179**

SEM images of carbon foam prepared from a mesophase pitch at 280 and 300°C under a pressure of 6.8 MPa and a releasing time of 5 s [539].

for structural composite materials, as an alternative to carbon fibers, the preparation of a large-sized carbon foam with high thermal conductivity from mesophase pitch by a new, less time-consuming process has been extensively studied [535–541]. The graphitized foams have the bulk density of 0.2–0.6 g/cm$^3$, average pore diameter of either 275–350 µm or 60–90 µm and relatively high graphitization degree. The bulk thermal conductivity depends on the bulk density of the foams and varied from 40 to 150 W/m.K. SEM images of carbon foam prepared from a mesophase pitch are shown in Fig. 3.179 [539]. Softening point Ts of the

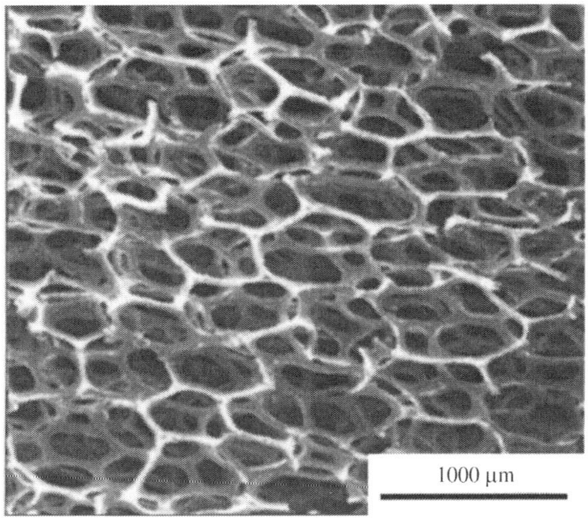


**FIGURE 3.180**

Carbon foam prepared from poly(urethane) foam, Reticulated Vitreous Carbon (RVC).

starting pitch influences on bulk density of the resultant carbon foam, as well as pressure during foaming [540]. The foams prepared from the pitch with low $T_s$ show better crystallite orientation and less number of microcracks in both ligaments and junctions, and less shrinkage during graphitization than those from the pitch with high $T_s$ [541].

Carbon foams with honeycomb-type open cell structure, as shown in Fig. 3.180, have been developed by carbonization of poly(urethane) (PU) foam by impregnation of carbon precursors and called reticulated vitreous carbon (RVC) [542,543]. This RVC foam is useful as optically transparent electrode for simultaneous spectroscopic and electrochemical measurements, which is applicable in corrosive solutions at elevated temperatures [544]. Various applications as a flow-through electrode have been proposed: flow injection analysis in amperometric and coulombic modes [545], flowing potential coulometry [546], monitoring the acidity of solutions [547], etc.

Carbon foams have also been prepared by impregnating poly(amide acid) into poly(urethane) foams as template followed by imidization and carbonization [548]. The carbon foams are used as adsorbents of atmospheric humidity and also a support for photocatalyst anatase-type $TiO_2$. In Fig. 3.181, the changes in pore appearance with impregnation of polyimide and carbonization at 1000°C are shown.

Graphite foams have been prepared by compression of exfoliated graphite, of which bulk density and porosity can be controlled by changing pressure. Exfoliation of graphite residue compound of sulfuric acid under compression can

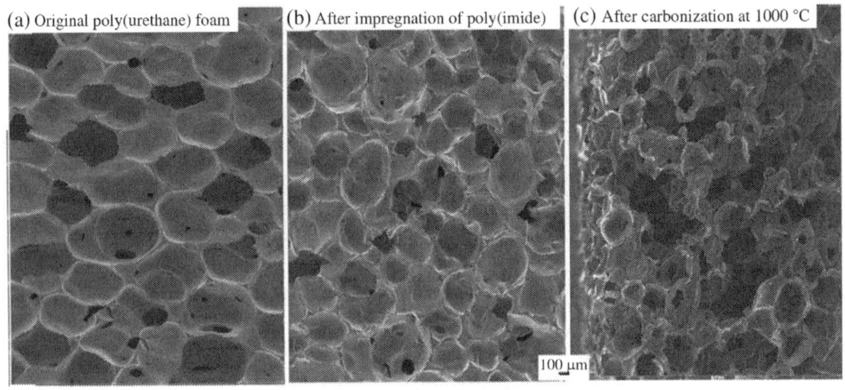


**FIGURE 3.181**

Changes in appearance of pores in the foams [548].

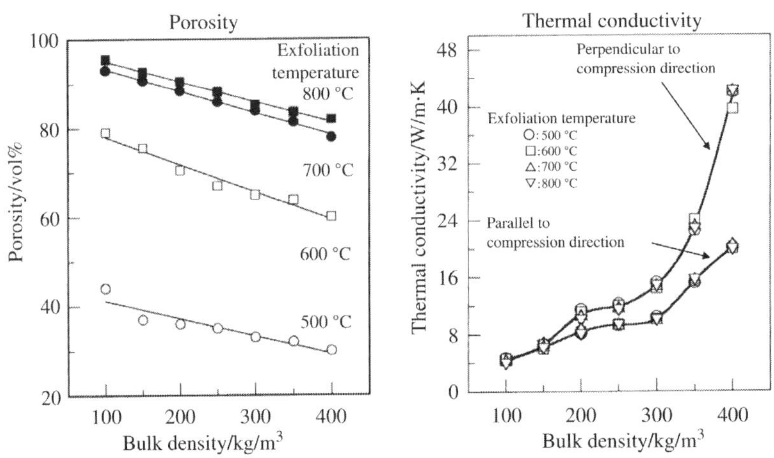


**FIGURE 3.182**

Porosity and thermal conductivity against bulk density on EG-foams [549].

give graphite foam [549]. In Fig. 3.182, porosity and thermal conductivity are plotted against bulk density on the foams. Porosity of the foam depends strongly on exfoliation temperature; exfoliation at the lower temperature results in the lower porosity due to incomplete exfoliation. Thermal conductivity of the foam depends strongly on bulk density of the foam, but not on exfoliation temperature because thermal energy passes through the foam walls consisting of graphite crystals, although it shows a marked anisotropy.

# 3.7 Carbon-based composites

## 3.7.1 Carbon-based composites

Many industrial carbon products are composites, consisting of different kinds of carbon components. In general, different combinations of various materials are employed in order to design the functions, one of these materials being carbon in the present case. In Fig. 3.183, the concept for the construction of carbon-based composites is summarized. For the combination of materials, three factors have to be considered, what materials, what shapes and what sizes. For the design of functions of carbon-based composites, three cases may be classified, hybridization of fundamental functions of each constituent materials, improvement in performance of carbon materials and obtaining of new functions.

According to this concept of the composites, most carbon materials are included into carbon/carbon composites: the polycrystalline graphite blocks are the composites between the particles of filler coke, three-dimensional irregular shape with either millimeter or micrometer size, and the carbons derived from a binder pitch (Section 3.1). The intercalation compounds are the composites in nanometer scale between host carbon and intercalates aiming to obtain different new functions (Section 3.8). Various porous carbons may be considered to be the composites between carbon matrix and nanometer-sized pores (Section 3.6). The composites between anatase-type $TiO_2$ and porous carbon, i.e., $TiO_2$-mounted

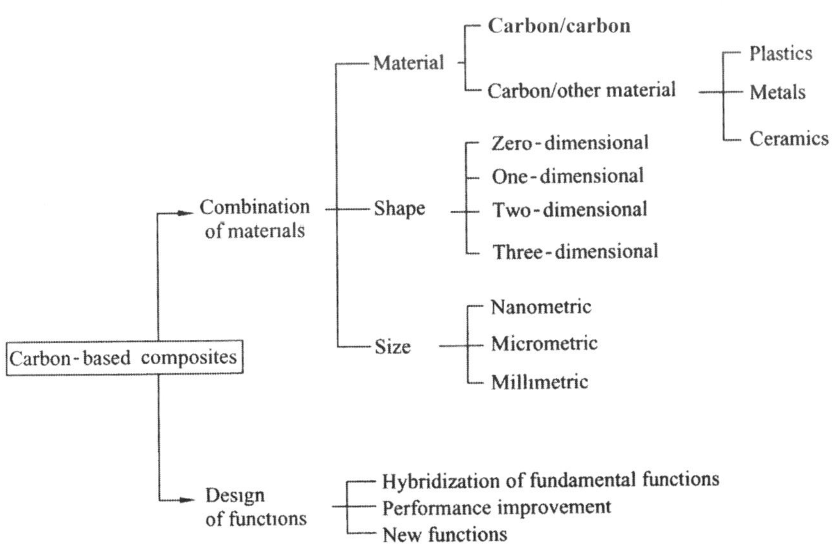


**FIGURE 3.183**

Concept of carbon-based composites.

activated carbons and carbon-coated $TiO_2$, have been developed to hybridize the photoactivity and adsorptivity (Section 3.10).

Among monolithic carbon materials, carbon fibers are the most important component for various composites with different matrix, such as carbons, plastics, ceramics, including concrete, and metals. In the present chapter, the description is focused mainly to carbon-fiber-reinforced composites, because other composites, such as polycrystalline graphite, porous carbons, intercalation compounds, etc., are described in individual chapters.

### 3.7.2 Carbon/carbon composites

Two major components of carbon/carbon composites (C/Cs) are the filler, usually coke particles, carbon blacks, carbon fibers, etc., and the matrix carbons, both of which can have different structures and nanotextures. Based on combinations of these two components with different structures and textures, the properties of the C/Cs can be significantly altered. By the selection of processing conditions, additional variations of C/Cs are obtained. When coke particles were used as fillers with a binder pitch, however, the resultant polycrystalline blocks were usually not called carbon/carbon composites, being called polycrystalline graphite in the present book. The names 'carbon/carbon composites', therefore, have customarily been used for the composites of which fillers are carbon fibers.

Carbon fibers give a wide range of variation in properties by selecting the precursors and the fiber forms, either strand, yarn or chopped. In the C/Cs, the matrix is also carbon, whose purpose is to assist in utilizing the high strength of the fibers, for example, by maintaining the space and geometries between yarns/fibers in order to transfer mechanical and heat loads between them. Variations of the structure and nanotexture of the matrix carbon are also important in determining the properties of C/Cs, such as thermal conductivity, fracture behavior, strength and toughness. For example, the fracture behavior in a composite is mainly controlled by the propagation of cracks within the matrix.

By controlling the architecture of the yarns of the filler and also the structure and nanotexture of the matrix, C/Cs can have the highest specific strength and modulus among any known high-temperature materials, including bulk graphite blocks. This is possible because of the low specific density and the high strengths of the carbon fibers, especially above 1500 to 2000°C. Although C/Cs is also known to have unique thermal properties, such as zero coefficient of thermal expansion near room temperature and negative values below room temperature. The various problems on the interaction between filler fibers and matrix carbon were discussed from the viewpoint of stress graphitization [550], using numerous examples, to illustrate how the thermal and mechanical properties of C/Cs could be extensively altered.

In Fig. 3.184, the relation between tensile strength and Young's modulus at room temperature was shown on various carbon materials. The strength and modulus of C/Cs are almost comparable with carbon fibers of general purpose grade

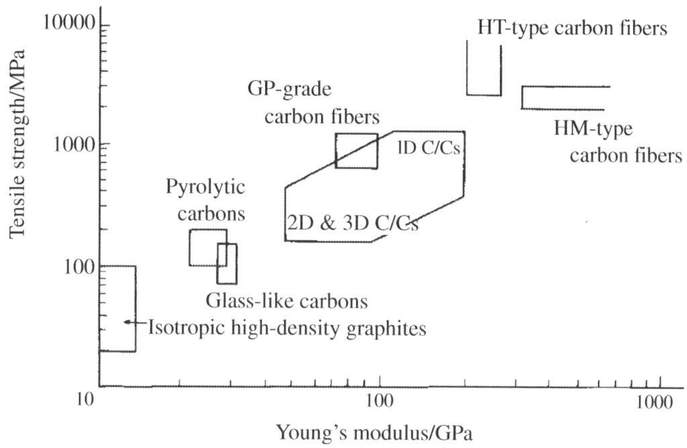


**FIGURE 3.184**

Relation between tensile strength and Young's modulus for various carbon materials.

**FIGURE 3.185**

Products of C/Cs.

(GP-grade), being much inferior to high-performance grade (HP-grade) carbon fibers, but it has to be emphasized that they have the highest values of strength and modulus among the bulky carbon materials. From the practical view point of performance as structural materials, they are much superior to carbon fibers themselves because C/Cs can be supplied in different sizes and shapes. Some of the products of C/Cs, bolts, nuts and heating elements, are shown in Fig. 3.185.

The production processes of C/Cs are shown in Fig. 3.186. The forming process is done mainly by impregnation, but also filament winding in the case of continuous fibers passing through the precursor of matrix carbon (pitches or resins) and injection molding of a mixture of matrix precursor with chopped carbon fibers are applied. The densification process is essential to fill the pores formed during carbonization of these precursors in the matrix, in order to get a high performance of C/Cs, and is carried out either by chemical vapor infiltration of some hydrocarbon gases (CVI) or impregnation of organic precursor (pitch or resin).

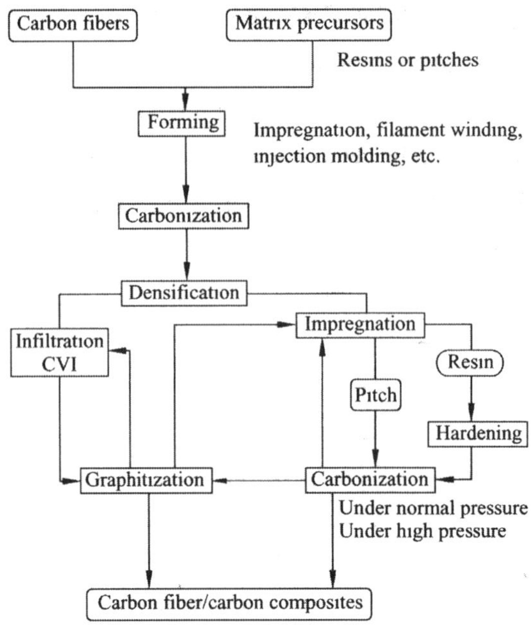


**FIGURE 3.186**

Flow chart for the production of C/Cs.

The repetition of the cycle of impregnation of pitch and its carbonization is effective to decrease the content of open pore and, as a consequence, to increase the density and strength. The results are shown as a relation between bending strength and open porosity in Fig. 3.187a [551]. The insertion of the heat treatment up to high temperatures such as 2500°C is recommended to get high density of C/Cs. In Fig. 3.187b, bulk density of C/Cs is plotted against impregnation times as a function of heat treatment temperature [552]. The heat treatment at 2600°C is efficient to have high density.

The selection of precursor for carbon matrix is also an important factor for getting high performance of the resultant C/Cs. In Fig. 3.188, bending strength of C/Cs with various matrix precursors is plotted against heat treatment temperature (HTT) [553]. Strength increases with increasing HTT, but absolute strength value depends strongly on the matrix.

The structure and texture developments in the matrix have been discussed by emphasizing the mechanical interactions between the filler carbon fiber and the impregnated matrix, because the precursor impregnated shows generally large volume shrinkage during carbonization, though the filler carbon fibers do not because they have already been carbonized. In Fig. 3.189, the change in polarized-light micrograph of the cross-section perpendicular to the carbon fibers with HTT is shown on the composite prepared from the matrix of glass-like

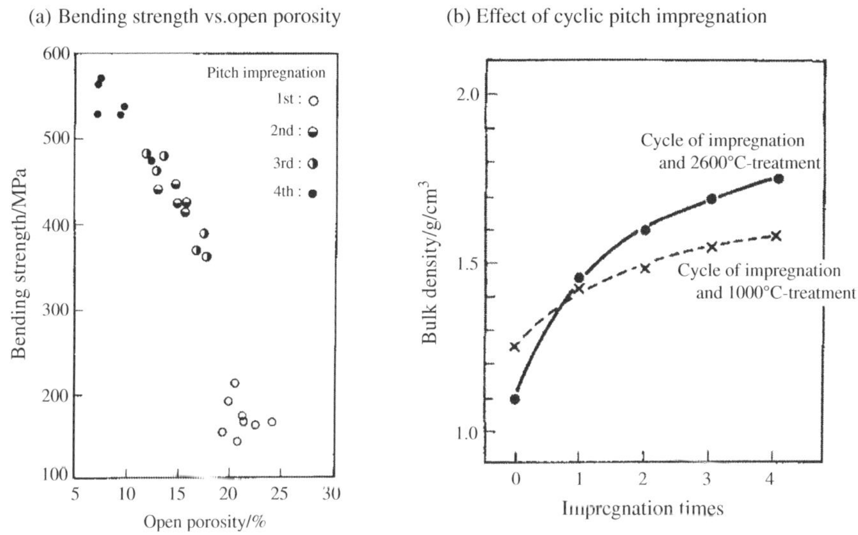


**FIGURE 3.187**

Effect of pitch impregnation on bending strength (a) [551] and bulk density (b) [552].

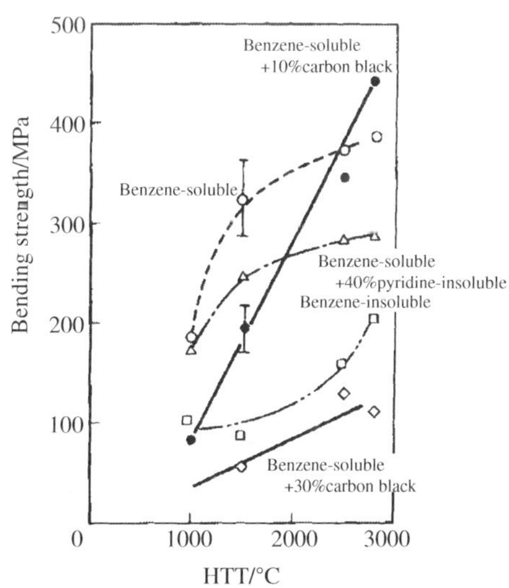


**FIGURE 3.188**

Dependence of bending strength of C/Cs on HTT for various matrix precursors [553].

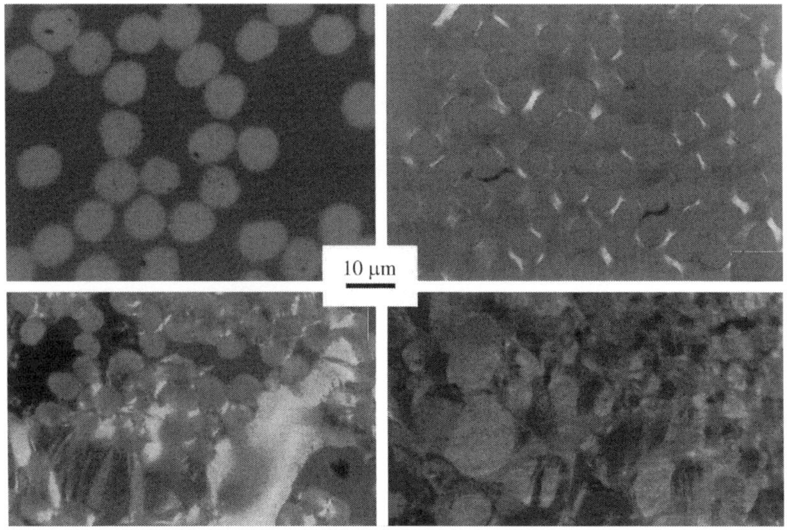


**FIGURE 3.189**

Polarized-light micrographs of the cross-section of a composite prepared from the filler PAN-based carbon fiber and the matrix glass-like carbon [554].

carbon [554]. The glass-like carbon by itself looks isotropic under an optical microscope, even after being heat-treated at high temperatures. Also the filler PAN-based carbon fibers themselves appear isotropic in their cross-section perpendicular to the fiber axes. When a composite has been prepared from these two carbons, however, a marked texture change occurs. After the heat treatment at 1000°C, anisotropic regions are developed preferentially at the fiber/matrix interfaces (Fig. 3.189b), though the composite cured at 110°C is isotropic (Fig. 3.189a). In these anisotropic regions, small carbon layers align parallel to the fiber surface and to the fiber axis from the analysis of pleochroism under polarized-light microscopy. These anisotropic regions are greatly enlarged by the high-temperature treatment (Fig. 3.189c), where the fibers with isotropic nature are distinctly seen. After heat treatment at 2800°C, almost the whole matrix becomes anisotropic, as shown in Fig. 3.189d, and isotropic areas due to carbon fiber cross-sections are difficult to recognize, though there are some. In Fig. 3.190, an SEM image is shown for the boundary between the filler carbon fiber and the matrix formed from a thermosetting resin, which usually gives glass-like carbon, after the heat treatment at 2800°C [555]. The orientation of layer planes around the carbon fibers is clearly seen. Graphitization of matrix carbon is also shown by X-ray diffraction measurement. This structural change at the boundaries between the filler carbon fiber and the matrix carbon in C/Cs is supposed to be due to the relaxation of accumulated stress during carbonization and

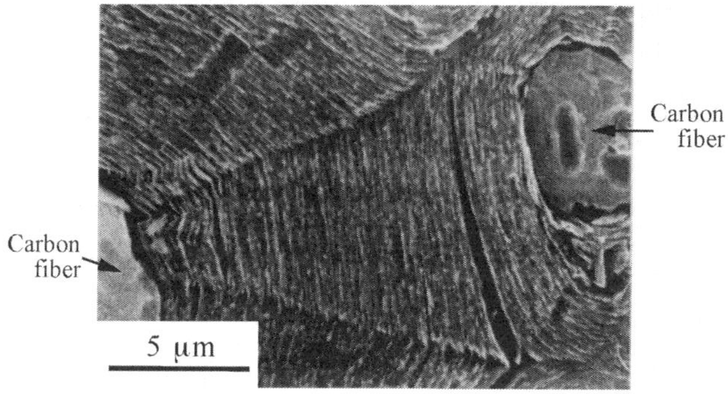


**FIGURE 3.190**

SEM image of the boundary between carbon fiber and matrix carbon after high temperature treatment [555].

graphitization associated with a large shrinkage of matrix, and discussed in the relation with stress graphitization [550].

Since the application of C/Cs is mostly as the structural components particularly at high temperatures, mechanical and thermal properties have been intensively studied. Mechanical strength of carbon materials is known to increase with increasing temperature up to around 2500°C [20,556,557], but their fracture behavior is brittle. By making composites with carbon fibers, therefore, it is expected to modify their fracture mode to ductile, in other words, to give toughness, in addition to increase their breaking strength.

It has been known that the fracture mode is quite different between glass-like carbon and graphite; in the former usually a single crack propagates through the specimen and gives an instant macroscopic breaking, but in the latter multiple cracks with varying lengths give rather ductile fracture behavior. In C/Cs, the matrix texture has a great influence of their fracture behavior. In the composites with glass-like carbon matrix, a crack usually occurs at the fiber–matrix boundary and travels unimpeded through the fibers and surrounding matrix. In a graphitic matrix, on the other hand, multiple cracks occur at the filler–matrix boundary and travel primarily along the lamellar structure of the matrix. Failure usually takes place when these cracks join together to form a critical crack size. The difference of fracture modes makes a big difference to the mechanical properties of the C/Cs. Also the strain characteristics of C/Cs are very different, strain being low for the matrix with glass-like carbon but relatively high for that with graphitic structure.

The effect of textures on mechanical and thermal properties has been studied on various C/Cs. In Fig. 3.191a relative toughness of the C/Cs, which have been prepared from a three-dimensionally woven PAN-based carbon fibers by the densification with a pitch, is plotted against HTT [558]. The toughness value shows a maximum at HTT of 2400°C and then decreases at 2500 and 2750°C. This result

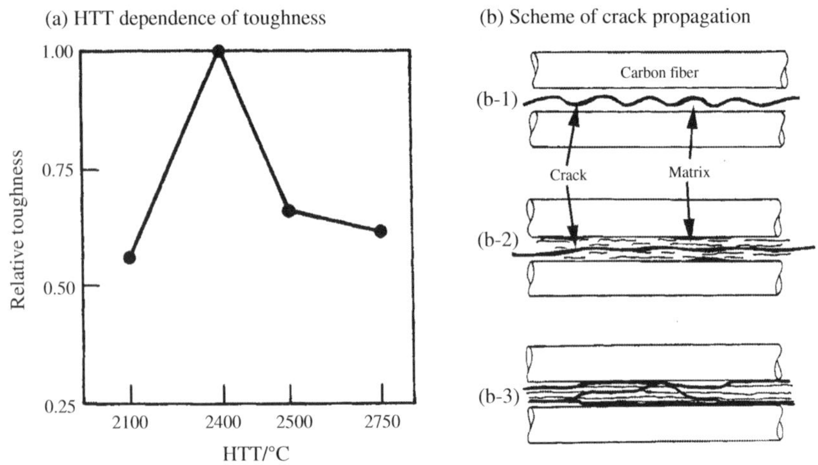


**FIGURE 3.191**

HTT dependence of relative toughness of C/Cs and the scheme of crack propagation [558].

is explained by the change in the mode of crack propagation through the matrix in C/Cs, as illustrated in Fig. 3.191b. In glass-like carbon matrix (Fig. 3.191b-1), single crack propagates along the matrix with high modulus and isotropic glass-like nanotexture. Since not much energy is dissipated during propagation of this crack, the toughness of C/Cs is low. The matrix, which was heat-treated at 2400°C and composed of two regions of glass-like carbon and graphite, has anisotropic modulus values and facilitates the formation of multiple small cracks (Fig. 3.191b-2), which absorb more energy, and, as a consequence, the toughness of this type of matrix nanotexture becomes higher than the previous one. As the matrix nanotexture becomes highly graphitic with longer lamellae and very anisotropic modulus after the heat treatment above 2500°C, however, many large cracks can easily propagate along the matrix (Fig. 3.191b-3) and less energy is expended. Therefore, the toughness of the composite with highly graphitic nanotexture becomes smaller. The change in modulus of the matrix in C/Cs with heat treatment temperature has been experimentally shown [559].

The selection of the proper condition of heat treatment has been shown to be very important for C/Cs with glass-like carbon matrix to get proper mechanical properties. Fig. 3.192 shows the relations between load and load-point displacement during 4-point bending test on the C/Cs with different HTTs [560], in which carbon fibers are aligned along the longitudinal direction of the C/Cs and perpendicular to the loading direction.

After the initial elastic deformation segment (the portion of linear increase in load with displacement), the composite heat-treated above 2500°C has a plastic-like displacement, as shown by the portions of A–B and C–D of the curves in

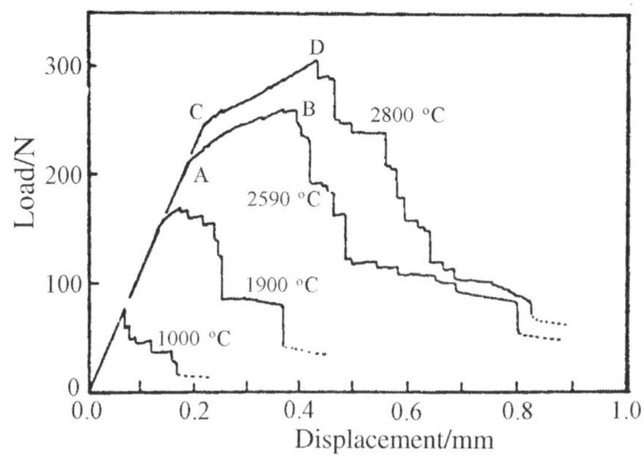


**FIGURE 3.192**

Load-displacement curves during 4-point bending test on C/Cs as a function of HTT [560].

Fig. 3.192, which occurs before the formation of macroscopic cracking in the specimen. By combining with microscopic observation of texture of the matrix in the composites, this plastic-like displacement is attributed from the formation of lamellar texture in the matrix, as shown in Fig. 3.190, where microscopic-sized cracks are formed, thereby causing stress relaxation and a redistribution of the local mechanical stresses. This microcracking decreases the modulus $E$ and increases its plastic deformation. Ultimately, the microcracks join together to form critical size macrocrack and to cause failure, as occurs at B and D points of the curves, with jagged steps of declining strength. On the other hand, the composite heat-treated at low temperatures as 1000°C does not have such plastic-like portion.

The texture change in the matrix with heat treatment affects sensitively on mechanical behavior, which has been shown by using C/Cs composed from two-dimensionally aligned PAN-based carbon fibers and phenolic resin [561]. In the C/C heat-treated at 2480°C, failure occurs in a brittle manner. After additional heat treatment at about 2500°C, however, the failure point increases by 45% with more ductile type of failure. By much longer heat treatment at about 2500°C a plastic mode of failure is more clearly observed and a gradual increase in deflection up to a much larger value. SEM examinations of the matrix surrounding the fibers show that the 2500°C-annealed C/Cs are supposed to contain a larger amount of the graphitic component than the as-prepared.

In Fig. 3.193a, the relations between load and displacement are compared on two C/Cs and polycrystalline graphite block (isotropic high-density graphite) [562,563]. One of the C/Cs has been prepared by carbonization of carbon fiber/phenol resin composite with two-directional alignment of fibers followed by the repetition of pitch impregnation (2D-C/C), and another C/C by densification of

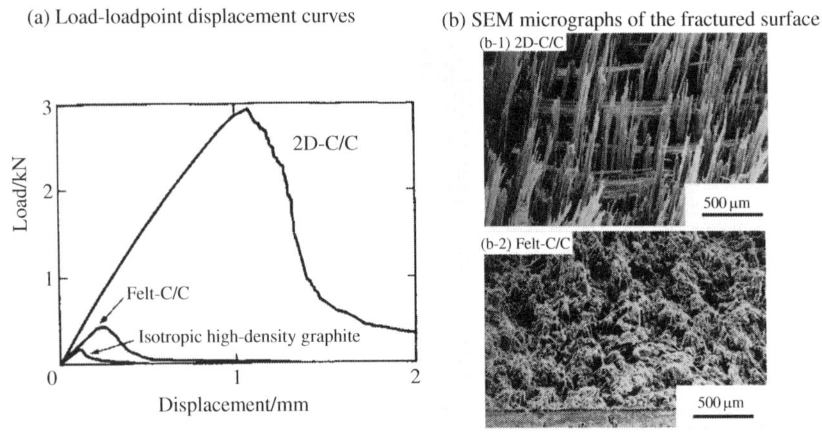


**FIGURE 3.193**

Fracture behaviors of C/Cs [562].

carbon fiber felt with CVI (felt-C/C). The 2D-C/C shows much higher strength and larger strain than felt-C/C and polycrystalline graphite. On the fractured surface of the 2D-C/C, marked pull-out of carbon fibers is seen, but no carbon fiber pull-out is observed on the felt-C/C, as shown in Fig. 3.193b, this being the reason why the increase in strength is not pronounced in the felt-C/C. The difference among these three C/Cs in mechanical behavior has been discussed by measuring energy to break (work of fracture); the present 2D-C/C has a large value at about 6000 $J/m^2$, but felt-C/C only 275 $J/m^2$ and polycrystalline graphite block much lower at 80 $J/m^2$ [562].

From the detailed measurements of fracture behavior on different C/Cs, the process to fracture has been discussed [563]. Fig. 3.194 illustrates schematically the process, showing a single filament of carbon fiber. With increasing load, the matrix breaks first because it has much lower strength than carbon fiber, but the fiber does not break at the load level, which the crack reaches to fiber, because of its high strength. Further increase in load causes the breaking at the interface between fiber and matrix, because this part is usually much weaker than fiber itself, and then the crack propagates beyond the fiber. Finally, carbon fiber itself breaks somewhere and then macroscopic fracture of the composite occurs. Since breaking of carbon fiber occurs not necessarily at the same place as the crack in the matrix arrives, it is well understood that fiber pull-out is observed. In this scheme, two processes, fiber bridging due to interface cracking and fiber pull-out, are characteristic for C/Cs and spend a large part of energy for fracture. This suggests the importance to control the structure of the interface between matrix carbon and filler carbon fiber, in addition to the control of the structure and texture of matrix carbon and also the selection of filler carbon fibers.

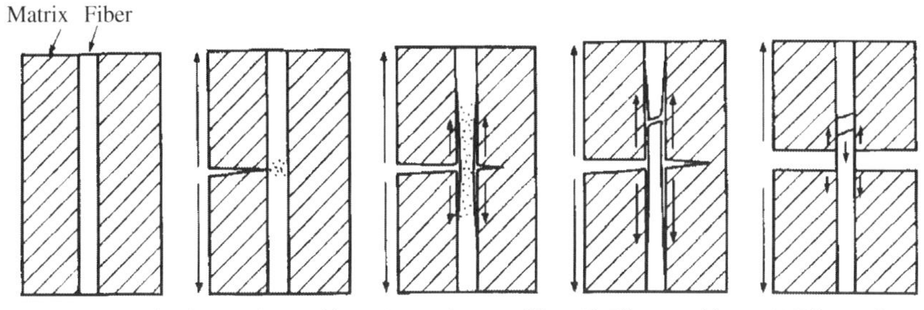


**FIGURE 3.194**

Propagation process of a crack in C/C [563].

The interface between the fiber and its surrounding matrix also has a great influence on the thermal properties of C/Cs, as well as the mechanical properties discussed above. The details of the structure of these interfaces were studied by the use of high-resolution transmission microscopy [564].

Along the fiber axis, a shear force is supposed to be developed with increasing temperature, because the thermal expansion coefficient of the fibers is negative but that of the matrix is positive. In certain cases, therefore, this shear stress at the filler–matrix interface can be large enough to break the bonds and then the matrix dominates the thermal expansion of C/Cs [565]. In the usual situation, however, only a fraction of the bonds is broken in the C/Cs during thermal cycling. Accordingly, the loads due to thermal stress are redistributed or relaxed, so that the expansion/contraction of the C/C occurs incrementally during thermal cycling. In Fig. 3.195, the length of thin wall C/C tube was continuously measured, with the precision of 0.2 $\mu\varepsilon$ units, during the heating/cooling cycle between 121°C and room temperature.

The saw-tooth-like pattern is considered to be probably indicative of the bond breaking at filler–matrix interfaces. As temperature increases, the negative slope of saw-teeth is due to the axial contraction of the fiber, whereas the positive slope is caused by the expansion of the matrix, as is schematically illustrated in the figure. The non-uniform nature of the saw-tooth-like pattern attributes to the variability of the shear stresses that are created or destroyed during heating/cooling cycle of C/C [566].

Nanotexture of the matrix and the state of filler–matrix bonding determine also thermo-mechanical properties of C/Cs. There is an optimum combination of disordered glass-like carbon and oriented graphitic regions that enables the maximum thermal-mechanical properties to be obtained for specific engineering applications. One of the most powerful methods of obtaining these optimum matrix textures is the use of the stress graphitization effect [550].

C/Cs are one of the important materials for aerospace systems, because of their high specific strength and Young's modulus (strength and Young's modulus

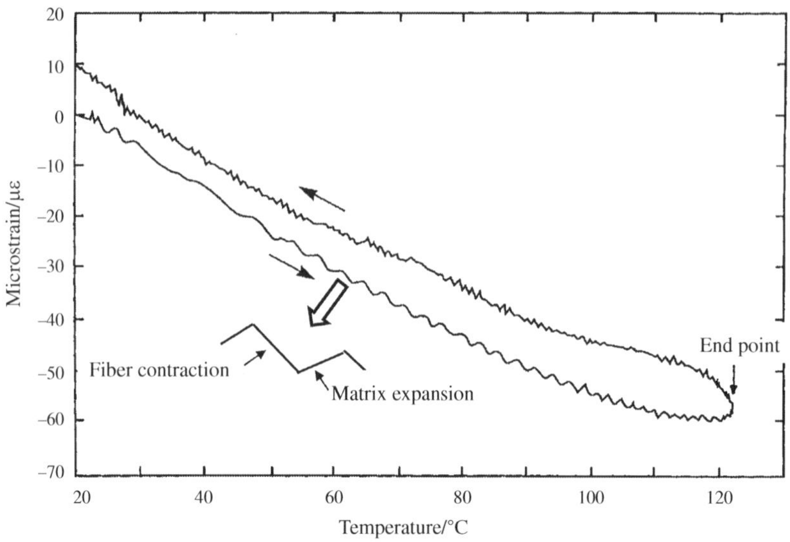


**FIGURE 3.195**

Deformation in the length of a thin-wall tube of C/C during a cycle of heating to 121°C and cooling to room temperature [565].

per unit weight) and also of high thermal shock resistance. The different parts of space carriers, such as nose cone, tipfin, radar and frap have been made from C/Cs. During re-entry into the atmosphere, these parts are expected to be exposed in very severe circumstance, to temperatures as high as 1700°C in oxidative atmosphere, very strong shock wave by hypersonic fluid, etc. In order to overcome these problems, different ideas have been proposed, such as coating of SiC on the surface.

## 3.7.3 Carbon/plastics composites

### a. Carbon-fiber-reinforced plastics

Carbon-fiber-reinforced plastics (CFRPs) are very important structural materials in different fields of industries, not only for aerospace but also for sport and leisure goods. In Table 3.33, the applications of CFRP and its merits to be used are summarized. Practical problems related to their production and applications were discussed from the view point of engineering [567b]. In addition to high strength and modulus of carbon fibers, different merits are obtained by using these CFRPs.

A large amount of carbon fibers, mostly PAN-based ones, are used in rackets for badminton and tennis, golf clubs and also fishing rods. The usage of carbon fibers has given a feeling of high grade to the customers in the beginning, but now it is common in many kinds of sporting goods. The application to airplanes

**Table 3.33** Applications of Carbon Fiber Reinforced Plastics (CFRPs) and their Merits

| Applications | Merits |
|---|---|
| **Sporting goods** | |
| Golf club, rackets for tennis and badminton | Lightweight, fashionable |
| Fishing rod | Lightweight, fashionable |
| Skiing | Lightweight, fashionable |
| Sailboat spar | Lightweight, high rigidity, anticorrosion |
| **Aerospace** | |
| Aircraft structure, wing, fuselage | Lightweight, high fatigue strength |
| Satellite body | Lightweight, high rigidity, size stability, High thermal diffusivity |
| **Automobiles** | |
| Shaft | Lightweight, high speed rotation |
| Wheel, bonnet | Lightweight, high fatigue resistance |
| Parts for engine | Lightweight for the reduction of inertia |
| **Industry** | |
| Robots | Lightweight, less inertia |
| Chemical plants | Anticorrosion, less electrostatic charging |
| Medical instruments | Transparency for X-rays, high rigidity |
| Rotary blades | Light weight, high speed rotation |
| **Construction** | |
| Piers of high ways | Lightweight, high strength |
| Chimneys | Lightweight, high rigidity |

has also been developed quickly in the field of military, but now even business jets and also jumbo jets, like Boeing 787, are produced by using the CFRPs to reduce their weights. The application to automobiles is not yet as wide, mainly because of the high price of carbon fibers. However, low weight body is strongly required in order to save energy and also to reduce waste gas, so that the usage of CFRPs is expecting to be essential. To the parts of robots and other industrial equipment, CFRPs are frequently used.

Prepreg sheets of carbon fibers and strands of carbon fibers were used for the reinforcement of highways in Japan, because of an unexpected increase in traffic in big cities and also a high possibility of having a big earthquake in Japan. Particularly after the large earthquake in Kobe area, Japan (January 17, 1995), the urgent reinforcement of piers of highways and pillars in buildings was pointed out. The reinforcement applications of carbon fibers to highways are shown schematically in Fig. 3.196. Not only the increase in endurance of the piers for earthquakes and for the expansion of the width of the road, but also the increase in the resistance to erosion due to salt in the wind from the sea was the reason.

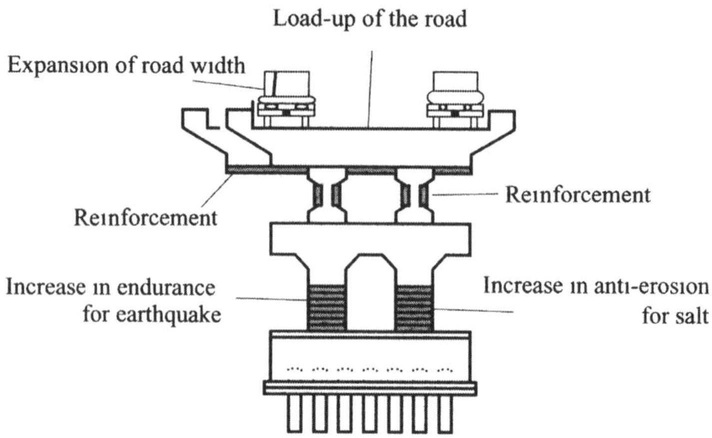


**FIGURE 3.196**

Applications of carbon fibers for reinforcement of highways.

*(Courtesy of Mitsubishi Rayon Co., Ltd.)*

**FIGURE 3.197**

Snapshot of the winding process of the prepreg sheets of carbon fibers onto a pier.

*(Courtesy of Mitsubishi Rayon Co., Ltd.)*

A snapshot of the practical winding process of the prepreg sheets of carbon fibers onto a pier for its reinforcement is shown in Fig. 3.197. The tapes of carbon fiber prepregs are firstly glued on the surface of either a pier or a pillar along its axis, after finishing the surface of concrete, and then either the same prepreg tapes or carbon fiber strands of 1200–2000 fibers are wound up perpendicular to

**FIGURE 3.198**

Reinforcement of a chimney by winding carbon fiber strands.

*(Courtesy of Mitsubishi Rayon Co., Ltd.)*

the pier or pillar axis. Since most highway piers have squared cross-section, as shown in Fig. 3.197, special attention is required at these corners not to break carbon fibers. Many new technologies have been introduced for these repairs, a remote-controlled system to check the corrosion of large concrete constructions, a technique passing electrical current in order to harden the plastic matrix after forming, etc.

The same process is applied to the chimney, which has been eroded by highly acidic exhausts, because of the advantages of lightweight and easy repair from the outside of the chimney. More than 20 cases of repairing the chimneys of factories by using carbon fiber strands have been reported in Japan, the biggest chimney repaired being 100 m high. Fig. 3.198 shows an example of chimney reinforcement by carbon fiber strands using remote-controlled equipment.

The use of the rod of CFRPs has been tried for reinforcing concrete, instead of steel wires. The rods consisting of a few strands of CFRP are tested for reinforcing a small concrete bridge, which is directly exposed to erosive sea wind from the Japan Sea and is necessary to be reconstructed at least once a year because of quick erosion of steel wires in concrete [567a]. By using rods of carbon fibers, no defects in the bridge have been observed even after 6 years. The same system has successfully been used in a pedestrian bridge, which crosses a large traffic road since 1989.

## b. Plastics reinforced by carbon nanotubes and carbon nanofibers

Reinforcement of plastics by using carbon nanotubes (CNTs) and carbon nanofibers (CNFs) has been tried in a number of research papers, but prominent success for reinforcement is very rare. A significant enhancement in thermal conductivity of a carbon fiber/phenolic resin composite has been obtained when highly crystalline multi-walled carbon nanotubes (MWCNTs) are additionally mixed [568].

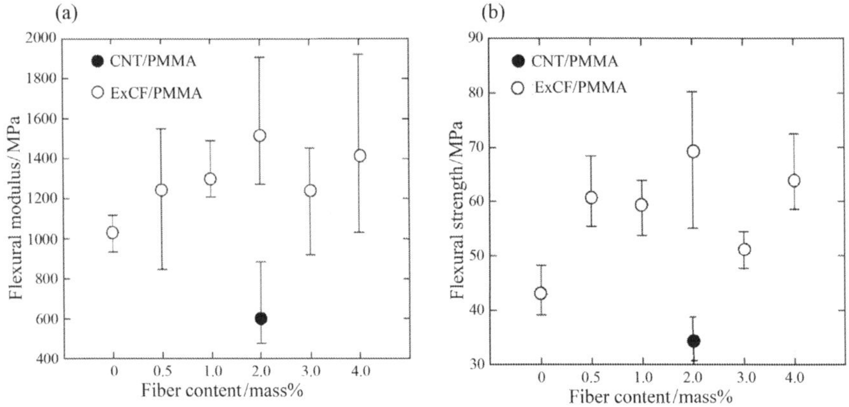


**FIGURE 3.199**

Dependences of flexural modulus (a) and strength (b) of ExCF/PMMA and CNT/PMMA composites on fiber content [570].

MWCNTs (7 mass%) dispersed homogeneously in a phenolic resin seems to act as an effective thermal bridge between adjacent carbon fibers and result in an enhancement of the thermal conductivity from 250 to 393 W/m K. MWCNTs-reinforcement of fluorinated rubbers developed a new application as durable sealants for the probing and production of oil in deeper wells [569]. MWCNT/rubber composite can satisfactorily work as oil sealant at extremely high temperatures and pressures (e.g., 260°C and 239 MPa).

Composites of exfoliated carbon fibers (ExCFs), which have been prepared by exfoliation of mesophase-pitch-based carbon fibers after converting graphite oxides, are able to reinforce more effectively than carbon fibers and even than CNTs, as shown for the composites with poly(methyl methacrylate) (PMMA) in Fig. 3.199 [570]. ExCF/PMMA composites with fiber content more than 0.5 mass% gave much higher flexural modulus and strength than CNT/PMMA. ExCFs consist of the filaments with sub-μm diameter and do not show significant aggregation in the composite, which may cause an effective reinforcement, more effective than CNTs which tend to form aggregation in the composites. Marked reinforcement by using ExCFs has also been demonstrated on Nylon-6 matrix [571].

## 3.7.4 Carbon/ceramics composites

### a. Carbon fiber reinforced concrete

Reinforcement of concrete by mixing GP-grade chopped carbon fibers led to prominent success and applied to various constructions [572,573]. The first application of carbon-fiber-reinforced concrete (CFRC) was in the Al Shaheed monument in Bagdad, Iraq, consisting of twin domes with the height of about 40 m and

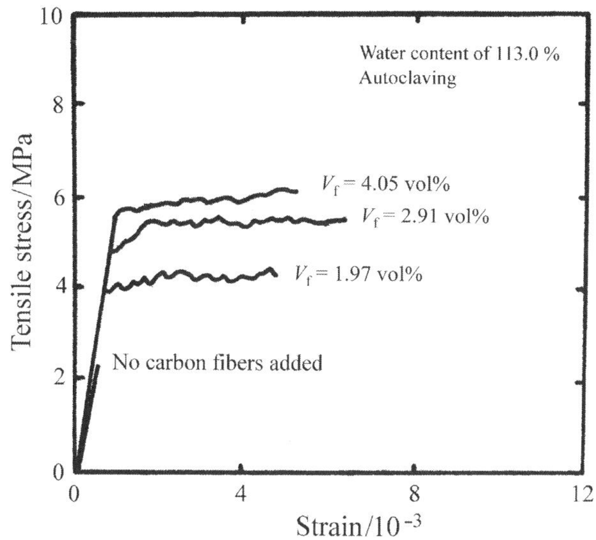


**FIGURE 3.200**

Stress—strain curves for CFRCs with different $V_f$ [574].

the bottom diameter of about 45 m (Fig. 1.7a). The steel skeleton and foundation of the Al Shaheed monument was constructed according to the original plan to cover the domes with gold-plated copper plates. However, the replacement of copper plates by porcelain tiles with Turkish blue color was decided after finishing the steel skeleton. Therefore, the cladding tile panels were necessary to have a weight of less than 60 $kg/cm^2$. In addition to being lightweight, enough mechanical strength and also durability for severe weather conditions in Baghdad, high temperature and low humidity in summer and below zero temperature in winter, were required. Only CFRC satisfies all these quality requirements [574]. In this dome, a large amount of CFRC panels (about 10 000 $m^2$) with very delicate curvature have been used, all of which are produced in Japan via autoclave curing, and transported to Iraq. This success leads to the application of this CFRC to the office building, Ark Tower with 37 stories, in Tokyo (Fig. 1.8b), as curtain walls (32,000 $m^2$), which gives a remarkable reduction of wall weight, easy transportation using a small lift and, as a consequence, shortening of the construction period [575]. Their prominent successes led to further applications of CFRCs in different components of different constructions. Many fundamental and engineering studies on CFRCs have been carried out by using PAN-based and pitch-based carbon fibers [573].

The addition of only 2 vol% of GP-grade isotropic-pitch-based carbon fibers into the cement mortar makes tensile strength double, as shown by stress—strain curves for CFRCs with different volume fractions of carbon fibers ($V_f$) in Fig. 3.200. Not only a marked increase in tensile strength but also a change in fracture mode from

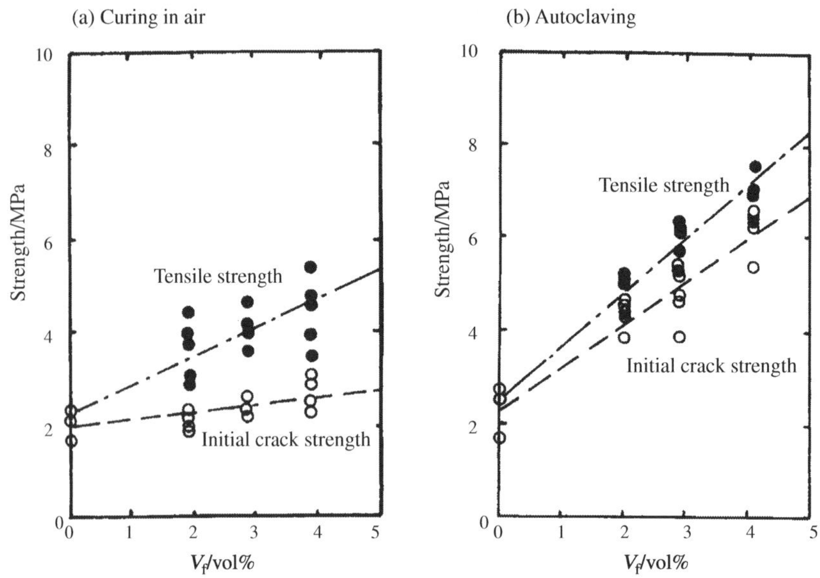


**FIGURE 3.201**

Dependences of the strengths on $V_f$ for CFRCs prepared under different curing conditions [574].

brittle to ductile by the addition of carbon fibers is clearly seen. Fig. 3.201 shows the dependences of initial crack strength, which is defined by the first break on the stress–strain curve, and ultimate fractured strength in tension on carbon fiber content for CFRCs cured in air and autoclave. Prominent increases in tensile strength with the increase in $V_f$ and also by autoclaving are seen. High resistance of carbon fibers to alkali solutions makes autoclave treatment of CFRCs possible. Various durability tests of the CFRCs have been performed before applying them to the Al Shaheed domes [574]. Keeping in hot water at 75°C and repeated cycles between freezing at −18°C and thrawing at 10°C are confirmed not to give any changes in mechanical properties and dimensions under severe weather conditions, like in Baghdad.

The fracture behaviors under tensile and bending stresses are studied in detail on concrete reinforced by unidirectionally aligned continuous GP-grade pitch-based carbon fibers [574]. The typical tensile stress–strain curves along the fiber direction are shown in Fig. 3.202. Cracks occur between the points A and B and most of them run perpendicular to the loading direction, in other words, perpendicular to the fiber direction. With the increase in strain from point A, the number of cracks gradually increases and so their spacing decreases. Beyond point B, no additional formation of cracks is observed. In this region, all the load seems to be supported by the carbon fibers, and so the slope of the stress–strain curve between points B and C is roughly the same as that calculated from Young's modulus and

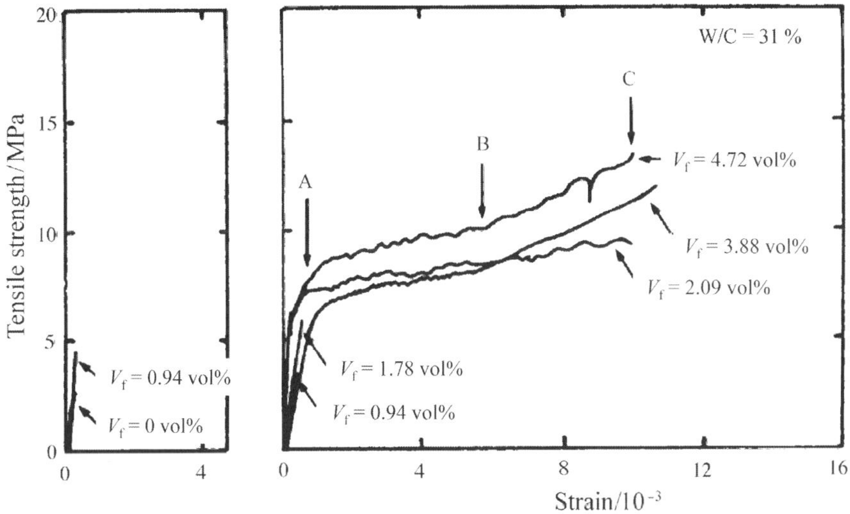


**FIGURE 3.202**

Stress–strain curves of the concrete plates reinforced by unidirectionally aligned continuous carbon fibers with different $V_f$ under tensile stress [574].

the content of carbon fibers. In Fig. 3.203, the dependence of tensile strength on $V_f$ is shown for two water/cement ratios. Above 1 vol%, the tensile strength of the CFRCs increases rapidly with increasing fiber content.

Non-woven mats of carbon fibers have also been used to reinforce concrete from Portland and alumina cements [576,577]. In Fig. 3.204, bending strength of the CFRCs is plotted as a function of the number of carbon fiber mats laid up. Bending strength of CFRC reaches 72 MPa by the lay-up of 25 sheets ($V_f$ of 4.4 vol%) which is about 18 times higher than without carbon fibers. By using carbon fiber mats and alumina cement, very thin plate of CFRC, as thin as 1 mm thickness, which showed marked bending deflection. The surface treatment of carbon fibers by a low-temperature oxygen plasma helps the wetting of the carbon fiber surfaces with cement paste and consequently to increase the strength of the resultant CFRCs [576].

CFRCs are effective for electromagnetic shielding in modern buildings, in order to block out electrical and/or magnetic noises from the outside and to keep the steady operation of building's functions by computers. By the addition of only one sheet of carbon fiber non-woven mat (carbon fiber content of 0.5 vol%), the shielding effect of 15 to 45 dB is obtained over the whole range of frequency of electrical field [577]. The increase in number of the mats increases the shielding effect. CFRCs have an additional advantage, i.e., the possibility to detect the fracture, even localized fracture, from the change in electrical resistivity of filler carbon fibers [578]. This principle has already been applied to practical buildings,

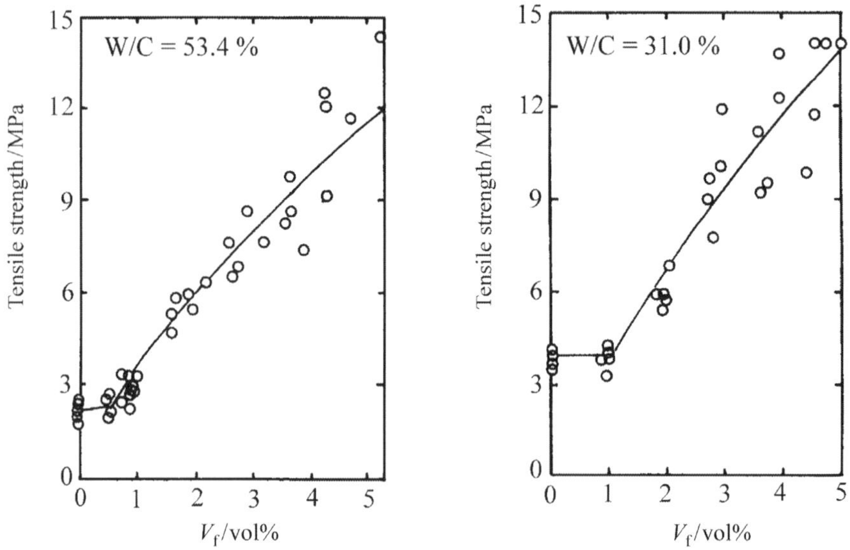


**FIGURE 3.203**

Dependences of tensile strength of CFRCs on $V_f$ for two different water/cement ratios (W/C) [574].

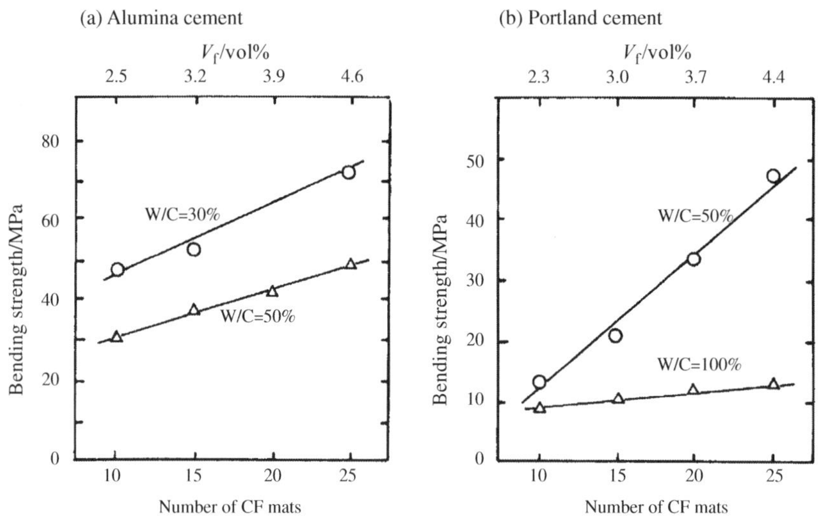


**FIGURE 3.204**

Bending strength of the concrete reinforced by the lay-up of carbon fiber mats [576].

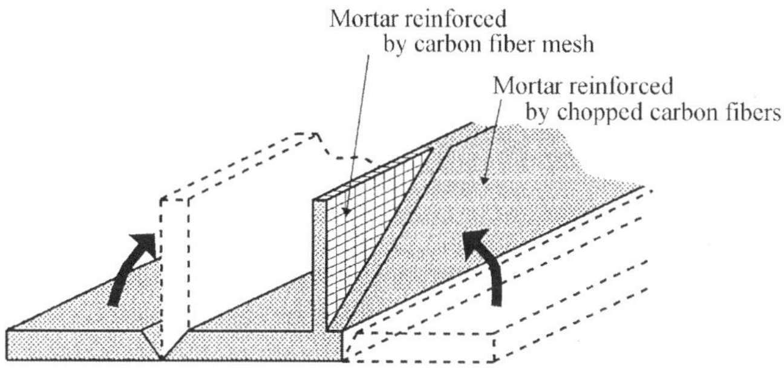


**FIGURE 3.205**

Frame of carbon fiber reinforced mortar for concrete placing.

*(Courtesy of Toho Rayon Co., Ltd.)*

not only to detect partial fracture due to fatigue but also to monitor an intentional breaking.

CFRCs have been used in different parts of buildings, curtain walls, water-proof roofs, floors with chemical stability and low level of noise, etc., and now they are used not only in buildings but also in the construction of bridges, tunnels, etc. They have also been used as a main component of free access floor panel due to the following advantages: light weight, high strength, high dimensional stability, good heat insulation and also low walking noise level, in comparison with the aluminum die-cast floor commonly used.

Thin mortar plate reinforced by carbon fiber has been used as the frames for concrete placing, as shown in Fig. 3.205, with many advantages; easy forming by bending at the site of construction, easy transportation and storage because of light weight, thin thickness, chemical and mechanical stabilities, etc. CFRCs have also been used for tunneling with an automatic drilling tool to strengthen the wall of the tunnel. It is used successfully at a depth of about 73 m.

### b. Ceramics reinforced by carbon fibers and carbon nanotubes

The main purposes to make composites of ceramics with carbon fibers (CFRCer's) are to give them toughness, to avoid a catastrophic breaking of ceramic parts, as in the case of carbon-fiber-reinforced concretes (CFRCs). An example of fracture behavior observed on one of CFRCers, carbon fiber/$Si_3N_4$ is shown in Fig. 3.206, where carbon fiber bundles are aligned in uni-direction [579]. With increasing load, deflection increases gradually in the beginning, i.e., cracks propagate gradually in the composite and no catastrophic breaking occurs. Electrical resistivity change of the composite ($\Delta R$) shows a good correspondence to load-deflection curve, i.e., crack propagation. Even saw-tooth-like small changes in a load-deflection curve, which are supposed to reflect partial breaking of carbon fibers, show good correspondence to resistivity changes $\Delta R$ (the arrows in the figure). This result indicates a possibility to detect partial fracture in this

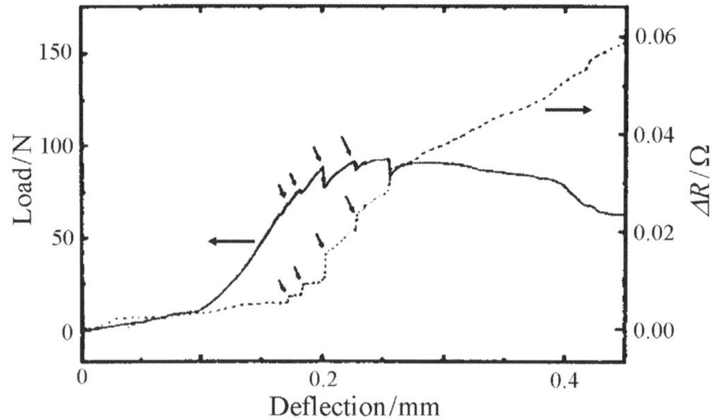


**FIGURE 3.206**

Changes in load and electrical resistance $\Delta R$ with deflection for a composite of $Si_3N_4$ with 40 vol% carbon fibers.

*(Courtesy of Dr. K. Matsuo of NGK Spark Plug Co., Ltd.)*

composite, which may give an advantage to use these composites to structural components, as CFRCs mentioned above.

To produce composites from a combination of carbon and ceramics, chemical and physical compatibility between these two components is the most important factor to be controlled. Most ceramics are produced at high temperatures as 1600–1800°C. At such high temperatures, chemical interaction of the matrix ceramics with the filler carbon fibers has to be taken into consideration. Physical compatibility between the filler carbon fibers and the matrix ceramics is caused from the stress at the boundary between two component materials mainly due to a difference in their thermal expansion during the cooling process from high treatment temperatures. Therefore, a combination between carbon and ceramics with a small difference in thermal expansion coefficient, in other words, ceramics with similar thermal expansion coefficient to carbon are preferable.

Single-wall carbon nanotube (SWCNT)/$Al_2O_3$ composites have been fabricated by spark plasma sintering of the mixture of $\gamma$-$Al_2O_3$ with 5.7 vol% SWCNTs at the temperature as low as 1150°C under the pressure of 63 MPa [580]. Matrix $Al_2O_3$ is transferred to $\alpha$-phase with a size of about 200 nm after sintering. A marked toughness increase to 9.7 MPa.m$^{1/2}$ is achieved for the SWCNT content of 10 vol%, nearly three times that of pure alumina. SWCNT/$Al_2O_3$ composites have also been prepared by hot-pressing at 1300°C and 43 MPa from $MgAl_2O_3$ foam after depositing carbon nanotubes (CNTs) via catalytic CVD [581]. The composites give marked increase in electrical conductivity with increasing CNT content to about 1 vol%, from $10^{-10}$ to $4 \times 10^{-3}$ S/cm. For multi-walled carbon nanotube (MWCNT)/$Al_2O_3$ composite shows a perfect plastic deformation at 1400°C [582].

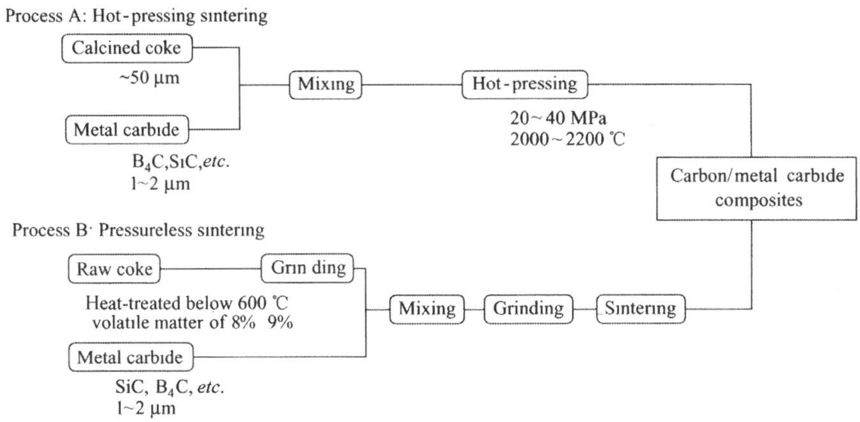


**FIGURE 3.207**

Two processes for the production of carbon/metal carbide composites.

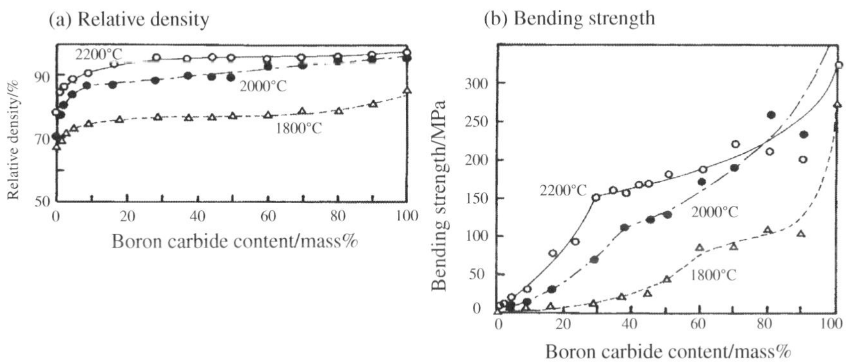


**FIGURE 3.208**

Dependences of relative density and bending strength of carbon/metal carbide composites on boron carbide content [583].

## c. Carbon/metal carbide composites

The preparation of the composites of carbon with metal carbides, mainly boron carbide, has been studied extensively [583]. Two processes are employed: hot-pressing using calcined coke and pressureless sintering by using raw coke, as shown in Fig. 3.207.

In Fig. 3.208a and b, bulk density and bending strength of the composites are plotted against the mixing ratio of boron carbide $B_4C$ as a function of hot-pressing temperature [583]. Densification and strengthening are observed above 2000°C, which is supposed to be caused by the diffusion of boron atoms into carbon particles. The composites thus prepared showed high resistance to oxidation in air at 800°C and also in sulfuric acid with 10% nitric acid. The formation of a thin layer of boron oxide on the surface of the composites in the very beginning of oxidation is supposed to act as a protective layer for further oxidation. The composites are also resistant in molten metals, such as aluminum and copper [583].

The addition of a third component of metal carbide, such as NbC, TaC and TiC, is effective in increasing the strength and thermal shock resistance of the composites [583]. They have a much higher strength than graphite materials up to 1700°C, as shown in Fig. 3.209. In these composites, the metal carbides added change to borides such as NbB, $TiB_2$ and $TaB_2$, after hot-pressing.

When the raw coke powder, which has been manufactured at about 500°C and contains about 8.5 mass% volatile matters, is mixed with carbide powder, as $B_4C$, SiC, etc., and ground in a mortar for a long time, it is possible to be sintered under atmospheric pressure (Process B: pressureless sintering in Fig. 3.207) [2]. Small balls with a size of 1–5 mm, in which the carbide particles are well dispersed in raw coke, are sintered into the composites at 1000 to 2400°C without

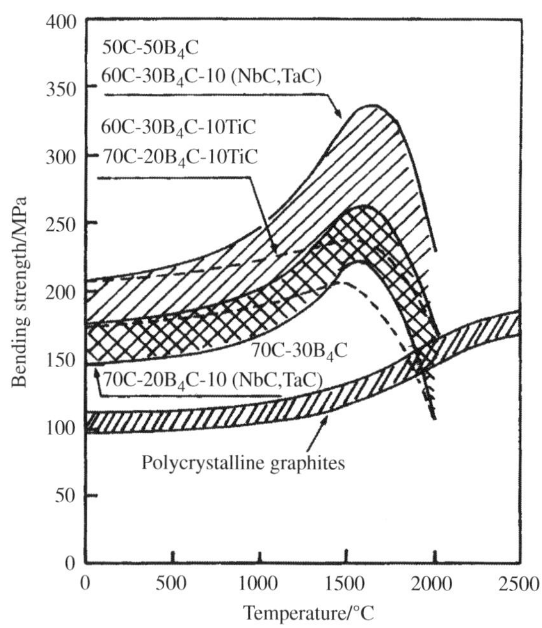


**FIGURE 3.209**

Temperature dependences of bending strength of various metal carbide/carbon composites in comparison with polycrystalline graphite blocks [583]. Here, 60C-30$B_4$C-10TiC, for example, stands for the composite prepared from a mixture of 60 mass% raw coke, 30 mass% $B_4C$ and 10 mass% TiC.

any pressure, in which small carbide particles are dispersed uniformly in a carbon matrix [577]. The addition of $B_4C$ is very effective for the increase in bending strength. Boron is substitutionally incorporated into carbon hexagonal networks up to about 3 at% [584,585]. Associated with this formation of solid solution with boron, the crystallinity of carbon matrix is improved [586–588] and also oxidation resistance increases. The composites with other carbide have also been studied [573,589]. Pressureless sintering is advantageous for the preparation of large-sized and complex-shaped products. In addition, these composites are machinable, particularly by electric discharge machining. This pressureless-sintered $C/B_4C/SiC$ composite has industrially been produced and commercialized.

### d. Coating of ceramic films on carbon materials

Ceramics coating has mostly been done in order to improve the oxidation resistance of carbon materials at high temperatures. There has been a proposal [590] on the fundamental concept to get sufficient oxidation resistance of carbon materials by forming multiplayers as oxide/silica glass/refractory oxide/refractory carbide/carbon. However, it is very difficult in practice to find a good combination to satisfy the physical compatibility between these materials, i.e., matching in their thermal expansion coefficients at each boundary.

Boron oxide has been used to coat carbon materials by different techniques, of which the principal role is to form a layer of boron-containing glass on the surface and avoid the contact between oxygen gas and carbon at high temperatures [591]. Antioxidation is performed to be effective at the temperature range of 600–1000°C, but no oxidation resistance is obtained at temperatures below 600°C because of no formation of the glass layer. Above 1000°C, antioxidation is not expected because of vaporization of boron oxide. In order to get high performance of boron oxide coating, the addition of silica has been tried to modify the property of the glassy phase [583]. Coating of some oxides, which has been used as frits for porcelains, is also reported [592].

The coating of either zircon ($ZrSiO_4$) or mullite ($3Al_2O_3$ $2SiO_2$) on carbon materials is performed by dipping the substrate carbons into the precursor solutions which are prepared from $Zr(OC_4H_9)_4$, $Si(OC_2H_5)_4$ and $Al(NO_3)_3$ by selecting an appropriate procedure [593]. The sol films on the carbon substrates thus prepared are kept at room temperature to gelation. The gel films on the substrates are heated at high temperatures in Ar flow to crystallize to either zircon or mullite.

The deposition of silicon carbide SiC on the surface of carbon materials is done by CVD method and its effectiveness for oxidation resistance has been reported [594,595]. In this case, the outermost layer changes to silica glass, the same as boron oxide. Impregnation of molten silicon into carbon material at 1450°C results in the formation of SiC concentration gradient in carbon matrix, by selecting an appropriate ratio of silicon to physical surface area of carbon material [596]. The coating of either zircon or mullite on the outermost surface of the carbon materials with a SiC concentration gradient is effective to have a good

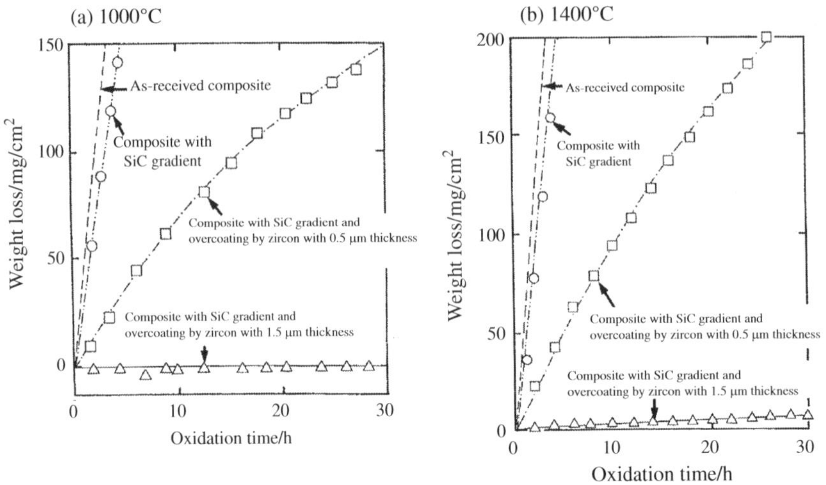


**FIGURE 3.210**

Weight loss of zircon-coated C/Cs with SiC concentration gradient during oxidation in an air flow of 200 mL/min [593].

oxidation resistance at high temperatures as 1400°C [593,597]. In Fig. 3.210, the changes in weight loss per unit area at 1000 and 1400°C with oxidation time in an air flow of 200 mL/min are compared among the starting C/C, that with SiC gradient, those with SiC gradient and over-coated zircon film of the thickness of 0.5 and 1.5 μm. The formation of SiC gradient in the C/C is not effective for the protection from oxidation at 1400°C. For an improvement in oxidation resistance of carbon materials at high temperatures of 1400°C, a coupling of SiC concentration gradient and over-coating of zircon film with a thickness of 1.5 μm is clearly shown to be effective. For this composite, the oxidation loss is negligibly small at 1000°C and only a few mg per square centimeter of physical surface area of the composite at 1400°C after 30 hours. From the XRD measurement, the surface oxidized at 1000°C for 30 hours gave only diffraction peaks of zircon. By oxidation at 1400°C, however, small diffraction peaks of tetragonal-$ZrO_2$ and α-$SiO_2$ are detected in addition of those of zircon, which are supposed to be formed by the phase separation of zircon at the contact with substrate carbon under high temperatures.

On the mullite-coated carbon materials, rapid heating up to 1400°C and quenching down to room temperature are repeated in an air flow, and weight loss of the composites is measured at room temperature. No weight loss is detected even after five times repetition of rapid heating and quenching, as shown in Fig. 3.211. From XRD, only the phase of mullite is observed on all quenched specimens, indicating no decomposition of mullite phase. Under SEM, no cracks are

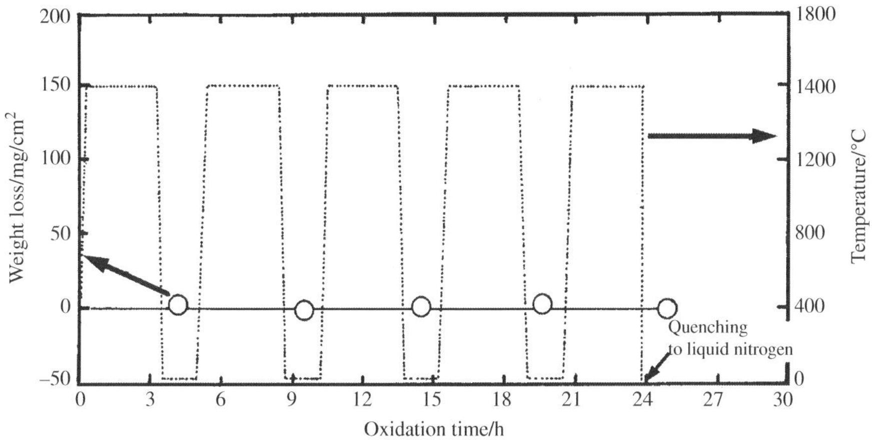


**FIGURE 3.211**

Weight loss of mullite-coated isotropic high-density graphite block with SiC concentration gradient by rapid heating to 1400°C and quenching to room temperature.

*(Courtesy of Prof. O Yamamoto of Akita Univ.)*

observed on the specimen surfaces after quenching to room temperature and only a few after quenching to liquid nitrogen temperature.

## 3.7.5 Carbon/metal composites

Carbon is well known to make interstitial alloys with iron, which resulted in various kinds of iron, from cast iron with high carbon content to steel with low content. Supersaturated carbon in molten iron is precipitated as graphite. When this supersaturated carbon is precipitated as spherical particles, cast iron having high mechanical performance is obtained.

The preparation of carbon-fiber-reinforced Al has been tried in order to obtain metals with both lightweight and high mechanical properties, but no success has been obtained, because the reaction between carbon and Al results in the formation of aluminum carbide $Al_4C_3$, which is very reactive with moisture even at room temperature. MWCNT/Al composites are successfully prepared [598]. MWCNTs with an average diameter of 13 nm and a length of 10–50 μm are mixed with Al powder containing a small amount of magnesium in an elastomer (natural rubber), and then molded into a block, followed by heat treatment at 800°C. Stress–strain curves in Fig. 3.212 shows great improvement in mechanical properties of Al by the addition of only 1.6 vol% MWCNTs. No melting of the composite is observed even at a temperature far above the melting point of Al. The preparation of MWCNT/elastomer composites is the key factor for the reinforcement, which results in high and uniform dispersion of MWCNTs, and also in improvement of their wettability without the formation of carbide.

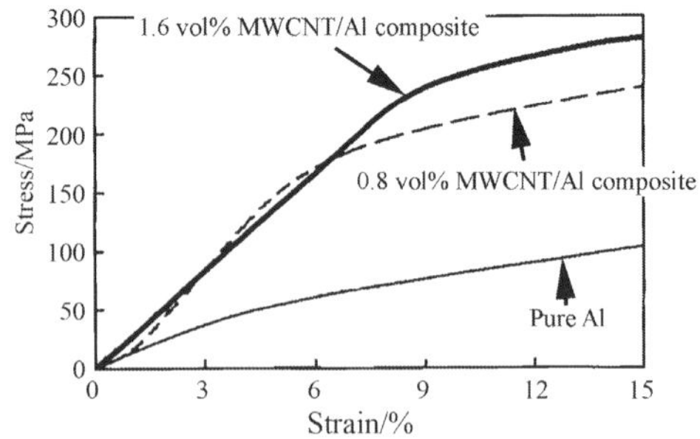


**FIGURE 3.212**

Stress–strain curves in compression for aluminum reinforced by MWCNTs.

*(Courtesy of Dr. T. Noguchi of Nissin Kogyo Co., Ltd.)*

# 3.8 Intercalation compounds

## 3.8.1 Possible applications

The intercalation compounds of graphite have been attracting attention since the first finding of the compounds with sulfuric acid [599]. The drastic changes in color occur from black for the original graphite to golden for the reaction product with potassium vapor to the composition of $KC_8$ having the first stage structure and to dark blue for the compound of $KC_{24}$ having the second stage structure. In 1977, high electrical conductance, comparable to metallic copper, was reported on graphite intercalation compound with $AsF_5$ [600] and then a number of researchers started to work on various intercalation compounds of different carbon materials with different intercalates, though the practical application of these high conductivity compounds was not realized. The intercalation/deintercalation reaction of lithium ions into graphite has successfully been applied to electrochemical energy storage [601], though the first application is obtained by using non-graphitizing carbons, which develops new rechargeable batteries and opens a huge market of carbon materials as the anode, not only graphite but also different carbon materials, such as mesocarbon microbeads. The insertion of alkali metal ions into the tetrahedral and octahedral interstices of cubic closest packing of $C_{60}$ spheres [602], some people calling this phenomenon intercalation, attracted attention because of superconductivity of the resultant compounds.

The intercalation phenomena into various carbon materials have been reviewed in different books and reviews [603–606]. A comprehensive review by focusing mainly on physical properties of intercalation compounds in the relation to new and interesting applications has been published [606]. Structural characteristics and synthesis of intercalation compounds of carbon materials are reviewed in Section 2.8.2.

With respect to the applications of intercalation compounds of carbon materials in graphite family, their functions have been discussed by dividing them into

**Table 3.34** Possible Applications of Intercalation Compounds and their Basic Processes

| Function for Applications | Intercalates | Basic Process |
|---|---|---|
| Highly conductive materials | $AsF_5$, $SbF_5$, $HNO_3$, $CuCl_2$, $FeCl_3$, $F_2$, K–Bi, residual $Br_2$ | Combination between host graphite and intercalates |
| Electrode materials in batteries | | |
| Primary battery | $(CF)_n$, $(C_2F)_n$, graphite oxide, $F_2$, $CoCl_2$, $TiF_4$ | Efficiency enhancement of function of intercalates |
| Secondary battery | Li, $H_2SO_4$, $Ni(OH)_2$, Mn $(OH)_2$ | Intercalation and deintercalation processes |
| Thermocell | $Br_2$, $HNO_3$ | Intercalation and deintercalation processes |
| Catalysts for organic synthesis | Li, K, K–Hg, K–$FeCl_3$, $SbF_5$, $Br_2$, $H_2SO_4$, $HNO_3$, etc. | Efficiency enhancement of function of intercalates |
| Materials for storage of gases and isotope-separation of hydrogen | K, Cs, Rb | Creation of functional space in intercalation compounds |
| Others | | |
| Exfoliation of graphite | $H_2SO_4$, $HNO_3$, $FeCl_3$, K-THF, Na-THF | Intercalation and deintercalation processes |
| Thermal energy storage | $MnCl_2$–$NH_3$ | Intercalation and deintercalation processes |
| Electrochromic | Li-DMSO | Intercalation and deintercalation processes |

five categories, as summarized in Table 3.34 with representative intercalates and the basic process for each function [607].

The high electrical conductivity of some intercalation compounds of graphite is originated from the electronic structure change due to charge transfer between host carbon and intercalate, and therefore depends strongly on the combination between host carbons and intercalates. High electrical conductivity, higher than that of metallic copper, has been found on the intercalation compounds of graphite with $AsF_5$ and $SbF_5$ prepared from highly graphitized carbon materials. Electrochemical and catalytic functions come from intercalates themselves by enhancing their performances through intercalation into the carbon gallery. The recent development of lithium ion rechargeable batteries is principally based on the intercalation and deintercalation reactions into carbon at the anode. Space formed in intercalation compounds plays an important role in their function of adsorption and storage of gases. A large amount of hydrogen is adsorbed into intercalation compounds of alkali metals with different selectivity of hydrogen isotopes. Different hydrocarbon gases are also able to be intercalated together with alkali metals, their oligomerization in the carbon gallery being confirmed.

In this chapter, intercalation compounds of carbon are explained on the basis of their functions summarized in Table 3.34.

### 3.8.2 High conductivity function

Intercalation compounds of graphite with $AsF_5$ and $SbF_5$ have been found to have a high electrical conductivity, in the order of $10^8$ S/m, higher than that of metallic copper [600]. Some other intercalation compounds with $CuCl_2$ and $FeCl_3$, and also with F in an ionic nature of bonding, have a rather high conductivity of $10^7$ S/m. In Table 3.35, the highest values of electrical conductivity reported on various couples between intercalates and host carbon materials are summarized together with some comments. For practical application as electrical conductors, not only their absolute value of electrical conductivity but also their light weight, stability and morphology have to be taken into account.

Electrical conductivity of intercalation compounds, $\sigma_i$, depends strongly on that of the host carbon materials, of which structure and crystallinity are often evaluated by using galvanomagnetic properties, and so $\sigma_i$ is often studied in the relation of electrical conductivity of host carbon $\sigma_h$. In Fig. 3.213, the relation between $\sigma_i$ and $\sigma_h$ for the compounds with $AsF_5$ prepared from three kinds of carbon fibers, vapor-grown carbon fibers (VGCFs), PAN-based carbon fibers and pitch-based carbon fibers with different heat treatment temperature (HTT) [608]. $\sigma_i$ depends strongly on host carbon fibers: in order to realize high conductivity, close to $10^7-10^8$ S/m, the host has to be selected as VGCF, particularly heat-treated at high temperature as 3300°C. For PAN-based and pitch-based carbon fibers, $AsF_5$-intercalation does not promote electrical conduction. Highly oriented pyrolytic graphite (HOPG) has been frequently used as the host, but has not always led to conductivity values as high as $10^8$ S/m [609]. VGCFs and vapor-deposited carbon films, both having been heat-treated at a high temperature above

**Table 3.35** Highest Values of $\sigma_i$ of Various Intercalation Compounds Reported

| Intercalate/Host | The Highest Conductivity (S/m) | Comments |
|---|---|---|
| **Pentafluoride** | | |
| $AsF_5$/HOPG or VGCF | $1 \times 10^8$ | Higher than metallic copper |
| **Metal chlorides** | | |
| $CuCl_2$/MPCF | $7.8 \times 10^6$ | Creditable values of specific conductivity, comparable with copper |
| $FeCl_3$/VGCF | $1.4 \times 10^7$ | |
| **Fluorine** | | |
| $F_2$/HOPG and various CFs | $2 \times 10^7$ | Stable in air |
| **Donnor couple** | | |
| (K + Bi)/HOPG | $2 \times 10^7$ | Stable in air |
| **Residue halogen** | | |
| $Br_2$ or ICl/MPCFs | $1 \times 10^6$ | Vary stable under severe conditions |

*HOPG: highly oriented pyrolytic graphite; VGCF: vapor-grown carbon fibers; MPCF: mesophase-pitch-based carbon fibers; CF: carbon fibers.*

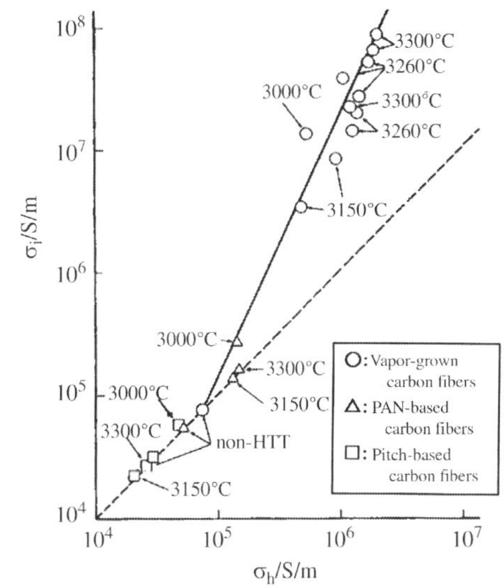


**FIGURE 3.213**

Relation between $\sigma_i$ and $\sigma_h$ for the intercalation of $AsF_5$ into different carbon fibers with different HTTs [608].

3000°C, are successfully used as host graphite, resulting in $\sigma_i$ as high as $10^8$ S/m [610,611]. The corrosive and poisonous nature of pentafluorides, $AsF_5$ and $SbF_5$, as well as the lack of stability in air of the compounds, gave serious obstacles to practical applications. Copper and Teflon tubes have been proposed to be used as protective sheaths [600,612]. It has been reported that the conductivity of the compounds with $AsF_5$ prepared from graphitized VGCFs is stabilized at a value around $10^7$ S/m after exposure to air and the sample stays for more than 2 years at the same value of conductivity [611].

Although the absolute values of $\sigma_i$ for the compounds with various transition metal chlorides, such as $FeCl_3$ and $CuCl_2$, are not very high, they might give creditable values comparable to that of copper if their low specific weight is taken into consideration (specific conductivity) [613–616]. For the compounds with these metal chlorides, the higher $\sigma_h$ gives the higher $\sigma_i$, but $\sigma_i$ is almost 10 times larger than $\sigma_h$, irrespective of the crystallinity of the host and the kind of transition metal chloride, as shown in Fig. 3.214.

The stability of the compounds with metal chlorides has been studied; high stability of the compounds with $NiCl_2$ in air and also in a number of organic solvents [617], and no detectable decomposition of the compounds of $NiCl_2$-$FeCl_3$ in boiling water [618]. On the other hand, instability has been reported on the same kinds of compounds [619]. Full understanding of the reasons behind the stability

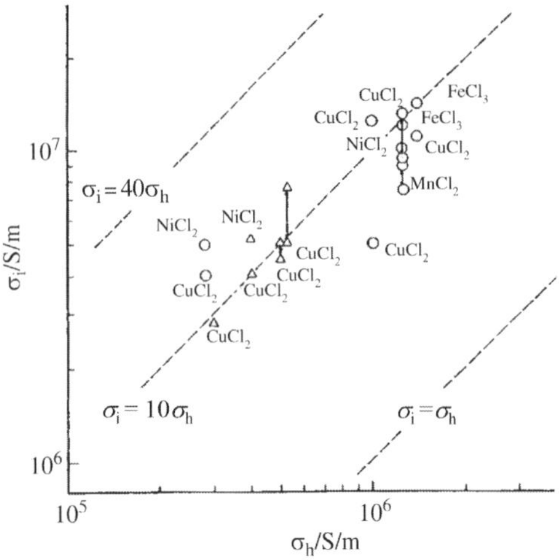


**FIGURE 3.214**

Relation between $\sigma_i$ and $\sigma_h$ for the intercalation of transition metal chlorides.

of these compounds with metal chlorides in an aggressive reagent, as boiling water, is still needed.

Intercalation compounds with fluorine, in which most of the fluorine atoms are considered to exist as ions, in other words, to have ionic interaction with graphite layers, were found to be synthesized under the coexistence of some metal fluorides at relatively low temperatures, much lower than that for the synthesis of the covalent compounds, graphite fluorides $(CF)_n$ and $(C_2F)_n$ [620,621]. These 'ionic' compounds with fluorine have relatively high $\sigma_i$, which greatly depends on the host carbon materials and also slightly on the coexisting metal fluorides during synthesis. The compounds prepared from HOPG and graphitized VGCF yielded the conductivity as high as $10^7$ S/m [622,623]. In Fig. 3.215, $\sigma_i$ of the compounds with ionic fluorine prepared from different carbon fibers is plotted against the content of fluorine. Using graphitized VGCFs, high $\sigma_i$ as $10^7$ S/m is obtained. The stability of these compounds with ionic fluorine is supposed to be responsible for the presence of partial fluorine atoms covalently bonded at the edge of the carbon layers [623].

Most donor-type intercalation compounds of graphite, such as the compounds with potassium in stage-1 structure $KC_8$, also have high $\sigma_i$, but their marked reactivity with humidity in air interferes with their practical applications not only for electrical conductors but also for other functional materials. Ternary compounds, in which triple layers of potassium-bismuth-potassium are formed in graphite gallery, are found to be relatively stable and to have relatively high $\sigma_i$ [624,625].

Emphasizing the stability for aerospace applications, residual compounds with bromine prepared from carbon fibers have been studied extensively, the electrical

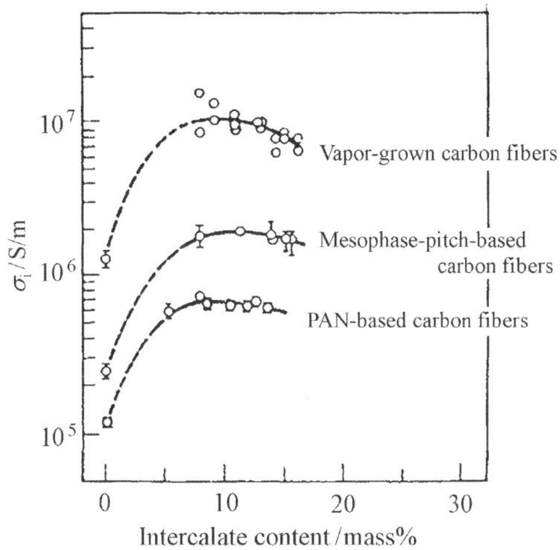


**FIGURE 3.215**

$\sigma_i$ for the compounds with ionic fluorine prepared from different graphitized carbon fibers.

*(Courtesy of Prof. T. Nakajima of Aichi Inst. Tech.)*

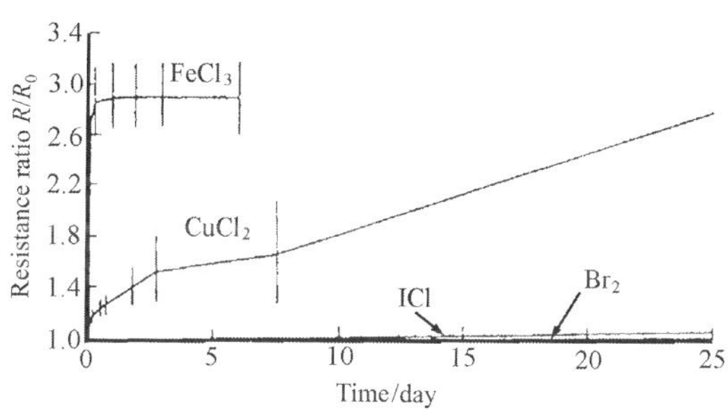


**FIGURE 3.216**

Changes in relative resistance $R/R_0$ of different residue compounds with time under the condition of 100% humidity at 60°C.

*(Courtesy of Dr. J. R. Gaier of NASA, USA.)*

conductivity of which reached the order of $10^6$ S/m [619,626]. This residue compound is stable under high humidity and a high vacuum. In Fig. 3.216, the stability of electrical resistance in a severe condition of 100% humidity at 60°C is compared among different residue compounds of $FeCl_3$, $CuCl_2$, ICl and $Br_2$, showing high stability of the latter two residue compounds.

On the intercalation compounds of graphite, superconductivity has been reported, as summarized in Table 3.36. The compounds containing alkali metals, $KC_8$, $KC_4Hg$, etc., have been mostly studied, but critical temperature $T_c$ reported is lower than 2 K [627–631]. The compounds, $LiC_2$, $KC_4$ and $CsC_4$, synthesized

**Table 3.36** Interlayer Distance $d_i$ and Critical Temperature $T_c$ of Intercalation Compounds

| Compound | $d_i$ (nm) | $T_c$ (K) | References |
|---|---|---|---|
| $KC_8$ | 0.535 | 0.15 | [29, 30] |
| $KC_4Hg$ | 1.015 | 0.73 | [31] |
| $KC_8Hg$ | 1.351 | 1.93 | [32] |
| $RbC_8Hg$ | 1.421 | 1.46 | [32] |
| $LiC_2$ | | 1.9 | [33] |
| $KC_4$ | | 5.5 | [34] |
| $CsC_4$ | | 6 | [34] |
| $YbC_6$ | 0.457 | 6.5 | [35] |
| $CaC_6$ | 0.451 | 11.5 | [35–38] |
| $Li_3Ca_2C_6$ | 0.97 | 11.15 | [39] |

under a high pressure show relatively high $T_c$, 1.9, 5.5 and 6 K, respectively [632]. The compound YbC6 synthesized via vapor transport method (two-bulb method) showed relatively high $T_c$ of 6.5 K [633]. The $T_c$ for the compounds containing Ca, $CaC_6$ and also $Li_3Ca_2C_6$, is a little higher than 11 K [633–637].

$CaC_6$ has been synthesized by the two-bulb method [633,635] and also by immersing in molten Li–Ca alloy [634]. Electronic structure of $CaC_6$ synthesized in Li–Ca alloy is studied in detail [638,639]. The compound $Li_3Ca_2C_6$ is also synthesized via liquid Li–Ca alloy [637]. Synthesis conditions for superconductive compounds in Ca–Li alloys have been studied in detail [640]. High crystallinity of the host graphite is a key to attaining superconductivity [641].

### 3.8.3 Electrochemical functions

Electrode reactions studied by using the intercalation compounds of various carbon materials are summarized in Table 3.37.

Graphite fluoride $(CF)_n$, a covalent compound with fluorine, has been successfully used as an electrode material for lithium primary batteries [642]. The electrode reaction is expressed by Eq. 3.12 in Table 3.37. It provides many advantages to the primary batteries, such as light weight, small size, relatively high voltage (about 2.5 V), high energy density (285 Wh/kg), high preservability and small self-discharge [643]. In Fig. 3.217, discharge curves of lithium-$(CF)_n$ primary cells are shown for various discharge conditions [643]. Even after storage at 20°C for 5 years, no marked change in discharge performance was detected (Fig. 3.217a) and also even at as low temperature as −20°C a relatively high discharge capacity was obtained. In Fig. 3.218, the structure of a commercially available cell of pin-type is illustrated.

**Table 3.37** Electrochemical Reactions Used in the Electrodes of Intercalation Compounds

| General Formular | Practical Examples | | Batteries |
|---|---|---|---|
| $C_nA + B^+ + e^- \leftrightarrow nC + AB$ | $(CF)_n + nLi^+ + e^- \leftrightarrow nC + LiF$ | (3.12) | Primary batteries |
| $nC + A^+ + e^- \leftrightarrow CnA$ | $nC + Li^+ + e^- \leftrightarrow C_nLi$ | (3.13) | Lithium ion rechargeable batteries |
| $nC + A^- \leftrightarrow C_nA + e^-$ | $nC + H_2SO_4 + HSO_4^- \leftrightarrow C_nH_2SO_4\ nHSO_4 + e^-$<br>$nC + ClO_4^- \leftrightarrow C_nClO_4 + e^-$ | (3.14)<br>(3.15) | Secondary batteries |
| $C_nA + B^- \leftrightarrow C_nAB + e^-$ | $C_nNi(OH)_2 + OH^- \leftrightarrow C_nNiOOH + e^- + H_2O$ | (3.16) | Secondary batteries |

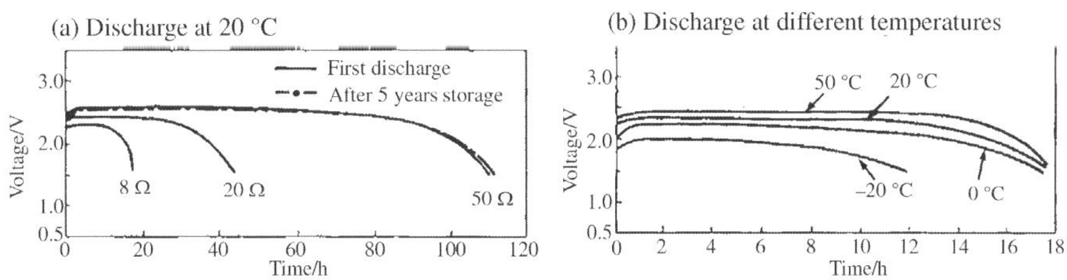


**FIGURE 3.217**

Discharge curves of lithium-$(CF)_n$ primary cell under different conditions [643].

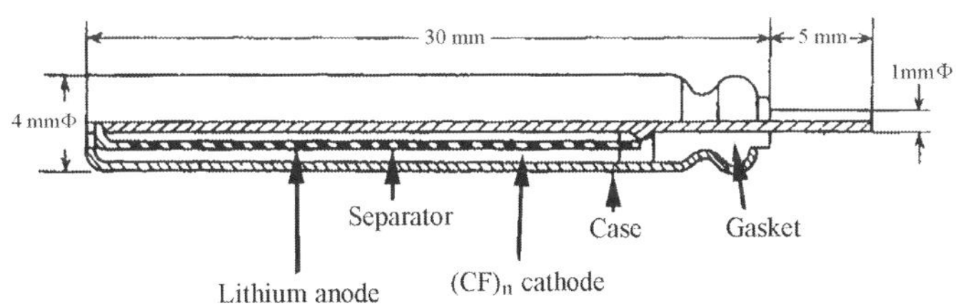


**FIGURE 3.218**

Commercially available pin-type primary cell [643].

The compound $(C_2F)_n$ gives higher potential than $(CF)_n$ [644]. Also, an intercalation compound with fluorine synthesized in the existence of $MgF_2$ (ionic compound with fluorine) is also used as electrode of lithium primary cell [645]. In Fig. 3.219, discharge curves of lithium primary cells with different compounds with fluorine are compared.

Another covalent compound, graphite oxide with a composition of $C_4O(OH)$, gives a high energy density (1200 Wh/kg), although the open-circuit voltage is slightly low (2.2 V) [646].

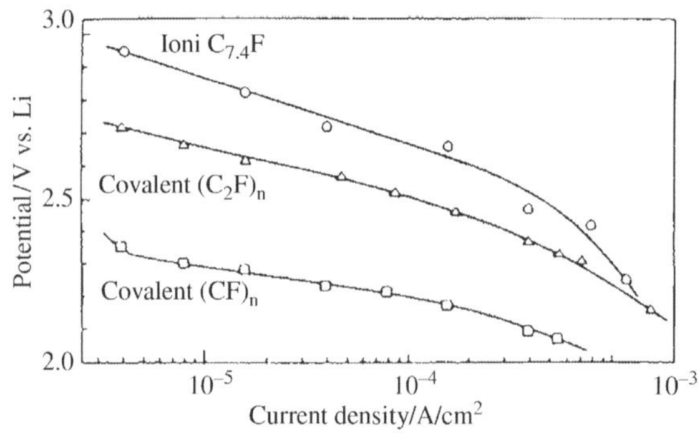


**FIGURE 3.219**

Discharge characteristics of lithium primary cells using different intercalation compounds with fluorine [645].

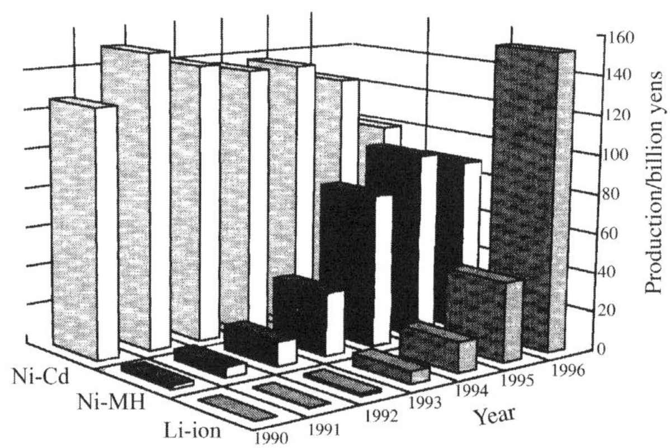


**FIGURE 3.220**

Production of rechargeable batteries in Japan.

A great success has been achieved by the use of carbon anode in lithium ion rechargeable batteries [601]. Not only graphite but also various carbon materials have been used as an anode for the batteries, the principal electrode reaction being supposed to be intercalation and de-intercalation of lithium ions (Eq. 3.13 in Table 3.37). These lithium ion rechargeable batteries are commonly used in cellular phones and mobile computers. Industrial production has accelerated markedly since 1996, and has become greater than other rechargeable batteries, Ni–Cd and Ni–metal hydride batteries in Japan, as shown in Fig. 3.220. Since the composition of stage-1 intercalation compounds of graphite with Li is $LiC_6$, the theoretical capacity of batteries using graphite as anode is 372 mAh/g. However, practical capacity values for most carbon anodes are lower than

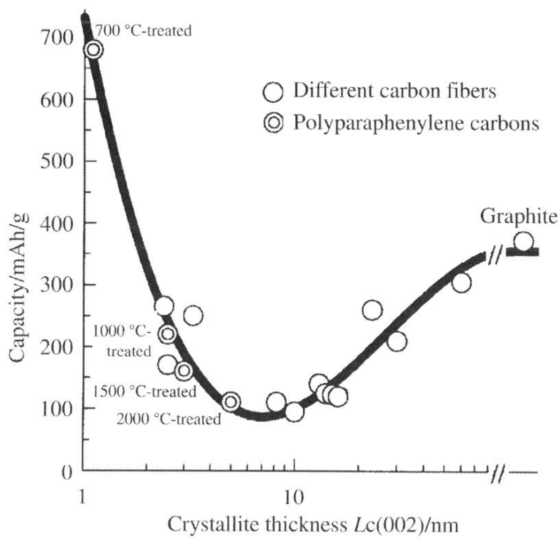


**FIGURE 3.221**

Dependence of anode capacity on crystallite thickness $L_c(002)$ of different carbon materials.

*(Courtesy of Prof. Endo of Shinshu University.)*

theoretical values, and depend strongly on the crystallinity of the host carbon materials. In Fig. 3.221, the capacity of batteries with various carbon fibers as the anode are plotted against crystallite thickness $L_c(002)$, as a measure of crystallinity of the host carbon materials, showing a minimum at certain crystallinity [647,648]. A similar dependence of battery capacity on the crystallinity of anode carbon materials is reported on various carbon materials [649]. In non-graphitizing carbons, the development of a crystalline structure is poor even after heat treatment at temperatures as high as 3000°C, and so low capacity was reported, which also revealed that heat treatment temperature cannot be a parameter to evaluate the anode capacity for carbon materials.

However, a very high capacity, much higher than the theoretical value, has been found on some low-temperature-treated carbons. A value higher than 1000 mAh/g has been reported [650–652]. The data on carbons derived from polyparaphenylene, one of non-graphitizing carbons, are plotted against $Lc(002)$ in Fig. 3.221. On these non-graphitizing carbons, the capacity decreases with increasing crystallite thickness $Lc(002)$ by high-temperature treatment, as shown on polyparaphenylene carbons in the figure. These carbon materials usually show a very high irreversible capacity, which become a high barrier to their practical use as anodes. The reason why these low-temperature-treated non-graphitizing carbons can accept such a large amount of lithium ions is still controversial, including whether all of these lithium ions are intercalated into the carbon gallery. Three models in Fig. 3.222 have been proposed for why non-graphitizing carbons

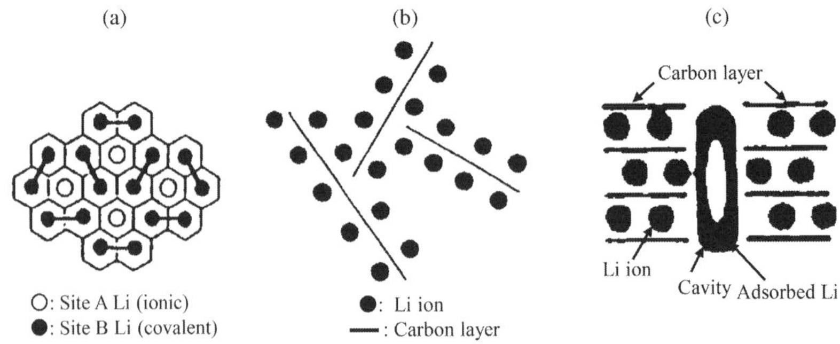


**FIGURE 3.222**

Models proposed to interpret a high anode capacity of low-temperature-treated carbon materials [651–653].

can contain a large amount of lithium ions. In the model (a), all lithium ions are registered to hexagons of carbon layer and the theoretical composition is $LiC_2$ (theoretical capacity of 1120 mAh/g) [651]. In the model (b), lithium ions are located on both sides of single carbon layers, which are the main structural units in these low-temperature carbons [653], and the model (c) suggests a simultaneous occurrence of both intercalation into the gallery of small carbon layers and capillary condensation in the pores (cavities) existed in these carbons [652].

A great prospect for these lithium ion rechargeable batteries is also anticipated to extend their application to electric vehicles and electricity storage for its leveling. The development of lithium ion rechargeable batteries has been reviewed by focusing on their researches, industrial production and applications [654]. The carbon materials for the anode of these lithium ion rechargeable batteries are also discussed in Section 3.9.1 as one of the tools for energy storage.

A variety of trials have been reported for the applications of various intercalation/deintercalation reactions to rechargeable batteries. The intercalation reactions of the perchlorate ion $ClO_4^-$ (Eq. 3.15 in Table 3.37) and bisulfate ions $HSO_{4-}$ into graphite electrodes (Eq. 3.14 in Table 3.37) were attempted for use as electrode reactions [655,656], but the repeated charge–discharge cycles were found to lead to the disintegration of the graphite electrode. However, a composite electrode of natural graphite (80 mass%) with polypropylene (20 mass%) was reported to withstand over 2500 cycles of charge–discharge in sulfuric and hydrofluoric acids [657]. Alkali rechargeable battery using carbon fibers and NiOOH as electrode materials with 30% KOH electrolyte solution were constructed [658]. Although cell performance was not so high, open-circuit voltage of 1.80 V and energy density of 14–22 Wh/kg, its cyclic performance was reported to be high, about 20% reduction of battery capacity after 500 cycles of charge–discharge.

Repetition of charging by chemical oxidation and electrochemical discharging was reported on a combination of graphite electrodes and $H_2SO_4$ electrolyte

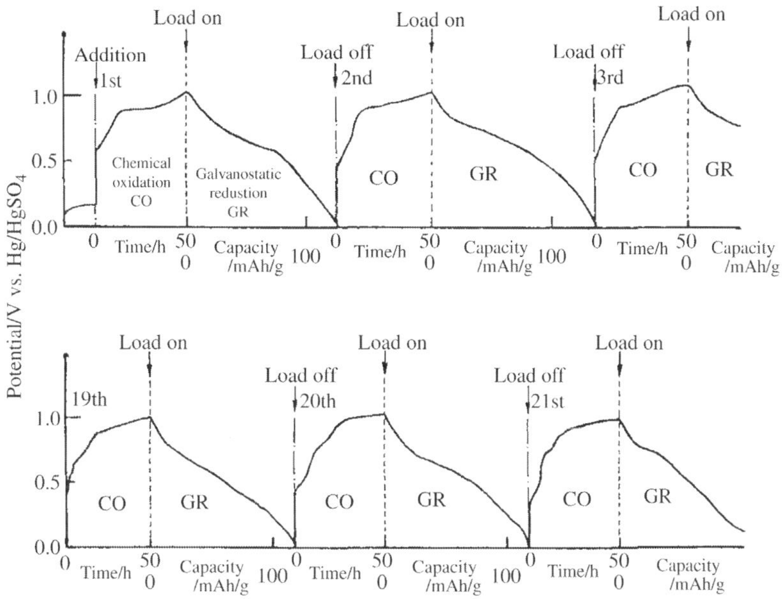


**FIGURE 3.223**

Potential changes with the repetition of chemical oxidation with nitric acid and galvanostatic reduction with a current of 4 μA for the combination of the electrode of natural graphite and $H_2SO_4$ electrolyte.

[659,660]. In Fig. 3.223, the potential changes with the repetition of chemical oxidation by nitric acid (intercalation of sulfuric acid, charging) and galvanostatic reduction with the current of 4 μA (deintercalation of sulfuric acid, discharging) are shown. The discharge capacity in every cycle of galvanostatic reduction is about 120 mAh/g. No change in the potential curves during chemical oxidation and galvanostatic reduction are observed even after the 21-s cycle.

The transformation of $NiCl_2$ to $Ni(OH)_2$ in the graphite gallery may offer an effective usage of nickel hydroxide for Ni–Cd cells [661,662]. A simple secondary battery composed of carbon fiber anode and 30% KOH aqueous electrolyte was also proposed [663].

A kind of concentration cell, a thermocell, using either Br-saturated KBr solution or $HNO_3$ has been proposed, which can convert a small temperature difference between two electrodes to electric power [664,665]. In Fig. 3.224, the structure of one of the thermocells used and the scheme of reactions at the electrodes is illustrated. The reaction of $Br_2$ molecules to $Br^-$ ions at the cathode and the reverse reaction at the anode are supposed to occur through the intermediate formation of intercalation compound with Br. The cell performance depends strongly on the construction of the cell, such as the lengths of the bridge between cathode and anode where $Br_2$ gas and $Br^-$ are diffusing. By using graphitized

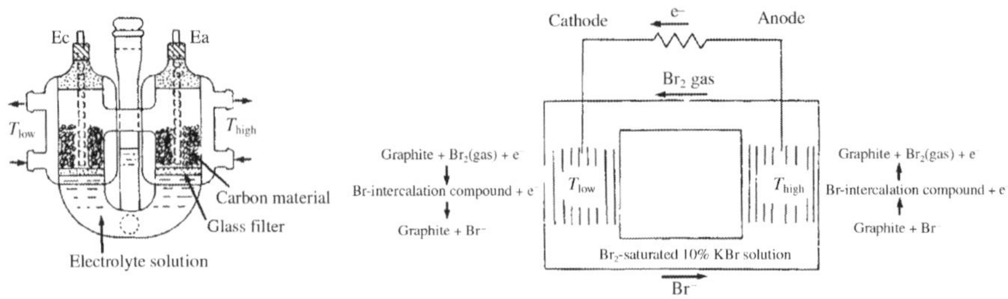


**FIGURE 3.224**

A thermocell for the conversion of a small temperature difference between two electrodes to electric power.

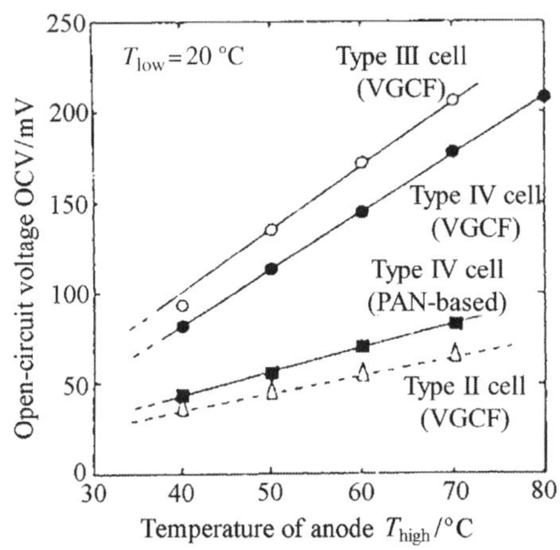


**FIGURE 3.225**

Dependences of open-circuit voltage of the thermocells on the temperature of the anode $T_{high}$ [667]. Different types of cell with VGCFs and PAN-based carbon fibers are used.

VGCFs as the electrode, an open circuit voltage of 200 mV and short circuit current of 10 mA/cm$^2$ were obtained from the temperature difference of about 80°C [666,667]. Dependences of open-circuit voltage on temperature of the anode, i.e., temperature difference between two electrodes, are shown for the cells with a slightly different construction in Fig. 3.225. By using PAN-based carbon fibers and nitric acid, electric power was continuously obtained from waste hot water for 2 years without any maintenance action [668a].

### 3.8.4 Catalytic functions

Various intercalation compounds, i.e., various carbon materials with various intercalates, have been tested for use in the synthesis process of different organic compounds. The role of the compounds is divided into two cases, the catalyst for accelerating the reactions and the support of reagents in carbon gallery [668b,c].

In the former case, two cases have to be differentiated: the compounds that act as catalysts in a strict meaning and do not show any change in structure, and those that help the formation of reaction intermediates but decompose during the reaction. Intercalation compounds with alkali metals are the effective catalysts for the polymerization of various hydrocarbons, such as ethylene and styrene, not only to increase the yield of polymerization but also to improve the selectivity. Published results of polymerization reactions of some hydrocarbons using the compounds with alkali metals are summarized in Table 3.38 [669].

For the polymerization of isoprene and butadiene, the use of $LiC_6$ and $KC_8$ as a catalyst in a solvent of cyclohexane is effective, the polymerized products being yielded in a relatively high percentage. In the case of isoprene, $KC_{24}$ gives a little higher yield than $LiC_6$, which is explained by the intercalation of monomer into

**Table 3.38** Catalytic Polymerization of Hydrocarbons Using Alkali Metal-Intercalation Compounds

| Monomer | Intercalation Compounds | Yield (mass %) | Composition of Polymers | Conditions |
|---|---|---|---|---|
| Ethylene | $KC_8$ | – | *trans,* monoolefine | 200°C, 68 bar, 21 h, iso-octene |
| Butadiene | $KC_8$ | 40–80 | 51%: 1–2, 49%: *trans* 1–4, | 30°C, 15 h, cyclohexane |
| | $KC_{12}$ | 51.3 | 90%: 1–4, 10%: 1–2 | 15°C, 100 h, cyclohexane |
| | $KC_{12}$ | 21.5 | 88%: 1–4, 12%: 1–2 | 15°C, 100 h, toluene |
| | $KC_{24}$ | – | 66%: 1–2, 35%: *trans* 1–4 | 75°C, 12 h, cyclohexane |
| Isoprene | $KC_8$<br>$LiC_{12}$<br>$KC_{37}$ | 90<br>80<br>95 | 60%: 3–4, 5%: 1–2, 35%: *trans* 1–4<br>44%: *cis* 1–4, 21%: 3–4, 35%: *trans* 1–4<br>43%: 1–4, 57%: 3–4 | 25°C, 12 h, *n*-heptane<br>15°C, 168 h, cyclohexane<br>15°C, 76 h, cyclohexane |
| Methyl-methacrylate | $LiC_{12}$<br>$KC_8 + KC_{24}$ | 80<br>44 | 4%: *iso*, 63.6%: *syndio*, 32.4%: *atactic*<br>0.5%: *Iso*, 50.5%: *syndio*, 49%: *atactic* | −63°C, 48 h, DME<br>23°C, 4 h, DME |

the graphite gallery widened by the intercalation of alkali metal ions, K ion being larger than Li ion [670]. In Fig. 3.226, molecular weight distributions are shown for the caprolactone polymerized by using different intercalation compounds of graphite [671]. The catalysts $LiC_{12}$ and $KC_8Hg$ give a simple distribution in molecular weight and the average molecular weight of the polymer obtained by using the latter was pretty high as $1.14 \times 10^5$.

The synthesis of ammonia from nitrogen and hydrogen gases is accelerated by the presence of graphite intercalation compounds with metal halogenides, as shown in Table 3.39 [672a]. Ternary intercalation compound with K and $FeCl_3$ gives a several hundred times higher formation rate than binary compounds with either K or $FeCl_3$ [672b].

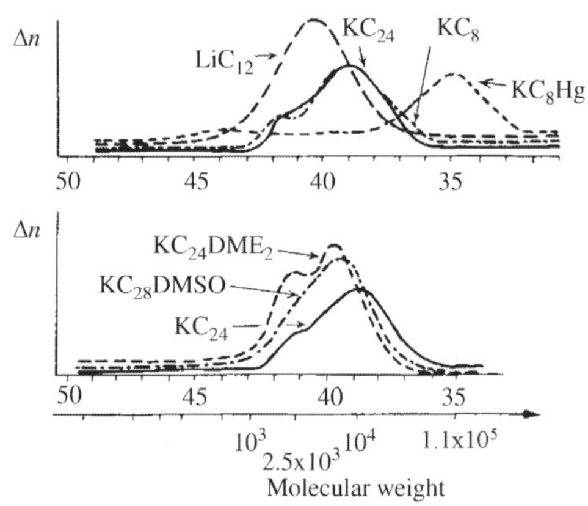


**FIGURE 3.226**

Molecular weight distribution of caprolactone polymerized by using different intercalation compounds of graphite [671].

**Table 3.39** Synthesis of Ammonia in the Existence of Graphite Intercalation Compounds [672a]

| Graphite Intercalation Compound | | Gas Pressure (Torr) | | Temperature (°C) | Ammonia Formed (cc/h) | Conversion Ratio After 10 h Reaction (%) |
|---|---|---|---|---|---|---|
| | Composition | $N_2$ | $H_2$ | | | |
| Graphite-$FeCl_3$-K | 2 : 0.2 : 2 | 100 | 300 | 350 | 10.6 | 90 |
| | | 100 | 300 | 300 | 5.1 | 55 |
| | | 230 | 350 | 30 | 0.0 | 1.5 |
| Graphite-K | 2 : 1 | 80 | 240 | 306 | 0.008 | 0.1 |
| Graphite-$FeCl_3$ | 2 : 0.2 | 100 | 300 | 300 | 0 | 0 |

The intercalation compounds with K are known to have a high catalytic activity for the reactions concerning hydrogen; hydrogenation of benzene to cyclohexane, dehydrogenation of benzene to biphenyl, etc. Exchange of hydrogen with deuterium is markedly accelerated in the presence of $KC_8$ and $KC_{24}$ [673]. Graphite intercalation compounds are effective for isomerization of various hydrocarbon molecules [665,674].

An example of the cases where the reagents are kept in graphite gallery as intercalation compounds is Br, intercalation compounds of Br being a good reagent for bromination of hydrocarbons with a high selectivity [675]. Intercalation compounds with $SbCl_2$ are able to exchange bromine in organic molecules by chlorine [676].

### 3.8.5 Gas adsorption and storage

Intercalation compounds of graphite with K can react with hydrogen gas to form the compounds with two different structures; $KC_8H_{2/3}$ formed from stage-1 $KC_8$ at room temperature [677] and $KC_{24}(H_2)_{1.9\sim2.1}$ from stage-2 $KC_{24}$ at around liquid nitrogen temperature [678–680]. Some characteristics of these two compounds are summarized in Table 3.40.

In $KC_8H_{2/3}$, intercalates form K-H-K triple layer along the normal to carbon hexagonal layer with the stage-2 structure (Fig. 2.180a). Since $KC_8$ adsorbs hydrogen preferentially to form $KC_8H_{2/3}$, hydrogen is concentrated in the compound rather than other isotopes, deuterium and tritium.

From the viewpoint of hydrogen storage, $KC_{24}(H_2)_2$ has certain advantages, including a small change in size accompanied by the adsorption/desorption of hydrogen, complete reversibility of adsorption and desorption of hydrogen, simple evacuation or heating for desorption, and only a small effect by contaminants such as oxygen. However, some disadvantages are also pointed out, such as the storage capacity being slightly smaller than a hydrogen storage alloy $LaNi_5H_6$, 15.5 L(NTP) per 100 g for $LNi_5H_6$ but 13.7 L per 100 g for $KC_{24}(H_2)_{1.9}$, and the need for as low a temperature as 77 K for the hydrogen storage into $KC_{24}$.

**Table 3.40** Ternary Graphite Intercalation Compounds with Potassium and Hydrogen

| | High-Temperature Form | Low-Temperature Form |
|---|---|---|
| Composition | $KC_8H_{2/3}$ | $KC_{24}(H_2)_2$ |
| Starting GIC | $KC_8$ (stage-1) | $KC_{24}$ (stage-2) |
| Reaction temperature | Room temperature | 77 K |
| Structure | Stage-2 K–H–K triple layer | Stage-2 $H_2$ molecules between K |
| Formation energy | 15–17 kcal $mol^{-1}$ | 2.2–2.9 kcal $mol^{-1}$ |
| Isotope effect | Concentration of H | Concentration of D |
| Exchange | Formation of HD | No formation of HD |

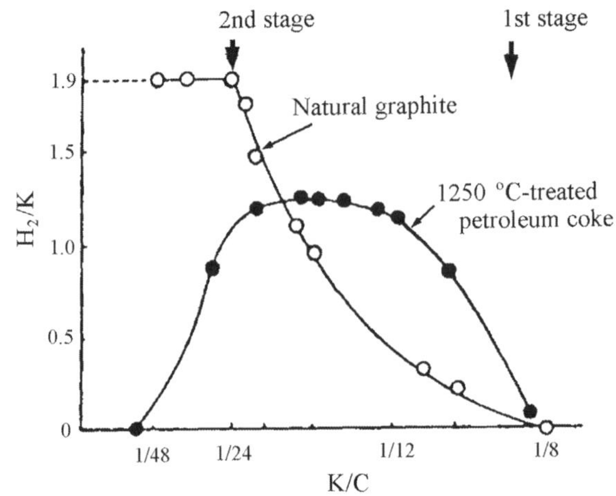


**FIGURE 3.227**

Hydrogen adsorption in the intercalation compounds with K prepared from different hosts [680].

Adsorption of hydrogen into intercalation compounds with K depends strongly on the structure of the host carbon materials. In Fig. 3.227, the adsorbed amount of hydrogen per K, $H_2/K$, is plotted against the composition of adsorbent compounds K/C for two different host carbon materials [680]. When natural graphite is host, the compounds with higher stage structure than 2 ($K/C < 1/24$) can adsorb the maximum amount of hydrogen, i.e., $H_2/K = 1.9$, but the compounds with the composition in between stage-2 and -1 ($K/C = 1/24-1/8$) $H_2/K$ decreases gradually, the change being explained by the coexistence of stage-2 ($KC_{24}$) with stage-1 ($KC_8$). For the compound prepared from petroleum coke heat-treated up to 1250°C, however, higher adsorption of hydrogen in the composition range of $1/24-1/8$ is obtained.

The use of the intercalation compounds of graphite with alkali metals as materials for cryosorption has been proposed [681]. In Fig. 3.228, isotherms of the equilibrium pressure $P_{eq}$ vs. the volume of hydrogen gas adsorbed per gram of intercalation compounds, $V_{ads}$, are shown for the graphite intercalation compounds with three alkali metals and a molecular sieve MS-5A. Even under as low a pressure as 1 Torr, the intercalation compounds with Cs and Rb could adsorb a relatively large amount of hydrogen, about 10 times larger than that with K and MS-5A. Table 3.41 shows the equilibrium pressure $P_{eq}$ of hydrogen and deuterium gases at two fractional adsorptions, $\Phi$, of 0.2 and 0.5 for four adsorbents. Into the intercalation compound with Cs, a half of maximum adsorption of hydrogen ($\Phi = 0.5$) can be attained at as low a pressure as 0.32 Torr, although the compounds with K and MS-5A needs 32 and 83 Torr, respectively. The equilibrium pressure of adsorption depends also on hydrogen isotopes, $H_2$ and $D_2$.

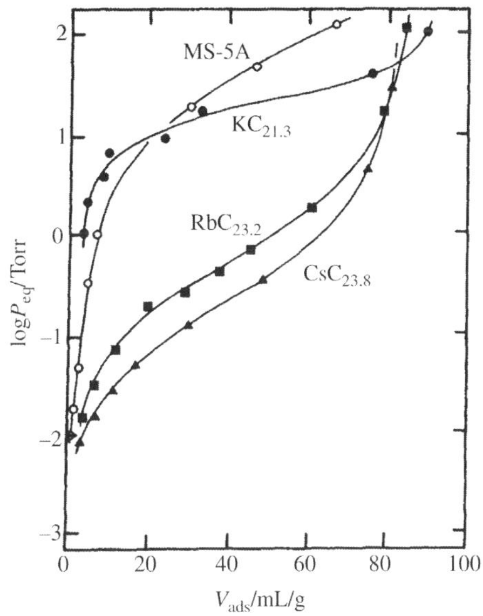


**FIGURE 3.228**

Adsorption isotherms of hydrogen at 77 K [681].

**Table 3.41** Equilibrium Pressure $P_{eq}$ at Fractional Adsorption $\Phi$ of Hydrogen and Deuterium [681]

| Adsorbent | | Gas | Equilibrium Pressure $P_{eq}$ (Torr) | |
|---|---|---|---|---|
| | | | $\Phi = 0.2$ | $\Phi = 0.5$ |
| Graphite intercalation compounds | with Cs | $H_2$ | 0.05 | 0.32 |
| | | $D_2$ | 0.04 | 0.25 |
| | with Rb | $H_2$ | 0.16 | 0.71 |
| | | $D_2$ | 0.05 | 0.20 |
| | with K | $H_2$ | 8.6 | 32 |
| Molecular sieve MS-5A | | $H_2$ | 11 | 83 |

For the separation of hydrogen isotopes, the structure and texture of the host carbon materials play a decisive role [682–684]. In Table 3.42, separation coefficients of hydrogen isotopes are listed for the intercalation compounds with K prepared from different host carbon materials, comparing with a molecular sieve MS-5A. High separation coefficient for $H_2$ to $D_2$ and to HT is obtained on the intercalation compounds with K, much higher than MS-5A. The compound $KC_{12}$

**Table 3.42** Separation Coefficients and Adsorption Volume of Hydrogen Isotopes for the Intercalation Compounds with K and a Molecular Sieve MS-5A

| Host | Intercalation Compound | Separation Coefficient | | Adsorption Volume (mL of $H_2$, NTP) |
|---|---|---|---|---|
| | | $H_2/D_2$ | $H_2/HT$ | |
| Petroleum coke heat-treated at 1500°C | $KC_{12}$ | 8.7 | 5.0 | 92 |
| | $KC_{24}$ | 5.7 | 3.7 | 111 |
| Petroleum coke heat-treated at 2300°C | $KC_{12}$ | 7.1 | 4.5 | 72 |
| | $KC_{24}$ | 5.7 | 3.7 | 120 |
| Graphite sheet | $KC_{22}$ | 5.9 | – | 119 |
| Molecular sieve MS-5A | | 2.6–2.8 | 1.7 | 102 |

*(Courtesy of Prof. N. Akuzawa of Tokyo Nat. Coll. Tech.)*

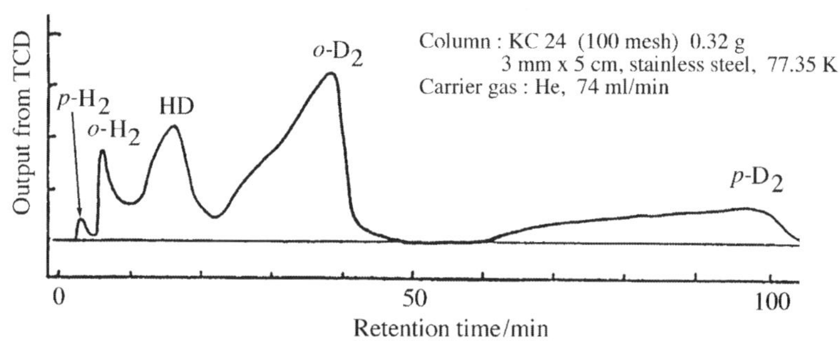


**FIGURE 3.229**

Gas chromatogram of hydrogen gas using a column of $KC_{24}$.

*(Courtesy of Prof. Terai of Tokyo Univ.)*

prepared from 1500°C-treated petroleum coke gives higher coefficients than the same petroleum coke heat-treated at 2300°C and graphite sheet manufactured from natural graphite.

It has been shown experimentally that these intercalation compounds with K are an effective column material for gas chromatography of hydrogen isotopes, only a 5 cm column of $KC_{24}$ giving clear separation of $p$-$H_2$, $o$-$H_2$, HD, $o$-$D_2$ and $p$-$D_2$ around 77 K, as shown in Fig. 3.229 [685], much better separation than a 5 m long column of alumina.

Intercalation compounds with alkali metals are able to adsorb different kinds of hydrocarbon gases [686]. In Table 3.43, the saturated adsorption of hydrocarbons, $C_mH_n$/Cs, and the reversibility of the adsorption process are summarized on stage-2 $CsC_{24}$. In the case of $n$-alkanes, the adsorbed amount decreases with

**Table 3.43** Adsorption of Various Hydrocarbon Gases into Stage-2 Graphite Intercalation Compound with Cs, $CsC_{24}$ [686]

| Hydrocarbon $C_mH_n$ | Vaporization Point (K) | Adsorption Temperature (K) | Saturated Adsorption $C_mH_n$/Cs | Reversibility of Adsorption |
|---|---|---|---|---|
| $CH_4$ | 111.7 | 90–181 | 1.2 | Reversible |
| $C_2H_6$ | 184.5 | 194–252 | 0.9 | Reversible |
| $C_3H_8$ | 231.1 | 252–287 | 0.69 | Reversible |
| n-$C_4H_{10}$ | 272.7 | 275–326 | 0.55 | Reversible |
| n-$C_5H_{12}$ | 309.2 | 300–327 | 0.54 | Reversible |
| n-$C_6H_{14}$ | 341.9 | 327–353 | 0.9 (0.44) | Reversible |
| $C_2H_4$ | 169.5 | 194–323 | 2.1 (1.1) | Reversible (< 200 K) Irreversible (> 273 K) |
| $C_3H_6$ | 225.4 | 253–273 | 0.8 | Reversible |
| $C_2H_2$ | 189.2 | 194–273 | 0.65 | Irreversible |

increasing molecular size and all of these adsorption processes are reversible, easily desorbed by simple cryopumping. In contrast, some alkenes and alkynes containing double and triple C–C bonds are irreversibly adsorbed into $CsC_{24}$. The isotherms of ethylene adsorption at different temperatures are shown in Fig. 3.230 [688]. Below 250 K, adsorption of ethylene $C_2H_4$ occurs in two steps, the value $C_2H_4$/Cs of around 1.1 at the first plateau and about 2.1 at the second. At 194 K, the adsorption of $C_2H_4$ is completely reversible, as in the case of *n*-alkanes. Above 273 K, however, only the first plateau is observed and the adsorption becomes irreversible. This irreversible adsorption is explained by the oligomerization of $C_2H_4$ molecules in the graphite gallery due to some catalytic action of $Cs^+$, on the basis of gas chromatography/mass spectrometry (GC/MS) analysis of toluene extracts from the compound and in situ calorimetry of the adsorption process [688].

### 3.8.6 Other functions

Abrupt heating of either graphite intercalation compounds or their residue compounds induces a marked exfoliation of the host graphite, preferentially perpendicular to its layers, mainly due to the rapid decomposition of intercalates to gaseous species in the graphite gallery. Flexible graphite sheets, which are prepared from exfoliated graphite by roll-forming without any binder, are currently in demand for use as gaskets, packing and thermal insulators at high temperatures [689,690]. The exfoliation process and the pore structure in the exfoliated graphite are described in Sections 2.7.3b and 3.6.3. The preparation, properties and

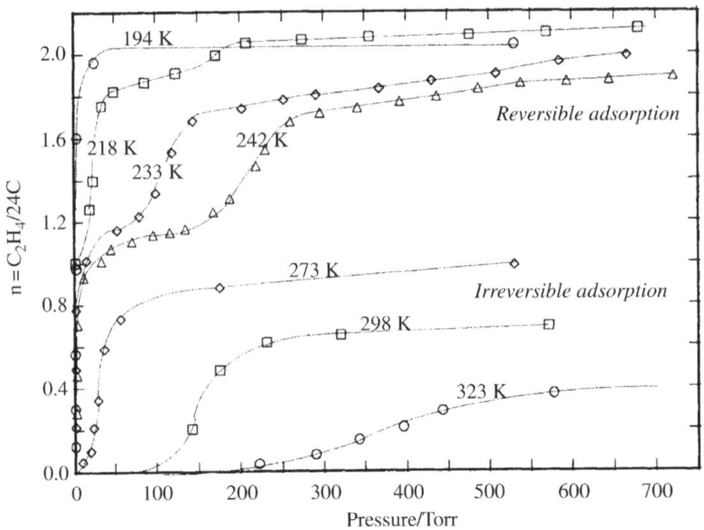


**FIGURE 3.230**

Adsorption isotherms of ethylene $C_2H_4$ into stage-2 $CsC_{24}$ at different temperatures [687].

applications of flexible graphite sheets prepared from exfoliated graphite were reviewed in Section 3.2.5. Also, exfoliation of natural graphite flakes is attracting attention as one of the routes to prepare graphenes (Section 3.5.3b).

In most industrial production of exfoliated graphite, residue compounds have been used, which are obtained from the graphite intercalation compounds with sulfuric acid by washing. In the process of this production, different problems have to be overcome; the necessity of using concentrated sulfuric acid for the compounds' preparation by either chemical oxidation or electrochemical oxidation, the formation of poisonous oxide gases $SO_x$ and $NO_x$ during exfoliation, and the possibility of erosion of metals by a trace of sulfur remaining in the resultant graphite sheets. Therefore, different intercalates, such as $FeCl_3$ and Na-THF, have been tested to prepare exfoliated graphite, as summarized in Table 3.44 with some comments for merits and demerits of the process [305,691–694]. The synthesis of the graphite intercalation compounds with formic acid through electrochemical oxidation and their exfoliation by rapid heating are proposed to prepare the exfoliated graphite without any sulfur [693].

Exfoliated graphite can sorb a large amount of heavy oil very quickly, more than 80 g of heavy oil per 1 g of exfoliated graphite within 1 min, which seems to be promising for the recovery of heavy oil spilled in water [695–697], as described in the Section 3.10.2.

A system for the storage of thermal energy has been proposed, of which the scheme is shown in Fig. 3.231a [698]. Thermal energy is stored during deintercalation of ammonia molecules from ternary intercalation compounds with metal

**Table 3.44** Intercalates Used for the Exfoliation of Graphite

| Intercalates Used | Comments |
|---|---|
| $H_2SO_4$ with $HNO_3$ or electrolysis | Industrially applied process |
| | Necessity of concentrated acid and formation of $SO_x$ and $NO_x$ |
| $FeCl_3$ | Possibility of using the hydrated $FeCl_3$ |
| | Formation of metallic Fe particles |
| Na-THF or K-THF | Intercalation at room temperature and no possibility to include sulfu. |
| | Dispersion of fine alkali metal particles in exfoliated graphite |
| Co-THF | Intercalation at room temperature |
| | Dispersion of fine Co metal particles, magnetic |
| $LiClO_4$-propylene carbonate | Spontaneous exfoliation in air at room temperature |
| $SbCl_5$ | Applied for exfoliation of graphitized vapor-grown carbon fibers |
| Formic acid | No possibility to include sulfur |
| | Necessity of electrochemical intercalation |

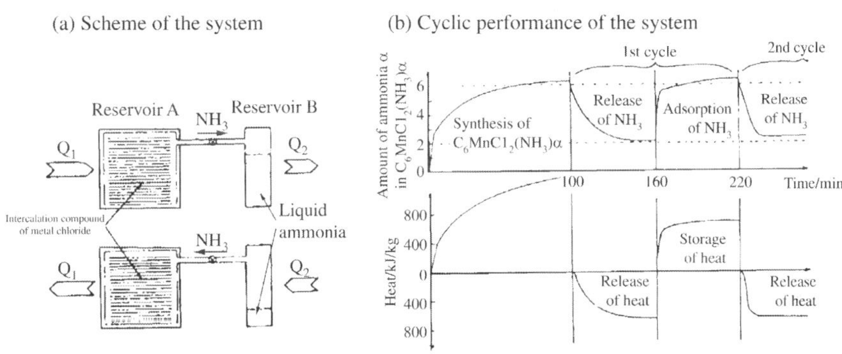


**FIGURE 3.231**

Scheme and an example of performance for thermal energy storage/emission cycle by using graphite intercalation compounds with metal chloride.

*(Courtesy of Prof. Ph. Touzain of Univ. of Grenoble.)*

chloride and $NH_3$, and released by a reverse reaction, i.e., formation of ternary intercalation compounds by the intercalation of $NH_3$ and de-intercalation to binary compounds with metal chloride. An example of the performance for the cycle of thermal energy storage/release is shown for the intercalation compound with $MnCl_2$ in Fig. 3.231b. A thermal capacity of about 200 Wh per 1 kg of the compound is reported.

A reversible change in color due to intercalation and deintercalation of Li, black in the host graphite and blue in the intercalation compound with Li, in the organic electrolyte has been tested for the use in electrochromic devices [699,700].

## 3.9 Carbon materials for energy storage

### 3.9.1 Rechargeable batteries

The fundamental electrochemical reactions in lithium ion rechargeable batteries are intercalation/deintercalation of lithium ions at both electrodes, as expressed by Eq. 3.13 in Section 3.8.3 and schematically illustrated in Fig. 1.8, in which lithium ions are extracted from cathode material, either ordered NaCl-type $LiCoO_2$ or spinel-type $LiMn_2O_4$, and inserted into graphite anode during discharging, and vice versa during charging. Therefore, the theoretical capacity of the anode is 372 mAh/g, because of the formation of the stage-1 intercalation compound with lithium, $LiC_6$.

The intercalation of lithium ions into graphite gallery is demonstrated to occur accompanied by stage structure change, as shown by its charge–discharge curve in Fig. 3.232 [701]. This charge–discharge curve shows clearly the presence of potential plateaux, corresponding to the stepwise formation of stage structures; the stage-2 structure occurs at the potential around 0.13 V and the stage-1 at around 0.09 V. At a high potential range down to about 0.21 V, no marked

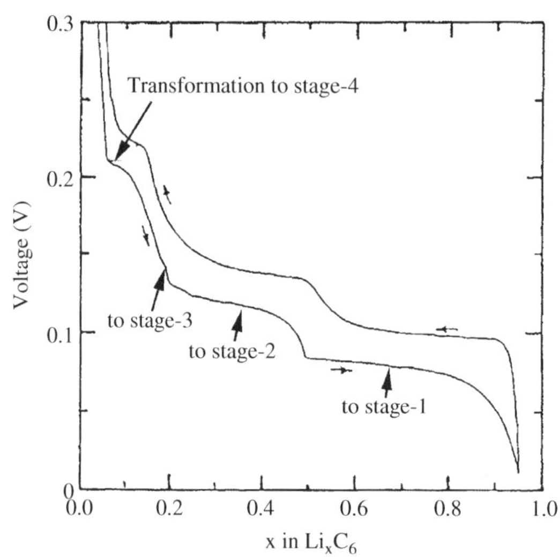


**FIGURE 3.232**

Charge-discharge curves for the anode of natural graphite [701].

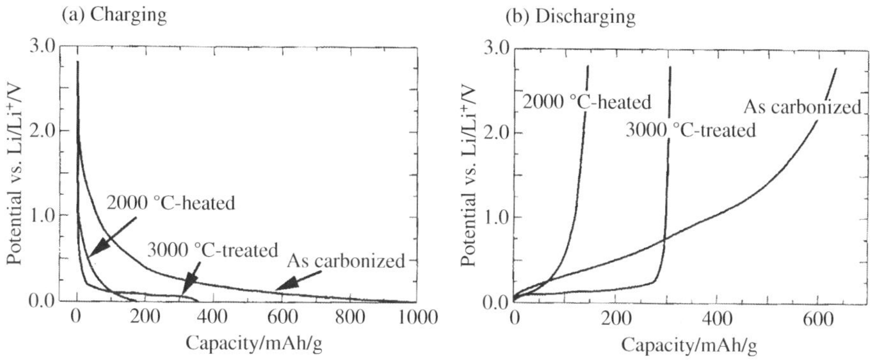


**FIGURE 3.233**

Charge and discharge behaviors for the coke with different HTTs [705].

plateau are observed, but a slight and gradual shift of 002 diffraction line of graphite where the stage-4 structure is clearly observed.

The charge–discharge curves for the anodes have been studied on various carbon materials, most of them being reviewed in different books [702–704]. In Fig. 3.233, charge and discharge curves observed on a coke with different heat treatment temperatures (HTTs) are shown [705]. This result shows clearly that anode performance of carbon materials in lithium ion batteries strongly depends on the structure of carbon. For well-graphitized carbon with high crystallinity, as 3000°C-treated coke in Fig. 3.233, rather flat discharge and charge plateaux are observed at a potential near zero volt, discharge capacity is very close to the theoretical value of 372 mAh/g for graphite, and charge capacity is only a little different from discharge capacity, i.e., relatively high Coulombic efficiency. For the carbon with poor crystallinity, i.e., as-carbonized coke in Fig. 3.233, however, potential changes gradually in both discharging and charging processes, no plateaux being detected. For 2000°C-treated coke, the capacity for discharging process is very small and only a small plateau is observed in the charging process.

A similar change in the charging process with different HTTs has been reported on mesocarbon microbeads (MCMB). The effect of high-temperature treatment on charging behavior of MCMB is shown in Fig. 3.234 [706]. With increasing HTT above 2000°C, plateau in discharging process becomes clear and longer, i.e., increase in discharge capacity. At low HTT below 2000°C, charge and discharge curves for MCMBs change also with HTT, as shown in Fig. 3.235 [707], discharge capacity determined at the potential up to 2 V increasing with decreasing HTT.

For different carbon materials derived from non-graphitizing carbons, having poor crystallinity, results similar to low-temperature-treated MCMB have been reported, i.e., large irreversible capacity and a discharge capacity larger than theoretical value, as can be seen by the comparison between charge and discharge curves at a HTT in Fig. 3.235. In Fig. 3.236, charge–discharge behavior is shown

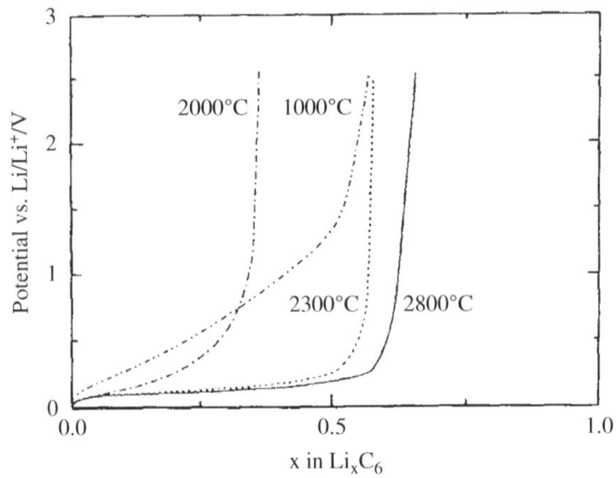


**FIGURE 3.234**

Discharge curves for MCMBs heat-treated at high temperatures [706].

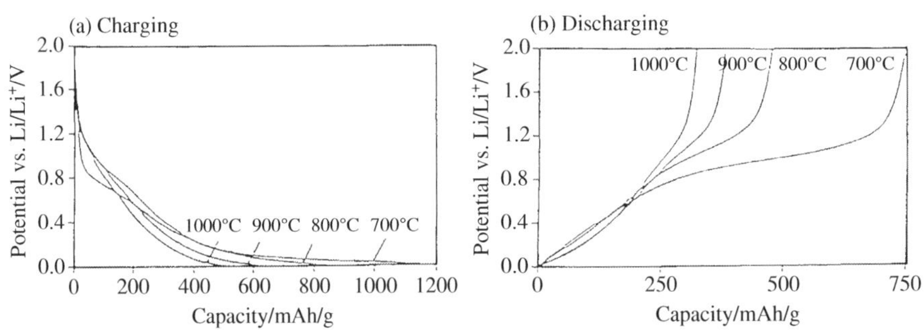


**FIGURE 3.235**

Charge and discharge curves for MCMBs heat-treated at low temperatures [707].

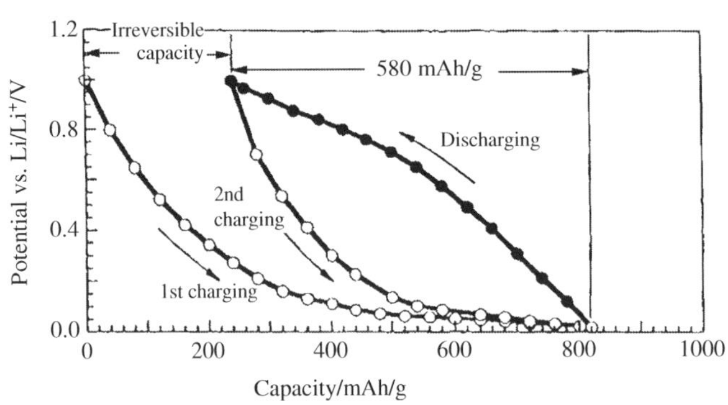


**FIGURE 3.236**

Charge-discharge curves for polyacene carbonized at about 700°C [708].

by plotting the discharge curve against the same abscissa to charging for the carbon derived from polyacene [708]; gradual change of potential in both discharging and charging processes, a large irreversible capacity as 230 mAh/g, and a high reversible capacity as 580 mAh/g, higher than the theoretical one.

Anode capacity has been studied on various carbon materials by differentiating graphitizing and non-graphitizing carbons with different HTTs [653,709]. Non-graphitizing carbons give high capacity, higher than the theoretical one, at low HTT below 1000°C and their capacity decreases with increasing HTT, keeping low capacity even after 3000°C treatment. Capacity of graphitizing carbons, on the other hand, increases with increasing HTT above 2000°C and approaches the theoretical one [653]. Increase in HTT accelerates the growth of crystallites in carbon, depending strongly on the nanotexture of carbon materials, as explained in Section 2.5. Since crystallite growth is strongly depressed, the non-graphitizing carbon has a small *Lc*(002) even after heat treatment at 3000°C, being comparable to low-temperature-treated graphitizing carbons. A plot of capacity against *Lc* (002) for mesophase-pitch-based carbon fibers (graphitizing) and the carbon derived from polyparaphenylene (non-graphitizing) with different HTTs shows a unique dependence of capacity on *Lc*(002), as shown in Fig. 3.221 [709]. Fig. 3.237 shows the same conclusion by plotting the amount of Li intercalated into carbon anode, $x$ in $Li_xC_6$, corresponding to capacity against degree of graphitization $P_1$ for various carbon materials [648,710], revealing that intercalated lithium becomes a minimum at small $P_1$, around 0.1−0.2. With increasing $P_1$, in

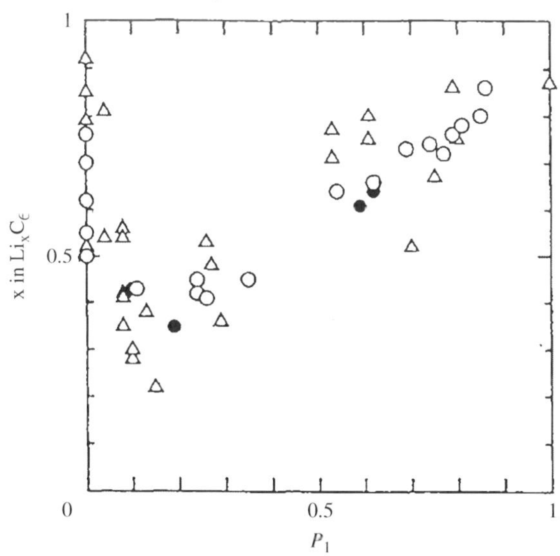


**FIGURE 3.237**

Dependences of capacity (measured as x in $Li_xC_6$) on $P_1$ [648].

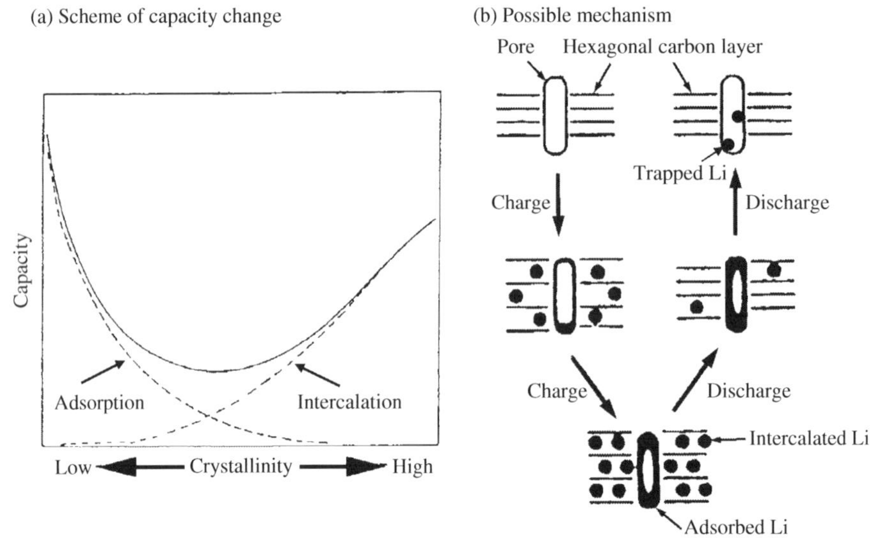


**FIGURE 3.238**

Scheme for the change in capacity with crystallinity of anode carbon (a) and possible mechanism for Li storage (b).

other words, increasing crystallinity intercalated lithium increases gradually, but it increases abruptly with decreasing $P_1$ below 0.1.

In Fig. 3.238a, the scheme for the change in capacity with crystallinity, where the position of capacity minimum depends on the parameters to evaluate the crystallinity, by $Lc(002)$ at around 10 nm (Fig. 3.221) and by P1 at around 0.1–0.2 (Fig. 3.237). This scheme suggests that two mechanisms work for the storage of Li in anode carbon, one increasing and the other decreasing with improving crystallinity. The former, capacity increase with increasing crystallinity, is due to the intercalation of $Li^+$ into the graphite gallery. The amount of intercalated $Li^+$ increases with the improvement of crystallinity, in other words, the growth of crystallites increasing the accommodation places for $Li^+$. Though quick increase in capacity with decreasing crystallinity (decreasing $P_1$ below 0.2) has not been understood completely, it is supposed to be due to adsorption of $Li^+$ into nanopores. Most carbons, particularly the carbons with random nanotexture and poor crystallinity, contain a large amount of nanopores, which possibly adsorb $Li^+$. Some of the adsorbed Li may not be able to be recovered from nanopores by discharging, i.e., trapped lithium, which causes irreversible capacity, as illustrated in Fig. 3.238b.

The species intercalated into graphite gallery are lithium ions, although they are solvated by solvent molecules in the electrolyte solution. The intercalation occurs from the edge of crystallite, and so solvated lithium ions have to be decomposed into lithium ions and solvent molecules; the former can intercalate

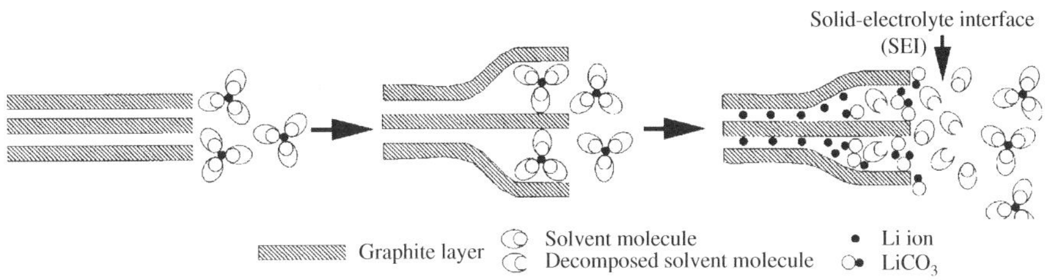


**FIGURE 3.239**

Illustration of the formation of SEI during intercalation of lithium ions [711].

into graphite gallery and the latter remains at the edge surface to form a solid–electrolyte interface (SEI), as schematically demonstrated in Fig. 3.239 [711]. The formation of SEI is so important to get high performance of batteries, i.e., high capacity and good cyclability, as to select the anode carbon material and electrolyte. The analysis of the composition of this SEI, however, showed the presence of a certain amount of $LiCO_3$ in addition to electrolyte molecules, such as ethylene carbonate (EC), etc. [712].

Cokes heat-treated up to 2200°C have a long cycle life, longer than natural graphite, and excellent pulse charge/discharge characteristics, although discharge capacity is relatively low [713]. With high charge/discharge rates, more than 3C, the coke heat-treated at 2200°C was reported to have higher discharge capacity than natural graphite.

Doping of boron into graphite structure increases the capacity of batteries due to the up-ward shifting of the potential for the lithium intercalation/deintercalation reaction [714,715]. Not only an increase in discharge capacity and a decrease in irreversible capacity but also improvement of cyclability by boron-doping have been observed on the graphitized mesophase-pitch-based carbon fibers [716].

By the addition of a small amount of high-temperature-treated vapor-grown carbon fibers (VGCFs) with small diameter of about 0.2 μm, a pronounced improvement on Coulombic efficiency of the battery is reported [717], as shown in Fig. 3.240. This improvement of cyclability is due to the formation of the network of carbon fibers, which can absorb and retain certain amount of electrolyte and can provide sufficient resiliency and compressibility to the electrode. Into the electrode materials of lead-acid rechargeable batteries, the addition of a small amount of graphitized VGCFs improves the performance of the batteries [718]. The only 1.5 mass% addition of graphitized VGCFs into the positive electrode results in a drastic decrease in resistivity, about 7 orders of magnitude decrease. When graphitized VGCFs are added into the negative electrode by 0.5–1.0 mass%, cyclic performance is greatly improved. These marked effects of the addition of graphitized VGCFs into the electrodes are owing to low electrical resistivity and mechanical strength of graphitized VGCFs.

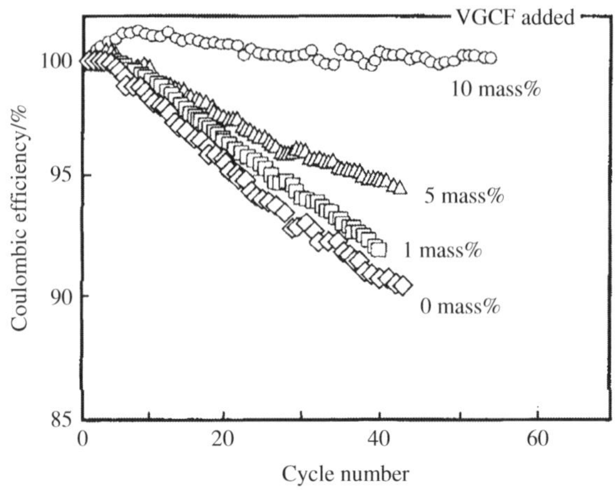


**FIGURE 3.240**

Cycle performance for the carbon anode added different amounts of high-temperature-treated VGCFs [717].

Coating of electrode materials has been shown to be effective to improve battery performances. Carbon coating of graphite results in an increase in reversible capacity, decrease of irreversible capacity, and improvement of cyclic performance [64–66]. Carbon coating of oxide materials, $Li_4Ti_5O_{12}$ for anode and $LiFePO_4$ for cathode, is essential for achieving high reversible capacity and stable charge/discharge performance, because of easy control of charge state of cations, keeping particle size of oxides small, improving electrical conductivity of the electrode sheets and improving wettability of oxides with organic electrolytes [719–721]. Carbon coating of different materials, including electrode materials, is reviewed [722].

Since the invention of the Lclanche cell in 1864, various carbon materials have been used in various primary batteries. In a lithium primary battery, graphite fluoride $(CF)_n$, which is one of the intercalation compounds of graphite and consists of covalent bonds between carbon atoms in graphite layers and fluorine (Section 3.8.3), is successfully used as an electrode material. In other types of primary batteries, such as alkaline batteries, polycrystalline graphite rods have also been used. Not only in lithium ion batteries, but also in various secondary batteries, different carbon materials are used; carbon fabrics for sodium-sulfur cells, electrodes of carbon fibers and bipolar plates for zinc-bromine cells, carbon fiber cloths with high surface area for redox flow cells, etc. In Table 3.45, carbon materials used in primary and secondary batteries are listed, together with active materials and electrolytes.

Graphite materials are important structural components in different types of fuel cell. A separator for a fuel cell has to be gas-impervious because of the separation of two fuel gases, such as oxygen and hydrogen, and so be made of

**Table 3.45** Carbon Materials Used in Primary and Secondary Batteries

| Battery | | Potential (V) | Active Materials | | Electrolyte | Carbon Materials Used |
|---|---|---|---|---|---|---|
| | | | Cathode | Anode | | |
| PRIMARY | Manganese | 1.5 | $MnO_2$ | Zn | $ZnCl_2$, $NH_4Cl$ | Carbon rod (collector), CB (conductive) |
| | Alkaline | 1.5 | $MnO_2$ | Zn | KOH | Graphite (collector) |
| | Siver oxid | 1.00 | $Ag_2O$ | Zn | KOH | Graphite (collector) |
| | Mercury | 1.35 | HgO | Zn | KOH or NaOH | Graphite (collector) |
| | Air | 1.4 | Air ($O_2$) | Zn | $NH_4Cl$ or KOH | AC (electrode) |
| | Lithium | 3 | $(CF)_n$ | Li | | $(CF)_n$ (electrode) |
| | | 3 | $MnO_2$ | Li | | Graphite, CB (collector) |
| | | 3.6 | $SOCl_2$ | Li | $SOCl_2 + LiCl + AlCl_4$ | Carbon rod (collector) CB (conductor) |
| SECONDARY | Nickel-Cadmium | 1.3 | NiOOH | Cd | KOH | Graphite (collector) |
| | Air | 0.2–1.2 | Air ($O_2$) | Zn or Fe | KOH | |
| | Sodium-sulfur | 1.0–2.1 | S | Na | $\beta$-$Al_2O_3$ | CF (catalyst support) |
| | Zinc-bromine | 1.8 | Br | Zn | $ZnBr_2$ | CF cloth, CFRP (electrode) |
| | Redox flow | 2.1 | Fe ion | Cr ion | HCl, HBr | CF felt (electrode) |
| | Lithium ion | 3.4–3.7 | Li-oxide | Graphite | Organic electrolyte | Graphite (electrode), CB (conductive) |

*CB: carbon black, CF: carbon fiber, $(CF)_n$: graphite fluoride.*

glass-like carbon. The plate formed by a phenol resin is carbonized under controlled processes. The electrodes for collecting electricity are also made of graphite materials and the porous carbon materials, such as felt of carbon fibers, have been used as suscepter for minute particles of platinum catalyst.

## 3.9.2 Electrochemical capacitors

### a. Construction and characteristics of electrochemical capacitors

Electric double-layer capacitors (EDLCs) attracted attention as one of the devices for the storage of electrical energy. It is based on the formation of electric double-layers on the surface of electrodes, as schematically shown in Fig. 3.241, where cations and anions of an electrolyte form Helmholtz layers on the surface of both electrodes. In its construction of cell and also in its performance, it may be located between batteries, e.g., lithium ion rechargeable batteries mentioned in the previous section, and ferroelectric capacitors (condensers). In Table 3.46, three devices are compared. In performance, EDLCs have relatively low energy density in comparison with batteries, but can be charged and discharged quickly, much faster than batteries; in batteries electrochemical reactions, intercalation and deintercalation of lithium ions, determines the rate of charging and discharging, but in EDLCs the electric double-layers are quickly formed on the surfaces of the electrodes. From performance, EDLCs are very similar to ferroelectric capacitors, low energy density but quick charge/discharge process.

The advantages of EDLCs are summarized as follows; (1) possible to charge/discharge quickly, (2) long cycle life because of no chemical reactions, just adsorption/desorption of electrolyte ions, (3) high efficiency for charge/discharge cycle, (4) no heavy metals used and consequently environment friendly, and (5) possible to discharge in high current density. On the basis of these advantages

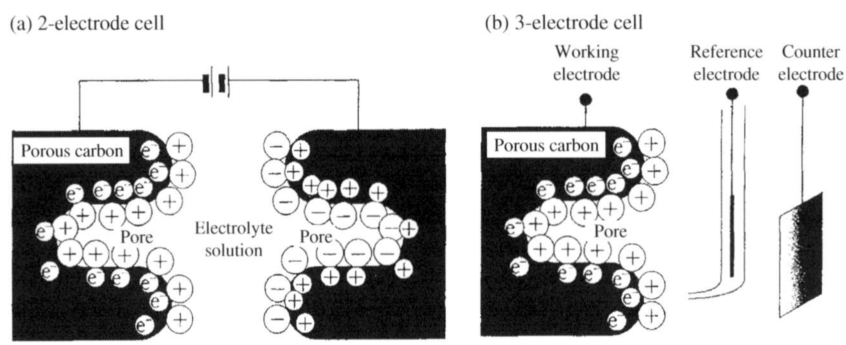


**FIGURE 3.241**

Schematic illustrations of the formation of electric double layers on the surface of carbon electrodes in 2-electrode and 3-electrode cells.

**Table 3.46** Three Devices for Electrical Energy Storage

| | | Batteries | Electric Double-Layer Capacitors | Ferroelectric Capacitors (condensers) |
|---|---|---|---|---|
| Construction | Functional materials | Electrolyte | | Ferroelectrics |
| | Fundamental process | Electrochemical reaction | Adsorption | Polarization |
| Performance | Energy density | High | Low | |
| | Charge/discharge | Slow | Fast | |
| | Current density | High | Small | |
| | Life | Limited | Long | |

and, in addition, because of their lightness, low cost and maintenance-free operation, they are now used as memory backup power sources for various electronic devices [415,723,724]. There have been many investigations for EDLCs to attain high energy density in order to use them in electric vehicle drivelines.

Various electrolytes have been used, which can be classified as aqueous and non-aqueous solutions. In the former, $H_2SO_4$ and KOH are often used as electrolyte. In the latter, propylene carbonate (PC) solution of either $(C_2H_5)_4NBF_4$ ($Et_4NBF_4$), $(C_2H_5)_4PBF_4$ ($Et_4PBF_4$) or $LiClO_4$ are frequently used. In order to test the EDLC performance of electrode materials, two set-ups have been employed, one consists of two carbon electrodes separated by a separator (Fig. 3.241a) and the other has only one carbon electrode coupled with a counter electrode, usually platinum or nickel metal, and a reference electrode, such as $Hg/HgSO_4$ (Fig. 3.241b). The capacitance of the electrode material was determined from either charge/discharge or cyclic voltammogram (CV) curves.

In order to characterize carbon materials as the electrode of EDLC, their specific capacitance was determined by using an electrolyte, either aqueous or non-aqueous electrolyte, with a cell, either 2-electrode or 3-electrode type, calculating either from charge/discharge curves or CV curve. The value of capacitance strongly depends on what electrolyte, which cell and also which curve are used, in other words, not only the value of capacitance but also the electrolyte, the cell and the measurement method have to be specified. The capacitance value measured by the 2-electrode type cell is 1/4 of that by the 3-electrode type cell [725]. The calculated capacitance from charge/discharge curves is usually a little different from that calculated from the CV curve.

Carbon materials for electrochemical capacitors have been reviewed by focusing on their preparation processes and their structure and texture [726].

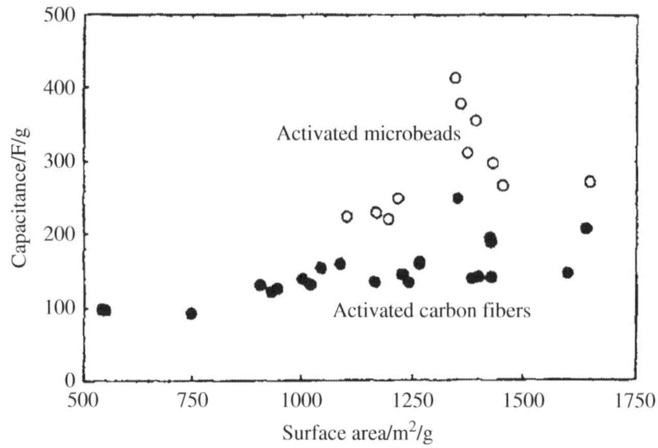


**FIGURE 3.242**

EDLC capacitance vs. total surface area measured by DFT for activated carbon microbeads and activated carbon fibers [727].

### b. In aqueous electrolytes

For a number of activated carbon microbeads and activated carbon fibers, capacitance of EDLC has been measured in 30 mass% KOH aqueous electrolyte [727]. In Fig. 3.242, the capacitance measured is plotted against total surface area determined by the DFT method. Activated carbon microbeads have higher capacitance than activated carbon fibers, although both of them have relatively high surface area as 1700 $m^2/g$, and it is difficult to deduce a common relation between capacitance and surface area.

In Fig. 3.243, the capacitance measured on various activated carbon fibers and activated carbons in 30 mass% $H_2SO_4$ aqueous solution is plotted against BET surface area $S_{BET}$ [728]. In this plot, any defined relation between capacitance and $S_{BET}$ cannot be concluded. It may say that high surface area is desirable for high capacitance. However, the experimental point for the carbon derived from poly(vinylidene chloride) (PVDC) at 700°C in $N_2$ atmosphere (no activation process has been applied) shows a high capacitance (about 250 F/g), although $S_{BET}$ is rather low (about 700 $m^2/g$) [728]. The pore size distribution on this PVDC-derived carbon showed that most pores in this carbon had a size less than 0.9 nm.

Defluorination of poly(tetrafluoroethylene) (PTFE) gives porous carbons, of which microporous and mesoporous surface areas are almost the same, and irradiation of $\gamma$-rays increases microporous surface area [729] (Section 3.6.2c). Theses PTFE-derived carbons gave very high capacitance of 200–250 F/g in 1 M $H_2SO_4$ electrolyte [440,729]. The capacitance values are compared to the carbon fibers activated in steam in Fig. 3.244 [729]. The PTFE-derived carbons gave much higher capacitance than activated carbon fibers and in PTFE-derived carbons defluorination reagent, either Li or K, seems to give an influence on the

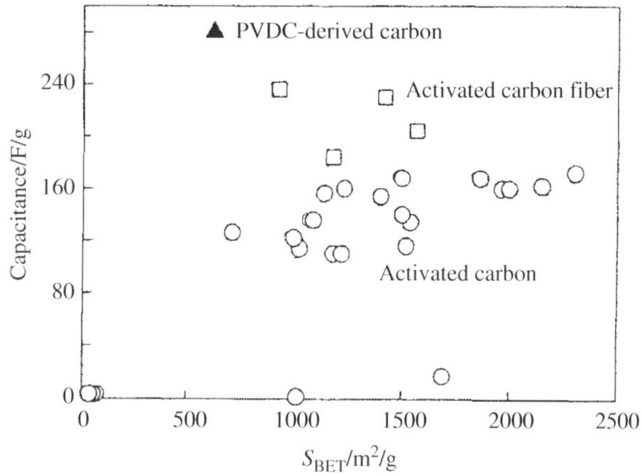


**FIGURE 3.243**

Specific capacitance of activated carbon fibers, activated carbons and PVDC-derived carbon plotted against $S_{BET}$ [728].

capacitance value. It has to be pointed out that a marked increase in $S_{BET}$ of the carbons by γ-ray irradiation, as shown in Fig. 3.244, seems not to be effective to increase the specific capacitance.

The observed capacitance $C_{obs}$ has been analyzed by dividing into two parts, capacitance due to the surface of micropores (microporous surface area $S_{micro}$), and that due to the surface of larger pores (external surface area $S_{ext}$), as follows;

$$C_{obs} = C_{ext} \times S_{ext} + C_{micro} \times S_{micro}, \tag{3.17}$$

where $C_{micro}$ and $C_{ext}$ are the capacitance per unit $S_{micro}$ and per unit $S_{ext}$, respectively [727]. Equation (3.17) is rewritten to

$$C_{obs}/S_{ext} = C_{ext} + C_{micro}(S_{micro}/S_{ext}), \tag{3.18}$$

which suggests the linear relation between ($C_{obs}/S_{ext}$) and ($S_{micro}/S_{ext}$). In Fig. 3.245a and b, the relation is shown for activated carbon microbeads and fibers separately.

For two carbon materials, linear but different relations are obtained. Based on Eq. (3.18), $C_{micro}$ and $C_{ext}$ are calculated as the slope and the intercept of the linear relations. For activated carbon microbeads, $C_{ext} = 74\ \mu F/cm^2$ and $C_{micro} = 19.5\ \mu F/cm^2$, and for activated carbon fibers $C_{ext} = 7.5\ \mu F/cm^2$ and $C_{micro} = 15.5\ \mu F/cm^2$ are obtained.

A similar analysis has been carried out using $S_{micro}$ and mesoporous surface areas $S_{meso}$ determined from adsorption/desorption isotherms of benzene vapor at 25°C on coal-derived activated carbons [439]. $C_{micro}$ was 10.1 and 9.8 $\mu F/cm^2$ and $C_{meso}$ was 9.1 and 23.1 $\mu F/cm^2$ in 6 mol/L KOH and 1 mol/L $H_2SO_4$ electrolyte solutions, respectively.

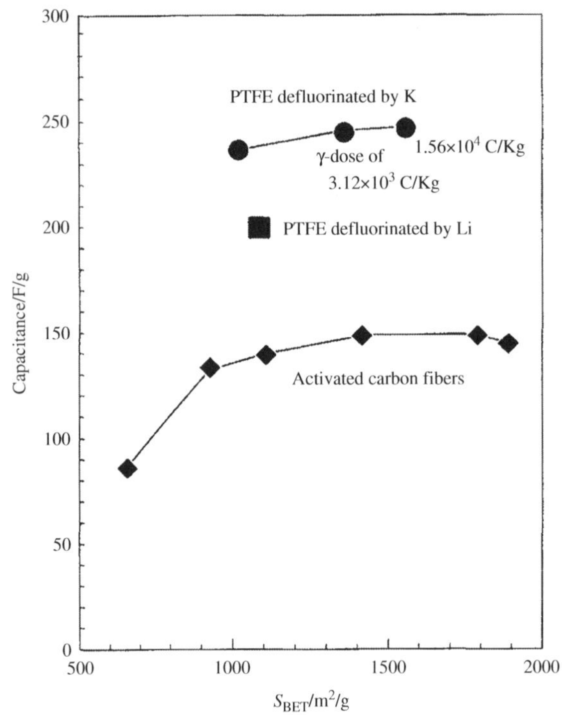


**FIGURE 3.244**

Dependences of capacitance on $S_{BET}$ for the carbons derived from PTFE and activated carbon fibers [728].

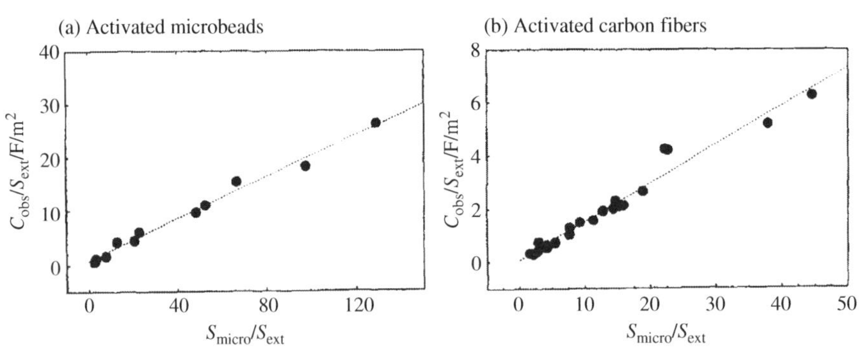


**FIGURE 3.245**

Relations between ($C_{obs}/S_{ext}$) and ($S_{micro}/S_{ext}$) in aqueous $H_2SO_4$ [727].

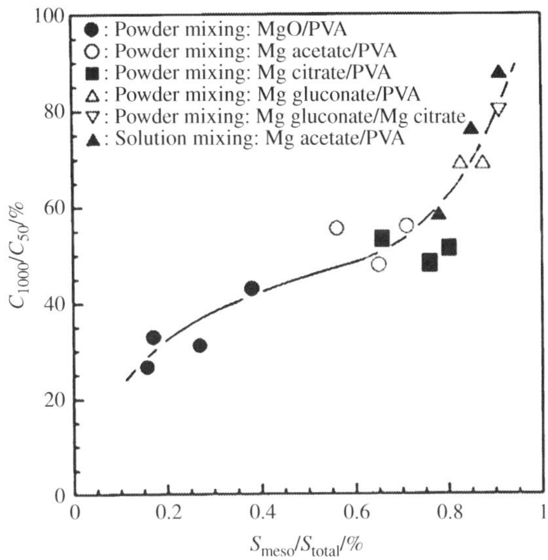


**FIGURE 3.246**

Relation between capacitance ratio $C_{1000}/C_{50}$ (a measure of rate performance) and $S_{meso}/S_{total}$ for the MgO-templated mesoporous carbons [731].

The capacitance values obtained for micropore surface $C_{micro}$ for different carbons in these papers [439,727] are in the range of 10–20 μF/cm$^2$, which are almost the same as the capacitance assigned for graphitic surface [730]. The capacitance values $C_{ext}$ and $C_{meso}$ are largely different for carbon materials used.

Mesopores are important to have high rate performance in EDLCs. In Fig. 3.246, rate performance evaluated by the ratio of capacitance with a current density of 1000 mA/g to that with 50 mA/g, $C_{1000}/C_{50}$, is plotted against the ratio $S_{meso}$ to total surface area $S_{total}$ for mesoporous carbons prepared using MgO template [731,732]. With increasing $S_{meso}/S_{total}$, i.e., increasing concentration of mesopores, rate performance of symmetric EDLC evaluated by $C_{1000}/C_{50}$ is markedly improved.

In Fig. 3.247a, capacitance values in 1 mol/L $H_2SO_4$ aqueous solution is plotted against $S_{BET}$ for exfoliated carbon fibers, which have been prepared from 3000°C-treated mesophase-pitch-based carbon fibers [496]. In the figure, activated carbon fibers are shown for comparison. Exfoliated carbon fibers do not have high $S_{BET}$, but show relatively high capacitance. By simple oxidation in air (activation), both $S_{BET}$ and capacitance increased. The relation between capacitance and $S_{BET}$ does not depend on intercalates used for exfoliation, either $H_2SO_4$ or $HNO_3$, and this relation is extended by the activation with air oxidation. On exfoliated carbon fibers, huge capacitance in high concentration $H_2SO_4$ electrolyte has been reported [733,734]. By using 18 mol/L $H_2SO_4$ electrolyte, huge capacitance as high as 555 F/g is obtained [733]. As shown in Fig. 3.247b,

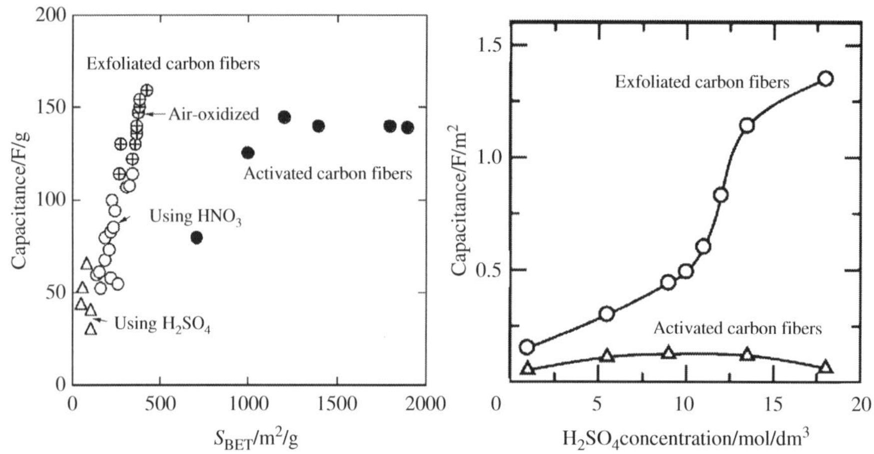


**FIGURE 3.247**

Dependences of gravimetric capacitance on $S_{BET}$ and capacitance per unit surface on $H_2SO_4$ concentration for exfoliated carbon fibers, in comparison with activated carbon fibers.

*(Courtesy of Prof. M. Toyoda of Oita Univ.)*

capacitance per unit surface area increases with increasing $H_2SO_4$ concentration above 10 mol/L and reaches 1.4 F/m$^2$ (gravimetric capacitance of 450 F/g) [734]. This huge capacitance is supposedly caused by the appearance of a large pseudo-capacitance due to oxygen-containing functional groups formed on the layer edges of exfoliated carbon fibers, and intercalation/deintercalation of $SO_4^{2-}$ into the crystallites of the pristine carbon fibers, in addition to the formation of electric double-layers on the surface of the fibers [735].

These results showed that EDLC capacitance cannot be explained simply as a function of $S_{BET}$. More detailed investigations have to be carried out by taking into account the pore sizes in the electrode carbons and also the size of electrolyte ions which are solvated.

### c. Non-aqueous electrolyte

In Fig. 3.248a and b, specific capacitance values in non-aqueous 1 mol/L $LiClO_4$/propylene carbonate (PC) solution and in 1 mol/L $H_2SO_4$ aqueous solution are shown as a function of $S_{BET}$ for a number of activated carbons and carbon fibers [736]. The correspondence between these two capacitance values is shown in Fig. 3.248c. The carbons with $S_{BET}$ of around 1000 m$^2$/g have a capacitance value of around 120 F/g in aqueous solution, but negligibly small capacitance in non-aqueous solution, which has been explained by the difference in ion sizes, $SO_4^{2-}(H_2O)_{12}$ ion of 0.533 nm but $Li^+(PC)_4$ ion of 1.19 nm, in the relation with pore size distribution determined from transmission electron microscope images.

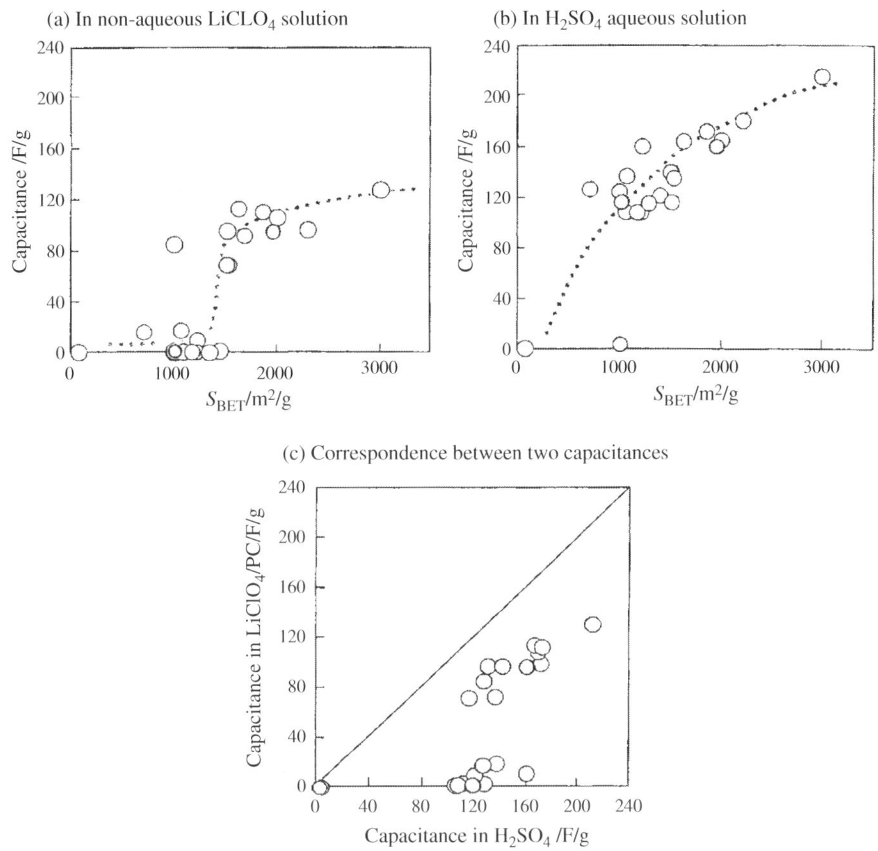


**FIGURE 3.248**

Relations between capacitances measured in non-aqueous $LiClO_4$/PC solution and aqueous $H_2SO_4$ solution as a function of $S_{BET}$ [736].

On PVDC-derived carbons, capacitance has been measured in 1 mol/L $Et_4NBF_4$/PC [737]. The capacitance is shown as a function of carbonized temperature in Fig. 3.249 for two series of PVDC-derived carbons, one is just carbonized in $N_2$ atmosphere and the other is activated by using 200 mass% KOH. The results show that KOH activation is essential to get high capacitance in non-aqueous electrolyte $Et_4NBF_4$. However, the same carbons gave relatively high capacitance in aqueous electrolyte $H_2SO_4$ without KOH activation [737]. This is explained by the difference in pore size distribution in relation to the size of electrolyte ions, electrolyte ions in $Et_4NBF_4$/PC solution being much bigger than those in $H_2SO_4$ aqueous solution.

For various carbons activated by KOH at a temperature between 700–950°C, capacitance in 1 mol/L $LiClO_3$/PC is shown as a function of $S_{BET}$ of carbon

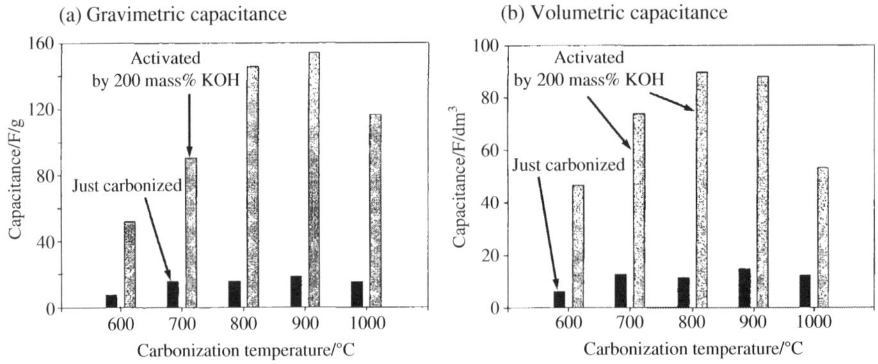


**FIGURE 3.249**

Capacitances in 1 mol/L $Et_4NBF_4$/PC for two series of PVDC-derived carbons carbonized at different temperatures [737].

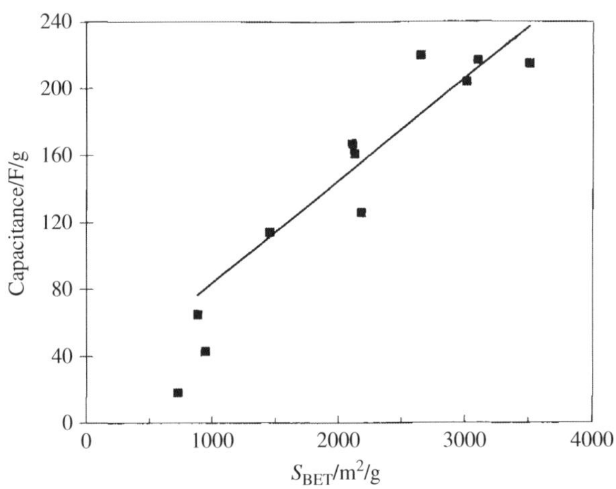


**FIGURE 3.250**

Dependence of capacitance of activated carbons in 1 M $LiClO_3$/PC solution on $S_{BET}$ [738].

electrodes in Fig. 3.250 [738]. An abrupt decrease in capacitance below $S_{BET}$ of 1000 $m^2/g$, similar to Fig. 3.248a, is explained by accessibility of micropores for solvated electrolyte ions, in other words, the presence of non-accessible micropores (ion-sieving effect) [738,739].

The values of capacitance observed are considered to be an averaged value between the capacitance derived from the cation adsorption/desorption and that from anion adsorption/desorption [740]. In Fig. 3.251, cation and anion capacitances are separately determined for the samples used in Fig. 3.250. In low $S_{BET}$

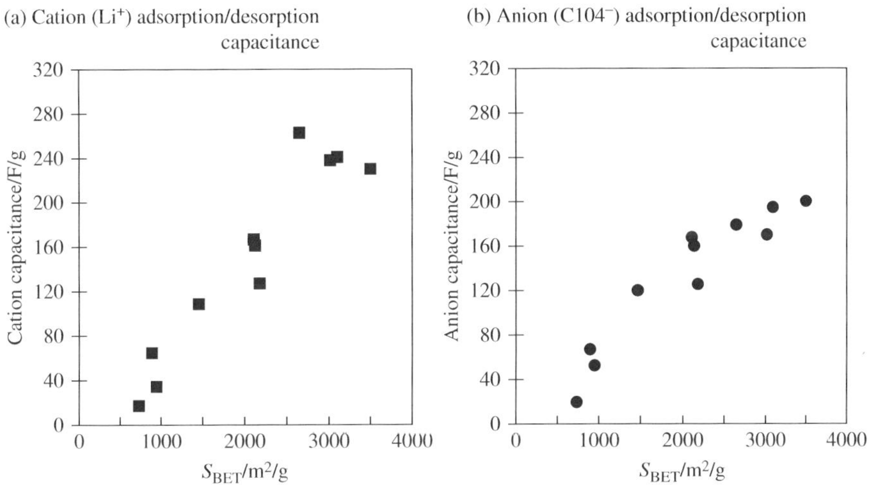


**FIGURE 3.251**

Cation and anion capacitances *vs.* $S_{BET}$ [738].

range, cation capacitance is lower than anion capacitance, which may be explained by the ion-sieving effect, larger solvated $Li^+$ ions (0.82 nm) being sieved more marked than smaller $ClO_4^-$ ions (0.52 nm). In high $S_{BET}$ range, however, cation capacitance is larger than anion capacitance, which may be explained by the presence of oxygenated surface functional groups on the surface of activated carbons [738].

Confirmation of Eq. (3.17) has been done by using non-aqueous electrolyte solution $Et_3MeNBF_4$/PC on air-activated carbon spheres [741,742] and various activated carbons [743]. By increasing the current density from 100 to 1000 mA/g, the contribution of $S_{micro}$, i.e., $C_{micro}$, becomes smaller, as shown for activated carbons in Fig. 3.252a [743]. Fig. 3.252b shows the dependences of $C_{micro}$ and $C_{ext}$ on current density during charge/discharge processes. $C_{micro}$ decreases with increasing current density, which can be explained by the fact that adsorption of electrolyte ions into micropores becomes difficult with increasing current density. $C_{micro}$ in $Et_3MeNBF_4$/PC is smaller than that in $H_2SO_4$, caused by the difference in size of electrolyte ions. The values of $C_{ext}/C_{micro}$ are in a range of 2.4–2.8 in aqueous $H_2SO_4$, but 4.3–7.2 in $Et_3MeNBF_4$/PC, suggesting a smaller contribution of $S_{micro}$ to observed capacitance in $Et_3MeNBF_4$/PC.

Single-wall carbon nanotubes (SWCNTs) have a high surface area at 2630 $m^2/g$ theoretically and homogeneous pores with the size of about 1 nm if nanotubes are open at the edges. Therefore, they are thought to be homogeneously microporous and so many works to apply them to EDLC electrode have been reported [744–748]. However, the capacitance values reported are so scattered and most of them are not as high as expected. On SWCNTs, it is interesting that their

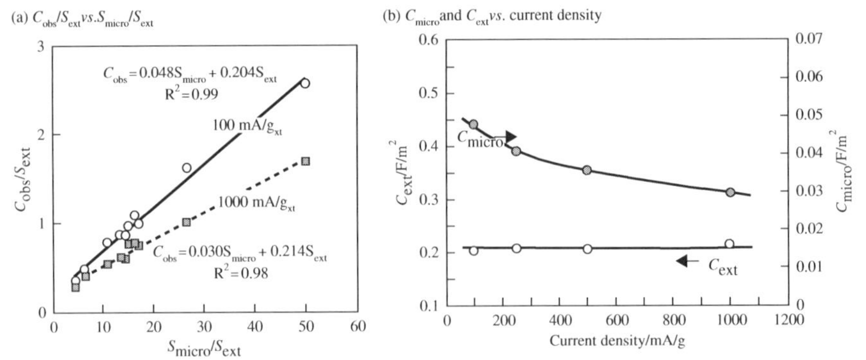


**FIGURE 3.252**

Relation between ($C_{obs}/S_{ext}$) and ($S_{micro}/S_{ext}$), and dependence of $C_{micro}$ and $C_{ext}$ on current density in non-aqueous $Et_3MeNBF_4$/PC electrolyte for various activated carbons [743].

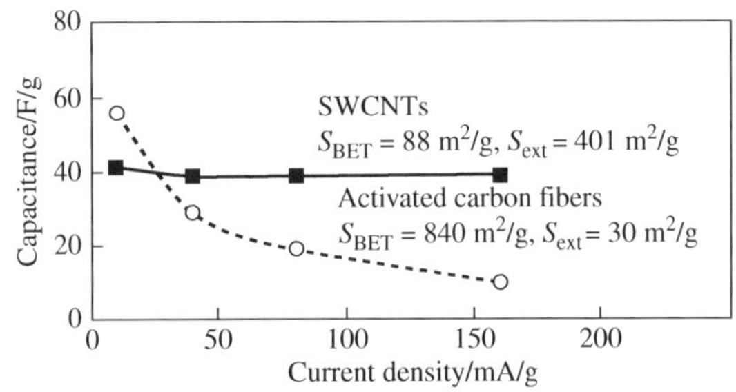


**FIGURE 3.253**

Dependence in capacitance in 1 mol/L $LiClO_4$/PC solution on current density for SWCNTs [748].

capacitance did not change with increasing current density at least up to 160 mA/g although activated carbon fibers with a similar $S_{BET}$ show a drastic decrease in capacitance, as shown in Fig. 3.253. This result is explained by the presence of a large external surface area of SWCNTs [748].

## d. Pseudocapacitance

Capacitors with higher energy density are strongly desired in order to apply them to electric vehicle drivelines. Composite electrode materials consisting of porous carbon and fine metal particles have been proposed as electrode materials of capacitors, the latter giving some capacitance due to redox reaction with electrolyte. Functional groups, which are formed at the surface of carbon particles and often contain oxygen and nitrogen, give some capacitance due to the redox reaction. The capacitance caused by redox reaction of fine metallic particles and

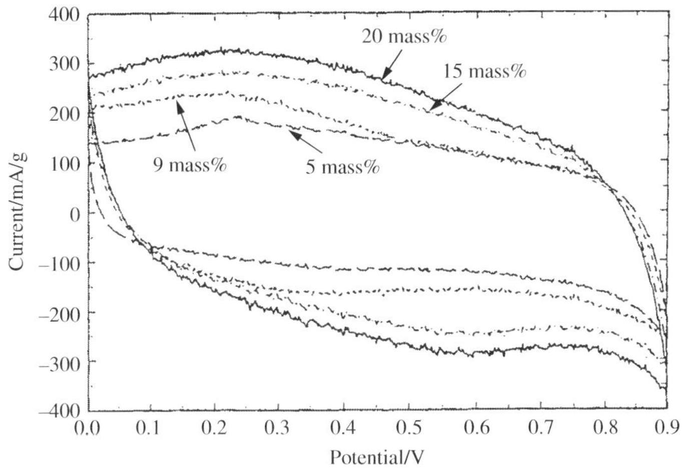


**FIGURE 3.254**

Cyclic voltammograms for the activated carbons with different $RuO_2$ loading and oxidized at 100°C [750].

surface functional groups with electrolyte is called pseudocapacitance [749], which is often coexistent with EDLC capacitance.

Loading of ruthenium oxide $RuO_2$ onto activated carbon has been carried out by electroless deposition of Ru metal in a $RuCl_3$ aqueous solution and then heating at 100°C in an oxygen atmosphere [750]. Specific capacitance obtained in 1 mol/L $H_2SO_4$ by a 3-electrode cell is 260 F/g for 20 mass% $RuO_2$-loaded activated carbon, though the pristine activated carbon has 98 F/g. Cyclic voltammograms are shown for the activated carbon with different $RuO_2$ loading in Fig. 3.254. With decreasing $RuO_2$ content, capacitance decreases. It is very interesting that there is no peak corresponding to redox reaction of $RuO_2$. Oxidation temperature after Ru deposition also gives a certain influence on capacitance [750].

Similar capacitance increase by $RuO_2$ loading is reported on various activated carbons by the hydrolysis of $RuCl_3$ [751]. Loading of $MoO_3$ on activated carbon increases the capacitance from 132 to 177 F/g in 1 mol/L NaOH electrolyte solution [752].

Functional groups on the surface of carbon materials often contain oxygen, such as -COOH, = CO, and others, depend strongly on the precursors and the preparation conditions. Some of these acidic functional groups are electrochemically active, and contribute to the capacitance of EDLCs. The functional groups containing oxygen are classified into two groups by temperature programmed desorption (TPD), the functional groups desorbed as CO (CO desorbing complexes) and those desorbed as $CO_2$ ($CO_2$-desorbing complexes); the former are supposed to assist the formation of electric double layers, but the latter to give a

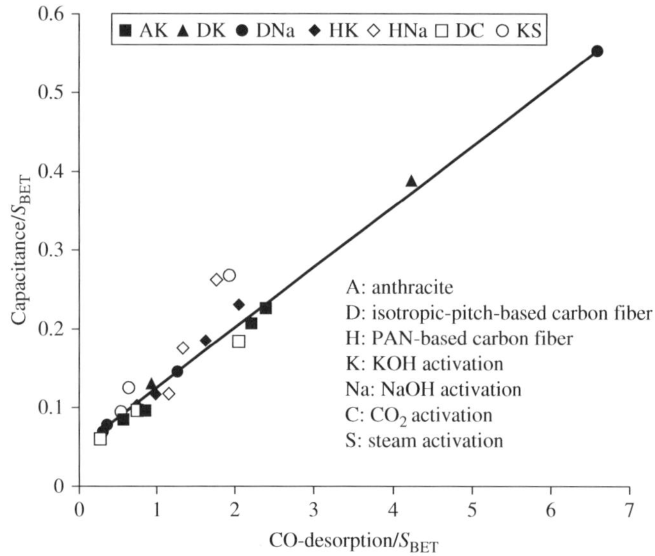


**FIGURE 3.255**

Relation between capacitance and CO-desorption after normalized by $S_{BET}$ on three carbon materials under different activation conditions [757].

negative effect [753]. The observed capacitance has been discussed in relation to the amounts of CO- and $CO_2$-sorbing complexes on various carbons [753–758]. In Fig. 3.255, a linear relation between capacitance and CO-desorption after normalizing by $S_{BET}$ is shown on three kinds of carbon materials with activation under different conditions [757]. The capacitance with zero current density, which is estimated by the extrapolation of the dependence of observed capacitance on current density, increases linearly with the increasing total amount of oxygen-containing functional groups for activated carbon fibers and activated carbon derived from a petroleum coke [759].

Huge capacitance as 450–555 F/g observed on exfoliated carbon fibers in 18 mol/L $H_2SO_4$ aqueous solution is caused by the appearance of a large pseudo-capacitance [733], as mentioned above. In Fig. 3.256, TPD curves of CO- and $CO_2$-desorbing complexes from 3000°C-treated PAN-based carbon fibers after electrolysis in 13 mol/L $HNO_3$ and after exfoliation at 1000°C are shown [735]. During electrolysis oxidation of pristine carbon fibers occurs by the association with intercalation of nitric acid and so the carbon fibers contain large amounts of CO- and $CO_2$-desorbing complexes. After exfoliation of these electrolyzed carbon fibers, $CO_2$-desorbing complexes are reduced to negligibly small amounts, but CO-desorbing complexes, quinone, ether, etc., even increase, which seems to be formed during high-temperature exfoliation. The formation of a large amount of CO-desorbing complexes and almost negligible $CO_2$-desorbing complexes is

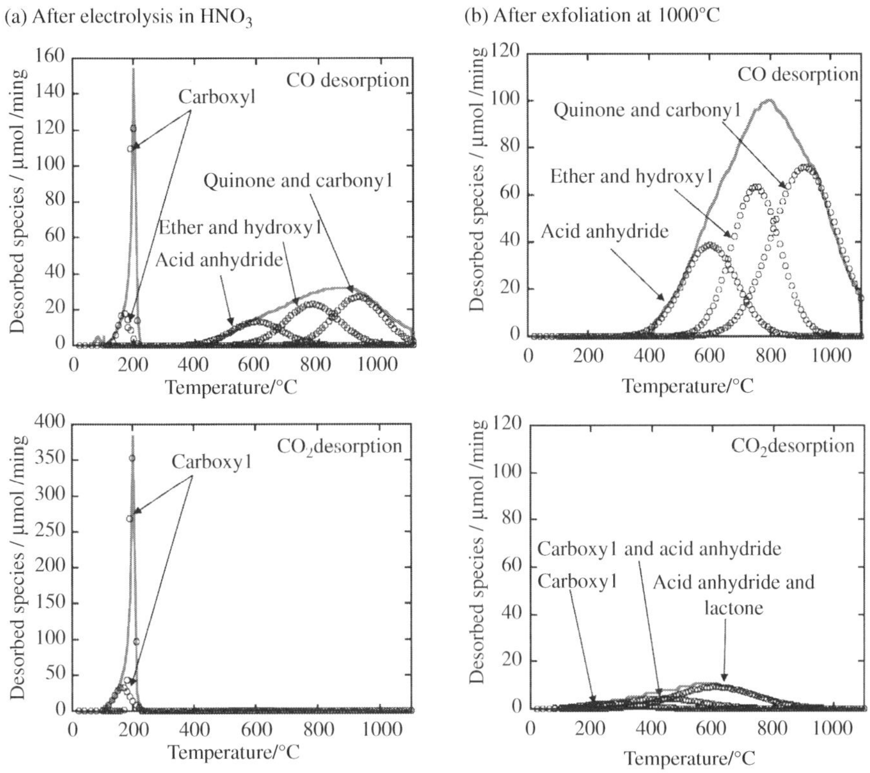


**FIGURE 3.256**

TPD curves of CO- and $CO_2$-desorbing complexes from 3000°C-treated PAN-based carbon fibers after electrolysis (a) and exfoliation (b) [745].

responsible for increasing pseudocapacitance, resulting in huge capacitance observed.

Doping of nitrogen into carbon materials also gives pseudocapacitance [760–764]. N-doped carbon has been synthesized from various hydrocarbons containing nitrogen by using different templates; fluorine expandable mica [760,762], mesoporous silica [761], MgO [765], etc. Flaky N-doped carbon prepared by mica-template gives a capacitance of 100–180 F/g in 1 mol/L $H_2SO_4$ electrolyte, which corresponds to the capacitance per unit $S_{BET}$ of 1.2–2.2 F/m$^2$, more than 10 times higher than that of activated carbons, mainly due to low $S_{BET}$ [760]. Microporous carbons prepared from brown coal by carbonization at 700°C, followed by activation with KOH and N-enriching by heating with urea, have high $S_{BET}$, as high as 2700–3200 m$^2$/g, and contain nitrogen of 5.0 mass% and oxygen of 5.4 mass%, which give high capacitance of 222–227 F/g [764]. MgO templated carbons from poly(vinylpyrrolidone) (PVP), polyacrylamide (PAA) and

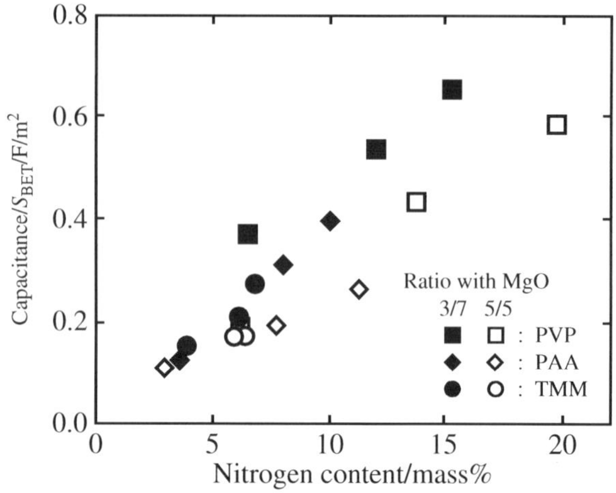


**FIGURE 3.257**

Dependence of gravimetric capacitance per unit $S_{BET}$ on nitrogen content for N-doped MgO-templated carbons [765].

trimethylolmelamine (TMM) give the capacitance per unit $S_{BET}$ increasing linearly with increasing nitrogen content, as shown in Fig. 3.257 [765]. Carbon foam prepared via carbonization of melamine foam at 800°C give a capacitance of 222 F/g in 1 mol/L $H_2SO_4$ with current density of 1000 mA/g [743].

### e. Hybrid capacitors

Hybrid capacitors consisting of different storage mechanisms have been proposed, electric double-layer formation at the cathode and faradaic charge-transfer reaction with $Li^+$ in the electrolyte at the anode [511,766,767]. The cell composed of a low-temperature-treated carbon at the anode and activated carbon at the cathode in a non-aqueous solution of lithium containing electrolyte gives a high-energy density, where the intercalation/deintercalation of $Li^+$ at the anode, occurring in lithium-ion rechargeable batteries (LIBs), and the formation of electric double-layers at the cathode are supposed to occur. The cell having such construction can have an extended potential range and is called a lithium-ion capacitor (LIC). In order to extend the potential range, however, a preliminary charging process using auxiliary Li electrode in the cell, i.e., doping of Li into the carbon of the anode, is necessary, of which energy seems to correspond to the irreversible capacity for non-graphitizing carbon in LIBs, as explained in the previous section. Energy and power densities became 1.14 and 2.3 times higher by extending the potential range from 2.5–4.3 to 1.5–4.3 V. The use of graphite as the anode is also proposed [768].

Different materials for the anode have been proposed; amorphous $MnO_2$· has often been selected [769–773]. A hybrid capacitor constructed from the cathode of an amorphous hydrous $MnO_2$ and the anode of an activated carbon are tested in 0.1 mol/L aqueous solutions of nitrates of alkaline earth metals, $Mg^{2+}$, $Ca^{2+}$ and $Ba^{2+}$, giving rectangular CVs in a potential window from 0 to 2 V, and high cyclability over 5000 cycles with a current density of 300 mA/g [772,773]. A couple of NiO cathode and mesoporous carbon anode in 6 mol/L KOH electrolyte is also proposed [774], and of activated carbon cathode and $Li_4Ti_5O_{12}$ anode in non-aqueous electrolyte [775].

Hybrid capacitors have been studied by aiming at improvement in performance, but they are not yet forwarded to the practical application mainly because the performance improvement is not marked enough and/or because the cost of the electrode materials increases.

## f. Commercially available cells

Electric double-layer capacitors are commercialized and used in various electronic devices. In Fig. 3.258, the construction of the cell with relatively large capacity, 1300 F, is schematically shown, together with its charge and discharge characteristics [776]. Anode and cathode of carbon materials were stacked with each other, separated by non-woven fabrics, and each unit cell is connected in parallel by Al collectors. The size of the cell is $150 \times 70 \times 22$ $mm^3$ and total weight is 0.38 kg. This cell can be charged to 2.5 V within 15 s by a current of 200 A or within 30 s by 100 A. By discharge, cell voltage decreases linearly, discharging from 2.5 to 1.0 V for 200 s by 10 A, for 30 s by 50 A and for 15 s by 100 A.

In Fig. 3.259, change in capacitance with cycling with a current of 50 A at 45°C and effect of temperature on discharge energy density are shown for 1300 F cell. Even after $4 \times 10^5$ cycles of charge/discharge with a current of 50 A, the decrease in capacitance of the cell is less than 15%. With decreasing temperature, energy density of the cell decreases mostly because of the increase in internal resistance associated with the increase in the resistance of electrolyte solution.

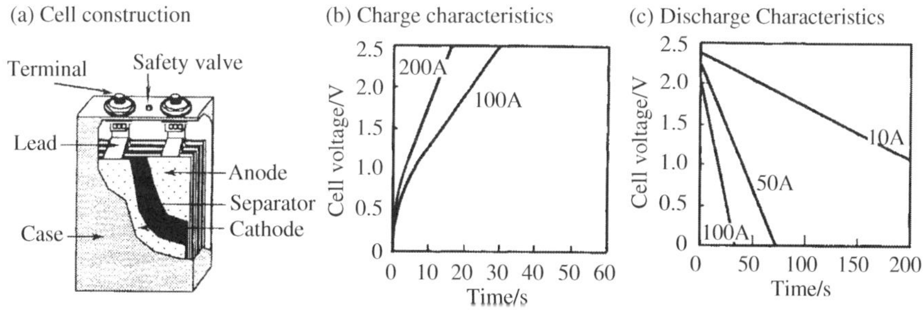


**FIGURE 3.258**

Scheme of construction and charge-discharge characteristics of 1300 F cell [776].

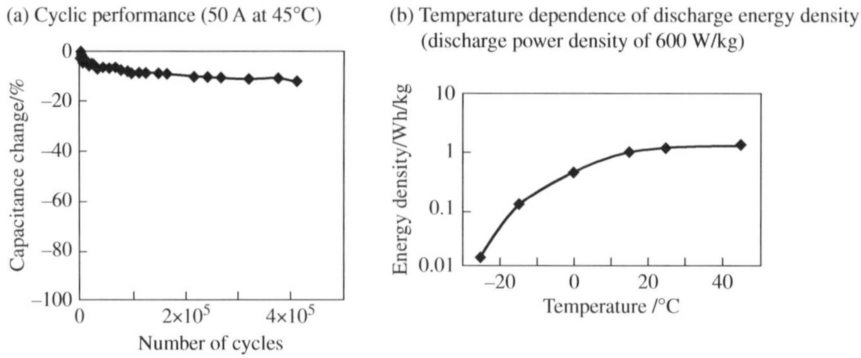


**FIGURE 3.259**

Cycle performance and temperature effect of 1300 F cell [776].

### 3.9.3 Storage of hydrogen gas

Hydrogen gas is an ideal alternative to fossil fuels from the global environmental point of view. It gives only $H_2O$ by combustion and does not give any pollutants, such as CO, $SO_x$, $NO_x$ as well as soot-like particles. Its storage, however, is an important problem for its transportation and use [777]. There are essentially four ways to store hydrogen gas; liquefaction, compression, formation of metal hydrides and adsorption into porous materials. Hydrogen liquefaction gives high densities, but needs low temperature as 21 K, being expensive and energy consuming. Compression of hydrogen gas requires a container with large volume and high strength, i.e., a heavy container, in order to get the density similar to liquid hydrogen. The large volume and heavy weight of the tank and the risks to using high pressures of hydrogen for compressed hydrogen inhibit its application to vehicles. The formation of metal hydrides does not need high pressure of hydrogen gas, but the weight of the storage system, i.e., tank plus metallic compounds, becomes heavy. In addition, heating of metal hydrides is essential to recover hydrogen gas. Therefore, metal hydrides are thought not to be the best way to store hydrogen.

Hydrogen adsorption into porous materials, particularly in carbon materials, attracted attention after the paper which reported enormous amount, more than 60 mass%, of adsorption into carbon nanofibers [778]. Different carbon materials, not only microporous activated carbons and activated carbon fibers but also carbon nanotubes and carbon nanofibers, and also these carbon materials doped by alkali metals have been studied for hydrogen storage, some giving striking results but some being discouraging. In Table 3.47, some data on hydrogen storage reported in the literatures are listed. The hydrogen plan of the U.S. Department of Energy (DOE) has set a benchmark for hydrogen storage of 6.0 mass% (45 g/L) for the year 2010 and 9.0 mass% (81 g/L) for 2015 in the application to

Table 3.47 Gravimetric Storage of Hydrogen in Various Carbon Materials Reported

| Carbon Material | Hydrogen Storage | | | Reference |
|---|---|---|---|---|
| | T (K) | P (MPa) | Stored (Mass%) | |
| SWCNTs (low purity) | 133 | 0.04 | 5–10 | [790] |
| SWCNTs (high purity) | 300 | 0.04 | 6.5 | [791] |
| SWCNTs | 300 | 10 | 4.2 | [792] |
| SWCNTs (high purity) | 80 | 7 | 8.25 | [793] |
| SWCNTs (high purity) | 77 | 0.1 | 1.8 | [795] |
| CNFs (tubular) | | | 11.26 | |
| CNFs (herringbone) | 298 | 12 | 67.55 | [778] |
| CNFs (platelet) | | | 53.68 | |
| Graphite | | | 4.52 | |
| CNFs | 298 | 12 | 18–35 | [796] |
| CNFs | 300 | 11 | 10–13 | [797] |
| | | | 3.4–5.7 | [798] |
| CNFs | 298–773 | | 0.4 | |
| Li-doped CNFs | ~473–673 | | 20.0 | |
| Li-doped graphite | ~473–673 | | 14.0 | |
| K-doped CNFs | <313 | | 14.0 | |
| K-doped graphite | <313 | 0.1 | 5.0 | [799] |
| Li- & K-doped CNFs | 293 | 0.1 | 12–1.8 | [800] |

automobile fuel cell system [789]. For carbon materials studied, the hydrogen storage data are scattered in a wide range.

SWCNTs were firstly reported to adsorb 5–10 mass% hydrogen at room temperature [790], but later this value was corrected to be about 6.5 mass% because SWCNTs used contained a certain amount of metallic alloys [791]. Also the value of 4.2 mass% was reported for SWCNTs at room temperature under about 10 MPa, of which about 78% could be desorbed under reduced pressure at room temperature [792]. At a low temperature of 80 K, the value of 8.25 mass% was reported as the highest under about 4 MPa pressure [793]. Storage of about 1.8 mass% at room temperature under 0.1 MPa pressure was reported by careful experiment [794]. Electrochemical storage of hydrogen was also tried in KOH electrolyte at room temperature and hydrogen storage into multi-walled carbon nanotubes (MWCNTs) was reported to be about 14 mass% [795]. Also for CNFs prepared under different conditions, various experimental data were reported. At first, extremely high value of hydrogen adsorption of more than 60 mass% at room temperature was reported for CNFs with herringbone nanotexture [778], but later it was corrected to 18–35 mass% [796]. CNFs prepared by CVD using iron catalysts was reported to store hydrogen up to 13 mass% at room temperature and 11 MPa pressure [797], but also corrected to be 5.7 mass% [798]. Doping of

**Table 3.48** Theoretical Prediction of Hydrogen Storage into Carbon Materials

| Simulation | Results | Reference |
|---|---|---|
| Grand canonical Monte Carlo | Close to DOE value on (9,9) nanotubes at 77K, 5 MPa | [791] |
| Self-consistent charge density-functional-based tight-binding | 14.3 mass% for (10,10) nanotubes | [794] |
| Geometrical calculation | 4.0 mass% for (10,10) nanotubes | [806] |
| Grand canonical Monte Carlo | 11.24 mass% at 77 K, 10 MPa | [807] |
| Charge density-functional-based tight binding | 14 mass% for (10,10) nanotubes | [808] |

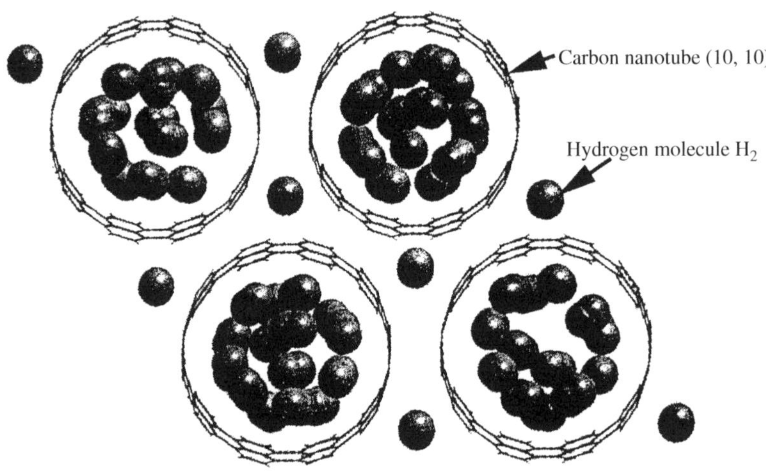


**FIGURE 3.260**

Configuration of hydrogen molecules in the rope of carbon nanotubes [805].

alkali metals into these CNFs was reported to increase the hydrogen storage [799], but another paper reported much lower values [800].

The results of theoretical calculation for hydrogen storage into these carbon materials are also scattered in a wide range. Table 3.48 shows some of the results reported. From a simple geometrical argument, close packing of $H_2$ molecules into the core of the (10,10) SWCNTs gave 3.3 mass% of hydrogen and 0.7 mass% hydrogen could be placed in the space among the tubes, leading to the total adsorption of 4.0 mass%, as shown in Fig. 3.260 [805]. The predominant part of hydrogen adsorption observed on sonicated SWCNTs was attributed to the impurity alloys coming from the vibrator [806].

Hydrogen storage in activated carbons and activated carbon fibers has also been studied. In Fig. 3.261, adsorption isotherms of hydrogen gas into various

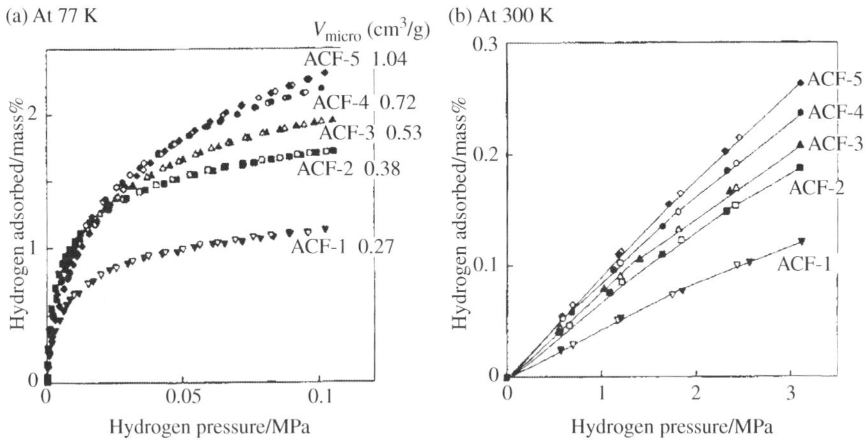


**FIGURE 3.261**

Hydrogen adsorption isotherms into the activated carbon fibers (ACFs) with different $V_{micro}$.

*(Courtesy of Dr. H. Takagi of AIST, Japan)*

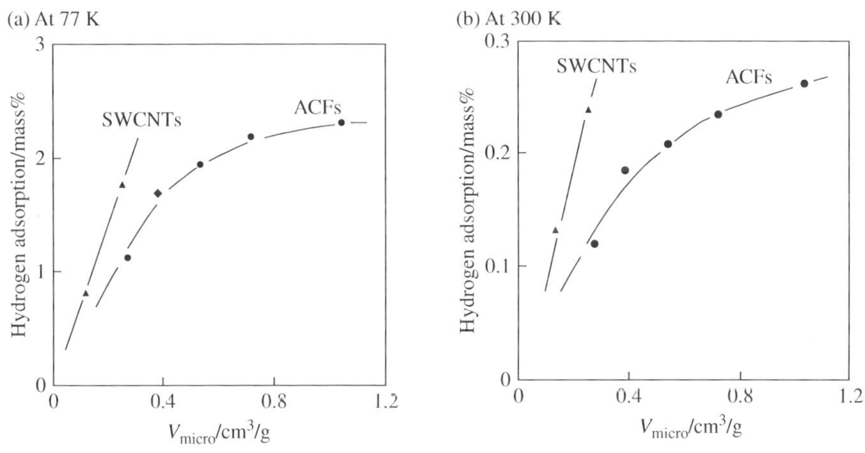


**FIGURE 3.262**

Dependences of hydrogen adsorption on $V_{micro}$ of activated carbon fibers.

*(ACFs) and SWCNTs (courtesy of Dr. H. Takagi of AIST, Japan)*

activated carbon fibers with different $V_{micro}$ are shown at two temperatures [795]. At 77 and 303 K, hydrogen adsorbed increased with increasing pressure, depending on $V_{micro}$. In Fig. 3.262, hydrogen uptakes measured on activated carbon fibers in Fig. 3.261, together with SWCNTs, are plotted against $V_{micro}$ of absorbents [796]. The hydrogen adsorbed into activated carbon fibers increases with

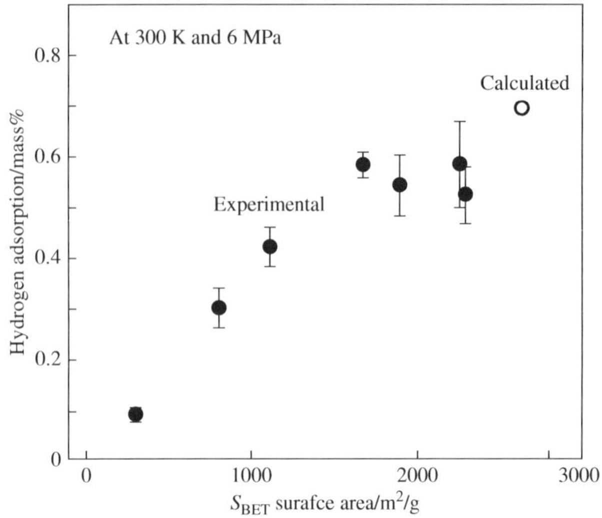


**FIGURE 3.263**

Hydrogen adsorption of different activated carbon fibers against $S_{BET}$ with calculated value for ideal graphite plane [808].

increasing $V_{micro}$, but tends to saturate to about 2.2 mass% at 77 K under 0.1 MPa pressure. Hydrogen adsorbed at 303 K under 3 MPa reaches close to 0.3 mass%. A similar $V_{micro}$ dependence of hydrogen adsorption is reported under the pressure of 10 and 70 MPa using commercially available activated carbon fibers [807]. The dependence of hydrogen adsorption, which is measured at 300 K under 6 MPa pressure, on $S_{BET}$ is shown in Fig. 3.263 [808]. It shows also a tendency to saturate with increasing $S_{BET}$ and to agree with the value calculated by grand canonical Monte Carlo simulation (about 0.7 mass% at 300 K and 6 MPa for $S_{BET}$ of 2600 m$^2$/g).

As seen in Fig. 3.262, the dependence of hydrogen adsorption on $V_{micro}$ is different for SWCNTs than for ACFs [795]. If we take into consideration that theoretical surface area of SWCNTs is 2630 m$^2$/g and that the present ACFs with $V_{micro}$ of 1.0 cm$^3$/g has $S_{micro}$ of about 2200 m$^2$/g, CNTs may have a possibility to have high hydrogen storage [795]. The results shown in Fig. 3.262 may suggest that not only pore size but also pore shape influence hydrogen adsorption.

Theoretical calculation has been carried out for slit-shaped and tubular pores with different sizes. In Fig. 3.264, the calculated hydrogen storage is plotted against hydrogen pressure at 77 and 300 K for slit-shaped and tubular pores with two different sizes [808]. Calculated hydrogen capacity for tubular pores is smaller than that for slit-shaped pores. Even for slit-shaped pores, hydrogen storage capacity at room temperature is not so high, less than 0.1 mass% under 0.1 MPa and about 1.2 mass% under 10 MPa.

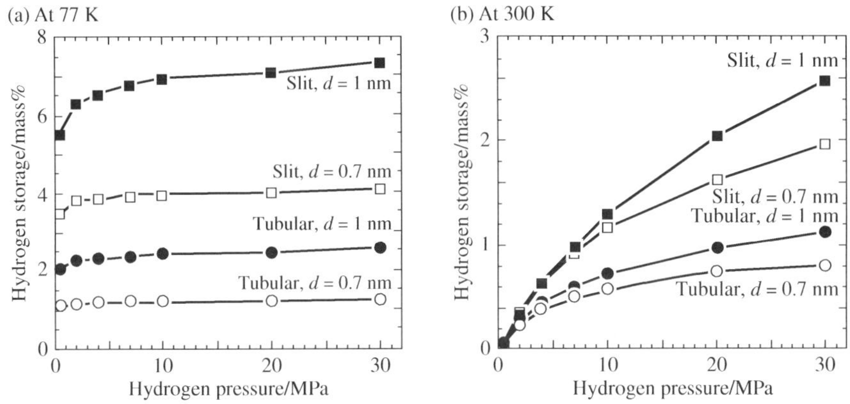


**FIGURE 3.264**

Calculated hydrogen storage capacity at two temperatures for different pore shapes and sizes [808].

Graphite intercalation compounds of alkali metals are known to adsorb hydrogen into graphite gallery to form two kinds of ternary compounds of $KC_{24}H_2$ and $KC_8H_{2/3}$ (Section 3.8.5). Intercalation of hydrogen to form $KC_{24}$ $H_2$ occurs at a low temperature around 77 K and deintercalation at room temperature, of which gravimetric storage capacity corresponds to 1.2 mass%. Ternary intercalation compound $KC_8H_{2/3}$ can form at room temperature by simple contact of stage-1 intercalation compound $KC_8$ with hydrogen gas, but its hydrogen content is only 0.4 mass%.

Hydrogen storage explained above is due to physical adsorption of hydrogen molecules. Chemical adsorption of hydrogen atoms has also been studied by different techniques. Using Ni catalyst, buckminsterfullerene $C_{60}$ can be hydrogenized to $C_{60}H_{36}$, which corresponds to the hydrogen content of 4.8 mass% [809]. In these cases, however, hydrogen is chemically bonded with a carbon atom and so a certain amount of energy is expected to be required for both the formation and destruction of these chemical bonds.

Milling of natural graphite in hydrogen atmosphere gives hydrogen-adsorbed graphite [810]. The hydrogen content of ground graphite increases with increasing milling time and reached 7.4 mass% after 80 h, as shown in Fig. 3.265a, although its structure examined by X-ray diffraction becomes amorphous. For desorption, it is necessary to heat above 600°C, as shown by the thermal desorption mass spectra in Fig. 3.265b [779]. Departure of hydrogen molecules occurs in two steps, around 700 and 1000 K. At the temperature range of the first step, the species with mass number of 16 and 28 ($CH_4$ and $C_2H_6$, respectively) are detected, as shown in Fig. 3.265b, but their total amount is less than 1% of the species with mass number of 2 ($H_2$ molecules).

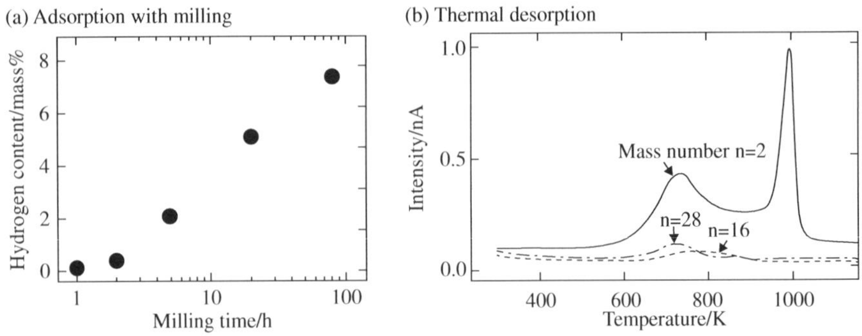


**FIGURE 3.265**

Hydrogen adsorption into natural graphite by milling in hydrogen atmosphere and desorption by heating [779].

Through electrodecomposition of 6 mol/L KOH aqueous solution at room temperature, hydrogen storage has been carried out on activated carbon electrode [780]. The release of 1.5 mass% hydrogen was observed at about −0.5 V with a well-defined potential plateau.

Hydrogen storage into carbon materials, carbon nanotubes, carbon nanofibers, activated carbons and activated carbon fibers, is still controversial on both experiments and theoretical analyses. Two kinds of purities have to be exactly defined to be clarified on the adsorbent carbon materials in order to have reliable data; purity in kind of carbon materials and impurities of foreign atoms. It has frequently been pointed out that it is pretty difficult to get really pure single-wall carbon nanotubes.

### 3.9.4 Storage of methane gas

For the storage of methane and natural gas, there are essentially three ways: liquefaction, compression and adsorption into porous materials. Natural gas can be liquefied at a temperature below 112 K (−161°C), but its energy density (the energy of combustion per unit volume) is only 23 MJ/L, which is considerably smaller than that of diesel oil of 37 MJ/L. Natural gas can be compressed to 20 MPa at room temperature, but has much smaller energy density at less than 10 MJ/L and in addition heavy storage tank is needed. Therefore, the use of adsorbent materials, such as activated carbons, porous silica and porous polymers, is an alternative way to store methane and natural gas [781]. Adsorption of natural gas into adsorbents is possible at relative low pressure as 3.5 MPa at room temperature. For the application to vehicles, the US Department of Energy (DOE) has set a tentative target for methane storage of 180 v/v under the pressure of 3.5 MPa at 25°C [782].

For a number of commercially available activated carbons, methane adsorption has been measured under 3.5 MPa at 25°C [801]. Dependence of methane

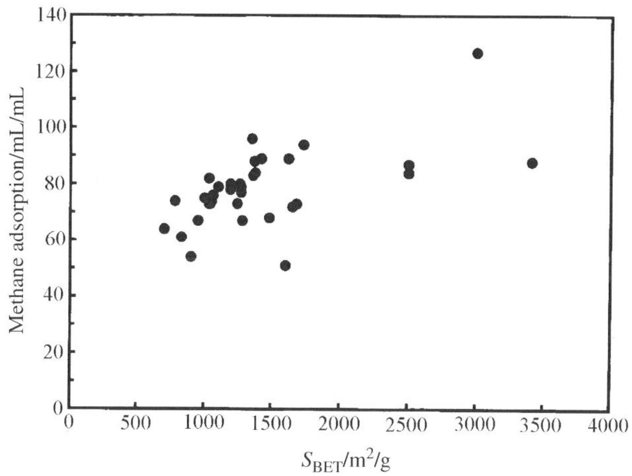


**FIGURE 3.266**

Relation between methane adsorption and $S_{BET}$ for commercially available activated carbons [801].

adsorption on $S_{BET}$ is not defined uniquely, as shown in Fig. 3.266, qualitatively large $S_{BET}$ being desired to attain a large adsorption of methane. This result suggests two points; (1) more detailed characterization of pores in carbon materials is necessary to relate it with methane adsorption and (2) the surface nature of carbons has to be taken into consideration.

For methane adsorption, theoretical predictions have been presented by different methods [783–786]. Methane is adsorbed exclusively into the micropores and capillary condensation in the mesopores does not occur [783]. By assuming the slit-shaped pores composed of hexagonal carbon layers, methane adsorption is simulated on the basis of grand canonical Monte Carlo method [784]. For three different pore sizes of 0.76, 1.14 and 1.52 nm, density of methane adsorbed into these pores is calculated as a function of hydrogen pressure at 0°C, as shown in Fig. 3.267. In the figure, the dependence of density on pressure for methane gas was also plotted. This calculation shows that the density of methane adsorbed into pores at low pressures is much higher than its gas state, which shows the advantage of adsorption storage of methane. It shows also that the smaller pores can adsorb methane at the lower pressure, but under high pressures the larger pores can absorb the larger amount of methane. In the figure, theoretical supply of methane gas (i.e., adsorption under 3.5 MPa minus that under atmospheric pressure) was shown. From these calculations, the slit-shaped pores with the size of 1.14 nm are desired to store methane, its supply of 166 g/L corresponding to 249 mL/mL, much larger than DOE benchmark. Another calculation by a similar method shows that 1.14 nm pores are the most suitable for methane adsorption and the methane supply is 195 mL/mL [785]. The calculation by changing not

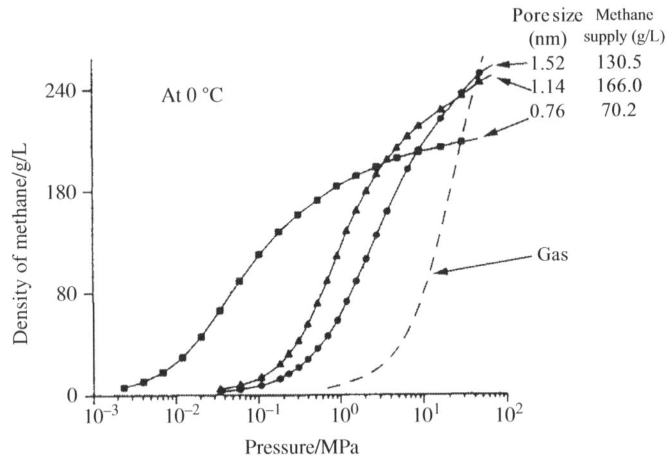


**FIGURE 3.267**

Theoretical adsorption isotherms for methane into different sizes of slit-shaped pore [784]. The relation between density and pressure for methane gas is shown for comparison.

only pore size but also number of hexagonal carbon layers stacked and their interlayer spacing give a similar result; for 1.12 nm pores composed of single layers the methane supply is 153 mL/mL [786].

These theoretical calculations suggested that micropores with the size around 1.14 nm responsible for methane adsorption, and other pores (mesopores, macropores and spaces among the particles) do not contribute methane storage. In other words, monolithic carbon materials with a large micropore volume and a high bulk density are suitable for methane storage.

In Fig. 3.268, methane adsorption under 3.5 MPa at 25°C is plotted against $V_{\mathrm{micro}}$ of the carbons activated by phosphoric acid and followed by heat treatment below or above 620°C [787]. The carbons heat-treated above 620°C give higher methane adsorption than the carbons heat-treated below 620°C. The result is explained by the elimination of oxygen-containing surface functional groups by the heat treatment above 620°C. It has experimentally been demonstrated that nitric acid treatment of activated carbons decreases methane adsorption and hydrogen treatment recovers methane adsorption [788].

Single-wall carbon nanohorns synthesized by laser-ablation of graphite at room temperature (Section 3.5.5) give the methane storage of 160 mL/mL under 3.5 MPa pressure at 30°C [811]. In order to get this value of storage, the nanohorns have to be opened by heating in $O_2$ at 420°C for 10 min, followed by overnight treatment in $H_2O_2$. The residues are heat-treated in argon at 900°C to remove the oxygen-containing functionalities, which are produced during the opening process (oxidation). The carbon nanohorns thus treated are mechanically compressed at 50 MPa under vacuum for 10 min. The pellet with a bulk density

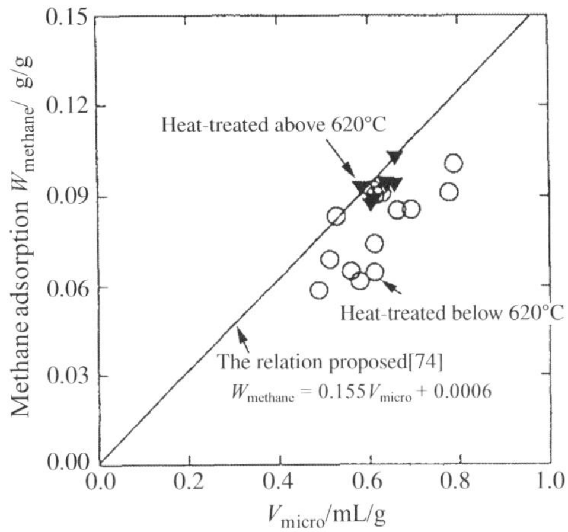


**FIGURE 3.268**

Relation between methane adsorption and $V_{micro}$ for the carbons activated by phosphoric acid and following heat treatment at different temperatures [787].

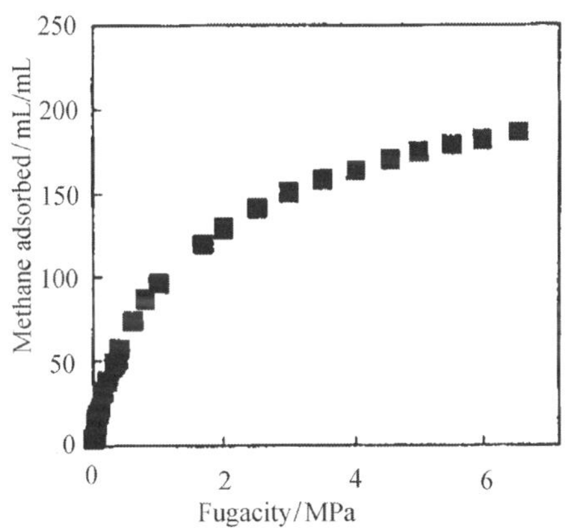


**FIGURE 3.269**

Methane adsorption isotherm of compressed single-wall carbon nanohorns at 30°C [811].

of 0.97 $g/cm^3$, which is obtained after nine cycles of crushing the pellet and successive compressing, shows 160 mL/mL storage ability. In Fig. 3.269 is shown methane adsorption isotherm for the compressed single-wall carbon nanohorns with 0.9 $g/cm^3$ density. This pressure dependence of methane adsorption for the compressed single-wall carbon nanohorns shows a good agreement with those calculated theoretically for CNTs [811].

A noticeable enhancement of methane adsorption by the pre-adsorbed water in micropores has been found by using steam-activated carbon fibers and

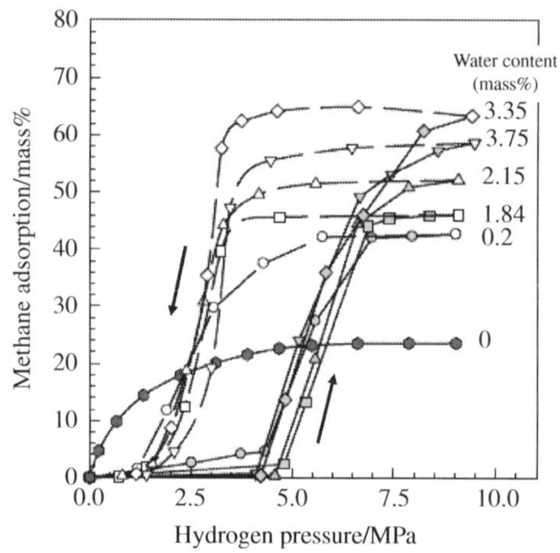


**FIGURE 3.270**

Adsorption/desorption isotherms of methane for activated carbons with different water contents [812].

KOH-activated pitch-based carbon [397,802–804,812]. For activated carbon fibers with $S_{BET}$ of 1800 $m^2/g$, methane adsorption increases to about 180 mg/g when the fractional filling of micropores by the pre-adsorbed water is 0.40 [397]. In Fig. 3.270, change in saturated adsorption of methane with water content is shown by using an activated carbon as adsorption/desorption isotherm [812]. Gravimetric methane adsorption increases to about 65 mass%, volumetric adsorption to about 200 mL/mL, at water content of around 3 mass%. In micropores, adsorbed methane forms stable methane-water clathrate (methane hydrate) with the composition of $CH_4.2H_2O$.

# 3.10 Carbon materials for environment remediation

## 3.10.1 Carbon/anatase composites

### a. Composites of carbon with $TiO_2$

Titanium dioxide $TiO_2$ has different crystalline phases, brookite, anatase and rutile. One of these phases, anatase, is active under the irradiation of ultraviolet (UV) rays, to be called 'photoactive', and one of the important photocatalysts, especially for the detoxification of water and air [813]. Many works have been focused on the preparation as well as on the modification of $TiO_2$, in the relation to its applications. Various composite materials of $SiO_2/TiO_2$, $ZrO_2/TiO_2$, zeolite/$TiO_2$ and $TiO_2$-coated polystyrene spheres were proposed in order to realize a high performance for photocatalysis, some of them being practically used

[814–822]. The composites with activated carbon have also been studied actively.

For the composites between photocatalyst $TiO_2$ and carbon, three cases have been reported, $TiO_2$-mounted activated carbon, carbon-doped $TiO_2$ and carbon-coated $TiO_2$. $TiO_2$-mounted activated carbon has been studied and certain successes are obtained for the coupling of photoactivity of anatase-type $TiO_2$ with adsorptivity of activated carbon, even though adsorptivity of carbon decreases by mounting fine particles of $TiO_2$. The mechanical mixture of anatase-type $TiO_2$ and activated carbon powder have been reported to give high performance for the decomposition of different model organic contaminants. Carbon-doped anatase-type $TiO_2$ have been prepared and reported to have visible-light sensitivity. Carbon-embedded and carbon-modified anatase photocatalysts are also prepared. Carbon-coated $TiO_2$ shows enhanced photoactivity, which is different from $TiO_2$-mounted carbons in the positional relation between two components ($TiO_2$ and carbon) and also different from carbon-doped anatase in carbon content and color. The importance of carbon materials in photocatalysis has been extensively reviewed [823].

### b. Anatase-mounted activated carbons

Composites between photoactive $TiO_2$ and different activated carbons have been studied. The increase in adsorptivity of the catalysts has certain benefits in the photocatalytic process, because the adsorbed molecules can easily react with OH radicals formed on $TiO_2$ surface, which are very strong oxidizers and taking part in the decomposition of organic matters. Therefore, high surface area of catalyst is very important as well as high dispersion of $TiO_2$ particles. However, a high degree of particle dispersion is not convenient for the separation of catalysts from water. In this respect, increase in accessibility of organic matters to the active radicals formed during excitation of $TiO_2$ seems to be interesting. Therefore, $TiO_2$-mounting on activated carbon surface has often been employed through various processes [824–830].

In Fig. 3.271, changes in relative concentration, $c/c_0$, of methylene blue (MB, $C_{16}H_{18}ClN_3S$) in the solution with time are shown for oxidized carbon spheres before and after $TiO_2$ mounting. Carbon spheres before $TiO_2$-mounting show only a small amount of adsorption, which seems to be saturated after soaking in the solution for first 2 h in the dark. After the saturation of adsorption in the dark, UV irradiation is performed by repeating on and off processes of UV light. On the $TiO_2$-mounted carbon spheres, concentration of MB in the solution decreases only during UV irradiation, proving that photocatalytic decomposition of MB surely occurs in the present system. The decomposition rate is almost the same in each irradiation period. The oxidized carbon spheres without anatase mounting adsorb only a small amount of MB and no effect of UV irradiation is observed. The adsorbed amount of MB for carbon spheres is a little larger than those after $TiO_2$-mounting, being consistent with the difference in BET surface area ($S_{BET}$).

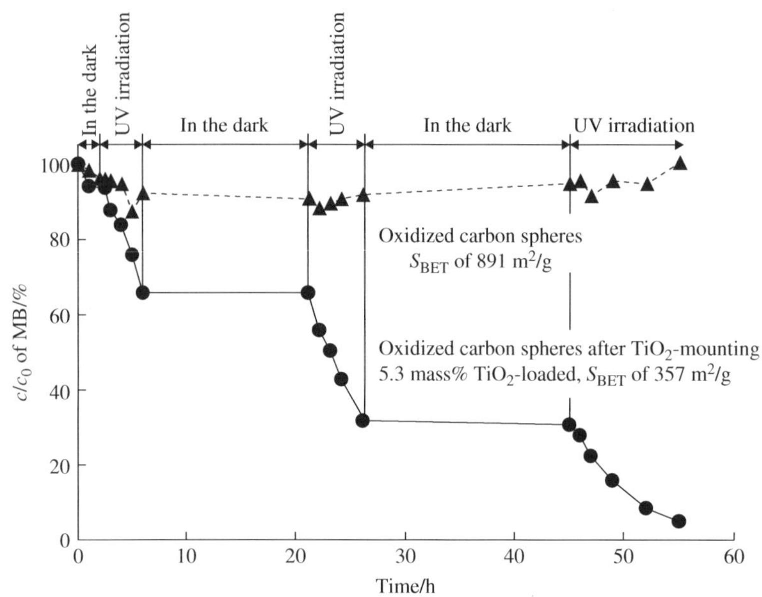


**FIGURE 3.271**

Changes in $c/c_0$ of MB with UV-irradiation time for oxidized carbon spheres before and after $TiO_2$-mounting [830].

In Fig. 3.272, change in logarithmic relative concentration, $\ln(c/c_0)$, of iminoctadine triacetates (IT, $C_{24}H_{53}N_7O_6$), which is one of agriculture chemicals often used in Japan, in its 100 ppm aqueous solution with irradiation time of UV is shown for an activated carbon with different amounts of anatase mounted [829]. Photodecomposition of IT proceeds slowly, and can be repeated with almost the same rate by renewing the IT solution. The decomposition rate of IT depends strongly on the amount of anatase mounted, revealing that there is an optimum mounting of anatase to attain maximum decomposition rate.

$TiO_2$ mounting onto carbon precursor is also tried, in which two processes, annealing of $TiO_2$ mounted and carbonization of the precursor, proceed simultaneously during the heat treatment at high temperatures and the resultant anatase-mounted carbons are studied on their structural characteristics, constituent crystalline phases of $TiO_2$, $S_{BET}$, and adsorption and photoactivity for phenol (Ph, $C_6H_6O$) [829].

A suspended mixture of $TiO_2$ particles (Degussa, P-25) with an activated carbon powder in aqueous solution of a contaminant gives higher rate constant, about 2.5 times higher, for the decomposition of phenol 4-chlorophenol and 2,4-dichlorophenoxyacetic acid [827,831]. As shown in Fig. 3.273a, the suspended mixtures of $TiO_2$ with activated carbon show marked adsorption of phenol in the dark and one of mixture gives faster removal of phenol under UV irradiation, as re-plotted by $\ln(c/c_0)$ vs. irradiation time in Fig. 3.273b.

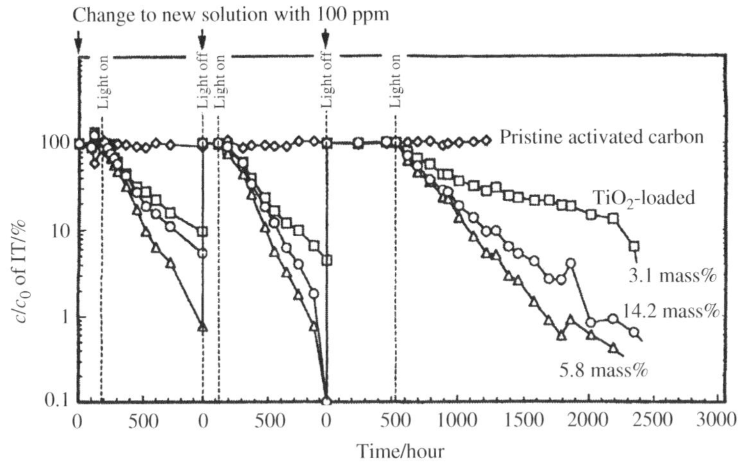


**FIGURE 3.272**

Changes in $c/c_0$ of IT with UV-irradiation time for $TiO_2$-mounted and pristine activated carbon [830].

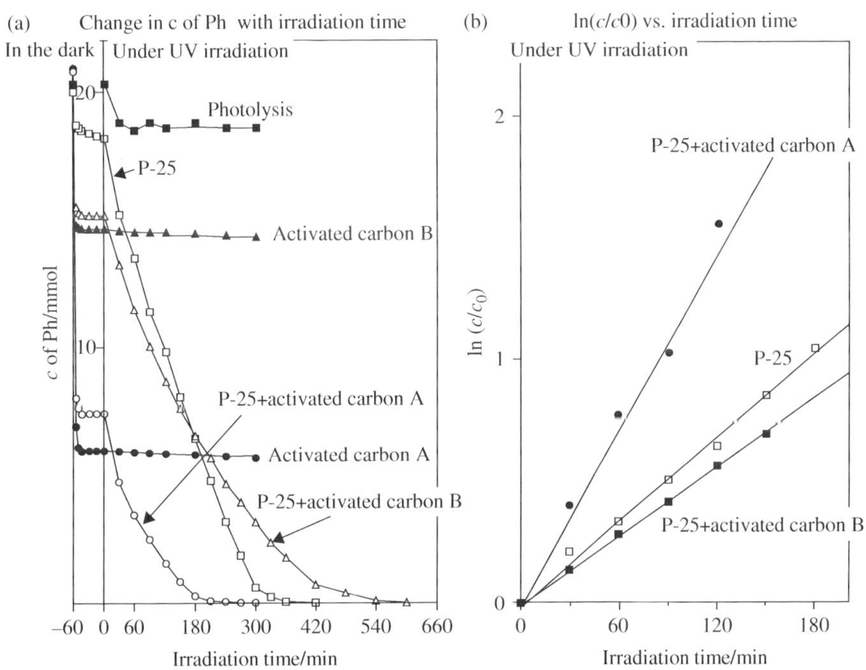


**FIGURE 3.273**

Dependence of concentration of Ph with irradiation time for the mixture of $TiO_2$ with activated carbon [827].

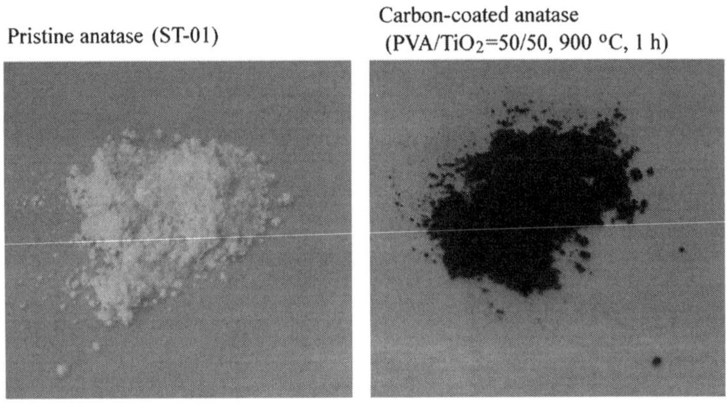


**FIGURE 3.274**

Appearance of anatase powder before and after carbon coating.

### c. Carbon-coated anatase

Another composite between anatase-type $TiO_2$ and carbon, i.e., carbon-coated anatase, have photocatalytic activity as anatase itself [832], even though it becomes black. In a series of works on carbon-coated anatase, it has clearly been shown that it has certain advantages; stabilization of anatase phase, consequently high crystallinity of anatase phase, high photoactivity, high adsorptivity and inert for organic binder for its fixation [501,832–841].

In Fig. 3.274, the appearance of the carbon-coated anatase is shown with the pristine anatase powder (Ishihara-Sangyo, ST-01). The process of carbon coating is very simple; the powder mixture of a pristine anatase and a carbon precursor, poly(vinyl alcohol) PVA, is heated at a temperature above 400°C in an inert atmosphere. If the mixing ratio of anatase to PVA is less than 50 to 50 in mass, no agglomeration of particles is observed [814], as can be seen in Fig. 3.274.

In Fig. 3.275, changes in X-ray diffraction pattern with heat treatment temperature (HTT) are shown for the pristine ST-01 and the mixture of ST-01 with PVA in 50/50 mass ratio. By heat treatment to 500°C, diffraction peaks of ST-01 were sharpened, showing crystallinity improvement, but after 700°C almost half of anatase is transformed to rutile structure, which has almost no photoactivity, and the transformation to rutile is completed at 900°C. For ST-01/PVA mixture, on the other hand, rutile phase appeared only in a small amount even above 900°C. Therefore, phase transformation from anatase to rutile is largely suppressed by carbon coating.

In Fig. 3.276, diffuse reflectance spectra are compared on anatase and carbon-coated anatase with different carbon contents [838]. With increasing carbon content, which has been supposed to correspond to the increase in the thickness of carbon layer coated on anatase particles, absorbance in the wavelength ranging from about 400 to 850 nm increased. Even for the sample with a high carbon

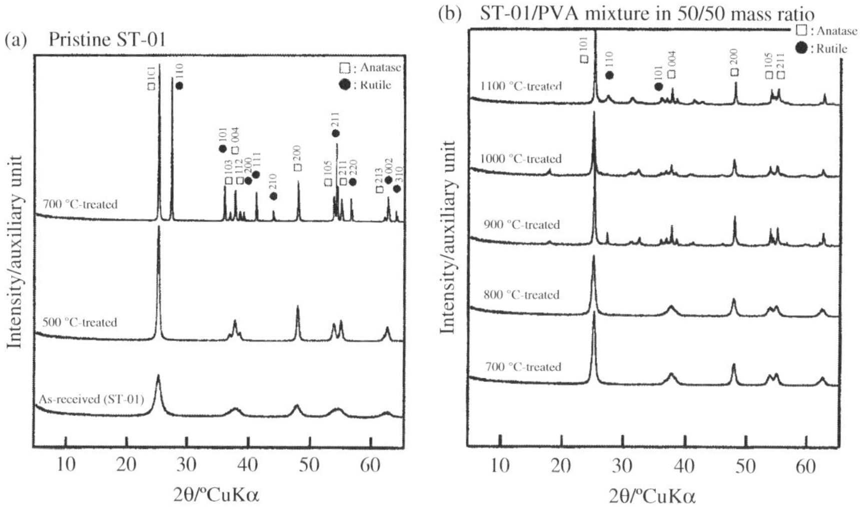


**FIGURE 3.275**

Changes in XRD pattern with HTT [833].

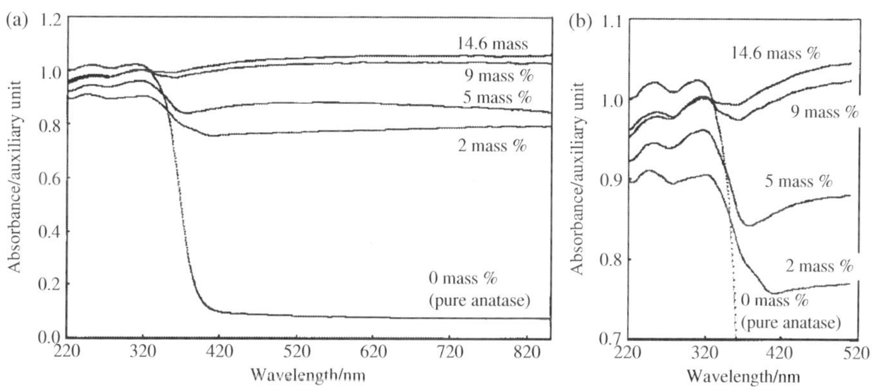


**FIGURE 3.276**

Diffuse reflectance spectra of anatase and carbon-coated anatase with different carbon contents [838].

content (14.6 mass%), however, absorption edge for anatase crystals, as seen for pure anatase, is recognized clearly, as shown by the enlarged figure in Fig. 3.276b. This suggests that anatase particles are activated by UV rays even after carbon-coating, though UV rays reach on the surface of anatase particle are weakened.

Under TEM, coated carbon gives weak contrast in comparison with anatase $TiO_2$ particle and sometimes $TiO_2$ shows Moire fringes because of high

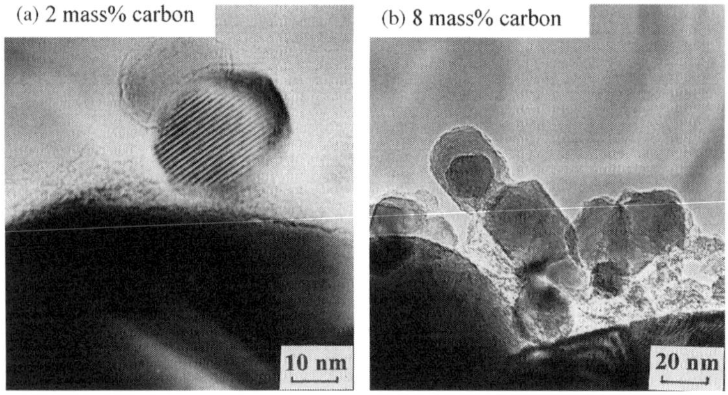


**FIGURE 3.277**

TEM images of the particles of carbon-coated anatase with different carbon contents [833].

crystallinity. Therefore, we can recognize that $TiO_2$ particles are coated by carbon, as shown in Fig. 3.277. The carbon-coated anatase shows relatively large $S_{BET}$ as 40–170 $m^2/g$, depending on carbon content, suggesting the formation of porous carbon. The carbon-coated anatase prepared at 700°C from the anatase/PVA mixture with a mixing ratio of 50/50, for example, gives apparent $S_{BET}$ of 130 $m^2/g$ and carbon content of 14.6 mass%, which leads to the calculated surface area of about 700 $m^2/g$ for carbon layer by taking into consideration of the fact that the pristine anatase ST-01 has $S_{BET}$ of 31 $m^2/g$ after 700°C treatment.

High surface area usually gives high adsorptivity, which can be an advantage for the removal of contaminants. In Fig. 3.278, adsorptivity for four contaminants, two azodyes (MB and reactive black 5, RB5), Ph and IT, is plotted against $S_{BET}$ for the carbon-coated anatase prepared at 700 and 900°C. Although adsorption into carbon layer of carbon-coated anatase depends strongly on the contaminant, MB is adsorbed in a large amount but RB5 and Ph are in relatively small amounts, adsorption is closely related to apparent $S_{BET}$ for each contaminant.

Photocatalytic activity of carbon-coated anatase has been measured through the determination of rate constant for the decomposition reaction of MB, RB5, Ph and IT [837–840]. The rate constant $k$ is determined from the slope of the linear relation between logarithm of relative concentration of a contaminant remained in the solution, $\ln(c/c_0)$, and irradiation time $t$. In Fig. 3.279, the plots of $\ln(c/c_0)$ against $t$ for the decomposition of MB and IT are shown for carbon-coated anatase photocatalysts, which are prepared at 900°C and have different carbon contents, in comparison with the pristine anatase ST-01. Carbon-coated anatase can decompose MB more rapidly, but IT more slowly than the pristine ST-01.

On anatase, its photocatalytic activity depends strongly on its crystallinity, which is evaluated by the full width at half maximum intensity (*FWHM*) of 101

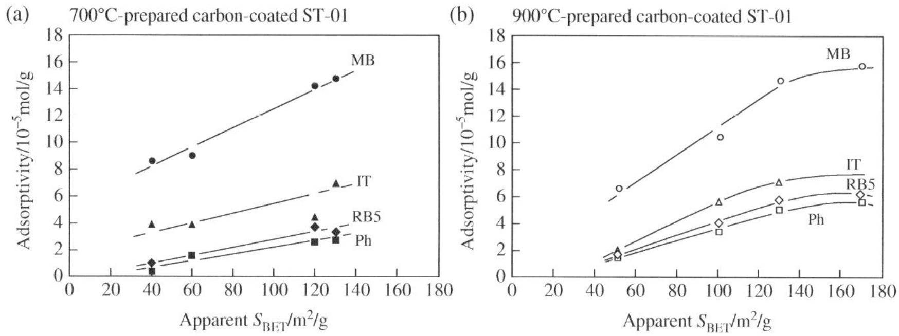


**FIGURE 3.278**

Dependences of the adsorptivity for MB, RB5, Ph and IT on apparent $S_{BET}$ of carbon-coated anatase prepared at 700 and 900°C [840].

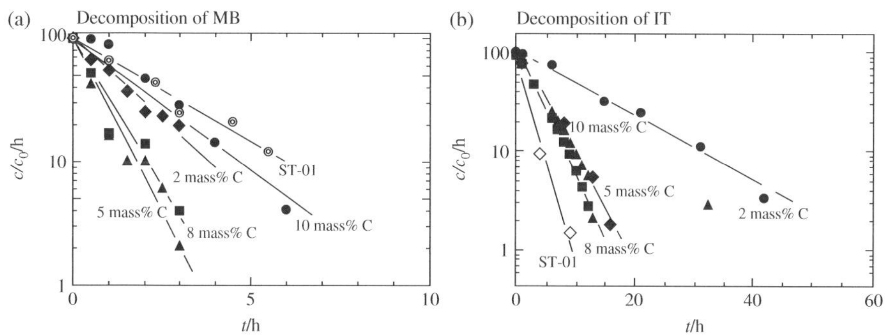


**FIGURE 3.279**

Plots of logarithm of $c/c_0$ against $t$ for carbon-coated anatase with different carbon contents and the pristine anatase ST-01 [841].

diffraction line of anatase phase [842,843]. For carbon-coated anatase, the rate constant $k$ for the decomposition of four contaminants is plotted against *FWHM* of anatase in Fig. 3.280. Although absolute values of $k$ are different for four contaminants, the dependence of $k$ on *FWHM* is the same, maximum rate constant being obtained at around 0.6° in 2θ (CuKα).

For the decomposition of contaminants on carbon-coated anatase, the molecules of contaminant have to pass through 4 steps, (1) diffusion to the surface of the particle of carbon-coated anatase in the solution, (2) adsorption into carbon layer, (3) diffusion in carbon layer to the surface of anatase particle and then (4) decomposition by photoactivity of anatase. Although the photoactivity of anatase particles is governed by its crystallinity [842,843], as shown in Fig. 3.280, carbon layer on the anatase particles is supposed to also play an important role on

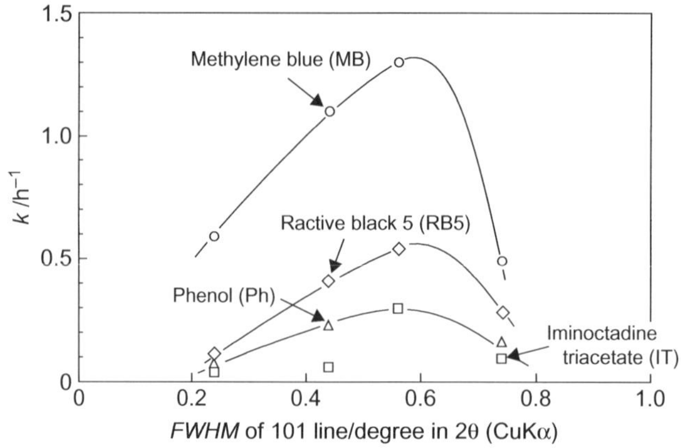


**FIGURE 3.280**

Relations between *k* for the decomposition of four contaminants and *FWHM* of 101 diffraction line of anatase [842].

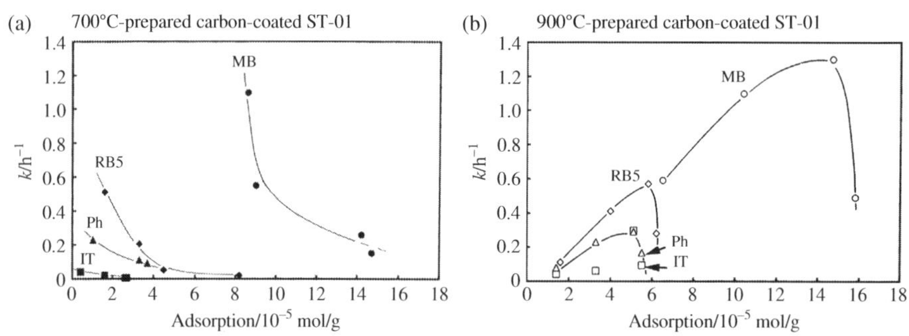


**FIGURE 3.281**

Relations between *k* for the decomposition of four contaminants and adsorption of carbon-coated anatase [501].

the photoactivity of carbon-coated anatase. In Fig. 3.281, k is plotted against adsorptivity for carbon-coated anatase prepared at 700 and 900°C. Between *k* and adsorptivity, a certain relation is observed, which is common for four contaminants used. This result suggests that *k* observed on carbon-coated anatase is also governed by the adsorptivity of coated carbon layer. The relation is different for preparation temperatures, 700 and 900°C, because the nature of carbon depends strongly on the carbonization conditions.

Carbon coating on anatase particles works effectively to prevent the interaction with binder resins. For fixing photocatalyst particles, various inorganic supporting materials have been proposed [814–822] to avoid the usage of organic binder. In Fig. 3.282, weight loss under UV irradiation with time is compared

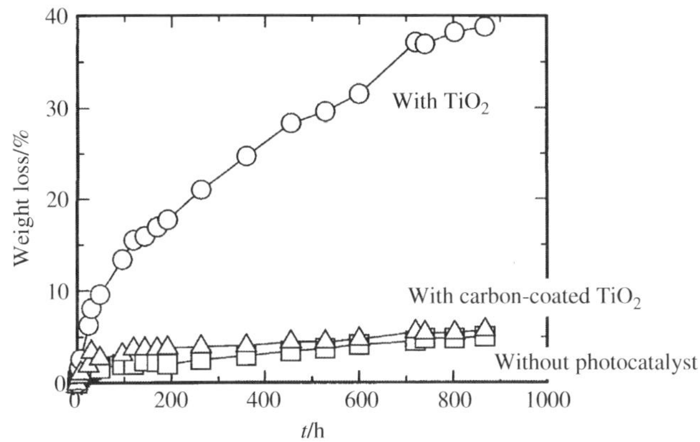


**FIGURE 3.282**

Weight loss of photocatalyst-dispersed acryl resin under UV irradiation [833].

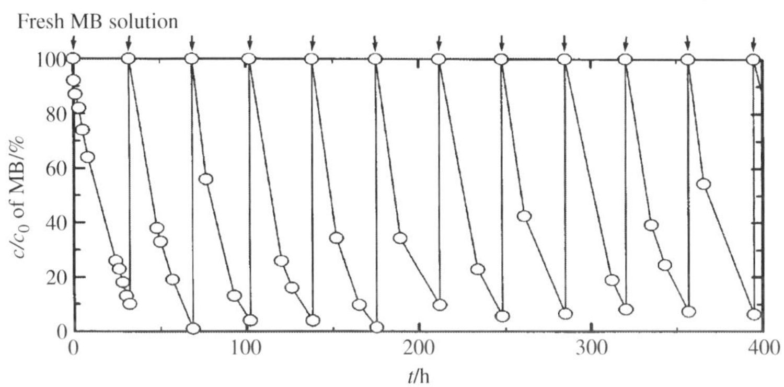


**FIGURE 3.283**

Performance for cyclic use of carbon-coated anatase fixed by acryl resin in $2.94 \times 10^{-5}$ mol/L MB solution [833].

among three acryl resin films, the film without any photocatalysts, the film containing anatase-type $TiO_2$ (ST-01) and the film containing carbon-coated anatase [833]. The weight loss increases quite rapidly for the film containing $TiO_2$ and reaches 39% loss after 870 h, which is mainly due to the photocatalytic decomposition of the resin by $TiO_2$. However, the film containing carbon-coated $TiO_2$ shows only 5% loss even after 870 h irradiation, which is almost the same as the acryl resin film without any photocatalyst.

In Fig. 3.283, cycling performance of carbon-coated anatase fixed with acryl resin is shown for the decomposition of MB in aqueous solution [833]. This result proves clearly that cyclic use of carbon-coated anatase is possible without any

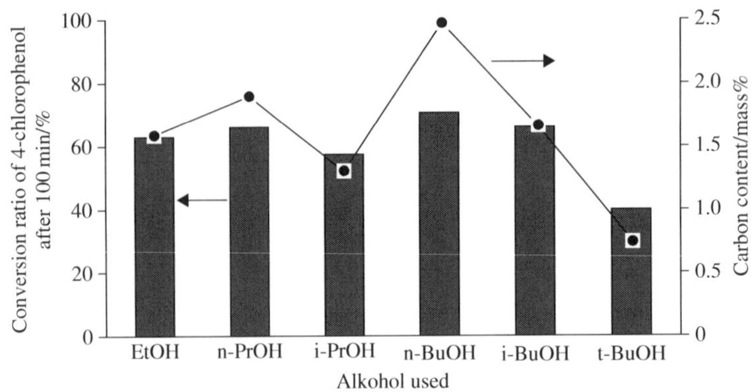


**FIGURE 3.284**

Photoactivity under visible light and carbon content of carbon-embedded anatase prepared from $Ti(O\text{-}iBu)_4$ alkoxide in various alcohols [844].

pronounced reduction in photocatalytic activity. Even in higher concentrations of MB ($1.47 \times 10^{-4}$ mol/L) a good cycling performance was confirmed.

### d. Carbon-doped anatase

By the oxidation of titanium carbide TiC at 350°C for 36 h in air and at 600°C for 5 h in $O_2$, carbon-doped anatase has been prepared, which shows photoactivity under visible light [844]. The carbon content is 0.32 mass%. Carbon is supposed to substitute oxygen in anatase crystal from XPS analysis and the particles of the photocatalyst are yellowish.

Doping of carbon into anatase has been carried out by heating the gels prepared from an alkoxide, titanium *iso*-butoxide, $Ti(O\text{-}iBu)_4$, at 250°C in air, in which a part of carbon is substituting oxygen in anatase but another part precipitate as coke-like particles in anatase particles (carbon-embedded anatase) [845]. In Fig. 3.284, conversion ratio of 4-chlorophenol by carbon-embedded anatase after 100 min and their carbon content are shown for the alcohols used during gel preparation. Rather good correspondence between photoactivity under visible light and carbon content are observed.

Surface of anatase-type $TiO_2$ (Police, Tytanpole A11) has been modified by heating in the vapor of n-hexane at a temperature ranging from 400–500°C [846]. After these heat treatments, $TiO_2$ particles are confirmed to keep anatase structure. Carbon content is less than 1 mass% (carbon-modified anatase), much less than carbon-coated anatase described above. In Fig. 3.285a, decomposition fraction of Ph under UV irradiation is plotted against carbon content as a function of HTT. On the carbon-modified anatase, turbidity of its slurry after 5 h irradiation and 10 min sedimentation is measured by using a turbidimeter, and plotted against carbon content in Fig. 3.285b. With increasing carbon content, photoactivity of carbon-modified anatase decreases but their turbidity decreases also.

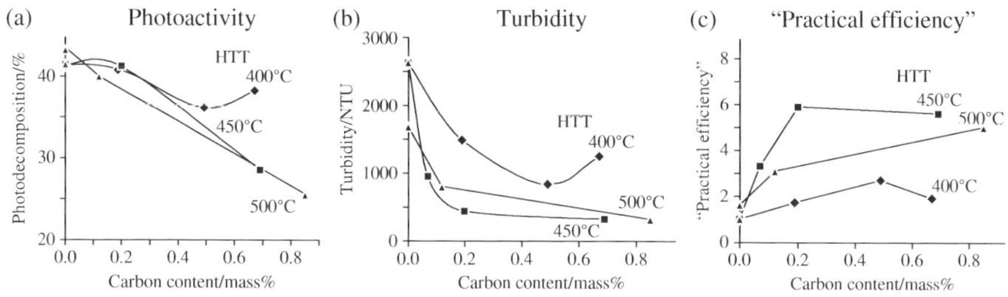


**FIGURE 3.285**

Changes in photoactivity for Ph decomposition (a), turbidity (b) and practical efficiency (c) with carbon content for carbon-modified anatase with different HTTs [846].

When the photocatalyst powders are used for purification of water, easy separation of photocatalyst particles from water, in other words, low turbidity of photocatalyst slurry, is desired. If the practical efficiency of the photocatalysts is calculated from decomposition fraction after 5 h irradiation, $A$, by the following formula;

$$\text{Practical efficiency} = (A_{\text{catalyst}}/A_{\text{reference}})(1/B),$$

where $A_{\text{catalyst}}$ is the decomposition fraction of carbon-modified anatase, $A_{\text{reference}}$ is that of pristine anatase and $B$ is turbidity. This practical efficiency calculated for the carbon-modified anatase is shown in Fig. 3.285c. Carbon-modified anatase has high practical efficiency for photocatalytic oxidation of Ph, higher than the pristine anatase without carbon modification, mainly due to marked improvement of turbidity of the slurry, though photoactivity of the catalysts is reduced by carbon modification.

## 3.10.2 Carbon materials for sorption of viscous fluids

### a. Sorption of heavy oils

Heavy oil spills result in not only a great deal of damage to the global environment but also a great loss of energy resources. Exfoliated graphite was firstly claimed to be able to sorb a heavy oil in a patent proposal in 1979 [847]. A Chinese group studied on sorption of heavy oils into exfoliated graphite paying particular attention to its preparation conditions [848]. It was demonstrated by a Japanese group that porous carbon materials with macropores, not only exfoliated graphite but also carbonized fir fibers and carbon fiber felts, were able to sorb a large amount of heavy oil very quickly, more than 80 g of heavy oil per 1 g of carbon materials within 1 min [849]. However, microporous carbon materials, such as activated carbons, have only small sorption capacity, as small as a few g/g. Dependences of sorption capacity on bulk density of carbon materials, temperature of sorption and viscosity of heavy oils, recovery of sorbed heavy oil from carbon materials and also recycling of both heavy oils and carbon materials have been explored on different carbon materials, exfoliated graphite [849–854], carbonized

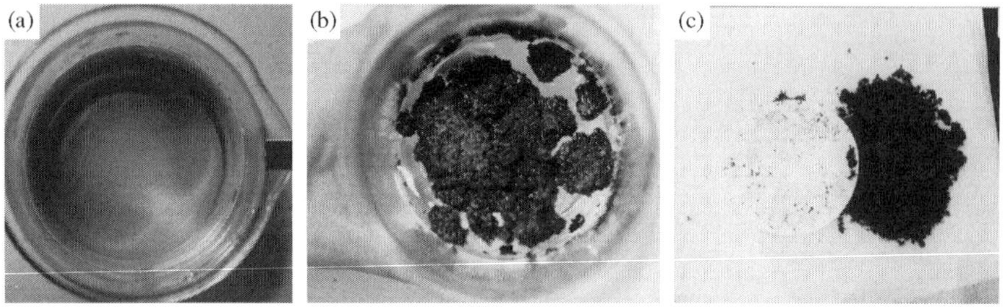


**FIGURE 3.286**

Photographs of A-grade heavy oil with exfoliated graphite. (a) Heavy oil floating on water, (b) 1 min after addition of a lump of exfoliated graphite, (c) after recovery of oil-sorbed exfoliated graphite by filtration.

fir fibers which are fibrous components extracted from natural fir trees [855,856] and carbon fiber felts [857]. Some trials for practical applications by using exfoliated graphite packed into plastic bags have also been carried out [858,859]. Fundamental sorption kinetics of heavy oils into carbons have been studied by using so-called wicking method [857,860,861]. These experimental results were reviewed in different journals [696,862–865].

Typical change in appearance is shown in the case of sorption of heavy oil floating on water by exfoliated graphite in Fig. 3.286. A-grade heavy oil floating on water (Fig. 3.286a) is completely sorbed into exfoliated graphite added (Fig. 3.286b). The characteristic brown color of A-grade heavy oil disappears within 1 minute after the addition of exfoliated graphite, if the amount of heavy oil is less than the sorption capacity of exfoliated graphite. The exfoliated graphite sorbed oil loses its luster and looks a deep black (Fig. 3.286b). Exfoliated graphite sorbed heavy oil can be easily separated from water by filtration. After taking out exfoliated graphite, no contamination appears in water and also in the used white filter paper, as shown in Fig. 3.286c. When the amount of heavy oil is a little larger than the sorption capacity of exfoliated graphite, the periphery of the lump of exfoliated graphite is trimmed by a thin oil layer. When largely excess oil presents, the exfoliated graphite lump looks wet and the brown color of oil remains on water.

In Fig. 3.287, dependences of sorption capacity on bulk density of three carbon sorbents, exfoliated graphite, carbonized fir fibers and carbon fiber felts, are shown for A-grade heavy oil, of which viscosity is 0.004 Pa.s. In Fig. 3.287b and 17c, representative data for exfoliated graphite and fir fibers are shown for making comparison easier. The sorption capacity of exfoliated graphite and carbonized fir fibers with low bulk density is surprisingly high; 1 g of exfoliated graphite with bulk density of 7 kg/m$^3$ can sorb more than 80 g of A-grade heavy oil, which is much higher than that of conventional sorbents, such as polyurethane foam (10–30 g/g). The sorption capacity decreases markedly with increasing bulk

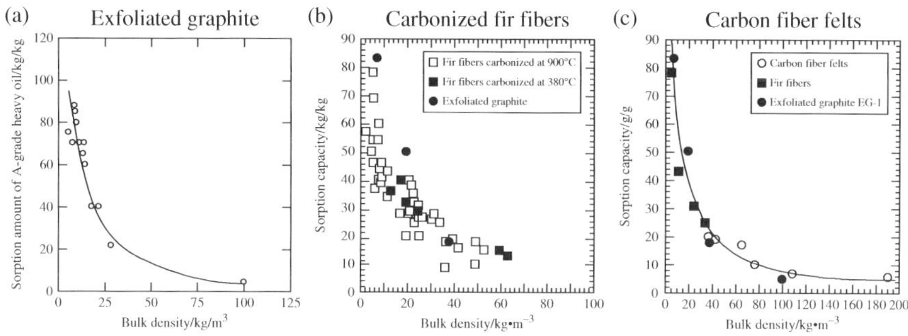


**FIGURE 3.287**

Dependences of sorption capacity for A-grade heavy oil on bulk density of carbon sorbents.

density. The sorption capacity of carbonized fir fibers is a little inferior to exfoliated graphite in the low bulk density region. Carbon fiber felts have rather high bulk density and so sorption capacity is not high, but its dependence on bulk density is almost the same as those observed on exfoliated graphite and carbonized fir fibers, as shown in Fig. 3.287c.

Correspondence between sorption capacity and porosity measured by different methods has been discussed and also pore structure in exfoliated graphite has been studied by different methods (Section 2.7.3b) [687,866–872]. Sorption capacity for A-grade heavy oil shows certain correspondence to pore volume measured by using new dilatometer of mercury porosimetry [866], as shown in Fig. 3.288, which is expected to attribute mostly to spaces among the particles [867]. Volume of large spaces among the worm-like particles of exfoliated graphite determined by using paraffin impregnation method is about 75% of the total volume of the lump and about 68% of sorbed heavy oil is thought to be kept in these spaces [868].

In Fig. 3.289, bulk density dependences of sorption capacity for B-grade heavy oil with the viscosity of 0.27 Pa s and C-grade heavy oil with 0.35 Pa s are shown on exfoliated graphite and carbonized fibers. Sorption capacity for viscous C grade oil is relatively low and a marked decrease is observed with increasing bulk density, a more pronounced decrease than for less viscous A-grade. No sorption is detected on dense exfoliated graphite, but on carbonized fir fibers certain capacity can be observed. Sorption capacity of exfoliated graphite for B-grade heavy oil, which is usually a mixture of less viscous A-grade oil and viscous C-grade oil, is a little larger than C-grade.

In Fig. 3.290a, the dependences of sorption capacity of exfoliated graphite with bulk density of about 7 $kg/m^3$ on temperature of oils are shown for four heavy oils. Even exfoliated graphite cannot sorb viscous C-grade oil below 15°C. On the other hand, a large amount of less viscous A-grade oil is sorbed into the exfoliated graphite even at 0°C. Crude and B-grade oils show intermediate

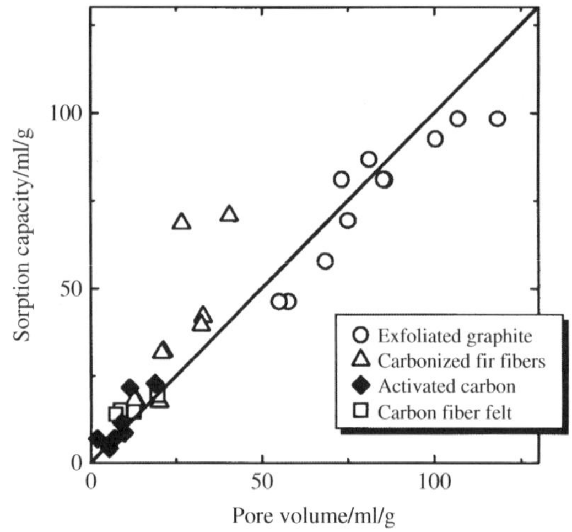


**FIGURE 3.288**

Relation between sorption capacity of different carbon sorbents for A-grade heavy oil and pore volume measured by using a new dilatometer for mercury porosimetry [865].

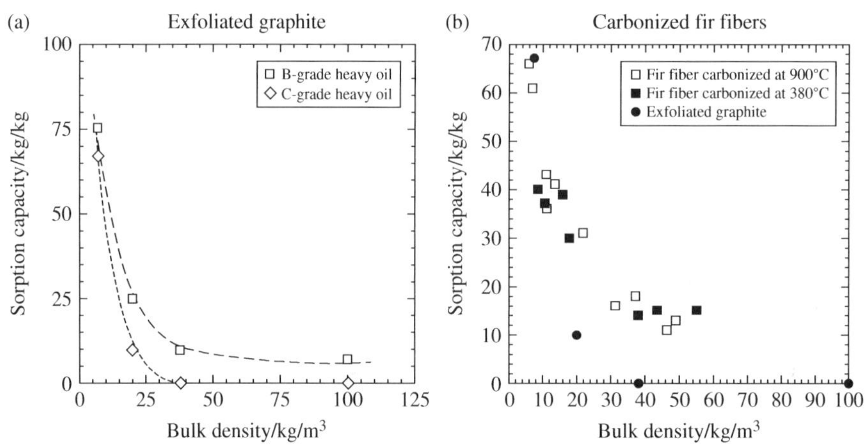


**FIGURE 3.289**

Dependences of sorption capacity of exfoliated graphite (a) and carbonized fir fibers (b) for C-grade heavy oil on bulk density of carbon sorbents.

dependences on temperature. In Fig. 3.290b, sorption capacities measured on the oils at different temperatures are re-plotted against viscosity of oils at different temperatures. The dependence of sorption capacity on viscosity of oil is divided into two regions, one for A-grade and crude oils and the other for B- and C-grade oils, which correspond to the viscosity below 0.01 and above 0.1 Pa s,

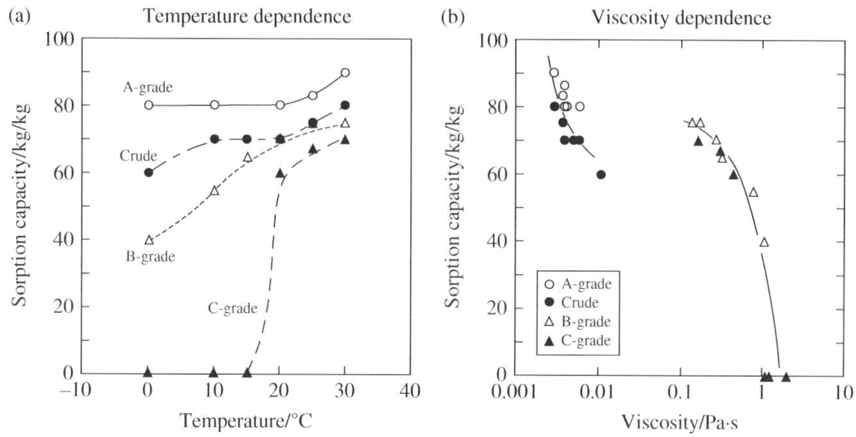


**FIGURE 3.290**

Dependences of sorption capacity of exfoliated graphite with a bulk density of 7 kg/m$^3$ on temperature of sorption and viscosity of heavy oils.

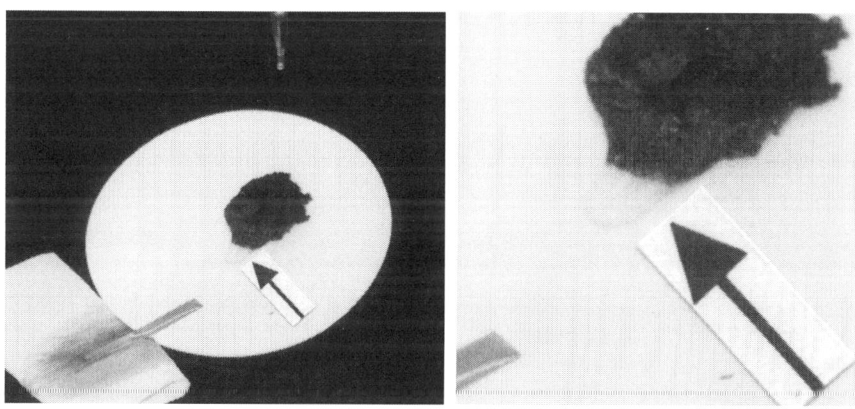

**FIGURE 3.291**

Water coming out from water-saturated exfoliated graphite by dropping A-grade heavy oil on its top.

respectively. In both regions, sorption capacity of the exfoliated graphite depends strongly on viscosity of oils.

When A-grade heavy oil is dropped onto one end of the lump of the water-saturated exfoliated graphite (sorbed about 1.8 kg of water per 1 kg of exfoliated graphite), water comes out from another end of the lump [696], as shown in Fig. 3.291. By continuing the dropping of heavy oil, water comes out continuously and finally heavy oil appears to come out. This experimental fact reveals

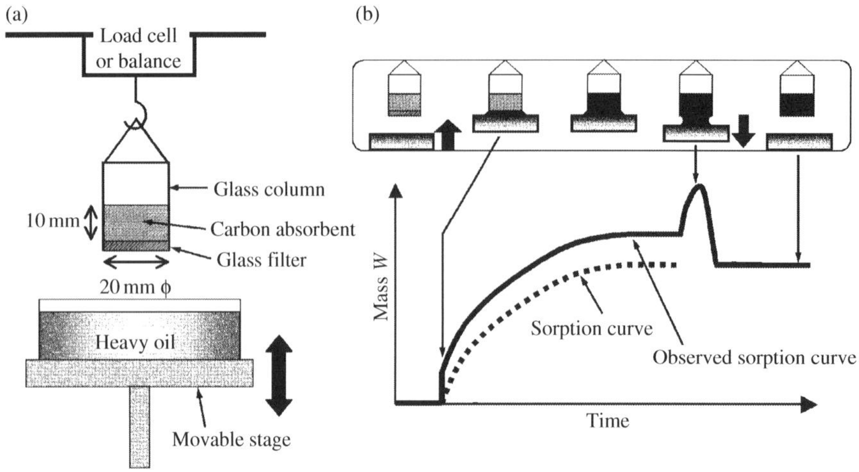


**FIGURE 3.292**

Scheme of measurement system for wicking method and sorption curve.

that water sorbed into exfoliated graphite can be replaced by heavy oil, in other words, exfoliated graphite sorbs preferentially heavy oil rather than water. On the other hand, a poly(propylene) mat can sorb about 4 kg of water per 1 kg of mat, but no water is recognized to come out from the mat by dropping A-grade heavy oil.

### b. Kinetics of heavy oil sorption

Kinetics of sorption into carbon materials for different oils has been studied by using the so-called wicking method [873]. The system for the measurement is schematically shown in Fig. 3.292a. Mass increase by capillary suction of oils from the bottom into carbon sorbents, either exfoliated graphite and carbonized fir fibers packed into a glass tube with a cross-sectional area of 3–10 $mm^2$ with different densities or carbon fiber felts cut into similar cross sectional area, is measured at room temperature as a function of time. The change in mass of carbon sorbents due to the sorption of oils is plotted against time (sorption curve), as shown schematically in Fig. 3.292b.

Some of the sorption curves for A- and C-grade heavy oils are shown on exfoliated graphite with different bulk densities in Fig. 3.293. Initial slope of sorption curves depends strongly on both bulk density of exfoliated graphite and viscosity of heavy oil. Gradual suction of A-grade heavy oil is observed by exfoliated graphite column with a bulk density of 7 $kg/m^3$. On the column with high densities of more than 20 $kg/m^3$, however, very rapid suction and reaching saturation within 10 s are observed (Fig. 3.293a). The initial slope of the curves increased with increasing bulk density of the column, but saturated mass increase shows a maximum at a bulk density of around 12 $kg/m^3$. Very similar dependence of sorption curve on the bulk density of exfoliated graphite column is observed for much

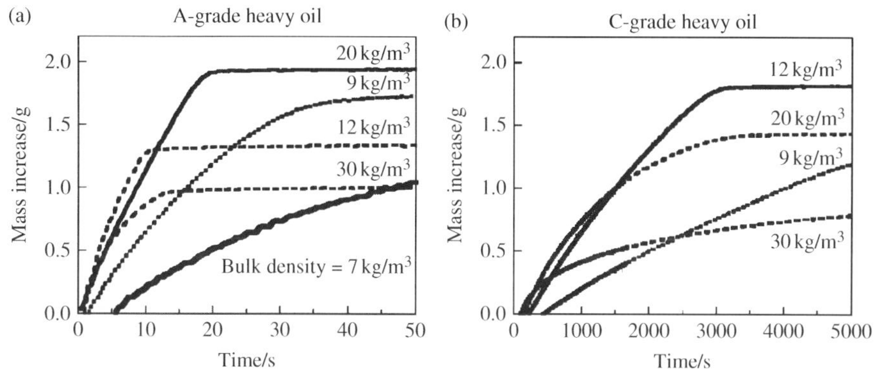


**FIGURE 3.293**

Sorption curves observed for heavy oil for exfoliated graphite with different bulk densities [55]. (a) A-grade heavy oil, (b) C-grade heavy oil.

viscous C-grade heavy oil, but it needs a much longer time than A-grade heavy oil to reach saturation (Fig. 3.293b).

The sorption curve before reaching saturation is well approximated by the equation:

$$m_s = K_s\, t^{1/2} + B, \tag{3.19}$$

where $m_s$ is mass increase per cross-sectional area of the sorbent column, $t$ time, $K_s$ sorptivity or liquid sorption coefficient and $B$ a constant. $K_s$ (kg/m$^2$ s$^{1/2}$) is a measure of sorption rate. This equation is theoretically derived for the sorption of a liquid into cylindrical capillary of a porous body [874], and successfully applied to porous ceramics [875,876].

In Fig. 3.294, the plots of $m_s$ vs. $t^{1/2}$ for A-grade heavy oil are shown on three carbon sorbents, exfoliated graphite, carbonized fir fibers and carbon fiber felts, respectively, with different bulk densities [864]. The initial slope of these curves, i.e., sorptivity $K_s$, depends strongly on the carbon sorbent and also its bulk density. The dependences of sorptivity $K_s$ on bulk density of three carbon sorbents are shown for A-grade heavy oil in Fig. 3.295. $K_s$ of carbonized fir fiber increases drastically with increasing bulk density in the region from 7 to 20 kg/m$^3$, and the increase in $K_s$ seems to be saturated at around 5.5 kg/m$^2$ s$^{1/2}$ when the fiber lump is densified above 30 kg/m$^3$. In the case of the carbon fiber felts, which have bulk density higher than 50 kg/m$^3$, the value of $K_s$ is approximately constant at about 5.5 kg/m$^2$ s$^{1/2}$. For exfoliated graphite, the maximum $K_s$ is a half smaller than those for the others, showing at 16 kg/m$^3$. Sorptivity $K_s$, is independent of the column height, though the saturated value $m_s$ is different.

For viscous C-grade heavy oil, the value of sorptivity $K_s$ is so small, about 0.2 kg/m$^2$ s$^{1/2}$, that its slight dependence on bulk density of two carbon sorbents, carbonized fir fibers and carbon fiber felts, has been observed.

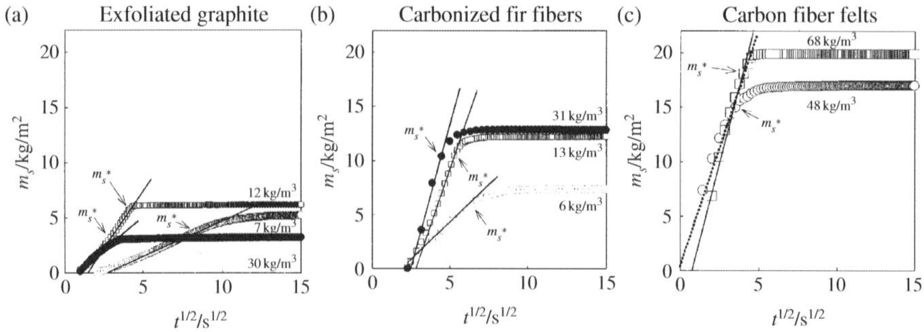


**FIGURE 3.294**

Sorption curves of carbon sorbents with different bulk densities for A-grade heavy oil [864].

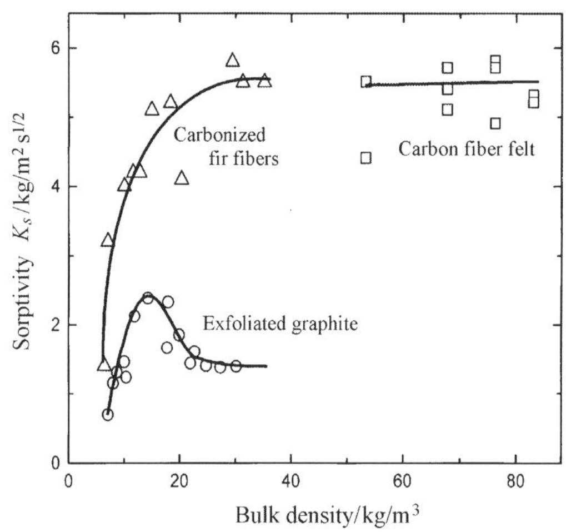


**FIGURE 3.295**

Dependences of sorptivity $K_S$ on bulk density of carbon sorbents [865].

Sorptivity or liquid sorption coefficient $K_s$ is theoretically expressed as follows [874]:

$$K_s = \left[ d_1 \sqrt{\frac{\gamma}{\mu}} \right] \left[ \frac{\varepsilon^*}{\lambda} \sqrt{r_o} \right] \left[ \sqrt{\frac{\cos\theta}{2}} \right] \tag{3.20}$$

where $d_1$ is density of liquid, $\gamma$ surface tension, $\mu$ viscosity, $\varepsilon^*$ effective sorption porosity of porous materials, $\lambda$ average tortuosity factor of the capillaries ($\lambda > 1$), $r_0$ average pore radius, and $\theta$ contact angle of interface between liquid and pore wall. Thus, the first blanket in Eq. (3.20) is the parameters based on the sorbates, the second one is variables for sorbents, and the third is the parameter for the interface between sorbate and sorbent. In order to understand the dependences of sorptivity $K_s$ on bulk density of carbon sorbents for A-grade heavy oil, the first blanket in Eq. (3.20) for the sorbate, A-grade heavy oil, must be a constant, and also the third blanket can be assumed to be a constant because the contact angle $\theta$ between A-grade heavy oil and pore wall of carbon is supposed not to be so much different among carbon materials used. Therefore, sorptivity $K_s$ on bulk density of carbon sorbents is tried to explain from these three parameters of the carbon sorbent, effective sorption porosity $\varepsilon^*$, average tortuosity factor $\lambda$ and average pore radius $r_0$ [864].

### c. Recovery of heavy oil and cycling of carbon sorbents

Because of the importance of heavy oils as energy resources, their recovery from sorbents is so important that recovered heavy oils have to be usable as energy resources and also recycling of the sorbent carbons is strongly desired. From this point of view, cyclic performance of carbon sorbents has been examined by different processes; filtration under suction, washing by solvent, centrifugation, etc.

Less viscous oils, as A-grade and crude oils, can be recovered from all carbon sorbents by the filtration under suction using a simple system as shown in Fig. 3.296. The changes in amounts of sorbed and recovered oils with cycling are shown on A-grade oil in Fig. 3.297. On exfoliated graphite, about a half of sorbed oil can be recovered in each sorption and recovery cycle. The remaining oils in an exfoliated graphite lump disturb the further sorption and so sorption capacity of the lump decreases roughly a half for each cycle. This decrease in sorption capacity is reasonably supposed to be due to the trapping of the oils at small crevice-like pores and also elliptic pores inside the particles (Section 2.7.3b).

Performance of a sorption/recovery cycle for carbonized fir fibers by filtration under suction is much better than exfoliated graphite, as shown in Fig. 3.297b. In each cycle about 80% of sorbed oil is recovered and so the decrease in sorption

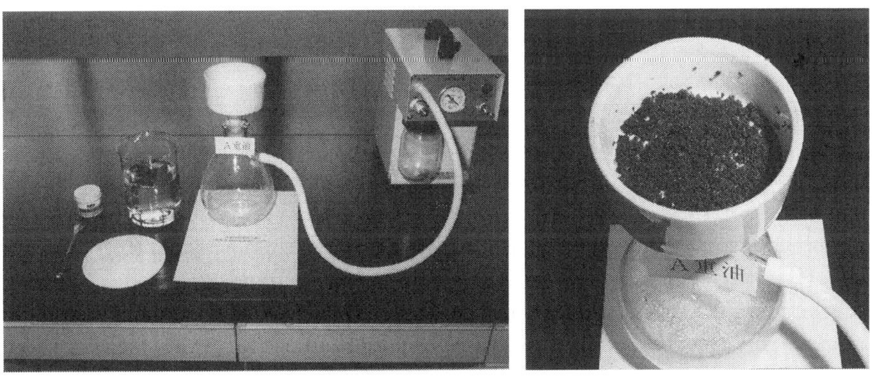


**FIGURE 3.296**

Filtration under suction of heavy oil-sorbed exfoliated graphite for the recovery of heavy oil.

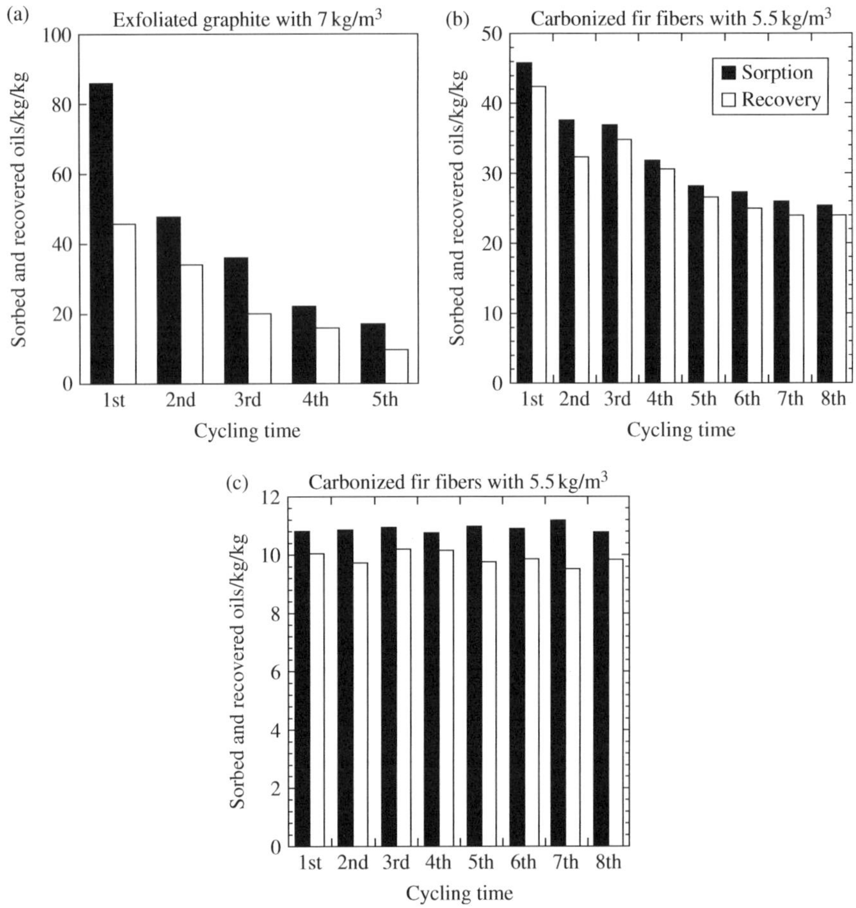


**FIGURE 3.297**

Cycle performances of carbon sorbents for A-grade heavy oil by filtration under suction.

capacity with cycling is much slower, after 8 cycles sorption capacity becoming about 60% of that of first cycle. When the fiber lump with a high bulk density is used, the decrease in sorption capacity with cycling is much less, though absolute value of sorption capacity is less. Though sorption capacity cannot be high, cycle performance of carbon fiber felts is excellent. By filtration under suction, about 90% of A-grade heavy oil sorbed could be recovered and no reduction in sorption capacity is observed even after 8 cycles, as shown in Fig. 3.297c.

Viscous oils cannot be recovered from either exfoliated graphite or carbonized fir fibers by filtration even under strong suction. Sorbed heavy oils can be recovered by washing with a solvent, such as n-hexane, but the exfoliated graphite after washing cannot be reused as sorbent for heavy oil, mainly because of the

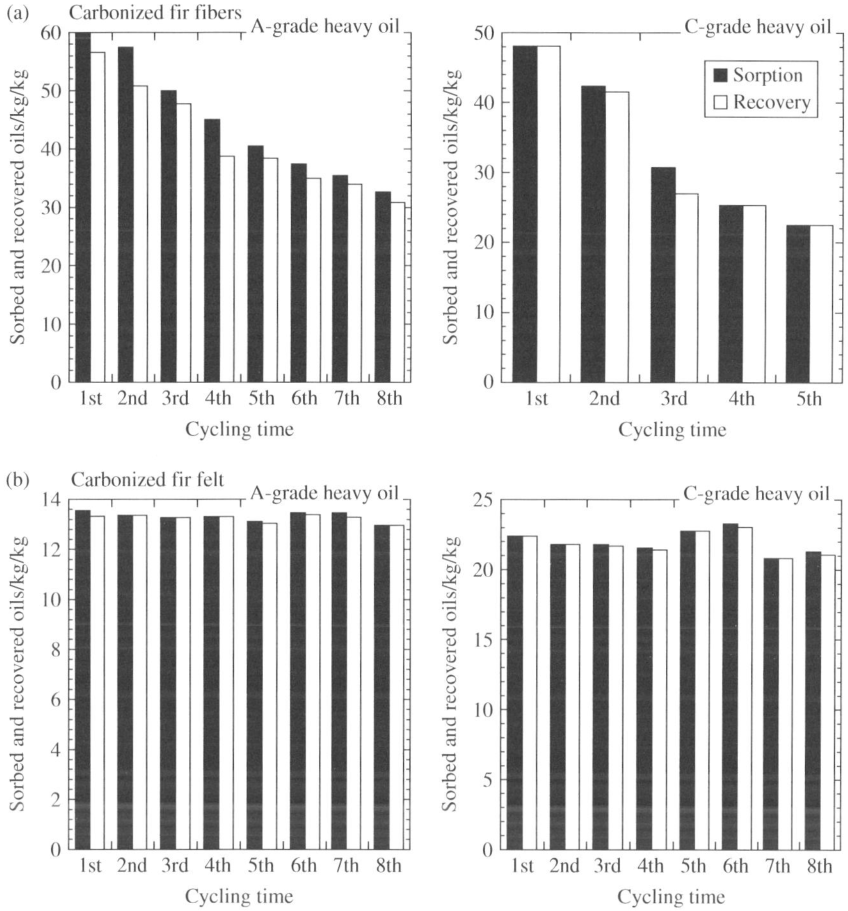


**FIGURE 3.298**

Cycle performance of carbonized fir fibers and carbon fiber felt for A- and C-grade heavy oils by washing with n-hexane.

destruction of bulky texture of exfoliated graphite. From fir fibers and carbon fiber felts, however, the oils, even viscous C-grade oil, can be washed out by using a solvent. In Fig. 3.298, cycle performances of carbonized fir fibers and carbon fiber felts for A- and C-grade heavy oils by washing with n-hexane are shown. On carbon fiber felts, almost 100% recovery and excellent cyclability by washing with n-hexane are obtained on both A- and C-grade heavy oils (Fig. 3.298b). In the case of C-grade heavy oil, less viscous A-grade oil can be used as a solvent.

In the case of carbon fiber felts consisting of PAN-based carbon fibers, even centrifuging at 3800 rpm can be applied without any reduction of sorption

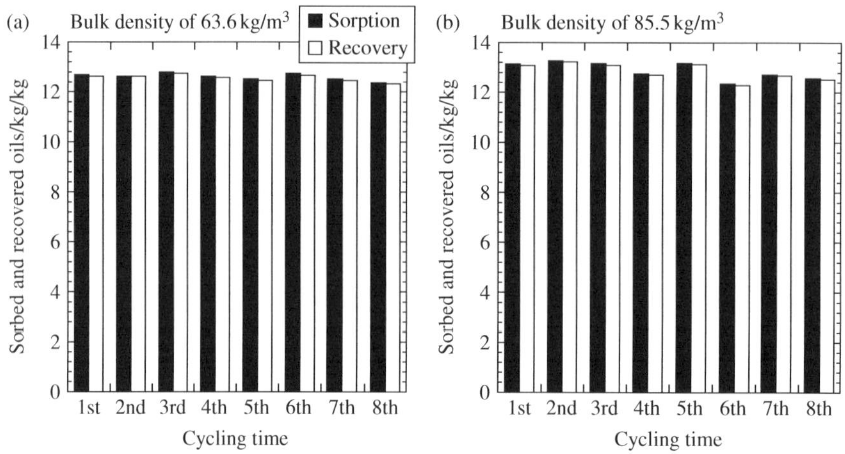


**FIGURE 3.299**

Cycle performance of carbon fiber felts by centrifuging with 3800 rpm.

**Table 3.49** Fractions of Aromatic Hydrocarbons and Averaged Molecular Weight Values Measured by FD-MS Analysis on Crude and C-grade Oils

| Sample Heavy Oil | | $F_{arom}$ (%) | Mn | Mw | Mz | Mw/Mn | Mz/Mw |
|---|---|---|---|---|---|---|---|
| A-grade heavy oil | Original | 4.0 | 258 | 274 | – | 1.06 | – |
| | Recovered | 4.2 | 259 | 274 | – | 1.06 | – |
| Crude oil | Original | 4.9 | 645 | 869 | 1102 | 1.35 | 1.27 |
| | Recovered | 4.5 | 672 | 915 | 1147 | 1.36 | 1.25 |
| C-grade heavy oil | Original | 5.4 | 1071 | 1768 | 2428 | 1.65 | 1.37 |
| | Recovered | 4.6 | 1207 | 1839 | 2393 | 1.52 | 1.30 |

*FD-MS: Field-desorption mass spectrometry; $F_{arom}$: Fraction of aromatic hydrocarbon; Mn : Number-averaged molecular weight; Mw : Weight-averaged molecular weight; Mz : Z averaged molecular weight.*

capacity during 8 cycles (Fig. 3.299). Pretty good cyclability is obtained even by squeezing the felt of PAN-based carbon fibers.

On recovered oils, different analyses on chemical composition, hydrocarbon contents and molecular weights have been carried out. No appreciable difference is detected between original and recovered oils. In Table 3.49, the results on fraction of aromatic hydrocarbons and different molecular weights are summarized on A-grade, crude and C-grade oils, revealing that recovered oils have no problem for all applications.

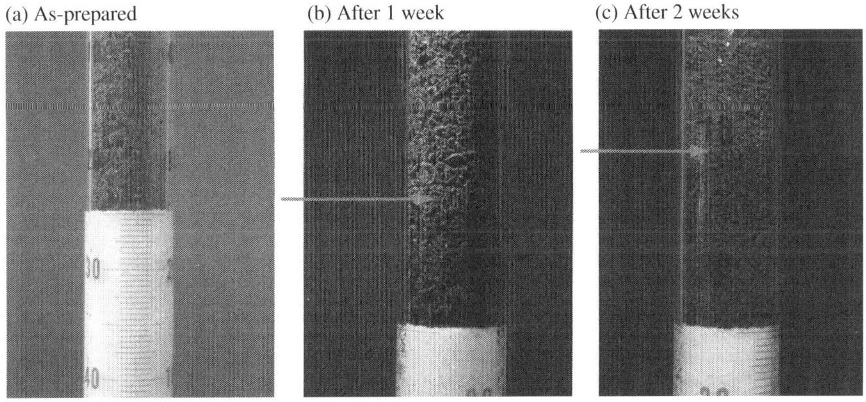


**FIGURE 3.300**

Experimental set-up and appearance of heavy oil climbing into exfoliated graphite with time [877].

### d. Recovery of heavy oil from contaminated sand

Capillary suction of heavy oil from the mixtures of model sand with defined particle sizes, alumina powders, with A-grade heavy oil has been performed [877]. Experimental set-up is as follows: in a glass cylinder with a diameter of 20 mm a mixture of sand with heavy oil (contaminated sand) is placed at the lower part and the upper part is exfoliated graphite with different packed densities, as shown in Fig. 3.300a. Upon heavy oil climbing, exfoliated graphite contacted with contaminated sand looks to be wetted, and changes its color to much darker black than the original, as shown in Fig. 3.300b and c. The change in the height of the wet part of exfoliated graphite is measured with time.

In Fig. 3.301a, changes in the height of climbed heavy oil in exfoliated graphite (bulk density of about 10 kg/m$^3$) with time are shown on the model sand, α-alumina ($\alpha$-$Al_2O_3$), with different particle sizes. With increasing time, the height of climbed heavy oil increases gradually and looks to saturate at a certain height. After 14 days, therefore, the height is thought to reach the saturation, though it increases slightly in the case of the particle size less than 303 μm. Saturated height of climbed oil is plotted against particle size of alumina sand in Fig. 3.301b. The climbing rate and saturated height of heavy oil depend strongly on the particle size of sand, as shown in Fig. 3.301a. From the sand with particle size of about 425 μm, heavy oil can climb up to the highest position in exfoliated graphite with a packed density of about 10 kg/m$^3$, in other words, the largest amount of heavy oil can be pumped up into exfoliated graphite. From the sand with a small particle size, like 175 μm, sorption of heavy oil is very slow, and so climbing height is rather low and increases gradually, even after 14 days. With a little larger particle size of sand, 303 μm, the saturated height of heavy oil is

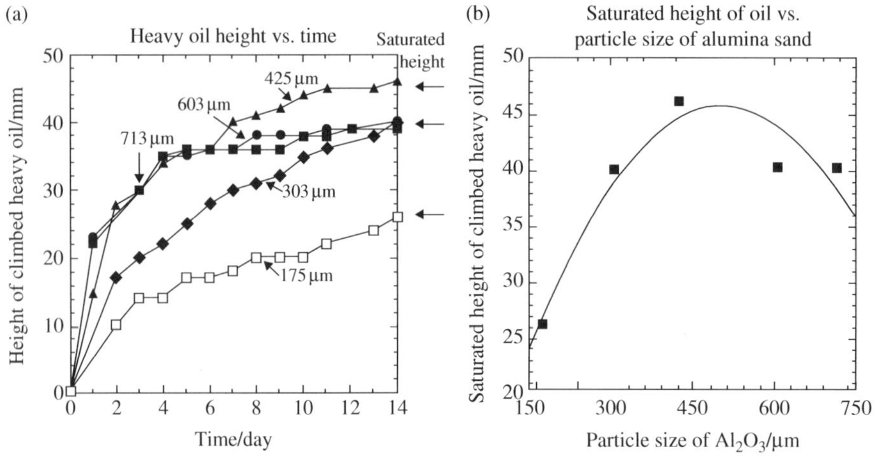


**FIGURE 3.301**

Changes in the height of climbed heavy oil in exfoliated graphite with time and dependence of saturated height of heavy oil on the particles size of sand [877].

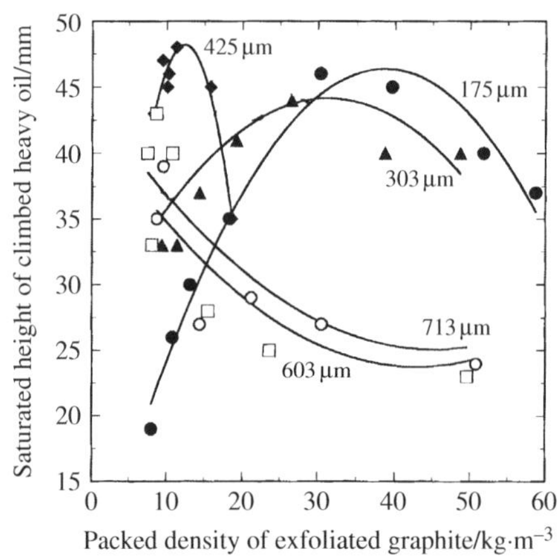


**FIGURE 3.302**

Saturated height of climbed oil as functions of packed density of exfoliated graphite sorbent and particle size of sand particles [877].

almost the same as the cases of particle size of 603 and 713 μm, but takes longer time to reach the saturation.

In Fig. 3.302, the saturated value of the height of climbed heavy oil from the sand with different particle sizes is plotted against packed density of exfoliated graphite used. For each particle size of alumina powder, there is a packed density to give a maximum saturated height of climbed heavy oil, i.e., to sorb a maximum amount of heavy oil. From alumina powder with the particle size larger than 425 μm, the maximum saturated height cannot be obtained in the present work, because the packed density of the exfoliated graphite cannot be less than 7 $kg/m^3$. It is seen clearly, however, that the packed density of exfoliated graphite to give the maximum saturated height shifts to lower value with increasing the particle size of alumina sand.

In order to have efficient recovery of heavy oil, it is important to have an appropriate combination between average size of sand particles and packed density of exfoliated graphite. For coarse sands, which have large pores to keep heavy oil, low packed density of exfoliated graphite where relatively large pores are formed among worm-like particles is required. Pumping of heavy oil into exfoliated graphite occurs as a balance in attractive force of capillary between exfoliated graphite and sand, the former being supposed to be stronger than the latter because of the hydrophobic nature of the surface of graphite. Climbing of heavy oil into exfoliated graphite from contaminated sand in the present positional relation, exfoliated graphite being in the upper part of contaminated sand, has to overcome the gravity.

For sea sand, which was sampled from the seashore where the accident of the tanker Nakhodka happened in 1997 and had an average particle size of 236 μm, the results consistent with those obtained on model sands were obtained on a saturated amount of climbed oil [877].

### e. Sorption of various oils

For various oils with a wide range of viscosity, as tabulated in Table 3.50, sorption capacity and sorption kinetics have been studied by using the exfoliated graphite with a bulk density of about 7 $kg/m^3$ [878].

**Table 3.50** Density and Viscosity of Oils Used

| Oil | Density ($kg/m^3$) | Viscosity (Pa · s) | Oil | Density ($kg/m^3$) | Viscosity (Pa · s) |
|---|---|---|---|---|---|
| Kerosene | 788.3 | 0.001 | Safflower oil | 914.0 | 0.069 |
| Light oil | 823.2 | 0.001 | 2-cycle motor oil | 858.3 | 0.118 |
| A-grade heavy oil | 852.3 | 0.004 | 4-cycle motor oil | 876.1 | 0.126 |
| Crude oil | 867.7 | 0.007 | Diesel oil | 877.7 | 0.127 |
| Mineral oil | 842.5 | 0.033 | B-grade heavy oil | 902.7 | 0.160 |
| Grape seed oil | 919.9 | 0.052 | C-grade heavy oil | 925.5 | 0.350 |
| Salad oil | 918.3 | 0.056 | | | |

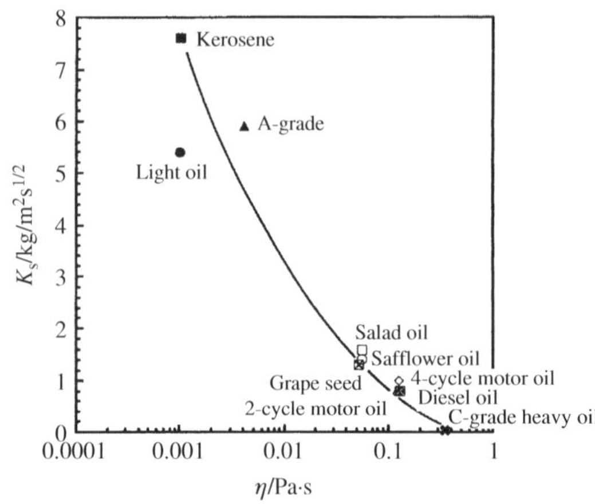


**FIGURE 3.303**

Dependence of $K_S$ for the exfoliated graphite with the bulk density of about 7 kg/m$^3$ on $\eta$ of oils [878].

The sorption rate depended strongly on viscosity. In Fig. 3.303, sorptivity $K_s$ is plotted against viscosity $\eta$ of oils in logarithmic scale. $K_s$ value shows a strong dependence on $\eta$, the oil with the higher viscosity being sorbed into a column of exfoliated graphite with the slower rate.

## f. Sorption of biomedical fluids

Exfoliated graphite has a high sorption capacity for biomedical molecules with large molecular size and weak polarity [879]. Sorption performances of exfoliated graphite with different bulk densities is studied on several kinds of biomedical molecules, ovalbumin, serum albumin, bovine serum albumin (BSA), lysine and herring sperm DNA [861]. The dependence of sorption capacity of exfoliated graphite for BSA on its bulk density is compared to heavy oil and a gasoline in Fig. 3.304. Changes in sorbed amount of BSA into a lump of exfoliated graphite with a low bulk density with time show that sorbed amount increases quickly in the first 20 minutes, but into a high density lump sorption turns quickly to saturation at a low amount. The lower-density lump has the larger sorption capacity, because of its larger pore volume for capillary condensation, but takes a longer time to be saturated.

The characteristics of carbon materials, such as low weight, chemical inertness, excellent compatibility with the human body [880] and also bacteriostasis have led to the new application as medical dressing in preventing a traumatized surface from infection, using large sorption capacity of exfoliated graphite for medical fluids [881].

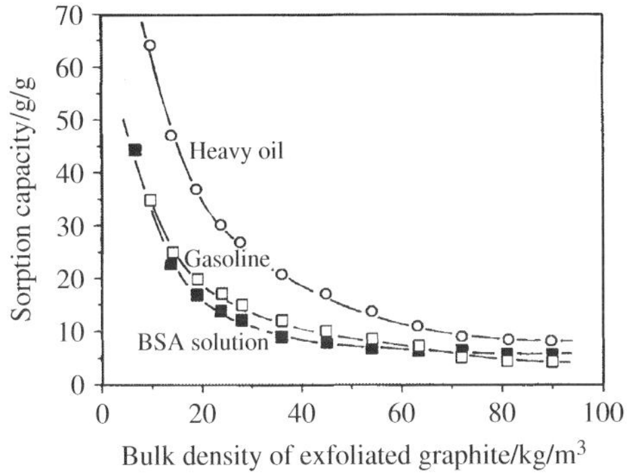


**FIGURE 3.304**

Dependence of sorption capacity for bovine serum albumin (BSA) on its bulk density of exfoliated graphite sorbent, in comparison with heavy oil and engine oils [861].

## g. Discussion on sorption of viscous fluids into macroporous carbon materials

So far, mats of some polymers, such as poly(propylene) and poly(urethane), have been used for the sorption of spilled oil. Their maximum sorption capacity is about 10–30 g of heavy oil per 1 g of polymer [882]. However, they sorb water, as well as heavy oil, and show no special selectivity for heavy oils. Therefore, the effective sorption capacity of the polymer mats for heavy oils floating on water must be lower than the figures mentioned above. Some natural sorbents prepared from cotton fibers, milkweed flosses and kenaf plants have been reported to have rather high sorption capacity and certain potential for oil recovery and sorbent re-usability [882–887]. Sorption capacity of macroporous carbon materials, exfoliated graphite and carbonized fir fibers, is very high in comparison with these materials. Preferential sorption of oils is an advantage of carbon materials in addition to their high sorption capacity.

It is interesting to point out that most materials, which have been either used or tested for sorption of heavy oils, are composed from fibrous particles, as explained above. Carbon materials, which have interesting performance for heavy oil sorption, are also fibrous, worm-like particles in exfoliated graphite having also fibrous morphology. The reason for this is not clear yet, but easy formation of large spaces with appropriate size for heavy oil sorption, associated with easy deformation of fiber networks to give appropriate morphology to keep oils, might be one of the reasons.

For large sorption capacity of carbon materials, large spaces among fibrous particles are reasonably supposed to be responsible. In the case of a lump of exfoliated graphite, there are at least three kinds of pores, large spaces among entangled worm-like particles with fibrous morphology, crevice-like pores on the surface of worm-like particles and elliptic pores inside of the particles. Large spaces among the particles is about 75% of total volume of a lump of exfoliated graphite and about 70% of sorbed heavy oil is thought to be kept in these spaces [868]. These large spaces among the particles can be easily destroyed by a slight compression, and, as a consequence, the sorption capacity is reduced correspondingly, as shown in Figs 3.287 and 3.289. However, another two pores, crevice-like pores on the surface of particles and elliptic pores inside the particles, also have important roles for heavy oil sorption. Observation under optical microscope showed that the oil climbed up through the edge tip of crevices formed on the surface of particles at the beginning of pumping of heavy oil into a lump of exfoliated graphite [696]. This complicated pore structure in exfoliated graphite lump results in rather strong holding of sorbed heavy oils, which does not move to filter paper during filtration to recover from water surface, even though sorption rate is low in comparison with carbon fiber felts which have a smooth surface.

The same discussion on heavy oil sorption into carbonized fir fibers is assumed, which have similar pore structure, large spaces among the fibers of fir plants, small pore inside the fibers and also rough surface of the fibers. In carbon fiber felts, however, only inter-particle pores exist, which results in small sorption capacity, but high sorption rate and high recovery ratio.

Hydrophobic (oleophilic) nature of the surface of carbon materials is a factor governing heavy oil sorption, particularly for preferential sorption of heavy oils.

## 3.10.3 Carbon fibers for environment remediation

It has been found that a large amount of microorganisms in water cling to carbon fibers in a short time [888]. When PAN-based carbon fibers in strand are hung in a solution containing revitalized dirt with 5% drainage (biomedical oxygen demand BOD of 1000 ppm), microorganisms in the dirt are clung quickly to carbon fibers, forming a large ball as shown in Fig. 3.305a. A strand of carbon fibers of 0.25 g became 440 g by clinging of microorganisms after 7 days. However, other organic fibers, such as cotton, nylon and polyethylene, cannot cling to microorganisms in such a large amount. Only for five minutes in this dirt, a large number of colony are cultured on PAN-based carbon fibers, more than ten times the number of other organic fibers, indicating that some bacteria prefer to be with carbon fibers, in other words, carbon fibers allure some bacteria. In order to have such a large amount of clung dirt, HT-type carbon fibers without sizing agents on their surface are preferable to HM-type ones and aeration of the solution is essential. During soaking of carbon fibers for a long time, carbon fibers act as substrate for organism metabolism and are supposed to be able to construct a small ecosystem. A number of practical applications of carbon fibers for water purification in sea, rivers and ponds have been developed [889].

In Fig. 3.305b, the test performed in the beginning of the research is shown, in which carbon fibers were hung in a pond with an area of 25 $m^2$ and a depth of 1 m. After a half hour, the water was not yet clear, but clearness of the water in the pond was markedly improved after 1.5 h with a slight aeration.

**FIGURE 3.305**

Clinging of microorganisms from the dirt onto PAN-based carbon fiber strands and purification of water in a pond by using carbon fibers.

*(Courtesy of Prof. A. Kojima of Gunma Nat. Coll. Tech.)*

Purification of environmental water by carbon fibers has been tested in various ponds and small lakes, and also in rivers. In the river Nakagawa in Gunma Prefecture, Japan, the river water was passed through a bypass with a cross-section of $50 \times 50\ cm^2$, where the strands of carbon fibers were hanged, with a rate of 30–50 L/min. In Fig. 3.306, change in biochemical oxygen demand (BOD) during the year 1999 is compared between the water sampled at the entrance and the outlet of the bypass. By passing through carbon fibers, BOD of water reduces by more than 50%. Clearness of water is also improved. This technique has been practically used to purify the sewerage in Coolgardie, Australia.

On Lake Haruna in Gunma Prefecture, Japan, carbon fiber strands were successfully used in order to increase the number of smelts. Lake Haruna was famous as a fishing place of smelts in winter through a hole on the ice, but the number of fish reduced greatly around 1995. In 1997, carbon fibers were tried to hang as the floats, some of them being shown in Fig. 3.307, at one place on the lake. On carbon fibers, many spawns were found and many young smelts were observed, as shown in Fig. 3.308. After various trials of the shape of carbon fiber

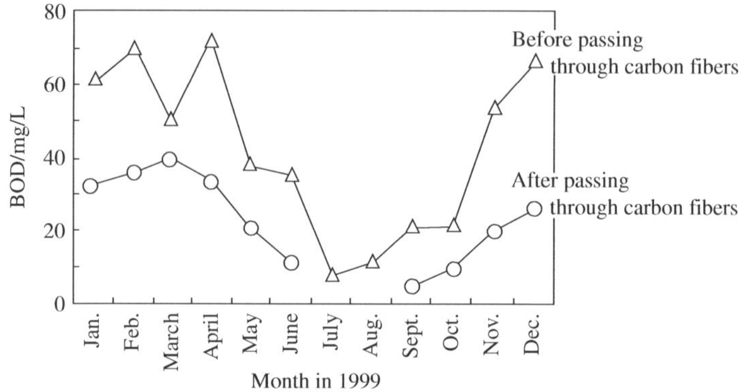


**FIGURE 3.306**

Changes in BOD of the river water during the year 1999.

*(Courtesy of Prof. A. Kojima of Gunma Nat. Coll. Tech.)*

**FIGURE 3.307**

Various types of the floats tried in Lake Haruna.

*(Courtesy of Prof. A. Kojima of Gunma Nat. Coll. Tech.)*

bundles, now a good fishery of smelts in winter has been recovered by setting artificial algae of floating carbon fibers in 9 places in the periphery of Lake Haruna [889].

By placing similar carbon fiber floats in the sea, rapid fixation of marine microorganisms onto carbon fiber surfaces was observed [889]. In Table 3.51,

**FIGURE 3.308**

Photos of carbon fibers in the lake.

*(Courtesy of Prof. A. Kojima of Gunma Nat. Coll. Tech.)*

**Table 3.51** Changes in Varieties of Marine Organisms on Carbon Fibers

| After Dipping of Carbon Fiber Float | Marine Organisms Observed |
|---|---|
| 1 day | Adhesive materials |
| 5 days | Diatom, zoo-plankton |
| 7 days | Diatom, zoo-plankton, *Caprella* and small shrimps (*Maera serratipalma*, etc.) |
| 14 days | Small shrimps, *Protohydroides elegans*, *Caprella* and ascidian |
| 1 month | Small shrimps ascidian, *Protohydroides elegans* and *Hydrozoa* |
| 2 months | Small shrimps, hydrozoa, barnache and fishes |
| 3 months | Small shrimps, barnache, spomges and fishes |

*(Courtesy of Prof. M. Shiraishi of Tokai University.)*

changes in varieties of marine organisms fixed or living on carbon fibers with time are summarized. A food chain is supposed to be established among bacteria, algae, zoo-plankton, small animals such as small shrimps (*Maer serratipalma* and *Photis reinhardi*) and fishes.

Carbon fibers have been found to accelerate the oxidation of Mn(II) ions in water by the manganese-oxidizing fungus [890]. In the culture medium containing nutrition, such as peptone, yeast extract and glucose, a manganese-oxidizing fungus, which has been isolated from a hot spring in Japan, can oxidize the dissolved Mn(II) ions to precipitate as mostly the solid $MnO_2$. This oxidizing process by the fungus is accelerated in the presence of carbon fibers, as shown in Fig. 3.309. The presence of carbon fibers in the culture medium is favorable for funguses, as shown in Fig. 3.310a where funguses are entangled with carbon

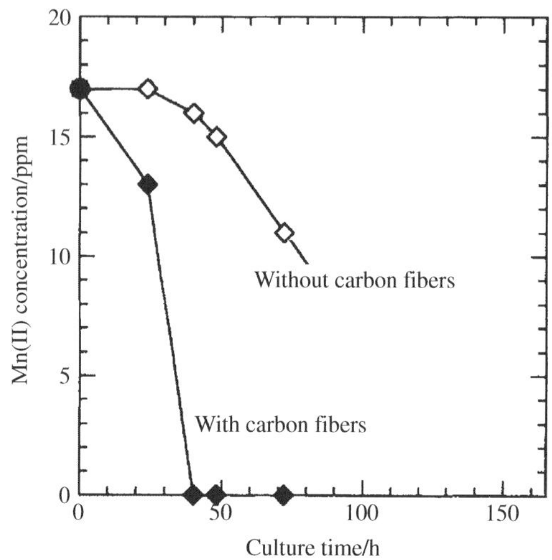


**FIGURE 3.309**

Oxidation of Mn(II) ions by the fungus either with or without carbon fibers in the medium containing peptone, yeast extract and glucose (0.05 g/L each).

*(Courtesy of Prof. H. Konno of Hokkaido Univ.)*

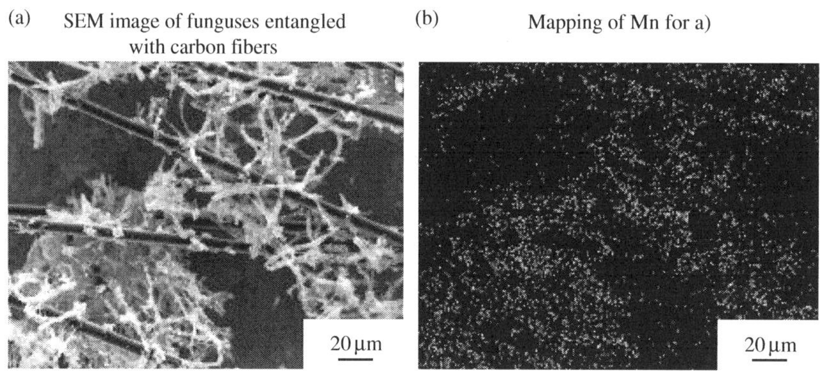


**FIGURE 3.310**

Manganese-oxidizing funguses with carbon fibers.

*(Courtesy of Prof. H. Konno of Hokkaido Univ.)*

fibers, and the elemental mapping shows that Mn is mainly associated with funguses, as shown in Fig. 3.310b. Without the funguses (i.e., in the sterilized medium), no oxidation of Mn(II) ions was observed. This process for oxidizing dissolved Mn(II) ions using the fungus may be effective to immobilize the dissolved manganese ions in water coming from some manganese mines by overcoming the disadvantage of a slow rate of biological treatment.

## References


[1] K. Matsuo, S. Hori, K. Fukada, TANSO (55) (1968) 114 [in Japanese]
[2] I. Ogawa, H. Yoshida, K. Kobayashi, J. Mater. Sci. 16 (1981) 2181.
[3] Y. Hishiyama, Y. Kaburagi, TANSO (98) (1979) 89 [in Japanese]
[4] A. Oberlin, J. Terriere, J. Microsc. 14 (1972) 1.
[5] T. Yokono, Y. Sanada, TANSO (105) (1981) 73 [in Japanese]
[6] M. Sakai, M. Inagaki, Reorozi Gakkaishi 15 (1987) 76 [in Japanese]
[7] Y. Yamada, H. Honda, H. Tanaka, et al., Sekiyu Gakkaishi 18 (1975) 765 [in Japanese]
[8] M. Sakai, M. Yoshihara, M. Inagaki, Carbon 19 (1981) 83.
[9] S.H. Chen, S.L. Eilenberg, R.J. Diefendorf, Int. Symp. Carbon, Tokyo (1982) 42.
[10] M. Sakai, K. Sasaki, M. Inagaki, Carbon 21 (1983) 593.
[11] M. Sakai, T. Sogabe, H. Kitagawa, et al., Carbon 21 (1983) 601.
[12] M. Sakai, M. Inagaki, Carbon 19 (1981) 37.
[13] M. Sakai, T. Kida, M. Inagaki, J. Mat. Sci. 19 (1984) 2651.
[14] M. Sakai, TANSO (106) (1981) 121 [in Japanese]
[15] M. Sakai, Carbon 17 (1979) 139.
[16] M. Sakai, Carbon 17 (1979) 145.
[17] Y. Sato, T. Kitano, M. Inagaki, et al., Carbon 28 (1990) 143.
[18] K. Oshida, N. Ekinaga, M. Endo, et al., TANSO (173) (1996) 142 [in Japanese]
[19] J.F. Andrew, J. Okada, D.C. Wobschall, Proc. 4th Biennial. Conf. Carbon (1960) 559.
[20] C. Malmstrom, R. Keen, L. Green, J. Appl. Phys. 22 (1951) 593.
[21] H.E. Martens, L.D. Jaffe, J.E. Jepson, Proc. 3rd Conf. Carbon. (1958) 529.
[22] M. Inagaki, H. Takeuchi, F. Yamanaka, et al., Kogyo Kagaku Kaishi 66 (1963) 169 [in Japanese]
[23] H.E. Martens, W.V. Kotlensky, Proc. 5th Conf. Carbon 2 (1963) 617.
[24] L. Green, J. Appl. Phys. 20 (1953) 289.
[25] P. Wagner, A.R. Driesner, J. Appl. Phys. 30 (1959) 148.
[26] P. Wagner, A.R. Driesner, L.A. Haskin, J. Appl. Phys. 30 (1959) 152.
[27] S. Sato, H. Awaji, H. Akuzawa, Carbon 16 (1978) 95.
[28] S. Sato, H. Awaji, H. Akuzawa, Carbon 16 (1987) 103.
[29] M. Sakai, K. Urashima, M. Inagaki, J. Am. Ceram Soc. 66 (1983) 868.
[30] M. Sakai, K. Urashima, M. Inagaki, Zairyo 32 (1983) 1260 [in Japanese]
[31] M. Sakai, K. Urashima, N. Mitani, et al., TANSO 120 (1985) 48 [in Japanese]
[32] M. Sakai, Y. Goto, M. Inagaki, Yogyo Kyoukaishi 94 (1986) 163 [in Japanese]
[33] T. Oku, S. Ishiyama, M. Eto, et al., Nihon Seramikkusu Kyoukaishi 96 (1988) 773 [in Japanese]
[34] M. Sakai, J. Yoshimura, Y. Goto, et al., J. Am. Ceram Soc. 71 (1988) 609.

[35] M. Sakai, M. Inagaki, J. Am, Ceram. Soc. 72 (1989) 388.
[36] M. Sakai, T. Miyajima, M. Inagaki, Composites Sci. Tech. 40 (1991) 231.
[37] A. Bailey, B. Yates, J. Appl. Phys. 41 (1970) 5088.
[38] J. Harrison, High Temp-High Press 9 (1977) 211.
[39] K. Miyazaki, Hitech Carbon Mater. 64 (1987) [in Japanese]
[40] K. Miyazaki, K. Kobayashi, H. Honada, TANSO 91 (1977) 121 [in Japanese]
[41] S. Mrozowski, Proc. 1st and 2nd Conf. Carbon (1956) 31.
[42] A.W. Smith, N.W. Rasor, Phys. Rev. 104 (1956) 885.
[43] W.W. Tyler, A.C. Wilson Jr, Phys. Rev. 89 (1953) 870.
[44] C.P. Jaminson, S. Mrozowski, Proc. 1st and 2nd Conf. Carbon (1956) 155.
[45] C.A. Klein, M.G. Holland, Phys. Rev. 136 (1964) A576.
[46] Y. Hishiyama, A. Ono, Carbon 19 (1981) 441.
[47] T. Takezawa, T. Tsuzuku, A. Ono, et al., Phil. Mag. 19 (1971) 623.
[48] K. Sugihara, J. Phys. Soc. Jpn. 29 (1970) 1465.
[49] K. Matsuo, T. Hoshikawa, T. Sakashita, Zairyo 23 (1973) 464.
[50] L. Zhou, F.Y. Kang, X.L. Li, et al., J. Phy. Chem. Solids 69 (2008) 1265.
[51] P.J. Wyllie, O.F. Tuttle, Nature 183 (1959) 770.
[52] P.R. Buseck, B.J. Huang, Geochim. Cosmochim. Acta 1985 (2003) 49.
[53] M. Inagaki, K. Fujita, Y. Takeuchi, et al., Carbon 39 (6) (2001) 921.
[54] H. Konno, K. Fujita, H. Habazaki, et al., TANSO (No.203) (2002) 113.
[55] T. Noda, H. Kato, Carbon 3 (1965) 289.
[56] T. Noda, Carbon 6 (1968) 125.
[57] K. Kamiya, M. Mizutani, T. Noda, et al., Bull. Chem. Soc. Jpn. 41 (1968) 2169.
[58] O. Beyssac, F. Brunet, J.-P. Petitet, et al., Eur. J. Miner. 15 (2003) 937.
[59] T. Noda, S. Hirano, K. Amanuma, et al., Bull. Chem. Soc. Jpn. 41 (1968) 1245.
[60] S. Hirano, H. Saito, M. Inagaki, Bull. Chem. Soc. Jpn. 43 (1970) 2599.
[61] S. Hirano, M. Inagaki, H. Saito, Bull. Chem. Soc. Jpn. 43 (1970) 2624.
[62] S. Hirano, M. Inagaki, H. Saito, Carbon 17 (1979) 395.
[63] L. Zhou, F.Y. Kang, Y.P. Zheng, et al., Electrochim. Acta 54 (2009) 3930.
[64] H. Nozaki, K. Nagaoka, K. Hoshi, et al., J. Power Sources 194 (2009) 486–493.
[65] N. Ohta, K. Nagaoka, K. Hoshi, et al., J. Power Sources 194 (2009) 985–990.
[66] K. Hoshi, N. Ohta, K. Nagaoka, et al., TANSO (240) (2009) 213–220.
[67a] K. Shen, PhD Thesis, Tsinghua University, 2013.
[67b] S. Liu, C.R. Loper Jr., Carbon 29 (1991) 547.
[67c] S. Liu, C.R. Loper Jr., Carbon 29 (1991) 1119.
[67d] W. Wang, K.M. Thomas, R.M. Poultney, et al., Carbon 33 (1995) 1525.
[68] A. Yoshida, Y. Hishiyama, TANSO (130) (1987) 110 [in Japanese]
[69] A. Yoshida, Y. Hishiyama, J. Mater. Res. 7 (1992) 1400.
[70] S. Liu, C.R. Loper Jr., Carbon 29 (1991) 547.
[71] Y. Hishiyama, Y. Kaburagi, M. Inagaki, Chem. Phys. Carbon 32 (1991) 1.
[72] M. Inagaki, Y. Hishiyama, New Carbon Mater. Gilhoudo Shuppan, Tokyo 1994 [in Japanese].
[73] Y. Hishiyama, TANSO (130) (1987) 18 [in Japanese]
[74] Y. Hishiyama, A. Ono, Carbon 23 (1985) 445.
[75] K. Sugihara, Y. Hishiyama, A. Ono, Phys. Rev. B. 34 (1986) 4298.
[76] S.B. Austerman, S.M. Myron, W. Wagner, Carbon 5 (1967) 549.
[77] T. Noda, Y. Sumiyoshi, N. Ito, Carbon 6 (1968) 813.
[78] C. Baraniecki, P.H. Pinchbec, B. Pickerin, Carbon 7 (1969) 213.

[79] I. Minkoff, I. Einbinder, Nature 194 (1962) 765.
[80] P. Ching, Nature 192 (1961) 864.
[81] H.J.C. Tulloch, D.A. Young, Nature 211 (1966) 730.
[82] L.M. Foster, G. Long, H.C. Stumpf, Amer Miner. 43 (1958) 285.
[83] M. Inagaki, K. Taoka, T. Noda, Carbon 9 (1971) 94.
[84] T. Ishii, Denki Kagaku 35 (1967) 688.
[85] J. Bokros, Chem. Phys. Carbon 5 (1969) 3.
[86] M. Inagaki, TANSO (34) (1963) 18 [in Japanese]
[87] A.W. Moor, Chem. Phys. Carbon 10 (1973) 69.
[88] A.W. Moor, Chem. Phys. Carbon 17 (1982) 233.
[89] M. Inagaki, S. Harada, T. Sato, et al., Carbon 27 (1989) 253.
[90] M. Inagaki, Y. Hishiyama, T. Takeichi, et al., Chem. Phys. Carbon 26 (1999) 246.
[91] M. Inagaki, N. Ohta, Y. Hishiyama, Carbon 61 (2013) 1.
[92] M. Inagaki, M. Sato, T. Takeichi, et al., Carbon 30 (1992) 903.
[93a] C. Bourgerrette, A. Oberlin, M. Inagaki, J. Mater. Res. 10 (1995) 1024.
[93b] C. Bourgerrette, A. Oberlin, M. Inagaki, J. Mater. Res. 7 (1992) 1158.
[94] Y. Hishiyama, K. Igarashi, I. Kanaoka, et al., Carbon 35 (1997) 657.
[95] H. Konno, T. Nakahashi, M. Inagaki, Carbon 35 (1997) 669.
[96] T. Nakahashi, H. Konno, Y. Hishiyama, Carbon 36 (1998) 1021.
[97] H. Hatori, Y. Yamada, M. Shiraishi, J. Appl. Polym. Sci. 57 (1995) 871.
[98] M. Inagaki, K. Sakamoto, Y. Hishiyama, J. Mater. Res. 6 (1991) 1108.
[99] S. Isoda, H. Shimada, M. Kochi, H. Kambe, J. Polym. Sci., Polym. Phys. 19 (1981) 1293.
[100] Y. Nagata, N. Nakama, K. Saito, Polym. Preprint, Jpn. 40 (1991) 4304.
[101] Y. Hishiyama, M. Nakamura, Y. Nagata, et al., Carbon 32 (1994) 645.
[102] Y. Kaburagi, A. Yoshida, Y. Hishiyama, et al., TANSO 19 (1995) [in Japanese]
[103] Y. Hishiyama, A. Yoshida, M. Inagaki, Carbon 36 (1998) 1113.
[104] Y. Hishiyama, A. Yoshida, Y. Kaburagi, Carbon 31 (1993) 1265.
[105] H. Hatori, Y. Yamada, M. Shiraishi, Carbon 30 (1992) 305.
[106a] M. Inagaki, Y. Hishiyama, Y. Kaburagi, Carbon 32 (1994) 637.
[106b] Y. Hishiyama, A. Yoshida, Y. Kaburagi, et al., Carbon 30 (1992) 333.
[107] M. Murakami, N. Nishi, K. Nakamura, et al., Carbon 30 (1992) 255.
[108] M. Murakami, T. Hoshi, N. Nishiki, Housyako 6 (1993) 43 [in Japanese]
[109] T. Ohnishi, I. Murase, T. Noguchi, et al., Synth. Met. (1986) 207.
[110] M. Murakami, K. Watanabe, S. Yoshimura, Appl. Phys. Lett. 48 (1986) 1594.
[111] M. Murakami, S. Yoshimura, Synth. Met. 18 (1987) 509.
[112] M. Shioya, K. Shinotani, A. Takaku, J. Mater. Sci. 34 (1999) 6015.
[113] J. Yamashita, M. Shioya, H. Hatori, et al., TANSO (245) (2010) 196 [in Japanese]
[114] Y. Hishiyama, S. Yasuda, A. Yoshida, et al., J. Mater. Sci. 23 (1988) 3272.
[115] H. Hatori, Y. Yamada, M. Shiraishi, Carbon 30 (1992) 763.
[116] Y. Kaburagi, T. Kimura, A. Yoshida, et al., TANSO (253) (2012) 106.
[117] Y. Hishiyama, A. Yoshida, Y. Kaburagi, TANSO (254) (2012) 176.
[118] D.D.L. Chung, J. Mater. Sci. 22 (1987) 4190.
[119] G. Furdin, Fuel 77 (1998) 479.
[120] M. Inagaki, M. Toyoda, F. Kang, Chem. Phys. Carbon 29 (2004) 1.
[121] M.B. Dowell, R.A. Howard, Carbon 24 (1986) 311.
[122] Y. Leng, J. Gu, W. Cao, T.Y. Zhang, Carbon 36 (1998) 875.
[123] R.A. Howard, Eng. Design Manual, Petrunich PS, Vol. 1 (1987).

[124] GRAFTECH Inc., GRAFOIL Flexible Graphite of Material Safety Data Sheet, MSDS No. 0001 (2002).
[125] R.A. Reynolds III, R.A. Greinke, Carbon 39 (2001) 479.
[126] X. Luo, D.D.L. Chung, Carbon 38 (2000) 1510.
[127] X. Luo, D.D.L. Chung, Carbon 39 (2001) 615.
[128] M. Inagaki, Y. Kaburagi, Y. Hishiyama, Adv. Eng. Mater. DOI: 10.1002/adem.201300418.
[129] D.D.L. Chung, J. Mater. Eng. Perf. 9 (2000) 161.
[130] D.D.L. Chung, Carbon 39 (2001) 279.
[131] A. Oberlin, G. Terriere, J. Microsc. 14 (1972) 1.
[132] A. Oberlin, G. Terriere, L.J. Boulmier, TANSO (80) (1975) 29.
[133] A. Oberlin, G. Terriere, L.J. Boulmier, TANSO (82) (1975) 153.
[134] A. Oberlin, Chem. Phys. Carbon 22 (1989) 1.
[135] G.M. Jenkins, K. Kawamura, Nature 231 (1971) 175.
[136] M. Shiraishi, Tansozairyou Nyuumon (1984) 29 [in Japanese]
[137] T. Noda, M. Inagaki, S. Yamada, J. Non-Cryst. Solids 1 (1969) 285.
[138] A. Yoshida, Y. Kaburagi, Y. Hishiyama, Carbon 29 (1991) 1107.
[139] Y. Hishiyama, Y. Kaburagi, D.F. Baker, et al., TANSO (128) (1987) 18 [in Japanese]
[140] Y. Hishiyama, Y. Kaburagi, A. Yoshida, et al., Carbon 31 (1993) 773.
[141] Y. Tanabe, J. Yamanaka, K. Hoshi, et al., Carbon 39 (2001) 2347.
[142] Y. Kaburagi, K. Hosoya, A. Yoshida, et al., Carbon 43 (2005) 2817.
[143] Y. Kaburagi, Y. Kaitou, E. Shindo, et al., TANSO (257) (2013) 110.
[144] T. Yamaguchi, Carbon 1 (1963) 47.
[145] T. Yamaguchi, Carbon 1 (1963) 535.
[146] D.F. Baker, R.H. Bragg, J. Non-Cryst. Solids 53 (1983) 57.
[147] Y. Kaburagi, Y. Hishiyama, R.H. Bragg, Phil Mag. B. 54 (1986) 381.
[148] Japanese Patent 412,380 (1963); British Patent 1.033,277 (1966).
[149] H.W. Davidson, H.H.W. Losty, GEC J. 30 (1963) 22.
[150] British Patent 860,342 (1961); 889,351 (1962).
[151] E. Fitzer, W. Schaffer, S. Yamada, Carbon 7 (1969) 643.
[152] Y. Hishiyama, Y. Kaburagi, A. Yoshida, Sciences and New Applications of Carbon Fibers, Toyohashi Univ. Tech. 21 (1984).
[153] J.B. Donnet, T.K. Wang, S. Rebouillat, et al., Carbon Fibers, 3rd Edition, Marcel Dekker, 1998.
[154] H. Ogawa, Nihon Kagaku Kaishi 1994 (1994) 560 [in Japanese]
[155] H. Ogawa, K. Saito, Carbon 33 (1995) 783.
[156] O.P. Bahl, Carbon Fibers, 3rd Edition, Marcel Dekker, 19981
[157] W. Watt, W. Johnson, Appl. Polm. Symp. (9) (1969) 215.
[158] M. Inagaki, Y. Hishiyama, New Carbon Mater., Gihoudo Shuppan, 1994 [in Japanese]
[159] A. Oberlin, M. Guiggon, in: Bunsell (Ed.), Fiber Reinforcement for Composite Materials, Elsevier, 1988.
[160] A. Oberlin, Carbon 22 (1984) 521.
[161] Y. Hishiyama, Y. Kaburagi, M. Inagaki, Chemistry and Physics of Carbon, Vol. 23, Marcel Dekker, 19911
[162] K. Sato, F. Sato, T. Tomioka, TANSO (157) (1993) 107 [in Japanese]
[163] J.B. Barr, S. Chwastiak, R. Didchenko, et al., Appl. Polym. Symp. (29) (1976) 161.

[164] F.F. Nazem, Fuel 59 (1980) 851.
[165] F.F. Nazem, Carbon 20 (1982) 345.
[166] H. Fujimaki, S. Otani, Ceramics 11 (1976) 612 [in Japanese]
[167] M. Inagaki, N. Iwashita, Y. Hishiyama, et al., TANSO (147) (1991) 57.
[168] Y. Tanabe, E. Yasuda, K. Yamaguchi, et al., TANSO (147) (1991) 66.
[169] J.G. Lavin, D.R. Boyington, J. Lahijani, et al., Carbon 31 (1993) 1001.
[170] T. Koyama, M. Endo, Jpn. J. Appl. Phys. 13 (1974) 1175.
[171] M. Ishioka, Y. Hishiyama, M. Inagaki, TANSO (169) (1995) 218 [in Japanese]
[172] M. Endo, M. Shikata, Ouyou Butsuri 54 (1985) 507 [in Japanese]
[173] M. Egashira, H. Katsuki, K. Hayashi, et al., Sekiyu Gakkaishi 26 (1983) 247 [in Japanese]
[174] M. Ishioka, T. Okada, K. Matsubara, Carbon 31 (1993) 123.
[175] T. Masuda, S.R. Mukai, K. Hashimoto, Carbon 31 (1993) 783.
[176] S.R. Mukai, T. Masuda, Y. Fujikata, et al., Chem. Eng. Sci. 49 (1994) 4909.
[177] S.R. Mukai, T. Masuda, Y. Fujikata, et al., Carbon 33 (1995) 733.
[178] S.R. Mukai, T. Masuda, T. Harada, et al., Carbon 34 (1996) 645.
[179] S.R. Mukai, T. Masuda, Y. Matsuzawa, et al., Chem. Eng. Sci. 53 (1998) 439–448.
[180] S.R. Mukai, Y. Rikima, R. Furukawa, et al., Ind. Eng. Chem. Res. (2013)in press
[181] M. Ishioka, T. Okada, K. Matsubara, Carbon 30 (1992) 975.
[182] M. Ishioka, T. Okada, K. Matsubara, Carbon 30 (1992) 859.
[183] M. Ishioka, T. Okada, K. Matsubara, M. Endo, Carbon 30 (1992) 865.
[184] A. Oberlin, M. Endo, T. Koyama, J. Cryst. Growth 32 (1976) 335.
[185] G.G. Tibbetts, M.G. Devour, F.J. Rodda, Carbon 25 (1986) 367.
[186] M. Endo, T. Koyama, Y. Hishiyama, J. Appl. Phys. 15 (1976) 2073.
[187] N. Iwashita, M. Inagaki, Y. Sawada, Rep. Osaka Res. Inst. 44 (1993) 115 [in Japanese]
[188] A. Yoshida, Y. Hishiyama, M. Ishioka, et al., TANSO (168) (1995) 169 [in Japanese]
[189] M. Ishioka, T. Okada, K. Matsubara, et al., J. Mater. Res. 1993 (1866) 8.
[190] A. Oya, N. Kasahara, Carbon 38 (2000) 11414.
[191] N. Kasahara, S. Shiraishi, A. Oya, Carbon 41 (2003) 1654.
[192] K. Okabe, S. Shiraishi, A. Oya, Carbon 42 (2004) 667.
[193] Y. Kaburagi, M. Ohoyama, Y. Yamaguchi, et al., Carbon 50 (2012) 4757.
[194] Y. Kaburagi, M. Ohoyama, Y. Yamaguchi, et al., TANSO (255) (2012) 225.
[195] S. Motojima, M. Kawaguchi, K. Nozaki, et al., Appl. Phys. Lett. 56 (1990) 321.
[196] S. Motojima, M. Kawaguchi, K. Nozaki, et al., Carbon 29 (1991) 379.
[197] S. Motojima, I. Hasegawa, S. Asakura, et al., Carbon 33 (1995) 1167.
[198] X. Chen, T. Saito, M. Kusunoki, et al., J. Mater. Res. 14 (1999) 4329.
[199] X. Chen, W. In-Hwang, S. Shimada, et al., J. Mater. Res. 15 (2000) 808.
[200] Helical materials and their applications, Mater. Integr. 17 (2004) 1 [in Japanese]
[201] S. Iijima, Nature 354 (1991) 56.
[202] S. Iijima, T. Ichihashi, Nature 363 (1993) 603.
[203] D.S. Bethune, C.H. Kiang, M.S. deVries, et al., Nature 363 (1993) 605.
[204] S. Iijima, T. Ichihashi, Y. Ando, Nature 358 (1992) 220.
[205] R. Bacon, J. Appl. Phys. 31 (1960) 283.
[206] M. Inagaki, K. Kaneko, T. Nishizawa, Carbon 42 (2004) 1401
[207] R. Saito, M. Fujita, G. Dresselhaus, et al., J. Appl. Phys. Lett. 60 (1992) 2204.
[208] R. Saito, G. Dresselhaus, M.S. Dresselhaus, J. Appl. Phys. 73 (1993) 494.

[209] A. Jorio, R. Saito, J.H. Hafner, et al., Phys. Rev. Lett. 60 (2001) 2204.
[210] S. Maruyama, R. Kojima, Y. Miyauchi, et al., Chem. Phys. Lett. 360 (2002) 229.
[211] Y. Murakami, S. Chiasji, Y. Miyauchi, et al., Chem. Phys. Lett. 385 (2004) 298.
[212] M.S. Dresselhaus, G. Dresselhaus, P.C. Eklund, Science of Fullerenes and Carbon Nanotubes, Academic Press, 1996.
[213] T.W. Ebbessen (Ed.), Carbon Nanotubes: Preparation and Properties, CRC Press, 1997.
[214] R. Saito, G. Dresselhaus, M.S. Dresselhaus, Physical Properties of Carbon Nanotubes, Imperial College Press, 1998.
[215] Y.A. Kim, H. Muramatsu, T. Hayashi, et al., Int. Symp. Nanocarbons (2004) 158.
[216] E.W. Wang, P.E. Shechan, C.M. Lieber, Science 277 (1997) 1971.
[217] S.P. Frank, P. Ponchard, Z. Wang, et al., Science 280 (1998) 1744.
[218] M. Kociak, A.Y. Kasumov, S. Gueron, et al., Phys. Rev. Lett. 86 (2001) 2416.
[219] P. Kim, L. Shi, A. Majundar, et al., Phys. Rev. Lett. 87 (2001) 215502.
[220] R.H. Baughman, A.A. Zakhidov, W.A. de Heer, Science 297 (2002) 787.
[221] D.A. Walters, L.M. Ericson, M.J. Casavant, et al., Appl. Phys. Lett. 74 (1999) 3803.
[222] P.-X. Hou, C. Liu, H.-M. Cheng, Carbon 46 (2008) 2003.
[223] N. Izard, S. Kazaoui, K. Hata, et al., Appl. Phys. Lett. 92 (2008) 243112.
[224] T. Tanaka, Y. Urabe, D. Nishide, et al., Appl. Phys. Expr. 2 (2009) 125002.
[225] O. Tanaike, O. Kimizuka, N. Yoshizawa, et al., Electrochem. Commun. 11 (2009) 1441.
[226] H.W. Zhu, C.L. Xu, D.H. Wu, et al., Science 296 (2002) 884.
[227] Y. Li, I.A. Kinloch, A.H. Windle, Science 306 (2004) 276.
[228] K. Jiang, Q. Li, S. Fan, Nature 419 (2002) 801.
[229] M. Zhang, S. Fang, A.A. Zakhidov, et al., Science 309 (2005) 1215.
[230] K. Hata, D.N. Futaba, K. Mizuno, et al., Science 306 (2004) 1362.
[231] D.N. Futaba, K. Hata, T. Yamada, et al., Nat. Mater. 5 (2006) 987.
[232] T. Hiraoka, T. Yamada, K. Hata, et al., J. Am. Chem. Soc. 128 (2006) 13338.
[233] O. Kimizuka, O. Tanaike, J. Yamashita, et al., Carbon 46 (2008) 1999.
[234] O. Tanaike, D.N. Futaba, K. Hata, et al., Carbon Lett. 10 (2009) 90.
[235] T. Hiraoka, A. Izadi-Najafabadi, T. Yamada, et al., Adv. Funct. Mater. 20 (2010) 422.
[236] B. Liu, W. Ren, L. Gao, et al., J. Am. Chem. Soc. 131 (2009) 2082.
[237] M. Endo, H. Muramatsu, T. Hayashi, et al., Nature 433 (2005) 476.
[238] Y.A. Kim, H. Muramatsu, T. Hayashi, et al., Chem. Vap. Deposition 12 (2006) 327.
[239] M. Endo, T. Hayashi, Y.A. Kim, et al., Jpn. J. Appl. Phys. 45 (2006) 4883.
[240] M. Endo, Y.A. Kim, T. Hayashi, et al., Carbon 39 (2001) 1287.
[241] Q. Lin, J.N. Harb, J. Electrochem. Soc. 151 (2004) 1115.
[242] X. Li, F. Kang, W. Shen, Carbon 44 (2006) 1334.
[243] K. Sheem, Y.H. Lee, H.S. Lim, J. Power Sources 158 (2006) 1425.
[244] C. Sotowa, G. Origi, M. Takeuchi, et al., Chem. Sus. Chem. 1 (2008) 911.
[245a] C. Liu, Y. Chen, C.-Z. Wu, et al., Carbon 48 (2010) 452.
[245b] M.S. Kim, N.M. Rodriguez, R.T.K. Baker, J. Catal 131 (1991) 60.
[246] N.M. Rodriguez, J. Mater. Res. 8 (1993) 3233.
[247] R.T.K. Baker, S. Terry, P.S. Harris, Nature 253 (1975) 37.
[248] M. Audier, A. Oberlin, M. Oberlin, et al., Carbon 19 (1981) 217.
[249] S. Motojima, S. Ueno, T. Hattori, et al., J. Cryst. Growth 96 (1989) 383.

[250] Y. Soneda, M. Makino, Carbon 38 (2000) 478.
[251] H. Konno, S. Sato, H. Habazaki, et al., Carbon 42 (2004) 2756.
[252] M. Endo, Y.A. Kim, T. Hayashi, et al., Appl. Phys. Lett. 80 (2002) 1267.
[253] M. Endo, Y.A. Kim, T. Hayashi, et al., Carbon 41 (2003) 1941.
[254] M. Endo, Y.A. Kim, M. Ezaka, et al., Nano. Lett. 3 (2003) 723.
[255] T. Kyotani, L. Tsai, A. Tomita, Chem. Mater. 7 (1995) 1427.
[256] T. Kyotani, L. Tsai, A. Tomita, Chem. Mater. 8 (1996) 2109.
[257] D. Hulicova, F. Sato, K. Okabe, et al., Carbon 39 (2001) 1438.
[258] D. Hulicova, K. Hosoi, S. Kuroda, et al., Adv. Mater. 14 (2002) 452.
[259] Y. Zhang, M.N. Gamo, K. Nakagawa, et al., J. Mater. Res. 17 (2002) 2457.
[260] M. Kusunoki, T. Suzuki, C. Honjo, et al., Chem. Phys. Lett. 366 (2002) 458.
[261] M. Kusunoki, M. Rokkaku, T. Suzuki, Appl. Phys. Lett. 18 (1997) 2620.
[262] M. Kusunoki, J. Shibata, M. Rokkaku, et al., Jpn. J. Appl. Phys. 37 (1998) L605.
[263] M. Kusunoki, T. Suzuki, K. Kaneko, et al., Phil. Mag. Lett. 79 (1999) 153.
[264] Y. Zhang, M. Nishitani-Gamo, C. Xiao, et al., Jpn. J. Appl. Phys. 41 (2002) 408.
[265] Y. Saito, S. Uemura, Carbon 38 (2000) 169.
[266] J.M. Bonard, M. Croci, C. Klinke, et al., Carbon 2002 (1715) 40.
[267] Y. Nakayama, H. Nishijima, S. Akita, et al., J. Vac. Sci. Technol. B. 18 (2000) 661.
[268] H. Kuramochi, K. Ando, Y. Shikakura, et al., Nanotech 15 (2004) 1126.
[269] V.M. Tsefrikas, L.T. Scott, Chem. Rev. 106 (2006) 4868.
[270] L.T. Scott, M.M. Boorum, B.J. McMahon, et al., Science 295 (2002) 1500.
[271] L.T. Scott, Angw. Chem. Int. Ed. 43 (2006) 4994.
[272] F. Diederich, Y. Rubin, Angw. Chem. Int. Ed. 31 (1992) 1101.
[273] R.W. Saalfrank, Nature 383 (1996) 124.
[274] D.S. Bethune, R.D. Johnson, J.R. Salem, et al., Nature 366 (1993) 123.
[275] Y. Rubin, Chem. Eur. J. 3 (1997) 1009.
[276] O. Vostrowsky, A. Hirsch, Chem. Rev. 106 (2006) 5191.
[277] F. Hauke, O. Vostrowsky, A. Hirsch, et al., Chem. Eur. J. 12 (2006) 4813.
[278] S. Liu, S. Sun, J Organomet. Chem. 599 (2000) 74.
[279] M. Saunders, H.A. Jimenez-Vazquez, R.J. Cross, et al., Science 259 (1993) 1428.
[280] M. Saunders, H.A. Jimenez-Vazquez, R.J. Cross, et al., J. Am. Chem. Soc. 116 (1994) 2193.
[281] R.J. Cross, A. Khong, M. Saunders, J. Org. Chem. 68 (2003) 8281.
[282] G. Schick, T. Jarrosson, Y. Rubin, Angew. Chem. Int. Ed. 38 (1999) 2360.
[283] Y. Murata, M. Murata, K. Komatsu, Chem. Eur. J. 9 (2003) 1600.
[284] Y. Murata, M. Murata, K. Komatsu, J. Am. Chem. Soc. 125 (2003) 7152.
[285] Y. Rubin, T. Jarrosson, G.-W. Wang, et al., Angew. Chem. Int. Ed. 40 (2001) 1543.
[286] K. Komatsu, M. Murata, Y. Murata, Science 307 (2005) 238.
[287] K. Tanigaki, J. Phys. Chem. Solids 54 (1993) 1645.
[288] A. Takeda, Y. Yokoyama, S. Ito, T. Miyazaki, et al., Chem. Commun. (2006) 912.
[289] M. Mikawa, H. Kato, M. Okumura, et al., Bioconjug. Chem. 12 (2001) 510.
[290] H. Kato, Y. Kanazawa, M. Okumura, et al., J. Am. Chem. Soc. 125 (2003) 4391.
[291] H.P. Boehm, R. Setton, E. Stumpp, Carbon 24 (1986) 241.
[292] C. Soldano, A. Mahmood, E. Dujardin, Carbon 48 (2010) 2127.
[293] M.J. Allen, V.C. Tung, R.B. Kaner, Chem. Rev. 110 (2010) 132.
[294] M. Terronesa, A.R. Botello Méndez, J. Campos Delgado, et al., Nano Today 5 (2010) 351.
[295] M. Inagaki, Y.A. Kim, M. Endo, J. Mater. Chem. 21 (2011) 3280.

[296] Y. Ohashi, T. Koizumi, T. Yoshikawa, et al., TANSO (180) (1997) 235.
[297] Y. Ohashi, T. Hironaka, T. Kubo, et al., TANSO (195) (2000) 410.
[298] K.S. Novoselov, A.K. Geim, S.V. Morozov, et al., Science 306 (2004) 666.
[299] Y. Zhang, J.P. Small, W.V. Pontius, et al., Appl. Phys. Lett. 86 (2005) 073104.
[300] X. Lu, H. Huang, N. Nemchuk, et al., Appl. Phys. Lett. 75 (1999) 193.
[301] K.S. Novoselov, D. Jiang, F. Schedin, et al., PNAS 102 (2005) 10451.
[302] J.C. Meyer, A.K. Geim, M.I. Katsnelson, et al., Nature 446 (2007) 60.
[303] W. Hummers, R. Offeman, J. Am. Chem. Soc. 80 (1958) 1339.
[304] B.C. Brodie, Ann. Chim. Phys. 59 (1860) 466.
[305] F. Kang, Y. Leng, T.Y. Zhang, Carbon 35 (1997) 1089.
[306] W. Schloz, H.P. Boehm, Z .Anorg. Allg. Chem. 369 (1969) 327.
[307] R. Yazami, P.H. Touzain, et al., Rev. Chim. Minerale 22 (1985) 398.
[308] T. Nakajima, Y. Matsuo, Carbon 32 (1994) 469.
[309] O. Akhavan, Carbon 48 (2010) 509.
[310] S. Stankovich, R.D. Piner, et al., J. Mater. Chem. 16 (2005) 155.
[311] S. Stankovich, et al., Carbon 44 (2006) 3342.
[312] N.A. Kotov, Nature 442 (2006) 254.
[313] S. Stankovich, D.A. Dikin, G.H.B. Dommett, et al., Nature 442 (2006) 282.
[314] Z.-S. Wu, W. Ren, L. Gao, et al., Carbon 47 (2009) 493.
[315] S.J. Wang, Y. Geng, Q. Zheng, J.K. Kim, Carbon 48 (2010) 1815.
[316] W. Lv, D.M. Tang, Y.B. He, et al., ACS Nano 3 (2009) 3730.
[317a] C. Zhang, W. Lv, X. Xie, et al., Carbon 62 (2013) 11.
[317b] F. Su, C. You, Y.B. He, et al., J. Mater. Chem. 20 (2010) 9644.
[317c] F. Su, Y.B. He, B. Li, et al., Nano Energy 1 (2012) 429.
[318] H. Shioyama, J. Mater. Sci. Lett. 20 (2001) 499.
[319] L.M. Viculis, J.J. Mack, R.B. Kaner, Science 299 (2003) 1361.
[320] D. Guo, T. Kondo, T. Machida, et al., Nat. Commun. 3 (2012) 1068.
[321] T.A. Land, T. Michely, R.J. Behm, et al., Surf. Sci. 264 (1992) 261.
[322] L. Papagno, L. Caputi, Phys. Rev. B. 29 (1984) 1483.
[323] H.J. Park, J. Meyer, S. Roth, et al., Carbon 48 (2010) 1088.
[324] X. Li, W. Cai, J. An, et al., Science 324 (2009) 1312.
[325] S. Bae, H. Kim, Y. Lee, et al., Nat. Nanotech. 5 (2010) 574.
[326] A. Nagashima, K. Nuka, H. Itoh, et al., Surf. Sci. 291 (1993) 93.
[327] C. Berger, Z. Song, T. Li, et al., Bull. Am. Phys. Soc. 49 (A17) (2004) 8.
[328] C. Berger, Z. Song, T. Li, J. Phys. Chem. B. 108 (2004) 19912.
[329] X. Yang, X. Douy, A. Rouhanipour, et al., J. Am. Chem. Soc. 130 (2008) 4216.
[330] C.D. Simpson, J.D. Brand, A.J. Berresheim, et al., Chem. Eur. J. 8 (2002) 1424.
[331] B.E. Hamaoui, L. Zhi, W. Pisula, et al., Chem. Commun. (2007) 2384.
[332] L. Zhi, J. Wu, J. Li, et al., Angew. Chem. Int. Ed. 44 (2005) 2120.
[333] J.O. Sofo, A.S. Chaudhari, G.D. Barber, Phys. Rev. B. 75 (2007) 153401.
[334] M.H.F. Sluiter, Y. Kawazoe, Phys Rev B 68 (2003) 085410.
[335] D.C. Elias, R.R. Nair, T.M.G. Mohiuddin, et al., Science 323 (2009) 610.
[336] A.M. Ilyin, N.R. Guseinov, I.A. Tsyganov, et al., Physica E 43 (2011) 1262.
[337] H.L. Poh, F. Sanek, Z. Sofer, et al., Nanoscale 4 (2012) 7006.
[338] H.L. Poh, Z. Sofer, M. Pumera, Electrochem. Commun. 25 (2012) 58.
[339] J. Zhou, Q. Wang, Q. Sun, et al., Nano Lett. 9 (2009) 3867.
[340] J. Zhou, M.M. Wu, X. Zhou, et al., Appl. Phys. Lett. 95 (2009) 103108.
[341] A.V. Talyzin, S. Luzan, I.V. Anoshkin, et al., ACS Nano 5 (2011) 5132.

[342] M. Jaiswal, CHYX Lim, Q. Bao, et al., ACS Nano 5 (2011) 888.
[343] K.J. Jeon, Z. Lee, E. Pollak, et al., ACS Nano 5 (2011) 1042.
[344] R.R. Nair, W. Ren, R. Jalil, et al., Small 6 (2010) 2877.
[345] J.T. Robinson, J.S. Burgess, C.E. Junkermeier, et al., Nano Lett. 10 (2010) 3001.
[346] X. Hong, S.H. Cheng, C. Herding, et al., Phys. Rev. B. 83 (2011) 085410.
[347] R. Zbořil, F. Karlický, A.B. Bourlinos, et al., Small 6 (2010) 2885.
[348] H. Chang, J. Cheng, X. Liu, et al., Chem. Eur. J. 17 (2011) 8896.
[349] P. Gong, Z. Wang, J. Wang, et al., J. Mater. Chem. 22 (2012) 16950.
[350] Z. Wang, J. Wang, Z. Li, et al., Carbon 50 (2012) 5403.
[351] A. Markevich, R. Jones, P.R. Briddon, Phys. Rev. B. 84 (2011) 115439.
[352] S. Tang, S. Zhang, J. Phys. Chem. C. 115 (2011) 16644.
[353] O. Leenaerts, H. Sahin, B. Partons, et al., Phys. Rev. B. 88 (2013) 035434.
[354] L.F. Wang, T.B. Ma, Y.Z. Hu, et al., J. Phys. Chem. C. 117 (2013) 12520.
[355] D.R. Dreyer, S. Park, C.W. Bielawski, et al., Chem. Soc. Rev. 39 (2010) 228.
[356] T. Szabó, O. Berkesi, P. Eorgó, et al., Chem. Mater. 18 (2006) 2740.
[357] S. Kim, S. Zhou, Y. Hu, et al., Nat. Mater. 11 (2012) 544.
[358] K.A. Mkhoyan, A.W. Contryman, J. Silcox, et al., Nano Lett. 9 (2009) 1058.
[359] Y. Zhu, S. Murali, W. Cai, et al., Adv. Mater. 22 (2010) 3906.
[360] J. Zhao, S. Pei, W. Ren, et al., ACS Nano 4 (2010) 5245.
[361] S. Pei, H.M. Cheng, Carbon 50 (2012) 3210.
[362] J.W. Burress, S. Gadipelli, J. Ford, et al., Angew. Chem. Int. Ed. 49 (2010) 8902.
[363] C.G. Navarro, R.T. Weitz, A.M. Bittner, et al., Nano Lett. 7 (2007) 3499.
[364] P. Johari, V.B. Shenoy, ACS Nano 5 (2011) 7640.
[365] I. Jung, D.A. Dikin, R.D. Piner, et al., Nano Lett. 8 (2008) 4283.
[366] E. Morales-Narváez, A. Merkoci, Adv. Mater. 24 (2012) 3298.
[367] Y. Cao, Y. Chong, H. Shen, et al., J. Mater. Chem. B. 1 (2013) 5602.
[368] R.H. Baughman, H. Eckhardt, M.J. Kertesz, J. Chem. Phys. 87 (1987) 6687–6699.
[369] N. Narita, S. Nagagi, S. Suzuki, et al., Phys. Rev. B. 58 (1998) 11009.
[370] M. Long, L. Tang, D. Wang, et al., ACS Nano 5 (2011) 2593.
[371] D. Malko, C. Neiss, F. Vines, et al., Phys. Rev. Lett. 108 (2012) 086804.
[372] J. Cao, C.P. Tang, S.-J. Xiong, Physica. B. 407 (2012) 4387–4390.
[373] J. He, S. Ying, P. Zhou, et al., J. Phys. Chem. C. 116 (2012) 26313–26321.
[374] T. Ouyang, Y. Chen, L.-M. Liu, et al., Phys. Rev. B. 85 (2012) 235436.
[375] A.L. Ivanovskii, Prog. Solid State Chem. 41 (2013) 1.
[376] M.M. Haley, S.C. Brand, J.J. Pak, Angew. Chem. Int. Ed. 36 (1997) 836.
[377] W.B. Wan, M.M. Haley, J. Org. Chem. 66 (2001) 3893.
[378] J.A. Marsden, M.M. Haley, J. Org. Chem. 70 (2005) 10213.
[379] W.B. Wan, S.C. Brand, J.J. Pak, et al., Chem. Eur. J. 6 (2000) 2044.
[380] J.M. Kehoe, J.H. Kiley, J.J. English, et al., Org. Lett. 2 (2000) 969.
[381] G. Li, Y. Li, H. Liu, et al., Chem. Commun. 46 (2010) 3256.
[382] X. Qian, Z. Ning, Y. Li, et al., Dalton Trans. 41 (2012) 730.
[383] G. Li, Y. Li, X. Qian, et al., J. Phys. Chem. C. 115 (2011) 2611.
[384] S. Wang, L. Yi, J.E. Halpert, et al., Small 8 (2012) 265.
[385] N. Yang, Y. Liu, H. Wen, et al., ACS Nano 7 (2013) 1504.
[386] H. Zhang, M. Zhao, X. He, et al., J. Phys. Chem. C. 115 (2011) 8845.
[387] H. Zhang, Y. Xia, H. Bu, et al., J. Appl. Phys. 113 (2013) 044309.
[388] C. Li, J. Li, F. Wu, et al., J. Phys. Chem. C. 115 (2011) 23221.
[389] H.J. Hwang, Y. Kwon, H. Lee, J. Phys. Chem. C. 116 (2012) 20220.

[390] Y.S. Wang, P.F. Yuan, M. Li, et al., J. Solid State Chem. 197 (2013) 323.
[391] Y. Jiao, A. Du, M. Hankel, et al., Chem. Commun. 47 (2011) 11843.
[392] S. Iijima, M. Yudasaka, R. Yamada, et al., Chem. Phys. Lett. 309 (1999) 165.
[393] S. Bandow, F. Kokai, K. Takahashi, et al., Chem. Phys. Lett. 321 (2000) 514.
[394] E. Bekyarova, K. Kankeo, D. Kasuya, et al., Langmuir 18 (2002) 4138.
[395] E. Bekyarova, K. Kankeo, M. Yudasaka, et al., J. Phys. Chem. 107 (2003) 4479.
[396] E. Bekyarova, K. Kaneko, M. Yudasaka, et al., Adv. Mater. 14 (2002) 973.
[397] E. Bekyarova, K. Murata, M. Yudasaka, et al., J. Phys. Chem. 107 (2003) 4681.
[398] K. Murata, K. Hirahara, M. Yudasaka, et al., J. Phys. Chem. B. 106 (2002) 12668.
[399] J. Miyawaki, M. Yudasaka, H. Imai, et al., Adv. Mater. 18 (2006) 1010.
[400] M. Kyotani, S. Matsushita, T. Nagai, et al., J. Am. Chem. Soc. 130 (2008) 10880.
[401] S. Matsushita, M. Kyotani, K. Akagi, J. Am. Chem. Soc. 133 (2011) 17977.
[402] S. Matsushita, M. Kyotani, K. Akagi, TANSO (258) (2013) 201 [in Japanese]
[403] Y. Sanada, M. Suzuki, K. Fujimoto, Kasseitann (Activated Carbons), Koudannsya Sci., 1992.
[404] F. Rodriguez-Reinoso, A. Linares-Salino, Chem. Phys. Carbon 21 (1995) 1.
[405] J.W. Patrick, Ed, Porosity in Carbons: Characterization and Applications, Edward Arnold, London (1995).
[406] H. Marsh, E.A. Heintz, F. Rodriguez-Reinoso, Introduction to Carbon Technologies, University of Alicante (1997).
[407] S. Morooka, K. Kusakabe, Mater. Res. Soc. Bull. 24 (1999) 25.
[408] T. Kyotani, Carbon 38 (2000) 269.
[409] C.Y. Wang, M. Inagaki, Carbon 37 (1999) 147.
[410] S. Kisamori, S. Kawano, I. Mochida, Chem. Lett. (1993) 1899.
[411] I. Mochida, Y. Korsi, M. Shirahama, et al., Carbon 38 (2000) 227.
[412] Y. Abe, R. Imamura, S. Yoshida, et al., TANSO (172) (1996) 11 [in Japanese]
[413] A. Oya, J. Odor. Res. Eng. 28 (1997) 57.
[414] A. Yoshida, I. Tanahashi, Y. Takauchi, et al., IEEE CHEM-10 1 (1987) 100.
[415] A. Nishino, TANSO (132) (1988) 57 [in Japanese]
[416] Nissan Diesel Co., <http//www.nissandiesel.co.jp/low-env/s-capacitor.htm/>.
[417] T. Kyotani, T. Nagai, S. Inoue, et al., Chem. Mater. 9 (1997) 609.
[418] Z. Ma, T. Kyotani, A. Tomita, Carbon 40 (2002) 2367.
[419] T. Kyotani, A. Tomita, J. Jpn. Petrol. Inst. 45 (2002) 261.
[420] T. Ohba, K. Kaneko, J. Phys. Chem. B. 106 (2002) 7171.
[421] H. Hatori, Y. Yamada, M. Shiraishi, et al., TANSO (167) (1995) 94 [in Japanese]
[422] H. Hatori, Y. Yamada, M. Shiraishi, Carbon 30 (1992) 303.
[423] H. Hatori, H. Takagi, Y. Yamada, Carbon 42 (2004) 1169.
[424] L.J. Wang, F.C.N. Hong, Microp. Mesop Mater. 77 (2005) 167–174.
[425] Y. Sakata, A. Muto, Uddin MdA, et al., Sep. Purif. Technol. 40 (1999) 97–100.
[426] T.A. Centeno, A.B. Fuertes, Carbon 38 (2000) 1067–1073.
[427] A. Merritt, R. Rajagopalan, H.C. Foley, Carbon 45 (2007) 1267–1278.
[428] C.W. Jones, W.J. Koros, Carbon 32 (1994) 1419–1425.
[429] V.C. Geiszler, W.J. Koros, Ind. Eng. Chem. Res. 35 (1996) 2999–3003.
[430] H. Suda, K. Haraya, J. Phys. Chem. B. 101 (1997) 3988–3994.
[431] A.B. Fuertes, T.A. Centeno, Microp. Mesop Mater. 26 (1998) 23–26.
[432] M. Ogawa, Y. Nakano, J. Membr. Sci. 162 (1999) 189–198.
[433] A. Singh-Ghosal, W.J. Koros, J. Membr. Sci. 174 (2000) 177–188.
[434] Q. Zeng, D. Wu, C. Zou, et al., Chem. Commun. 46 (2010) 5927.

[435] C. Zou, D. Wu, M. Li, et al., J. Mater. Chem. 20 (2010) 731.
[436] J. Jansta, F.P. Dousek, V. Patzelova, Carbon 13 (1975) 377.
[437] F.P. Dousek, J. Jansta, J. Baldrian, Carbon 18 (1980) 13.
[438] L. Kavan, P. Novak, F.P. Dousek, Electrochim. Acta 1988 (1605) 11.
[439] S. Shiraishi, H. Kurihara, H. Tsubota, et al., Electrochem. Solid-State Lett. 4 (2001) A5.
[440] Y. Yamada, O. Tanaike, T.T. Liang, et al., Electrochem. Solid-State Lett. 5 (2002) A283.
[441] O. Tanaike, N. Yoshizawa, H. Hatori, et al., Carbon 40 (2002) 457.
[442] O. Tanaike, N. Yoshizawa, H. Hatori, et al., Mol. Cryst Liq. Cryst. 388 (2002) 459.
[443] O. Tanaike, H. Hatori, Y. Yamada, et al., Carbon 41 (2003) 1759.
[444] T.T. Liang, Y. Yamada, N. Yoshizawa, et al., Chem. Mater. 13 (2001) 2933.
[445] N. Miyajima, O. Tanaike, T. Matsumura, et al., Chem. Lett. 41 (2012) 53.
[446] Y. Gogotsi, A. Nikitini, H. Ye, et al., Nat. Mater. 2 (2003) 591.
[447] R.K. Dash, A. Nikitin, Y. Gogotsi, Microp. Mesop Mater. 72 (2004) 203.
[448] A. Jaenes, L. Permann, M. Arulepp, et al., Electrochem. Commun. 6 (2004) 313.
[449] J. Chmiola, G. Yushin, R.K. Dash, et al., Electrochem. Solid-State Lett. 8 (2005) A357.
[450] R.K. Dash, G. Yushin, Y. Gogotsi, Microp. Mesop Mater. 86 (2005) 50.
[451] M. Rose, E. Kockrick, I. Senkovska, et al., Carbon 48 (2010) 403.
[452] E. Lust, A. Janes, M. Arulepp, J. Electroanal. Chem. 562 (2004) 33.
[453] A. Jaenes, L. Permann, P. Nigu, et al., Surf. Sci. 560 (2004) 145.
[454] J. Chmiola, G. Yushin, Y. Gogotsi, et al., Science 2006 (1760) 313.
[455] J. Chmiola, G. Yushin, R. Dash, et al., J. Power Sources 158 (2006) 765.
[456] A. Jaenes, E. Lust, J. Electrochem. Soc. 153 (2006) A113.
[457] R. Dash, J. Chmiola, G. Yushin, et al., Carbon 44 (2006) 2489.
[458] J. Leis, M. Arulepp, A. Kuura, et al., Carbon 44 (2006) 2122.
[459] H. Wang, Q. Gao, Carbon 47 (2009) 820.
[460] Y. Gogotsi, R.K. Dash, G. Yushin, et al., J. Am. Chem. Soc. 127 (2005) 16006.
[461] G. Yushin, R. Dash, J. Jagiello, et al., Adv. Funct. Mater. 16 (2006) 2288.
[462] M. Oschatz, E. Kockrick, M. Rose, et al., Carbon 48 (2010) 3987.
[463] R.W. Pekala, F.M. Kong, Polym. Prpts 30 (1989) 221.
[464] R.W. Pekala, C.T. Alviso, F.M. Kong, et al., J. Non-Cryst Solids 145 (1992) 90.
[465] A.W.P. Fung, G.A.M. Reynolds, Z.H. Wang, et al., J. Non-Cryst Solids 186 (1995) 200.
[466] G.A.M. Reynolds, A.W.P. Fung, Z.H. Wang, et al., J. Non-Cryst Solids 188 (1995) 27.
[467] G. Reichenauer, A. Emmerling, J. Fricke, et al., J. Non-Cryst Solids 225 (1998) 210.
[468] Y. Hanzawa, K. Kaneko, N. Yoshizawa, et al., Adsorption 4 (1998) 187.
[469] Y. Hanzawa, K. Kaneko, Langmuir 26 (1996) 6167.
[470] K. Kaneko, Y. Hanzawa, T. Iiyama, et al., Adsorption 5 (1999) 7.
[471] K. Kaneko, C.K. Yang, T. Ohkubo, et al., Adsorpt. Sci. Technol. (2000) 1.
[472] D. Mowla, D.D. Do, K. Kaneko, Chem. Phys. Carbon 28 (2003) 229.
[473] T. Ohba, Y. Hanzawa, T. Kimura, et al., Characterization of Porous Materials, Princeton (2003).
[474] E. Bekyarova, K. Kaneko, Adv. Mater. 2000 (1625) 12.
[475] E. Bekyarova, K. Kaneko, J. Coll. Interface Sci. 238 (2001) 357.
[476] Y. Hanzawa, H. Hatori, N. Yoshizawa, et al., Carbon 40 (2002) 575.

[477] H. Tamon, H. Ishizaki, M. Mikami, et al., Carbon 35 (1997) 791.
[478] H. Tamon, H. Ishizaki, T. Araki, et al., Carbon 36 (1998) 1257.
[479] H. Tamon, H. Ishizaki, Carbon 36 (1998) 1397.
[480] H. Tamon, H. Ishizaki, T. Yamamoto, et al., Carbon 37 (1999) 2049.
[481] H. Tamon, H. Ishizaka, T. Yamamoto, et al., Carbon 38 (2000) 1099.
[482] Y. Yamamoto, T. Nishimura, T. Suzuki, et al., J. Non-Cryst Solids 288 (2001) 46.
[483] T. Yamamoto, T. Nishimura, T. Suzuki, et al., Carbon 39 (2001) 2374.
[484] T. Yamamoto, T. Nishimura, T. Suzuki, et al., Drying Tech. 19 (2001) 1319.
[485] Y. Tao, H. Kanoh, K. Kaneko, J. Am. Chem. Soc. 125 (2003) 6044.
[486] Y. Tao, H. Kanoh, K. Kaneko, J. Phys. Chem. 107 (2003) 10971.
[487] R. Ryoo, S.H. Joo, M. Kruk, M. Jaroniec, Adv. Mater. 13 (2001) 677.
[488] J. Lee, S. Han, T. Hyeon, J. Mater. Chem. 14 (2004) 478.
[489] S. Tanaka, N. Nishiyama, Y. Egashira, K. Ueyama, Chem. Commun. 2005 (2005) 2125.
[490] F.Q. Zhang, Y. Meng, D. Gu, et al., J. Am. Chem. Soc. 127 (2005) 13508.
[491] C.D. Liang, S. Dai, J. Am. Chem. Soc. 128 (2006) 5316.
[492] Y. Meng, D. Gu, F. Zhang, et al., Chem. Mater. 18 (2006) 4447.
[493] F. Zhang, Y. Meng, D. Gu, et al., Chem. Mater. 18 (2006) 5279.
[494] J. Jin, N. Nishiyama, Y. Egashira, et al., Microp. Mesop Mater. 118 (2009) 218.
[495] T. Morishita, T. Tsumura, M. Toyoda, et al., Carbon 49 (2010) 2690.
[496] T. Morishita, L. Wang, T. Tsumura, et al., TANSO (242) (2010) 60 [in Japanese]
[497] J. Lee, S. Yoon, T. Hyeon, et al., Chem. Communn. (1999) 2177.
[498] R. Ryoo, S.H. Joo, S. Sun, J. Phys. Chem. B. 103 (2004) 7743.
[499] S. Yoon, J. Lee, T. Hyeon, et al., J. Electrochem. Soc. 147 (2000) 122.
[500] S. Jun, S.H. Joo, R. Ryoo, et al., J. Am. Chem. Soc. 122 (2000) 10712.
[501] M. Inagaki, S. Kobayashi, F. Kojin, et al., Carbon 42 (2004) 3153.
[502] T. Morishita, R. Suzuki, T. Nishikawa, et al., TANSO (219) (2005) 226 [in Japanese]
[503] T. Morishita, R. Suzuki, T. Tsumura, et al., TANSO (223) (2006) 220 [in Japanese]
[504] T. Morishita, Y. Soneda, T. Tsumura, M. Inagaki, Carbon 44 (2006) 2360.
[505] T. Morishita, K. Ishihara, M. Kato, T. Tsumura, M. Inagaki, TANSO (226) (2007) 19.
[506] M. Inagaki, M. Kato, T. Morishita, K. Morita, K. Mizuuchi, Carbon 45 (2007) 1121.
[507] W. Liu, C. Wang, J. Wang, et al., *Chin. Sci. Bull.* 58 (2013) 992.
[508] T. Morishita, K. Ishihara, M. Kato, M. Inagaki, Carbon 45 (2007) 209.
[509] J. Przepiorski, J. Karolczyk, K. Takeda, et al., Ind. Eng. Chem. Res. 48 (2009) 7110.
[510] J. Przepiorski, A. Czyzewski, J. Kapica, et al., Pol. J. Chem. Technol. 4 (2011) 42.
[511] H. Konno, H. Onishi, N. Yoshizawa, et al., J. Power Sources 195 (2010) 667.
[512] M.-J. Jung, J.S. Im, E. Jeong, et al., Carbon Lett. 10 (2009) 217.
[513] W.F. Zhang, Z.-H. Huang, G.P. Cao, et al., J. Power Sources 204 (2012) 230.
[514] W.F. Zhang, Z.-H. Huang, C.J. Zhou, et al., J. Mater. Chem. 22 (2012) 7158.
[515] W.F. Zhang, Z.-H. Huang, G.P. Cao, et al., J. Phys. Chem. Solids 73 (2012) 1428.
[516] S. Han, T. Hyeon, Chem. Commun. (1999) 1955.
[517] Z. Li, M. Jaroniec, J. Am. Chem. Soc. 123 (2001) 9208.
[518] J. Pang, X. Li, D. Wang, et al., Adv. Mater. 16 (2004) 884.

[519] A.A. Zakhidov, R.H. Baughman, Z. Iqbal, et al., Science 282 (1998) 897.
[520] D. Wu, Z. Li, Y. Liang, et al., Carbon 47 (2009) 916.
[521] Y. Liang, F. Liang, D. Wu, et al., Phys. Chem. Chem. Phys. 13 (2011) 8852.
[522] S. Li, Y. Liang, D. Wu, et al., Carbon 48 (2010) 839.
[523] S.-W. Woo, K. Dokko, H. Nakano, et al., J. Mater. Chem. 18 (2008) 1674.
[524] Y. Liang, D. Wu, R. Fu, Langmuir 25 (2009) 7783.
[525] D. Wu, C.M. Hui, H. Dong, et al., Macromolecules 44 (2011) 5846.
[526] D. Wu, H. Dong, J. Pietrasik, et al., Chem. Mater. 23 (2011) 2024.
[527] M. Inagaki, J.M.D. Tascon, Activated Carbon Surfaces in Environmental Remediation (2006) 49.
[528] M. Inagaki, H. Orikasa, T. Morishita, RSC Adv. 1 (2011) 1620.
[529] T. Takeichi, M. Zuo, A. Ito, High Perform. Polym. 11 (1999) 1.
[530] T. Takeichi, Y. Yamazaki, A. Ito, et al., J. Photopolym. Sci. Technol. 12 (1999) 203.
[531] T. Takeichi, Y. Yamazaki, M. Zuo, et al., Carbon 39 (2001) 257.
[532] T. Takeichi, Y. Yamazaki, T. Fukui, et al., TANSO (195) (2000) 388.
[533] S. Koyama, H. Haniu, S. Tanaka, et al., Mol. Cryst. Liq. Cryst. 388 (2002) 587.
[534] P.C. Cowlard, J.C. Lewis, J. Mater. Sci. 2 (1967) 507.
[535] J. Klett, R. Hardy, E. Rimine, et al., Carbon 38 (2000) 953.
[536] J.W. Klett, A.D. McMillan, N.C. Gallego, et al., Carbon 42 (2004) 1841.
[537] S. Li, Q. Guo, Y. Song, et al., Carbon 45 (2007) 2843.
[538] J. Song, Q. Guo, Y. Zhong, et al., New Carbon Mater. 27 (2012) 27.
[539] A. Eksilioglu, N. Gencay, M.F. Yardim, et al., J. Mater. Sci. 41 (2006) 2743.
[540] M. Wang, C.-Y. Wang, T.-Q. Li, et al., Carbon 46 (2008) 84.
[541] S. Li, Y. Tian, Y. Zhong, et al., Carbon 49 (2011) 618.
[542] J. Wang, Electrochim. Acta 26 (1981) 1721.
[543] A.G. Chakovskoi, C.E. Hunt, G. Forceberg, et al., J. Vac. Sci. Tech. B. 21 (2003) 571.
[544] V.E. Norvell, G. Mamantov, Anal. Chem. 49 (1977) 1470.
[545] A.N. Strohl, D.J. Curran, Anal. Chem. 51 (1979) 1045.
[546] A.N. Strohl, D.J. Curran, Anal. Chem. 51 (1979) 1050.
[547] A.N. Strohl, D.J. Curran, Anal. Chim. Acta. 108 (1979) 379.
[548] M. Inagaki, T. Morishita, A. Kuno, et al., Carbon 42 (2004) 497.
[549] J.H. Han, K.W. Cho, K.-H. Lee, et al., Carbon 36 (1998) 1801.
[550] M. Inagaki, R.A. Meyer, Chem. Phys. Carbon 26 (1999) 246.
[551] Y. Tanabe, E. Yasuda, M. Takabatake, et al., Rep. RLEMIT 15 (1990) 83.
[552] S. Furukawa, S. Otani, A. Kojima, et al., TANSO (141) (1990) 23 [in Japanese]
[553] S. Kimura, K. Yasuda, E. Yasuda, et al., TANSO (128) (1987) 30 [in Japanese]
[554] Y. Hishiyama, M. Inagaki, S. Kimura, et al., Carbon 12 (1974) 249.
[555] E. Yasuda, S. Kimura, V. Shibusa, Trans. Jap. Soc. Compo. Mater. 6 (1980) 14 [in Japanese]
[556] H.E. Martens, L.D. Jaffe, J.E. Jepson, Proc. 3rd Conf. Carbon (1958) 529.
[557] M. Inagaki, H. Takeuchi, F. Yamanaka, et al., Kogyou Kagaku Kaishi 66 (1963) 169.
[558] R.A. Meyer, Proc. Carbon Conf., Baden-Baden (1986).
[559] F.A. Feldman, S.R. Gyetvay, Aerospace Report TOR-0086(6728-020)-1 (1986).
[560] S. Kimura, E. Yasuda, Zairyo Kagaku 20 (1983) 36 [in Japanese]
[561] M. Buechler, G.F. Hawkins, R.A. Meyer, Aerospace Rpt TOR-0084(4622–01) (1984).

[562] M. Sakai, T. Miyajima, M. Inagaki, Compos. Sci. Tech. 40 (1991) 231.
[563] T. Miyajima, M. Sakai, Fract. Mech. Ceram. 9 (1992) 83.
[564] R. Pleger, W. Braue, R.A. Meyer, Proc. 20th Biennial Conf. Carbon (1991) 399.
[565] R.A. Meyer, Proc. 19th Biennial Conf. Carbon (1989) 332.
[566] R.A. Meyer, G.W. Henderson, 39th Int. SAMPE Symp. (1994) 311.
[567a] K. Katawaki, L. Nishizaki, Pre-stressed Concr. 30 (1988) 31.
[567b] Plastics and Rubber Institute, Carbon fibers Technology, Uses and Prospects, Noyes Publications, New Jersey, USA, 1986.
[568] Y.A. Kim, S. Kamio, T. Tajiri, et al., Appl. Phys. Lett. 90 (2007) 093125.
[569] M. Endo, T. Noguchi, M. Ito, et al., Adv. Funct. Mater. 18 (2008) 3403–3409.
[570] M. Toyoda, R. Kohara, T. Tsumura, et al., Mat. Sci. Eng. B. 161 (2009) 202.
[571] M. Toyoda, R. Kohara, S. Yoshinaga, et al., TANSO (238) (2009) 92 [in Japanese]
[572] M. Inagaki, L'Ind. Ceramique (846) (1990) 122.
[573] M. Inagaki, Carbon 29 (1991) 287.
[574] S. Akihama, T. Suenaga, T. Banno, KICT Rep. (53) (1984).
[575] S. Akihama, Sen-i Gakkai Shi 44 (1988) 125 [in Japanese]
[576] A. Kojima, S. Otani, K. Nakamura, et al., TANSO (139) (1989) 9 [in Japanese]
[577] A. Kojima, S. Otani, Y. Sakaniwa, et al., TANSO (141) (1990) 23 [in Japanese]
[578] N. Muto, H. Yanagida, M. Miyayama, et al., J. Ceram. Soc. Jpn. 100 (1992) 585.
[579] T. Mitsuoka, H. Matsubara, M. Takada, et al., Trans. Mater. Res. Soc. Jpn. 14A (1994) 421.
[580] G.-D. Zhan, J.D. Kuntz, J. Wan, et al., Nat. Mater. 2 (2003) 38.
[581] S. Rul, F. Lefevre-schlick, E. Capria, et al., Acta Mater. 52 (2004) 1061.
[582] M. Estili, Y. Sakka, W.-W. Wu, et al., J. Am. Ceram. Soc. 2013 (1904) 96.
[583] K. Kobayashi, K. Miyazaki, I. Ogawa, et al., Mater. Des. 9 (1988) 10.
[584] C.E. Lowell, J. Am. Ceram. Soc. 50 (1967) 142.
[585] Y. Hishiyama, Y. Kaburagi, K. Kobayashi, et al., Mol. Cryst. Liq. Cryst. 310 (1998) 279.
[586] K. Miyazaki, K. Kobayashi, H. Honda, TANSO (91) (1977) 121 [in Japanese]
[587] T. Sogabe, K. Nakajima, M. Inagaki, J. Mater. Sci. 31 (1996) 6469.
[588] T. Sogabe, O. Okada, K. Kuroda, et al., Carbon 35 (1997) 67.
[589] I. Ogawa, K. Kobayashi, S. Nishikawa, J. Mater. Sci. 23 (1988) 1363.
[590] J.R. Strife, E. Sheehan, Ceram. Bull. 67 (1988) 369.
[591] P. Ehburger, P. Baranne, J. Lahaye, Carbon 24 (1986) 495.
[592] Y. Tanabe, E. Yasuda, S. Kimura, et al., TANSO (131) (1987) 181 [in Japanese]
[593] O. Yamamoto, T. Sasamoto, M. Inagaki, Carbon 33 (1995) 359.
[594] D.W. McKee, Carbon 24 (1986) 737.
[595] T.D. Nixon, J. Cawley, J. Am. Ceram. Soc. 75 (1992) 703.
[596] O. Yamamoto, K. Imai, T. Sasamoto, et al., J. Eur. Ceram. Soc. 12 (1993) 435.
[597] O. Yamamoto, T. Sasamoto, M. Inagaki, Nihon Seramic Kyokai Shi 102 (1994) 165.
[598] T. Noguchi, A. Magario, S. Fukazawa, et al., Mater. Trans. 45 (2004) 602.
[599] P. Schaffaeutl, J. Prakt. Chem. 21 (1841) 155.
[600] F.L. Vogel, J. Mater. Sci. 12 (1977) 982.
[601] G. Pistoia, Lithium Batteries, Elsevier, 1994483
[602] A.F. Hebard, M.J. Rossinsky, R.C. Haddon, et al., Nature 358 (1991) 600.
[603] M.S. Dresselhaus, Ed Intercalation in Layered Materials NATO ASI series B: Physics, 148 (1986).

[604] A.P. Legrand, Flandrois S, Ed Chemical Physics of Intercalation NATO ASI series B: Physics 172 (1987).
[605] H. Zabel, S.A. Solin, Graphite Intercalation Comp. I & II (1990 & 1992).
[606] T. Enoki, M. Suzuki, M. Endo, Graphite Intercalation Compounds and Applications, Oxford Univ. Press, 2003.
[607] M. Inagaki, J. Mater. Res. 4 (1989) 1560.
[608] M. Endo, M. Inagaki, M.S. Dresselhaus, Adv. Carbon II. Graphite Intercalation Comp., Carbon Soc. Jpn. (1990) 129 [in Japanese]
[609] M. Inagaki, Hyoumen 20 (1982) 130 [in Japanese]
[610] J. Shioya, Y. Yamaguchi, H. Matsubara, et al., Nihon Kagaku Kai Shi (1986) 238 [in Japanese]
[611] H. Matsubara, Y. Yamaguchi, J. Shioya, et al., Synth. Met. 18 (1987) 503.
[612] I.L. Kalnin, H.A. Goldberg, Synth. Met. 8 (1983) 277.
[613] H. Oshima, J.A. Woollam, J. Appl. Phys. 53 (1982) 9220.
[614] H. Oshima, J.A. Woollam, A. Yavrouian, et al., Synth. Met. 5 (1983) 113.
[615] T.C. Chieu, M.S. Dresselhaus, M. Endo, Phys. Rev. B. 26 (1982) 5867.
[616] T.C. Chieu, G. Timp, M.S. Dresselhaus, et al., Synth. Met. 27 (1983) 3686.
[617] S. Flandrois, J.M. Masson, J.C. Rouillon, et al., Synth. Met. 3 (1981) 1.
[618] M. Inagaki, Z.D. Wang, Y. Okamoto, et al., Synth. Met. 20 (1987) 9.
[619] J.R. Gaier, D.A. Jaworske, Synth. Met. 12 (1985) 525.
[620] T. Nakajima, M. Kawaguchi, N. Watanabe, Naturforsch 36 (1981) 1419.
[621] T. Nakajima, M. Kawaguchi, N. Watanabe, Carbon 20 (1982) 287.
[622] T. Nakajima, M. Kawaguchi, N. Watanabe, Synth. Met. 7 (1983) 117.
[623] T. Nakajima, N. Watanabe, I. Kameda, et al., Carbon 24 (1985) 343.
[624] E. McRae, J.F. Mareche, A. Bendriss-Rerhrhaye, et al., Ann. Phys. C-2 11 (1986) 13.
[625] P. Lagrange, A. BendrisspRerhrhaye, Carbon 26 (1988) 283.
[626] J.R. Gaier, M.E. Slabe, N. Shaffer, Carbon 26 (1988) 381.
[627] N.B. Hannay, T.H. Geballe, B.T. Matthias, et al., Phys. Rev. Lett. 14 (1965) 225.
[628] Y. Koike, S. Tanuma, H. Suematsu, et al., J. Phys. Chem. Solids 41 (1980) 1111.
[629] A. Chaiken, M.S. Dresselhaus, T.P. Orlando, et al., Phys. Rev. B. 41 (1990) 71.
[630] M.G. Alexander, M.P. Goshorn, D. Guérard, et al., Solid State Commun. 38 (1981) 103.
[631] I.T. Belash, A.D. Bronnikov, O.V. Zharikov, et al., Solid State Commun. 69 (1989) 921.
[632] V.A. Nalimova, V.V. Avdeev, K.N. Semenenko, Mater. Sci. Forum 91–93 (1992) 11.
[633] T.E. Weller, M. Ellerby, S.S. Saxena, et al., Nat. Phys. 1 (2005) 39.
[634] N. Emery, C. Hérold, M. d'Astuto, et al., Phys. Rev. Lett. 95 (2005) 087003.
[635] R. Xie, D. Rosenmann, A. Rydh, et al., Physica. C. 439 (2006) 43.
[636] N. Emery, C. Hérold, J.-F. Marêché, et al., Sci. Technol. Adv. Mater. 9 (2008) 044102.
[637] N. Emery, C. Hérold, J.F. Marêché, et al., J. Solid State Chem. 179 (2006) 1289.
[638] H. Okazaki, R. Yoshida, K. Iwai, et al., Phys. Rev. B. 80 (2009) 035420.
[639] H. Okazaki, R. Yoshida, K. Iwai, et al., Physica. C. 469 (2009) 1041.
[640] N. Emery, S. Pruvost, C. Herold, et al., J. Phys. Chem. Solids 67 (2006) 1137.
[641] M. Toyoda, A. Takenaka, Y. Takano, et al., TANSO 233 (2008) 148.
[642] R. Okazaki, A. Aoki, K. Tsubaki, et al., Nat. Tech. Rep. 24 (1978) 281.
[643] H. Touhara, H. Fujimoto, N. Watanabe, et al., Solid State Ionics 14 (1984) 163.
[644] Y. Kita, N. Watanabe, Y. Fujii, J. Am. Chem. Soc. 101 (1979) 3832.

[645] T. Nakajima, Kagaku 38 (1983) 268 [in Japanese]
[646] R. Yazami, P.H. Touzain, J. Power Sources 9 (1983) 365.
[647] M. Endo, H. Nakamura, A. Emori, et al., TANSO (150) (1991) 319 [in Japanese]
[648] M. Endo, Y. Nishimura, T. Takahashi, et al., J. Phys. Chem. Solids 57 (1996) 725.
[649] J.R. Dahn, A.K. Sleigh, H. Shi, et al., Lithium Batteries G. Pistoria Ed, Elsevier, Amsterdam (1994).
[650] S. Yata, H. Kinoshita, M. Komori, et al., Synth. Met. 62 (1994) 153.
[651] K. Sato, M. Noguchi, A. Demachi, et al., Science 264 (1994) 256.
[652] A. Mabuchi, K. Tokumitu, H. Fujimoto, et al., J. Electrochem. Soc. 142 (1995) 1041.
[653] J.R. Dahn, 22nd Biennial Conf. on Carbon San Diego, 1995
[654] M. Endo, C. Kim, K. Nishimura, et al., Carbon 38 (2000) 183.
[655] R. Fujii, Denki Kagaku 41 (1973) 52 [in Japanese]
[656] J.S. Dunning, W.H. Tiedemen, L. Hsuch, et al., Solid State Ionics 14 (1984) 163.
[657] F. Beck, H. Krohn, Synth. Met. 7 (1983) 193.
[658] S. Otani, K. Matsumoto, Y. Mogi, Denki-Kagaku 50 (1982) 684 [in Japanese]
[659] N. Iwashita, M. Inagaki, Solid State Ionics 70–71 (1994) 425.
[660] M. Inagaki, N. Iwashita, J. Power Sources 52 (1994) 69.
[661] S. Flandrois, J.M. Masson, C.J. Rouillon, et al., Synth. Met. 3 (1981) 1.
[662] N. Iwashita, M. Inagaki, Nihon Kagaku Kai Shi (1992) 1414 [in Japanese]
[663] S. Otani, H.I. Phung, T. Kubota, et al., Denki Kagaku 44 (1976) 27 [in Japanese]
[664] M. Inagaki, A. Matsumoto, M. Sakai, et al., Nihon Kagaku Kai Shi 1983 (1983) 309 [in Japanese]
[665] J.M. Lalancette, R. Roussel, Can. J. Chem. 54 (1976) 3541.
[666] M. Endo, T. Mori, T. Koyama, et al., Ouyou Buturi 49 (1980) 563 [in Japanese]
[667] M. Endo, M. Inagaki, Synth. Met. 7 (1983) 203.
[668a] M. Inagaki, E. Itoh, A. Tanaka, Synth. Met. 35 (1990) 383.
[668b] J. Gole, J. Mater, Sci. Eng. 31 (1977) 309.
[668c] R. Setton, J. Mater, Sci. Eng. 31 (1977) 303.
[669] M. Inagaki, TANSO (133) (1988) 127 [in Japanese]
[670] G. Merle, J. Polym. Sci. 15 (1977) 2067.
[671] I. Rashkov, I. Panayotow, I. Gitsov, Polym. Bull. 4 (1981) 97.
[672a] M. Ichikawa, T. Kondo, K. Kawase, et al., JCS Chem. Comm. (1972) 176.
[672b] M.E. Vol'pin, Z. Anorg, Allg. Chem. 428 (1977) 231.
[673] M. Ichikawa, K. Kawase, K. Tamaru, JCS Chem. Comm. 1980 (1980) 1088.
[674] H. Imamura, S. Tsuchiya, Z. Phys. Chem. 125 (1981) 251.
[675] H.B. Kagan, Chemtech 1976 (1976) 510.
[676] R. Setton, J. Mat. Sci. Eng. 31 (1977) 303.
[677] G. Furdin, P. Lagrange, A. Herold, CR Acad. Sci. Paris 283 (1976) C-563.
[678] K. Watanabe, M. Soma, T. Onishi, et al., Nature 233 (1971) 160.
[679] K. Watanabe, T. Kondow, M. Soma, et al., Proc. Roy. Soc. A. 333 (1973) 51.
[680] P. Lagrange, A. Herold, CR Acad. Sci. Paris 281 (1975) C-381.
[681] N. Akuzawa, K. Katano, Y. Ohmura, et al., TANSO (133) (1988) 100 [in Japanese]
[682] T. Terai, Y. Takahashi, J. Nucl. Sci. Tech. 18 (1981) 643.
[683] T. Trai, Y. Takahashi, Carbon 22 (1984) 91.
[684] N. Akuzawa, T. Amemiya, et al., Science and New Applications of Carbon Fibers, Toyohashi Univ., (1984) 113.
[685] T. Terai, Ph D Thesis, Tokyo Univ., (1983) 206.

[686] Y. Takahashi, TANSO (160) (1993) 301 [in Japanese]
[687] M. Inagaki, N. Saji, F. Kang, et al., TANSO (215) (2004) 258.
[688] H. Pilliere, Y. Takahashi, T. Yoneoka, et al., Synth. Met. 59 (1993) 191.
[689] R. Fujii, T. Dohi, in: N. Watanabe (Ed.), Graphite Intercalation Compounds, Kindai-Hensyu-Sha, 1986, p. 281. Chapter 5, [in Japanese]
[690] M. Inagaki, M. Toyoda, F. Kang, in: L.A. Radovic (Ed.), Chemistry and Physics of Carbon, 29, Dekker, 2004, p. 1.
[691] D. Berger, J. Maire, J. Mater. Sci. Eng. 31 (1977) 335.
[692] M. Inagaki, K. Muramatsu, Y. Maeda, Synth. Met. 8 (1983) 335.
[693] F. Kang, Y. Leng, T.Y. Zhang, Carbon 35 (1997) 1089.
[694] H. Jimenez-Gonzalez, J.S. Speck, G. Roth, et al., Carbon 24 (1986) 627.
[695] M. Toyoda, M. Inagaki, Carbon 38 (2000) 199.
[696] M. Inagaki, M. Toyoda, N. Iwashita, et al., Carbon Sci Korea 12 (2001) 1.
[697] M. Inagaki, M. Toyoda, N. Iwashita, et al., TANSO (201) (2002) 16 [in Japanese]
[698] P.h. Touzain, J. Michel, P. Blum, Synth. Met. 8 (1983) 313.
[699] P. Pfluger, H.V. Kunzi, H.J. Guentherodt, Appl. Phys. Lett. 35 (1979) 771.
[700] K. Yoshino, H. Ueno, TANSO (136) (1989) 29 [in Japanese]
[701] J.R. Dahn, R. Fong, M.J. Spoon, Phys. Rev. B. 42 (1990) 6424.
[702] G. Pistoia (Ed.), Lithium Batteries New Materials, Developments and Perspectives, Elsevier, 1994.
[703] M. Inagaki (Ed.), Anode Carbon Materials for Lithium Ion Secondary Batteries, Carbon Society of Japan, 1996 [in Japanese]
[704] T. Enoki, M. Suzuki, M. Endo, GICs and Batteries in Graphite Intercalation Compounds, Oxford Univ. Press, 2003336
[705] J.R. Dahn, T. Zheng, Y. Liu, J.S. Xue, Science 270 (1995) 590.
[706] K. Tatsumi, N. Iwashita, H. Sakaebe, et al., J. Electrochem. Soc. 142 (1995) 716.
[707] A. Mabuchi, K. Tokumitsu, H. Fujimoto, et al., J. Electrochem. Soc. 142 (1995) 1041.
[708] S. Yata, H. Kinoshita, M. Komori, et al., Synth. Met. 62 (1994) 153.
[709] M. Endo, Y. Nishimura, T. Takahashi, et al., J. Phys. Chem. Solids 57 (1996) 725.
[710] M. Noguchi, K. Miyashita, M. Endo, Tanso (155) (1992) 315.
[711] J.O. Besenhard, M. Winter, J. Yang, et al., J. Power Sources 54 (1995) 228.
[712] F. Fong, K. Sacken, J.R. Dahn, J. Electrochem. Soc. 1990 (2009) 137.
[713] H. Fujimoto, J. Power Sources 195 (2010) 5019.
[714] J.R. Dahn, J.N. Reimers, A.K. Sleigh, T. Tiedje, Phys. Rev. B. 45 (1992) 3773.
[715] B.M. Way, J.R. Dahn, J. Electrochem. Soc. 141 (1994) 907.
[716] Y. Nishimura, T. Takahashi, T. Tanaki, et al., TANSO (172) (1996) 89 [in Japanese]
[717] K. Nishimura, Y.A. Kim, T. Matushita, et al., J. Mater. Res. 15 (2000) 1303.
[718] M. Endo, Y.A. Kim, T. Hayashi, et al., Carbon 39 (2001) 1287.
[719] G. Liang, L. Wang, X. Ou, et al., J. Power Sources 184 (2008) 538–543.
[720] Z.Y. Chen, H.L. Zhu, S. Ji, et al., Solid State Ionics 179 (2008) 1810–1815.
[721] A.V. Murugan, T. Muraliganth, A. Manthiram, J. Electrochem. Soc. 156 (2009) A79–A83.
[722] M. Inagaki, Carbon 50 (2012) 3247.
[723] S. Sarangapani, P. Lessner, J. Forchione, et al., J. Power Sources 29 (1990) 355.
[724] B.E. Conway, Electrochemical Supercapacitors, Kluwer Academic/Plenum Publishers, 1999.

[725] Q. Deyang, S. Hang, J. Power Sources 74 (1998) 99.
[726] M. Inagaki, H. Konno, O. Tanaike, J. Power Sources 195 (2010) 7880–7903.
[727] H. Shi, Electrochem. Acta 41 (1996) 1633.
[728] M. Endo, Y.J. Kim, T. Takeda, et al., J. Electrochem. Soc. 148 (2001) A1135.
[729] Y. Yamada, O. Tanaike, T.T. Liang, et al., Electrochem. Solid-State Lett. 5 (2002) A283.
[730] K. Kinoshita, Carbon: Electrochemical and Physicochemical Properties, Wiley and Sons (1988).
[731] T. Morishita, L. Wang, T. Tsumura, et al., TANSO (242) (2010) 60 [in Japanese]
[732] T. Morishita, Y. Soneda, T. Tsumura, et al., Carbon 44 (2006) 2360.
[733] Y. Soneda, M. Toyoda, K. Hashiya, et al., Carbon 41 (2003) 2680.
[734] Y. Soneda, M. Toyoda, Y. Tani, et al., J. Phys. Chem. Solids 65 (2004) 219.
[735] H. Hara, T. Kinumoto, T. Tsumura, et al., TANSO (255) (2012) 245.
[736] M. Endo, T. Maeda, T. Takeda, et al., J. Electrochem. Soc. 148 (2001) A910.
[737] Y.J. Kim, Y. Horie, S. Ozaki, et al., Carbon 42 (2004) 1491.
[738] D. Lozano-Castello, D. Cazorla-Amoros, A. Linares-Solano, et al., Carbon 41 (2003) 1765.
[739] G. Salitra, A. Soffer, L. Eliad, et al., J. Electrochem. Soc. 147 (2000) 2486.
[740] S. Shiraishi, H. Kurihara, L. Shi, et al., J. Electrochem. Soc. 149 (2002) A855.
[741] L. Wang, M. Fujita, M. Inagaki, Electrochim. Acta 51 (2006) 4096.
[742] L. Wang, M. Toyoda, M. Inagaki, Ads. Sci. Technol. 26 (2008) 491.
[743] L. Wang, M. Inagaki, M. Toyoda, TANSO (240) (2009) 230.
[744] C. Liu, J. Bard, F. Wudl, et al., Electrochem. Solid-State Lett. 2 (1999) 577.
[745] J.N. Barisci, G.G. Wallace, R.H. Baughman, Electrochim. Acta 46 (2000) 509.
[746] E. Franckowiak, K. Jurewicz, S. Delpeux, et al., J. Power Sources 97–98 (2001) 822.
[747] K.A. An, W.S. Kim, Y.S. Park, et al., Adv. Mat. 13 (2001) 497.
[748] S. Shiraishi, H. Kurihara, K. Pkabe, et al., Electrochem. Comm. 4 (2002) 593.
[749] B.E. Conway, V. Nirss, J. Wojtowicz, J. Power Sources 66 (1997) 1.
[750] M. Ramani, B.S. Haran, R.E. White, et al., J. Power Sources 93 (2001) 209.
[751] Y. Sato, K. Yomogida, T. Nanaumi, et al., Electrochem. Solid-State Lett. 3 (2000) 113.
[752] Y. Takasu, T. Ohnuma, W. Sugimoto, et al., Electrochemistry 67 (1999) 1187.
[753] Y. Nian, H. Teng, J. Electroanal. Chem. 540 (2003) 1197.
[754] D. Qu, J. Power Sources 109 (2002) 403.
[755] C. Hsieh, H. Teng, Carbon 40 (2002) 667.
[756] Y.R. Nian, H. Teng, J. Electrochem. Soc. 149 (2002) A1008.
[757] M.J. Bleda-Martinez, J.A. Macia-Agullo, D. Lozano-Castello, et al., Carbon 43 (2005) 2677.
[758] V. Ruiz, C. Blanco, E. Raymundo-Pinero, et al., Electrochim. Acta 52 (2007) 4969.
[759] H. Oda, A. Yamashita, S. Minoura, et al., J. Power Sources 158 (2006) 1510.
[760] M. Kodama, J. Yamashita, Y. Soneda, et al., Mater. Sci. Eng. B. 108 (2004) 156.
[761] M. Kodama, J. Yamashita, Y. Soneda, et al., Chem. Lett. 35 (2006) 680.
[762] D. Hulicova, M. Kodama, H. Hatori, Chem. Mater. 18 (2006) 2318–2326.
[763] M. Kodama, J. Yamashita, Y. Soneda, et al., Carbon 45 (2007) 1105–1107.
[764] K. Jurewicz, R. Pietrzak, P. Nowicki, et al., Electrochim. Acta 53 (2008) 5469.
[765] H. Konno, H. Onishi, N. Yoshizawa, et al., J. Power Sources 195 (2010) 667.
[766] T. Aida, K. Yamada, M. Morita, Electrochem. Solid-State Lett. 9 (2006) A534.
[767] T. Aida, I. Murayama, K. Yamada, et al., J. Power Sources 166 (2007) 462.
[768] V. Khomenko, E. Raymundo-Pinero, F. Beguin, J. Power Sources 177 (2008) 643.

[769] M.S. Hong, S.H. Lee, S.W. Kim, Electrochem. Solid-State Lett. 5 (2002) A227.
[770] T. Brousse, P.L. Taberna, O. Corsnier, et al., J. Power Sources 173 (2007) 633.
[771] V. Khomenko, E. Raymundo-Pinero, F. Beguin, J. Power Sources 153 (2006) 183.
[772] C. Xu, H. Du, B. Li, et al., J. Electrochem. Soc. 156 (2009) A73.
[773] C. Xu, H. Du, B. Li, et al., J. Electrochem. Soc. 156 (2009) A435.
[774] D.W. Wang, F. Li, H.M. Cheng, J. Power Sources 185 (2008) 1563.
[775] G. Amatucci, F. Badway, A.D. Pasquier, et al., J. Electrochem. Soc. 148 (2001) A930.
[776] T. Morimoto, TANSO (189) (1999) 188 [in Japanese]
[777] A. Appleby, J. Int. Hydrogen Energy 19 (1997) 175.
[778] A. Chambers, C. Park, T.K. Baker, et al., J. Phys. Chem. B. 102 (1998) 4253.
[779] S. Orimo, T. Matsushima, H. Fujii, et al., J. Appl. Phys. 90 (2001) 1545.
[780] K. Jurewicz, E. Franckowiak, F. Beguin, Electrochem. Solid-State Lett. 4 (2001) A27.
[781] T.L. Cook, C. Komodromos, D.F. Quinn, et al., Carbon Materials for Advanced Technologies. Burchell TD, Ed., Pergamon, Amsterdam (1999) 269.
[782] T. Burchell, M. Rogers, SAE Tech. Pap. Ser. (2000) 01−2205.
[783] V. Menon, S. Komarneni, J. Porous Mater. 5 (1998) 43.
[784] R.F. Cracknell, P. Gordon, K.E. Gubbins, J. Phys. Chem. 97 (1993) 494.
[785] K.R. Matrange, A.L. Myers, E.D. Glandt, Chem. Eng. Sci. 47 (1992) 1569.
[786] B. McEnaney, T.J. Mays, X. Chen, Fuel 77 (1998) 557.
[787] J.A.F. MacDonald, D.F. Quinn, Carbon 34 (1996) 1103.
[788] S.S. Barton, M.J.B. Evans, J.A.F. Macdonald, Carbon 29 (1997) 1099.
[789] FY 2006 Annual Progress Report for the DOE Hydrogen Program, (2006) http://www.hydrogen.energy.gov/annual_progress.html.
[790] A.C. Dillon, K.M. Jones, T.A. Bekkedahl, et al., Nature 386 (1997) 377.
[791] A.C. Dillon, T. Gennett, J.L. Alleman, et al., Proc. 2000 DOE/NREL Hydrogen Program Review, NRE + /CP-570-28890, (2000).
[792] C. Liu, Y.Y. Fan, M. Liu, et al., Science 286 (1999) 1127.
[793] Y. Ye, C.C. Ahn, C. Witham, et al., Appl. Phys. Lett. 74 (1999) 2307.
[794] H. Takagi, H. Hatori, Y. Soneda, et al., Mater. Sci. Eng. B. 108 (2004) 143.
[795] S.M. Lee, K.S. Park, Y.C. Choi, et al., Synth. Met. 113 (2000) 209.
[796] C. Park, P.E. Anderson, A. Chambers, et al., J. Phys. Chem. B. 103 (1999) 10572.
[797] Y.Y. Fan, B. Liao, M. Liu, et al., Carbon 37 (1999) 1649.
[798] H.M. Cheng, C. Liu, Y.Y. Fan, et al., Metallkd 91 (2000) 306.
[799] P. Chen, X. Wu, J. Lin, et al., Science 285 (1999) 91.
[800] R.T. Yang, Carbon 38 (2000) 623.
[801] N.D. Parkyns, D.F. Quinn, Porosity of Carbons, Parick JW, Ed, Edward Arnold, London (1995) 291.
[802] L. Zhou, M. Li, Y. Sun, et al., Carbon 39 (2001) 773.
[803] M.Y. Liang, G.J. Chen, C.Y. Sun, et al., J. Phys. Chem. B. 109 (2005) 19034.
[804] L.J. Yan, G.J. Chen, W.X. Pang, et al., J. Phys. Chem. B. 109 (2005) 6025.
[805] M.S. Dresselhaus, K.A. Williams, P.C. Eklund, MRS Bull. (1999) 45.
[806] M. Hirshner, Appl. Phys. A. 72 (2001) 129.
[807] M.A. de la Casa-Lillo, F. Lamari-Darkrim, et al., J. Phys. Chem. B. 106 (2002) 10930.
[808] M. Rzepka, P. Lamp, M.A. de la Casa Lillo, J. Phys. Chem. B. 102 (1998) 10894.
[809] T. Ozaki, Sci. Carbon Precursors Mater. Des. 1 (1999) 97 [in Japanese]
[810] S. Orimo, G. Majer, T. Fukunaga, et al., Appl. Phys. Lett. 75 (1999) 3093.

[811] E. Bekyarova, K. Murata, M. Yudasaka, et al., J. Phys. Chem. 107 (2003) 4681.
[812] J. Liu, Y. Zhou, Y. Sun, et al., Carbon 49 (2011) 3731.
[813] Ollis DF, Al-Ekabi H, Photocatalytic Purification and Treatment of Water and Air Elsevier, Amsterdam (1993).
[814] H. Kominami, M. Kohno, Y. Matsunaga, et al., J. Am. Ceram. Soc. 84 (2001) 1035.
[815] H. Chen, J. Shi, W. Zhang, et al., Chem. Mater. 13 (2001) 1035.
[816] H. Chun, W. Yizhong, T. Hongxiao, Appl. Catal. B. 35 (2001) 95.
[817] A. Imhof, Langmuir 17 (2001) 3579.
[818] Z. Ding, G.Q. Lu, P.F. Greenfield, Colloid. Interf. Sci. 232 (2000) 1.
[819] M. Hirano, C. Nakahara, K. Ota, et al., J. Am. Ceram. Soc. 85 (2002) 1333.
[820] J. Chen, L. Eberlein, C.H. Langford, J, Photochem. Photobiol. A. 148 (2002) 183.
[821] A.D. Modestov, O. Lev, J. Photochem. Photobiol. A. 112 (1998) 261.
[822] J.-M. Herrman, H. Tahiri, Y. Ait-Ichou, et al., Appl. Catal. B. 13 (1997) 219.
[823] M. Toyoda, T. Tsumura, B. Tryba, et al., Chem. Phys. Carbon 31 (2012) 172.
[824] T. Torimoto, S. Itoh, S. Kuwabara, et al., Environ. Sci. Technol. 30 (1996) 1275.
[825] J. Prepiorski, N. Yashizawa, Y. Yamada, J. Mater. Sci. 36 (2001) 4249.
[826] Hildenbrand K Lettmann Ch, H. Kisch, et al., Appl. Catal. B. 32 (2001) 215.
[827] J. Matos, J. Laine, J.-M. Herrmann, J. Catal. 200 (2001) 10.
[828] B. Tryba, A.W. Morawski, M. Inagaki, Appl. Catal. B. 41 (2003) 427.
[829] B. Tryba, A.W. Morawski, M. Inagaki, Appl. Catal. B: Environ. 46 (2003) 203.
[830] M. Toyoda, Y. Nanbu, T. Kito, et al., Desalination 159 (2003) 273.
[831] J.M. Herrmann, J. Matos, J. Disdier, et al., Catal. Today 54 (1999) 255.
[832] I. Izumi, K. Kuroda, Y. Onishi, et al., Mizushori Gizyutu 42 (2001) 461 [in Japanese]
[833] T. Tsumura, N. Kojitani, I. Izumi, et al., J. Mater. Chem. 12 (2002) 1391.
[834] M. Toyoda, T. Tsumura, Y. Nanbu, et al., Mizukannkyo Gakkaishi 4 (2003) 209 [in Japanese]
[835] T. Tsumura, N. Kojitani, H. Umemura, et al., Appl. Surf. Sci. 196 (2002) 429.
[836] M. Inagaki, Y. Hirose, T. Matsunaga, et al., Carbon 41 (2003) 2619.
[837] B. Tryba, T. Tsumura, M. Junas, et al., Appl. Catal. B. 50 (2004) 177.
[838] B. Tryba, A.W. Morawski, T. Tsumura, et al., J. Photochem. Photobio. A. 167 (2004) 127.
[839] M. Toyoda, Y. Yoshikawa, T. Tsumura, et al., Appl. Catal. B. 51 (2004) 247.
[840] M. Toyoda, B. Tryba, E. Ito, et al., Mizukannkyo Gakkaishi 29 (2006) 9 [in Japanese]
[841] M. Inagaki, S. Kobayashi, F. Kojin, et al., Carbon 42 (2004) 3153.
[842] M. Toyoda, Y. Nanbu, Y. Nakazawa, et al., Appl. Catal. B. 49 (2004) 227.
[843] M. Inagaki, T. Imai, T. Yoshikawa, et al., Appl. Catal. B. 51 (2004) 247.
[844] H. Irie, Y. Watanabe, K. Hashimoto, Chem. Lett. 32 (2003) 772.
[845] C. Lettmann, K. Hildenbrand, H. Kisch, et al., Appl. Catal. B. 32 (2001) 215.
[846] M. Janus, B. Tryba, M. Inagaki, et al., Appl. Catal. B. 52 (2004) 61.
[847] Fujiraito Ind. Co., Ltd., Jpn. Pat. Propos. (95333) (1979).
[848] N.Z. Cao, W.C. Shen, S.Z. Wen, et al., Abstr. Carbon 96 (1996) 114.
[849] M. Toyoda, J. Aizawa, M. Inagaki, Desalination 115 (1998) 199.
[850] M. Toyoda, K. Moriya, M. Inagaki, TANSO (187) (1999) 96 [in Japanese]
[851] M. Toyoda, K. Moriya, J. Aizawa, et al., Desalination 128 (2000) 205.
[852] M. Inagaki, H. Konno, M. Toyoda, et al., Desalination 128 (2000) 213.
[853] B. Tryba, R.J. Kalenczuk, F. Kang, et al., Mol. Cryst. Liq. Cryst. 340 (2000) 113.
[854] B. Tryba, A.W. Morawski, R.J. Kalenczuk, et al., Spill. Sci. Tech. Bull. 8 (2003) 569.
[855] M. Inagaki, A. Kawahara, H. Konno, Carbon 40 (2002) 105.

[856] M. Inagaki, A. Kawahara, T. Hayashi, Res. Rep. Aichi. Inst. Tech. (36) (2001) 69 [in Japanese]
[857] M. Inagaki, A. Kawahara, N. Iwashita, et al., Carbon 240 (2000) 1487.
[858] M. Inagaki, K. Shibata, S. Setoh, et al., Desalination 128 (2000) 219.
[859] M. Toyoda, N. Dogawa, T. Seki, et al., TANSO (199) (2001) 166 [in Japanese]
[860] M. Toyoda, M. Inagaki, Spill Sci. Tech. Bull. 8 (2003) 467.
[861] F. Kang, Y.P. Zheng, H. Zhao, et al., New Carbon Mater. 18 (2003) 161.
[862] M. Toyoda, M. Inagaki, Carbon 38 (2000) 199.
[863] Y. Nishi, G. Dai, N. Iwashita, et al., Mater. Sci. Res. Intl. 8 (2002) 243.
[864] Y. Nishi, N. Iwashita, Y. Sawada, et al., Water Res. 36 (2002) 5029.
[865] M. Toyoda, N. Iwashita, M. Inagaki, Chem. Phys. Carbon 30 (2007) 177.
[866] Y. Nishi, N. Iwashita, M. Inagaki, TANSO (201) (2002) 31 [in Japanese]
[867] M. Toyoda, Y. Nishi, N. Iwashita, et al., Desalination 151 (2002) 139.
[868] Y.P. Zheng, H.N. Wang, F.Y. Kang, et al., Carbon 42 (2004) 2603.
[869] M. Inagaki, T. Suwa, Carbon 39 (2001) 915.
[870] F. Kang, Y.P. Zheng, H.N. Wang, et al., Carbon 40 (2002) 1575.
[871] M. Inagaki, R. Tashiro, M. Toyoda, et al., J. Ceram. Soc. Jpn. 112 (2004) S1513.
[872] M. Inagaki, R. Tashiro, Y. Washino, et al., J. Phys. Chem. Solids 65 (2004) 133.
[873] R. Aggarwal, Carbon 15 (1977) 291.
[874] E.W. Washburn, Phys. Rev. 17 (1921) 273.
[875] V. Beltran, A. Escardino, C. Feliu, et al., Br. Ceram. Trans. J. 87 (1988) 64.
[876] V. Beltran, A. Barba, M.D. Rodrigo, et al., Br. Ceram. Trans. J. 88 (1989) 219.
[877] M. Inagaki, A. Kawahara, H. Konno, Desalination 17 (2004) 77.
[878] M. Inagaki, F.Y. Kang, M. Toyoda, Chem. Phys. Carbon 29 (2004) 1.
[879] W. Shen, S. Wen, N. Cao, et al., Carbon 37 (1999) 351.
[880] J. Bokros, L.D. LaGrange, F.J. Shoen, Chem. Phys. Carbon 9 (1972) 103.
[881] N.Z. Cao, W.C. Shen, S.Z. Wen, et al., Eur. Carbon Conf. (1996) 258.
[882] J. Drelich, J. Hupka, B. Gutkowski, Chem. Prot. Environ. (1988) 207.
[883] H.M. Chol, R.M. Cloud, Environ. Sci. Technol. 26 (1992) 772.
[884] R.F. Johnson, T.G. Manjrekar, J.E. Halligan, Environ. Sci. Technol. 7 (1973) 439.
[885] H. Yamamoto, Cellulose Commun. 5 (1998) 148.
[886] N. Miyata, Sen'i Gakkaishi 55 (1999) 576 [in Japanese]
[887] K. Umehara, S. Nakamura, M. Saito, 27th Symp. Chem. Treat. Woods (1997) 49 [in Japanese]
[888] A. Kojima, S. Otani, International Workshop on Advanced Materials for Functional Manifestation of Frontier and Environmental Consciousness, Tokyo (1997)
[889] M. Shiraishi, Carbon Alloys -Novel Concepts Develop Carbon Sci. Technol. (2003) 515.
[890] K. Sasaki, H. Konno, M. Endo, et al., Biotech. Bioeng. 85 (2004) 489.

# Index

*Note*: Page numbers followed by "*f*" and "*t*" refer to figures and tables, respectively.

## C

## D



## E

## R



## S



## T

Edwards Brothers Malloy
Ann Arbor MI. USA
June 19, 2014